INTERNATIONAL UNION OF GEODESY AND GEOPHYSICS
INTERNATIONAL ASSOCIATION OF GEODESY
COMMISSION ON RECENT CRUSTAL MOVEMENTS

DEUTSCHE AKADEMIE DER WISSENSCHAFTEN ZU BERLIN

Klasse für Bergbau, Hüttenwesen und Montangeologie

INTERNATIONALE UNION FÜR GEODÄSIE UND GEOPHYSIK
INTERNATIONALE ASSOZIATION FÜR GEODÄSIE

Kommission für rezente Erdkrustenbewegungen

I. Internationales Symposium über rezente Erdkrustenbewegungen vom 21. bis 26. Mai 1962 in Leipzig, DDR

Herausgegeben von

OTTO MEISSER,

Freiberg/Sa.

AKADEMIE-VERLAG · BERLIN

1962

DEUTSCHE AKADEMIE DER WISSENSCHAFTEN ZU BERLIN
Klasse für Bergbau, Hüttenwesen und Montangeologie

INTERNATIONAL UNION OF GEODESY AND GEOPHYSICS
INTERNATIONAL ASSOCIATION OF GEODESY

Commission on recent crustal movements

1st International Symposium on Rezent Crustal Movements
Leipzig, GDR, May 21st to 26th, 1962

AKADEMIE-VERLAG · BERLIN
1962

Vorgelegt von Hrn. MEISSER in der Klassensitzung vom 7. Juni 1962

Zum Druck genehmigt am gleichen Tage, ausgegeben am 20. Dezember 1962

Erschienen im Akademie-Verlag GmbH, Berlin W 8, Leipziger Straße 3—4

Lizenznummer: 202 · 100/761/62
Kartengenehmigung: MdI der DDR Nr. 7921/62
Offset-Druck: Nationales Druckhaus, Berlin
Bestellnummer: 2001/62/III/2 · E S: 18 F 2

V O R W O R T

Die Klasse für Bergbau, Hüttenwesen und Montangeologie bei der Deutschen Akademie der Wissenschaften zu Berlin hat gemäß Resolution X des ersten Internationalen Symposiums über rezente Erdkrustenbewegungen vom 21. bis 26. Mai 1962 in Leipzig die Veröffentlichung aller Materialien dieses Symposiums übernommen.

Das Redaktionskollegium

O. Meißer, Freiberg (Sa)
J.D. Boulanger, Moskau
J.A. Mescherikov, Moskau
H. Peschel, Dresden
H. Philipps, Potsdam
K. Reicheneder, Potsdam
S. Miyamura, Tokyo

hat die Herausgabe außerordentlich mit unterstützt.

Neben dem Dank an die Herren Autoren und Herren Diskussionsredner für eine schnelle Erledigung aller Rückfragen möchte ich folgenden Mitarbeitern des Instituts für Angewandte Geophysik der Bergakademie Freiberg für weitgehende Unterstützung bei der Anfertigung des druckfertigen Manuskriptes noch besonders danken: Dr. L. Hiersemann, Dr. H.-G. Thon, G. Schmidt, H. Arnold, Frl. B. Laue, Frl. L. Piontek, Frl. U. Sauerstein, Frl. M. Thieme, ebenso dem Geodätischen Institut Potsdam wie dem Geodätischen Institut der TU Dresden für Korrekturlesen und der Sprachabteilung der Bergakademie Freiberg.

Für eine zügige Herstellung gebührt dem Akademieverlag eine lobende Anerkennung.

O. Meißer

Freiberg, den 1.10. 1962 — Vorsitzender des Redaktionskollegiums

F i n a l   R e m a r k

J. A. M e s c h e r i k o v (Moskau)

President of Comm. on Recent Crustal Movements

Ladies and Gentlemen,

Dear Colleagues,

We have finished the work of this first International Symposium on Recent Movements of the Earth's Crust. I hope that it will be of use to all of us. Here we had the opportunity of taking notice of the latest results of the research work we are all interested in. We have exchanged our opinions and views which is very important for the development of science.

I am very glad of the unanimous support of the project of a world-wide collaboration on recent crustal movements.

I may hope that the realization of this project will be a valuable contribution to the geo-sciences, - to geodesy, geophysics, geology, and physical geography.

I am very glad of the friendly atmosphere and of the fact that the full mutual understanding of the scientists of various countries and various special fields manifested itself in this symposium.

This is to be welcomed very much. It lays the foundations for further success.

The Symposium could be only successful because of the excellent organization. We express our deep gratitude to the German Academy of Sciences in Berlin and its President Prof. Dr., Dr. honoris causa W. Hartke. With a feeling of admiration and gratitude we shall always remember the ever-lasting unselfish work of Prof. Dr. Meißer, the Chairman of the Organizing Committee. Our best thanks to all the collaborators of the Organizing Committee, too, and to the translators for their hard and extremely important work.

I very heartily thank all the participants in this symposium who represent the different countries and continents and I hope that we all shall often meet again.

TABLE OF CONTENTS

LECTURES

P r o b l e m s   o f   p r a c t i c a l   i n t e r e s t

1st International Symposium on Recent Crustal Movements

Leipzig, GDR, 21st to 26th May, 1962

At the invitation of the German Academy of Sciences in Berlin and under the honorary chairmanship of its president Prof. Dr., Dr., h.c., W. Hartke a symposium was held in Leipzig from 21st to 26th May 1962 in order to discuss the problems of the recent movements of the earth's crust on a wider basis and to develop a suitable programme for further advancement. The organization had been carried out by a committee consisting of Prof. O. Meißer (Chairman), Prof. J.D. Boulanger, Dr. J.A. Mescherikov and Dr. S. Miyamura. Ninety-four scientists from eighteen different countries attended; a list of their names and adresses is given below. This was the first symposium of the Commission on Recent Crustal Movements.

After an informal gathering of participants who had arrived by Monday, May 21st evening, the opening session was held on Tuesday, May 22nd in the festival hall of the Old Town Hall at Leipzig. The chairmen of the opening session were: O. Meißer, J. D. Boulanger, J. J. Levallois, T. Sorgenfrei, St. Krynski, J. A. Mescherikov, S. Miyamura, R. Tomaschek. The scientists were welcomend by Prof. Dr. phil. O. Meißer, Sekretär der Klasse für Bergbau, Hüttenwesen und Montangeologie der Deutschen Akademie der Wissenschaften zu Berlin, as the representative of the president of the German Academy of Sciences, Prof.Dr. Dr. h.c. W. Hartke. Prof. J. D. Boulanger replied on behalf of the International Association of Geodesy, of which he is a vice-president. Dr. J. A. Mescherikov, the president of the Commission on Recent Crustal Movements then outlined, after his welcoming speech, the fundamental problems of recent movements of the earth's crust.

6 scientific conferences were held in the Parkgaststätte Leipzig-Markkleeberg with the following lecture programme. All lectures and discussions were simultanously interpreted in English, French, German and Russian.

Lecture Programme

Tuesday, May 22nd

1st scientific conference with the following chairmen: J. A. Mescherikov, T. J. Kukkamäki, L. Asplund

General questions and the task of the Commission on Recent Crustal Movements

Boulanger, J.D.; Mescherikov, J.A. (Moscow):
Some additions to the program of the international researches of the crustal movements

Kukkamäki, T. J. (Helsinki):
The task of the Commission on Recent Crustal Movements

Geodetic problems concerning the recent movements of the earth's crust

A. World-wide problems and methodical problems

Zakatov, P.S. (Moscow); read by Kondrashkov (Moscow):
On the application of the method of relevelling in a study of recent vertical movements of the earth's crust

Wyrzykowski, T. (Warsaw):
Problemy proektirovanija setej povtornogo precizionogo nivelirovanija dla issledovanija sovremennych vertikal'nych dvizenij zemnoj kory
(Probleme der Entwurfsbearbeitung von Netzen wiederholter Präzisionsnivellements für die Bedürfnisse von Untersuchungen gegenwärtiger Vertikalbewegungen der Erdkruste)

Wyrzykowski, T. (Warsaw):
The methods of geodetic determination of the recent vertical movements of the earth's crust

Hermanowski, A. (Warsaw):
Stricte méthode de compensation des déplacements verticaux de l'écorce terrestre déterminés par les mesurages périodiques de nivellement de précision

Böhm, J. (Prague):
Das Vertrauensintervall bei den gemessenen Erdkrustenbewegungen

Hradilek, L. (Prague):
Methode zur Bestimmung der rezenten Bewegungen der Erkruste in Hochgebirgsgebieten

Izotov, A. A. (Moscow):
On the geodetic methods of studying the movements of the earth's crust

Setunskaya, L.E.; Finko, H.A.; Martynova, Z.I. (Moscow):
Some results of field geological inspection of precise levelling lines

B. Regional and local problems

Pevnev, A.A. (Moscow):
Vertical crustal movements of the Baskunchak Salt Dome

Kazanchian, P.P. (Yerewan):
Some results of the study of vertical movements in Armenia

Homoródi, L. (Budapest):
Untersuchungen der rezenten Erdkrustenbewegungen in Ungarn

Kruis, B. (Prague):
Erforschung vertikaler Erdkrustenbewegungen in der Tschechoslowakei

Wednesday, May 23rd

2nd scientific conference with the following chairmen: J. D. Boulanger, O. Simonsen, G. W. Lennon, L. Cahierre.

B. Regional and local problems (continued)

Toperczer, M. (Vienna):
Über die Möglichkeit, in Österreich rezente Erdkrustenbewegungen aus dem Präzisions-

nivellement abzuleiten

Niewiarowski, I. (Warsaw):

Entwurf eines Netzes für wiederholte Präzisionsnivellements für Untersuchungen vertikaler Erdkrustenbewegungen im polnischen Raum

Niewiarowski, I.; Wyrzykowski, T. (Warsaw):

Die Bestimmung gegenwärtiger vertikaler Erdkrustenbewegungen im polnischen Raum durch Vergleich von Ergebnissen wiederholter Präzisionsnivellements

Simonsen, O. (Copenhagen):

Remarks on secular movements within Denmark

Gudelis, V.K. (Vilnius); read by Lesis, J.P. (Vilnius):

Results of investigations of the latest and recent vertical displacements of the earth's crust in the Baltic Region and some problems of methodical character

Zhelnin, G.A. (Tartu); read by Mescherikov (Moscow):

Some results on neotectonic movements of the earth's crust in the Estonian SSR obtained by the method of repeated levellings

Waterlevel observations concerning the recent movements of the earth's crust

Weise, H. (Potsdam):

Ein Beitrag zur wahrscheinlichen Senkung der deutschen Ostseeküste, abgeleitet aus 100jährigen Wasserstandsbeobachtungen

Ionin, A.S.; Kaplin, P.A.; Medvedev, V.S. (Moscow); read by Kazmin, J.B. (Moscow):

On the methods of determining the present day tectonic movements of the seacoast

Miehlke, O. (Warnemünde):

Zur Methodik der Berechnung des statischen Luftdruckeffektes auf den Wasserstand abgeschlossener Meeresbecken

Siniagina, M.I. (Moscow); read by Shokin (Moscow):

Some researches on the velocity of eustatic changes of the world ocean level

Lennon, G.W. (Birkenhead/Cheshire):

Tidal observations as a measure of movements of the earth's crust

Wednesday, May 23rd

3rd scientific conference with the following chairmen: O. Meißer, E. Nishimura, T. Sorgenfrei, R. Tomaschek

Earth tide instruments for the study of recent crustal movements

Tomaschek, R. (Breitbrunn):

Crustal movements as revealed by measurements with horizontal pendulum and gravimeter in the course of earthtide

Boulanger, J.D. (Moscow):

On secular gravity changes

Ostrovsky, A.Ye.; Bakhrushin, A.B.; Mironova, L.I. (Moscow); read by Latyniny, L.A. (Moscow):

The test of tectonic deformation measurements with the tiltmeter in Kondara (near Dushanbe) 1957-1959

R e c e n t m o v e m e n t s o f t h e e a r t h ' s c r u s t r e l a t e d w i t h e a r t h q u a k e s

Nishimura, E. (Kyoto):
Anomalous tilting movement of the ground observed before destructive earthquakes

Miyamura, S. (Tokyo):
Types of crustal movements accompanied with earthquakes

Sugimura, A. (Tokyo); read by Miyamura (Tokyo):
Two examples of recent and pleistocene crustal movements in Japan

Pavoni, M. (Zürich):
Rotierende Felder in der Erdkruste

Zátopek, A. (Prague):
Junge Erdkrustenbewegungen und die Erdbebentätigkeit auf dem Gebiet der Tschechoslowakei

Tàrczy-Hornoch, A. (Sopron):
Über die Epizentrumberechnung am Erdellipsoid und Rückschlüsse auf die Fortpflanzungsgeschwindigkeiten

Kraus, E.C. (Munich); read by Hiersemann, L. (Freiberg):
Über Bewegungstendenzen in der Erdrinde. Ein Beitrag zur Seismotektonik

G e o l o g i c a l a n d g e o m o r p h o l o g i c a l p r o b l e m s c o n c e r n i n g t h e r e c e n t m o v e m e n t s o f t h e e a r t h ' s c r u s t

A. General problems

Mescherikov, J.A. (Moscow):
Secular crustal movements: some results and tasks of researches

Hristov, K.; Galabov, J. (Sofia):
Mitteilung über eine vorläufige Untersuchung über die neuen vertikalen Bewegungen der Erdkruste in Bulgarien

Setunskaja, L.E. (Moscow):
Ein Versuch der Analyse von Längsprofilen der Flüsse zwecks Untersuchung der rezenten Erdkrustenbewegungen

Friday, May 25th

4th scientific conference with the following chairmen: J. Galabov, T. Sorgenfrei, St. Pawlowski

B. Regional problems

Gerasimov, I.P. (Moscow); read by Marinich, A. (Kiev):
The study of recent tectonic movements of the earth's crust in the territory of the USSR

Gorelov, S.K.; Martynova, Z.I.; Matskova, V.A. (Moscow):
Study of recent crustal movements in the region of repeated precise levellings in the Ternopol-Kursk area

Rozhdestvensky, A.P.; Zhourenko, J.M. (Ufa):
Recent vertical movements of the earth's crust in the Volga-Ural region and

methods of their study

Finko, E.A.; Ranrsman, E.I. (Moscow); read by Rastvorova, V. (Moscow):
Geomorphological interpretation of the data provided by repeated levelling in the low-mountain areas of Kazakhstan (Arys-Alma Ata-Semipalatinsk)

Doumitrashko, N.V.; Lilienberg, D.A. (Moscow):
Tectonique actuelle du Caucase

Sokolovsky, I.L.(Kiev); read by Aristarkhova, L. (Moscow):
On the inheritance of the contemporary movements of the earth's crust on the territory of the Ukrainian SSR from older movements

Grave, M.K.; Evzerov, V.J. (Moscow):
Latest and present tectonic movements in the central part of the Kola Peninsula

Rosa, B. (Torun); read by Mojski, J.E. (Warsaw):
Über die Bedeutung der Neotektonik in der morphologischen Entwicklung der polnischen Küste in der Nachlitorinazeit

Jaranoff, D. (Sofia):
Structure tectonique et mouvements récents de la croûte terrestre en Bulgarie

Pécsi, M. (Budapest):
Ausmaß der holozänen Krustenbewegungen in Ungarn

Vogt, H. (Strasbourg):
Quartäre Tektonik am Ostabfall der Vogesen: Der Stand der Forschung

Dubourdieu, G. (Paris); read by Destombes, J.P. (Paris):
Le rapprochement de l'Europe et de l'Afrique depuis le miocène supérieur

Friday, May 25th

5th scientific conference with the following chairmen: J.D. Boulanger, St. Krynski, K. Neubert

Recent movements of the earth's crust and continental drift in connection with the deep structure

Stoyko, A. (Paris); read by Cahierre, L. (Paris):
Le Service International de l'Heure et le déplacement des continents

Stoyko, N. (Paris); read by Cahierre, L. (Paris):
La variation des latitudes et leur relation avec le mouvement de l'écorce terrestre

Scheglov, V.P. (Tashkent):
Continental drift in the light of modern astronomical data and prospects of further researches in this field

Vening-Meinesz, F.A. (Amersfoort); read by Kukkamäki, T.J. (Helsinki):
The importance of determining the relative horizontal movements in continental and island-arc areas

Lustikh, E.N.; Magnitsky, V.A. (Moscow):
Vertical movements of the earth's crust, temporal changes of gravity field and shift of subcrustal material

Egyed, L. (Budapest):
Some geophysical problems in connection with recent and past crustal movements

Barta, G. (Budapest):
Recommendation for detecting and surveying the secular variation of the earth's gravity field (The connection between the triaxiality of the earth and the eccentricity of the geomagnetic field)

Fanselau, G. (Potsdam):
Geomagnetische Information über die Bewegung der Erdkruste

P r o b l e m s   o f   p r a c t i c a l   i n t e r e s t

Gough, D.I.; Gough, W.I. (Salisbury):
Geophysical investigations at Lake Kariba

Kowalczyk, Z. (Cracow):
Scheme of a complex investigation method of tectonic movements with special regard to mining areas

Neubert, K. (Freiberg):
Bewegungsvorgänge der Erdkruste als Folge bergbaulicher Einwirkungen

Niewiarowski, I. (Warsaw):
Vertikalbewegungen längs hauptsächlicher Linien des Präzisionsnivellements im Górnóśląskie Zagłebie (Oberschlesisches Kohlenbecken) im Zeitraum 1955 bis 1958

Broggi, J.A. (Lima):
About the measurements of active faults

Broggi, J.A. (Lima):
Condiciones del pronóstico sísmico para que sea eficaz

Lauterbach, R. (Leipzig):
Submeridionale Tiefenbrüche Mitteleuropas als Gebiete fossiler wie rezenter Krustenbewegungen

Radecki, J. (Warsaw):
Über einige Grundsätze der Bestimmung von Unterschieden geographischer Längen

Schoeneich, K. (Warsaw):
Očerk po neotektonike pol'ši (Preliminary outline of neotectonics in Poland)

Svoboda, K. (Prague):
Zur Frage der Erdkrustenbewegungen in dem karpatischen Teil der ČSSR

Wyrzykowski, T. (Warsaw):
Influence des mouvements systématiques verticaux de l'écorce terrestre sur les résultats et la détermination des grandeurs d'erreurs de nivellements de précision

Saturday, May 26th

6th scientific conference with the following chairmen: J.A. Mescherikov, S. Miyamura, T. Sorgenfrei, St. Krynski, J.D. Boulanger

Summarizing discussions, final remarks of the Vicepresident of the IAG, Prof. J.D. Boulanger, final remarks of the president of the Commission on Recent Crustal Movements, Dr. J.A. Mescherikov.

The symposium was attended by scientists from the following 18 countries: Austria, Bulgaria, Czechoslowakia, Denmark, England, Federal Republic of Germany, Finland, France, German Democratic Republic, Hungary, Italy, Japan, Poland, Rhodesia, Romania, Switzerland and USSR.

The symposium was dedicated to the centenary of the foundation of the "Internationale Erdmessung (International Association of Geodesy) in Germany in 1862.

R e s o l u t i o n

The following resolutions were adopted:

A c k n o w l e d g e m e n t s

I. The participants in the symposium express their deep gratitude to the German Academy of Sciences in Berlin and to its President, Professor Dr.Dr. h.c. W. Hartke, the Honorary President of the International Symposium on Recent Movements of the Earth's Crust taking place in Leipzig,

and to the Committee of Geodesy and Geophysics in the German Democratic Republic and its President Professor Dr. Philipps for having been given the opportunity of meeting in Leipzig to discuss and to debate present day scientific problems.

II. The participants in the symposium wish to express their gratitude and great appreciation to the Organinzing Committee of the German Academy of Sciences in Berlin for all it had done for the benefit of the Symposium on Recent Movements of the Earth's Crust in Leipzig.

The participants in the Symposium sincerely thank the Chairman of the Organizing Committee Professor Dr. Meißer and all his collaborators in this organizing committee for the admirable preparations which they have made for the work of the Symposium and at the same time they express their gratitude for the friendly and hearty welcome.

III. a) The participants in the Symposium are deeply satisfied with the creation of a permanent Commission on Recent Crustal Movements in the framework of the IUGG (International Union of Geodesy and Geophysics). They express their acknowledgement to the Couneil of the International Association of Geodesy for having supported the activity of this commission in establishing international cooperation of scientists in the field of neotectonics.

The participants in the Symposium appreciate the contacts, which have now been initiated with the International Union of Geological Sciences, since efforts of geodesists, geophysicists, and geologists are necessary for the solution of the problems of recent crustal movements.

b) The participants in the Symposium express their appreciation to the special study group No. 22, and to the Permanent Service for Mean Sea Level for the work already carried out and emphasize its importance at present and for the future. In particular the Symposium gives its support to the Permanent Service for Mean Sea Level in its efforts to increase the world coverage of tidal stations.

T h e I n t e r n a t i o n a l R e s e a r c h P r o g r a m m e

IV. The participants in the symposium emphasize the scientific and practical significance of research on present tectonic movements and the prospects of work along these lines.

They support the project of international cooperation in research on recent movements of the earth's crust which was established by Professor J.D. Boulanger and Dr. J.A. Mescherikov and which includes:

a) The creation of maps of recent crustal movements by means of geodetic, oceanographic, geophysical, geological, and geomorphological data. Cooperation with the International Geographical Union in the field of geomorphology is envisaged.

b) The foundation of international cooperation for the observation of crustal movements (creation of special local polygons for the study of recent vertical and horizontal movements of the earth's crust).

c) The creation of a close cooperation between the International Union of Geodesy and Geophysics (IUGG) and the International Astronomical Union (IAU) for the study of continental drift in member countries.

It is recommended that working groups for the study of recent crustal movements should be organized by the National Committees of the International Union of Geodesy and Geophysics (IUGG) in order to facilitate collaboration among scientists concerned in this field.

V. The International Commission on Recent Crustal Movements (President J.A. Mescherikov and the Secretary S. Miyamura) is requested to send the project approved by the present Symposium regarding international research on recent crustal movements incorporating the results of this Symposium) to the XIIIth General Assembly of the IUGG in 1963 in order to have this project, together with the proposals for its realization, approved by the Assembly.

VI. In order to systematize and coordinate international research according to the above programme, it is proposed to form special study groups for research on recent crustal movements of different continents and regions within the IAG.

It is proposed to start with the formation of special study groups for

a) a map of recent crustal movements in Eastern Europe

b) a map of recent crustal movements in Western Europe

c) recent crustal movements in the Pacific Area

d) horizontal movements of the crust (continental drift)

e) local polygons for the study of recent crustal movements and the standardisation of observations

and to submit proposals for the membership of these study groups.

VII. It is desirable that the methods introduced by earth tide measurements and long period seismic observations, which give entirely new possibilities of ascertaining proper movements of the earth's crust, be improved by increasing both the density of stations and, so far as possible, the number of instruments at each station.

VIII. The participants in the Symposium note with satisfaction the preparation of an extensive bibliography undertaken in Japan by the Japanese National Committee for the IUGG in the field of crustal movements, under the direction of Dr. Miyamura, and request all member countries of the IUGG to carry out such work in the future - they ask the Commission of the International Geodetic Bibliography, IAG, to include this field within its scope.

Exchange of Research Material

IX. Considering that great attention should be paid to the regular international excahnge of materials and data on recent crustal movements, this symposium believes it necessary:

- to use for this purpose the World Data Centres A and B.
- to put "Recent Crustal Movement" into the register of the materials available for exchange in conformity with the proposals by the International Geophysical Committee.

P u b l i c a t i o n   o f   M a t e r i a l s

X. The participants in the Symposium express their gratitude to the German Academy of Sciences in Berlin for preparing the publication of the material of this Symposium for presentation at the XIIIth General Assembly of the IUGG.

They request the following colleagues to form the editorial board, namely:

Prof. Meißer (chairman)
Prof. Boulanger
Dr. Mescherikov
Dr. Miyamura
Prof. Peschel
Prof. Philipps
Prof. Reicheneder

N e x t   S y m p o s i u m

XI. The Commission requests the I.A.G. to sponsor another symposium in 1965 on the same subject, and to take the necessary steps with the IUGG for its preparation.

Р Е З О Л Ю Ц И Я

Следующие резолюции были приняты:

БЛАГОДАРНОСТИ

I. Участники симпозиума выражают свою глубокую признательность

Немецкой Академии наук в Берлине и ее президенту, профессору-доктору, почетному доктору В. Хартке, почетному председателю международного симпозиума по современным движениям земной коры в Лейпциге,

Комитету по геодезии и геофизике Германской Демократической Республики и его президенту профессору-доктору Филиппсу за предоставленную им возможность встретиться в Лейпциге для обсуждения актуальных научных проблем.

II. Участники симпозиума выражают свою самую искреннюю благодарность Организационному комитету Симпозиума по современным движениям земной коры при Немецкой Академии наук в Берлине. Участники симпозиума сердечно благодарят председателя Организационного комитета профессора-доктора О. Майссера и всех его сотрудников по Организационному комитету за прекрасные условия для работы симпозиума, за дружеский, теплый прием.

III. а) Участники симпозиума выражают глубокое удовлетворение созданием постоянной Комиссии по современным движениям земной коры в рамках Международного геодезического и геофизического союза. Они выражают свою признательность руководству Международной ассоциации геодезии за поддержку начинаний Комиссии в налаживании международного сотрудничества ученых в области неотектоники.

Участники симпозиума приветствуют налаживающийся контакт с Международным геологическим союзом, поскольку решение комплексной проблемы современных движений земной коры требует совместных усилий геодезистов, геофизиков и геологов.

б) Симпозиум выражает признательность Специальной исследовательской группе № 22 МАГ и Постоянной службе среднего уровня моря за выполненную уже работу и выражает надежду, что эта важная работа будет продолжена. Симпозиум особенно поддерживает усилия Постоянной службы среднего уровня моря, направленные к расширению мировой сети уровнемерных станций.

МЕЖДУНАРОДНАЯ ПРОГРАММА ИССЛЕДОВАНИЙ

IУ. Участники симпозиума отмечают важное научное и практическое значение и перспективность исследований современных тектонических движений.

Симпозиум поддерживает проект международного сотрудничества в области изучения современных движений земной коры, разработанный Ю. Д. Буланже и Ю. А. Мещеряковым и предусматривающий

а) Составление сводных карт современных движений земной коры на основе геодезических, океанографических, геофизических, геологических и геоморфологических данных, в связи с чем весьма желательной представляется кооперация с Международным географическим союзом в области геоморфологии;

б) Международное сотрудничество в организации наблюдений над ходом движений земной коры (создание специальных полигонов для изучения местных вертикальных и горизонтальных движений земной коры);

в) Тесное сотрудничество Международного геодезического и геофизического союза с Международным астрономическим союзом в деле изучения больших горизонтальных перемещений материков.

В целях усиления сотрудничества ученых разных специальностей в исследовании современных движений земной коры участники симпозиума рекомендуют создавать специальные рабочие группы при Национальных комитетах стран-участниц Международного геодезического и геофизического союза.

У. Международной комиссии по современным движениям земной коры (президент - Ю. А. Мещеряков, секретарь - С. Миямура) поручается представить XIII Генеральной Ассамблее МГГС в 1963 г. одобренный настоящим симпозиумом проект международных исследований современных движений земной коры (с учетом результатов его обсуждения на данном симпозиуме) для окончательного утверждения проекта и вынесения рекомендаций по его осуществлению.

УI. Для общего методического руководства и координации международных исследований по указанной выше программе представляется необходимым создать в рамках МАГ специальные исследовательские группы по изучению современных движений земной коры различных территорий. Симпозиум предлагает начать с образования следующих рабочих групп:

а) Группа по составлению карты современных движений земной коры Восточной Европы;
б) Группа по составлению карты современных движений земной коры Западной Европы;
в) Группа по изучению современных движений земной коры Тихоокеанской области;
г) Группа по изучению горизонтальных перемещений материков;
д) Группа по полигонам для изучения современных движений земной коры и унификации методов наблюдений.

Представляется необходимым в ближайшее время подготовить предложения о составе этих исследовательских групп.

УII. Желательно, чтобы методы, применяемые при изучении приливов в твердой Земле и при сейсмологических исследованиях, которые открывают совершенно новые возможности при изучении современных движений земной коры, нашли более широкое применение, для чего необходимо расширение сети существующих станций, а также количества приборов на каждой станции.

УIII. Участники симпозиума с удовлетворением отмечают опыт составления сводной библиографии по современным движениям земной коры, охватывающей все работы, выполненные в Японии, подготовленной к настоящему симпозиуму д-ром Миямурой и опубликованной Национальным комитетом МГГС Японии. Они обращаются ко всем странам-участницам МГГС с просьбой проводить аналогичную работу. Участники симпозиума просят Комиссию по международной геодезической библиографии включить в свое рассмотрение современные движения земной коры.

ОБМЕН МАТЕРИАЛАМИ ИССЛЕДОВАНИЙ

IX. Придавая большое значение регулярному международному обмену материалами по изучению современных движений земной коры, участники симпозиума признают необходимым:

использовать для этой цели мировые центры данных А и Б;

включить раздел "Современные движения земной коры" в список материалов, подлежащих обмену, в соответствии с рекомендациями Международного геофизического комитета.

ПУБЛИКАЦИЯ МАТЕРИАЛОВ

X. Участники симпозиума с благодарностью отмечают решение Немецкой Академии наук в Берлине о публикации материалов настоящего симпозиума в виде отдельного тома, подготовливаемого к XII Генеральной Ассамблее МГГС.

Участники симпозиума просят взять на себя труд по подготовке указанного тома редакционную коллегию в следующем составе:

проф. Майссер (председатель);
проф. Буланже;
д-р Мещеряков;
д-р Миямура;
проф. Пешель;
проф. Филиппс;
проф. Райхенедер

СЛЕДУЮЩИЙ СИМПОЗИУМ

XI. Просить Международную ассоциацию геодезии провести следующий симпозиум по современным движениям земной коры в 1965 г. и принять необходимые меры для подготовки его в рамках МГГС.

R é s o l u t i o n

I. Les participants du Symposium expriment leur profonde reconnaissance:

A l'Académie des Sciences de Berlin et à son Président, le Professeur Dr.Dr. W. Hartke, Président honoraire de Symposium International sur les mouvements récents de la croûte terrestre, tenu à Leipzig.

Au Comité de Géodésie et Géophysique de la République Démocratique Allemande ainsi qu'à son Président le Prof. Dr. Philipps grâce à qui ils ont eu la possibilité de se réunir à Leipzig pour discuter et débattre les problèmes scientifiques actuels.

II. Les participants du Symposium expriment leur gratitude au Comité de l'Académie des Sciences de Berlin chargé de l'organisation du Symposium sur les mouvements récents de la croûte terrestre à Leipzig.

Ils remercient cordialement le Président du Comité d'organisation, le Prof. Dr. Meißer et tous ses collaborateurs pour leur remarquable préparation des travaux du Symposium et expriment leur gratitude pour leur accueil si amical et si cordial.

III. a) Les participants du Symposium expriment leur satisfaction de la création d'une commission permanente sur les mouvements récents de la croûte terrestre au sein de l'Union Géodésique et Géophysique Internationale. Ils remercient la Présidence de l'Association Internationale de Géodésie d'avoir donné son appui aux activités de cette commission en établissant une coopération internationale des savants dans le domaine de la néo-tectonique.

Ils apprécient les contacts qui ont été pris avec l'Union Internationale des Sciences Géologique, étant donné que les efforts des Géodésiens et Géophysiciens et des Géologues sont nécessaires pour résoudre les problèms complexes des mouvements récents de la croûte.

b) Le Symposium exprime ses remerciements au Groupe Spécial d'Etudes n° 22 ainsi qu'au Service Permanent du niveau moyen des mers pour le travail déjà accompli dont il souligne l'importance actuelle et future. En particulier, le Symposium donne tout son appui au Service Permanent du niveau moyen des mers dans se ses efforts pour augmenter dans le monde entier la densité des marégraphes.

P r o g r a m m e   i n t e r n a t i o n a l   d e   r e c h e r c h e s

IV. Les participants du Symposium soulignent l'importance scientifique et pratique de la recherche des mouvements tectoniques actuels ainsi que les perspectives d'avenir du travail dans cette direction.

Ils approuvent le projet de coopération internationale établi par le Prof. J.D. Boulanger et le Dr. J.A. Mescherikov dans le domaine de recherches des mouvments récents de l'écorce terrestre qui comporte:

a) L'établissement de cartes des mouvements récents de la croûte tirées de documents géodésiques, océanographiques, géophysiques, et géomorphologiques. On compte amorcer une coopération avec l'Union Géographique Internationale dans le domaine de la géomorphologie.

b) l'établissement d'une coopération internationale pour l'observation des mouvements de la croûte (création de polygones locaux pour l'étude des mouvements récents de la croûte dans le sens vertical et horizontal).

c) la création d'une coopération étroite entre l'Union Internationale de Géodésie et de Géophysique et l'Union Astronomique Internationale pour l'étude de la dérive des continents parmi les pays membres de ces Unions.

d) l'organisation de groupesde travail pour l'étude des mouvements récents de la croûte au sein des Comités Nationaux de l'U.G.G.I. dans le but de faciliter la collaboration parmi les savants qui s'occupent de la question des mouvements récents de la croûte.

V. La Commission Internationale sur les mouvements récents de la croûte terrestre (Président J.A. Mescherikov, Secrétaire S. Miyamura) est priée d'expédier le projet de recherches internationales approuvé par la Symposium (Compte-tenu des résultats de ce Symposium) à la 13e Assemblée Générale de l'U.G.G.I. en 1963 pour y faire approuver les propositions visant à le réaliser.

VI. En vue de systématiser et de coordonner la recherche internationale conformément au programme ci-dessus, il est proposé de former des groupes spéciaux d'études au sein de l'Association Internationale de Géodésie, pour la recherche des mouvements récents de la croûte.

On propose la création des groupes suivants:

a) carte des mouvements récents de la croûte dans l'Europe de l'Est

b) carte des mouvements récents de la croûte dans l'Europe de l'Ouest

c) carte des mouvements récents de la croûte dans la zone du Pacifique

d) mouvements horizontaux de la croûte (dérive des continents)

e) polygones locaux pour l'étude des mouvements récents et uniformisation des observations.

On propose également de préparer des propositions pour désigner les membres de ces groupes d'études.

VII. Il est désirable que les méthodes introduites dans les mesures des marées terrestres et les observations des longues périodes séismiques, qui fournissent des possibilités entièrement nouvelles pour déterminer les mouvements propres de la croûte terrestre, soient améliorées en augmentant la densité des stations ainsi que, dans toute la mesure du possible, le nombre des instruments de chaque station.

VIII. Les participants du Symposium notent avec satisfaction le préparation d'une Bibliographie complète couvrant le domaine des mouvements de la croûte, entreprise pour le compte de l'U.G.G.I., par le Comité National Japonais de Géodésie et Géophysique sous la direction du Dr. Miyamura, et prient tous les pays membres de l'U.G.G.I. de procéder désormais à de tels travaux. Ils demandent à la Commission Géodésique Internationale de Bibliographie de l'.A.I.G. d'inclure ce programme dans son domaine.

E c h a n g e   d e   m a t é r i e l   d e   r e c h e r c h e

IX. Considérant que la plus grande attention doit être attachée à un échange régulier de la documentation et des résultats des travaux sur les mouvements récents de la croûte, le Symposium estime nécessaire:

- d'utiliser dans ce but les Centres mondiaux de données A et B
- d'inclure les données relatives aux mouvements récents parmi celles qui peuvent figurer dans les échanges, conformément aux propositions faites par le Comité Internaional de Géophysique.

P u b l i c a t i o n   d e s   r é s u l t a t s

X. Les participants du Symposium expriment leur reconnaissance à l'Académie des Sciences de Berlin pour la préparation de la publication de ce Symposium pour le présenter à la 13e Assemblée Générale de l'U.G.G.I. Ils prient

Mr. le Prof. Meißer (Président)
Mr. le Prof. Boulanger
Mr. le Dr. Mescherikov
Mr. le Dr. Miyamura
Mr. le Prof. Peschel
Mr. le Prof. Philipps
Mr. le Prof. Reicheneder

de bien vouloir partie du Comité d'Edition.

P r o c h a i n   S y m p o s i u m

XI. La Commission demande à l'Association Internationale de Géodésie de patronner en 1965 un autre Symposium sur le même sujet et d'entreprendre auprès de l'U.G.G.I. les démarches que nécessitent sa préparation.

R e s o l u t i o n

Die folgende Resolution ist angenommen worden:

A n e r k e n n u n g e n

I. Die Teilnehmer am Symposium drücken ihre tiefe Dankbarkeit aus der Deutschen Akademie der Wissenschaften zu Berlin und ihrem Präsidenten Dr.Dr. h.c. W. Hartke, dem Ehrenpräsidenten des internationalen Symposiumsüber rezenter Erdkrustenbewegungen in Leipzig,

und dem Komitee für Geodäsie und Geophysik in der Deutschen Demokratischen Republik und seinem Präsidenten Prof.Dr. Philipps, dafür, daß die Möglichkeit geschaffen wurde, in Leipzig zu tagen und aktuelle wissenschaftliche Probleme zu diskutieren.

II. Die Teilnehmer an dem Symposium wünschen, ihre Dankbarkeit und volle Anerkennung dem Organisationskomitee für das Symposium über rezente Erdkrustenbewegungen in Leipzig bei der Deutschen Akademie der Wissenschaften in Berlin auszudrücken. Die Teilnehmer an dem Symposium danken herzlich dem Vorsitzenden des Organisationskomitees, Prof.Dr. Meißer, und allen seinen Mitarbeitern in diesem Organisationskomitee für die ausgezeichneten Vorbereitungen, die für die Arbeit des Symposiums geschaffen wurden und gleichzeitig bringen sie ihre Dankbarkeit für die freundliche und herzliche Aufnahme zum Ausdruck.

III. a) Die Teilnehmer am Symposium sind tief befriedigt über die Bildung einer permanenten Kommission für rezente Erdkrustenbewegungen innerhalb der IUGG (Internationalen Union für Geodäsie und Geophysik). Sie drücken der Leitung der Internationalen Assoziation für Geodäsie ihre Anerkennung dafür aus, daß sie die Aktivität dieser Kommission bei der Schaffung einer internationalen Zusammenarbeit der Wissenschaftler auf dem Gebiet der Neotektonik unterstützt hat. Mit großer Befriedigung vermerkten die Teilnehmer des Symposiums die Kontakte, die sich jetzt mit der Internationalen Union der Geologischen Wissenschaften angebahnt haben, da die Anstrengungen von Geodäten, Geophysikern und Geologen für die Lösung der Probleme rezenter Krustenbewegungen notwendig sind.

b) Die Teilnehmer an dem Symposium drücken ihre Dankbarkeit der speziellen Studiengruppe Nr. 22 - und dem Permanent Service for Mean Sea Level für die schon geleistete Arbeit aus und unterstreichen die Bedeutung dieser Arbeit jetzt und für die Zukunft. Besonders gibt das Symposium seine Unterstützung dem Permanent Service for Mean Sea Level und seinen Anstrengungen, die Verteilung von Gezeitenstationen auf der Welt zu verdichten.

I n t e r n a t i o n a l e   F o r s c h u n g s p r o g r a m m e

IV. Die Teilnehmer am Symposium betonen die wissenschaftliche und praktische Bedeutung der Erforschung der rezenten tektonischen Bewegungen und die Perspektive der Arbeit in dieser Richtung.

Sie unterstützen das Projekt einer internationalen Zusammenarbeit bei der Erforschung der rezenten Erdkrustenbewegungen, das von Professor J.D. Boulanger und Dr. J.A.

Mescherikov erarbeitet wurde und folgendes vorsieht:

a) die Schaffung von Karten rezenter Erdkrustenbewegungen mit Hilfe geodätischer, ozeanographischer, geophysikalischer, geologischer und geomorphologischer Unterlagen. Die Zusammenarbeit mit der Internationalen Geographischen Union auf dem Gebiete der Geomorphologie wird vorgesehen.

b) die Schaffung einer internationalen Zusammenarbeit für die Beobachtungen von Krustenbewegungen (Schaffung spezieller lokaler Polygonzüge für die Untersuchung vertikaler und horizontaler rezenter Erdkrustenbewegungen).

c) die Schaffung einer engen Zusammenarbeit zwischen der Internationalen Union für Geodäsie und Geophysik (IUGG) und der Internationalen Astronomischen Union (IAU) für die Untersuchung der Kontinentalverschiebung in den Mitgliedländern.

Es wird empfohlen, Arbeitsgruppen für das Studium der rezenten Krustenbewegungen in den Nationalen Komitees der Internationalen Union für Geodäsie und Geophysik (IUGG) zu organisieren, um die Zusammenarbeit der Wissenschaftler, die auf dem Gebiet der rezenten Krustenbewegungen arbeiten, zu erleichtern.

V. Die Internationale Kommission für rezente Erdkrustenbewegungen (Präsident J.A. Mescherikov, Sekretär S. Miyamura) wird gebeten, das von dem gegenwärtigen Symposium angenommene Projekt über die internationalen Forschungen auf dem Gebiet der rezenten Erdkrustenbewegungen (unter Berücksichtigung der Ergebnisse dieses Symposiums) an die 13. Generalversammlung der IUGG im Jahre 1963 zu senden, um dieses Projekt mit den Vorschlägen zu seiner Realisierung durch die Generalversammlung billigen zu lassen.

VI. Um die internationale Forschung entsprechend dem obigen Programm zu systematisieren und zu koordinieren, wird vorgeschlagen, spezielle Studiengruppen für die Erforschung der rezenten Krustenbewegungen verschiedener Kontinente und Regionen innerhalb der IAG zu bilden.

Es wird vorgeschlagen, zu beginnen mit der Bildung spezieller Studiengruppen für

a) eine Karte rezenter Krustenbewegungen in Ost-Europa

b) einer Karte rezenter Krustenbewegungen in West-Europa

c) die rezenten Krustenbewegungen im pazifischen Gebiet

d) die horizontalen Krustenbewegungen (Kontinentalverschiebung)

e) die lokalen Polygonzüge für die Untersuchung rezenter Krustenbewegungen und die Standardisierung der Beobachtungen

und Vorschläge für die Mitgliedschaft dieser Studiengruppen zu unterbreiten.

VII. Es ist wünschenswert, daß die durch die Erdgezeitenmessungen und Beobachtungen langperiodischer Erdbebenwellen eingeführten Methoden, die völlig neue Möglichkeiten zur Feststellung von Eigenbewegungen der Erdkruste liefern, sowohl durch eine Verdichtung des Stationsnetzes als auch wenn möglich durch eine Erhöhung der Anzahl der Instrumente je Station verbessert werden.

VIII. Die Teilnehmer am Symposium stellen mit Befriedigung die Vorbereitung einer umfassenden Bibliographie fest, die in Japan durch das japanische Nationalkomitee für die IUGG auf dem Gebiet der Krustenbewegungen unter der Leitung von Dr. Miyamura getroffen wurde, und leiten sie an alle Mitgliedsländer der IUGG mit der Bitte weiter, solche Arbeiten in Zukunft auszuführen, - sie bitten die Kommission für Internationale Geodätische Bibliographie, IAG, diese Punkte in ihr Arbeitsgebiet aufzunehmen.

Austausch von Forschungsmaterial

IX. Bei Berücksichtigung der Tatsache, daß dem regelmäßigen internationalen Austausch von Materialien und Unterlagen bezüglich der rezenten Krustenbewegungen große Aufmerksamkeit geschenkt werden muß, hält es das Symposium für angebracht,

für diesen Zweck die Weltzentren A und B zum Sammeln wissenschaftlicher Daten zu verwenden,

die "Rezente Krustenbewegung" in das Verzeichnis von Materialien aufzunehmen, die für einen Austausch vorgesehen sind, und dies in Zusammenhang mit den Vorschlägen zu tun, die von dem Geophysikalischen Komitee gemacht wurden.

Veröffentlichung von Materialien

X. Die Teilnehmer am Symposium drücken der Deutschen Akademie der Wissenschaften zu Berlin ihre Dankbarkeit dafür aus, daß sie die Veröffentlichung der Materialien dieses Symposiums zur Vorlage bei der 13. Generalversammlung der IUGG vorbereitet hat.

Sie bitten die folgenden Kollegen, das Redaktionskollegium zu bilden:

Prof. Meißer (Vorsitzender)
Prof. Boulanger
Dr. Mescherikov
Dr. Miyamura
Prof. Peschel
Prof. Philipps
Prof. Reicheneder

Nächstes Symposium

XI. Die Kommission bittet die IAG, ein weiteres Symposium über dasselbe Thema zu fördern und die notwendigen Schritte bei der IUGG für seine Vorbereitung zu unternehmen.

Scientific Excursions

Thursday, May 24th

to

1. the Karl-Schwarzschild-Observatory of the German Academy of Sciences in Berlin at Tautenburg near Jena and the astronomical department of the nationally-owned Carl-Zeiß-Works at Jena.

to

2. the Karl-Schwarzschild-Observatory of the German Academy of Sciences in Berlin at Tautenburg near Jena and the National Places of Worship at Weimar.

to

3. the Institute of Geodesy of the German Academy of Sciences in Berlin at Potsdam and the Park of the Castle of Sanssouci.

L i s t   o f   P a r t i c i p a n t s

A u s t r i a

Prof. Dr. M. Toperczer, Wien XIV, Geophys. Abtlg. an der Zentralanstalt für Meteorologie und Geodynamik, Hohe Warte 38

B u l g a r i a

Prof. J.S. Gâlâbov, Sofia, Geophysical Inst. Acad. of Sci., Benkovski 3

Prof. W. Hristov, Sofia, Acad. of Sci.

Prof. Dr. D. Jaranoff, Sofia, Geological Inst. Acad. of Sci., Skoboleff 4

C z e c h o s l o v a k i a

Prof. Dr. J. Böhm, Praha 1, Nat. Com., Sect. Geodesy, Husova 5

Dr. L. Hradilek, Praha 2, Naturwiss. Fak. Universität, Albertov 6

Dr. A. Kopecky, Praha, Geologisches Zentralamt

Ing. B. Kruis, Praha 1, Forschungsinstitut f. Geodäsie, Topographie, Kartographie, Tr. Politických Vězňu 12

Dr. A. Matejka, Praha, Geologisches Zentralamt

Prof. Dr. A. Zátopek, Praha 2, Inst. f. Geophysik der Karls-Universität, Ke Karlovu 3

D e n m a r k

Dr. O. Simonsen, Copenhagen, Geodetic Inst. of Denmark, Proviantgaarden

Prof. Dr. T. Sorgenfrei, Generalsekretär der Int. Geol. Un., Copenhagen, Ostervoldgade 7

E n g l a n d

Dr. G.W. Lennon, Birkenhead/Cheshire, Liverpool Tidal Inst. and Observatory

Dr. A.R. Robbins, Oxford, Department of Surveying and Geodesy, 10 South Parks Road

F e d e r a l   R e p u b l i c   o f   G e r m a n y

Prof. Dr. R. Tomaschek, Präsident der Permanenten Gezeitenkommission Breitbrunn/Chiemsee

F i n l a n d

Prof. Dr. T.J. Kukkamäki, Helsinki, Geodetic Inst., Bulevardi 40

F r a n c e

Prof. Dr. L. Cahierre, Paris 7e, Inst. Géographique National, 136 bis, Rue de Grenelle

Ing. J.D. Destombes, Paris 15e, Bureau des Rech. géol. et minière, 74, Rue de la Fédération

Ing. J.Levallois, Generalsekretär der AIG, Paris 9e, 19 Rue Auber

Dr. H. Vogt, Strasbourg, Inst. de Géographique, Poste 449

G e r m a n   D e m o c r a t i c   R e p u b l i c

Dr. K. Arnold, Potsdam, Geodätisches Inst., Telegraphenberg

Dr. H. Boddin, Leipzig C 1, VEB Geophysik, Martin-Luther-Ring 13

Prof. Dr. W. Buchheim, Freiberg/Sa., Inst. f. Theor. Physik und Geophysik, Akademiestr. 6

Prof. Dr. G. Fanselau, Potsdam, Geomagnetisches Inst., Telegraphenberg

Prof. Dr. G. Hollweg, Rektor der Bergakademie, Freiberg/Sa., Akademiestr. 6

Prof. Dr. O. Meißer, Freiberg/Sa., Inst. f. Angew. Geophysik, Nonnengasse 35

Dr. O. Miehlke, Warnemünde, Küstenschutzamt, Schillerstr. 10b

Prof. Dr. K. Neubert, Freiberg/Sa., Inst. f. Markscheidewesen und Bergschadenkunde, Prüferstr. 1

Prof. Dr. H. Peschel, Dresden A 27, Geodätisches Inst., George-Bähr-Straße

Prof. Dr. H. Philipps, Präs. d. Meteor. u. Hydrol. Dienstes d. DDR, Potsdam, Alfred-Wegener-Haus

Prof. Dr. K. Reicheneder, Potsdam, Geodätisches Inst., Telegraphenberg

Dipl.-Ing. D. Rotter, Berlin W 8, Otto-Nuschke-Str. 22/23

Prof. Dr. W. Rüger, Freiberg/Sa., Inst. f. Geodäsie, Akademiestr. 6

Prof. Dr. Sandig, Dresden A 27, Geodätisches Inst., George-Bähr-Straße

Dr. habil. W. Sponheuer, Jena, Inst. f. Bodendynamik und Erdbebenforschung, Burgweg 11

Dr. Schliephake, Leipzig C 1, Erich-Weinert-Str. 32

Dr. habil. G. Viete, Freiberg/Sa., Geologisches Inst., Bernhard-v.-Cotta-Str. 2

Prof. Dr. A. Watznauer, Freiberg/Sa., Geol. Inst., Bernhard-v.-Cotta-Str. 2

Dipl.-Ing. H. Weise, Potsdam, Geodätisches Inst., Telegraphenberg

Dipl.-Ing. A. Zappe, Berlin W 8, Mauerstr. 29-32

Prof. Dr. Zill, Dresden A 27, Lehrstuhl f. Vermessungskunde, Mommsenstr. 13

H u n g a r y

Dr. G. Barta, Budapest, Damjanich ut. 28/b

Prof. Dr. L. Egyed, Budapest VIII, Rol. Eötvös Univ., Muzeum-krt. 4/a

Dr. L. Homoródi, Budapest XI, Techn. Univ., Budafoki ut. 3

Dr. M. Pécsi, Budapest VII, Inst. of Geography, Muzeum-krt. 6-8

Prof. Dr.Dr. h.c. A. Tárczy-Hornoch, Sopron, MTA Geodéziai és Geofizikai Kutató Labóratoriumainak igazgatója, Postafiók 9

I t a l y

Prof. Dr. C. Morelli, Bari, Università degli Studi di Bari

J a p a n

Prof. Dr. S. Miyamura, Tokyo, Earthquake Res. Inst., Hongo

Prof. Dr. E. Nishimura, Kyoto, Geophysical Inst., Kyoto Univ.

P o l a n d

Dipl.-Ing. A. Hermanowski, Warsaw 35, Politechnika Wydzial Geodezji i Kartografii, Osiedle Przyjaiú

Prof. C. Kamela, Warsaw, Politechnika Wydzial Geodezji i Kartografii

Prof. Dr. Z. Kowalczyk, Kraków, al. Mickiewicza 30

Doz. St. Kryński, Warsaw, Instytut Geodezji i Kartografii, ul. Jasna 2/4

Dipl.-Ing. W. Milewski, Gdansk, Katedra Geodezji, Majakowskiego 11/12

Dr. J.E. Mojski, Warsaw, Instytut Geologiczny, Rakowiecka 4

Doz. J. Niewiarowski, Warsaw, Instytut Geodesji i Kartografii, ul. Jasna 2/4

Prof. M. Odlanicki-Poczobutt, Krakow, ul. 'sw. Anny 4 - 14

Prof. St. Pawłowski, Warsaw, Instytut Geologiczny, Rakowiecka 4

Doz. J. Skorupa, Warsaw, Instytut Geologiczny, Rakowiecka 4

Dipl.-Ing. B. Szmielew, Warsaw, ul. Jasna 2/4

Prof. Dr. J. Witkowski, Poznan, Observatorium Astronomiczne, ul. Sloneczna 36

Dipl.-Ing. T. Wyrzykowski, Warsaw, Instytut Geodesii i Kartografii, ul. Jasna 2/4

R h o d e s i a

Prof. Dr. D.J. Gough, Salisbury, Physics Department, University College, Private Bag 167H

R o m a n i a

Prof. Liteanu, Bukarest, Calia Griwitze 64

S w e d e n

fil. lic. L.-W. Åse, Stockholm Sö, Högbergsgatan 2

Prof. Dr. L. Asplund, Stockholm 8, Geographical Survey Office, Hantverkargatan 29

S w i t z e r l a n d

Dr. N. Pavoni, Zürich 6, Inst. f. Geophysik, Leonhardstraße 33

U S S R

Dr. L.B. Aristarkhova, Moscow, State Lomonosov University, Geographical Faculty

Prof. J.D. Boulanger, Moscow B-296, Molodezhnaya 3, Soviet Geophysical Committee

Prof. N.V. Dumitrashko, Moscow B-17, Staromonetny per. 29, Institute of Geography, USSR Acad. of Sci.

Dr. S.K. Gorelov, Moscow B-17, Staromonetny per. 29, Institute of Geography, USSR Acad. of Sci.

Dr. M.K. Grave, Moscow B-17, Staromonetny per. 29, Institute of Geography, USSR Acad. of Sci.

A.A. Izotov, Moscow, Gorokhovsky per. 4, Institute of Geodesy, Airsurvey and Cartography

P.P. Kazanchian, Yerewan, Acad. of Sci. of Armenian SSR

Dr. J.B. Kazmin, Moscow B-296, Molodezhnaya 3, Soviet Geophysical Committee

Prof. M.V. Karandeeva, Moscow, State Lomonosov University, Geographical Faculty

Dr. A.V. Kondrashkov, Moscow, Gorokhovsky per. 4, Institute of Geodesy, Airsurvey and Cartography

S. Kochetkov, Moscow B-296, Molodezhnaya 3, Soviet Geophysical Faculty

Dr. L. A. Latynina, Moscow, B. Gruzinskaya 10, Institute of the Physics of the Earth, USSR, Acad. of Sci.

Dr. I.P. Lesis, Vilnius, Acad. of Sci. of Lithuanian SSR

E.J. Magnitskaya, Moscow, B. Gruzinskaya 10, Institute of the Physics of the Earth, USSR Acad. of Sci.

Prof. V.A. Magnitsky, Moscow, State Lomonosov University, Institute of the Physics of the Earth

Prof. A.M. Marinich, Kiev, State University, Geographical Faculty

Dr. J.A. Mescherikov, Moscow B-17, Staromonetny 29, Institute of Geography, USSR Acad. of Sci.

Dr. G.D. Panasenko, Moscow B-296, Molodezhnaya 3, Soviet Geophysical Committee

A.K. Pevnev, Moscow B-296, Molodezhnaya 3, Soviet Geophysical Committee

Dr. V.A. Rastvorova, Moscow, B. Gruzinskaya 10, Institute of the Physics of the Earth, USSR Acad. of Sci.

Dr. A.P. Rozhdestvensky, Ufa, Geological Institute, Bashkirian Department, USSR Acad. of Sci.

Dr. L.N. Rykunov, Moscow, State Lomonosov University, Institute of the Physics of the Earth

Prof. V.P. Scheglov, Taschkent, Astronomical Observatory, Acad. of Sci. of the Uzbekian SSR

Dr. L.E. Setunskaya, Moscow B-17, Staromonetny per. 29, Institute of Geography, USSR Acad. of Sci.

Prof. P.F. Shokin, Moscow, Gorokhovsky per. 4, Institute of Geodesy, Airsurvey and Cartography

Dr. S.N. Scheglov, Moscow, B. Gruzinskaya 10, Institute of the Physics of the Earth

J. A. Mescherikov (President of the C.R.C.M.)

S. Miyamura (Secretary of the C.R.C.M.)

O. Meißer (Chairman of the Organizing Committee)

Begrüßungsansprache des Herrn Präsidenten der Deutschen Akademie der Wissenschaften zu Berlin, Prof. Dr. Dr. h. c. W. Hartke

verlesen von

Herrn Prof. Dr. O. M e i ß e r

Sekretar der Klasse für Bergbau, Hüttenwesen und Montangeologie bei der Deutschen Akademie der Wissenschaften zu Berlin

Sehr geehrter Herr Präsident der Internationalen geodätischen Assoziation

Sehr geehrter Herr Präsident der Kommission für rezente Erdkrustenbewegungen

Sehr geehrte Herren Akademiker und Herren Delegierte

Meine sehr geehrten Damen und Herren

Der Präsident der DAW zu Berlin, Herr Prof. Dr. Dr. h.c. W. Hartke, bedauert außerordentlich, daß er durch eine dringende Dienstreise ins Ausland verhindert ist, heute hier anwesend zu sein. Als Ehrenpräsident unseres Symposiums entbietet er Ihnen folgende Adresse, die ich die Ehre habe, in meiner Eigenschaft als Mitglied des Präsidiums der DAW und als Sekretar der Klasse für Bergbau, Hüttenwesen und Montangeologie zur Verlesung zu bringen.

Im Namen der Deutschen Akademie der Wissenschaften zu Berlin habe ich die Ehre, Sie zur Eröffnung des ersten Internationalen Symposiums über rezente Erdkrustenbewegungen herzlich willkommen zu heißen. Es war mir eine hohe Ehre, die angetragene Ehrenpräsidentschaft über Ihr Symposium zu übernehmen.

Die Anregung von Herrn Vizepräsident Prof. Boulanger und Ihres Präsidenten, Herrn Dr. Mescherikov, das von Ihnen 1960 während der IUGG-Tagung in Helsinki beschlossene erste Symposium über rezente Erdkrustenbewegungen in der DDR abzuhalten, wurde von der Deutschen Akademie der Wissenschaften begrüßt und führte zur Organisation der heutigen Tagung in Leipzig.

Die DAW hat seit jeher in geodätischer Hinsicht nach der astronomisch-geodätischen Seite auf breiter geophysikalischer Basis und nicht zuletzt auch auf dem Gebiete der Geologie traditionsgemäß die Geowissenschaften umfassend gepflegt, so daß in unserem Lande eine starke wissenschaftliche Aufgeschlossenheit für Ihre Probleme vorliegt. Verfügt doch die DAW in ihrer Forschungsgemeinschaft der naturwissenschaftlichen, technischen und medizinischen Institute auf dem geologisch-geophysikalischen und astronomischen Sektor über mehr als 15 einschlägige Institute und Einrichtungen, die neben den entsprechenden Instituten der Universitäten und Hochschulen und der staatlichen Einrichtungen personell und materiell an Ihren Problemen weitgehend interessiert sind.

Ihre Bestrebungen über die klassischen Verfahren der Geodäsie hinaus, die vertikalen und horizontalen Bodenbewegungen des gesamten zeitlichen Spektrums, von plötzlichen Erdbebenbewegungen bis zu säkularen Hebungen und Senkungen auch im geologischen Zeitmaß zu erfassen, kartenmäßig darzustellen und gegebenenfalls auch mit neuen Verfahren und Instrumenten zu beobachten, gehen weit über die Aufgaben der IUGG hinaus und bedürfen der verständnisvollen Mitarbeit und einer umfassenden Auswertung auch der wissenschaftlichen Union für Astronomie und Geologie.

Ich freue mich ganz besonders, daß unseren Einladungen zur komplexen Diskussion der rezenten Bodenbewegungen auf einer so umfassenden Grundlage durch die Teilnahme von hervorragenden Fachwissenschaftlern aus allen daran beteiligten Disziplinen Folge geleistet wurde.

Insbesondere begrüße ich es, daß auch die wirtschaftlichen Aspekte neben den wissenschaftlichen Grundlagen eine entsprechende Berücksichtigung im Vortragsprogramm finden. Hat doch unsere Akademie den Leibniz'schen Grundsatz "Theoria cum praxi".

In voller Aufgeschlossenheit für Ihre komplexen, weltweiten Probleme wünsche ich Ihrem Symposium wissenschaftlich einen vollen Erfolg für die weitere internationale Zusammenarbeit und hoffe darüber hinaus, daß Sie in kollegialer Verbundenheit während Ihrer Tagung in unserem Lande eine recht schöne Zeit haben.

Begrüßungsansprache

von

Herrn Prof. J.D. B o u l a n g e r

Vizepräsident der Internationalen Assoziation für Geodäsie

Hochgeehrter Herr Präsident!

hochgeehrte Kollegen!

Damen und Herren!

Wir haben uns daran gewöhnt, unseren Planeten "Erde" für fest und unerschütterlich zu halten. Wahrscheinlich hat sich ein großer Teil der Anwesenden öfters experimentell davon überzeugt. Man braucht nur einmal zu stolpern und zu fallen, um sich von der Festigkeit und Härte der Erde zu überzeugen. Jedoch ist unsere Erde kein absolut starrer Körper. Sie ist elastisch und kann in dieser Hinsicht mit einem Fußball verglichen werden. Äußere und innere Kräfte deformieren sie ständig und verändern ihre Form. Man kann behaupten, daß sich kein Punkt auf der Erde in Ruhe befindet. Das Problem ist nun, mit welcher Geschwindigkeit und in welche Richtung sich dieser Punkt bewegt. Meist ist diese Frage nicht leicht zu beantworten.

Die Menschheit hat ungefähr 70 Jahre gebraucht, um die Frage zu entscheiden, ob die Erde seitens der Pole etwas abgeplattet oder entlang der Polarachse gestreckt ist. Man brauchte mehr als 150 Jahre, um mit der Genauigkeit, die den Bedürfnissen der Gesellschaft entspricht, die Abmessungen der Polar- und Äquatorachsen festzustellen, das heißt, die Abmessungen sowie die die Form der Erde charakterisierenden Grundparameter zu bestimmen.

Bis in die letzte Zeit befassen sich Geodäten mit diesen Fragen, indem sie sowohl die Form und die Abmessungen der Erde als auch die Lage der geodätischen Punkte für unveränderlich halten.

Infolge der Erhöhung der Präzision dieser Messungen stellten sich Veränderungen in der Lage der geodätischen Punkte heraus. Wir Geodäten befinden uns an der Schwelle einer neuen Ära der Erforschung unseres Planeten. Wenn wir bisher die Form und Abmessungen der Erde zu bestimmen strebten, so haben wir nun die Aufgabe vor uns, die Veränderungen der Abmessungen und der Form der Erde zu erforschen. Jetzt sind wir gezwungen, bei Durchführung vieler Messungen zur Bestimmung der Lage eines Punktes auf der Erdoberfläche die vierte Koordinate - die Zeit - einzuführen. Dabei haben wir oft Schwierigkeiten, auf welche Weise man den Ausgangspunkt dieser Koordinate wählen soll.

In dieser hochansehnlichen Versammlung ist es gar nicht nötig zu beweisen, wie wichtig diese Forschungen sowohl für die Wissenschaft als auch für die praktische Tätigkeit des Menschen sind.

Das Problem der Erforschung der Formveränderung der Erde ist ein Problem globaler Art. Die Wissenschaftler eines Landes sind nicht in der Lage, es allein zu lösen, unabhängig davon, wie groß und mächtig dieses Land sei, und dieses Problem allein vorwärtszubringen. Seine Lösung bedarf der Beteiligung von Fachleuten aus vielen angrenzenden Wissenschaften: der Astronomen, Geophysiker, Geologen, Geomorphologen, Geographen, Physiker, Ozeanologen und anderen. Nur eine solche umfangreiche wissenschaftliche Zusammenarbeit bietet uns die Möglichkeit, richtige Wege zu beschreiten, unsere Kräfte zu sparen und das Ziel schnell zu erreichen.

Diese Gedanken veranlaßten uns, im Rahmen der Internationalen Assoziation für Geodäsie, die ein Mitglied der bedeutendsten internationalen Union ist, ein spezielles ständiges Organ zu bilden. Dieses Organ kann maßgebende Empfehlungen zur optimalen Lösung dieses Problems ausarbeiten.

Also sind diese Ideen realisierbar. Ein entsprechendes Organ ist gebildet, und wir haben das Vergnügen und die Möglichkeit, im Kreise hervorragender Spezialisten die Lösungsmethoden des uns alle bewegenden Problems zu besprechen.

Having the high honour and pleasure to speak here from this tribune I should like on behalf of the International Association for Geodesy to express a deep gratitude to our colleagues Prof. Tsuboi, Prof. Baeschlin and Dr. Mescherikov for their initiative and efforts connected with the creation of the Permanent Commission on studying the present movements of the Earth's crust.

I should like to stress that great care, constant aid and thorough understanding of the question of organising the Commission, which we felt all the time on the part of Mr. Levallois, the IAG Secretary General; I should like to express my deep and hearty gratitude to him.

Wir haben es der Gastfreundschaft der Deutschen Akademie der Wissenschaften zu Berlin und dem hochgeehrten Präsidenten, Professor, Doktor, Doktor honoris causa, Hartke zu danken, daß wir das Vergnügen und die Möglichkeit haben, hier in dieser schönen alten Stadt zu tagen. Gestatten Sie mir bitte, dem Herrn Präsidenten der Akademie im Namen der Internationalen Assoziation für Geodäsie und in meinem eigenen Namen den herzlichsten Dank auszusprechen.

Die Last der Organisation dieses Symposiums übernahm unser unermüdlicher, energievoller und hochgeehrter Professor Doktor Meißer. Gestatten Sie mir bitte, lieber Professor, Ihnen für Ihre große Arbeit bei der Vorbereitung dieses Symposiums herzlich zu danken.

Das bevorstehende Symposium verspricht sehr interessant zu werden. Die Gelehrten von ca. 20 Ländern haben Vorträge eingereicht, die das Problem der rezenten Erdkrustenbewegungen von verschiedenen Standpunkten aus behandeln. Wir werden diese Vorträge anhören, sie besprechen und die Empfehlungen ausarbeiten, die die Richtungen unserer Arbeit in den nächsten Jahren bestimmen können.

Zu meinem großen Bedauern muß ich feststellen, daß nicht alle Wissenschaftler, die im Rahmen des zu besprechenden Problems mitarbeiten wollten, an diesem Symposium teilnehmen können. Es liegt hier keine Schuld der Organisation des Symposiums vor. Die Wissenschaftler, die an dieser Arbeit teilnehmen wollten, sind auch nicht schuld daran. Diese Wissenschaftler waren nicht imstande zu kommen, weil Kräfte in der Welt existieren, welche die Festigung freundschaftlicher Beziehungen zwischen den Wissenschaftlern hindern. Wollen wir hoffen, daß diese Wolke unsere Bestrebungen zur Schaffung dieser für unsere Aufgaben so notwendigen Zusammenarbeit nicht abkühlen läßt.

Gestatten Sie mir, allen hier Anwesenden erfolgreiche Arbeit zu wünschen.

Begrüßungsansprache

von

Herrn Dr. J. A. Meschcherikov

Präsident der Kommission für rezente Erdkrustenbewegungen bei der I.A.G.

Meine Damen und Herren!

Liebe Kollegen!

Ich glaube, daß ich dem Wunsch aller Anwesenden entgegenkomme, wenn ich in meinen ersten Worten der Akademie der Wissenschaften zu Berlin und ihrem hochgeehrten Präsidenten Herrn Professor, Doktor, Doktor honoris causa, Hartke, unseren innigsten Dank für die Möglichkeit der Einberufung dieses Symposiums hier, in Leipzig, ausdrücke. Wir messen eine außerordentlich große Bedeutung der Tatsache bei, daß Herr Präsident, Professor Doktor Hartke, die Ehrenpräsidentschaft über das Symposium übernommen hat.

Wir alle kannten bereits im Vorbereitungsstadium die umfangreiche Arbeit unseres hochgeehrten Kollegen, Professor Doktor Meißer, der die schwierigen und komplizierten Aufgaben des Vorsitzenden des Organisationskomitees übernahm und einen großen Beitrag zur praktischen Vorbereitung sowie zur gesamten Organisation dieses wichtigen internationalen Symposiums leistete. Wir schätzen die Anstrengungen der Mitglieder und Mitarbeiter des Organisationskomitees und seines Vorsitzenden Professor Doktor Meißer sehr. Im Namen aller am Erfolg des Symposiums interessierten Wissenschaftler möchte ich den deutschen Kollegen unseren herzlichen Dank für die vorbildliche Vorbereitung, für die großartigen Arbeitsbedingungen, die für uns geschaffen wurden, und für das freundschaftliche Verhalten gegenüber uns allen aussprechen.

The present Symposium is convened under the auspices of the International Association of Geodesy - one of the oldest international scientific associations, which will celebrate its centenary this year. I am glad to welcome here Prof. J.D. Boulanger, the Vice-President of the International Association of Geodesy, who is the official representative of the IAG to this Symposium and who did a lot of work to ensure its success. I am also glad to see Mr. J.J. Levallois, Secretary-General of the International Association of Geodesy. With profound understanding of the trends and prospects of the development of geodesy Mr. Levallois from the very beginning showed his sympathies with the organisation of a special Commission for the Recent Crustal Movements within the IAG; and afterwards the Commission always felt his authoritative assistance. I heartily welcome Prof. Dr. T. Sorgenfrei, Secretary-General of the newly organised International Geological Union. The presence at this Symposium of the Secretaries-General of two international associations, the Geodetical and the Geological, seems significant and justifies the hope in successful development of the problems of recent crustal movements which are within the scope of the technical and the natural sciences.

I take great pleasure in welcoming at this Symposium our guests of honour Prof. F.A. Vening-Meinesz and Prof. R. Tomaschek, the distinguished scientists of our time, who made valuable contributions to the study of the problems that Mr. Ch.A. Whitten, President of the International Association of Geodesy, Dr. W. Markowitz and a number of other scientists from the U.S.A. whose remarkable researches and scientific interests are in the closest connection with the theme of the Symposium, could not come to Leipzig, not with the fault of the organisers of this Symposium, who provided all possible facilities to make this Symposium an imposing forum of scientists of all countries and of all continents.

I am very sorry to be forced to inform the honourable assembly that Professor Baeschlin is no longer with us.

This outstanding scientist was the chairman of the "Special Study Group on the Problems of Recent Crustal Movements" within the IUGG.

The commission now existing will continue these studies. The death of Prof. Baeschlin is a great loss to the sciences of geodesy and neotectonics.

Presidental address

by

J. A. M e s c h e r i k o v

President of the Commission on Recent Crustal Movements

Herr Präsident der Deutschen Akademie der Wissenschaften zu Berlin,
Ladies and Gentlemen,
Dear Colleagues,

The problem, for the sake of which we have gathered here from different parts of the earth, has drawn attention of inquisitive minds of all epochs. Even the philosophers of the ancient times have noticed that though the surface of the solid earth at first glance seems so stable, in reality it moves: some parts of the surface slowly rise, the other subside. Strabon and Aristotle already knew about the secular movements of the earth; Lucretius Carus in his poem "De rerum natura" gave a vivid description of this phenomenon and of the related changes in the earth's surface. Beginning with the first half of the 18th century systematic observations of the modern uplifting of the earth in Scandinavia are being carried out, initiated by Linnaeus and Celsius. Since then many distinguished scientists of different countries engaged in the studies of the secular crustal movements: Lomonosov, Lyell, Darwin, Leopold von Buch, Humboldt, Suess, Davis, Karpunsky, Obruchov, von Bubnoff, Gutenberg, Sauramo. These scientists have contributed much to our knowledge of these most interesting events in nature.

Despite the brilliant ideas and conjectures of eminent scientists the progress of knowledge about the secular movements of the earth's crust has been in general extremely slow. As a matter of fact, the observations of secular crustal movements, which occur with the intensity of several millimeters per year, are involved with great difficulties and need long-term and rather precise determinations.

The new stage in the study of crustal movements really began only recently almost before our eyes, and it is connected with considerable improvement of the precision of geodetic measurements and with the accumulation of materials of repeated measurements at the basic geodetic networks. These materials allowed to proceed from the identification of the character of movements in the separate parts of the earth to the compilation of summarising maps of the recent crustal movements on large territories. The obtained data in an altogether different way illumine the problem of the recent crustal movements, thus ranking it with the most up-to-date problems of modern science about the Earth - geodesy, geophysics, geology, geography.

As a geodesist by education, I am glad to note the exceptional role of geodetic data and of geodetic institutions in the definition and elaboration of the recent movements problem. This example shows, how the ancient science of geodesy is not exhausted as a region of research, but is growing and developing. Similar to a sturdy tree geodesy becomes more vigorous from year to year, spreading forth fresh and young branches. One of these branches, now breaking into leaves and bearing fruit, is the study of the recent crustal movements.

In the history of the organisation of international researches the first sprigs of that branch should be dated as far as 1954, when according to a suggestion by Prof. F.A. Vening-Meinesz a special research group on the recent crustal movements has been organised within the International Association of Geodesy of the IUGG. Headed by Prof. Baeschlin the Group accomplished valuable work. At the XII General IUGG Assembly in Helsinki a special symposium on recent crustal movements has been convened under the direction of Prof. Tsuboi and Prof. Boulanger. At this Assembly it was decided of expand these researches still further and to set up a permanent Commission on Recent Crustal Movements. At present, the Commission includes representatives from 28 countries adhering to the IUGG. The necessity for such an international center for the study of recent crustal movements is caused by the fact, that this universal, global process can be studied only by the joint efforts of scientists of manycountries.

It seems quite appropriate that the Commission has been organised within the framework of the International Union of Geodesy and Geophysics, which unites scientists not only from different countries, but also of different specialities, who study the structure and development of our planet from different aspects. I would like to stress the significance of this problem, which to a great extent has been introduced by geodetic works, for geodesy itself. The new approach to the problems of geodesy is connected with the problem of the recent crustal movements. The new trend in science, which is sometimes called "dynamic geodesy", or more exactly "kinematic geodesy", launches from the predicament that all geodetic measurement are made not on the stable but on the mobile substratum. In connection with the preparation of this Symposium, Dr. de Graaff-Hunter wrote me that until now geodesy took interest in the figure of the Earth; the new task of geodesy is the study of the process resulting in the changes of the figure of the Earth. The same idea has been expressed recently by Prof. Heiskanen in the IUGG Chronicle, where he wrote that a fourth coordinate is now emanated in geodesy - the time. It would be difficult to overestimate the significance of this new approach not only for the theory of geodesy, but also for practical application. Our century is the time when the requirements to the precision of the geodetic bases of constructions are extremely high, and often it is impossible to ignore the fact of gradual "aging" of geodetic coordinates.

Oceanographic sea-level observations have made and continue to make valuable contribution to the study of recent crustal movements. In their turn, the data on the secular variations of the level of the world ocean endow the theory and practice of physical oceanography.

The researches of geomorphologists, geologists,

g e o g r a p h e r s take their own courses in the study of the problem of recent crustal movements. The latest data about the history of the Earth's relief and related sedimentation lead to the conclusion that the last pages of the Earth's history were far from being the period of tectonic peace; on the contrary, they were abounding in complicated and violent tectonic events. A separate scientific trend was formed which studied the new crustal movements. Academician V.A. Obruchev called this branch of science n e o t e c t o n i c s . Profound study of neotectonic phenomena persuaded Academician I.P. Gerassimov to distinctly define the difference between the n e w (Neogene-Quarterly) movements and the r e c e n t t e c t o n i c m o v e m e n t s. The first are identified by geologo-morphological methods, the latter - by geodetic, oceanographic, astronomical observations. The improvement of the methods of the geomorphological analysis, the application of modern methods of sedimentation dating (radio-carbon method, etc.) allow to use the geologomorphological method in fixing tectonic events which occurred in different periods of the Holocene and which approach nearer and nearer to our epoch.

The determination of the new and the recent tectonic movements has an extremely great significance for the geologo-geographical sciences. It sets a reliable substantial basis for the understanding of processes which took place in the geological past and which had great influence on the formation of the structure and relief of the Earth, on the deposition of ores and minerals. The analysis of modern tectonics opens new vistas for physical oceanography. I began my work as a geographer and a geomorphologer with a man who discovered the last large ridge on the surface of the continents; I mean Prof. S.V. Obruchev and the Czersky Ridge in Eastern Siberia. The time of geographical discoveries on the Earth is now passing away. It does not mean, though, that the tasks of geography are fulfilled. A new and not less absorbing task becomes evident - the study of recent processes that change the face of the Earth. Does a certain mountain ridge continue to rise, or probably, the movements has changed to general subsiding of this part of the crust? How the recent tectonic movements are reflected in the processes of erosion and accumulation, in the changes of the whole outlook of a landscape? The study of the recent tectonic movements provides a basis for the analysis of recent processes which form the landscape. This study is important not only for enriching our knowledge, but also for applied purposes.

The development of special g e o p h y s i c a l methods, especially the tiltmeter and gravimetrical, for the crustal movements determinations has a promising future. The application in practice of more "ductile", than repeated levelling, methods of research shall ensure their successful use for the evaluation of the stability of constructions sites, for reconnaisance of ores, and in the first place of such mobile substances as oil, gas, thermal waters, magmatic melts. Continuous observations of the variations of crustal movements shall provide a closer approach to the forecasting of destructive earthquakes. Together with the gravimetrical and magnetic data and with the results of seismic soundings the data about the recent crustal movements will give a better understanding not only of the structure of deep layers of our planet but also of the processes that are going on in these regions: the deconsolidation, and condensation of matter, convection currents of subcrustal masses, etc. Thus we shall come closer to finding the causes of tectonic movements, which is the most important theorectical problem of the physics of the Earth and of geology.

In our days the work of the researcher demands profound specialisation. This profoundness is the strength of modern researcher, but at the same time it is dangerous. The great poet, who lived not far from here in Weimar, aimed his irony at a scientist, who

"... hat er die Teile in seiner Hand
fehlt, leider! nur das geistige Band."

The problem of recent crustal movements is just such a problem that must be worked out by different specialists and with different methods. We should not forget, though, that the p h e n o m e n o n of secular crustal movements is a complicated and manifold, but integrated process. A single specialist is unable to embrace all the aspects of this process. Therefore, our aim now is to unite specialists for joint work in all the stages of research. This is the necessary condition and token of success.

In the extensive programm of international cooperation in the field of the study of recent crustal movements, presented to this Symposium by Prof. Boulanger and me, a plan is proposed for joint or parallel work of geodesists, oceanographers, geophysicists geologists and geographers in the generalisation of the existing data in the form of m a p s o f r e c e n t m o v e m e n t s of large territories, in the o b s e r v a t i o n s o f t h e c h a n g e s o f s e c u l a r m o v e m e n t s on special polygones, the network of which should cover all continents. Joint efforts are also necessary for the successful interpretation of the problem of c o n t i n e n t a l d r i f t s , whih intrigues not only specialists but also wide communities of educated people. The application of different and unconnected methods of recent crustal determinations shall render not only mutual control of results, but also would allow to understand the p r o c e s s of recent crustal movements in all its complexity, with all its many connections and relationships with other processes in nature. The presence of scientists of different specialities at this Symposium allows to hope in gaining success in the field of research.

As you know, the study of the recent crustal movements has not been planned in the program of the International Geophysical Year. And it is most satisfactory to note that these researches are now taking a consopicuous place in the new U p p e r M a n t l e P r o j e c t . I think that the data about the recent crustal movements shall prove to be an important contribution to the Upper Mantle Project: the picture of the deep layers structure of the Earth will appear as a feature of development. The data about the recent movements would render dynamic aspect to the geophysical materials about the structure of the Earth's interior and to the results of deep drilling. It will become possible to contemplate as mutually connected the processes underway in a large section of the Earth -- from its upper surface to the upper mantle.

A hundred years ago here in Germany the foundation was laid for the international cooperation in geodesy and geophysics which at our time yielded results astonishing in significance and scope. I wish to hope that here at this Symposium a sound basis is established for a large cycle of extremely inspiring and promising researches which shall assign still more importance to geodesy and geophysics in the life of people and in their strife for a better life on this planet.

Some additions to the program of the international researches of crustal movements

von

J. D. Boulanger und J. A. Mescherikov (Moskau)

The study of recent crustal movements becomes now one of the most vital problems in the geosciences. Data on recent tectonic movements are necessary in the solution of a number of fundamental theoretical problems in geology, geophysics and physical geography. These data can serve to establish the nature and causes of crustal movements - both slow (secular) and rapid (seismic) ones. In a complex with other geological and geophysical information, observations over recent deformations of the Earth's surface will permit to study in more detail the structure of the Earth and the processes, which take place in the crust and the subcrustal layer - in the so-called upper mantle of the Earth. A record of the data on recent movements is important for a correct appraisal of many processes taking place not only at depth but also on its very surface - erosion, accumulation of sediments, bogging, etc. Along with its purely scientific significance, the analysis of recent tectonic movements is of a practical interest, which will, undoubtedly, increase with the perfection of devices and methods used in the determination of tectonic deformations.

And yet, it is necessary to point out that the present state of our knowledge of recent crustal movements is far from being perfect. For the time being we do not know anything about the character of recent crustal movements throughout the vast expanses of our planet. The methods and implements used in the study of these movements are not sufficiently elaborated and not standardized. The studies proceeding in different countries are not properly coordinated and there is not enough exchange of information obtained.

Getting ready for this symposium and thinking over its aims and purposes, the authors came to the conclusion that it would not be enough if we here only summarize the results of the work already done in the study of recent tectonic movements. It is necessary also to accept a definite agreed upon program for the work in future, a program, which would guarantee a fundamental change in the organization of studies on recent crustal movements on a global scale. For this reason the authors have prepared in 1961 a note "On the development of international cooperation in the study of recent crustal movements".

In a short period of time this program became quite widely known among the specialists in this line. Owing to the initiative of our colleague Prof. O. Meißer, who made so much to provide a successful work of this symposium, our note has been printed as a small pamphlet and distributed together with information circulars on the Leipzig symposium. We were also happy to see our note published in the "IUGG Newsletter" magazine, published by the General Secretary of IUGG M. G. R. Laclavère.

We are greatly satisfied to state that our suggestions have been commented upon by many scientists of various countries. Letters with remarks or additions have been received by us from many people and among them from Honorary President of the International Association of Geodesy (AIG) Dr. J. de Graaff-Hunter (Great Britain), President of AIG Mr. C.A. Whitten (U.S.A.), Secretary General of AIG M. J. Levallois

(France), Director of the Permanent Service on the Mean Sea Level Mr. Rossiter (Great Britain), Director of the International Center on Earth Tides, Dr. Melchior (Belgium), President of the Commission on European Levelling Dr. Simonsen (Denmark) and many others.

At a special session of the Soviet Geophysical Committee in Moscow devoted to the problems of crustal movements valuable comments have been made by Prof. Magnitsky, Prof. Scheglov, Prof. Gzovsky, Dr. Pellinen. We would like to express our sincerest thanks to all those scientists who let us have their comments and ideas regarding this draft of a program.

Despite the remarks and additions, the draft as a whole has been unanimously and positively accepted. Many specialists had indicated that the draft can serve as a basis for discussion and acceptance of definite decisions by the Leipzig symposium.

We assume, there is no need to dwell in detail on all the points of the draft. We would like to mention their main aims and to inform you about the most important additions, which came in during the preparation for the symposium.

We would like to point out that our suggestions were aimed to direct the efforts of the researchers in three main lines:

1. Compilation of a world map of recent crustal movements;
2. Organization of an international observation service over the progress of secular crustal movements;
3. Study of possible horizontal movements of continents.

Point 1 of the program assumes, first of all, the compilation of velocity maps for recent vertical crustal movements on the continents of Europe and North America on the basis of repeated levelling results. Dr. de Graaff-Hunter and some other scientists recommended to stress the great significance of oceanographic (levelgauge) data for the fulfillment of this project. In reply to our address Mr. Rossiter, Director of the Permanent Service on the Mean Sea Level, expressed a readiness to cooperate with the Commission on Recent Crustal Movements along the following lines: a) interpretation of level-gauge data; b) detalization and perfection of observation methods at existing stations; c) recommendations on improving the geographical distribution of levelgauge stations. It is, apparently, necessary to reflect in the draft the necessity to obtain a full record of all level-gauge data and to use them in the compilation of maps on secular movements of continents. It would be also advisable to summarize level-gauge observations for the compilation of a world map of vertical movements on the coasts of seas and oceans.

At the present time great success has been achieved in working out geological and geomorphological methods for the establishment of crustal movements, which took place during periods very close to our times, by observing young relief forms (sea and river terraces, flood plains, longitudinal river profiles) and associated sediments; in these determinations exact methods of absolute dating (C-14 etc.) are used. Very promising is, apparently, the method of comparing topographic surveys and airsurveys data for various years, which permits to establish changes in the landscape caused by secular uplifts or subsidences of the Earth's crust. It seems advisable, parallel to the compilation of maps of secular movements according to geodetic and oceanographic data, organize a map compilation also by another independent method - a geological and geomorphological method. Dr. Miyamura raised the question of correlating marine terraces and establishing their deformations. This work done parallel with the compilation of a world map of vertical movements on the coasts, would, obviously, be most interesting by itself and would serve as a valuable check of geodetic and level-gauge data.

Point 2 schedules the creation of a world network of polygons, which would effect a systematic study of vertical and horizontal movements of the Earth's crust by means of repeated levellings, triangulation, linear measurements, tiltmeter and level-gauge observations. It is also necessary to add to the complex of geodetic and geophysical researches a series of geological and geomorphological observations over recent processes of erosion and sedimentation in the areas of the polygons. A quantitative characteristic of recent erosion and accumulation processes will permit to appraise the role of the isostatic factor in the formation of recent uplifts and subsidences of the crust and will help to penetrate into the mechanism of relations between endogenetic and exogenetic processes. An effective way of solving these problems might be a systematic repeated air photosurvey.

Point 3 was based on the realization of the program of astronomical observations on the existing latitudinal stations and on special longitudinal stations, which should be located approximately along three meridians crossing: a) North and South America, b) Europe and Africa and c) Asia and Australia. Dr. Bartha suggested to organize a network of points along the equator to carry out observations intended to eastablish secular changes in gravity. These observations expected to last for a long time, are to determine general deformations of the Earth, which is vital for an interpretation of possible horizontal displacements of separate segments of the crust established by the astronomic method. It was also pointed out bow necessary it is to study the question of using not only astronomical determinations, but also radio and long-range light determinations for the establishment of horizontal displacements of continents and big islands.

The fulfillment of each point demands the solution of a number of special problems - first of all of a methodological character. Thus, for project 1 it is necessary to agree on the scales, content and legends of generalized maps of recent crustal movements; to agree on general principles of a joint treatment of geodetic and level-gauge data. On project 2 it is necessary to work out common principles of locating a network of stations (polygons), to fix the methods and terms of observations. The same refers, actually, to project 3. There is no doubt that the solution of these problems will demand a discussion. Thus, we have already discussed by correspondence with Dr. Melchior the problem of the principles according to which tiltmeters are to be placed, of the needed precision of the tiltmeters to establish secular crustal movements, on the possibility of joining tiltmeter and level-gauge stations. We hope that this useful discussion will be continued here and will result in working out definite methodological recommendations.

To work out one common plan and to coordinate methods of research it seems advisable to create several working groups within the Commission on recent crustal movements of the AIG. The beginning of their activity should be associated with this symposium; in future these groups should coordinate researches along specific lines and discuss the results of current work. It seems necessary to organize the following groups:

1) Group on the compilation of a map of secular crustal movements in Eastern Europe;
2) Group for the compilation of a map of secular crustal movements in Western Europe;
3) Group for the compilation of a map of secular crustal movements in Northern America;
4) Group on the world network of polygons for observations over secular crustal movements;
5) Group for the study of continental drift;
6) Group for the determination of devices necessary in the establishment of recent crustal movements.

The question should be discussed regarding the compilation of a map secular movements on the coasts of seas. If this work will be undertaken by the special researchgroup No. 22 of AIG (Mean sea level) of the Commission on coastal oceanography of IUGG formed at the XI General Assembly of IUGG - there is no necessity to create anv special group within the Commission of recent movements.

The realization of a program of international researches demands an organized mutual exchange of observation material. The best decision of this problem , seems to be, the utilization of existing International centers of data A and B.

The program of international researches agreed upon at this symposium should be presented by the Commission on recent crustal movements to the XIII General Assembly of IUGG in Berkely for approval and for the compilation of corresponding recommendations to the National Committees of the countries participating in the IUGG. It is also necessary to get directly in touch with the IAU to coordinate astronomical work on the determination of continental drift. It is possible to expect that at the next symposium of the Commission on recent crustal movements, which it would be advisable to hold within the period between the XIII and XIV General Assemblies of the IUGG, we will be able to discuss some results of the work carried out according to the program.

It should be, however, clearly realized that we stand now at the very beginning of a long route of investigations. Secular crustal movements - as their name makes it clear, - demand a long and systematic study. The President of AIG, Mr. C.A. Whitten, who, unfortunately, is not present at this symposium, was quite right when he wrote to J.A. Mescherikov that this is a task of many years and that it is the duty of our generation to provide the collection of data, which will be used by our successors.

In conclusion it is also necessary to stress that the program suggested is far from embracing all the aspects of the study of recent crustal movements.

Actually, the program is mainly expected to include a study of recent crustal movements within the territories of continents, while it is quite obvious that identical, if not stronger movements of the solid surface of our planet, are also taking place at the ocean floor. We can only hope that the work on such a program by itself will serve as an impulse for an extension and detalization of researches. It is still difficult for us now to form an idea of the vast and maybe unexpected results and consequences, which might result from detailed studies of crustal movements. One can only be quite confident that this work will provide a lot for our theory and will be most helpful for humanity in mastering the riches of the Earth and the forces of Nature.

D i s k u s s i o n

J. Witkowski (Poznan):

Ich möchte eine Frage über die Wahl der Lage der Längen- und Breitenstationen für die Bestimmung der Lageänderungen der Kontinentalschollen stellen. Die Wahl muß mit der Struktur der Erdkruste rechnen, speziell, wenn solche Fragen ins Auge gefaßt werden wie die möglichen Drehungsphänomene der Elemente der Erde.

J.D. Boulanger (Moskau):

In order to solve the problem of continent drift it is necessary to use the existing and the newly organised latitude-longitude stations. A special Working Group should be nominated to consider the program of work of these stations and their distribution. This work should be carried out in close contact with the International Astronomical Union.

The task of the Commission on Recent Crustal Movements

von

T. J. Kukkamäki (Helsinki)

Summary

Several individual investigations on different fields of research are listed. The organizing of these investigations is not the task of the CRCM. To propose international cooperation and to point out the investigations, to compile the results of the geodetic, geophysical and geological studies concerned with the recent crustal movement and to offer the collected material and to stress the special problems the CRCM fulfills the most important aspects of its task.

Zusammenfassung

Es werden mehrere Einzeluntersuchungen aus verschiedenen Forschungsbereichen aufgeführt. Die Organisation dieser Untersuchungen ist nicht Aufgabe der CRCM. Eine internationale Zusammenarbeit vorzuschlagen und die Untersuchungen auszuwerten, die Ergebnisse geodätischer, geophysikalischer und geologischer Forschungen im Zusammenhang mit den rezenten Krustenbewegungen zu vergleichen, das gesammelte Material darzubieten und die speziellen Probleme der CRCM zu betonen, all dies ergibt die bedeutsamen Aspekte ihrer Aufgabe.

The paper "On the development of international cooperation in the study of recent crustal movements" prepared by J.D. Boulanger and J.A. Mescherikov, makes suggestion of the task and domain of this symposium and the Commission on Recent Crustal Movement. As examples there are given the establishment of a map showing the hitherto collected knowledge of the secular vertical movements, the continuous observation of the crustal movement in some special areas and the organized observation of the longitude. These are good examples of important tasks which, when performed, will essentially help the general study of the recent and the present crustal movement.

For the plans of the work of the Commission on Recent Crustal Movement, it is necessary at first to list all possible investigations which may contribute to the work of the Commission, and to try to coordinate these heterogeneous efforts into one structure around this central problem.

The researches in question belong to many different international organizations, such as most of the associations of the International Union of Geodesy and Geophysics, the International Astronomical Union and the recently established International Union of Geological Sciences. It is natural to treat these investigations in their respective contextures.

The International Association of Geodesy shows a great activity in this field, and all of its sections have worked and will work on this problem.

The most concrete and succesful program has been the determination of the secular vertical movement through relevellings, as indicated in the paper of Boulanger and Mescherikov. The results achieved up till now are unfortunately scattered and uneven. Thus the suggested map will be filled only for the part of some isolated limited areas, such as Finland, Denmark, The Netherlands, Hungary and parts of the Soviet Union. Even here the accuracy of the values depends largely on the length of interval time and the accuracy of the levellings, especially on that of the older one. Thus it is necessary that the accuracy is clearly indicated in the map in order to avoid misinterpretation of the circumstances.

Relevellings are largely used in Japan, India and USA for the determination of sudden crustal movements in connection with earthquakes. Comparison of the levelling before the earthquake and the levelling after the earthquake, gives valuable information on the deformation occurring because of earthquakes.

Continuously repeated levellings of some specially selected lines have been performed in Spain and Germany, as recommended by the Levelling Section of the IAG, in order to determine the possible daily and yearly pulsation of the earth's upper layer. The results have been negative until now.

Closely related with the last mentioned efforts is the work of the Permanent Commission on Tides of the Earth's Crust. The extensive recordings with tiltmeter, and the strain measurements collected and worked by this Commission, give information on the strength of the earth's crust and must contribute to the study of recent crustal movements in an essential way.

Due to its smaller relative accuracy triangulation has been less successful than levelling in detecting the crustal movements. Some beautiful examples can be seen, however, such as the reobservation of the triangulation net across San Andreas Fault in California. However, no systematical observation over larger areas can be planned yet with the present triangulation and trilateration techniques.

Attempts have been made to detect the recent drifting of continents by a comparison of latitude and longitude obtained in this century with the values obtained in the last century. The large shifts obtained at first for Greenland, have later turned out to have been caused by the insufficient accuracy of the older observations. It has been supposed that the international latitude station Mizusawa in Japan has shifted continuously 0.2 m/y to the south, but even this case is not evident. Thus until now it has not been possible to show any drift of the continents by the aid of astronomical position observations.

The relation of the work of the CRCM with the gravimetrical section of IAG is not very close. Not considering gravity recordings already mentioned, the gravity anomaly, however, seems to be correlated with the crustal movements.

The planning of the investigations outside the domain of geodesy is beyond the knowledge of the writer of this paper, and it is to be entrusted to geophysicists and geologists. Only some investigations which are most clearly related to the recent crustal movement, can be listed here.

Within the IUGG, most associations make important contributions toward settling this problem. The Association of Seismology and Physics of the Earth's Interior shows the weak zones and lines in the earth's crust. The areas of the earthquake foci and the areas with active crustal movements show correlation with each other and help the investigation of both reciprocically.

The observations of the water level were the first to draw attention, many centuries ago, to the changing of the mutual position of the water level and the solid ground.

Even nowadays the recordings carried out by oceanographers and hydrologists give better information than any other observations, on the crustal movements in the greater part of the world.

Outside the IUGG, the most important contribution is given by the investigation made in the field of the youngest scientific union, the International Union of Geological Sciences. Notwithstanding the fact that geology actually does not deal with recent phenomena, it is possible, by the aid of geological findings, to link the scattered pieces of geodetic and geophysical results with each other and to extend these results beyond the limited areas investigated geodetically and geophysically. The postglacial clay layer countings offer a good example of how the landuplift results of mareograph recordings and of relevellings have been extended to the interior of countries where no relevellings are available.

Bearing in mind the multitude of the programmes, of which several examples are given above, it is evident that the CRCM cannot go into detail in organization and techniques. Numerous investigations have already been organized on an international basis and many observation series are going on. However, some of the investigations, which are important for the study of crustal movement, are scattered over separate areas without coordination with each other, and there are many interesting areas in different parts of the world where a proper geodetic, geophysical or geological investigation would be of great value. To propose international cooperation in different fields and to point out the areas where investigations would be desirable in order to solve the recent crustal movement is an important part of the work of this commission.

On the other hand, individual researches connected with the crustal movement solve special problems, but because of their dissimilarity and the fact that they are scattered over many different fields, it is difficult to obtain a total picture of any kind of the crustal movement. It is therefore important to compile the results of the geodetic, geophysical and geological studies in any way concerned with the recent crustal movement. By offering to scientists the material thus collected and by stressing the special problems, the CRCM fulfills the most important aspects of its task.

Diskussion

J.J. Levallois (Paris):

Les questions soulevées par les rapports du Prof. Boulanger et Kukkamäki sont de la plus haute importance. Il est nécessaire de prendre tous les contacts avec l' Union Astrnonomique Internationale et l' Union Internationale de Géologie. C'est pourquoi, il faut réfléchir à cette indispensable liaison et à l'organisation de détail à l'intérieur de l'Association. Il faut donc peut-être attendre un peu plus longtemps pour pouvoir proposer une organisation rationnelle.

On the application of the method of relevelling in a study of recent vertical movements of the earth's crust

von

P. S. Zakatov (Moskau)

S u m m a r y

The report reviews different types of recent vertical movements of the land surface and the use of relevelling methods for their determination.

The difficulties are indicated to organize studies for the establishment of recent vertical movements of the Earth's crust and of glacio-isostatic crustal movements involving vast territories. A brief analysis is given of the compilation of maps of the velocity of recent movements in various countries.

Z u s a m m e n f a s s u n g

Der Bericht behandelt die verschiedenen Typen rezenter Vertikalbewegungen der Landoberfläche und die Anwendung von Wiederholungsnivellements für ihre Bestimmung.

Die Schwierigkeiten werden aufgezeigt mit dem Ziel, Untersuchungen zu veranlassen für die Feststellung der rezenten Vertikalbewegungen der Erdkruste und der glazio-isostatischen Krustenbewegungen einschließlich größerer Territorien. Eine kurze Analyse von Darstellungen der Geschwindigkeit rezenter Bewegungen in verschiedenen Ländern wird gegeben.

At the present time the study of recent vertical movements of the Earth's crust is a big problem embracing researches on various types of recent movements as well as an investigation into the causes of these movements.

The main types of movements recognized at the present time are:

1. Intermittent and sudden movements alternating with more or less lengthy periods of quiet. Such movements, are, mainly, characteristic of seismically active areas.

2. Oscillation movements of the crust with comparatively low velocities of the movements. Researches of geodesists and geomorphologists in various countries have proved that these movements are ubiquitous, are developed in all geological areas located on different continents of the globe.

3. Glacio-isostatic crustal movements.

4. Displacements of the upper layers of the land surface under the influence of various phenomena not related directly to tectonic crustal movements, but changing the position of bench marks with time. These phenomena include: changes in the level of ground waters, effect of soil freezing and thawing, karst and landslide phenomena, subsidences of the land surface caused by the activity of man.

Owing to a great number of causes, which result in movements of the land surface, the organization of work on the study of recent vertical movements of the Earth's crust is very complicated and inevitably necessitates its subdivision into separate lines of research. It is also vital to have a very good idea what sort of movement is being studied and plan the work subsequent data processing of relevelling according to the specific type of movement, which has been investigated.

In studying sudden Saltatory crustal movements it is recommended to do repeated relevellings at short time intervals on relatively limited stretches of the land surface. Such a kind of study of crustal deformations is developed in seismic areas of Japan. A great number of levelling profiles from 6 to 300 km long has been created there. To a smaller extent similar work is being also carried out on the territory of the Soviet Union.

The study of oscillation movements of the Earth's crust and of glacio-isostatic movements extending to considerable territories should be carried out simultaneously. It is advisable to use for this purpose networks of relevelling connected with level gauge installations on the sea coasts. However, until the present time levelling networks of great extent have not been specially developed for use in the study of recent vertical movements of the Earth's crust. Usually state levelling networks have been used, their relevelling having served to renew on data the elevation control at sufficiently lengthy time intervals.

To study local displacements of the upper layers of the land surface, for instance, within the territories of large cities, in the areas of an intense activity of man, in karst and landslide regions, special networks of relevelling lines are created. They cover rather small areas and refer to one certain bench mark (or a group of bench marks) placed either on a rocky basement or at a much greater depth than other levelling marks. Studies of this kind are reviewed in Small's report (USA) read at the 12th Assembly of IUGG and are effected by Soviet geodesist M.S. Uspensky.

Most complicated both in its organization and in its geological and geomorphological interpretation is the study of recent oscillation and glacio-isostatic movements of the crust.

The difficulty in studying this problem lies first of all in the necessity to investigate the laws governing the manifestation of such movements on a vast territory, if possible not limited by the boundaries of one state but embracing the territory of individual continents. In this case it is necessary to use relevelling lines of great extent, where systematic errors of levelling can be considerably reflected. At the present time,unfortunately, they are not yet properly studied. We lack also sufficiently reliable methods of appraising the quality of levelling. These circumstances make the use of the method of relevelling in the study of recent oscillation movements much more difficult than in researches on the crustal movements of other types. This means that it is necessary to work out special methods of using data of relevelling and combine them with other independent methods (oceanographic, geological-geomorphological, etc.)

On using relevelling data for the compilation of maps of the velocity of recent movements, the fundamental premises accepted in all countries as the first approximation, is a hypothesis that the velocity of recent vertical movements of the crust in any point of its territory is constant and continuous throughout the time interval between the dates of the first and last levelling, the data of which are used for the study of recent tectonic movements. The main difference in processing relevelling data for the study of recent movements in different countries consists in the choice of the reference surface for the compilation of velocity maps of these movements.

In different countries at the present time there is no single opinion regarding the choice of the reference surface for the compilation of these maps.

There are three approaches to the choice of the reference surface:

1) as a reference surface the mean sea level is accepted, the position of which is conventionally accepted as constant in time (USSR);

2) as a reference surface a level surface is accepted conditionally, which does not

change its position in time and passes through a point of the country (Hungary, Poland, etc.). Such a choice of reference surface takes place on the territory of countries with a comparatively small area;

3) as a reference surface the surface of the mean level of the World ocean is accepted, which changes its position in time; consequently, velocity values of recent movements obtained on the basis of level observations are corrected according to the data on eustatic changes in the level of the World ocean (Great Britain, Finland).

In our opinion, one of the foremost tasks of the coming international symposium is the choice of a reference surface for the compilation of velocity maps of recent movements.

Special investigation of this problem was made at the Moscow Institute of Geodesy, Airphotosurvey and Cartography. In these researches results of sea-level observations have been used given in the Publications of the International Association of Physical Oceanography. It has been established that for all points used in the processing, velocity values of recent movements deducted from data on sea-level observations during fifty years and more are nearly constant and differ within the limits of $\pm$ 0.2 — $\pm$ 0.3 mm/y. Consequently, the position of the surface of the mean level of the World ocean established for a fifty-year period can be regarded as constant, considering that eustatic factors should change the data obtained by the same amount. However, not all points of sea-level observations located on the coasts of seas and oceans have such a duration of sea-level observations. Thus, in certain cases it is necessary to use sea-level posts with a shorter period of observations. According to our investigations, errors in velocity determination for recent movements deducted on the basis of sea-level observations during 25 - 30 years rest within the limits of $\pm$ 0.6 — $\pm$ 1.2 mm/y, there being appreciable velocity fluctuations as compared with their values during certain periods. German researcher H. Valentin, who used data of sea-level observations for the compilation of a velocity map of movements on the territory of Great Britain, also indicates the presence of short-period errors in the series of sea-level observations lasting 20 - 25 years. Apparently, on the basis of this fact E. Kääriainen (Finland) thinks that the velocity of recent movements is determined by sea-level observations three times rougher than by relevelling data.

On the basis of the abovesaid, it seems that, considering it is time now to start a study of recent movements on a scale of entire continents, within the bounds of international cooperation the reference surface should be the position of the surface mean levels of seas washing the investigated territory and established during a certain period of time.

Such a choice of reference surface has the following advantages:

a) it permits to determine so-called "absolute" values of the velocity of recent movements comparable with velocity values determined in other parts of Europe and even on other continents, and

b) lessens the error of velocity determinations by a multiple tying to the datum sea-level surface.

The question of the choice of the reference surface is also directly related to the question of choosing the method of adjustment for the final velocity values calculated according to the data of relevelling.

In the USSR velocity values calculated for the entire network by the usual formula:

$$v = \frac{h_2 - h_1}{t}$$

have been adjusted by polygons and between the points of sea-level observations on the sea coasts.

Special researches effected in the Moscow Institute of Geodesy, Aerophotosurvey and Cartography indicated that owing to great expanses of the territory of the USSR both longitudinally and latitudinally, an adjustment of the computed values of velocity between the points of sea-level (oceanographic) observations lessens the effect of systematic errors in levelling and permits to increase the reliability of the determined velocity values of recent movements.

In other countries of comparatively small areas (Finland, Bulgaria and the Baltic republics of the Soviet Union), where there is a network of closed level lines, the adjustment of calculated velocity values is done according to the principle of adjusting a free network.

In some countries the determination of final velocity values is done by already adjusted elevations of the first and second level network. However, such a method of determining final velocity values can result sometimes in considerable distortions due to these adjustment corrections.

One of the main differences in the methods of processing relevelling data for the computation of final velocity values in different countries is due consideration of difference in the time of levelling. We have two typical ways of processing relevelling data in USSR and in Finland: In the USSR in compiling the velocity map of recent movements there is a strict stipulation of the hypothesis on the permanency of velocity values of recent movements during the interval between the first and repeated levelling. In Finland, apparently, in order to lessen the influence of the time difference between the first and second levelling and to increase the equiprecision of the final velocity values of recent movements, the adjustment of calculated velocity values is effected by a method of approximations. At first approximate velocity values are determined on the basis of computed differences in heights, then they are brought down to one time period, which serves as a mean date for the fulfilment of each used level network. After the first and second levelling have been brought down to one time period, the height differences of the first and second levelling are adjusted again and then velocity values of recent movements are again determined according to the same formula, which, however, looks now:

$$Vok = \frac{h'_2 - h'_1}{44}$$

where h'_2 and h'_1 are values of adjusted height differences and 44 years the time elapsed between the mean dates of the first and second levelling.

Values of finally adjusted velocities are compared with the approximate values and, if these values do not come closely enough, the process of velocity determination is repeated until the values of final velocity will coincide sufficiently. An analysis of the method of processing used in Finland shows that the equiprecision in the determinations of adjusted velocities according to the investigations of Finnish geodesists is only a seeming one.

The last difference in the compilation of velocity maps of recent movements of the crust is the method of a visual presentation of the results.

There are two different ways of representing the velocity of recent movements on a territory under study:

1) plotting on the maps lines of equal velocity (isolines) in mm/y, mm/10 years, in cm/100 years, and

2) plotting the differences in the heights of points on the land surface (amplitudes of movements) for a certain time interval.

At the present time the main task in a further development of studies of recent vertical movements of the land surface by relevelling is an accumulation of factual material throughout the globe. That is why it is important to discuss the organization of geodetic work for the study of various kinds of recent movements, the methods of processing the results of these researches and the ways of interpreting their results for the establishment of shifts in the land surface created by various causes.Recommendations, which will be accepted, will speed up the compilation of a sufficiently reliable map for very vast territories that will enable to realize sooner the laws governing the distribution and the causes of origin of recent vertical movements.

Diskussion

L. Egyed (Budapest):

Bei dem Referenzniveau sollten wir auch die Veränderung polarer Eismassen berücksichtigen. In längeren geologischen Zeiten nimmt das mittlere Meeresniveau ab.

Problemy proektirovanija setej povtornogo precizionogo nivelirovanija dla issledovanija sovremennych vertikal'nych dvizenij zemnoj kory

(Probleme der Entwurfsbearbeitung von Netzen wiederholter Präzisionsnivellements für die Bedürfnisse von Untersuchungen gegenwärtiger Vertikalbewegungen der Erdkruste)

von

T. W y r z y k o w s k i (Warschau)

S u m m a r y

The projected canvas of precision levellings intended for the investigations of actual vertical movements of the earth's crust has to meet special requirements. Elements influencing the form and structure of the canvas are dealt with in this lecture (incorporation of existing levelling lines, consideration of the structural conditions of a given area, and others). Moreover the selection of the reference level (determination of relative and absolute motions), the stabilization of points of the canvas of investigation, and the distribution in space of fundamental bench marks are discussed. The real survey as well as the time intervals between the repeated levelling operations are also treated, time intervals which result from an accuracy assumed for measuring the velocity of vertical movements of the earth's crust.

Z u s a m m e n f a s s u n g

Das für Zwecke von Untersuchungen gegenwärtiger vertikaler Erdkrustenbewegungen projektierte Netz von Präzisionsnivellements muß besonderen Ansprüchen genügen. Im Referat werden Elemente behandelt, die die Gestalt und Struktur des Netzes beeinflussen (Einbeziehung bestehender Nivellementslinien, Berücksichtigung der tektonischen Bildung eines gegebenen Raums und andere). Weiterhin werden die Auswahl der Bezugsniveaus (Bestimmung von relativen und absoluten Bewegungen), die Stabilisierung von Punkten des Untersuchungsnetzes und die räumliche Verteilung von fundamentalen Höhenfestpunkten besprochen. Auch die Netzvermessung sowie die Zeitintervalle zwischen den wiederholten Vermessungen von Nivellementsnetzen, die sich aus einer für die Bestimmung der Geschwindigkeit vertikaler Erdkrustenbewegungen angenommenen Genauigkeit ergeben, wurden behandelt.

Р е з ю м е

Сеть прецизионного нивелирования, проектируемая для исследования современных вертикальных движений земной коры должна отвечать специальным требованиям. В докладе обсуждаются: элементы влияющие на форму и структуру сети (включение существующих линии нивелирования, учёт тектоники территории и другие), избрание исходной поверхности (определение движений относительных и абсолютных), закрепление пунктов исследовательской сети и размещение фундаментальных реперов. Обсуждается также нивелирование сети и, следующую из заложенной точности определения скорости вертикальных движений земной коры, частоту повторяемых измерений сети нивелирования.

Доклад на эту тему был уже опубликован в Трудах Института Геодезии и Картографии ("Zagadnienia projektowania sieci powtarzanej niwelacji precyzyjnej dla potrzeb badania współczesnych pionowych ruchów skurupy ziemskiej" Prace Instytutu Geodezji i Kartografii, tom VIII, zeszyt 2 (18), Warszawa 1961).

I. Вступление

Почти все использованные до сих пор для исследования современных вертикальных движений земной коры материалы повторных прецизионных нивелирований имели, в некотором смысле, случайный характер. Ни проект сети и трассы ее отдельных линии, ни первоначальное и повторное измерение не учитывали надобностей этих исследований. Вследствие этого возможность использования материалов нивелирования была в значительной степени ограничена, а получаемые на основании этих данных результаты определения движений земной коры были отягчены относительно большой ошибкой. Можно легко добиться уменьшения этой ошибки, если проект сети и повторяемое каждый раз ее нивелирование будут отвечать определенным, частично новым - по отношению к применяемым до сих пор - требованиям.

Доклад занимается именно теми характерными элементами служащей для исследования движений земной коры сети нивелирования, которые должны быть принимаемы во внимание при целесообразном проектировании такой сети и ее измерении.

Проектируя такую сеть и имея в виду ее использование для общих исследований движений на большой территории, например на территории целой страны, следует стремиться к осуществлению следующих основных условий:

а) сеть должна иметь возможно однородную структуру и должна обнимать всю исследуемую территорию,

б) линии всей сети должны пересекать важные с точки зрения геологического и тектонического строения структурные единицы исследуемой территории,

в) измерение целой сети должно производиться каждый раз в течение короткого периода времени и с возможно высокой точностью.

В нижеследующем мы рассмотрим отдельные элементы такой сети, которую в дальнейшем будем называть исследовательской сетью.

2. Элементы влияющие на форму и структуру сети

В начале следует обратить внимание на факт, что проектирование исследовательской сети будет, как правило, относиться к территориям, на которых уже существуют сети прецизионного нивелирования. Этот факт в большой степени ограничивает свободу проектирования, принуждая, из-за экономических соображений, использовать существующие уже линии нивелирования, одновременно давая возможность сравнения результатов давнейших нивелирований с будущими и создавая условия для первого, приблизительного определения скорости современных вертикальных движений земной коры.

Имея в виду те же самые экономические соображения проект исследовательской сети должен быть так составлен, чтобы линии этой сети могли быть целесообразно использованы в государственной сети прецизионного нивелирования, имеющей характер главной высотной основы для дальнейших измерений хозяйственного характера.

Соотношение исследовательской сети к основной государственной сети прецизионного нивелирования будет следствием величины территории, обнимаемой исследовательской сетью.

В случае исследовательской сети местного характера, когда исследование ведется на относительно небольшой территории, например интересной с точки зрения геологии, эта сеть должна состоять из существующих уже на этой территории линий прецизионного нивелирования и новопроектируемых, загущающих линий, что давало бы возможность получить более полное определение движений, но как правило, относительных.

Если исследовательская сеть обнимала бы территорию всей страны, могла бы возникнуть целесообразность перепроектирования существующей сети прецизионного нивелирования.

При отборе существующих линий и проектировании новых следует учитывать геологическое строение и тектонику территории. Исходя из предпосылки, что в районах отдельных структурных единиц могут проявляться характерные для них, но разные скорости вертикальных движений, линии нивелирования следует так проектировать, чтобы они соединяли и одновременно пересекали эти структурные единицы, давая наиболее характеристические профили, определяющие высотные изменения. Следовательно, направление линии нивелирования должно быть перпендикулярным к краям тектонических структур, к осям синклиналей и антиклиналей.

При проектировании исследовательской сети необходимым является близкое сотрудничество с геологами, геоморфологами и географами.

Чтобы добиться возможно высокой точности, исследовательская сеть должна иметь однородную структуру, и в то же время состоять из такого числа линий, чтобы измерения целой сети могли быть исполнены в течение возможно короткого периода времени (см. § 5.).

Исследовательская сеть должна содержать, по возможности, все фундаментальные вековые реперы существующей сети. Эти реперы могут быть использованы как исходные, опорные пункты при локальных исследованиях.

Чтобы иметь возможность определения абсолютных скоростей движений относительно среднего уровня моря, в проект сети должны быть включены специальные линии нивелирования, связывающие сеть с нулями уровенных постов и мареографических станций. Если мареографы и уровенные посты разбросаны вдоль длинного побережья моря, следует запроектировать линию нивелирования вдоль побережья, связывающую мареографы и уровенные станции между собой.

Исследовательская сеть, обнимающая территории ряда стран, территориально бо́льшая, чем выше обсуждаемая, ставит перед проектом высшие требования, но дает возможность получить более полную картину движений, более правильную геологическую интерпретацию, а также увеличивает возможность лучшей связи с средним уровнем моря или даже нескольких морей.

Совместная исследовательская сеть должна обнимать всю исследуемую территорию и состоять из связанных между собой сетей нивелирования, в которых загущение линиями, период и точность измерений, способ закрепления реперов и расстояния между ними должны быть согласованы так, чтобы объединенная сеть гарантировала однородную, ранее заложенную, по возможности высокую, но практически осягаемую точность определения движений земной коры.

В этом случае линии связывающие нивелирования отдельных стран должны также обеспечивать однородную структуру сети и устранять возможность обнижения точности определения величины движений при переходе одной сети к другой.

Уже проектируя исследовательскую сеть следует иметь в виду ее уравнивание; будет ли это однородная сеть или предвидится ее загущение линиями низшего класса, что будет влиять на отбор линии и ход измерения.

3. Исходная поверхность вертикальных движений земной коры

Значение выбора соответствующей исходной поверхности (горизонта относимости) растет вместе с увеличением исследовательской сети. Для большой сети обнимающей, например, поверхность страны, выбор исходного пункта (исходной поверхности) имеет большое значение, так как сильно влияет на разложение ошибок определения - возникающих из ошибок измерения - и на возможность сравнения определенных относительных величин движений земной коры, с их абсолютными величинами. Исходный пункт должен быть расположен центрально относительно сети и находиться в районе, относительно которого можно предполагать, что там не выступают движения локального характера, которые влияли бы на получаемую картину движений.

Чтобы устранить такую возможность целесообразным является определять абсолютные скорости движений, отнесенные к среднему уровню моря, поскольку положение исследуемой территории и многолетние наблюдения уровня моря на уровенных постах делают это возможным.

Отнесение определяемых скоростей движений к среднему уровню моря имеет следующие достоинства:

а) позволяет определять скорости движений, которые, в некоторой степени, могут быть сравниваемы с абсолютными скоростями движений, независимо определенными в других районах континентов,

б) уменьшает величины ошибок определения, так как дает возможность многократной и в разных точках связи с исходной поверхностью.

Это второе достоинство выступает особенно тогда, когда определение вертикальных движений земной коры ведется на очень большой территории (например на части континента), когда исследовательская сеть связана с большим числом отдаленных от себя точек, определяющих принятую исходную поверхность.

Так как при определении движений земной коры очень важным является постоянство исходной поверхности, а не ее высота, то как исходные можно принять несколько разных, но постоянных поверхностей. Из этого следует возможность отнесения определяемых движений к средним уровням нескольких, смывающих данную территорию, морей.

В исследовательскую сеть должны быть включены, путём привязки нулей, уровнемерные посты и мареографы, по возможности равномерно размещенные вдоль морских побережьи. При отборе уровнемерных станций следует руководиться длиной периода веденных на них постоянных наблюдений уровня моря, а также учитывать их положение и местные влияния (конфигурации побережья, влияние ветра, близость устья реки и другие).

Так как принимаем средний уровень моря (не учитывая изменений эвстатического характера) как неизменный, то v ср. ур. м. = 0, и следовательно, определенные на основании многолетних наблюдений, систематические изменения среднего уровня моря можно считать следствием изменений нуля поста, т. е. следствием вертикальных движений данного района побережья.

Для контроля постоянства закрепления уровнемерной рейки, следует периодически, прецизионным нивелированием проверять превышения между нулем рейки и двумя-тремя контрольными реперами, расположенными на небольшом расстоянии от береговой линии.

Определенные, исходя из неизменного среднего уровня моря, и проверенные скорости движений нулей уровнемерных постов являются исходными величинами (v_0) для определения скоростей вертикальных движений в целой исследовательской сети.

4. Закрепление пунктов исследовательской сети

Фундаментальные реперы

Как уже выше было сказано, сеть прецизионного нивелирования для исследования современных вертикальных движений земной коры будет состоять преимущественно из уже существующих линий, а новые проектируемые линии нивелирования должны одновременно отвечать условиям для линии государственного нивелирования. Ввиду этого следует обсудить, в какой степени применяемое до сих пор закрепление и густота реперов на линиях нивелирования отвечают условиям исследовательской сети.

Проблема исследования движений на большой территории обыкновенно не ставит существующим сетям прецизионного нивелирования общего условия загущения линии добавочными реперами. Существующее загущение реперов - отвечающее техническим и хозяйственным надобностям, при общих исследованиях является достаточным.

Закрепление новых реперов может быть однако в некоторых случаях желательным, имея в виду целесообразность их загущения в районах, в которых выступают большие и сложные движения, или в районах, где существующее закрепление реперов по причине климатических и геоморфологических условий не гарантировало достаточной стабильности реперов. Новозакладываемые реперы должны быть или земными реперами, заложенными ниже уровня замерзания грунта, или стенными реперами на фундаментальных постройках.

Было бы очень желательным, чтобы места закладки новых реперов, а также существующие реперы были обследованы геоморфологически, что позволило бы при будущей обработке данных повторного нивелирования иметь в своем распоряжении более достоверный, с добавочной характеристикой материал, а также уменьшило бы число исключаемых реперов, проявляющих движения определенно локального характера.

Размещение фундаментальных (вековых) реперов должно отвечать не только их нормальной роли в сети нивелирования, но и учитывать геологическое строение и тектонику исследуемой территории. Фундаментальные реперы должны быть закреплены на каждой отдельной тектонической структурной единице.

5. Измерение исследовательской сети

Скорости современных вертикальных движений земной коры преимущественно не превышают величины I см/год.

Разности между скоростями вертикальных движений соседних территорий тоже соответственно малы, вследствие чего разности превышений можно констатировать - и то по истечении большого промежутка времени - только между отдаленными между собой на несколько километров пунктами. Ввиду этого решительную роль имеет точность нивелирования. Снижение точности ниже некоторой границы обесценивает повторное нивелирование как метод исследования вертикальных движений.

Сеть повторного нивелирования для исследования современных вертикальных движений земной коры должна характеризоваться прежде всего высокой точностью и коротким периодом времени ее измерения, причем эта вторая черта способствует получению определяемых величин движений с большей точностью.

Учитывая вышесказанное, исследовательская сеть - особенно при исследовании больших территорий - должна состоять из линии прецизионного нивелирования I класса, а в случае объединенной исследовательской сети ряда стран, очень желательно, чтобы измерение сетей каждой из стран было исполнено с одинаковой точностью и по возможности, соблюдая одинаковую методику. Все составляющие сети должны быть наблюдаемы в одном для всех сетей периоде времени, причем этот период не должен превышать нескольких лет.

Было бы полезным для сокращения периода времени нивелирования, применение прецизионных автоматических нивелиров.

Является очень желательным, чтобы каждое повторное нивелирование отдельных линий сети производилось в одном и том-же времени года, что в некоторой степени элиминировало бы из определяемых величин движений земной коры влияние периодических изменений положения репера, вызываемых сезонными изменениями температуры и влажности грунта.

Одновременно с нивелированием всей сети - в случае определения абсолютных скоростей движений - должно быть произведено связывающее нивелирование нулей уровнемерных станций и мареографов, особенно тех, показания которых будут использованы для определения среднего уровня моря, т. е. исходной поверхности определяемых движений.

6. Минимальный интервал времени между очередными повторными нивелированиями исследовательской сети

Геодезический метод исследования современных вертикальных движений земной коры путем повторного нивелирования, кроме несомненных достоинств, имеет и серьезный недостаток. Именно, первое независимое определение скорости движений земной коры на исследуемой территории получаем лишь только после истечения многолетнего периода времени.

Причины этого следующие:

а) из-за экономических причин повторное нивелирование больших сетей производится через большие интервалы времени (десятки лет),

б) сравнение повторных нивелирований, разделенных меньшим интервалом времени было бы неценным по причине невыгодного в таком случае соотношения между ошибкой определения и определяемой величиной.

В несовершенстве метода содержится и другой его недостаток. В действительности повторное нивелирование дает величины вертикальных перемещений знаков за период времени между повторными нивелированиями, а не скорости их движений. Значения скорости получаем только принимая постоянство скорости в этом периоде, то есть принимая $V_n = const$.

Хотя эта предпосылка обоснована и является единственной возможной в существующих условиях, в ней кроется источник ошибки. Соблюдая точность, правильнее было бы говорить о определении средних скоростей.

Имея в виду факт, подчеркнутый выше в пункте б), следует определить такой минимальный интервал времени между повторными нивелированиями, который давал бы уже возможность использования этих измерений для исследования вертикальных движений.

Периодичность повторных нивелирований исследовательской сети можно с большим приближением определить исходя из формулы:

$$V'_n = \frac{H''_n - H'_n}{T'' - T'} = \frac{\Delta H_n}{\Delta T},$$

где:

$H''_n - H'_n = \Delta H_n$	разность высот n-того пункта сети, определенная из двух нивелирований сети,
$T'' - T' = \Delta T$	средний интервал времени между двумя нивелированиями сети,
V'_n	относительная скорость n-того пункта сети определенная относительно исходного уровня (пункта) нивелирования.

Получаем следующую зависимость между интервалом времени и ошибками отдельных величин:

$$\Delta T = \pm \frac{1}{m_{V'_n}} \cdot \sqrt{V'^2_n \cdot m^2_{\Delta T} + m^2_{\Delta H_n}}$$

Элементами имеющими наибольшее влияние на величину минимального интервала времени между нивелированиями на территориях, на которых не проявляются большие скорости движения земной коры, как например для территории Польши, и для случая, когда нивелирование сети может быть исполнено в течение короткого периода времени, являются следующие величины:

а) ошибка определения разности высоты n-того пункта ($m_{\Delta H_n}$), зависящая от предполагаемой точности обоих нивелирований, от протяжения сети, а особенно от расстояния (D) данного пункта от исходного пункта сети.

б) принятая, допустимая величина средней ошибки определения относительных скоростей движения земной коры ($m_{V'}$).

Из вышесказанного следует, что величину минимального интервала времени (ΔT_{min}) можно определить, пользуясь упрощенной формулой, подставляя наибольшее ожидаемое значение средней ошибки определения разности высот пункта сети ($m^{max}_{\Delta H}$) и принимая допустимую величину ошибки определения скорости движений ($m^{dop.}_{V'}$).

Получаем:
$$\Delta T_{min} \approx \left| \frac{m^{max}_{H}}{m^{dop}_{V'}} \right|$$

Если желаем определить абсолютные скорости вертикальных движений земной коры (V), то следует учитывать также скорость движения исходного пункта (пунктов) - (V_o) и ошибку ее определения (m_{V_o}). Принимая допустимую величину ошибки определения абсолютных скоростей движения земной коры (m_V^{dop}), получим допустимую величину ошибки относительной скорости движения:

$$m_{V'}^{dop} = \pm \sqrt{(m_V^{dop})^2 - m_{V_0}^2}$$

Чтобы наглядно представить вышесказанное, представим величины минимальных интервалов времени между очередными наблюдениями сети нивелирования, вычисленными при предпосылке, что средние квадратические ошибки определяемых относительных скоростей вертикальных движений не могут (для наиболее невыгодно расположенного пункта сети) превышать величины: $m_{V'}^{dop} = \pm$ I мм/год. Средние квадратические ошибки определения разности высот пункта вычислено по формуле:

$$m_{\Delta H} = \pm \sqrt{\eta^2 + \sigma^2 \cdot L_m} \cdot \sqrt{2D}$$

при предпосылке, что:

$$\eta = \pm 0{,}5 \text{ мм} \cdot \sqrt{\text{км}}, \quad \sigma = \pm 0{,}04 \text{ мм/км},$$

L_m = 50 км - средняя длина отрезка линии нивелирования, для которой можно принять постоянство систематической ошибки.

(Вышеподанные величины являются величинами порядка величин ошибок измерения современных сетей прецизионного нивелирования, в том числе и польской сети).

Получаем следующие, приблизительные - так как не учитываем влияния уравнивания сети - величины минимальных интервалов времени между нивелированиями сети, в зависимости от наибольшего расстояния (D) пунктов сети от исходного пункта нивелирования.

Максимальное расстояние (D) километры	I00	200	400	600	800	I000
Минимальный интервал времены (ΔT_{min}) годы	8	II	I6	20	23	27

Вполне понятно, что если примем величину допускаемой ошибки определения скорости движения n-кратно большей, то получим n-кратно меньший интервал времени. Однако принятая величина допускаемой ошибки определения скорости должна быть решительно меньше известных, или ожидаемых величин скоростей современных вертикальных движений земной коры на исследуемой территории, так как только в таком случае результаты определения скоростей будут иметь полную ценность.

The methods of geodetic determination of the recent vertical movements of the earth's crust

von

T. W y r z y k o w s k i (Warschau)

S u m m a r y

The recent vertical movements of the Earth's crust on the great areas are determined with geodetic methods by comparison of the results of reiterated precise levelling. There are several methods of using the levelling data. The choice of one of these methods depends on such things as: the quality of data of the second levelling, the accuracy to be obtained, the possibilities of eliminating subjectiv influences etc.

This paper deals with different methods of using the levelling data, their advantages and disadvantages, giving as well the limitations of their use.

Z u s a m m e n f a s s u n g

Die gegnwärtigen Vertikalbewegungen der Erdkruste von großen Gebieten werden mit Hilfe geodätischer Methoden durch den Vergleich der Ergebnisse von wiederholten Präzisionsnivellements bestimmt. Es gibt mehrere Methoden, diese Nivellementselemente zu verwenden. Die Wahl einer dieser Methoden hängt von folgenden Faktoren ab: Der Qualität der Daten des Wiederholungsnivellements, der erhaltenen Genauigkeit, der Möglichkeit der Ausschaltung von subjektiven Einflüssen usw.

Diese Arbeit berichtet über verschiedene Methoden für die Verwendung der Nivellementsdaten, ihre Vorzüge und Nachteile, und gibt auch ihre Verwendungsgrenzen an.

1. I n t r o d u c t i o n

The geodetic method of determination of the recent vertical movements of the Earth's crust by comparison of the results of successive precise levellings enables to determine the value of vertical displacement of a given point in the time interval between the measurements. Dividing the value of this displacement by the interval of time, we obtain the average (mean) velocity of the movement of point in this time interval.

Basing on data of reiterated levellings, collected till now, one can come to the conclusion, that, if the time interval is long enough - twenty or thirty years for instance - the mean velocity of the movement will be then quite reliable, because the short - period fluctuations of movements are thus eliminated.

There are several methods of use the reiterated levellings for determination of the recent vertical movements of the Earth's crust on large areas. The choise of one of them depends of such factors, as: the quality of the results of successive levellings, the accuracy in determination of the movements to be obtained and of elemination of subjectiv influences.

We shall describe here these methods, their advantages and disadvantages, giving characteristic concerning limitations of their use, as well.

2. The determination of the vertical movements by comparison the heights of points, determined from the independend adjustments of the two nets of precise levelling

This method consists in the comparison the heights of points from independent adjustment of two levelling nets, for determination of the vertical movements of the Earth's crust. The heights of points have to be computed in the same systems of heights and referred to the same reference level.

The velocity of movement of the point (n), common to both nets, in regard to the reference level, which we admit to be constant in the time between the measurements, can be expressed, as:

$$V'_n = \frac{H''_n - H'_n}{T'' - T'} = \frac{\Delta H_n}{\Delta T_{śr}} , \qquad (1)$$

where $\Delta T_{śr}$ is the mean value of the time interval between the measurements of the nets.

If the reference level (point) of the movements of the Earth's crust differs from the reference level (point) of the levelling, we shall have the following relation:

$$V_n = V_0 + V'_n \qquad (2)$$

According to the fact, whether we admit for the reference point of the movement to be determined the velocity $V_0 = 0$, or we determine it with reference to the mean sea level, we shall obtain for all the points the relative (V') or absolute (V) velocities.

The mean error of determination of the relative velocity of the vertical movements can be expressed by the following approximate formula:

$$m_{V'_n} = \pm \frac{1}{\Delta T_{śr}} \sqrt{V'^2 \cdot m^2_{\Delta T} + m^2_{\Delta H_n}} \qquad (3)$$

where: V' = the average value of velocity of movements of the Earth's crust between the reference point of the levelling and the given point,

$m_{\Delta T}$ = the average value of differencies between the real intervals of time separating the consecutive measurements of individual lines and the mean interval $\Delta T_{śr}$,

$m_{\Delta H_n}$ = the mean error of determination of the height difference of the nth point.

It results from this formula, that the high accuracy of the determination can be obtained, if a relatively long time interval separates the consecutive measurements and if the measurements of nets are made quickly (little value of $m_{\Delta T}$).

The advantages of the method of comparison of the adjusted heights are the following:

a) if the lines of reiterated levellings do not form the net of circuits, that is when the adoption of any other geodetic method of determination of movements is impossible, this method will enable yet to obtain the homogenous (to a certain degree) determination, referred to the common reference level [4],

b) it is possible to use this method in the case, when only the data from catalogue (the heights of points) without the field materials are available,

c) by using the adjusted values of heights in this method, the use is made of superabundant conditions of both levellings,

d) the method does not require much work, being based on the data obtained from the previous made adjustments of the levelling nets.

The weak features of this method are:

a) the uncertainty of the estimation of the error of ΔH determination, because the height differences in each net were separately adjusted, using conditions: (p'v'v') = minimum and (p"v"v") = minimum, do not connect one with the other,

b) the influence of the movement of points during the measurements of the net on the results are not eliminated in this method. As a result of this influence systematic errors are introduced to the results of adjustment [6],

c) the use of the mean interval of time ($\Delta T_{śr}$) induces an additional source of error to the determination of the movements.

For the diminuation of the influence of the two last factors mentionned above, the method of succeeding approximations can be used, but then the field data of the two compared nets are necessary [2]. This method consists in the use of velocities of the movements of the nodal points of nets (obtained from the first comparison of heights), for the reduction of the height differences of individual lines of these nets to the corresponding epoches [6], which are different for both nets. The data of both nets, computed in such a manner, are then adjusted for the second time. The heights of points thus determined are the base for determination of the second approximation of velocities of movements of the Earth's crust. The series of succeeding approximations is the more convergent, the smaller is the error of the levelling measurements, as compared with the values of reductions we have introduced.

3. Determination of the vertical movements by means of adjustment of differences of height differences determined from two measurements of the precision levelling nets, or by the common adjustment of differences of height differences determined from two measurements of nets

On the contrary to the former method, the lines of the second levelling have to form here a net of circuits. Thus this method can be used in the case, when the two different nets are almost identic, or when a special research net is established, on which the measurements are periodically repeated.

The two methods of equal theoretical value can be used here:

a) the joint adjustment of the measured differences of height differences, obtained from the two measurements of the net, with the condition (p'v'v') + (p"v"v") = min.,

b) the adjustment of the measured differences of height differences, obtained from the two measurements of the net, with the condition (pvv) = min., where v is the correction of the measured difference of height differences.

The values of unknowns, determined by the two methods, will be the same. It means, that the height differences (h''_i, h'_i), when adjusted according to the condition (p'v'v')+ + (p"v"v") = min., will satisfy with the differences of height differences (Δh_i), when adjusted according to the condition (pvv) = min., the equation: $h''_i - h'_i = \Delta h_i$. The characteristic of accuracy of the unknowns thus determined will be the same in both cases [3].

For the large nets it is more convenient to use the method of differences of height differences, because the number of unknowns and normal equations will be twice lesser.

This method of determination of the vertical displacements of points is more correct from the point of view of the compensating computation, as that described under point 2. As it is shown in the publication [3] (§ 44 of this publication is written by J. Gaździcki), the vertical displacement of points, when determined from the two separate adjustments of the two measurements of a levelling net and of their joint adjustment, will be equal only in this case, when the weights of the two measurements of the net are proportional. But the accuracy of determination of the displacements will be different in both cases.

According to the length of time intervals, in which each measurement of the net was made, it could be necessary to reject the supposition, that the point haven't moved at the time of measurement. If the computation of the variations of the heights of points by taking sums of the adjusted differences of height differences and the determination of velocities from the formula (1) have shown, that the variations of the height differences during the measurement of the net were remarkable, such, that the values of the reduction to the mean moment of each measurement have reached the values of accidental errors of this measurement, it would be necessary to introduce such reductions.

The introduction of reductions to the height differences of lines or to the conditions of closure [6] will enable to repeat the joint adjustment and to compute the second approximation of values of the vertical displacements of points and, as consequence - the velocities of movements. Thus the procedure will be here the same, as described under the point 2.

In the both methods (p.49, and p.50), described here for the determination of the vertical movements of the Earth's crust either separate adjustments of the two measurements of nets or the joint adjustment of both measurements were used.

To form the error equations by the method of indirect observations, the following relations between the heights of points and the differences of heights were used:

$$\begin{aligned} &\text{from the first measurement: } H'_{k_i} - H'_{P_i} = h'_i, \\ &\text{from the second measurement: } H''_{k_i} - H''_{P_i} = h''_i, \end{aligned} \qquad (4)$$

or for the height differences:

$$(H''_{k_i} - H'_{k_i}) - (H''_{P_i} - H'_{P_i}) = h''_i - h'_i = \Delta h_i \qquad (4a)$$

4. The determination of the velocity of the vertical movements by introducing the additional relation: $\Delta h = \Delta T \cdot \Delta V$ to the adjustment of differences of height differences, determined from the two measurements of a levelling net

As it was already said, the methods of determination of the recent movements of the Earth's crust, described above, are inaccurate, owing to the same existing of these movements; the vertical displacements of points, as determined by these methods, can't be referred to the strictly defined interval of time. It is especially significant for the cases, when the intervals of time between the repeated measurements of the same lines of a net much differ one from another.

Taking into account, that in the determination of the velocity of the actual vertical movements of the Earth's crust by geodetic methods, we are obliged to assume, the velocities of these movements to be constant in a given interval of time, it is possible to make use of this assumption already in setting conditions for the adjustment.

A. The method of indirect observations

Let's take the equation (4a), mentionned above, which we write here in a slightly changed form:

$$\Delta H_{K_i} - \Delta H_{P_i} = \Delta h_i \qquad (6)$$

Knowing the time interval between the measurements (ΔT_i), we can write:

$$\Delta T_i \cdot V_{K_i} - \Delta T_i \cdot V_{P_i} = \Delta h_i \qquad (7)$$

where V_{K_i} and V_{P_i} are respectively the velocities of movement of the end and initial point of the i - levelling line.

We divide the both sides of the equation (7) by ΔT_i and thus we obtain the equation of the velocity's increase between the end point of the levelling line:

$$V_K - V_P = \frac{\Delta h_i}{\Delta T_i} = \Delta V_i \,. \qquad (8)$$

The formulas (7) and (8) are different from these, we normally use to set the error equations in adjustment a levelling net. This is why we shall give the form of the error equations, which derives from the equations (7) and (8):

$$\Delta T_i \cdot dV_{K_i} - \Delta T_i \cdot dV_{P_i} + \Delta T_i (V^o_{K_i} - V^o_{P_i}) - \Delta h_{i_{obs.}} = v_{\Delta h_i} \,, \qquad (7a)$$

$$dV_{K_i} - dV_{P_i} + (V^o_{K_i} - V^o_{P_i}) - \frac{\Delta h_{i_{obs.}}}{\Delta T_i} = v_{\Delta V_i} \,, \qquad (8a)$$

where:

$V^o_{K_i}$, $V^o_{P_i}$ = the approximate values of velocities of the vertical movements of the end and initial point of the i - levelling line,

dV_{K_i} , dV_{P_i} = the corrections to the approximate values of the velocities,

$v_{\Delta h_i}$ = the correction to the observed difference of height differences of the i - line,

$v_{\Delta V_i}$ = the correction to the increase of velocity between the endpoints of the line expressed as: $\frac{\Delta h_{i_{obs.}}}{\Delta T_i}$

For the equations we give the weights, derived from the observations made in the net of repeated levelling:

for the equations of the type (7a): $p_i = \frac{K}{m^2_{\Delta h_i}}$, (9)

for the equations of the type (8a): $p'_i = \frac{K}{m^2_{\Delta V_i}} = \frac{K \cdot \Delta T_i^2}{m^2_{\Delta h_i}}$ (10)

After introducing the weights, the normal equations, resulting from the error equations of the type (7a) or (8a) will be identical. This is why, only the convenience of the computation will decide, whether we choose this or that type of equations.

After solving the normal equations and computation of the values dV_i, we can determine the mean error of unit weight observation, as:

$$m_0 = \pm\sqrt{\frac{(p \cdot v^2_{\Delta h})}{n}} = \pm\sqrt{\frac{(p' \cdot v^2_{\Delta V})}{n}} \,, \qquad (11)$$

where: n = the number of superabundant observations, and then the mean errors of the unknowns.

B. The method of the conditioned observations

Like in the method of indirect observations (see under A), we assume here, that the velocity of the movement of each point is constant, but we take any assumption as to the sign or magnitude of this movement. In this case, instead of the equation (5a) rather the following one is more suitable:

$$\Delta V_1 + \Delta V_2 + \Delta V_3 + \dots\dots\dots + \Delta V_n = 0 \quad , \qquad (12)$$

or:

$$\frac{\Delta h_1}{\Delta T_1} + \frac{\Delta h_2}{\Delta T_2} + \frac{\Delta h_3}{\Delta T_3} + \dots\dots\dots + \frac{\Delta h_n}{\Delta T_n} = 0 \; , \qquad (12a)$$

where: ΔV_i = increase of velocity of movement between the end point of the i - line of the circuit

Δh_i = the change of the height difference between the end points of the i - lines in the time interval between the two measurements (ΔT_i).

Here again, as before, we can give to the condition equations two formally different forms, viz.:

$$v_{\Delta V_1} + v_{\Delta V_2} + \dots\dots + v_{\Delta V_n} + (\Delta V_1 + \Delta V_2 + \dots\dots + \Delta V_n) = 0, \qquad (13)$$

or:

$$\frac{1}{\Delta T_1} \cdot v_{\Delta h_1} + \frac{1}{\Delta T_2} \cdot v_{\Delta h_2} + \dots\dots + \frac{1}{\Delta T_n} \cdot v_{\Delta h_n} + \left(\frac{\Delta h_1}{\Delta T_1} + \frac{\Delta h_2}{\Delta T_2} + \right.$$

$$\left. + \dots\dots + \frac{\Delta h_n}{\Delta T_n} \right) = 0, \qquad (13a)$$

where: $\Delta V_i = \frac{\Delta h_i}{\Delta T_i}$.

We give the weights to ΔV_i and Δh_i , according to the formulas (10) and (9).

The further procedure for the determination of adjusted values of ΔV_i , V_i , the mean error of the unit weight observation, and eventually the mean error of V_i after the adjustment don't require explanations.

When estimating the approximate error of determination of the velocity of movement of the net's point by this method, we can postpone the first term under the squareroot in the formula (3).

The consequences of the assumption, given at the beginning of this pragraph and of the procedure described here, are the following:

a) it is not necessary to use the method of approximations for the determination of the velocities of movement of points of the levelling net, because the values obtained from the adjustment are the definitive ones,

b) the velocities of movement of points are dermined from the strictly defined intervals of time,

c) for the sums of variations of height differences in a circuit there is not the false condition $\Sigma\Delta h = 0$ put on, and these sums don't fulfil the false assumption H_i - constant for the whole time of the measurements of the net, that means, for instance, that we don't assume the height of a nodal point, determined from the measurements of several lines measured in different times, has to be the same,

d) the corrections of the differences of height differences, obtained from the adjustment fulfil the condition (pvv) = min., but the sum of the differences of height differences in a circuit will not be zero, but it will be equal to the difference between the theoretical values of discrepancy of the circuit, computed for the two measurements of net, to be compared.

Thus, if the theoretical value of discrepancy in closure of the circuit due to the influence of movements of the Earth's crust, during the first (earlier) measurement of a net is Δ'_p [6] and that of the second (later) measurement is Δ''_p, then the theoretical value of the discrepancy of the adjusted differences of height differences in this circuit will be:

$$\Delta_p = \Delta''_p - \Delta'_p \quad , \tag{14}$$

which can be computed otherwise, as:

$$\Delta_p = \sum_{i=1}^{n} \Delta V_i \cdot \Delta T_i \tag{15}$$

The results obtained from (14) and (15) will give the control of the computations. For each circuit we shall have, after the adjustement:

$$\Sigma \Delta h_i = \Delta_p , \tag{16}$$

As we can see, the method described above under point 4, is the most suitable one for the determination of the recent movements of the Earth's crust; this method was already partly used in the practice [5]. It can be used as well for the comparison of the two partly overlapping levelling nets, or by the periodical measurements of a specially established research net and in all the cases, when we can reject the possibility of unexpected and sudden variations of the height of the point.

5. The connection of the determined movements of the Earth's crust to the reference level

As we have said above, the vertical movements of the Earth's crust and the velocities of these movements can be determined as relative, referred to an arbitrarily choosen point (reference level), or as absolute, that is, referred to the mean sea level, adequately determined.

In the first case, we choose normally one point of reference, which we consider to be constant ($V = 0$), thus the errors of the determined relative velocities are due to the errors in the levelling measurements only.

In the case of the connection of a net to one point, for which the vertical component of the velocity of movement ($V_0 = K$) was made from the long - period observations of the sea level, the determined values of the absolute movements in the whole net will be additionally influenced by the error of determination of the velocity of the initial point.

If the net of the second levelling is based on several points, the absolute velocities (V_0) of these points being determined (height marks near the mareographs), then the additional mathematical conditions will appear; these conditions have to be introduced to the adjustment of the net.

According to the magnitude of errors of the values V_0 in particular points, the procedure may be different. If these errors are small - in comparison with the errors of measurement of the net - one could regard the initial values as the true ones. If it is not the case, it would be better to regard the initial values V_0 as the observations with appropriate weights, thus these values will obtain corrections from the adjustment [1].

Bibliography

[1] Hausbrandt, S.: "Analiza porównawcza dokładności wielkotrójkątowych i małotrójkątowych sieci triangulacyjnych nawiązana do prac geodezyjnych

w Polsce" (Rozdz. XII) Prace Inst. Geod. i Kart., t. III, nr 1/5, Warszawa 1955

[2] Kukkamäki, T.J.: "The land uplift in Finland determined with two levellings as well as with water level observations", Bull. géod. Nr 36, Londyn 1955

[3] Lazzarini, T.: "Geodezyjne pomiary odkształceń i ich zastosowanie w budownictwie", Warszawa 1961

[4] Niewiarowski, J.; Wyrzykowski, T.: "Détermination des mouvements verticaux de l'écorce terrestre sur l'étendue de la Pologne par méthode de répétition des nivellements de précision", Bull. Géod. No 62, Paris 1961

[5] "Sowriemiennyje wiertikalnyje dwizenija ziemnoj kory na tierritorii zapadnoj połowiny jewropiejskoj czasti SSSR". Trudy CNIIGAiK, nr 123, Moskwa 1958

[6] Wyrzykowski, T.:"Influence des mouvements systématiques verticaux de l'écorce terrestre sur les résultats et la détermination des grandeurs d'érreurs de nivellement de précision". Comm. présentée à XII Assemblée UGGI, Helsinki, Warszawa 1960

D i s k u s s i o n

J. A. M e s c h e r i k o v (Moskau):

I would like to note the big principal significance of researches made by Dr. Wyrzykowski. In his researches of the values of level accuracy and methods of adjustment levelling network Dr. Wyrzykowski assumes that the earth's surface is not rigid (as it has been believed before in the study of these problems) but it is mobile. The consideration of the movements of the earth's crust leads Dr. Wyrzykowski to principally new important conclusions which deserve the closest attention of geodesists of other countries. The works in this direction should be continued and widened, because they are very important for the problems of construction of the basic geodetic networks and for the study of the recent movements of the earth's crust. In his further researches I would remind Dr. Wyrzykowski to give attention to the factors that indicate the changes of the sign and intensity of the crustal movements. It seems to me that Dr. Wyrzykowski, with depth and thoroughness inherent to him, can make valuable contribution to our knowledge of that problem. This would add the physical approach to the geometrical, i.e. with due consideration of the essence of the phenomena of crustal movements.

Stricte méthode de compensation des déplacements verticaux de l'écorce terrestre déterminés par les mesurages périodiques de nivellement de précision

von

A. Hermanowski (Warschau)

Summary

The method proposed is based on the simultaneous compensation of the initial and the actual measurings. The compensation is made by means of the method of conditioned observations with the application of the theory of compensations by groups and with the rejection of the condition of the stability of the reference points. The formulas and calculations are given in notations of Cracovian algebra. During the compensation the vertical displacement is referred not to a single point but to a fitting surface which is determined by coinciding points and which has approximately the same apparent displacement in the given area. After the compensation the fitting surface can be displaced parallel to any point. The mean errors of all fixed displacements are obtained by the compensation together with the errors of the displacements at the reference points. In order to increase the accuracy of the fixed displacements and to simplify the computations, the whole area ist divided into single parts, the size of which depends on the form of the canvas of the survey and on the presumed accuracy. The single parts of the periodical levelling of the canvas are referred by mutual parallel displacement of the levels to a common level and united by applying the theory of compensations by groups to form a uniform canvas.

Zusammenfassung

Die vorgeschlagene Methode beruht auf der gleichzeitigen Ausgleichung der Ausgangsmessungen und der aktuellen Messungen. Die Ausgleichung führt man nach der Methode der bedingten Beobachtungen mit Anwendung der Theorie von gruppenweisen Ausgleichungen und bei Ablehnung der Bedingung der Festigkeit der Bezugspunkte durch. Die Formeln und Berechnungen stellt man in Krakovianenalgebrabezeichnungen dar. Während der Ausgleichung bezieht man die vertikale Verschiebung nicht auf einen einzigen Punkt sondern auf eine sich anpassende Fläche, die durch Passpunkte bestimmt ist, und die ungefähr die gleiche scheinbare Verschiebung auf dem gegebenen Gebiet hat. Nach der Ausgleichung kann man die Passfläche parallel zu einem beliebigen Punkte verschieben. Die mittleren Fehler aller bestimmten Verschiebungen bekommt man bei der Ausgleichung zusammen mit den Fehlern der Verschiebungen auf den Bezugspunkten. Um die Genauigkeit der bestimmten Verschiebungen zu vergrößern und die Berechnungen zu vereinfachen, wird das ganze Gebiet in einzelne Teile geteilt, deren Größe von der Form des Netzes und der vorausgesetzten Genauigkeit abhängt. Die einzelnen Teile des periodischen Nivellements des Netzes werden durch gegenseitige parallele Verschiebungen der Niveauflächen auf ein gemeinsames Niveau gebracht und bei Anwendung der Theorie der gruppenweisen Ausgleichungen zu einem einheitlichen Netz vereinigt.

A b r é g é

Au cas ou les réseaux de nivellement périodiques sont rattachés a un seul point, l'erreur moyenne des grandeurs déterminées croît très vite en mesure qu'on s'éloigne du point initial. Quand les recherches embrassent de grandes étendues du terrain on arrive a obtenir de grandes erreurs moyennes lesquelles, comparées aux déplacements relativement petits en hauteur, mettent en doute la réalité de toute détermination du mouvement des points de l'écorce terrestre situés a la lisière de l'étendue en question.

Sur une étendue de terrain limitée faisant partie des nivellements périodiques, il est possible de déterminer le déplacement relatif des points avec une grande exactitude. Si l'on pouvait trouver dans le terrain en question une série de points ayant les déplacements identiques (dans les limites d'exactitude de la détermination), il serait possible de procéder au calcul et à la compensation des déplacements verticaux en s'appuyant sur ces points.

Cette solution nous amène à la conception d'une surface défini par les points des déplacements égaux et au calcul subséquent des déplacements relatifs par rapport à cette surface. Cette surface de référence constitue alors un lieu géométrique des points aux déplacements égaux quoique non définis en grandeur. Le déplacement de la surface de référence n'étant pas définie, celle-ci peut être mue dans la direction verticale, ce qui rendra possible la réunion des points, subdivisés en groupes, en un ensemble rapporté à un seul point arbitraire de référence. Je suis d'avis qu'une telle solution va augmenter l'exactitude de détermination des déplacements et la rendra indépendante de la grandeur de l'étendue étudiée, au moins dans une large mesure.

La méthode proposée consiste en une compensation simultanée des mesurages initial et actuel. Sous le nom de mesurage initial nous entendons le nivellement de précision primaire, comme mesurage actuel - le mesurage secondaire effectué après une période t de temps.

La compensation est faite par la méthode conditionnée avec l'application de la théorie de compensation par groupes et avec l'abandon de la condition de stabilité des points de référence. Pour le calcul et les formules on a appliqué les symboles de l'algèbre cracovienne. Procédant à la compensation, les déplacements verticaux au lieu d'être rapportés à un seul point, sont référés à une surface d'ajustement β définie par les points d'ajustement ayant approximativement les mêmes déplacements apparents sur une étendue donnée. Après la compensation la surface d'ajustement peut être déplacée verticalement à un point arbitraire en principe, souvent présentant une signification particulière. La compensation finale permet de calculer toutes les erreurs moyennes de déplacements celles par rapport aux points de référence.

Dans le but de simplifier les calculs et d'augmenter l'exactitude de détermination des points, il est à recommander de diviser l'étendue soumis à l'étude en parties.

La grandeur des parties particulières dépend:

1) de la précision, fixée à l'avance, de détermination de la surface d'ajustement, laquelle à son tour dépend de l'exactitude demandée pour la définition de la vitesse de mouvement de l'écorce terrestre,
2) de la forme et grandeur des polygones des lignes de nivellement de précision.

Afin de simplifier les calculs, l'organisation des nivellements périodiques doit tendre aux mesurage des parties séparées ne surpassant pas le double d'intervalle de temps t entre la mesure initiale et actuelle.

Les parties particulières du réseau d'un nivellement périodique étant réduite à seul niveau de référence grace au déplacement de surfaces d'ajustement, seront réunies en un seul ensemble par l'application de la théorie de compensation par groupes.

Avant de procéder à la compensation des parties séparées il faut corriger les différences de hauteur des points considérés par les corrections tenant compte de l'erreur theorique de fermeture des anneaux du réseau de nivellement. L'omission de ces corrections n'aura pas une influence sur la grandeur de déplacement de points, mais fera augmenter l'erreur moyenne d'une observation typique.

Compensation et calcul des déplacements verticaux d'une partie séparée

Nous supposons qu'une partie séparée d'une nivellement périodique forme un ou plusiers polygones fermés. Pour simplifier, je ne présente qu'un cas d'un seul intervalle de temps t entre les mesures. Le cas général de plusieurs intervalles t est décrit dans mon travail (rapport) l'exemple I.

L'étape I comprend une compensation indépendante du mesurage initial (Etape I^A) et du mesurage actuel (Etape I^B) en tenant compte uniquement des conditions de fermeture des polygones de nivellement. Dans ce but nous procédons en considérant les équations de condition du type

$$h_1 + v_{I1} + h_2 + v_{I2} + h_3 + v_{I3} + \ldots\ldots\ldots + h_n + v_{In} = 0$$

En outre nous calculons les cracoviens $\bar{U}_{IA}$ et $\bar{U}_{IB}$ des coefficients $Q_i Q_j$ des différences de hauteur h et h' d'après la formule

$$\bar{U} = ff - (f\tau a)\ \left[(\tau a)^2\right]^{-1}\ (f\tau a)$$

Après avoir corrigé les dénivellations h et h' en partant d'un point arbitraire situé dans la partie milieu du réseau, on calcule les déplacements verticaux apparents. Dans le cas ou la partie isolée est pourvue des cheminements ouverts et les conditions de fermeture ne peuvent pas être établies, l'étape I sera limitée au calcul des déplacements apparents directement d'après les documents de campagne. Dans les dit cas les cracoviens $\bar{U}_{IA}$ et $\bar{U}_{IB}$ sont égaux aux cracoviens respectifs des inverses des poids de dénivellations h et h'. Parmi les points de l'ensemble du réseau, dont les déplacements apparents ont été calculés, on choisit un groupe de points aux déplacements à peu près égaux et répartis uniformément sur toute l'étendue autant que possible.

Bien entendu il n'y a que les points indépendants, c.à d. situés assez loin l'un de l'autre, seront choisis. On leur donnera le nom de "points d'ajustement". Le nombre de ces points et la différence admissible des grandeurs des déplacements apparents, seront déterminés d'après les résultats d'une analyse détaillée du degré de précision des réseaux de nivellement existant à l'heure actuelle ou en project. Le rapport présenté ne tient pas compte de ces grandeurs, vu manque des données respectives.

Les différences des déplacements verticaux aux points d'ajustement sont dues:

1) aux différences minimes de vitesse des mouvements de l'écorce terrestre,
2) aux erreurs de détermination des déplacements verticaux.

Je suppose de traiter les déplacements verticaux des points d'ajustement comme étant déterminés directement. Comme surface de référence, dite alors "surface d'ajustement", nous admetterons une surface β qui va satisfaire à la condition minimum de sommes des carrés des corrections aux grandeurs de déplacements verticaux, conjointement avec les autres conditions de compensation. Nous aurons alors:

$$[pvv] + [pv'v'] + [p\delta\delta] = \text{minimum}$$

La surface β présente un lieu géométrique des points d'un déplacement identique, égale à A.

Tenant compte de cette supposition nous obtiendrons comme grandeurs des déplacements verticaux sur les points d'ajustement:

$$\Delta_i = A + \delta_i\ ; \qquad A = 0; \qquad \Delta_i = \delta_i$$

(Afin de simplifier les calculs on admet A = 0; cette supposition n'exerce aucune influence sur la grandeur des déplacements déterminés)

Les poids des déplacements verticaux sur les points d'ajustement ont été déterminés en se basant sur l'analyse de grandeur et de distribution des erreurs vraies, obtenues par l'étude pratique du problème sur le réseau expérimental de la Chaire de la Géodésie des Mesurages Spéciaux de l'Ecole Polytechnique de Varsovie.

É t a p e II. A j u s t e m e n t m u t u e l d e s m e s u r a g e s i n i t i a l e t a c t u e l

Les équations de condition d'ajustement sont de la forme

$$\delta_i + \left[(h' + v'_I) + v'_{II}\right] - \left[(h + v_I) + v_{II}\right] - \delta_{i+1} = 0$$

Des équations de condition on passe aux équations normales

$$k(\tau a \bar{U}_I \tau a) + \omega = 0$$

$$\bar{U}_I = \begin{Bmatrix} \bar{U}_{IA} & & \\ & \bar{U}_{IB} & \\ & & \bar{U}_D \end{Bmatrix}$$

$$\bar{U}_D = \begin{Bmatrix} P^{-1}_{\Delta 1}. & & \\ & \cdot & \\ & & \cdot\ P^{-1}_{\Delta N} \end{Bmatrix}$$

$\frac{1}{i}$ étant l'inverse du poids de déplacement sur le point d'ajustement.

Les équations secondaires sont calculés d'après la formule

$$v_{II} = k\ (\bar{U}_I \tau a).$$

erreur moyenne d'une observation typique pour une partie isolée donnée sera

$$\mu_0^2 = \frac{-[k\omega]_{IA} - [k\omega]_{IB} - [k\omega]_{II}}{W_{IA} + W_{IB} + W_{II}} = \frac{A}{W}$$

Après l'introduction des corrections secondaires aux déplacements $h + v_I$ et $h' + v'_I$ on calcule les valeurs des déplacements verticaux par rapport à la surface β en partant d'un point arbitraire d'ajustement. En même temps il faut tenir compte du fait que le déplacement Δ_i sur le point i-ème d'ajustement est égale à une correction correspondante δ_i . Les déplacements calculés sont portés une carte et on procède alors à une analyse des grandeurs déterminées. Pour les calculs qui vont suivre nous marquons les points caractéristiques de grandeur du mouvement et les points nécessaires pour la jonction de la partie donnée du réseau aux autres.

Pour les déplacements sur les points choisis on calcule le cracovien $\bar{U}_{II}$ des coefficients de poids $Q_i Q_j$

$$\bar{U}_{II} = f\bar{U}_I f - (f\bar{U}_I\ \tau\ a)\cdot(\tau a\bar{U}_I \tau a)^{-1}\cdot (f\bar{U}_I \tau a)$$

A j u s t e m e n t d e s p a r t i e s p a r t i c u l i è r e s d e s n i v e l l e m e n t s p é r i o d i q u e s

Comme résultat de compensation mentionnée plus haut nous obtiendrons les déplacements verticaux des différentes parties du réseau calculé, suivant le cas, par rapport aux différentes surfaces d'ajustement et pour différentes périodes du laps de temps qui sépare les mesurages. En supposant l'uniformité du mouvement de l'écorce terrestre nous pouvons réduire les déplacements obtenus aux différentes parties à une seul période de temps entre les observations.

Les cracoviens $\bar{U}$ des coefficients de poids Q_iQ_j seront aussi réduits à la même période de temps que les déplacements en tirant profit du théorème suivant: "l'erreur moyenne du produit d'une grandeur mesurée avec une erreur moyenne $\pm$ m par une constante c est égal au produit $\pm$ cm".

Les déplacements calculés sont alors réduits à une seule surface d'ajustement β, arbitraire en principe.

Le déplacement de toutes les surfaces β au même niveau étant effectué, nous obtenos sur les points communs aux différentes parties du réseau les déplacements verticaux provenant des erreurs d'ajustement.

On fera disparaitres ces dernières divergences en effectuant une compensation jointive de toutes les parties du réseau.

Les équations de condition des parties particulières ont la forme

$$\Delta_i^{I} + \delta_i^{I} = \Delta_i^{II} + \delta_i^{II}$$

Dans le cas d'un point commun à N parties il y aura N-1 équations. Des équations de condition on passe aux équations normales

$$k(\tau a \bar{U} \tau a) + \omega = 0$$

ou

$$\bar{U} = \begin{Bmatrix} \bar{U}_I & & & \\ & \bar{U}_{II} & & \\ & & \ddots & \\ & & & \bar{U}_N \end{Bmatrix}$$

Après avoir introduit les corrections a tous les déplacements verticaux, on peut déplacer parallélement la surface commune β au point qu'on a admis comme initial pour le calcul de la vitesse du mouvement de l'écorce terrestre sur l'étendue donnée.

La grandeur de l'erreur moyenne d'une observation typique sera calculée d'après la formule

$$\mu_0 = \pm\sqrt{\frac{A_I + A_{II} + \cdot \cdot \cdot + A_N + D}{W_I + W_{II} + \cdot \cdot \cdot + W_N + W_D}}$$

ou A et W ont été calculés a la compensation des parties particulières, $D = -[k\omega]$ a été calculé a la dernière étape de compensation, et W_D exprime le nombre des conditions d'ajustement mutuel des réseaux particuliers.

La justesse de la conception présentée a été confirmée par les résultats d'un mesurage expérimental.

Les expériences pratiques tirant profit des déplacements connus des repères de nivellement, mesurés avec une haute précision, ont confirmé l'équité des principes admis a la base de connaissance des erreurs vraies.

Das Vertrauensintervall bei den gemessenen Erdkrustenbewegungen

von

J. Böhm (Prag)

S u m m a r y

The change of position of a stabilized mark is proved by means of geodetic measurements repeated after certain time intervals. This motion in space can be expressed by the components ΔX, ΔY, ΔZ.

We get real compensated values Δx, Δy, Δz of the components by the effect of the errors of measurement, even if no movement of the crust has taken place (thus even in the case of the hypothesis $\Delta X = 0$, $\Delta Y = 0$, $\Delta Z = 0$).

By means of the so-called "interval of confidence" the so-called critical limiting values can be calculated, if the found values Δx, Δy, Δz

a) are to be regarded only as the product of the errors of measurement (e.g. $\Delta x < m_{\Delta x}$);

b) allow a movement of the crust (e.g. $m_{\Delta x} < \Delta x < 2\ m_{\Delta x}$)

c) are evidence of such movements with practical reliableness (e.g. $\Delta x > 2\ m_{\Delta x}$).

The simple or double value of the real error ist taken as critical limit or the calculated mean error, if there is a sufficient number of observations. The influence of the accidental and the systematic errors, however, must be included in this value.

Z u s a m m e n f a s s u n g

Mittels der nach gewissen Zeitintervallen wiederholten geodätischen Messungen stellt man die Lageänderung einer stabilisierten Marke fest. Diese Bewegung im Raume kann mittels der Komponenten ΔX, ΔY, ΔZ ausgedrückt werden.

Durch die Wirkung der Messungsfehler bekommt man reale ausgeglichene Komponentenwerte Δx, Δy, Δz, auch wenn es zu keiner Krustenbewegung gekommen ist (also auch bei der Hypothese $\Delta X = 0$, $\Delta Y = 0$, $\Delta Z = 0$).

Mittels des s.g. Vertrauensintervalles kann man die s.g. kritischen Grenzwerte bestimmen, wenn die gefundenen Werte Δx, Δy, Δz

a) nur als Produkt der Meßfehler zu betrachten sind (z.B. $\Delta x < m_{\Delta x}$);

b) eine Erdkrustenbewegung zulassen (z.B. $m_{\Delta x} < \Delta x < 2 \cdot m_{\Delta x}$);

c) mit einer praktischen Sicherheit für solche Bewegung zeugen (z.B. $\Delta x > 2\ m_{\Delta x}$).

Man wählt als kritische Grenze den einfachen oder den doppelten Wert des tatsächlichen Fehlers, oder wenn eine genügende Zahl von Beobachtungen vorliegt, den berechneten mittleren Fehler. In diesem Wert muß aber auch der Einfluß der zufälligen und der systematischen Fehler mit inbegriffen sein.

Man kann eine Lageänderung einer stabilisierten Marke mittels der nach gewissem Zeitabstand wiederholten geodätischen Messungen feststellen. Dabei bezeichnen die Koordinaten

X_1 , Y_1 , Z_1 die w i r k l i c h e Raumlage der Marke bei der ersten Messung

X_2 , Y_2 , Z_2 dieselbe bei der zweiten (wiederholten) Messung.

Dann bedeuten die Koordinatenunterschiede

$$\Delta X = X_2 - X_1 \qquad \Delta Y = Y_2 - Y_1 \qquad \Delta Z = Z_2 - Z_1$$

die räumlichen Komponenten der w i r k l i c h e n Krustenbewegung im erwähnten Punkte.

Mittels der Messung stellt man niemals die wirklichen Koordinatenwerte fest, sondern durch Messung und Ausgleichung nach der Methode der kleinsten Quadrate bekommt man nur die sog. a u s g e g l i c h e n e n Koordinatenwerte

$$x_1 = [\alpha l]_1 \qquad y_1 = [\beta l]_1 \qquad z_1 = [\gamma l]_1$$
$$x_2 = [\alpha l]_2 \qquad y_2 = [\beta l]_2 \qquad z_2 = [\gamma l]_2$$

als lineare Funktionen der ursprünglichen direkten oder vermittelnden Beobachtungen (Längen, Winkel, Höhen, astronomische Größen). Das betrifft auch die ausgeglichenen Werte der gesuchten Komponenten

$$\Delta x = x_2 - x_1 \qquad \Delta y = y_2 - y_1 \qquad \Delta z = z_2 - z_1$$

Diese sind mit dem Einfluß der Beobachtungsfehler belastet, z.B.

$$\varepsilon_{\Delta x} = \Delta X - \Delta x = [\alpha\varepsilon]_2 - [\alpha\varepsilon]_1$$

Infolge der Zufälligkeit der Beobachtungsfehler wird der wirkliche Fehler $\varepsilon_{\Delta x}$ i m m e r einen realen Wert aufweisen, a u c h w e n n e s z u ü b e r h a u p t k e i n e r B e w e g u n g i n d e r e r w ä h n t e n A c h s e X g e k o m m e n i s t . Also auch im Falle $\Delta X = 0$ (keine Krustenbewegung) ergibt die Messung und Ausgleichung einen gewissen Wert Δx, der in diesem Falle dem Fehler gleicht und so n u r a l s P r o d u k t d e r B e o b a c h t u n g s f e h l e r e r s c h e i n t.

Man wird nie den wahren Fehlerwert $\varepsilon_{\Delta x}$ kennen; er kann aber mit praktischer Sicherheit gewisse Grenzen, die man als Vertrauensintervall bezeichnet, nicht übersteigen. Diese Grenzen bestimmt man mittels der mittleren Koordinatenfehler

$$m_{x,1} \;, \qquad m_{x,2} \;, \qquad m_{\Delta x} = \sqrt{m_{x,1}^2 + m_{x,2}^2}$$, usw., deren Schätzung aus der Ausgleichung hervorgeht

$$m_{x,1} = \left\{ m_0 \sqrt{[\alpha\alpha]} \right\}_1 = \left\{ m_0 \sqrt{Q_{11}} \right\}_1 \qquad \text{usw.}$$

Die Verteilung der Wahrscheinlichkeiten bei den ursprünglichen Fehlern ε kann man praktisch als normal betrachten – der Parameter der zutreffenden Gauß'schen Kurve ist nahe m_0 . Dann folgen auch die möglichen wirklichen Fehler in der gemessenen Erdkrustenbewegungskomponente $\varepsilon_{\Delta x} = \Delta X - \Delta x$ der normalen Verteilung mit dem Parameter nahe $m_{\Delta x}$ (bzw. $m_{\Delta y}$, $m_{\Delta z}$). Eine Voraussetzung der praktischen Gültigkeit ist die Berechnung der mittleren Fehler m_x aus einer genügenden Anzahl der überschüssigen Beobachtungen $(n - k) > 20$. Dann ist es möglich, die sog. Intervallschätzung nach der Tabelle der normalen Verteilung anzuwenden, z.B. für die Wahrscheinlichkeit P

$$P = P\left\{ -t\cdot m_{\Delta x} < \varepsilon_{\Delta x} < t\cdot m_{\Delta x} \right\} = \frac{1}{\sqrt{2\pi}} \int_{-t}^{t} e^{-\frac{t^2}{2}}\, dt = \Phi(t).$$

Das ist gleich

$$P\left\{ (\Delta x - t\cdot m_{\Delta x}) < \Delta X < (\Delta x + t\cdot m_{\Delta x}) \right\} = \Phi(t),$$

wo Δx den gemessenen (berechneten) und ΔX den wirklichen Wert der Bewegungskomponente darstellt. Es gibt für die verschieden gewählten Werte

$$t = 1,\ 2,\ 2{,}5\ ,\ 3$$

die zutreffenden Wahrscheinlichkeiten

$$P = 68\ \%,\ 95\ \%,\ 99\ \%,\ 997\ ‰\ .$$

Man kann danach folgende vereinfachte Regel bilden:

a) im Falle $\Delta x < m_{\Delta x}$ hat die Messung keine Krustenbewegung festgestellt (entweder es ist zu keiner Bewegung gekommen, oder die Messungsgenauigkeit genügte nicht, sie festzustellen);

b) bei $m_{\Delta x} < \Delta x < 2m_{\Delta x}$ kann man die Bewegung zulassen;

c) bei $\Delta x < 2m_{\Delta x}$ ist man davon praktisch überzeugt.

Dasselbe Intervall gilt für die anderen Komponenten Δy, Δz mit der Anwendung der mittleren Fehler $m_{\Delta y}$, $m_{\Delta z}$.

In den mittleren Fehlern $m_{\Delta x}$, $m_{\Delta y}$, $m_{\Delta z}$ sollen alle störenden Einflüsse - also außer den zufälligen auch systematische Fehler - inbegriffen sein.

A. Bei den astronomischen Ortsbestimmungen wird die eventuelle Änderung der geographischen Koordinaten φ, λ gemessen. Die mittleren Fehlerwerte beider Komponenten betragen (durchschnittlich und annähernd) bei den üblich gemessenen Laplace'schen Punkten

$$m_\varphi = 0{,}2'' \qquad\qquad m_\lambda = 0{,}5''$$

was der Lageänderung ca. 6 - 10 m entspricht. Man sollte bei der wiederholten Messung möglichst dieselbe Apparatur benutzen, um den Einfluß der systematischen Achsenfehler (besonders die ungenaue Bolzenausschleifung) auszuschließen. Dann genügt die durch die Werte m_φ, m_λ ausgedrückte sog. "innere" Genaugikeit der Methode. Nach die Methode kann man keine Erdkrustenbewegungen kleiner als 10 - 20 m durch astronomische Messung verläßlich feststellen. Nur astronomischen Observationen, wo die Koordinaten eine lange Zeit und mit verschiedenen Instrumenten gemessen werden, sinken die Werte der mittleren Fehler auf ca. 25 % , das Vertrauensintervall beträgt aber immer noch einige Meter. Diese Studien betrachten vor allem die Bewegung von Kontinenten (z.B. nach der Theorie von Wegener).

B. Fall der durch die Triangulierung überprüften Entfernung von zwei festen Punkten. Für die kurzen Entfernungen L = AB, wo die geprüften Punkte A, B zu derselben Landestriangulation gehören, kann man als Vertrauensintervall die mittleren Fehlerellipsen mit verdoppelten Ausmaßen wählen. So kann man auf kurze Entfernungen schon relative Lageänderungen von einigen Zentimentern feststellen.

Die weit entfernten Punkte muß man mit einer trigonometrischen Kette verbinden. Dort werden außer den Richtungen auch Azimute und Basislängen (in Abständen von 200 - 250 km) gemessen. Bei den mittleren Fehlern 0,3" in der Richtung, 0,4" im Azimut und

2.10^{-6} in der von einer geodätischen Basis abgeleiteten Ausgangsseite und nach der Ausgleichung einer einfachen Kette beträgt der mittlere Lagefehler

in der Richtung der Kettenachse $m_x = 0,6$ m/250 km bis 1,5 m/1000 km
in der Querrichtung $m_y = 0,8\ m_x$

Dabei macht sich der mittlere Richtungsfehler m_α mit $\sqrt{L^3}$ und der mittlere Basisfehler m_B mit $\sqrt{L}$ geltend; der relative mittlere Fehler m_L/L sinkt nämlich mit der Anzahl der eingelegten Basen.

Bei der Verdopplung der Kette sinken die Zahlwerte ungefähr um 30 %. Bei den direkten Längenmessungen der trigonometrischen Seiten mittels des Geodimeters wird der mittlere Fehler m_x, nicht aber m_y vermindert. Bei einer Kombination der Richtungs- und Längenmessungen kann man die Senkung der Werte auf

$m_x = 0,1$ m/250 km und ein weiteres Wachsen ungefähr mit $\sqrt{L}$
$m_y = 0,4$ m/250 km und ein weiteres Wachsen ungefähr mit L

schätzen.

Mittels der Triangulierung kann man also die Lageänderungen erst von mehr als einigen Zentimetern bis einigen Metern - je nach der Entfernung der zwei verbundenen Endpunkte - feststellen.

C. Für die Feststellung der vertikalen Bewegungskomponenten ΔZ eignet sich am besten das wiederholte genaue N i v e l l e m e n t . Das zutreffende Vertrauensintervall für die gefundene Höhenunterschiedsänderung $\Delta Z = (\Delta H_{AB})_2 - (\Delta H_{AB})_1$,

$$\text{beträgt} \quad \pm 2\ m = \pm 2 \sqrt{\tau_{1,L}^2 + \tau_{2,L}^2}\ \sqrt{L_{AB}}$$

wo τ_L der sog. gesamte mittlere Nivellementskilometerfehler ist, der den Einfluß sowohl der zufälligen (mit $\sqrt{L}$ wachsenden) als auch der systematischen Fehler ausdrückt. Die letzten wachsen auf kürzere Entfernungen proportional der die zwei Endpunkte A, B verbindenden Nivellementsstrecke $\overline{AB} = L$, aber für größere Entfernungen ($L > 50$ km) nehmen sie schon einen zufälligen Charakter an. Deswegen ändert sich τ_L nach der erwähnten Entfernung L und muß für jedes Nivellementsnetz nach den internationalen Formeln von Vignal berechnet werden. In den modernen Nivellements beträgt

$$\tau_L = \eta = 0,4\ \text{mm} \quad \text{für } L \leq 1\ \text{km}$$
$$\tau_L = 1,0\ \text{mm} \quad \text{für } L \geq 50\ \text{km}$$

Der erste Wert wird aus den Differenzen der doppelten Nivellierung in kurzen Strecken ($L < 1$ km) berechnet, der zweite aus den Schleifenschlußfehlern. Der Wert τ_L für 1 km < L < 50 km wird durch eine Interpolation ermittelt; man kann eine Exponential - oder parabolische Funktion dazu benutzen, z.B. nach Ing. K r u i s :

$$\tau_L^2 = \eta^2 + (\tau^2 - \eta^2)\ \frac{100\ L - L^2}{2500}$$

Für das heutige Nivellement gilt z.B. folgende Tafel

L km	0	5	10	15	20	30	40	50
τ_L mm	0,40	0,57	0,68	0,76	0,83	0,93	0,98	1,00

In den älteren Netzen beträgt der mittlere Fehler τ_L das mehrfache gegenüber den modernen; dann überwiegt der Wert τ_L des älteren Netzes. Auch bei dem z.B. fünffachen Werte kann man die Höhenänderungen von einigen Millimetern (bis L = 10 km) oder nur von einigen Zentimetern (bis L = 400 km) verläßlich feststellen.

Methode zur Bestimmung der rezenten Bewegungen der Erdkruste in Hochgebirgsgebieten

von

L. H r a d i l e k (Prag)

S u m m a r y

Recent movements of the earth's crust have not yét been calculated till now by exact methods in high mountain-chains, though recent orogenic movements are known to reach considerable dimensions.

A new method of trigonometrical altimetry has been used in Praha in which the deflections of the vertical as well as the changes of refraction are taken into consideration in each point of the canvas serving for the measuring of altitudes. A high degree of accuracy is reached by making the survey in a suitable manner and by analizing the measured values.

The verification of the method by levelling, by astronomical measurements of the deflection of the vertical, and by the altitudes calculated by means of lengths measured electrooptically has proved that an accuracy of altitude of $\pm$ 2 to $\pm$ 3 cm can be attained at bench marks and at points on peaks situated at a distance of 17 km from the bench marks of levelling.

It will be possible to calculate the values of recent movements of the earth's crust in high mountain-chains, too, by a suitable repetition of such measurements.

Z u s a m m e n f a s s u n g

In Hochgebirgsgebieten wurden bisher noch nicht rezente Bewegungen der Erdkruste durch exakte Methoden bestimmt, obwohl es bekannt ist, daß die jungen gebirgsbildenden Bewegungen beträchtliche Größen erreichen können.

Es wurde in Prag eine neue Methode trigonometrischer Höhenmessungen ausgearbeitet, bei der in jedem Höhennetzpunkt Rücksicht auf die Lotabweichungen sowie auch auf die Refraktionsänderungen genommen wurde. Durch eine geeignete Anlage der Messung und Bearbeitung der Meßwerte wird eine große Genauigkeit erreicht.

Die Nachprüfung der Methode durch das Nivellement, durch astronomische Bestimmungen der Lotabweichungen und durch die aus elektrooptisch gemessenen Längen ermittelten Höhen hat bewiesen, daß sich eine Höhengenauigkeit von $\pm$ 2 bis $\pm$ 3 cm bei den Festpunkten erreichen läßt und bei Punkten auf Berggipfeln, die bis 17 km von den Festpunkten des Nivellements entfernt sein können.

Es wird möglich sein, durch eine geeignete Wiederholung von derartigen Messungen die Werte der rezenten Erdkrustenbewegungen auch in Hochgebirgsgebieten zu bestimmen.

Das geometrische Nivellement , das für die Bestimmung der vertikalen rezenten Erdkrustenbewegungen die wichtigste geodätische Methode ist, kann man nicht in Hochgebirgsgebieten anwenden. Es ist deshalb nötig, die Genauigkeit der trigonometrischen Höhenbestimmung zu steigern.

Bei den bisher üblichen Methoden der präzisen trigonometrischen Höhenbestimmung blieb das Problem des Refraktionseinflusses ungelöst. Obwohl die Refraktion in Gebirgen weniger veränderlich ist als im ebenen Gelände, kann jedoch nicht angenommen werden, daß diese auf sämtlichen Netzpunkten konstant ist. Eine solche Voraussetzung könnte zu Fehlern führen, die alle anderen Meßfehler überbieten.

In Prag wurde in den letzten Jahren eine Methode der präzisen Höhenbestimmung ausgearbeitet, bei der sowohl die Lotabweichungen, als auch die Werte der Refraktionskoeffizienten auf jedem Netzpunkte bestimmt werden.

Aus den Ergebnissen der umfangreichen Refraktionsforschungen von A.A. Izotov und L.P. Pellinen in der UdSSR und J.J. Levallois und J. de Masson d'Autume in Frankreich geht hervor, daß der Hauptwert in der Formel für den Refraktionskoeffizienten für sämtliche von einem Standpunkt ausgehende Sichtlinien immer dann gleich ist, wenn auf einer Bergspitze oder von einem Beobachtungsgerüst mindestens 12 bis 15 m über der Erde gemessen wird.

Unter dieser Voraussetzung kann man bei geeignetem Ausgleichsverfahren die Refraktion von den Lotabweichungen trennen und auf jedem Netzpunkte die Refraktionskoeffizienten, Lotabweichungen und Höhen bestimmen. Auf einigen Netzpunkten ist es nötig, die Lotabweichungen astronomisch zu messen und zur Erreichung der nötigen Genauigkeit die Höhen von mindestens zwei oder drei Randpunkten des Netzes nivellitisch zu bestimmen.

Bei der in den Jahren 1958-1961 vorgenommenen Überprüfung der vorgeschlagenen Methode in der Hohen Tatra zeigte es sich, daß es möglich ist, die Höhen der über 2.000 m hohen Berggipfel mit einer Genauigkeit von 2cm zu bestimmen. Es ist nötig, die Netzpunkte in geeigneter Weise zu signalisieren und die von einem Standpunkt ausgehenden Höhenwinkel alle in rascher Aufeinanderfolge zu messen (mit einem Fehler kleiner als 1") und die Beobachtung etwa nach zwei Stunden zu wiederholen.

Bei dem Anschluß des Netzes an zwei 30 km voneinander entfernten Nivellementspunkten wurden folgende Ergebnisse erreicht:

1. Die Höhe des 17 km von den Ausgangspunkten des Netzes entfernt liegenden Punktes Zuberec war von dem nivellitisch bestimmten Kontrollwert um 1,5 cm verschieden.

2. Die Komponenten der Lotabweichungen wichen von den auf sechs Punkten astronomisch bestimmten Kontrollwerten im Mittel um 2,2" ab.

3. Die Refraktionskoeffizienten, die aus bei verschiedenen Beobachtungsbedingungen gemessenen Höhenwinkeln berechnet worden sind, unterschieden sich bis um die Hälfte ihres Wertes. Der endgültige Wert des Refraktionskoeffizienten auf dem Punkt Volovec, der während der Beobachtung mit einer leichten Schneeschicht bedeckt war, erwies sich als beinahe doppelt so groß wie auf den Nachbarpunkten. Die Voraussetzung einer konstanten Refraktion würde daher zu fehlerhaften Ergebnissen führen.

4. Es wurde auch die Methode der Höhenbestimmung aus schiefen, elektrooptisch gemessenen Längen ausgearbeitet und überprüft. Dabei wurden keine vertikalen Winkel benutzt. Die trigonometrischen Höhen der höchsten Punkte des Netzes unterschieden sich von den durch Nivellement bestimmten und den aus direkt gemessenen elektronischen Längen bestimmten Kontrollwerten im Mittel um 2 cm.

Die Ausnutzung direkt gemessener Längen, gemeinsam mit den gemessenen Horizontal- und Vertikalwinkeln führt zur räumlichen Netzausgleichung und an Hand wiederholter Messungen können dann die Horizontal- und Vertikalbewegungen der Erdkruste in Hochgebirgsgebieten bestimmt werden.

Eine weitere Entwicklung und praktische Geltendmachung werden noch die im untersuchten Gebiet durchgeführten gravimetrischen Messungen bieten.

Aus den Ergebnissen der verschiedenen Überprüfungen der vorgeschlagenen Methode geht hervor, daß es möglich ist, an Hand dieser Methode die Größe und gegebenenfalls auch die Veränderungen der rezenten Bewegungen auch in den Gebirgsgebieten geodätisch zu bestimmen und zu verfolgen.

L i t e r a t u r

Studia geoph. et geod. 2 (1958), 101; 3 (1959), 334; 4 (1960), 217

Studia geoph. et geod. 2 (1958), 206; 6 (1962) (im Druck)

Studia geoph. et geod. 4 (1960), 1; 5 (1961), 302

On the geodetic methods of studying the movements of the earth's crust

von

A. A. I z o t o v (Moskau)

S u m m a r y

The present paper explains that so far only the relative movements of the earth crust can be determined by relevelling and retriangulation or retrilateration. It describes methods for obtaining relative horizontal movements of the earth crust directly from the difference in re-measured angles or sides of the triangles; adjustment and the calculation of the coordinates of points being dispensed with. It is noted that the best effect can be obtained when the results of the re-measurement of both angles and lengths are used simultaneously. Triangulation or trilateration alone enable only those horizontal displacements to be revealed whose values exceed one three-hundredth part of the distance between the stations.

Z u s a m m e n f a s s u n g

Die vorliegende Arbeit erläutert, daß bisher nur Relativbewegungen der Erdkruste durch Wiederholungsnivellement und Wiederholungstriangulationen oder Wiederholungstrilaterationen bestimmt werden können. Sie beschreibt Methoden, durch die man die relativen Horizontalbewegungen der Erdkruste direkt aus der Differenz der mehrfach gemessenen Winkel oder Seiten der Dreiecke erhält; eine besondere Auswahl und Berechnung der Koordinaten der (Fix-) Punkte ist nicht notwendig. Es wird gezeigt, daß man das beste Resultat erhält, wenn die Ergebnisse der Mehrfachmessungen von Winkeln und Längen gleichzeitig ausgewertet werden. Triangulationen oder Trilaterationen allein ermöglichen nur Horizontalverschiebungen festzustellen, deren Werte 1/300 des Abstandes zwischen den Stationen überschreiten.

By determining the positions of certain points of the earth surface in a definite, invariable system of coordinates from geodetic measurements, these latter being repeated after some intervals of time, it is possible to obtain the components of the shifts of these points along the coordinate axes. If the geodetic movements have been performed with the high degree of accuracy desired, the shift of those points may be accounted for by the movements of the crust, provided the selection and marking of the stations were done in the proper way.

Though the geodetic methods of investigating the earth crust movements are well known at present, there are still some problems associated with them that are worthy of consideration.

From the theoretical point of view it is feasible to find the positions of geodetic stations and, consequently, the movements of the earth crust directly in the three-dimensional coordinate system.

However, the difficulties encountered in measuring zenith distances and determining the deviation of the plumb line still prevent this method of studying the movements of the earth crust from being used. At present, therefore, studies of vertical and horizontal movements in the crust are carried out separately and are based on two kinds of measurements that are entirely disconnected.

The best method to find out vertical movements of the crust is by high precision levelling. Only the observed differences in elevations of two adjacent points - dz - can be obtained directly by levelling. To find out the heights of points of the earth surface above the reference surface, the gravity values g must be observed all along the line of levelling. These data enable us to get a single-valued difference between the geopotential at a given point B and that at the initial point of levelling 0:

$$W_B - W_0 = - \int_{(OB)} g\,dz$$

Different principles may underlie the computation of the elevations of points of the earth surface above the reference surface, which results in various systems of elevations being used. In the USSR, for instance, the system of normal elevations is employed, the formula used being as follows:

$$H_B - H_0 = Z_B - Z_0 + \frac{1}{\gamma^B_m} \int_{(OB)} (\gamma_0 - \gamma^B_0)\,dz + \frac{1}{\gamma^B_m} \int_{(OB)} (g - \gamma)\,dz$$

In this formula γ_0 and γ are the normal values of gravity at the variable point on the reference surface and on the surface of the earth, while γ^B_m and γ^B_0 are those at the point on the surface of reference and at one half of its elevation.

The geopotential at the initial point of levelling may be found from, are measurements with observations of gravity included. However, the exact gravity potential is not yet known. That is why in high precision levelling it is generally assumed that

$$W_0 = C_0$$

$$H_0 = Z_0 = 0$$

It appears that absolute and relative displacements of the surface of the earth should be defined with a greater exactitude. By absolute vertical displacements of points on the earth's surface one should understand the displacements of those points along the plumb line -- or, strictly speaking, the line of force -- with respect to the centre of the mass of the Earth; while changes in elevation of these points with respect to the reference surface will be referred to in the following as relative vertical movements. It must be understood that absolute and relative displacements will coincide only when the position of the initial point of levelling remains fixed.

For studying absolute vertical movements of the earth crust it is necessary, first of all, to determine the change in the gravity potential at the initial point of levelling. But so far the values of the changes in the gravity potential furnished by measurements are not sufficiently accurate to meet the requirements.

So, the data furnished by relevelling enable us to investigate only the relative vertical movements of the earth crust.

In processing the results of relevelling it is generally assumed that during the period of time that elapsed between the first levelling and the second no changes occurred in the values of gravity. It is understood that in this case the vertical displacements of points of the surface of the earth are equal to the changes in their heights, as directly measured

$$\Delta H_B = \Delta Z_B$$

It must be clear that for the purpose of studying vertical movements of the earth crust in the constant gravitational field of the Earth any kind of system of heights, and any initial point of levelling may be chosen. Nevertheless, it is a matter of importance that the initial point of levelling be selected at the mean sea level and that it coincide with the zero of the mariograph. This will facilitate the determination of sea level difference, and the studies of relative vertical oscillations of land and sea.

The gravitational field of the earth does not by any means remain constant during long periods of time. That is why at the time of high precision levelling, high precision determination of gravity must also be carried out.

Horizontal movements of the Earth crust can be studied by means of triangulation. By measuring the base line and two angles in every triangle we shall find out the relative positions of the triangulation points. As a rule, in triangulation an excessive number of base lines and angles are measured to secure a good check and increase the accuracy of the results. When the above measurements in the same triangulation scheme are repeated after a certain interval of time, we obtain the data on the horizontal movements of the earth surface.

One has to differentiate between absolute and relative horizontal movements, which in the following will be divided into components of two coordinate axes. The shift of the points of the earth surface on their level surface with respect to the invariable grid of meridians and parallels will be called absolute horizontal movements of these points; while the definition of their relative movements will be restricted to the changes in their coordinates whose origin coincides with one of the points.

To reveal absolute horizontal movements of triangulation stations it is essential to obtain first the change in position of one point and the change in the azimuth of the line convecting it with an adjacent point. Undoubtedly, this can be done by astronomical methods, but the latter do not yet enable minute changes in longitude, latitude or azimuth to be observed.

So at the present state of knowledge we can only deal with the problem of investigating relative horizontal movements of the earth crust.

To find out the relative horizontal displacements of triangulation points within a restricted area the plane rectangular coordinate system will be employed. The origin of these coordinates will be made to coincide with the initial point of the triangulation base line, their axes being oriented by the base line azimuth. In the case under consideration it would be admissible to assume an arbitrary direction of the coordinate axes. It would therefore be advisable to make one of the coordinate axes coincide with the base line, i.e. to assume the base line azimuth to be equal to zero. The essential thing is that the directions of the coordinate axes should remain fixed.

Horizontal displacements of triangulation points cause variations in triangle angles. Let the change in angle at point i between the bearings on points k and n be expressed in terms of changes in the coordinates of these points. Then the following observation equation will result:

$$\delta u_{kin} = du_{kin} + (a_{in} - a_{ik})\, dx_i + (b_{in} - b_{ik})\, dy_i + a_{ik} dx_k + b_{ik} dy_k - a_{in} dx_n - b_{in} dy_n \qquad (I)$$

$$a_{ik} = -\frac{\rho''\sin\alpha_{ik}}{S_{ik}}; \qquad b_{ik} = +\frac{\rho''\cos\alpha_{ik}}{S_{ik}}$$

In this equation dx and dy are the changes in the coordinates, du is the change in the angle taking place during the interval between the first measurement and the successive one, and δu is the difference between the errors of the measurements of this angle, and S und α are the length and the azimuth of the triangle side.

The calculation and solution of the above observation equations are not dealt with here. In practice it would, certainly, be an advantage to derive them directly from the observed values of the angles. Using these formulas for studying the horizontal movements of the earth crust enables one to dispense with the adjustment of triangulation and with the calculation of coordinates.

On the other hand, displacements of triangulation stations cannot be fully revealed by the changes of triangle angles alone. To do this the change in the length and the azimuth of the base line due to the displacement of its terminal point with respect to its initial point O assumed to be fixed must also be allowed for. Changes in the base line length and azimuth are expressed in the observation equations given below that do not call for any explanation:

$$\delta\beta_{01} = d\beta_{01} - \cos\alpha_{01}\, dx_1 - \sin\alpha_{01}\, dy_1 \qquad \text{(XX)}$$

$$\delta\alpha_{01} = d\alpha_{01} - a_{01}dx_1 - b_{01}dy$$

These equations should be added to, and solved simultaneously with, the observation equations (X) resulting from the changes in the angles of the triangles.

If the differences of errors $\delta\beta$ and $\delta\alpha$ in two successive measurements of base line and azimuth are negligible their equations (XX) may be used as conditional equations. Then these conditional equations will enable us to find the changes in the coordinates of the terminal point of the base line and substitute them in the angle observation equations (X). This done, the co ordinates of all the other triangulation points will be obtained by solving the observation equations for the angles.

In case base line and azimuth fail to be measured with a considerably higher degree of accuracy than their changes due to the movements of the earth crust, it would be best to assume them simply to be constant. Then in the observation equations for angles (X) it should be assumed that

$$dx_I = dy_I = 0$$

By neglecting the changes in the length and azimuth of the base line the displacements of triangulation points will be in error according to

$$\delta x = x\frac{d\beta_{01}}{B_{01}} - y\frac{d\alpha''_{01}}{\rho''}$$

$$\delta y = y\frac{d\beta_{01}}{\beta_{01}} + x\frac{d\alpha''_{01}}{\rho''}$$

Naturally, these errors will be of a systematic character and may cause apparent horizontal displacements of triangulation points.

Electro-optical and radio range-finders that are available at present, e.g. the geodimeter and the tellurometer, make it possible to measure distances with a sufficiently high degree of accuracy. As a result, horizontal movements of the earth crust can be observed by the method of trilateration, i.e. by measuring directly the sides of triangles with the aid of the above instruments.

Let it be assumed that all the sides of triangles have been measured with sufficient accuracy at different times. The changes in length of lines will be described in terms of changes in the coordinates of their endpoints, one of trilateration points being taken for the fixed origin of the coordinates. It follows that for every line there will be obtained an observation equation that does not call for any explanation.

$$\delta S_{ik} = dS_{ik} + \cos \alpha_{ik}\, dx_i + \sin \alpha_{ik} dy_i - \cos \alpha_{ik} dx_k - \sin \alpha_{ik} dy_k \qquad \text{(XXX)}$$

It is from these equations that the components of horizontal displacements of trilateration points can be found, with both its adjustment and the calculation of coordinates dispensed with. To produce observation equations for lines (XXX) an approximate knowledge of their azimuths will suffice, though the trilateration initial azimuth both for the first and the succeeding observation must meet the highest possible requirements as to accuracy.

To get the best results in studying horizontal movements of the earth crust it is desirable to measure the triangle sides as well as the angles every time. Then angle equations (X) will have to be solved simultaneously with side equations (XXX). If necessary, respective weights may be assigned for either type of equations.

At present the mean square error of angular measurements in triangulation of high accuracy is in the neighbourhood of 0".6 - 0".7. It seems unlikely that the relative error of ranging by either the geodimeter or the tellurometer can be less than 1 : 300,000. This means that so far we are able to disclose only those horizontal displacements of the earth crust whose values exceed the above-mentioned errors of observation.

From the point of view of studying the movements of the earth crust, further improvement of the methods and instruments used is of the greatest importance. It is also essential to succeed in minimizing various errors that affect the results of geodetic measurements and are often the cause of important research being cramped.

D i s k u s s i o n

L. E g y e d (Budapest)

Die Bestimmung von absoluten Krustenbewegungen ist nur dann möglich, wenn man die Anziehungskraft als konstant in der Zeit annimmt. Das enthält zwei Hypothesen:

1. Der Gravitationskoeffizient ist konstant.
2. Der Durchmesser der Erde ist konstant.

Die erste ist bestritten von Dirac, der die Abnahme der Gravitationszahl mit der Zeit abgeleitet hat. Die zweite scheint mit den geologischen Beobachtungen in Widerspruch zu stehen.

Some results of field geological inspection of precise levelling lines

von

L. E. Setunskaya, E. A. Finko
Z. I. Martynova (Moskau)

Summary

The article covers the effect various engineering and geological factors produce on vertical displacement of bench-marks on the basis of the materials of field geological inspection of three relevelling lines totalling over 3,000 km (Semipalatinsk-Alma Ata, Alma Ata - Arys', Vologda - Perm); an appraisal is made of the possibilities to utilise the precise relevelling data for the study of crustal movement.

The 277 bench-marks we have surveyed are subdivided into three categories: 1) practically stable (the engineering and geological factors have no effect on the position of the bench-marks) - some 50 per cent of the total number of the surveyed bench-marks, 2) unstable (displacement of the bench-marks is due to engineering and geological factors) - 15-20 per cent, 3) doubtful (the relationship between the tectonic and engineering geological factors is not clear) - 30-35 per cent.

The ground bench-marks have proved to be most stable while the bench-marks located in the bank-seats of railway bridges were least reliable. The hydrogeological conditions are the predominant physico-geographical factor causing the displacement of the bench-marks.

Despite the high percentage of unstable and doubtful bench-marks no changes have been registered in the general nature of the curves of intensities of the recent tectonic movements after correction in accordance with the results of the field geological inspection. The curves are smoothed mainly due to the elimination of local bends associated with indications of unstable and doubtful bench-marks. None of the large bends of the curves have been brought about by the exogenic factors, and they are due only to tectonic changes.

Therefor, the use of the data on precise relevelling for the study of recent crustal movements is fully justified.

Zusammenfassung

Die Abhandlung betrachtet die verschiedenen technischen und geologischen Faktoren, die Vertikalverschiebungen von Fixpunkten bedingen, auf der Grundlage des Materials feldgeologischer Inspektion dreier Wiederholungsnivellements über insgesamt 3000 km (Semipalatinsk-Alma Ata, Alma Ata - Arys', Vologda - Perm); es wird eine Abschätzung der Möglichkeiten über die Verwendung der Daten des Präzisionsnivellements für das Studium der Kurstenbewegungen gegeben.

Die 277 von uns vermessenen Fixpunkte sind in drei Kategorien eingeteilt: 1. praktisch unveränderliche, (die technischen und geologischen Faktoren haben keinen Einfluß auf die Lage der Fixpunkte) etwa 50 % der Gesamtzahl der vermessenen Fixpunkte. 2. nicht stabile (Verrückung der Fixpunkte infolge technischer und geologischer Faktoren) 15 -20 %. 3. nicht zuverlässige (die Beziehung zwischen den tektonischen und ingenieur-geologischen Faktoren ist nicht klar) 30 - 35 %.

Die Grundfixpunkte haben sich als sehr stabil herausgestellt, während die in der Nähe von Eisenbahnbrücken befindlichen Fixpunkte weniger zuverlässig sind. Die hydrogeologische Situation ist der vorherrschende physikalisch-geographische Faktor, der die Verschiebung der Fixpunkte verursacht.

Trotz des hohen Prozentsatzes der nicht stabilen und unzuverlässigen Fixpunkte sind keine Änderungen in der allgemeinen Natur der Intensitätskurven der rezenten tektonischen Bewegungen nach der Korrektur registriert worden, in Übereinstimmung mit den Ergebnissen der feldgeologischen Inspektion. Die Kurven werden hauptsächlich geglättet infolge der Elimination der lokalen Krümmungen, die mit der Anzeige der nicht stabilen und unzuverlässigen Fixpunkte verbunden sind. Keine der großen Krümmungen der Kurven wird durch äußere Faktoren hervorgerufen, sie sind nur durch tektonische Änderungen bedingt.

Deshalb ist die Verwendung der Daten von Präzisionsnivellements für das Studium der rezenten Krustenbewegungen voll gerechtfertigt.

During the last decade a number of Soviet institutions (Moscow Institute of Engineers of Geodesy, Air Surveying and Cartography, Central Scientific Research Institute for Geodesy, Air Surveying and Cartography, Geographical Institute of the Academy of Sciences of the U.S.S.R., etc.) have been conducting a joint research into recent vertical movements of the earth's crust broadly utilising the data obtained from repeated precise levellings. These data have been subjected to a through preliminary analysis for the purpose of appraising the methods of levelling, its precision and the reliability of bench-marks.

We know that vertical displacements of bench-marks revealed by repeated levelling may be caused not only by tectonic factors but also by various factors of a non-tectonic character (physico-geographical features of the area where the bench-marks are located, the state of the structures in which they are laid or the state of the bench-marks themselves).

This article makes an attempt to evaluate the role various non-tectonic (engineering and geological) factors play in the vertical displacement of bench-marks on the basis of field geological inspection of three lines of repeated levellings over 3,000 km long.

The lines under examination are located in distinctly unlike physico-geographical and structural-tectonic conditions.

The lines from Semipalatinsk to Alma Ata and from Arys to Alma Ata run through the submountain and lowmountain districts of the Eastern and Southern Kazakhstan which belong to semi-desert and desert zones.

The Vologda-Perm line intersects the Russian Plain in a latitudinal direction and passes through the sub-zone of southern Taiga.

All the three lines of repeated levelling are located along the railways and the time interval between the first and second levellings was about 20 years.

There were two types of bench-marks subjected to repeated levelling between Semipalatinsk and Alma Ata and between Arys and Alma Ata: 1) bench-marks secured on walls and 2) bench-marks in the ground. Along the Vologda-Perm line all bench-marks are in the wall-type category.

The field geological inspection of bench-marks has been conducted according to a definite programme. We described the buildings with bench-marks and determined their strength and stability (fractures, misalignments, state of the foundation), checked on the degree of bench-mark preservation (strength of cementation), found the position of bench-marks relative to relief elements, studied the characteristics of soils and estimated the depth of underground waters and the course of present geomorphological processes.

As a result, all bench-marks were placed in three categories.

1. Practically stable bench-marks;

2. Unstable bench-marks;

3. Doubtful bench-marks.

The stable category included well preserved bench-marks in favourable natural conditions. These bench-marks are located in solid and undamaged structures.

The unstable category included:

a) destroyed or heavily damaged bench-marks located in structures in unsatisfactory condition (ruined, considerably damaged buildings)

b) bench-marks in unfavourable natural conditions.

Among the doubtful category were:

a) bench-marks slightly damaged or those laid in buildings with insignificant defects.

b) bench-marks in partially unfavourable natural conditions.

The bench-marks of the first category can be relied upon in studying vertical movements of the earth's crust if we assume that their displacement is caused only by the tectonic factor -- general upheavals or subsindences of large areas of the earth's crust. The bench-marks of the second category cannot be used since the effect of the tectonic factor in the change of their vertical position is noticeably distorted by unfavourable engineering and geological conditions. The approach to the third-category bench-marks must be very cautious because the role played by tectonic and non-tectonic factors in their displacement is not clear.

A subjective factor also comes to the fore in appraising the stability of bench-marks. However, as the investigations of various lines by the authors indicate, the subjective factor is of no decisive importance. As a rule, the bench-marks which were placed in the stable category by the authors produce no peaks -- sinking or swelling -- on the curves of the rates of recent movements not corrected in accordance with the field data. Unstable bench-marks usually cause local bends to appear on the curve. Fig. 1 shows graphs of the rates of tectonic movements on all the three lines. The aligned curve of the movement intensities has been obtained on the basis of the data provided by the field inspection with account taken of the stable bench-marks. The peaks are mainly due to unstable and doubtful bench-marks.

These peaks can be clearly seen in Fig. 2 which shows some portions of these graphs in a larger scale.

Altogether we have investigated 277 bench-marks, of which 114 along the Semipalatinsk-Alma Ata line, 112 between Arys and Alma Ata and 51 between Vologda and Perm. Of the total number of bench-marks subjected to repeated levelling on each line, 86 per cent have been investigated between Semipalatinsk and Alma Ata, 75 per cent between Arys and Alma Ata and 34 between Vologda and Perm.

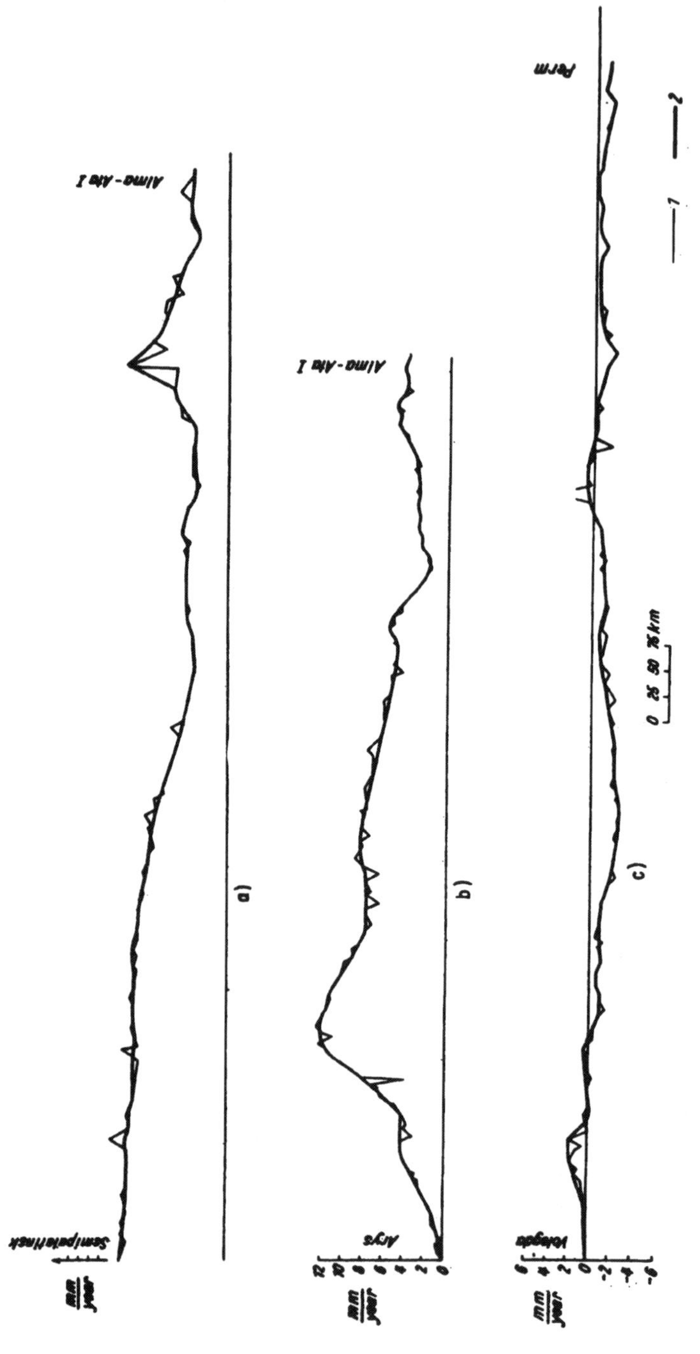

Fig. 1

Graphs of the rates of recent tectonic movements.

a) - along Semipalatinsk-Alma Ata line, b) - along Arys-Alma Ata line, c) - along Vologda-Perm line.

1. Curve of the rates of recent tectonic movements plotted by the geodetic data. 2. Curve of the rates of recent tectonic movements corrected (aligned) in accordance with the data of field geological inspection

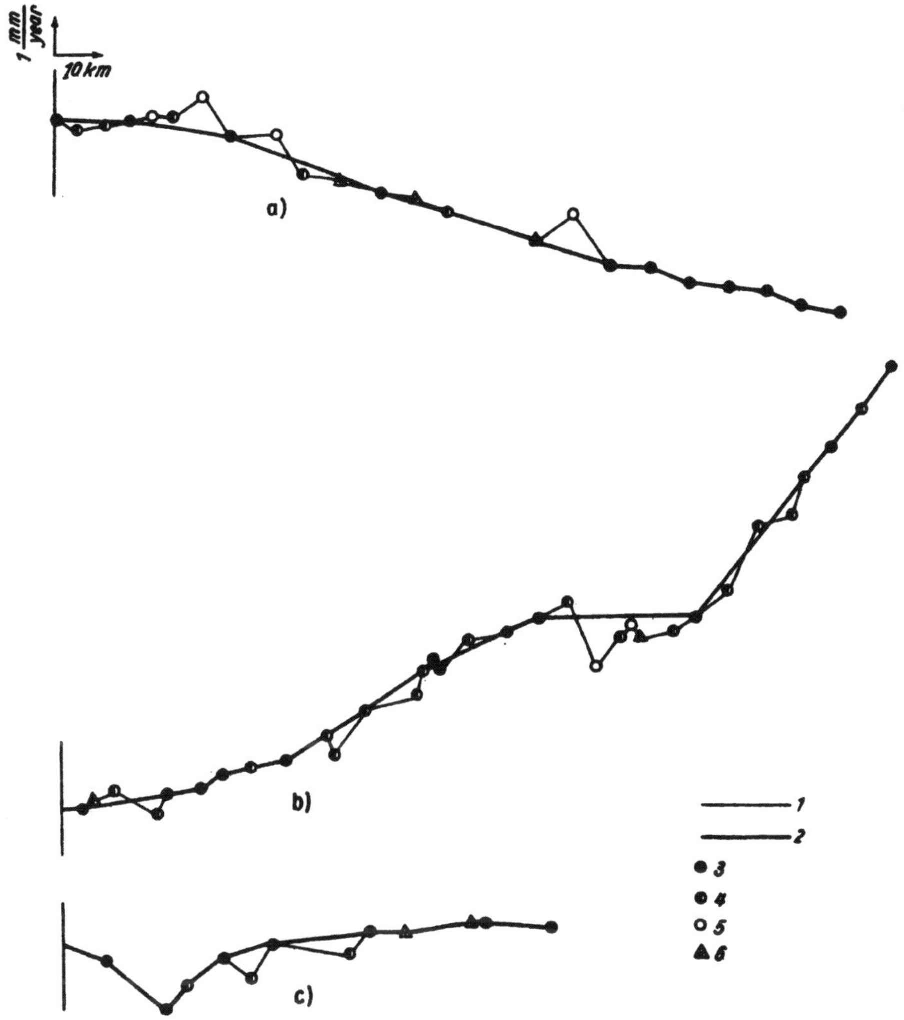

Fig. 2

Position of bench-marks with different degree of stability relative to the aligned curves of recent tectonic movements (Portions from graphs: a) - along Semipalatinsk-Alma Ata line, b) - along Arys-Alma Ata line, c) - along Vologda-Perm line).

1. Curve of the rate of recent tectonic movements plotted by the geodetic data. 2 . Curve of the rates of recent tectonic movements corrected (aligned) in accordance with the data of field geological inspection. 3. Stable bench-marks. 4. Doubtful bench-marks. 5. Unstable bench-marks. 6. Unstudied bench-marks.

The summary results of investigations are presented in Table 1.

Table 1

Line of repeated levelling	Total number of studied bench-marks	Location of bench-marks			
		bank-seats of railway bridges (on walls)	pump-houses (on walls)	buildings of railway stations, sidings, etc. (on walls)	in ground
Semipalatinsk Alma Ata	114	64 (56%)	5 (5%)	13 (11%)	32 (28%)
Arys Alma Ata	112	26 (23%)	2 (2%)	62 (65%)	22 (20%)
Vologda Perm	51	22 (43%)	26 (51%)	3 (6%)	-

The three categories of bench-mark stability are shown in Table 2.

Table 2

Line of repeated levelling	Total number of studied bench-marks	Stability category		
		1st category stable	2nd category unstable	3rd category doubtful
Semipalatinsk Alma Ata	114	52 (46%)	26 (23%)	36 (31%)
Arys Alma Ata	112	55 (49%)	17 (15%)	40 (36%)
Vologda Perm	51	28 (55%)	6 (12%)	17 (33%)

The Table shows that the relatively stable bench-marks make up nearly half of all the studied bench-marks.

The characteristics of unstable and doubtful bench-marks are given in Table 3 which shows the amount of bench-mark displacement under the action of non-tectonic factors.

Table 3

Line of repeated levelling	Unstable bench-marks			Doubtful bench-marks		
	Total number of bench-marks	Mean value of displacement mm/year (mm in 20 years)		Total number of bench-marks	Mean value of displacement mm/year (mm in 20 years)	
		swelling	sinking		swelling	sinking
Semipalatinsk Alma Ata	26	+0.5 (+10.0)	-0.3 (-6.0)	36	+0.2 (+4.0)	-0.2 (-4.0)
Arys Alma Ata	17	+0.4 (+8.0)	-0.6 (-12.0)	40	+0.3 (+6.0)	-0.4 (-8.0)
Vologda Perm	6	-	-0.7 (-14.0)	17	-	-0.3 (-6.0)

The results of analysis of the degree of stability of bench-marks depending on their type, and for bench-marks on walls depending on the structure in which they are laid, are shown diagrammatically in Fig. 3.

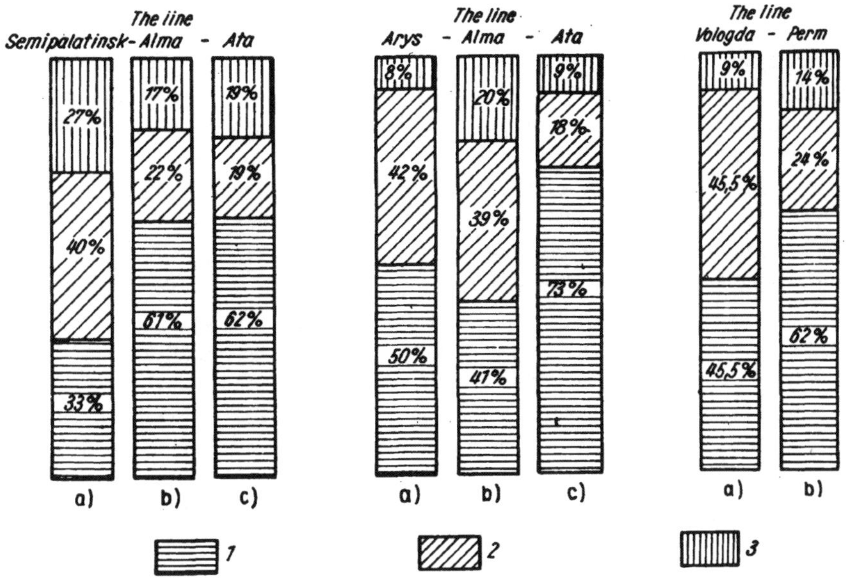

Fig. 3

Dependence of the degree of stability of bench-marks on their types and character of buildings in which they are laid.

a -- bench-marks in bank-seats of railway bridges; b -- bench-marks in the walls of water towers and various buildings near railway stations (railway stations, sidings, locomotive sheds, etc.), c -- ground bench-marks.

1 -- practically stable bench-marks, 2 -- doubtful bench-marks, 3 -- unstable bench-marks

The diagram shows most of the stable bench-marks were ground bench-marks (from 62 to 73%).

Of the wall-type bench-marks the more stable were those laid in various buildings near railway stations (water towers, railway stations, etc.) and less stable -- those arranged in bank-seats of railway bridges. Among the latter category the number of unstable and doubtful bench-marks came to 50-70%.

The poor stability of bench-marks in bank-seats of bridges can be explained by the fact that bridges are ordinarily in less favourable natural conditions than buildings situated near railway stations. For the most part, bridges rest on river flood-plains where the hydrogeological conditions do not favour the construction of engineering structures. Besides, they constantly suffer from passing railway transport which also deteriorates their stability. For this reason, bridges are the least reliable receptacles for bench-marks intended for subsequent study of recent crustal movements.

In appraising the role of various physico-geographical factors in the vertical displacement of bench-marks it is difficult to isolate the effect of individual factors in its "pure form". For all that, we have made an attempt to analyse the degree of bench-mark stability from the viewpoint of relief, lithologic composition of rock and the depth of underground waters. The results of analysis have shown that the hydrogeological conditions are the predominant factor affecting the vertical displacement of bench-marks. The majority of the bench-marks referred to the unstable and doubtful category are located in areas where underground waters are not deep. Almost all such bench-marks produce peaks on the curve of the rates of recent tectonic movements. Small depth of underground waters in combination with predominant clayey and loamy soils causes boggingup in the Vologda-Perm area. The conditions along the southern Kazakhstan lines with their high level of underground waters favour salinisation and development of salterns and salt bottoms. Both bog formation and salinisation are known to produce the most injurious effect on the engineering structures and bench-marks in the ground.

The lack of sufficient factual data does not permit us to give any definite characterisation of the mechanism of action of the hydrogeological conditions on the engineering structures and ground bench-marks. Our observations show that bogging-up causes most bench-marks to sink; in case of saline soils both sinking and swelling may take place. To arrive at a final solution of this problem special stationary observations must be conducted over the behaviour of bench-marks in conditions of different lithologic composition of the soils, various degree of their moistening, and particularly, in conditions of bog formation and salinisation.

Our investigations have shown that on the whole, despite the high percentage of unstable and doubtful bench-marks (some 50%) along the line, no changes were recorded in the general character of the curve of the rates of recent tectonic movements corrected according to the data of the field geological inspection of bench-marks. The curve is aligned mainly due to the elimination of local bends which, as a rule, are connected with the effect of the nontectonic factor. All large bends on the curve are not the result of the action of exogenetic factors and can be attributed to tectonic causes. Therefore, the use of the data provided by repeated levelling for the study of recent movements of the earth's crust is fully justified.

Field engineering and geological inspection of bench-marks is an inalienable element of these investigations. However, all bench-marks need not be investigated comprehensively. Attention should be focussed on bench-marks which produce peaks on the graph. The bench-marks must be investigated immediately after the second levelling has been performed and its results processed, because a long interval between the second levelling and the inspection of the bench-marks may give a distorted picture.

D i s k u s s i o n

L. A s p l u n d (Stockholm):

The question of stability of benchmarks certainly is a very important one. Of course it is important also to look from this aspect on the future benchmarks. Geodesists should cooperate with geologists and geotechnicians to develop bench-mark devises for different ground types, which are as stable as ever possible. In the future it would be better to use the specialists of geotechnic as consultants instead of inspectors.

S. M i y a m u r a (Tokyo):

I have one question of general character, namely what procedure is taken in law or in some other way in every country to keep safe geodetic marks, especially bench marks. For, in Japan recent development in construction works very often destroys them and it has become a serious problem.

Vertical crustal movements of the Baskunchak Salt Dome

von

A. K. Pevnev (Moskau)

Summary

The circuit located on the Baskunchak salt dome has been established for the purpose of obtaining quantitative data on recent movements of local structures of this type.

The circuit encircles Mt. Bolshoye Bogdo which is the most elevated part of the salt dome. Some bench-marks are located near Mt. Bolshoye Bogdo and on it, others stretch along the shore of Lake Baskunchak, still others are deliberately moved away from the mountain and the lake.

The initial levelling was completed in the autumn of 1951 and the first relevelling in the autumn of 1958. The following conclusions can be drawn from the results of these two levellings.

1. All bench-marks without exception situated on the shore of Lake Baskunchak sink on the average by 1 mm/year.

2. The bench-marks near and on Mt. Bolshoye Bogdo rise at a rate of about 0.5 mm/year.

3. The bench-marks removed from both the lake shore and the mountain show practically no change in their altitude.

These conclucions which presuppose a uniform vertical movements of the earth's crust are in full accord with the concept of the recent movements of the Baskunchak salt dome.

It should be noted that the rates of bench-mark movement have been calculated relative to one of the bench-marks farthest from the lake shore and the mountain.

The second levelling of the Baskunchak area was performed in the summer of 1961.

The comparison of the results of all the three levellings makes it possible to draw the following conclusions:

1. The main conclusion is that the general regularity of vertical movements in the Baskunchak area established by the first relevelling was confirmed by the 1961 levelling.

2. At the same time the relevelling produced slightly differing results which can apparently be ascribed to the seasonal surface fluctuations in the Baskunchak area.

Zusammenfassung

Am Baskunchak Salzdom ist eine Nivellementsschleife angelegt worden, um quantitative Daten über rezente Bewegungen lokaler Strukturen dieses Typs zu erhalten.

Die Schleife umschließt den Mt. Bolshoye Bogdo, den höchsten Teil des Salzdomes. Einige Höhenmarken befinden sich in der Nähe des Mt. Bolshoye Bogdo und darauf. Andere sind entlang der Küste des Sees Baskunchak angebracht, weitere sind absichtlich entfernt von dem Berg und dem See eingebracht worden.

Das erste Nivellement wurde im August 1951 abgeschlossen und das erste Wiederholungsnivellement im August 1958. Aus diesen 2 Nivellements können folgende Schlußfolgerungen gezogen werden.

1. Alle Höhenmarken, die an der Küste des Sees Baskunchak liegen, sinken ohne Ausnahme im Mittel um 1 mm/Jahr.

2. Die Höhenmarken in der Nähe und auf dem Mt. Bolshoye Bogdo heben sich um einen Betrag von ungefähr 0,5 mm/Jahr.

3. Die von beiden, der Seeküste und dem Berg entfernten Höhenmarken zeigen praktisch keine Höhenänderung.

Diese Schlußfolgerungen, die eine einheitliche Vertikalbewegung der Erdkruste voraussetzen, befinden sich in völliger Übereinstimmung mit der Vorstellung über die rezenten Bewegungen des Baskunchak Salzdomes.

Es muß bemerkt werden, daß die Beträge der Höhenmarkenbewegungen relativ zu einer der Höhenmarken berechnet wurden, die sich am weitesten von der Seeküste und dem Berg entfernt befindet.

Das 2. Nivellement des Baskunchak-Gebietes wurde im Sommer 1961 ausgeführt.

Der Vergleich der Ergebnisse all dieser drei Nivellements macht es möglich, folgende Schlußfolgerungen zu ziehen:

1. Die 1. Schlußfolgerung ist, daß die allgemeine Regelmäßigkeit der Vertikalbewegungen, die im Baskunchak-Gebiet durch das 1. Wiederholungsnivellement festgestellt worden sind, durch das Nivellement im Jahre 1961 bestätigt wurde.

2. Zur selben Zeit lieferte das Wiederholungsnivellement wenig verschiedene Ergebnisse, die wahrscheinlich auf die jahreszeitlichen Oberflächenschwankungen des Baskunchak-Gebietes zurückgeführt werden können.

The study of movements of local structures is by no means less important from the theoretical and practical point of view than the study of recent tectonic movements of entire regions. This prompted the creation in 1951 of two first special levelling grounds on the Baskunchak and Saikhin salt domes of the Caspian lowland for the purpose of obtaining quantitative data on the recent movements of salt domes. Extensive geological-geomorphological investigations that preceded the creation of these levelling grounds helped to reveal the manifestations of young movements of both domes. The two structures have different relief: the Saikhin dome is a small flat upheaval; at the same time the most elevated point of the Baskunchak dome -- Mt. Bolshoye Bogdo -- dominates over the surrounding steppeland by more than 100 metres. This mountain is situated on the south-west bank of Lake Baskunchak (Fig. 1). The Saikhin and Baskunchak levelling grounds were created by the joint effort of the Geographical Institute of the U.S.S.R. Academy of Sciences, the Central Scientific Research Institute for Geodesy, Air Surveying and Cartography and a Moscow aerogeodetic enterprise. In the autumn of 1958 the Geographical Institute of the USSR Academy of Sciences and the Moscow Institute of Engineers of Geodesy, Air Surveying and Cartography conducted the first repeated levelling in the Baskunchak area. In the summer of 1961 the Geographical Institute performed the second levelling in the same area.

The Baskunchak levelling ground is a closed type structure which encircles Mt. Bolshoye Bogdo. The grounds are provided with fourteen bench-marks, one of which is located on the mountain and is connected to the bench-mark on the bank of the lake with a hanging traverse. The area is about 50 km in perimeter. All the three levellings were conducted according to the 2nd class method but the sighting ray was limited to 50 m despite the fact that in case of 2nd class it can generally be 75 m. This limitation was aimed at increasing the accuracy of levelling.

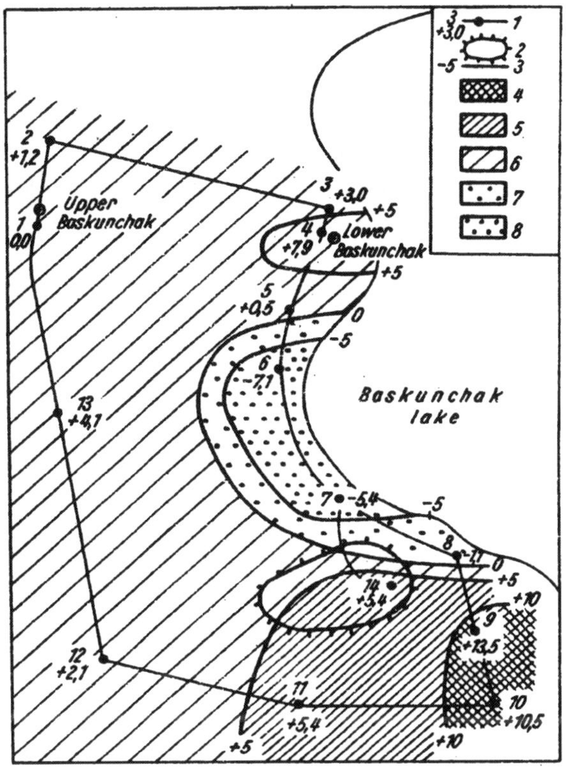

Fig. 1

Diagram of recent movements of the Baskunchak salt dome.

1 - line of repeated levelling, bench-marks (1, 2, ...) and amplitudes of their displacement (0.0 + 1.2) in mm from 1951 to 1961; 2 - contours of Mt. Bolshoye Bogdo; 3 - isometric lines of amplitude movements in mm (1951 - 1961); 4 - uplifts over + 10 mm; 5 - ditto, from + 10 mm to + 5 mm; 6 - ditto, from + 5 to 0 mm; 7 - subsidences from 0 to - 5 mm; 8 - ditto, over - 5 mm

Table 1 shows the values of accidental mean quadratic error for the levelling of various years per 1 km of traverse η and the values of discrepancies f. Due to a small levelling traverse it is very difficult to obtain the systematic error, but the small values of discrepancies prove its insignificance. Accuracy of levelling is likewise corrobated by the difference in the excess values obtained from forward and reverse

Table 1

Year	η mm	f mm
1951	± 0.8	- 7.8
1958	± 0.7	- 5.7
1961	± 0.6	- 1.8

traverses. These data for the levelling of 1961 are given in Table 2. This Table also

Table 2

Number of bench-mark	Distance between bench-marks	Excess				Difference between direct and reverse traverse	Permissible difference,
		direct		reverse			
	km	m		m		mm	mm
1	2.4	-	2.5313	+	2.5316	+ 0.3	± 4.6
2	7.6	-	37.6490	+	37.6468	- 2.2	± 8.3
3	1.1	+	4.4637	-	4.4642	- 0.5	± 3.1
4	2.0	+	0.0852	-	0.0843	+ 0.9	± 4.2
5	2.2	+	1.4797	-	1.4801	- 0.4	± 4.4
6	3.6	+	0.4733	-	0.4735	- 0.2	± 5.7
7	3.4	-	4.1913	+	4.1891	- 2.2	± 5.5
8	2.5	+	14.5747	-	14.5777	- 3.0	± 4.7
9	2.0	+	2.4429	-	2.4432	- 0.3	± 4.2
10	5.4	+	12.4222	-	12.4163	+ 5.9	± 7.0
11	5.4	+	5.2991	-	5.2970	+ 2.1	± 7.0
12	6.6	+	4.9818	-	4.9850	- 3.2	± 7.7
13	4.4	-	1.8542	+	1.8541	- 0.1	± 6.3
1							
7	3.2	+	108.3744	-	108.3771	- 2.7	± 5.4
14							

indicates the distances and excesses between the bench-marks. The rates of the bench-mark movements have been calculated by the formula

$$v = \frac{H_2 - H_1}{t}$$

where H_2 and H_1 are the data of the same bench-mark provided by the repeated and initial levelling respectively; t is the time interval between these levellings.

It should be noted that not the absolute but relative rates of bench-mark movement were computed in the Baskunchak area since one of the bench-marks whose position was assumed constant in time was used as a starting point in computing the bench-mark data.

This circumstance is of no essential value in studying recent movements of local structures if this initial starting point is selected regardless of the structure. In the Baskunchak area we were interested above all in the movements of the lake bank and Mt. Bolshoye Bogdo. In view of this, it was expedient to select an initial point removed farthest from the bank of the lake and the mountain. We used bench-mark 1 as such initial point.

Before describing the results of the second levelling we shall point out the main conclusions obtained from the first levelling. If we assume that the rates of bench-mark movements show us the true picture of the recent movements it can be concluded on the basis of the results provided by the 1958 levelling that the different sections of the earth's crust with the bench-marks are displaced differently in height. The area with the western boundary of the grounds (bench-marks 12, 13, 1, 2, 11) have a zero rate of movement, the bank of lake Baskunchak (bench-marks 3, 4, 5, 6, 7, 8) sinks at a rate of about 1 mm/year, Mt. Bolshoye Bogdo and the section of the steppeland east of it (bench-marks 14, 9, 10) rise at a rate of about 0.5 mm/year. These conclusions presuppose within the time interval between the two levellings the existance of a uniform vertical movement of those sections of the earth's crust where the grounds are located.

The rates of bench-mark movement according to the results of both repeated levellings are shown in Table 3. The graphs (Fig. 2 and 3) illustrate these tables. The curves in

Table 3

No. of bench-mark	Difference of bench-mark data			Rates of bench-mark movements			Possible mean quadratic errors for the time interval of 1951 - 1961	
	$H_{58}-H_{51}$ mm	$H_{61}-H_{51}$ mm	$H_{61}-H_{58}$ mm	1951-1958 mm/y	1951-1961 mm/y	1958-1961 mm/y	Difference of data, mm	Rate mm/year
1	0,0	0,0	0,0	0,00	0,00	0,00	0,0	0,00
2	+ 1,0	+ 1,2	+ 0,2	+ 0,14	+ 0,12	+ 0,07	± 1,5	± 0,15
3	- 7,1	+ 3,0	+ 10,1	- 1,02	+ 0,31	+ 3,68	± 3,1	± 0,32
4	- 4,4	+ 7,9	+ 12,3	- 0,63	+ 0,81	+ 4,47	± 3,3	± 0,34
5	- 5,9	+ 0,5	+ 6,4	- 0,84	+ 0,05	+ 2,33	± 3,6	± 0,37
6	- 11,8	- 7,1	+ 4,7	- 1,68	- 0,73	+ 1,71	± 3,9	± 0,40
7	- 10,0	- 5,4	+ 4,6	- 1,42	- 0,55	+ 1,67	± 4,3	± 0,44
8	- 5,7	- 1,1	+ 4,6	- 0,81	- 0,11	+ 1,67	± 4,8	± 0,49
9	+ 5,8	+ 13,5	+ 7,7	+ 0,83	+ 1,38	+ 2,80	± 4,9	± 0,50
10	+ 3,4	+ 10,5	+ 7,1	+ 0,49	+ 1,08	+ 2,58	± 4,8	± 0,49
11	+ 1,2	+ 5,4	+ 4,2	+ 0,18	+ 0,55	+ 1,53	± 4,1	± 0,42
12	+ 0,1	+ 2,1	+ 2,0	+ 0,02	+ 0,22	+ 0,73	± 3,3	± 0,34
13	+ 1,6	+ 4,1	+ 2,5	+ 0,23	+ 0,42	+ 0,91	± 2,1	± 0,21
14	+ 4,4	+ 5,4	+ 1,0	+ 0,64	+ 0,55	+ 0,36	± 4,8	± 0,49

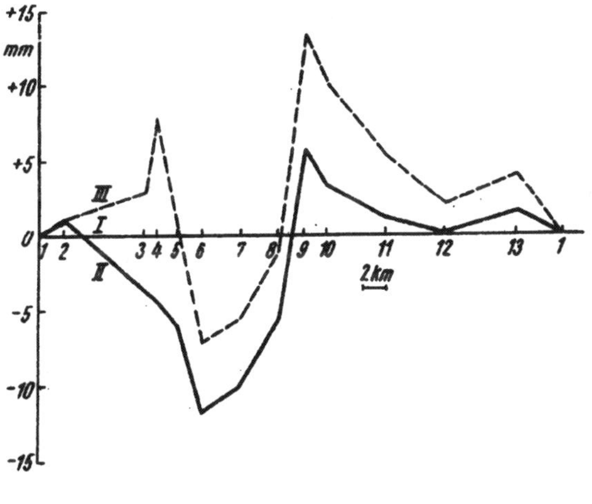

Fig. 2

Vertical movements of the Baskunchak salt dome according to the data of the repeated levellings of 1958 and 1961 (1, 2, ... are numbers of bench-marks)
I - 1951, II - 1958, III - 1961

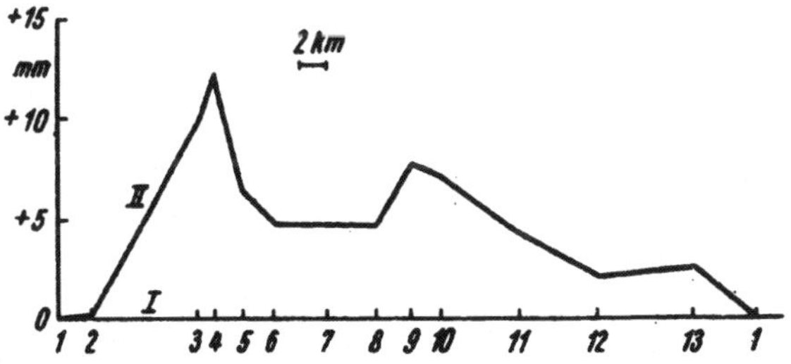

Fig. 3

Vertical movements of the Baskunchak salt dome within the time interval of 1958 to 1961(1, 2, ... are the numbers of bench-marks)
I - 1958, II - 1961

the figures were plotted according to the difference between the data of the bench-marks obtained from the levellings conducted in various years. In Fig. 2 the X-axis is an arbitrary level of the 1951 bench-mark levelling which is used to show the respective changes in the data of the identical bench-marks in 1958 and 1961. In Fig. 3 the X-axis is an arbitrary level of 1958 while the curve reflects the change in the data provided by the 1961 levelling. The X-axis also shows the distances between the bench-marks.

Analysis of these material leads us to the conclusion that:

1) the general tendency of vertical movements in the Baskunchak area established by the first levelling has been borne out by the second levelling. This can be clearly seen from the graph (Fig. 2) and is confirmed by the correlation factor of both curves (about 0.8);

2) however, despite this main conclusion, the results of the two repeated levellings essentially differ because of the change in the sign of the movement intensity of the bench-marks located on the bank of the lake. This is especially noticeable in the time interval of 1958 - 1961 during which the rates of movement of all the bench-marks are positive values. This fails to agree with the results of the first levelling which for the two different groups of bench-marks established inverse signs of the intensities of their movements.

The following causes responsible for this disagreement can be pointed out:

1. levelling errors;
2. instability of bench-marks;
3. seasonal variations of the section surface and, especially, of Lake Baskunchak bank.

These differences cannot be attributed to the levelling errors since these differences considerably exceed the possible levelling errors. Thus, for example, the mean possible quadratic error in the difference of the bench-mark 3 data computed on the basis of mean quadratic accidental errors of the 1958 and 1961 levellings comes to $\pm$ 2.9 mm but the actual difference between these data amounts to + 10.1 mm. A possible blunder is out of the question because each excess was determined twice and each time independently.

The instablity of the bench-marks can neither be the cause of this difference. The fact of the matter is that in 1951 the Central Scientific Research Institute for Geodesy, Air Surveying and Cartography set up in the Baskunchak area a special section for the study of the bench-mark behaviour in the soils of this district. The investigations carried out over many years under the guidance of the senior scientific worker M.S. Uspensky revealed that the bench-marks sink at a rate of about 0.3 mm/year. The 1958 and 1961 levellings showed that all the bench-marks on the bank of Lake Baskunchak rose in reference to the initial point which, if we remember all that has been said above, can by no means be explained by the instability of the bench-marks. Speaking about the stability of the bench-marks generally, we can hope that all the bench-marks in the ground should behave in about the same way since the conditions of their location (the soil and the depth of laying in) are very similar. The results of the first levelling which placed all the bench-marks in three groups with different rates of vertical movement are a forcible proof of this conclusion. Such subdivision of the bench-marks which is evidently due to the vertical crustal movements testifies to the fact that the bench-marks in the soil behave in a similar way.

It can be believed that the disagreements between the levelling data are brought about by the so-called seasonal variations of the crust. But the levellings of 1951 and 1958 were conducted in autumn (September, October) while the 1961 levelling was performed in summer (June). To answer this question a minimum of two levellings in summer and autumn must be conducted in the course of one year.

It should be stressed in conclusion that the results of levelling in the Baskunchak area show that in studying the recent movements of local structures not only levelling jobs should be undertaken. Suoh ground must be provided with dip-meter observation stations.

Discussion

T. Sorgenfrei (Kopenhagen):

1.) I ask for information regarding the exact location of the Salt Dome.

2.) I ask whether the survey was a single survey or whether it was a part of a larger survey of more than one Salt Dome.

3.) I ask the lecturer whether he had cooperated with geologists, and if it was possible to find geological evidences, others than those mentioned, of the recent movements of the Salt Dome.

A. K. Pevnev (Moskau):

The study is carried out in the Prekaspian depression on the Baskunchak salt dome; these studies were not carried out on other salt structures. Geodetic data about the rising of the dome are confirmed by the geological and geomorphological observations, and in particular by the data about the deformation of terraces.

Some results of the study of vertical movements in Armenia

von

P. Kazanchyan (Eriwan)

Summary

The study of the materials on the occasional precise levellings which have been conducted in Armenia since 1927 allowed us to conclude that the Leninakan Plateau (absolute elevation over 1,500 metres) rises by 3 mm per year while the north shore of Lake Sevan (absolute elevation about 2,000 metres) sinks by about 7 mm annually.

It is interesting to note that in Armenia with her peculiar geological structure and many active tectonic and seismic regions research into crustal movements can produce especially valuable results. Many of the regions under investigation may in future become very effective grounds for a permanent study of secular and seasonal movements of the Earth's crust. Our investigations, although carried out on a limited scale, show already that the rate of crustal movement in the given tectonic zone may be assumed constant. However, on the boundary between two tectonic zones there occur periodic subsidences and uplifts, i.e. the constancy of the rate of crustal movement is not maintained. A sharp change in these rates takes place after each earthquake. For this reason, these sections should be subjected to numerous precise levellings at brief intervals. To obtain a complete picture of local tectonic deformation, it would be desirable to study also horizontal movements. This has been corroborated by our investigations conducted since 1957 in a seismically active area near Leninakan where first class levellings are undertaken regularly every six months with the help of nine special benchmarks. Comparison of the results of precise levellings of all cycles with the data provided by the seismological station has enabled us also to locate the junction of two tectonic zones.

Zusammenfassung

Das Studium über Präzisionsnivellements, die in Armenien seit 1927 durchgeführt wurden, erlaubt den Schluß, daß das Leninakan Plateau (absolute Höhe über 1500 m) sich um 3 mm pro Jahr hebt, während das Nordufer des Sevansees (absolute Höhe ungefähr 2000 m) um ungefähr 7 mm pro Jahr absinkt.

Interessant ist, festzustellen, daß in Armenien, mit seiner besonderen geologischen Struktur und den vielen aktiven tektonischen und seismischen Gebieten, die Untersuchung der Krustenbewegung besonders unterschiedliche Ergebnisse liefern kann. Viele Untersuchungen dürften in Zukunft besonders Anlaß für das ständige Studium der säkularen und jahreszeitlichen Bewegungen der Erdkruste sein. Obgleich unsere Untersuchungen in begrenztem Umfang durchgeführt worden sind, zeigen sie schon, daß der Betrag der Krustenbewegungen in der betrachteten tektonischen Zone als konstant angenommen werden kann. An der Grenze zweier tektonischer Zonen treten jedoch periodische Senkungen und Hebungen auf, d.h. die Konstanz des Betrages der Krustenbewegung ist nicht gewährleistet. Eine plötzliche Betragsänderung findet nach jedem Erdbeben statt. Deshalb sollten in diesen Gebieten häufiger Präzisionsnivellements durchgeführt werden. Um ein vollständiges Bild von der lokalen tektonischen Deformation zu erhalten, ist es wünschenswert, auch die Horizontalbewegungen zu studieren. Dies ist durch unsere Untersuchungen bekräftigt worden, die seit 1957 in einem seismisch aktiven Gebiet nahe Leninakan durchgeführt wurden, wobei Nivellements 1. Klasse regelmäßig alle 6 Monate unter Benutzung von 9 Spezial-Fixpunkten zur Anwendung kamen. Ein Vergleich der Ergebnisse der Präzisionsnivellements aller Zyklen mit den von der seismologischen Station erhaltenen Daten ermöglichte uns, die Grenze zweier tektonischer Zonen zu lokalisieren.

Recent tectonic movements of the Earth's crust are of enormous scientific and practical interest which makes their study all the more important with each passing year.

Precise levellings repeated after 15-20 years can be used to determine the magnitude and rate of recent crustal movements, the rate being assumed constant. The available materials on precise levellings cannot fully satisfy the requirements presented by this problem. The precise levelling lines are usually laid for geodetic purposes along railways and motor-roads and frequently fail to intersect in the required direction large and small geological structures on whose boundaries the vertical crustal movements manifest themselves with special intensity. Apparently, these features should be taken into account when conducting special investigations and numerous repeated levellings within very brief time intervals undertaken in specially assigned areas should be employed for the study of local tectonic displacements.

Special research into recent movements of the Earth's crust has been recently undertaken in the Armenian SSR. Highly precise repeated levellings in Armenia with her complex geological structure and many active tectonic and seismic areas can produce extremely valuable results.

The levellings carried out in Armenia for geodetic purposes do not as yet meet the present-day requirements presented by the study of the oscillatory movements of the Earth's crust. For all that, the results of these repeated levellings have shown that in the period from 1927 to 1947 the Leninakan Plateau rose at a mean rate of about 3mm/year while the northern bank of Lake Sevan sank at a rate of 6.7 mm/year. When analysing the results of precise levellings account should be taken of the fact that at the junction of two tectonic zones there can occur distinctive variations of the Earth's crust, i.e. subsidences and upheavals, in the period between the repeated levellings, especially after earthquakes.

Hence, the rates of vertical movements obtained from the levellings repeated after 15-20 years will reflect only the remaining part of these variations. Regular observations should be arranged on the boundary between two adjacent tectonic zones to study the intensity of the crustal movements within very brief time intervals.

With a view to studying crustal movements taking place within brief time intervals or having a seasonal character special levelling jobs were undertaken in 1957 near Leninakan. For this purpose, we laid in nine specially designed basement bench-marks spaced at a distance of about one kilometre along the line of nearly eight kilometres curved in the middle at a right angle (side bearings 52° and 142°). Three months later, in October 1957, we conducted the levelling of the entire area in conformity with class I programme in favourable external conditions.

Subsequently, every six months we carried out precise levellings of this area in about the same external conditions and by the same observer who used the same tools.

The excess data we obtained for all the cycles are illustrated in Table 1.

Table 1

No. of bench-marks	Mean excess of forward and reverse traverse						
	October 1957 h_1	April 1958 h_2	October 1958 h_3	April 1959 h_4	October 1959 h_5	February 1960 h_6	April 1960 h_7
1	-5.6122	-5.6117	-5.6106	-5.6114	-5.6126	-5.6123	-5.6131
2	-9.5811	-9.5830	-9.5822	-9.5784	-9.5814	-9.5792	-9.5790
3	-6.7324	-6.7336	-6.7351	-6.7316	-6.7321	-6.7332	-6.7388
4	-6.0612	-6.0604	-6.0626	-6.0642	-6.0608	-6.0619	-6.0662
5	-2.6474	-2.6472	-2.6477	-2.6448	-2.6469	-2.6474	-2.6476
6	-4.5174	-4.5168	-4.5178	-4.5162	-4.5169	-4.5181	-4.5207
7	-1.8853	-1.8846	-1.8848	-1.8853	-1.8850	-1.8835	-1.8825
8, 9	+0.5488	+0.5488	-	-	+0.5496	-	-
Traverse disagreement in mm	-0.95	+0.2	+2.2	0	+0.2	+0.3	-1.1

Bench-mark No. 5 is situated in the middle of the area and is the vortex of the right angle. For this reason, it was taken as the initial point relative to which we computed the accumulations of the excess differences and the rates of the bench-mark displacements for all the seven cycles of observations. The results of these calculations are given in Table 2.

Table 2

No. of bench-marks	I diff.		II diff.		III diff.		IV diff.		V diff.		VI diff.		diff.	
	accumulation h_1-h_2	rate v_1	accumulation h_3-h_2	rate v_2	accumulation h_4-h_3	rate v_3	accumulation h_5-h_4	rate v_4	accumulation h_6-h_5	rate v_5	accumulation h_7-h_6	rate v_6	accumulation h_7-h_1	rate v
1	+1.8	+3.6	+1.8	+3.6	-4.9	-9.8	+1.3	+2.6	-0.3	-0.9	+10.5	+7.0	+10.2	+2.6
2	+2.3	+4.6	+2.9	+5.8	-5.7	-11.4	+0.1	+0.2	0	0	+ 9.7	+6.5	+ 9.3	+2.4
3	+0.4	+0.8	+3.7	+7.4	-1.9	-3.8	-2.9	-5.8	+2.2	+6.6	+ 9.9	+6.6	+11.4	+3.0
4	-0.8	-1.6	+2.2	+4.4	+1.6	+3.2	-3.4	-6.8	+1.1	+3.3	+ 4.3	+2.9	+ 5.0	+1.3
5	0	0	0	0	0	0	0	0	0	0	0	0	0	0
6	+0.2	+0.4	-0.5	-1.0	+2.9	+5.8	-2.1	-4.2	-0.5	+1.5	- 0.2	-0.1	- 0.2	0
7	+0.8	+1.6	-1.5	-3.0	+4.5	+9.0	-2.8	-5.6	-1.7	-5.1	- 2.8	-1.9	- 3.5	-0.9
8	+1.5	+3.0	-1.7	-3.4	+4.0	+8.0	-2.5	-5.0	-0.2	-0.6	- 1.8	-1.2	- 0.7	-0.2
9	+1.5	+3.0	-	-	-	-	-	-	-	-	-	-	-	-

During the period from October 1957 to August 1961 the seismic station in Leninakan recorded fifteen earthquakes with the force over two. Namely, on May 30 and 31, 1958 we recorded three earthquakes with the maximum intensity of 4.5; on June 26, 1958 - with intensity 4, on March 23, 1959 with intensity 2.5; on December 8-9, 1959 eight earthquakes with the maximum intensity of 6; on April 18, 1960 and on February 3, 1961 with intensity over 4.

Comparing these data with the data in Table 2 the following can be assumed:

The accumulations of the excess differences during the periods when there were no earthquakes differ from those registered during the periods with earthquakes.

During the periods without earthquakes the signs of the accumulations of excess differences were retained in most cases. The earthquakes violated this regularity.

The fact should be noted that the displacements of the benchmarks relative to the initial point change, now decreasing, now increasing. The rates of displacement of the bench-marks 3 and 4 changed with very high intensity which can be attributed to the alteration of the hydrological and structural conditions along the entire area under investigation. Under the bench-marks 1, 2 and 3 situated on one side of the right angle, the depth of ground waters was maintained at 2 to 3 metres. The depth of ground waters under the bench-mark 4 situated on the same side of the right angle and under all the other bench-marks sharply increased and reached sixteen metres. There is every ground to believe that the boundary between two heterogeneous geological structures passes exactly between the bench-marks 3 and 4.

Summarising all that has been said above we can point out that repeated levellings conducted after 15-20 years can be used only to determine the mean intensity, to be more precise the remaining value of the rates of recent crustal movements.

The true intensity of recent crustal movements requires for its determination regular and seasonal observations. We believe that the results of precise levellings can be used to mark off the junctions between various tectonic zones.

D i s k u s s i o n

S. M i y a m u r a (Tokyo):

Is there an active volcano near the area studied?

P. S. K a z a n c h y a n (Moskau):

All the fundamental benchmarks were set in bedrocks which are of volcanic origin. Therefore, the depth of their setting is about 4.5 m.

Untersuchungen der rezenten Erdkrustenbewegungen in Ungarn

von

L. Homoródi (Budapest)

Summary

In Hungarian territory three measurings of altitudes have hitherto been made, the results of which were used for the movement of the earth's crust, too. The Military Institute of Geography in Vienna carried out the first levelling between 1873 and 1897. The other two measurings of altitudes were conducted by the State Survey between 1921 and 1931, and between 1949 and 1954, respectively. The most extensive researches were based on the comparison of the results of the first and the second measurings of altitudes, the results of which have been published in the international literature, too.

The newest investigations have shown that considerable errors found their way into the levelling of the Military Institute of Geography so that its ascertainments can't be supposed to be real. The two modern measurings of altitudes have been used so far only for partial researches because each was carried out on another level. Thus only the compensated heights can be compared; these however, are influenced by the mode of compensation and the form of the canvas of the survey. The lecture mentions interesting final results and discusses further projects.

Zusammenfassung

Im Gebiet Ungarns wurden bisher drei Landeshöhenmessungen durchgeführt, deren Ergebnisse auch bei der Erdkrustenbewegung verwendet wurden. Das erste Nivellement hat zwischen 1873 und 1897 das Wiener Militär-Geographische Institut ausgeführt. Die anderen zwei Höhenmessungen wurden zwischen 1921 und 1931, bzw. zwischen 1949 und 1954 von dem Staatlichen Vermessungsdienst durchgeführt. Die ausgedehntesten Untersuchungen beruhten auf dem Vergleich der Ergebnisse der ersten und zweiten Höhenmessung, deren Resultate auch in der internationalen Literatur erschienen. Die neuesten Untersuchungen haben gezeigt, daß das Nivellement des Militär-Geographischen Instituts beträchtliche Fehler enthält, so daß man deren Feststellungen nicht für real halten kann. Die zwei neueren Höhenmessungen wurden bisher nur auf partielle Untersuchungen angewendet, weil jede auf einer anderen Linie geführt wurde. So kann man nur die ausgeglichenen Höhen vergleichen; diese werden aber durch die Ausgleichungsart und die Form des Netzes beeinflußt. Der Vortrag erwähnt interessante Endresultate und spricht von den weiteren Plänen.

Dieses Referat möchte kurzgefaßt diejenigen in Ungarn durchgeführten Untersuchungen darstellen, die auf Grund von Nivellementsangaben zum Nachweis der vertikalen, und auf Grund von Koordinaten trigonometrischer Punkte der horizontalen Erdkrustenbewegungen durchgeführt wurden.

Im Landesgebiet Ungarns wurde das erste Nivellement, das als Feinnivellement bezeichnet werden darf, im ersten Teil des 19. Jahrhunderts für Flußregulierungszwecke durchgeführt; das erste sich auf das ganze Landesterritorium erstreckende Nivellement wurde im wesentlichen von 1873 bis 1897 vom einstmaligen Wiener Militär-Geographischen Institut durchgeführt. Die in großem Maße erfolgte Zerstörung der Nivellementspunkte und die gesteigerten Genauigkeitsanforderungen haben die Entwicklung eines neuen Nivellementsnetzes erster Ordnung notwendig gemacht, das von 1921 bis 1931 durchgeführt wurde. Infolge der Ereignisse des zweiten Weltkrieges sind Punkte dieses neuen Nivellementsnetzes zerstört worden, so daß es notwendig wurde, das Netz vom Jahre 1949 an völlig zu erneuern. Diese Arbeit wurde im Jahre 1954 beendet.

Die Festpunkte des zu Flußregulierungszwecken dienenden Nivellements wurden leider nur mit provisorischer Vermarkung versehen; deshalb können die Ergebnisse dieses Nivellements zur Prüfung der Erdkrustenbewegung nicht verwendet werden. Die erste derartige Untersuchung hat die Höhenangaben des einstmaligen Militär-Geographischen Institutes und die des Nivellementes von 1921-31 gegenübergestellt. Die Ergebnisse dieses Vergleichs werden in Abb. 1 geschildert [1]. Die in der Abbildung dargestellten Differenzen zwischen

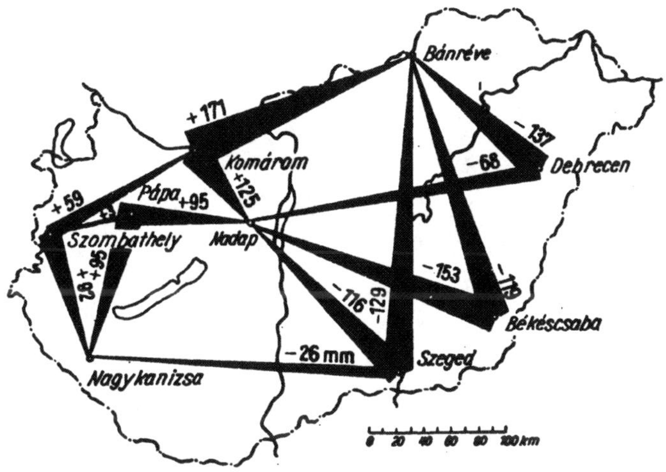

Abb. 1

Höhenunterschiede in mm zwischen Nivellementsergebnissen von 1873/97 und von 1921/31

den Höhen der Punkte des Nivellementsnetzes erregen den Anschein, daß das Territorium Ungarns um eine, ungefähr in SW-NO-Richtung liegende Achse derart kippt, daß die östlichen Landesteile sinken, die westlichen sich heben. Der Wert der jährlichen Niveauänderungen - stellenweise mehr als 3 mm/Jahr - hat zwar schon am Anfang hinsichtlich der Realität der Darstellung gewisse Zweifel erregt, und es wurde betont, daß an den Differenzen sowohl die Meßfehler, als auch der aus der Ausgleichung stammende Zwang beteiligt sind. Dennoch sind die Angaben in der Literatur mehr oder weniger als effektive Niveauänderungen angedeutet worden. Einige waren bestrebt, die Intensität der Bewegung so zu begründen, daß die in der Mitte des 19. Jahrhunderts durchgeführten Flußregulierungen das Wasserniveau des Bodens gesenkt hätten, was zum Austrocknen und zur Verdichtung der zur Oberfläche naheliegenden lehmigen Schichten führen sollte. Dieser Vorgang hätte natürlich seine Wirkung gegenüber dem Zeitpunkt der Flußregulierung ausgeübt und in der Zeit verschoben, auch im Zeitraume, der nach dem Nivellement des Militär-Geographischen Instituts folgte [2].

Das wiederholte Nivellement einzelner Linien des Militär-Geographischen Institutes und die gewissenhafte Kontrolle der früheren Rechenarbeiten haben es in der letzten Zeit bestätigt, daß der Grund der erwähnten Erscheinungen nur zum kleinsten Teil die effektive Niveauänderung sein kann. Im wesentlichen setzten sie sich aus Meßfehlern und aus dem Ausgleichungszwang zusammen. Es sind hier besonders die Komparationsfehler der Nivellierlatten und die Refraktionsfehler zu erwähnen [3]. Der Vergleich der Ergebnisse des Nivellements 1949–1955 mit den Ergebnissen des Nivellements 1921–1931, also der Vergleich zweier zeitgemäßer Nivellements zeigt klar, daß die früher ausgewiesenen Differenzen nicht geologischer Herkunft sind. Die Abbildungen 2 und 3 stellen an zwei Linien die Hö-

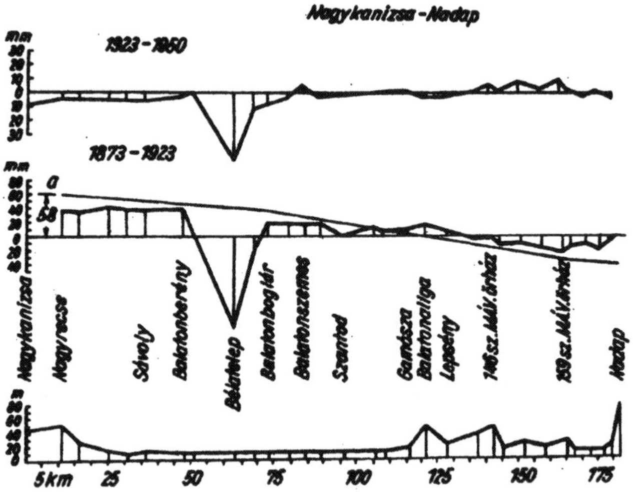

Abb. 2

Höhenunterschiede der Punkte der gleichen Nivellementslinie nach den Ergebnissen des Nivellements von 1873–1923 und von 1923 – 1950. (Unten das Längsprofil der Linie.)

henwertdifferenzen nach den Angaben von 1873–1923 und 1923–1950 einander gegenüber (an beiden Abbildungen stellt die untere Linie das Längsprofil des Geländes dar) [4]. Es ist ersichtlich, daß zwischen den zwei Ergebnissen nicht nur eine Größenordnungsdifferenz besteht, sondern auch die Vorzeichen verschieden sind. Übrigens ist es ja überhaupt nicht wahrscheinlich, daß binnen einer verhältnismäßig so kurzen Zeitspanne die frühere Bewegung von bedeutender Geschwindigkeit nicht nur gebremst wird, sondern sogar eine Phasenänderung erfolgen kann.

Der Umstand, daß die Fehler der Ergebnisse des Militär-Geographischen Institutes in erster Linie im Ausbleiben der Lattenkomparierungen und in Refraktionsauswirkungen gesucht wurden, haben die Vorstellung begründet, daß es unrichtig wäre, diese Angaben im Berg- und Hügelgelände zu benutzen. Bei unserem Tiefebenegebiet aber könnten sie zum Zwecke der Untersuchung der Niveauänderungen zum Ausgangspunkt werden. In nahezu flachem Gelände wird nämlich keine dieser Fehlerquellen zu einem beträchtlichen Fehlerwert führen. Auf Grund dieser Vorstellung versucht V. Scheffer nachzuweisen, daß die Niveauänderungen in Ungarn isostatischen Charakter haben [5]. Wenn er die oben erwähnten Niveauänderungen an einigen Linien mit den Schwerkraftanomalien gegenüberstellt, erhält er solche Korrelationen, die den isostatischen Charakter der Niveauänderungen zu bestätigen scheinen.

Den Niveauänderungs-Charakter der mittels der Angaben des Militär-Geographischen Institutes an der Ungarischen Tiefebene ausgewiesenen Höhendifferenzen unterstützt auch der interessante Zusammenhang der Quartär-Sedimentdicke mit dem Höhenunterschied, worauf

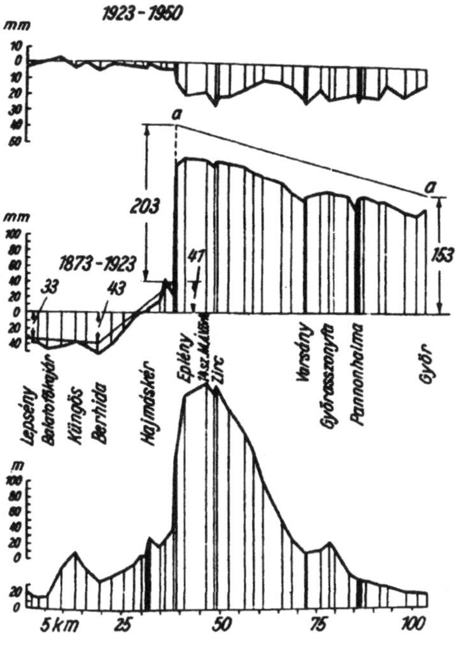

Abb. 3

Höhenunterschiede der Punkte der das Bakony-Gebirge durchschneidenden Nivellementslinie nach den Ergebnissen des Nivellements von 1873 - 1923 und von 1923-1950 (Unten das Längsprofil der Linie)

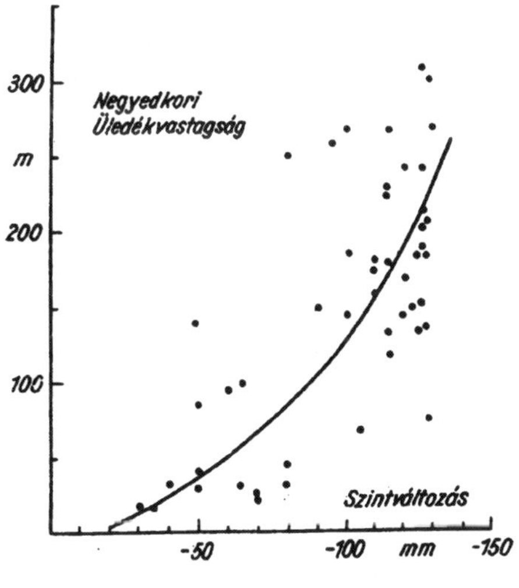

Abb. 4

Zusammenhang der Niveauänderung (Abszissen) und der Dicke des Quartärsediments (Ordinaten) nach Prof. Egyed

Prof. Egyed hingewiesen hat [6].

Der Zusammenhang wird in Abb. 4 geschildert, wo auf die Abszissenachse die Höhendifferenzen, an der Ordinatenachse die Dicken des Quartärsedimentes aufgetragen wurden.

Hinsichtlich des Charakters der Höhendifferenzen, die auf Grund der in den Jahren 1873 und 1921 begonnenen Nivellements ausgewiesen wurden, sind also die Meinungen recht verschieden. Entscheidend wichtig ist daher die Gegenüberstellung der Nivellements vom Jahre 1921 und 1949.

Dies ist das Thema der Abhandlung von L. Bendeffy [7], die durch Vergleich der Ergebnisse der Nivellements von 1921 und 1949 die transdanubischen Niveauänderungen und deren tektonische Ursachen analysiert. Als Ergebnis seiner Untersuchungen wird festgestellt, daß die Bewegungsverhältnisse Transdanubiens sehr variabel sind; das Ausmaß der Hebung und Senkung bleibt im allgemeinen unter 2 mm/Jahr. Die Blöcke, die im Mesozoikum eine stärkere Belastung erhielten, sinken, so in erster Linie das Bakonygebirge. Es ist interessant, daß hauptsächlich im südöstlichen Teil des Landes sich hebende und absinkende Zonen wechseln, was auf Grund der Abbildg.5 als Ergebnis eines, infolge des Sinkens des Bakonygebirges erfolgten Rutschens der aufeinander gehäuften Blöcke erklärt werden kann. Das schwarze Diagramm schildert das Ausmaß der Niveauänderung auf eine 10jährige Zeitspanne bezogen. Von einem solchen sinkenden bzw. sich hebenden Streifenpaar wird das Strombett der Donau an der Strecke Esztergom-Visegrád so geschnitten, daß an der oberen Strecke das Strombett sinkt, an der unteren sich erhebt (Abb. 6). Die Niveauänderung solchen Charakters wird an dieser Stelle durch die Prüfung der Stromschütte und durch archäologische Angaben unterstützt.

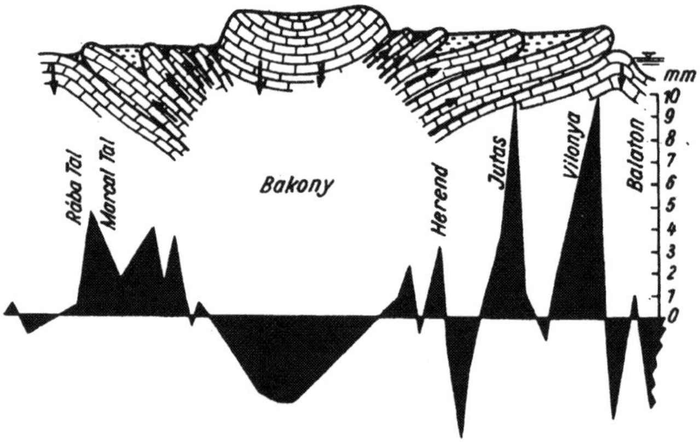

Abb. 5

Der Mechanismus der Senkung des Bakony-Gebirges nach Bendeffy (Unten das Diagramm der in dem NW-SO Längs-Profile gefundenen Höhenänderungen).

Die Ergänzung der zu Transdanubien gehörenden Untersuchungen mit ähnlicher Bearbeitung der östlichen Landesteile ist bis heute leider nicht erfolgt, hauptsächlich deshalb, weil wir zu der Überzeugung gekommen sind, daß eine Zeitspanne von 25 Jahren zur Ableitung zuverlässiger Angaben nicht genügt. Darum befaßten sich unsere Ingenieure vielmehr mit der Frage, mit welcher Wahrscheinlichkeit aus solchen Vergleichen auf eine Niveauänderung gefolgert werden kann und was sonstige Ursachen sein können, die eine Unterschiedlichkeit der Ergebnisse zweier Feinnivellements verursachen kann. Diese Frage erscheint auch deswegen interessant, weil aus technischen Gründen die Nivellementszüge erster Ordnung im Jahre 1921 und 1949 nicht an derselben Linie geführt wurden. Was wir also vergleichen können, das sind nur die Höhen der identischen Punkte nach der Aus-

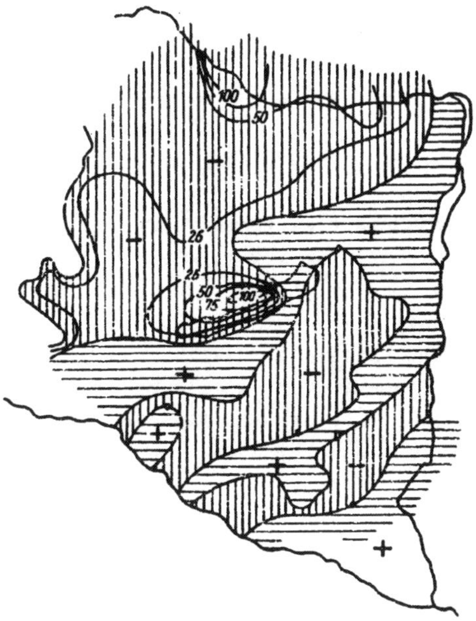

Abb. 6 Die Niveauänderung von Transdanubien nach Bendeffy. (Die senkrecht schraffierten Teile senken sich, die waagerecht schraffierten Teile heben sich. Die Ziffern zeigen die Niveauänderungen in mm/100 Jahr)

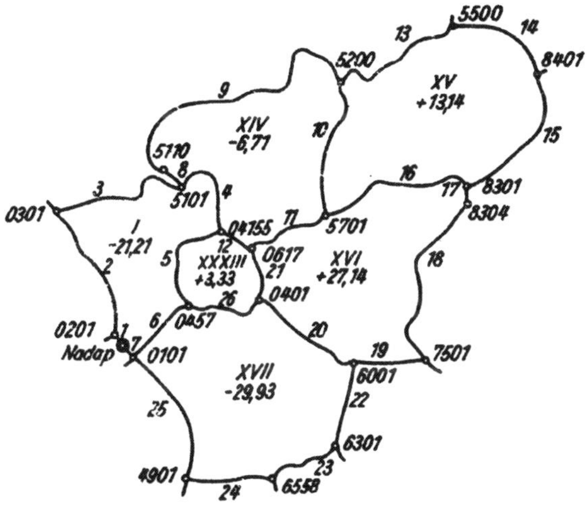

Abb. 7 Ein Teil des ungarischen Nivellementsnetzes, dessen Ausgleichung in verschiedenen Varianten durchgeführt wurde, um den Einfluß der Ausgleichung zu untersuchen

gleichung. So taucht die Frage auf, in welchem Unfange aus verschiedenen Netzen stammende Werte zusammen verwendet werden dürfen.

Um dies beurteilen zu können, haben wir den an Abb. 7 geschilderten Abschnitt des neuesten Nivellementsnetzes (vom Jahre 1949) auch in mehreren verschiedenen Variationen ausgeglichen, und zwar derart, daß einmal der eine, alsdann ein anderer Abschnitt ausgelassen wurde, als ob er nicht gemessen worden wäre. Es wurden also 6 Nivellementspolygone zu 4-5 Polygonen vereinigt. Die so erhaltenen Ergebnisse wurden dann mit den aus der gemeinsamen Ausgleichung der 6 Polygone erhaltenen Ergebnissen verglichen. Das Ergebnis war, daß unter den einzelnen Varianten Abweichungen von mehreren mm vorgekommen sind, sogar solche, welche die Größenordnung der Schlußfehler erreicht haben [8]. Hieraus wird gefolgert, daß eine bei den Niveauänderungen in Betracht kommende Differenz von einigen mm nur dann real nachgewiesen werden kann, wenn sich die verglichenen, im Zeitraum von 15-25 Jahren gemessenen Nivellements, streng auf dasselbe Netz beziehen. Wenn zwei verglichene Netze nicht dieselbe geometrische Form besitzen, so kann die Höhendifferenz der identischen Punkte mindestens in dem Maße aus der Verschiedenheit der Ausgleichungsfehlerverteilung stammen, wie die Größenordnungen der zu erwartenden Niveauänderungen sein würden.

Es soll noch erwähnt werden, daß durch eine neuere Ausgleichung des auf ungarischem Territorium befindlichen Netzteiles des militärischen Nivellements vom Jahre 1873 auch die Fortpflanzung der Wirkung der in den einzelnen Zügen begangenen Meßfehler geprüft wurde, wobei die Meßergebnisse der Nivellementszüge, die vom Militär-Geographischen Institut als fehlerhaft gefunden und neu gemessen wurden, durch neuere aus den Jahren 1897-1913, aus sorgfältigeren Messungen stammende Ergebnisse ersetzt wurden. Die Ergebnisse der ursprünglichen und der neuen Ausgleichung gegenübergestellt, galt es zu beweisen, daß sich die Meßfehler auch auf die unverändert gelassenen Linienteile bedeutend fortgepflanzt haben. Demnach müssen die sich an den Polygonzügen unserer Tiefebene ergebenden Höhenabweichungen zwischen den Ergebnissen älterer und neuerer Nivellements mit Vorbehalt behandelt werden, obgleich diese Züge der Tiefebene in sich selbst in großem Maße von den Fehlern, die aus der Vernachlässigung der Lattenkorrektionen und aus Refraktionswirkungen stammen, frei sind. Auf Grund dessen befaßte sich Bendeffy in mehreren Abhandlungen mit den Ursachen der Unterschiede zwischen nominalen und effektiven Höhen und mit den auf Grund der Unterschiede möglichen Folgerungen [4], [10].

In zwei Abhandlungen wurde versucht, die horizontalen Erdkrustenbewegungen auf Grund der Unterschiede der Koordinaten einer älteren und einer neueren Triangulation auszuweisen. Die eine bezog sich auf das Gebiet von Budapest, die andere auf das Gebiet des Mecsekgebirges. Der ersten Abhandlung [11] ist es infolge der größeren Anzahl der Punkte und ihrer Genauigkeit gelungen, gewisse Zusammenhänge zwischen der tektonisch zu erwartenden Bewegung der durch einzelne Verwerfungen abgegrenzten Blöcke und den Differenzen der Koordinaten auszuweisen (Abb. 8), die zweite wurde mit einem negativen Ergebnis abge-

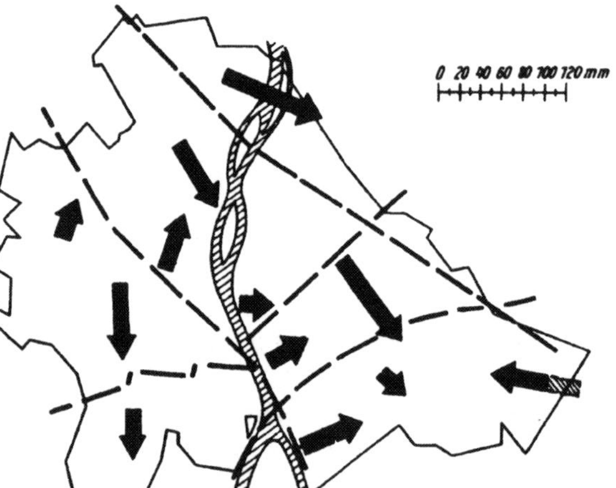

Abb. 8

Die Horizontalbewegungen des Gebietes von Budapest nach Bendeffy nach den Ergebnissen der Triangulationsarbeiten von 1878 und 1925. (Die Pfeile zeigen die durchschnittliche Richtung und Größe der Bewegungen, die Strichellinien die Bruchlinien.)

schlossen; denn auf die einzelnen, durch Verwerfungen abgegrenzten Blöcke fielen so wenige und so minderverläßliche Punkte, daß es zu einer zuverlässigen Folgerung keine Möglichkeit gab.

Im Zusammenhang mit der Bestimmung der horizontalen Erdkrustenbewegungen sind noch die an den fundamentalen Längenhauptpunkten seit 1939 systematisch durchgeführten Längenbestimmungen zu erwähnen. Eine sehr wichtige Erfahrung dieser Messungen, die auch bei anderen, ähnlichen Messungen zu beachten ist, war, daß bei Messungen solcher Art die Anwendung desselben Instruments und derselben Methode von grundlegender Bedeutung ist, und die Einschaltung eines neuen Beobachters auch nur nach einer mehrjährigen parallelen Meßarbeit erfolgen soll. Übrigens ist es nicht möglich, aus den an den Längenpunkten bisher durchgeführten Messungen eine Bewegung in einer Richtung zu folgern, es wären eher periodische Änderungen auszuweisen, diese müssen aber noch durch weitere Messungen unterstützt werden.

Schließlich sei es noch gestattet, zu erwähnen, daß wir aus unseren bisherigen Untersuchungen zu der Überzeugung gekommen sind, daß die zu praktischen Zwecken durchgeführten Nivellementsarbeiten zum Studieren der Erdkrustenbewegungen weniger geeignet sind; deshalb bereiten wir uns mit besonderer Umsicht zum Messen eines grundlegenden Nivellementsnetzes vor, das primär das Studium der Erdkrustenbewegungen zum Zwecke hat. Mit einer besonderen Vermarkungsmethode wollen wir die Stabilität der Punkte sichern, und wir sind bestrebt, die Beobachtung des ganzen Netzes möglichst in einer Sommerperiode zu beenden.

Im Interesse des Studiums der Erdkrustenbewegungen wollen wir in der Zukunft an mehreren Punkten geographische Ortsbestimmungen durchführen, an denen auch vor Jahrzehnten (an einigen nahezu 100 Jahre) Messungen durchgeführt wurden, und so ist es zu hoffen, daß trotz der unvermeidlichen Abweichung auf Grund anderer beobachtender Personen und anderer Instrumente, doch brauchbare Ergebnisse zustande kommen.

Literatur

[1] Gárdonyi, F.: A régi felsőrendű szintezési alappontok magasságainak változásai (Die Höhenänderungen der alten Feinnivellements-Festpunkte). Geodéziai Közlöny, VIII, 1932, Budapest

[2] Bikfalvy, B.: Az elsőrendű magassági jegyek, mint változásainak agrogeológiai okai (Agrogeologische Ursachen der Änderung der Höhenmarken). Geodéziai Közlöny, VIII, 1932, Budapest

[3] Bendeffy, L.: Szintezési alappontok időközi magassági változásainak meghatározása (Die Bestimmung der mittleren Höhenänderung der Nivellementsfestpunkte). Geofizikai Közlemények, IV, 1956, Budapest

[4] Bendeffy, L.: Szintezési munkálatok Magyarországon (Nivellementsarbeiten in Ungarn). Akadémiai Kiadó, 1958, Budapest

[5] Scheffer, V.: A magyarországi szintváltozások izosztatikus jellege és a szintezési alappontok magasságainak időbeni értékjavitási lehetősége (Der isostatische Charakter der Niveauänderungen Ungarns und die Möglichkeit der zeitlichen Wertkorrektion der Höhen der Nivellementspunkte). A Magyar Tudományos Akadémia Műszaki Tudományok Osztályának Közleményei, XIII, 1954, Budapest

[6] Egyed, L.: A kéregmozgások okai és a magyarországi kéregmozgások (Die Ursachen der Erdkrustenbewegungen und die Erkrustenbewegungen in Ungarn). Geofizikai Közlemények VI. 1957, Budapest

[7] Bendeffy, L.: Niveauänderungen im Raum von Transdanubien auf Grund zeitgemäßer Feineinwägung. Acta Technica Acad. Sc. Hungaricae XXIII, 1957, Budapest

[8] Németh, F.: Szintváltozások és szintezési adatok (Niveauänderungen und Nivellementsangaben). Geodézia és Kartográfia, 11, 1959, Budapest

[9] Niewiarowski, F.; Wyrzykowski, T.: Determination des mouvements verticaux de l'écorce terrestre sur l'étendue de la Pologne par la méthode de repetition des nivellements de precision. Bulletin Géodésique, No 62, 1961, Paris

[10] Bendeffy, L.: Szintezési alappontok időközi magasságváltozásnak meghatározása (Die Bestimmung der mittleren Höhenänderung der Nivellementsfestpunkte). Geofizikai Közlemények, V. 2. 1955, Budapest

[11] Bendeffy, L.: Vízszintes értelmű szekuláris mozgások Budapest területén (Horizontale säkulare Bewegungen in Budapest). Földméréstani Közlemények, 1953, Budapest

[12] Lukács, T.: A geodézia szerepe a vízszintes kéregmozgások vizsgálatában (Die Rolle der Geodäsie bei der Prüfung der horizontalen Erdkrustenbewegungen). Geodézia és Kartográfia, 1959, Budapest

D i s k u s s i o n

A. Z a t o p e k (Prag):

Ich war sehr glücklich zu hören, wie sich die Niveauänderungen in einem Gebiet verhalten, das an der Südgrenze der CSSR gelegen ist. Ich hätte in dieser Hinsicht zwei Fragen zu stellen und eine Bemerkung zu machen:

1) Wurden spezielle Messungen in den seismisch aktiven Gebieten gemacht?
2) Wurden in den Schleifen, deren Durchmessung ein längeres Zeitintervall erfordert hat, irgendwelche Einflüsse auf die Schlußfehler ausgeübt?

L. H o m o r ó d i (Budapest):

1) Die Gegend längs der Donau, wo eine Änderung im Vorzeichen der Niveauänderung stattfindet, ist seismisch aktiv. Auch südlich vom Plattensse liegt eine seismisch tätige Region.
2) Die Zeitspanne, während der die einzelnen Nivellementsarbeiten durchgeführt wurden, beträgt nicht mehr als 5 Jahre, was wenig Einfluß auf den Zeitfaktor hat.

A. Z a t o p e k (Prag):

Es ist ein Beispiel, wo die Notwendigkeit einer international organisierten Zusammenarbeit sehr klar auftritt.

L. H o m o r ó d i (Budapest):

Eine Zusammenarbeit zwischen der CSSR und Ungarn bei Problemen der Erdkrustenbewegungen ist sehr nützlich, sogar notwendig, und wir hoffen, daß sie auch verwirklicht wird.

J. A. M e s c h e r i k o v (Moskau):

Is there a map of recent crustal movements for the territory of Hungary?

Erforschung vertikaler Erdkrustenbewegungen in der Tschechoslowakei

von

B. K r u i s (Prag)

S u m m a r y

In Czechoslovakian territory three levellings are available for investigating the movements of the earth's crust:

a) the canvas of the Austrian - Hungarian survey (1873 - 1896)
b) the first Czech survey (1920 - 1938)
c) the modern unitary survey (1939 - 1960)

The Research Institute for Geodesy, Topography and Cartography in Praha has been charged with investigating the movements of the crust. Till now this research work has been made by means of the first two levellings. The results hitherto obtained are not entirely satisfying, because we meet with some difficulties when older measurements are used, e.g. an exceptionally small number of identical points and an insufficient accuracy of the older levelling. Nevertheless differences of height have been found out which are evidence of vertical movements (especially in the eastern part of the territory). Moreover an outline map of isobases was made. A sufficiently sharp boundary and a mutual movement of the two formations: the Bohemian massif and the Carpathian system can be recognized.

For further studies use will also be made of the modern survey, and the project of a levelling which is to be repeated for this purpose will be prepared. Besides, the motions in the plain of the Danube (subsidence 12 mm/5 years) as well as in the areas of coal basins and great industrial plants (especially barrages) have been investigated. The selection of the points and a method of measurement attaining the greatest possible accuracy have also been studied.

Z u s a m m e n f a s s u n g

Zur Bestimmung der Erdkrustenbewegungen auf dem Gebiete der Tschechoslowakei stehen drei Nivellements zur Verfügung:

a) österreichisch-ungarisches Nivellementsnetz (1873 - 1896)
b) das erste tschech. Nivellementsnetz (1920 - 1938)
c) das neueste "einheitliche" Nivellementsnetz (1939 - 1960)

Mit der Erforschung der Krustenbewegungen ist das Geodätische, Topographische und Kartographische Forschungsinstitut in Prag beauftragt worden. Bis heute wurde diese Forschung auf Grund der ersten zwei Nivellements durchgeführt. Die bisherigen Ergebnisse sind nicht völlig befriedigend, weil bei der Ausnützung älterer Messungen einige Schwierigkeiten auftraten, wie z.B. eine besonders kleine Anzahl von identischen Punkten und eine geringe Genauigkeit des älteren Netzes. Trotzdem wurden Höhendifferenzen festgestellt, die von vertikalen Bewegungen zeugen (besonders in dem östlichen Teil des Gebietes). Außerdem wurde eine Übersichtskarte von Isobasen zusammengestellt. Man sieht eine ausreichend scharfe Grenze und eine gegenseitige Bewegung folgender zwei Formationen: des Böhmischen Massivs und des Karpatensystems.

Zum weiteren Studium wird auch das neueste Netz ausgenützt und das Projekt des zu diesem Zwecke wiederholten Nivellements vorbereitet werden. Außerdem sind die Bewegungen in der Donauebene (Senkung 12 mm/5 Jahre) sowie in den Gebieten der Kohlenbecken und großen Ingenieurwerke (besonders Talsperren) verfolgt worden. Es wurde auch die Wahl der Punkte und eine die höchstmögliche Genauigkeit ergebende Meßmethode ausgearbeitet.

In der letzten Zeit macht sich in verschiedenen Zweigfächern der Geodäsie der Übergang von der statischen zur dynamischen Auffassung bemerkbar. Diese Entwicklung ist auch beim Nivellement zu beobachten, wobei dasselbe sogar eine der ersten Stellen einnimmt. Diese Stellung konnte das Nivellement mit Hilfe der ständig fortschreitenden Vervollkommnung des Festlegungsverfahrens, der Nivellierinstrumente, der Hilfsmittel und Methoden erreichen, was wesentlich zur gesteigerten Genauigkeit der Ergebnisse beiträgt. Das Nivellement wird somit zu einer geeigneten und verläßlichen Methode zur Ermittlung selbst kleiner Höhenänderungen bei verhältnismäßig weiten Entfernungen. Es ist also von großer Bedeutung bei der Erforschung vertikaler Erdkrustenbewegungen, welche durch geologische Einflüsse verursacht worden sind.

Ähnlich wie in anderen Staaten, widmet auch die Tschechoslowakei diesem Problem große Aufmerksamkeit. Die Erforschung vertikaler Bewegungen auf unserem Gebiet ist aber keine leichte Angelegenheit. Die Tschechoslowakei ist von allen Europa umgebenden Meeren bedeutend entfernt und hat infolgedessen keine Möglichkeit, die absoluten Bewegungen zu kontrollieren. Eine weitere Schwierigkeit bilden die reichlich zerbröckelten geologischen Schollen. Obzwar unser Territorium nur zu zwei geologischen Grundformationen, d.h. zur Böhmischen Masse und zu den Westkarpaten gehört, bildet es doch ein buntes Mosaik, da dieselben in eine Reihe von kleineren geologischen Einheiten aufgeteilt und von zahlreichen tektonischen Linien durchwoben sind.

Vor einer gewissen Zeit wurden einige Studien über Höhenänderungen der Nivellementpunkte veröffentlicht, welche zu Schlußfolgerungen über vertikale Bewegungen dienten. Diese Arbeiten behandelten nur kleinere Gebiete, wie z.B. das Gebiet von Groß-Prag, der Kleinen Donauebene u.a. Erst vor kurzer Zeit habe ich mir zur Aufgabe gemacht, alle Höhenänderungen der einzelnen Punkte auf unserem ganzen Territorium festzustellen und sie dann in der Karte der Isobasen zu verzeichnen. Diese Karte erschien im "Bulletin géodésique" Nr. 56. Es geschah ohne mein Wissen und ohne die diesbezüglichen Erklärungen.

Deswegen möchte ich vor allem ganz kurz ihre Entstehung und auch ihren Wert erläutern. Auf dem Territorium unserer Republik wurden bis dato schon drei Staats-Nivellementswerke errichtet. Ein Teil des österreichisch-ungarischen Nivellementsnetzes und das erste und zweite tschechoslowakische Nivellement. Selbstverständlich befaßte sich keines von ihnen ausschließlich mit dem gegebenen Problem. Ihr Hauptzweck ging dahin, ein Netz von Nivellementslinien und Nivellementspunkten für den Bedarf der technischen Öffentlichkeit zu bilden und erst in zweiter Reihe auch der Wissenschaft zu dienen. Deswegen müssen alle aus dem Vergleich der Höhen identischer Punkte gefolgerten Schlüsse als provisorisch erachtet werden.

Für meine Forschung benutzte ich einstweilen die ersten zwei Nivellements. Gegenwärtig setze ich meine Studien über Änderungen des zweiten und dritten Nivellements fort. Unser ältestes Netz ist das Feinnivellementsnetz, gegründet vom österreichisch-ungarischen Militär-geographischen Institut in Wien in den Jahren 1873-1896 für die damalige europäische Gradmessung. Dieses Netz wurde zu einer Zeit geschaffen, wo es noch keine vollkommenen Nivellierinstrumente und Verfahren gab, so daß es viele Mängel aufwies. Vor allem war das Mittelwasser des Adriatischen Meeres, welches als Bezugsniveau des ganzen österreichisch-ungarischen Nivellementsnetzes diente, nicht allzu zuverlässig festgelegt. Weiterhin waren die Linien und Punkte nicht genügend dicht. Dabei war der Großteil der Linien längs der Eisenbahnlinien geführt, und auch die Punkte waren auf Eisenbahnobjekten festgelegt. Anfangs widmete man keine Aufmerksamkeit den Längeänderungen des Lattenmeters. Das Zeitintervall zwischen zwei Messungen, d.h. zwischen der Messung hin und zurück, betrug bei vielen Linien einige Jahre, ja sogar Intervalle von 10-12 Jahren sind bekannt. Die Messungen wiesen auch grobe Fehler auf, welche erst bei der Errechnung der Polygonschlüsse festgestellt wurden. Einige grobe Fehler beseitigten die neuen Messungen, andere blieben jedoch im Netz, da es nicht gelungen ist, auch nur beiläufig die Stelle zu bezeichnen, wo der Fehler verursacht worden ist. Auch die Ausgleichungsart wies Mängel auf. Das Netz wurde nämlich in drei Teilen ausgeglichen. Zuerst wurde der westliche Teil ausgeglichen, dann der nordöstliche und endlich der südöstliche; dadurch wurden die Höhen etwas deformiert. Auch die Genauigkeit dieses Netzes ließ zu wünschen übrig, und nach den heutigen Kriterien würde sie in der Kategorie der Nivellements hoher Genauigkeit nicht bestehen. Nach meinen Berechnungen beträgt der wahrscheinliche Totalfehler τ, berechnet nach den internationalen Formeln vom Jahre 1948, 2,8 mm/km. Der einzige positive Faktor dieses Netzes war die Errichtung von 7 Grundnivellementspunkten an Stellen, welche sorgfältig auf Grund geologischer Beurteilungen gewählt und auf eine solche Art festgelegt wurden, welche selbst den an solche Punkte gestellten Ansprüchen entspricht. Sie wurden direkt auf einheitlichen und unversehrten Felsen festgelegt und durch Denkmäler vor Beschädigung geschützt. Zwei derselben befinden sich auf unserem Territorium. Das eine, noch in seinem Originalzustand, befindet sich in Südböhmen (bei Lišov) und das zweite, aus technischen Gründen um einige hundert Meter weiter versetzt, bei Strečno, im Nordwesten der Slowakei. Beide Punkte werden ihre Bedeutung für die Forschung der Bewegungen unserer geologischen Grundeinheiten beibehalten, da der eine in der Böhmischen Masse und der zweite in den Westkarpaten liegt. Verhältnismäßig gut waren die Punkte I. Ordnung, besonders diejenigen, welche auf Felsen und älteren, vollkommen gesetzten größeren Gebäuden festgelegt waren. Diese Art von Festlegungsverfahren der Grundnivellementspunkte wurde von den späteren tschechoslowakischen Nivellements übernommen. Heute haben wir im ganzen 20 neue Punkte und einige Hilfspunkte.

Quantitativ und qualitativ ist das zweite Nivellement besser, welches auf unserem Territorium in den Jahren 1920 - 1938 durchgeführt wurde. Es ist viel dichter als das vorhergehende, und es wurde nach den von der Internationalen geodätischen Assoziation genehmigten Richtlinien vorgenommen. Ein Hindernis dieses Netzes bildeten zwei Bezugsniveaus - das eine für den westlichen Teil mit dem Ausgangspunkt Lišov und das zweite für den östlichen Teil mit dem Ausgangspunkt Strečno. Auch bei diesem Netze wurde keine zusammenfassende Ausgleichung durchgeführt. Was seine Genauigkeit anbelangt, beträgt der wahrscheinliche Gesamtfehler $\tau = 0{,}9$ mm; daraus ergibt sich, daß dieses Netz beiläufig dreimal genauer ist als das vorhergehende.

Diese zwei Netze bildeten die Grundlage für die Zusammenstellung der Isobasenkarte. Ich betone, daß eine sorgfältigere Zusammenstellung durch die verhältnismäßig kleine Anzahl von identischen Punkten erschwert war, da vom ersten Netz nur die Punkte I. Ordnung benutzt werden konnten; außerdem deckten sich die Linien des zweiten Netzes in vielen Gebieten, besonders im westlichen Teil, nicht mit den Linien des älteren Netzes. Beide Netze wurden zusammenfassend ausgeglichen und die Höhen auf den Grundpunkt Lišov bezogen, und deswegen haben die errechneten Änderungen nur einen relativen Charakter. Die Karte weist wesentlich größere Unterschiede im östlichen Teil auf, und zwar im negativen Sinne. Manche Unterschiede betragen bis 35 mm. Selbst wenn wir die erwähnten Mängel der beiden Netze - besonders des österreichisch-ungarischen, in Erwägung ziehen, können die festgestellten Höhenunterschiede der identischen Punkte nicht nur auf Rechnung der Einflüsse der Nivellementsfehler gehen. Der gemeinsame wahrscheinliche Totalfehler τ, nach der Formel

$$\tau = \sqrt{\tau_1^2 + \tau_2^2}$$

errechnet, beträgt 3 mm/km. Für verschieden entfernte Punkte kann der Fehler nach der Formel

$$\tau = 3\,\text{mm} \sqrt{L\,\text{km}}$$

errechnet werden, wo L die Entfernung vom Ausgangspunkt bedeutet, in unserem Falle von Lišov. Für den östlichen Teil unserer Republik, d.h. für die Entfernung von 900 km, beträgt der Fehler 90 mm. Da die Unterschiede diesen Fehler mehrere Male übersteigen, können wir mit Recht annehmen, daß unsere beiden geologischen Grundformationen sich in gegenseitiger Bewegung befinden, jedoch nicht in einer solchen, wie man nach der Größe der Unterschiede urteilen könnte. Die festgestellten Unterschiede sind wahrscheinlich dreimal größer, als die wirklichen Unterschiede. Ungelöst bleibt die Frage, ob das Gebiet von Südböhmen steigt (wo Lišov liegt), oder ob das östliche Gebiet unserer Republik sinkt. Ein richtigeres Bild wird die Beendigung des zweiten Teiles der Forschung ergeben.

Außer der Forschung, welche vorwiegend wissenschaftlichen Charakter hat, werden bei uns Forschungen über die Bewegungen für praktische Zwecke durchgeführt, und zwar besonders dort, wo große Wasserwerke gebaut oder projektiert werden, und in den Bergbaugebieten. Im Rahmen der Ausnützung der Wasserkräfte wird ein großes Wasserwerk an der Donau projektiert. Das Projekt dieses Werkes stößt jedoch auf Schwierigkeiten, welche ihren Ursprung in den besonderen Verhältnissen der Kleinen Donauebene und besonders im Becken von Komárno haben. Laut den bisherigen Forschungen zeichnet sich dieses Gebiet durch die größte seismische Aktivität in unserer Republik aus. Es kommen hier von Zeit zu Zeit Erdbeben und schwankende Erdkrustenbewegungen vor. Einen Teil der kollektiven Forschung bildet die Erforschung vertikaler Bewegungen durch wiederholte Nivellements, durchgeführt auf den Linien des Grundnivellementsnetzes zwischen Bratislava und Sturovo, deren Länge im Jahre 1950 ungefähr 180 km betrug. Auf diesen Linien fand die erste Wiederholung des Nivellements im Jahre 1956 statt. Für die Vermessung wurden Nivellementspunkte nach dem bei uns üblichen Verfahren festgelegt. Um örtliche Höhenänderungen der Punkte auszuschalten, wurde das ganze Nivellement in Abschnitte von durchschnittlich 3 km aufgeteilt. Auf jedem Abschnitt wurde der Durchschnitt der Höhenänderungen aller Abschnittspunkte berechnet. Diese Durchschnitte wurden dann als Grundlage für die Forschung benutzt. Im Komárno-Gebiet wurde auf diese Weise eine Senkung von 12 mm festgestellt. Das Ergebnis wurde mit den Ergebnissen der geologischen und geophysikalischen Erforschungen verglichen, und der Vergleich ergab eine gute Übereinstimmung beider Ergebnisse. Zur weiteren Kontrolle werden die Ergebnisse des im diesem Jahre wiederholten Nivellements dienen.

Ein weiteres Forschungsgebiet bildet die Umgebung eines großen Wasserwerkes an der Moldau. Hier wird ein 68 km langer und nur höchstens 1 km breiter See entstehen. Auf einer Fläche von ungefähr 27 km^2 werden 720 Millionen m^3 Wasser angestaut Für die Erforschung der durch das Gewicht der Wassermengen hervorgerufenen Änderungen, wurden einige Nivellementslinien errichtet, welche senkrecht das Moldautal durchschneiden und nach Füllung des Beckens einige Mal neu gemessen werden. Um möglichst den Einfluß der Längenänderung des Lattenmeters auszuschließen, wurden die Linien so gewählt, daß sie auf einem ebenen oder nur wenig geneigten Gebiet laufen.

Das Ostrauer Kohlenbecken ist das erste Bergbaugebiet, wo methodisch die vertikalen Bewegungen des Beckens, hervorgerufen durch die Bergwerksarbeiten, erforscht werden. Das Becken wurde von einer Nivellementsschleife umgeben, in welcher ab 1949 alle 3 Jahre die Nivellements wiederholt werden. Nach den bisherigen Ergebnissen neigt sich das Bekken in der südöstlichen Richtung, wie dies auch die Senkungen beweisen, welche bis zu der slowakischen Grenze führen. Vom heurigen Jahr ab werden die wiederholten Nivellements in Zusammenarbeit mit dem polnischen geodätischen Dienst durchgeführt, um die Bewegung des ganzen schlesischen Kohlenbeckens festzustellen. Voriges Jahr wurde eine ähnliche Nivellementsschleife um das Kohlenbecken von Kladno errichtet.

Die Bedeutung der bisherigen Forschungen bewirkte, daß man nun systematisch die vertikalen Erdkrustenbewegungen durch wiederholte Nivellements im ganzen Gebiet unserer Republik verfolgt. Zu diesem Zweck wurde mit allen daran interessierten Institutionen ein Plan ausgearbeitet, welcher langsam nun realisiert wird. Der Plan teilt die Forschung in zwei selbständige, wenn auch eng zusammenhängende Teile.

Der erste Teil betrifft die Forschung der Bewegungen der geologischen Grundformationen. Diese Bewegungen können die Bildung der tektonischen Verhältnisse im großen Gebiet von Mitteleuropa wie auch Osteuropa, so auch der Tschechoslowakei beeinflussen. Das tschechoslowakische Territorium selbst weist zwei geologische Grundformationen - die Böhmische Masse und die Westkarpaten auf und grenzt gleichzeitig an einige weitere Formationen. Die Forschung auf diesem Gebiet setzt selbstverständlich eine enge internationale Mitarbeit voraus. Ein spezielles Netz von wiederholten Nivellements wurde in Erwägung gezogen. Bei der Planung dieses Netzes wurde mit einer solchen Dichte von Nivellementslinien gerechnet, welche in der möglichst kürzesten Frist vermessen werden können - d.h. binnen 2, höchstens 3 Jahren. Aus diesem, wie auch aus ökonomischen Gründen hat man nicht für nötig erachtet, die Nivellements auf allen Linien unseres Grundnetzes zu wiederholen, im Gegenteil, man sollte auch Linien der niedrigeren Ordnungen benutzen, falls ihre Lage für die Forschung geeigneter wäre. Der Plan wurde in Zusammenarbeit mit Fachleuten der Geologie und Geophysik ausgearbeitet. Die Nivellementslinien wurden nach den bisherigen Linien gewählt, damit anfangs die Ergebnisse der früheren Vermessung genutzt werden können.

Der zweite Teil befaßt sich mit dem Studium vertikaler Bewegungen kleinerer geologischer Einheiten, welche vorwiegend innerhalb der Grenzen der Tschechoslowakei stattfinden. Nur stellenweise, an den Grenzen der Nachbarstaaten, wird die internationale Mitarbeit in Erwägung gezogen werden müssen. Die in diesem Rahmen durchgeführten Forschungen haben einen mehr oder weniger praktischen Charakter, da sie außer der Wissenschaft auch Projekten, Bau und Sicherung großer sozialistischer Bauten dienen werden. Das konkrete Ziel dieser Aufgabe ist die Lösung einiger bis jetzt nocht nicht gelöster geologischer Fragen unseres Territoriums. Es soll dadurch eine eingehende geologische Erforschung ermöglicht werden. Die Nivellements werden in so große Gebiete eingeteilt, daß die diesbezüglichen Arbeiten innerhalb eines Jahres durchgeführt werden können.

Zur Durchführung dieser Arbeiten werden alle neuesten Erkenntnisse und Erfahrungen genützt. Eine besonders große Sorgfalt wird der Wahl der einzelnen Stellen und der Festlegung neuer Nivellementspunkte gewidmet, welche als Zeiger der Bewegungen dienen werden. Diese Punkte bezeichnen wir mit dem Namen Höhenindikationspunkte, damit dadurch möglichst genau ihre Funktion ausgedrückt wird. Wir teilen sie in zwei Kategorien ein a) Höhenindikationsfelder, b) Höhenindikationspunkte.

Unter einem Höhenindikationsfeld verstehen wir eine Gruppe von Nivellementspunkten, welche besonders auf einer Fläche von 1-2 km^2 festgelegt wurden. In einer jeden selbständigen geologischen Formation wurden einige Höhenindikationsfelder gewählt. Ihre Anzahl richtet sich nach dem Umfang der geologischen Einheit, sie darf aber nicht unter 3 sinken. Jedes Indikationsfeld besteht aus 3-6 Punkten in solcher Weise, daß bei gegenseitigem Vergleich der Höhenunterschiede der wiederholten Nivellements der geänderte Punkt eventuell ausgeschieden werden kann. In zwei benachbarten geologischen Einheiten werden die Höhenindikationsfelder in der Weise gewählt, daß die sie verbindenden Nivellementslinien möglichst kurz sind und die Grenzen der geologischen Einheiten senkrecht durchschneiden. Für die Höhenindikationsfelder werden vor allem die bisherigen

Grundnivellementspunkte benutzt, welche durch weitere Punkte ergänzt wurden.

Die Höhenindikationspunkte sind selbständige Nivellementspunkte, gesichert durch höchstens zwei Nachbarpunkte. Diese werden auf solchen Stellen gewählt, wo man die Bewegung benachbarter geologischer Schollen kleineren Umfanges erwarten kann, wobei auf jeder ihrer Grenzseite 2-3 Punkte gewählt werden. Auf dieselbe Art werden auch tektonische Linien gewählt. Weiterhin werden sie auf solchen Stellen gewählt, wo 3 oder mehrere Linien zusammentreffen, als in Knotenpunkten, und endlich auf allen Linien in der Entfernung von ungefähr je 6 km.

Zur Festlegung der Indikationspunkte werden vor allem solche Felsen benutzt, deren Gestein widerstandsfähig gegen Witterungseinflüsse ist. In Gebieten mit einem festen Untergrund, welchen bis 2 m unter die Oberfläche reicht, benutzt man die unterirdische Festlegung. Es ist ein Prisma aus widerstandsfähigem Material, für gewöhnlich ein Granitprisma von 25 x 25 x 110 cm, welches auf eine Betonplatte am Grund der Grube in einer Tiefe von 2-2,5 m aufgestellt wird.

Wenn der feste Untergrund tiefer als 2 m liegt, dann benutzt man eine Tieffestlegung, welche aus einem Stahlrohr von 10 cm Durchmesser besteht und bis auf den feste Untergrund reicht. In der Oberschicht wird dasselbe von einer Einfassung von 20 cm Durchmesser geschützt. Der Oberteil ist von einem Gußeisendeckel verdeckt. Eine Drehbohrmaschine mit Motorantrieb bohrt die nötige Bohrung in den Boden.

Die Vermessung wird mit den vollkommensten Nivellierinstrumenten der Gegenwart und mit Invarlatten vorgenommen. Die Länge des Lattenmeters wird im Laboratorium mittels eines Komparators festgestellt und dann weiter direkt im Feld mittels Vergleichsinvarmetern. Auf einigen längeren Linien wurden Versuche mit der Methode "unterbrochener Nivellements" unternommen. Bei Normalumständen wählt man eine Zielweite von 35 m und ist bedacht, daß die Ziellinie in ihrem Gesamtlauf wenigstens 0,8 m oberhalb des Bodens verläuft; nur bei kürzeren Ziellinien wird die Grenze bis auf 0,5 m erniedrigt. Die Höhenunterschiede, welche durch Messungen hin und zurück festgestellt wurden, werden auf jeder Strecke verglichen, und die Abweichung darf nicht die durch

$$1{,}5 \text{ mm} \sqrt{R_{km}}$$

ausgedrückte Grenze überschreiten. Für einen Abschnitt oder eine Linie länger als 50 km gilt

$$3{,}0 \text{ mm} \sqrt{I_{km}}$$

Bei der Ausgleichung wird derart vorgeschritten, daß zuerst das Hauptnetz der Linien, welche geologische Grundeinheiten verbinden, ausgeglichen wird, und diejenigen Linien, welche kleinere geologische Einheiten verbinden, werden selbständig in einem jeden Polygon des Hauptnetzes ausgeglichen.

Zum Schluß möchte ich nur noch erwähnen, daß das Projekt das Ergebnis einer musterhaften Zusammenarbeit der Geodäten, der Geologen und der Geophysiker ist. Die bisherigen Forschungen haben gezeigt, daß ohne eingehendere Kenntnisse und Forschungen in der Geologie und Geophysik die Nivellements ihre Aufgabe nicht erfüllen können. Die neuzeitlichen Nivellements können jedoch wesentlich die geologischen und geophysikalischen Erforschungen kontrollieren und genau festsetzen. Nur auf Grund eines Vergleichs der Ergebnisse aller Erforschungen können verläßliche Schlußfolgerungen zum Vorteil der Wissenschaft und Technik erzielt werden.

Über die Möglichkeit, in Österreich rezente Erdkrustenbewegungen aus dem Präzisionsnivellement abzuleiten

von

M. Toperczer (Wien)

Der größte Teil Österreichs gehört seinem geologischen Aufbau nach dem Bereich der alpinen Faltung an. Nördlich der Donau hat es Anteil an der geologisch viel älteren böhmischen Masse, im Osten berühren es die Ausläufer der ungarischen Tiefebene. Die zwar nicht sehr rege, doch keineswegs zu übersehende Bebentätigkeit weist darauf hin, daß der Schollenaufbau der Kruste im alpinen Bereich noch keineswegs zur Ruhe gekommen ist. Beben außerhalb des alpinen Bereichs sind äußerst selten. Am deutlichsten äußert sich dieser Gegensatz im Bereich des Wiener Beckens, wo die Donau ein habituelles Stoßgebiet scharf gegen Norden abgrenzt.

Die erhöhte seismische Aktivität des alpinen Bereiches läßt dort auch eine erhöhte Beweglichkeit der Kruste erwarten und damit auch das Auftreten langsam verlaufender Krustenbewegungen, die zur Herstellung und Aufrechterhaltung der isostatischen Kompensation erforderlich sind. Denn das "isostatische Gleichgewicht" ist im Gegensatz zum hydrostatischen Gleichgewicht kein statischer Zustand, sondern ein dynamisches Gleichgewicht großer innerer Spannungsfelder des Krustengefüges.

Die vertikale Beweglichkeit einzelner Krustenteile gegeneinander kann durch Nivellementsmessungen festgestellt und gemessen werden. In aktiveren Gebieten der Erdkruste genügen bei der heutigen Meßgenauigkeit relativ kurze Zeiträume, um Vertikalbewegungen zu erkennen und auch größenmäßig zu bestimmen.

Das erste, aus den Zeiten der Monarchie stammende Nivellement hatte seinen Bezugspunkt im Pegel des Maritimen Observatoriums Triest; aus technischen Gründen folgte es damals, in der Zeit vor dem ersten Weltkrieg, vorwiegend den Eisenbahnlinien. Auf dem Gebiete der heutigen Republik befand sich, eben wegen seiner tektonischen Bewegtheit kein einziger Hauptpunkt des damaligen Nivellements, dessen mittlere Fehler im Gebirgsland, also im Bereich größerer Höhenunterschiede bei 2 – 3 mm/km lag.

Nach dem Zerfall der Monarchie mußte zunächst das gesamte staatliche Vermessungswesen neu organisiert werden. Nivellementsmessungen wurden erst spät und nur in bescheidenem Umfang ausgeführt. Das am Ende der ersten Republik vorliegende Höhennetz war inhomogen und nicht widerspruchsfrei.

Nach Beendigung des zweiten Weltkrieges konnte das Bundesamt für Eich- und Vermessungswesen die Arbeiten am Höhennetz rasch wieder aufnehmen und nach modernen Gesichtspunkten fortführen. Aus verschiedenen Gründen erschien es am zweckmäßigsten, das geometrische Nivellement ganz neu aufzubauen. Eine große Zahl früherer Höhenfixpunkte war im Laufe der Zeit verloren gegangen, in technischer Beziehung war aber nunmehr die Straße weit wichtiger geworden als das schon bestehende Eisenbahnnetz, dessen Erweiterung nicht beabsichtigt war. Zur hypothesenfreien Reduktion der Nivellementsmessungen wurden laufend an fast allen Fixpunkten desselben auch Schweremessungen ausgeführt. Im Bereich des neuen Höhennetzes ist daher die Verteilung der Schwere bekannt und damit die Isostasie überprüfbar.

Der hier kurz geschilderte Entwicklungsgang der Höhenmessung in Österreich ergibt, daß nur in sehr wenigen Bereichen Streckenzüge früherer Nivellements sich mit den heutigen decken und damit zur Ableitung vertikaler Kurstenbewegungen verwendet werden können. Auch ist die Zahl der unberührt erhaltenen Höhenmarken aus früherer Zeit nur verhältnismäßig gering und dadurch werden die Vergleichsmöglichkeiten und damit die Möglichkeiten zur Bestimmung vertikaler Bewegungen weiter eingeschränkt.

Sichere Vergleichsmöglichkeiten finden sich noch am ehesten im Bereiche großer technischer Kunstbauten der Eisenbahnen, vor allem in den Tunnels. Als Beispiel sei hier die relative Krustenbewegung im Bereich des den Zentralalpenkamm durchquerenden Tauerntunnels angegeben. In ihm wurde ein Präzisionsnivellement im Jahre 1909 etwa gleichzeitig mit den Schweremessungen von Sterneck ausgeführt. Im Jahre 1949, gleich zu Beginn der Arbeiten am neuen Präzisionsnivellement wurden die Messungen durch den Tauerntunnel wiederholt, da er in diesem Teile der Zentralalpen die einzige Verbindung zwischen Nord- und Südseite darstellt, die für das geometrische Nivellement in Betracht kommt. Fest - punkte des alten Nivellements sind als gewachsene Steinpfeiler in Seitenkammern des Tunnels bis heute unverändert erhalten geblieben.

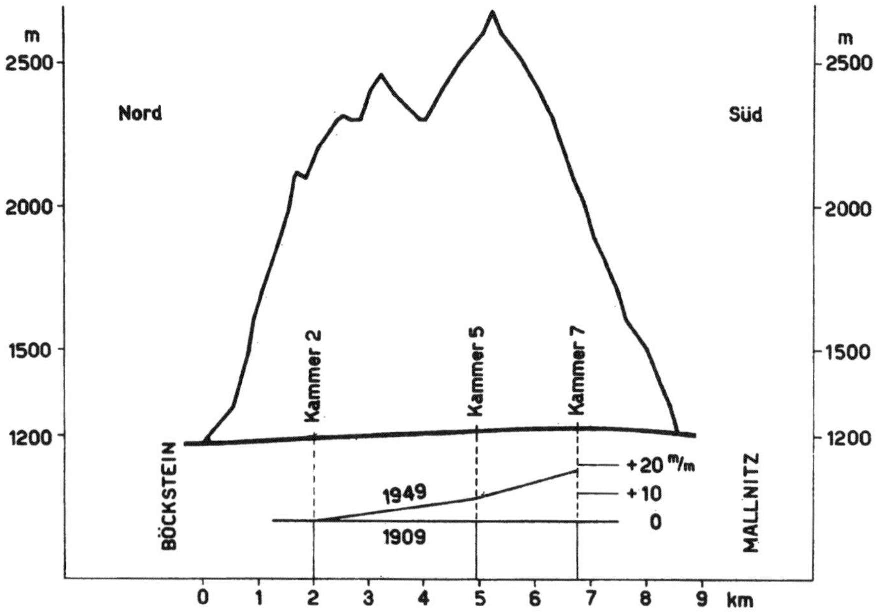

Abb. 1

Relative Vertikalbewegungen im Bereich des Tauerntunnels (Zentralalpen) 1909 bis 1949

In der Abb. 1 sind die relativen Höhenänderungen des Nivellements aus 1949 gegen die für 1909 erhaltenen Werte dargestellt. Dabei ist zunächst willkürlich der nördliche Ausgangspunkt bei Böckstein als unverändert angenommen worden. Die Darstellung entspricht also einer Hebung des Südteils gegen das Nordende mit einer Maximalamplitude von 19 mm, was einer Hebungsgeschwindigkeit von etwa 0.5 mm/Jahr auf der Südseite entspricht.

Die Abb. 2 zeigt den Verlauf der Schwerewerte innerhalb des gemessenen Profils, sowie der Bouguer-Schwerestörung einschließlich der topographischen Reduktion.

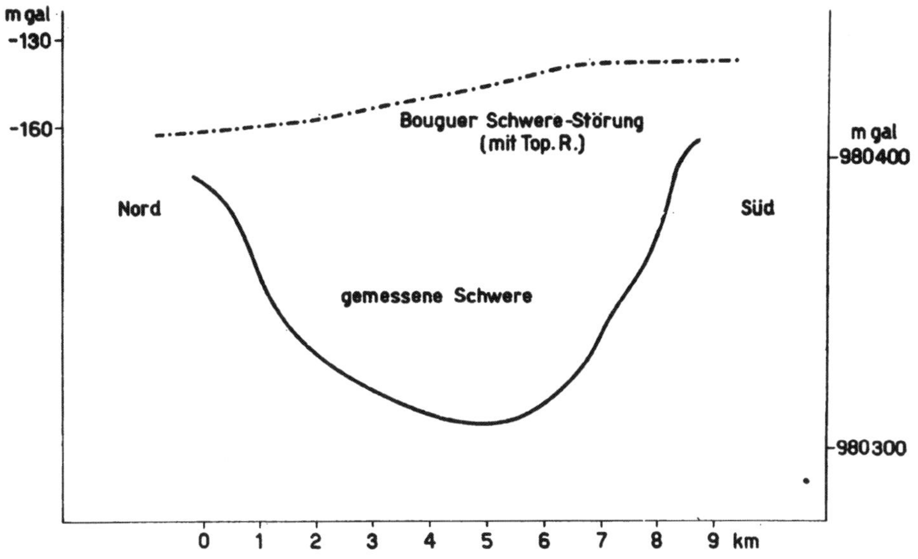

Abb. 2

Verlauf der Schwerkraft, gemessen an den Profilpunkten im Tauerntunnel, und der Bouguer-Schwerestörung

Eine Hebung des Südteils anzunehmen, wird dadurch noch gestützt, daß Vergleiche mit der alten Nivellementslinie, die bis ins Salzachtal am Nordfuß der Tauernkette sich erstrecken, eher für eine Hebung des südlichen Teiles als für eine Senkung des Nordfusses sprechen. Damit gibt das hier gebrachte Beispiel übrigens auch einen Einblick in die Größenordnung der Vertikalbewegungen in diesem Teil der Zentralalpen.

Es sei noch erwähnt, daß sich in der nächsten Umgebung des Tauerntunnels, nämlich in einem Stollen des Radhausberges bei Böckstein, die Gezeitenstation des Bundesamtes befindet, die auch Aufschluß über die Beanspruchung des Gebirges durch die Gezeiten geben kann.

Da nunmehr ein großer Teil des neuen Nivellementsnetzes fertiggestellt ist, werden die geplanten Wiederholungsmessungen auf verschiedenen Streckenzügen bereits in nächster Zukunft weitere Aufschlüsse über die rezenten Krustenbewegungen auf österreichischem Gebiet liefern.

D i s k u s s i o n

K. A r n o l d, (Potsdam):

Wenn Sie nach Süden hin ein Ansteigen des Gebirges festgestellt haben, so möchte ich fragen, ob dies mit dem Streben nach isostatischem Ausgleich in der Po-Ebene zusammenhängt.

M. T o p e r c z e r (Wien):

Die Bewegung des Gebirges ist sehr wahrscheinlich ein isostatischer Prozeß; ein schlüssiger Beweis kann erst in Verbindung mit den italienischen Messungen gegeben werden.

J. A. M e s c h e r i k o v (Moskau):

In connection with the project of compiling a summarising map of recent movements of the European continents I would like to know if there is a possibility of compiling a summarising map of recent crustal movement for the territory of Austria.

M. T o p e r c z e r (Wien):

Es gibt wohl eine seismotektonische Karte von Österreich, eine Bewegungskarte wollen wir aber erst fertigstellen, bis mehr zuverlässige Ergebnisse aus den Nivellementsmessungen vorliegen.

A. Z a t o p e k (Prag):

Ich möchte gern an Herrn Kollegen Toperczer 3 Fragen stellen:

1. Aus dem N-S-Profil der Höhenänderungen geht hervor, daß die Alpen sich im Süden schneller heben als im Norden. Wurden in diesem Zusammenhang spezielle Messungen am Übergang zur Böhmischen Masse angestellt?

M. T o p e r c z e r (Wien):

Es wurden zusätzlich zu den vorhandenen Nivellementsstrecken im Bereich der Böhmischen Masse noch Basisstrecken eingemessen, nach denen die relativen Bewegungen ermittelt werden.

A. Z a t o p e k (Prag):

2. Wurden irgendwelche Differenzen in der Hebungsgeschwindigkeit der Westalpen gegenüber den Ostalpen festgestellt?

M. T o p e r c z e r (Wien):

Auch hier können erst spätere Wiederholungsmessungen Aufschluß geben.

A. Z a t o p e k (Prag):

3. Beabsichtigen Sie in Österreich Messungen von Vertikalverschiebungen im Wiener Becken durchzuführen?

M. T o p e r c z e r (Wien):

In allen Bebengebieten werden ebenso wie in den ruhigen Bereichen eigene Teststrekken, bzw. Schleifen ausgewählt werden.

Entwurf eines Netzes für wiederholte Präzisionsnivellements für Untersuchungen vertikaler Erdkrustenbewegungen im polnischen Raum

von

J. N i e w i a r o w s k i (Warschau)

S u m m a r y

The line Kołobrzeg-Przemyśl divides the Polish territory into two large tectonic units. The crystalline Russian plateau constitutes the northern part of this area. The structure of those parts of this area situated in the south-west and in the central part of Poland results from the foldings originating from the Palaeozoic Age. The structures we meet with in the south of this area show characteristic formations dating from the epoch of Alpine foldings.

The scheme of the canvas of a repeated precision levelling includes the state canvas of the first class as well as five lines of the present precision levelling of the second class. The lines of the projected canvas result in profiles approximately perpendicular to the trend of the geological formations.

In the project the following connection with the canvases of the precision levellings of the neighbouring countries has been planned: 4 lines with the Soviet Union, 7 lines with Czechoslovakia, and 4 lines with the German Democratic Republic.

The description of a stabilization of the canvas is given and suggestions are made concerning the method and the accuracy of repeated surveys.

Z u s a m m e n f a s s u n g

Die Linie Kołobrzeg-Przemyśl trennt den polnischen Raum in zwei große tektonische Struktureinheiten. Der Nordteil dieses Raumes wird von der kristallinen Russischen Platte gebildet. Die Struktur der im Südwesten und in der Mittelpartie von Polen gelegenen Teile dieses Raumes ist das Ergebnis von Faltenbildungen, die aus dem Paläozoikum stammen. Die im Süden dieses Gebietes gelegenen Strukturen werden von Formationen eingenommen, die in der Epoche alpiner Faltenbildungen entstanden.

Im Entwurf des Netzes eines wiederholten Präzisionsnivellements sind das Staatsnetz I. Klasse sowie 5 Linien der gegenwärtigen Präzisionsnivellements II. Klasse inbegriffen. Die Linien des projektierten Netzes ergeben annähernd senkrecht zur Richtung der geologischen Strukturen verlaufende Profile.

Im Projekt ist folgende Verbindung mit den Netzen der Präzisionsnivellements der Nachbarländer vorgesehen: Mit der UdSSR - 4 Linien, mit der CSSR - 7 Linien und mit der DDR - 4 Linien.

Angegeben sind die Beschreibung einer Stabilisation des Netzes sowie Anregungen bezüglich der Methode und der Genauigkeit von wiederholten Netzvermessungen.

1. Einleitung. Geologische Charakteristik des polnischen Raums

Die vertikalen Bewegungen der Erdkruste in der Gegenwart sind hinsichtlich ihres Entstehens sowie ihres Verlaufs eine Folgeerscheinung von ehemals in der Erdkruste wirkenden Kräften. Ein Beispiel hierfür sind die in der vorletzten geologischen Epoche stattgefundenen Erscheinungen, als in der Eiszeit des Pleistozäns (Diluvium) riesige Massen von Festlandeis Teile von Kontinenten (fennoskandischer Schild, der Nordteil von Nordamerika) in die Unterlage des Sima eindrückten. Während des Alluviums hebt sich nach Schmelzen des Festlandeises in diesen Räumen die Erdkruste und strebt nach ihrer früheren Lage zurück. Es sind dies isostatische Bewegungen, die auf das Erreichen eines isostatistischen Gleichgewichts in der Erdkruste gerichtet sind. Diese Bewegungen treten in Skandinavien und im Nordteil des nordamerikanischen Festlands auch noch gegenwärtig auf.

Hinsichtlich ihrer Größe können die in vergangenen geologischen Perioden stattgefundenen tektonischen Prozesse sowie die durch dieselben bewirkten Bewegungen der Erdkruste nicht auf gegenwärtige Zeitabschnitte extrapoliert werden. Auf keinen Fall kann dies in bezug auf die Zeitabschnitte geschehen, während derer mittels geodätischer Verfahren durch wiederholte Präzisionsnivellements Untersuchungen gegenwärtiger Vertikalbewegungen der Erdkruste durchgeführt werden können, da diese Zeitabschnitte im Vergleich zum Zeitraum des Wirkens tektonischer Kräfte verschwindend kurz sind.

Um im polnischen Raum das Projekt eines Netzes wiederholter Präzisionsnivellements für die Untersuchungen gegenwärtiger rezenter Bewegungen der Erdkruste auf entsprechende Weise bearbeiten zu können, muß folgende Vorbedingung erfüllt werden. Man muß - wenigstens in allgemeinen Umrissen - die Tektonik des polnischen Raumes kennenlernen, d.h. also die Anordnung und die Geschichte der Felsformationen dieses Raumes.

Polen wird durch eine von Nordwest nach Südost verlaufende Linie (Kołobrzeg-Toruń-Przemyśl), die den Rand des osteuropäischen Plattengebietes bildet, in zwei große tektonisch-strukturelle Einheiten unterteilt. Der nordöstlich von dieser Linie gelegene Teil wird von der russischen Platte gebildet. Diese ist eine noch aus der Periode des Präkambriums stammende - aus kristallinen Felsen bestehende - zweischichtig gebaute Formation, die mit Restablagerungen jüngerer Epochen bedeckt ist. Im Nordosten Polens befindet sich die kristalline Unterlage dieser Platte in einer Tiefe von einigen hundert Metern und fällt stufenweise zum Rand des Plattengebietes hin ab, wobei sie bis auf etwa 10 km tief absinkt. Im Raum des westlichen Pomorze, der den Südabhang des skandinavischen (baltischen) Schildes bildet, befindet sich diese Unterlage in rund 2 km Tiefe (Gegend von Leba).

Komplizierter ist die Struktur der im Südwesten der Linie Kołobrzeg-Przemyśl gelegenen Gebiete. Sie ist das Ergebnis von Faltenbildungen, die im Paläozoikum stattgefunden haben, und das Entstehen der Sudety (Sudeten), der Góry Swietokrzyskie (Heiliges Kreuz-Gebirge im Bezirk Kielce) sowie der Faltengebiete Górny Slask (Oberschlesien) und des Südteils von Dolny Slask (Niederschlesien) bewirkten. Die Strukturen des Paläozoikums verschwinden nach Norden hin unter einer dicken Deckschicht von aus der mezozoischen Ära und jüngeren Epochen stammenden Ablagerungen.

In der Kreidezeit entstanden infolge des durch die Faltenbildung der Alpen verursachten und durch die große Fennosarmatische Platte (russische Platte) aufgehaltenen Druckes Kompressionen und Auffaltungen von Sedimenten in der sog. dänisch-polnischen Geosynklinale, wodurch die Entwicklung der Antiklinale Pomorze-Kujawy hervorgerufen wurde. Diese Antiklinale zieht sich von Nordwesten nach Südosten auf einer Breite von 30-80 km, bei leichtem Abfall an den Flügeln, durch den gesamten polnischen Raum (Kołobrzeg-Sanok).

Der Südteil Polens wird von zur Zeit alpiner Faltenbildungen entstandenen Strukturen wie der Tatra, den Pieniny und Karpaty gebildet. Es ist dies ein Teil der in der sich über Südeuropa und Südasien erstreckenden großen Geosynklinale entstandenen, mächtigen alpinen Gebirgsbildung. Innerhalb des polnischen Raumes können drei Zonen alpiner Gebirgsbildung ausgesondert werden. Es sind dies im Süden - die äußeren Karpaten (Tatry und Pieniny) sowie die äußeren flyschartigen Karpaten und im Norden - die Einsenkung der Vorkarpaten.

Auch das auf so vereinfachte Weise dargestellte Bild des geologischen Baus von Polen gestattet es, bereits allgemeine Richtlinien bestimmter Art für die Bearbeitung eines Projektes wiederholter Präzisionsnivellements aufzustellen. Es sollen demnach die Nivellementslinien möglichst senkrecht zur Richtung geologischer Strukturen verlaufen. Aus dem geologischen Bau Polens ergibt sich, daß ein derartiger grundsätzlicher Verlauf der Nivellementslinien für den nordöstlichen Raum (russische Platte), für die mittelpolnische Antiklinale und für die paläozoischen Faltengebiete (westliche und südwestliche Gebiete sowie ein Teil Mittelpolens) sich in Richtung von Nordosten nach Südwesten bewegt und im Süden Polens sich der Meridianrichtung nähert.

In einer weiteren, schon geodätischen Richtlinie ist die maximale Ausnutzung bereits bestehender und gemessener Linien des Präzisionsnivellements zu postulieren. Es soll dies so erfolgen, daß die erste zukünftige Nivellementsmessung nicht nur als Unterlage für zukünftige Untersuchungen diene, sondern bereits die Größenbestimmung von Vertikalbewegungen der Erdkruste durch einen Vergleich mit den Ergebnissen früherer, längs derselben Linien durchgeführter Nivellements zuließe.

2. Projekt eines Netzes wiederholter Präzisionsnivellements

Bei Berücksichtigung des Zieles und der Aufgaben für ein zu Untersuchungen rezenter Vertikalbewegungen der Erdkruste im polnischen Raum projektiertes Netz von Präzisionsnivellements sowie unter Innehaltung der angeführten Richtlinien, wurden die nachfolgenden, ein selbständiges Netz bildenden Linien des Präzisionsnivellements ausgewählt.

1. Alle in den Jahren 1952-1955 vermessenen, 8 geschlossene Polygone bildenden Linien des staatlichen Präzisionsnivellements I. Klasse,

2. die Linie des jetztigen staatlichen Präzisionsnivellements II. Klasse auf der Strecke Wieliczka-Przeworsk,

3. die beiden Linien Warszawa-Bialystock und Warszawa-Lublin, die längs der Strekken des polnischen Präzisionsnivellements der Jahre 1926-37 sowie längs der Strecken des derzeitigen Präzisionsnivellements II. Klasse verlaufen.

Das Projekt des Netzes wiederholter Präzisionsnivellements für Untersuchungen von Vertikalbewegungen der Erdkruste ist auf Abb. 1 dargestellt.

Die Durchführung erneuter Nivellements dieser Linien ermöglicht es, ein Bild von der Größe derzeitiger Vertikalbewegungen der Erdkruste längs der durch die wichtigsten und aktivsten tektonischen Strukturen Polens durchgehenden Querschnitte zu erhalten. Die für die erneute Führung des Präzisionsnivellements ausgewählten Linien verlaufen längs der Südküste des Baltischen Meers. Sie geben vier Querschnitte, und zwar durch die Antiklinale Pomorze-Kujawy und deren Verlängerung nach Südosten, im Süden gehen sie durch das Vorgebirge der Sudeten, durch die Flyschkarpaten und die Einsenkung der Vorkarpaten, gleichfalls ergeben sie einschließlich der für das Präzisionsnivellement ausgewählten Linien, die das polnische Nivellementsnetz mit dem Netz des Präzisionsnivellements der Tschechoslowakischen Sozialistischen Republik verbinden - Querschnitte durch Karpaten und Sudeten. Von den für die vorgesehene Wiederholung des Präzisionsnivellements ausgewählten Linien werden drei hauptsächliche tektonisch-strukturelle Einheiten Polens durchschnitten. Es sind dies die russische Platte im Nordosten, das Gebiet paläozoischer Faltenbildungen im Westen, im Südwesten und teilweise in Zentralpolen sowie das Gebiet alpiner Faltenbildungen im Süden.

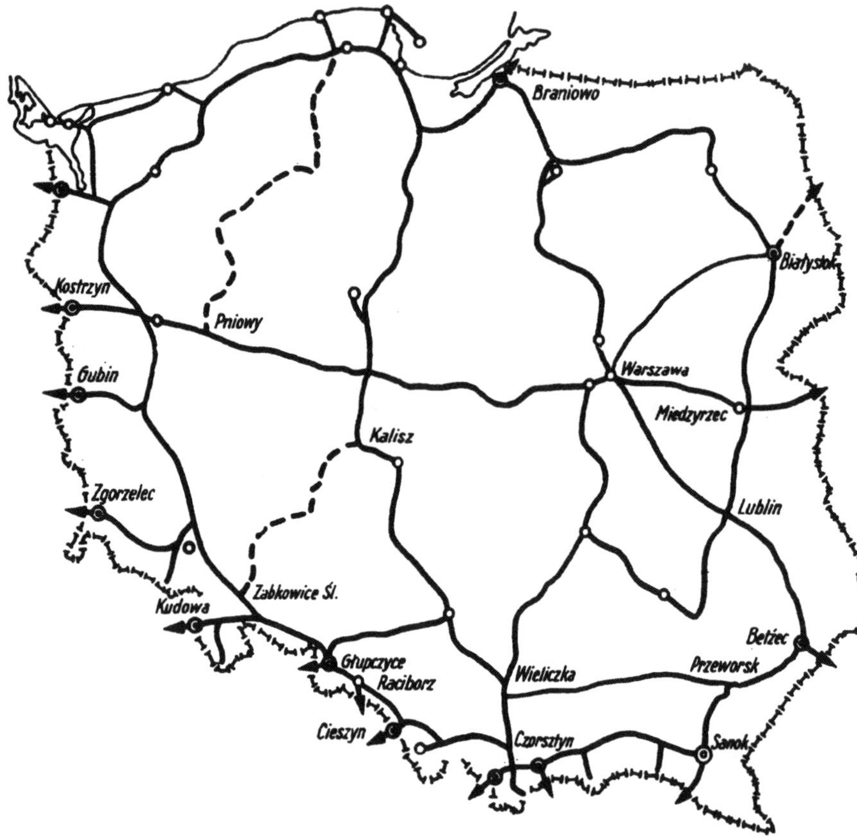

Abb. 1

Projekt des Netzes wiederholter Präzisionsnivellements für Untersuchungen von Vertikalbewegungen der Erdkruste in Polen

Im Projekt ist eine Veränderung der Struktur des staatlichen Nivellementsnetzes vorgesehen. Es sollen nämlich zwei große westliche Polygone des derzeitigen Netzes I. Klasse mit Linien des Präzisionsnivellements I. Klasse durchschnitten werden. Es wird weiterhin vorgeschlagen, das nordwestliche Polygon mittels der Nivellementslinie Lebork-Chojnice-Pniewy zu durchschneiden.

Das derzeitige südwestliche Polygon des Präzisionsnivellements I. Klasse soll - vorschlagsgemäß - mittels der Nivellementslinie Kalisz-Wrocław-Zabkowice Ślaskie durchschnitten werden.

Im südöstlichen Randgebiet des polnischen Raums ist kein säkularer Fundamentalpunkt vorhanden. Im Projekt des Präzisionsnivellements zur Untersuchung vertikaler Bewegungen ist die Anlage eines Fundamentalpunktes in der Gegend von Sanok an der Linie des Präzisionsnivellements I. Klasse im Raum der (Flysch) Karpaten vorgesehen.

Zielsetzung für die Wiederholung des Präzisionsnivellements zur Untersuchung rezenter Vertikalbewegungen der Erdkruste ist die Bestimmung der Geschwindigkeit dieser Bewegungen in mm/Jahr. Von vorrangiger Bedeutung ist in diesem Zusammenhang die Bestimmung des Bezugsniveaus der festzustellenden Bewegungen. Wenn man als Bezugsniveau die Höhenlage eines bestimmten Festpunktes im untersuchten Raum, z.B. der Marke in Toruń [2], annimmt, erhält man - bei der à priori zu treffenden Voraussetzung, daß dieser Bezugspunkt unveränderlich sei - die relativen Geschwindigkeiten von Vertikalbewegungen.

Um die absoluten Geschwindigkeiten von Vertikalbewegungen zu erhalten, ist als Bezugsniveaus das innerhalb langjähriger Beobachtung in ozeanographischen- und Pegelstationen erarbeitete mittlere Meeresniveau anzusetzen. In diesem Falle ist das auf einer gegebenen ozeanographischen Station ermittelte beziehungsweise das auf mehreren Stationen festgestellte mittlere Meeresniveau als unveränderlich anzusehen (wir nehmen dann die Geschwindigkeit seiner vertikalen Bewegung V = 0 mm/Jahr an - bei evtl. Berücksichtigung von sich aus eustatischen Veränderungen des Meeresniveaus ergebenden Korrekturen).

Alle ozeanographischen und Pegelstationen an der polnischen Ostseeküste sind in das derzeitige Netz des staatlichen Präzisionsnivellements eingeschaltet.

Die Kontrollmessungen von Höhenfestpunkten an Meeres- und Wasserstandspegeln auf den Verbindungslinien des Präzisionsnivellements I. Klasse sind in Zeitabständen von 5 Jahren zu wiederholen.

Die Kontrolle der Beständigkeit und Unveränderlichkeit hauptsächlicher Höhenfestpunkte an ozeanographischen und Pegelstationen sowie die Verknüpfung derselben mit den Nullpunkten von Pegellatten soll mindestens einmal im Jahre durchgeführt werden.

Durch die während vieler Jahre an ozeanographischen und Wasserpegelstationen angestellten Beobachtungen wird die Bestimmung der Schnelligkeit der Vertikalbewegungen von Meeresküsten ermöglicht. Darüber hinaus kann aber auf diese Weise auch der Verlauf (das Profil) des mittleren Meeresniveaus im Vergleich zu einer als Bezugsoberfläche angenommenen Niveaufläche festgestellt werden. Das Baltische Meer ist kein offenes Meer und die Lage seines mittleren Niveaus wird stark von anemobarischen Faktoren (Wind, atmosphärischer Druck) sowie durch Versalzung beeinflußt. Wie ozeanographische Untersuchungen beweisen, hebt sich das mittlere Niveau des Baltischen Meeres über die angenommene Niveaufläche mehr oder weniger regelmäßig von Westen nach Osten. Im Gesamtraum der polnischen Meeresküste hat diese Erhöhung des Meeresniveaus die Größenordnung 3 cm.

Sehr wichtige Fragen, die in einem Projekt für die Wiederholung von Nivellements zur Untersuchung gegenwärtiger Vertikalbewegungen der Erdkruste nicht außer Acht gelassen werden können, sind: innerhalb welcher Zeitabschnitte eine Wiederholung der Nivellements durchzuführen ist, und nach welcher Methode die Ergebnisse der Nivellements auszuarbeiten sind.

Die genannten Probleme sind in der Mitteilung von T. Wyrzykowski behandelt. Falls man für den polnischen Raum eine Bestimmungsgenauigkeit der Schnelligkeit von Vertikalbewegungen mit einer Größenordnung von ± 1 mm/Jahr erhalten will, dann soll der Zeitraum innerhalb wiederholter Präzisionsnivellements 15-20 Jahre betragen.

Die Durchführung des vorliegenden Projektes wird eine Bestimmung der Schnelligkeit gegenwärtiger Vertikalbewegungen längs der wichtigsten, in geologischer und tektonischer Hinsicht aktiven Profile im Raum von Polen ermöglichen.

Untersuchungen in zwischen den Hauptprofilen gelegenen Gebieten können in Anlehnung an Linien wiederholten Nivellements, bzw. in Anlehnung an Bezugspunkte zur Durchführung gelangen, die im letzteren Fall säkulare Fundamentalpunkte sein können.

3. Vorschläge für den Anschluß eines wiederholten Präzisionsnivellementsnetzes im polnischen Raum an die Netze der Nachbarländer

Die rezenten Vertikalbewegungen der Erdkruste, deren Entstehen einleitend in der angeführten geologischen Charakteristik des polnischen Raumes kurz behandelt worden war, treten auf allen Festlandsgebilden der Erdkruste in Erscheinung. Je größere Festlandsgebiete also durch verschiedene Verfahren und demnach auch durch geodätische Untersuchungen erfaßt werden können, umso vollständiger und geologisch verständlicher wird sich das so erhaltene Bild dieser Bewegungen gestalten.

Die Durchführung von gleichzeitigen Untersuchungen großer Festlandgebiete mittels des Verfahrens eines wiederholten Präzisionsnivellements im Raum mehrerer Länder ist möglich. Es kann dies durch Verbindung untereinander mittels Verbindungslinien der zur Untersuchung von Vertikalbewegungen bestimmten Nivellementsnetze der einzelnen Länder erfolgen. Diese Linien müssen dann derartig verlaufen, daß man bei gemeinsamer Behandlung von Netzen der einzelnen Länder ein großes Netz wiederholter Präzisionsnivellements möglichst einheitlicher Struktur erhalten könnte.

In Voraussicht derartiger Möglichkeiten wurden Vorschläge für Anknüpfungen erarbeitet, die u.E. geeignet wären, die Durchführung der Verbindung des polnischen Netzes wiederholter Präzisionsnivellements zu Untersuchungen von Vertikalbewegungen mit demselben Zweck dienenden Präzisionsnivellementsnetzen der Nachbarländer zu ermöglichen.

Für die Anknüpfung an das Präzisionsnivellementsnetz der Deutschen Demokratischen Republik kämen 4 von den Punkten Dołuje, Kostrzyn, Gubin und Zgorzelec ausgehende Linien in Frage.

Für die Verbindung mit dem Präzisionsnivellementsnetz der Tschechoslowakischen Sozialistischen Republik wären die von den Punkten Kłodzko, Racibórz, Cieszyn, Koniaków, Nowy Targ, Czorsztyn und Sanok ausgehenden Linien zu berücksichtigen.

Diese südlich von unseren Grenzen durch Präzisionsnivellementslinien verlängerten Linien würden Querschnitte durch die Struktur der Sudeten sowie durch die Depressionsbereiche von Śląsk, Tatry und Karpaty ergeben.

Zwecks Anknüpfung an das Netz der Sowjetunion sind als durch Netzstruktur und geologische Rücksichten bedingte, unentbehrliche Verbindungslinien 4 von den Punkten Braniewo, Sokółka, Miedzyrzec und Bełzec ausgehende Linien vorzuschlagen. Alle die im Gebiet benachbarter Länder liegenden, zur Untersuchung gegenwärtiger Vertikalbewegungen dienenden Netze und Verbindungs-Linien sollen Präzisionsnivellementslinien I. Klasse sein und sollen - nach Verbindung mit den Präzisionsnivellementslinien benachbarter Länder - ein geschlossenes Polygon bilden.

4. Netzstabilisierung

Das für Untersuchungen rezenter Vertikalbewegungen der Erdkruste in Polen als Netz wiederholten Präzisionsnivellements projektierte Nivellementsnetz enthält 16 bestehende säkulare Fundamentalpunkte sowie einen neuprojektierten Punkt dieser Art. Diese Punkte sind so verteilt, daß sie sich in allen tektonischen Einheiten Polens von grundsätzlicher Bedeutung befinden. Dem projektierten Netz des wiederholten Präzisionsnivellements stehen rund 700 unterirdische Höhenfestpunkte zur Verfügung; es sind dies Höhenfestpunkte der Typen II, III und IV [10], die in einer Tiefe von rund 1,7 m (Typ II in rund 2,5 m Tiefe) unter der Erdoberfläche, d.h. also unterhalb der Frostgrenze, angebracht sind.

Das eigentliche Höhenzeichen (Höhenfestpunkt) befindet sich nicht weniger als o,5 m unter der Erdoberfläche. Die Stabilisierung dieser Zeichen ist aus bewährtem Beton ausgeführt. Außer diesen Höhenzeichen verfügt das Netz über an bedeutenden Bauten befestigte Mauerbolzen sowie über Erdfestpunkte, so daß die durchschnittliche Dichte von Höhenfestzeichen im projektierten Netz - 1 Höhenfestpunkt auf weniger als 2 km Nivellementslinie - beträgt. Es wird beabsichtigt, vor der wiederholten Vermessung des gesamten für Untersuchungen derzeitiger Vertikalbewegungen benötigten Nivellementsnetzes eine Verdichtung der Stabilisierung mittels Höhenfestpunkten von Typ IV durchzuführen.

5. Methode und Meßgenauigkeit für ein Netz wiederholter Präzisionsnivellements

Die erneute Vermessung des für Untersuchungen rezenter Vertikalbewegungen der Erdkruste projektierten Netzes kann man für den Zeitraum der Jahre 1970–75 vorsehen. Diese Vermessung wäre in möglichst kurzer Zeit durchzuführen. Es ist anzunehmen, daß die Realisierung einer derartigen Aufgabe innerhalb von 3 Jahren in den Grenzen der in Polen gegebenen Möglichkeiten liegt. Falls die Untersuchungen Gebiete einer Reihe von Ländern zu umfassen hätten, dann sollten die Messungen der Vertikalbewegungen wenigstens innerhalb eines Gesamtnetzes gleichzeitig angestellt werden.

Mit Rücksicht auf die im Gebiet des Baus vom Präzisionsnivellementsinstrumenten einzuschlagenden Richtung in bezug auf Konstruktion und Leistungsfähigkeit, sowie im Hinblick auf die mögliche Einführung von automatischen Nivelliergeräten für Zwecke der Präzisionsnivellements im Verlauf eines Zeitraumes von etwa 10 Jahren, ist die beste für die Zukunft einzusetzende Meßmethode im Rahmen des Projektes nicht präzisiert worden. In bezug auf zukünftig wiederholte Präzisionsnivellements scheint die Bedingung gerechtfertigt zu sein, nach der die Größe des Gesamtfehlers auf 1 km doppelten Nivellements $\pm$ 0,5 mm nicht übersohreiten darf.

Zum Schluß wären noch Vorschläge anzuführen, die auf das Problem derjeningen Untersuchungen von Vertikalbewegungen bezugnehmen, welche bis zur Zeit einer erneuten Messung des Untersuchungsnetzes durchzuführen wären. Es handelt sich hierbei um:

1. Ausarbeitung und Festlegung des mittleren Niveaus für das Baltische Meer,
2. Kontrollmessungen aller säkularer Fundamentalpunkte einschließlich der nächstliegenden Höhenfestpunkte vom Typ IV auf Linien I. Klasse,
3. Arbeiten über Verfeinerung der Methoden der Präzisionenivellements mit dem Ziele, den Einfluß systematisch auftretender Fehler zu verringern.

Literatur

[1] Ksiazkiewicz, M.: Geologia Dynamiczna. Wyd. Geol., Warszawa 1959

[2] Niewiarowski, J.; Wyrzykowski, T.: Wyznaczenie współczesnych ruchów pionowych skorupy ziemskiej na obszarze Polski przez porównanie wyników powtarzanych niwelacji precyzyjnych. Prace IGiK., t. VIII, nr 1 (17), Warszawa 1961

[3] Passendorfer, E.: O budowie i ruchach skorupy ziemskiej, Wyd. Geol., Warszawa 1954

[4] Pozaryski, Wł.: Budowa geologiczna Polski. Wyd. Geol., Warszawa 1959

[5] Rudzki, M.: Fizyka Ziemi, Kraków 1909

[6] Witting, R.: Die Lage des Wasserspiegels des Baltischen Meeres. Warszawa 1930

[7] Tymczasowa instrukcja niwelacji precyzyjnej I. i II. klasy, CUGiK, Warszawa 1956

[8] Instrukcja po niwielirowaniju I, II, III i IV kłassow. Gieodiezizdat, Moskwa 1959

[9] Rapport sur les travaux géodésiques executés en Pologne de 1957 à 1960. Nivellement des précisions, presenté à la XII Assemblée de UGGI Helsinki, Warszawa 1960

[10] Sowriemiennyje wiertikalnyje dwizenija ziemnoj kory, Trudy CNIIKAiK, Nr 123, Gieodiezizdat, Moskwa 1958

Diskussion

J. A. Mescherikov (Moskau):

I would like to note the significance of the fact that the Geodetic Service of Poland, during elaboration of the project of the basic geodetic network, considers in detail the data on tectonics, geology of the Quaternary sediments, recent crustal movements. This example shows us that the data on recent crustal movements are very important in the construction of the basic geodetic networks. It seems valuable that the elaboration of the project of the basic levelling network was carried out in close cooperation of geodesist with geologists. This cooperation will ensure the success and provide for a high quality of geodetic works and for important scientific results. The methods of work of the Geodetic Service of Poland in the elaboration of the project of the levelling network should be seriously considered by the Geodetic Services of other countries.

I. N i e w i a r o w s k i (Warschau):

During the discussion Dr. Mescherikov has noted that the geological structure of the territory (tectonics) was considered in the project of the relevelling network for the study of recent vertical crustal movement in Poland. This method is of certain significance and the consideration of the geological structure shall yield accurate results.

Die Bestimmung gegenwärtiger vertikaler Erdkrustenbewegungen im polnischen Raum durch Vergleich von Ergebnissen wiederholter Präzisionsnivellements x)

von

J. Niewiarowski, T. Wyrzykowski (Warschau)

Summary

The velocities of present vertical movements of the earth's crust in Polish territory have been determined in the Institute of Geology and Cartography. Use was made of the following data of levelling:

1.) The present precise levelling net of the first class (~5500 km) dating from 1952-1955 as well as the precise levelling net of the second class (~10 000 km) dating from 1955-1958.

2.) The precise levelling net of the first order surveyed between 1926 and 1937.

3.) Former levellings carried out (after 1870) in the west and north of Poland.

The compensated altitudes of common points were compared, the velocity of motion being calculated by means of the formula $V = \frac{\Delta H}{\Delta T}$.

The relative velocities related to the point Toruń were determined. The mean error has the order of magnitude ± 1 mm/year in the marginal zone of the investigated area.

Because of the comparison methods being used necessarily, the determination of the velocities of motion is only a provisional one, yet it allows to draw the conclusion that the earth's crust is now subjected to minute vertical movements in Polish territory.

Zusammenfassung

Im Institut für Geodäsie und Kartographie wurden die Geschwindigkeiten gegenwärtiger Vertikalbewegungen der Erdkruste im polnischen Raum bestimmt. Benutzt wurden die nachfolgenden Nivellementsangaben:

1.) Das aus den Jahren 1952-1955 stammende, gegenwärtige Präzisionsnivellementsnetz I. Klasse (~5500 km) sowie das aus den Jahren 1955-1958 (~10 000 km) stammende Präzisionsnivellementsnetz II. Klasse.

2.) Das in den Jahren 1926-1937 vermessene Präzisionsnivellementsnetz I. Ordnung.

3.) Frühere (nach 1870) im Westen und Norden Polens durchgeführte Nivellements.

Verglichen wurden die ausgeglichenen Höhen gemeinsamer Punkte, wobei die Bewegungsgeschwindigkeit aus der Formel $V = \frac{\Delta H}{\Delta T}$ bestimmt wurde.

Bestimmt wurden die relativen, auf den Punkt Toruń bezogenen Geschwindigkeiten. Der mittlere Fehler im Randgebiet des untersuchten Raums hat die Größenordnung ± 1 mm/Jahr.

Die Bestimmung der Bewegungsgeschwindigkeiten trägt, wegen der notwendigerweise zur Anwendung gelangenden Vergleichsmethoden, provisorischen Charakter, gestattet es jedoch die Schlußfolgerung zu ziehen, daß im Raum von Polen die Erdkruste gegenwärtig geringen Vertikalbewegungen unterliegt.

x) Texte du rapport présenté à la XII Assemblée de l'U.G.G.I. à Helsinki en 1960, pourvu des quelques suppléments.

1. Introduction

Le développement continuel des sciences ayant pour objet l'étude de la Terre et l'accumulation croissante des documents d'observation ont permis de constater que les mouvements tectoniques actuels de l'écorce terrestre sont de l'ordre de grandeur qui les rend important au point de vue autant scientifique que pratique. Malgré leurs lenteurs, mais grâce à leurs continuité, les mouvements verticaux de l'écorce terrestre constituent un facteur qui peut rendre illusoires les résultats des nivellements de précision et, cas échéant, doit être pris en considération à l'éxécution de certains travaux du génie,(p.ex. à la construction des ports).

Les mouvements verticaux contemporains de l'écorce terrestre sont en élaboration par les différentes branches d'études de la Terre avec des méthods différentes. L'Institut de Géodésie et de Cartographie effectue ses recherches en appliquant la méthode de comparaison des résultats de nivellement de précision actuel avec ceux des nivellements anciens, sur une étendue définie du territoire. L'accord de l'ensemble des études, concernant les mouvements contemporains de l'écorce terrestre, exercées par les différentes branches de la science, appartient à l'Académie Polonaise des Sciences.

2. Matériaux de nivellement utilisés à l'étude des mouvements verticaux

2.1 Réseau de nivellement de précision de 1-re et 2-ième classe

Ce réseau, qui couvre toute étendue de la Pologne, a été mesuré pendant la période de 1952-1958. Les lignes de 1-re classe forment 8 polygones à l'intérieur desquels on a compensé le nivellement de 2-ième classe.

Ce réseau a été mesuré durant un laps de temps très court, comme on le voit d'après les dates d'exécution, ce que constitue son grand avantage. Il est caractérisé par des données suivantes:

1-re classe avant la compensation:
l'erreur moyenne accidentelle par km $|\eta| \leqslant 0{,}50$ mm,
l'erreur moyenne systématique par km $|\sigma| \leqslant 0{,}20$ mm,
l'erreur moyenne par km après la compensation

$$|m_{moyen}| \leqslant 0{,}80 \text{ mm},$$

et par analogie pour la 2-ième classe:

l'erreur moyenne accidentelle par km $|\eta| \leqslant 0{,}75$ mm,
l'erreur moyenne systématique par km $|\sigma| \leqslant 0{,}40$ mm,
l'erreur moyenne par km après la compensation

$$|m_{moyen}| \leqslant 1{,}50 \text{ mm}.$$

Le réseau de nivellement de précision des années 1952-1958 est calculé définitivement dans un système des altitudes normales. Aux différences de hauteur obtenues on a appliqué des corrections dues aux variations des anomalies de la pesanteur mesurées le long de lignes. Le niveau de référence est constitué par le niveau du marégraphe de Kronstadt. Anciennement le réseau de 1-re classe a été compensé dans un système des altitudes orthométriques normales, avec le point de départ à Toruń, Hôtel de Ville, et le niveau de référence "N.N.".

En principe les lignes du nivellement de précision longent les voies de communication principales du pays. Le réseau comprend environ 1500 repères souterrains établis au-dessous du niveau de congélation du sol.

2.2 Réseau de nivellement de précision de 1-re classe de Pologne mesuré dans les années 1926-1937

La longueur totale du réseau comporte environ 10000 km.

Le réseau est formé de 36 polygones fermés. Le niveau de référence est le niveau "Normal Null" de zéro du marégraphe à Amsterdam. Le point de départ est formé par le repère de 1-re classe à l'Hôtel de Ville de Toruń, dont l'altitude a été fixée d'après le catalogue allemand à + 50,518 m.

La précision de ce réseau est caractérisée par une erreur moyenne par km $m_k = \pm\ 0{,}75$ mm (par les mesures d'aller et de retour et par les différences de lignes particulières). L'erreur moyenne par km après la compensation de réseau est $M_k \pm\ 1{,}04$ mm.

Ce réseau, calculé dans le système des altitudes orthométriques, présente un réseau homogène de nivellement de précision qui peut être utilisé pour les comparaisons. La longueur de lignes communes des nivellements de précision des années 1952-1958 et 1926-1937, utilisées à l'étude des mouvements verticaux contemporains, comporte 2200 km. Malheureusement ces lignes ne forment pas des figures fermées.

2.3 Ancien nivellement allemand sur les territoires de l'Ouest et du Nord-Est de la Pologne

Ce sont les mesures de nivellement des années 1870 et postérieures, complétées ensuite par parties. Elles ne forment pas un réseau homogène. Le niveau de référence est le niveau du marégraphe d'Amsterdam. L'établissement de ce réseau, son hétérogénéité et précision sont donnés de façon détaillée par les sources allemandes (Planheft Großdeutsches Reich, Berlin 1944; "Handbuch der Vermessungskunde", Jordan, Eggert, Kneissl, III, Stuttgart 1956). La précision de ce nivellement ne correspond pas aux exigences qu'on présente aux nivellements contemporains. Les lignes de l'ancien nivellement allemand ont été utilisées à l'étude des mouvements verticaux sur une longueur de 1500 km environ.

3. Méthode de la détermination des mouvements verticaux de l'écorce terrestre

3.1 Le niveau de référence

Les mouvements de l'écorce terrestre sur l'étendue de la Pologne ont été déterminés sous forme de grandeurs relatives référées à l'ancien point fondamental du nivellement de Pologne à Toruń (Hôtel de Ville), qui servait du niveau de référence pour le nivellement de 1926-1937 et pour la compensation provisoire du nivellement de 1-re classe d'après la guerre. Il a été impossible de déterminer les valeurs absolues des mouvements de l'écorce terrestre par suite d'un temps trop court d'enregistrement des marégraphes, ou ce qui vient au même, par suite d'une erreur trop grande en détermination du niveau moyen de la mer.

3.2 Le choix de la méthode

Pour déterminer les grandeurs de mouvements verticaux de l'écorce terrestre on s'est servi de la méthode de comparaison des altitudes de points communs des deux réseaux de nivellement compensés et mesurés dans des périodes séparées par un intervalle de temps important. L'admission de cette méthode, approximative en quelque sorte, a été dictée par les considérations suivantes:

a) Les lignes communes des réseaux comparés ne formaient pas des polygones fermés et ne pouvaient donner une comparaison plausible que par application de cette méthode.

b) En comparant les altitudes compensées on tenait compte de leurs valeurs les plus probables, obtenues avec la considération de toutes les mesures surabondantes du réseau.

c) A peu près 60% de l'étendue du pays a été embrassé par la comparaison des deux réseaux de haute précision, dont un avait été mesuré dernièrement dans un temps très court, ce que garantissait que la majeure partie des résultats ne serait pas entachée de grandes erreurs.

3.3 Le mode d'utilisation des réseaux de nivellement

Sur les territoires du centre et de l'Est de la Pologne on a eu à comparer les altitudes des repères de réseau de nivellement de précision de 1-re classe mesuré en 1926-1937 avec celles de nivellement de 1-re et 2-ième classe mesuré en 1952-1958.

Une comparaison directe n'embrassa que les altitudes du réseau de 1926-1937 avec les altitudes du nivellement de 1-re classe d'après la guerre compensé provisoirement, puisque il n'y a que celles qui ont été rapportées au même niveau de référence et calculées dans un même système de hauteurs orthométriques normales.

La comparaison avec les altitudes du réseau de nivellement de 2-ième classe d'après la guerre, calculé dans un système de hauteurs normales à la base du réseau de 1-re classe compensé définitivement et rapporté au zéro du marégraphe de Kronstadt, n'a été effectuée qu'indirectement. Notamment aux altitudes de points du réseau de 2-ième classe on a ajouté les corrections de réduction à l'ancien niveau (de "Normal Null") et au système d'altitudes orthométriques normales. Pour calculer ces corrections on se servait entre autre de la comparaison des résultats de compensation provisoire et définitive du réseau de nivellement de 1-re classe.

Dans l'Ouest et le Nord-Est de la Pologne on a procédé à la comparaison du réseau d'après la guerre de 1-re et 2-ième classe avec les anciens réseaux allemands, calculés dans le soi-disant "vieux système" et rapportés au niveau de "N.N.".

3.4 Le mode de la détermination de la grandeur des mouvements de l'écorce terrestre

Pour déterminer la grandeur de mouvements des points particuliers on se servait de la formule

$V = \frac{\Delta H}{\Delta t}$, où V est la vitesse de mouvement en mm/an,

ΔH - la différence des altitudes du point, obtenue par la comparaison des deux réseaux,

Δt - l'intervalle de temps entre les mesures prises.

Au centre et à l'Est de la Pologne les intervalles de temps Δt ont été calculés pour chaque ligne séparément, étant compris dans les limites de 17 à 31 ans (22 en moyenne). Sur les territoires couverts par l'ancien réseau allemand, vu un mode particulier de mesure et de calcul de ce réseau, on a admis un intervalle de temps commun et approximatif d'une valeur $\Delta t = 70$ ans.

En tenant compte des grandeurs ΔH calculés pour les points particuliers, on définissait les sections de lignes du dernier nivellement ayant les mêmes valeurs moyennes de V aux intervalles de 0,5 mm/an. A la base des sections accusant les mouvements de valeur égale de l'écorce terrestre on a interpolé (en certains cas extrapolé) ces valeurs dans les espaces entre les lignes de nivellement. Ainsi on a obtenu une carte des mouvements de l'écorce terrestre pour presque toute étendue de la Pologne.

4. Résultats et la précision de la détermination des mouvements verticaux de l'écorce terrestre

Les résultats de la détermination montrent que les grandeurs relatives des mouvements de l'écorce terrestre sur le territoire de la Pologne sont peu différenciées et en général d'une petite valeur qui varie de -1 à + 1,5 mm par an. Vu que ces mouvements sont référés au point de Toruń qui n'accentue aucune variation par rapport à la côte Sud de la mer Baltique, que d'autre part cette côte n'accuse que des petits mouvements verticaux, en peut admettre que les mouvements relatifs déterminés diffèrent peu des valeurs absolues.

Malgré qu'on disposait d'un matériel de nivellement assez défectueux (une précision médiocre de l'ancien réseau allemand, un grand pourcentage de lignes de nivellements à un petit intervalle de temps entre les mesures, une quantité insuffisante de lignes de nivellement répété) on a obtenu une image présentant les mouvements de l'écorce terrestre de toute probabilité. Les étendues déterminées des mouvements verticaux de vitesse égale trouvent souvent une confirmation dans la structure tectonique de l'écorce. Cette convergence confirme la justesse de l'image obtenue en ce qui concerne la tendance générale de soulèvement ou d'abaissement des étendues en question.

Les valeurs numériques des mouvements verticaux de l'écorce terrestre sont entachées d'erreur assez considérable, ce qui a été d'ailleurs prévu et confirmé par les discordances aux limites du réseau. Pour les étendues du pays couvertes par les réseau du nivellement des années 1926-1937 et pour un intervalle de temps entre les deux mesures $\Delta t = 22$ ans, on a calculé l'erreur moyenne de détermination, en tenant compte des erreurs moyennes de nivellement et des distances au point de référence.

Distance au point de référence en km	100	200	400
Erreur moyenne de détermination du mouvement de l'écorce terrestre (en mm/an)	$\pm$ 0,5	$\pm$ 0,7	$\pm$ 1,0

Il est beaucoup plus difficile de définir la précision en détermination des mouvements de l'écorce sur les territoires de l'Ouest et du Nord-Est vu les traits caractéristiques de l'ancien réseau de nivellement allemand, mentionnés plus haut. On peut pourtant admettre qu'à cause d'un grand intervalle de temps ($\Delta t = 70$ ans) cette erreur est aussi de l'ordre de $\pm$ 1 mm/an.

En analysant les mouvements sur l'étendue de chacun des deux réseaux différents de nivellement anciens on constate des discordances aux limites qui font supposer l'existence d'une erreur en détermination des altitudes de l'ancien réseau allemand par rapport au point de référence à Toruń.

L'élaboration des mouvements verticaux de l'écorce presentée ici est la première qui embrasse presque toute étendue du pays. Les résultats pourtant ne peuvent être considérés qu'approximatifs vu les matériaux assez restreints et souvent accidentels dont on se servait à l'élaboration. C'est d'autant plus probable que les grandeurs des mouvements sont comprises souvent dans les limites assignées aux erreurs.

Néanmois, la détermination géodésique présentée ici permet, malgré son caractère provisoire, d'affirmer que le territoire de la Pologne est soumis à des mouvements de l'écorce terrestre d'un ordre minime. Cette thèse est d'ailleurs d'accord avec les opinions énoncées dans les autres domaines de la science.

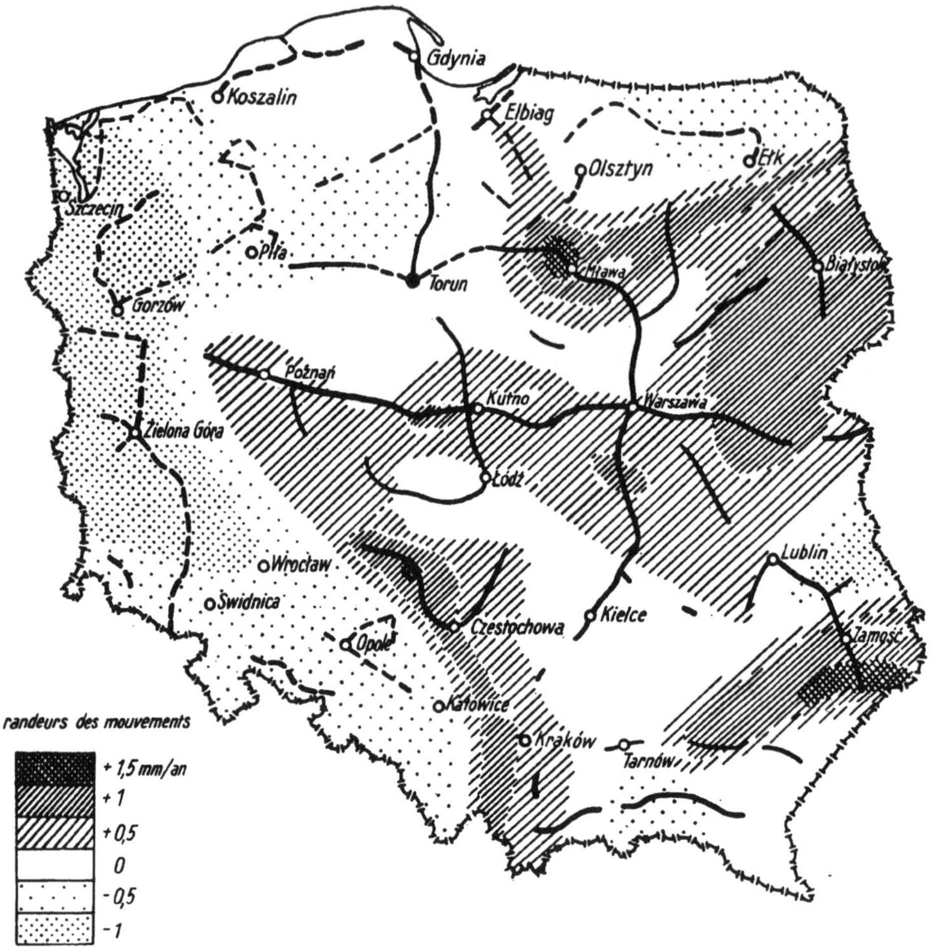

Fig. 1

Grandeurs des mouvements verticaux récents de l'écorce terrestre sur l'étendue de la Pologne

—— } Lignes de nivellement contemporain de précision de 1-re et 2-ième classe (1952-58)
—— } coincidant avec les lignes de nivellement de précision des années 1927-37

– – – } Lignes de nivellement contemporain de précision de 1-re et 2-ième classe
– – – } coincidant avec les lignes du nivellement ancien (mesurées après 1871)

◉ Niveau de référence des mouvements de l'écorce terrestre (repère Toruń - Hôtel de Ville)

Echelle 1 : 4000 000

Institut de Géodésie et de Cartographie, Varsovie 1960

La Pologne possède actuellement un réseau de marégraphes et indicateurs de niveau d'eau sur la côte de la mer Baltique, rattaché au réseau de nivellement de précision de l'État. On peut alors espérer que les observations marégraphiques permettront dans un temps proche de déterminer le niveau moyen de la mer. En outre il faut tenir compte du fait que le réseau de nivellement de precision mesuré dans la période 1952–1958 possède 1500 points environ établis au-dessous du niveau de congélation du sol. Une répétition de la mesure de ce réseau planée pour une période de 15 à 20 ans à partir d'aujourd'hui rendra sans doute possible l'établissement d'une image complète des mouvements verticaux absolus c.à d. rapportés au niveau de la mer pour l'étendue de toute la Pologne.

Les mouvements verticaux relatifs de l'écorce terrestre sur l'étendue de la Pologne sont présentés sur la carte ci-jointe.

D i s k u s s i o n

A. P. R o z h d e s t v e n s k i y (Ufa):

I would like to know the scale of the map compiled by you of recent crustal movements in Poland.

I. N i e w i a r o w s k i (Warschau):

The present map of recent vertical movements of the earth's crust in Poland has the scale 1 : 4 000 000.

Remarks on secular movements within Denmark

von

O. Simonsen (Kopenhagen)

Summary

The secular movements within the Danish islands are elucidated by the aid of:
(1) researches of raised shorelines
(2) tidal observations
(3) comparison between old and new levellings giving the differences $h_{new} - h_{old}$.

Information is given whether $h_{new} - h_{old}$ may be due to secular movement or to other causes. The variation of the values $h_{new} - h_{old}$ has been compared with the retreat of ice border lines in the late glacial period and also compared with Bouguer gravity anomalies.

Some remarks are given on the geological structure of the underground and finally some other remarks are given about levelling connections between the Danish islands and from them to Jutland and to Sweden.

Zusammenfassung

Die säkularen Bewegungen auf den dänischen Inseln sind aufgeklärt worden durch:
1. Untersuchungen der gehobenen Uferlinien
2. Tidenbeobachtungen
3. Vergleiche zwischen den alten und neuen Nivellements geben die Differenzen $h_n - h_a$

Es wird auseinandergesetzt, ob die Differenz $h_n - h_a$ durch säkulare Bewegungen bedingt ist oder andere Ursachen hat. Die Änderungen des Wertes $h_n - h_a$ sind mit dem Eisrückgang und den Bouguer-Schwereanomalien verglichen worden.

Es werden einige Bemerkungen zu Struktur des Untergrundes und zum Schluß auch einige Äußerungen über die Anschlüsse der Nivellements zwischen den dänischen Inseln und denen von Jütland und Schweden gemacht.

Raised shorelines and tidal observations

Even though Denmark is a small country the problem on secular movement within Denmark may be of some general interest because within Denmark both uplift and subsidence will be found, in other words the Zero-Isobase for movement of land relatively to the sea will cross through Denmark.

---

x) The text has been abridged from the stenalled report presented in Leipzig May 1962

G. Forchhammer already during the years 1835 - 1858 published several papers dealing with secular movement within Denmark. An important paper [13,1] translated into German [13,2] defines a line, the so-called F o r c h h a m m e r - L i n e running from the middle of Nissum Fjord (West Coast of Jutland) to a point 3 km South of the town Nyborg at the Great Belt and running still further against South-East and Forchhammer indicated that the areas to the North-East of this line is subject to an uplift. Forchhammer based his theory on raised shorelines and considered the uplift as belonging to the outer limit of the wellknown Scandinavian uplift with centre of uplift in the Northern part of Sweden.

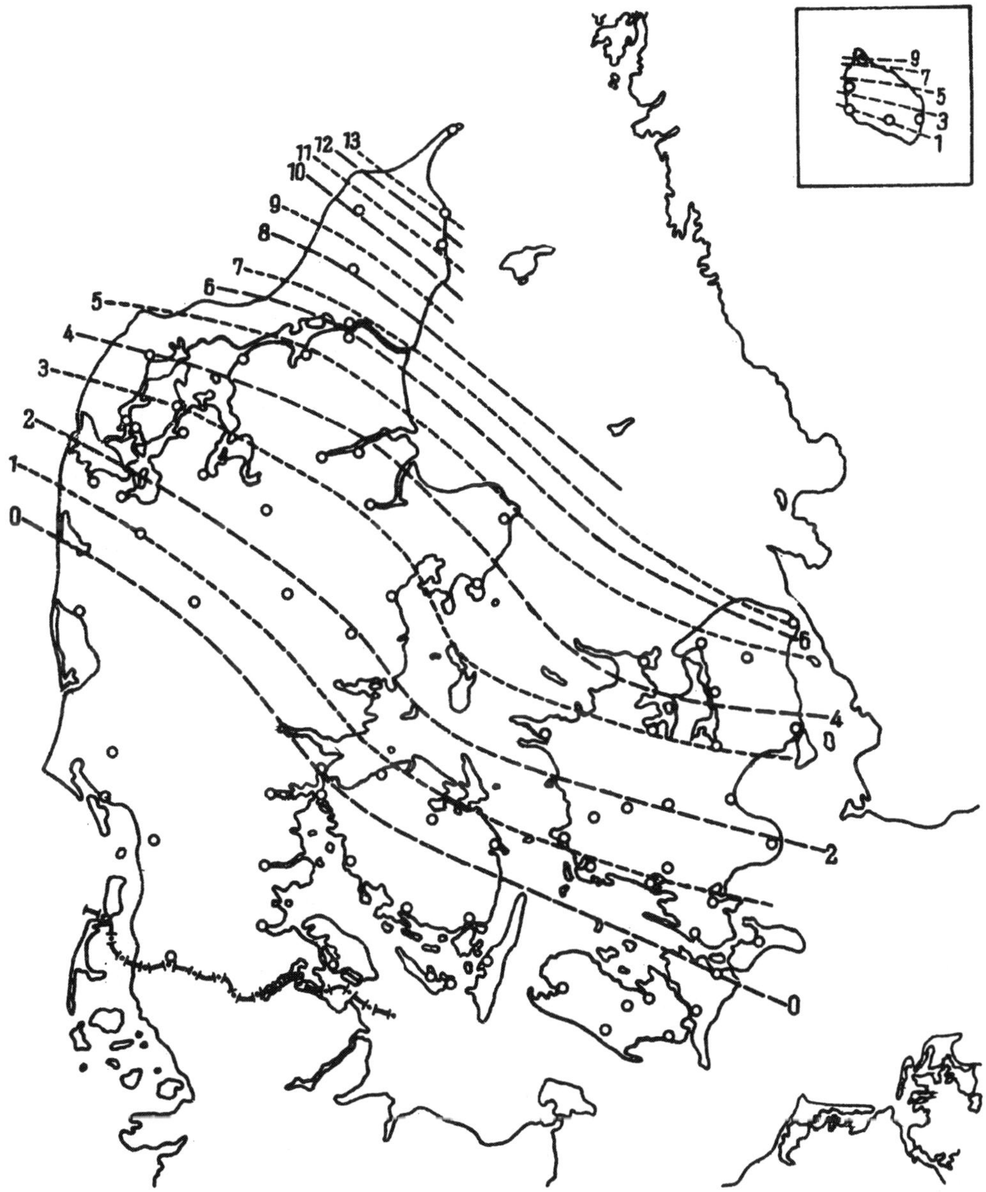

Fig. 1

GEOLOGICAL SURVEY OF DENMARK. GEOLOGIST ELLEN LOUISE MERTZ. (1924). Isobases of the uplift in the north-east Denmark after the maximum of the Litorina subsidence; equidistance 1 m

In this century the Danish geologist Mrs. E.L. Mertz [23] in 1924 has collected new investigations of land-uplift within Denmark based on observations of raised shore-lines. Mrs. Mertz published a map of isobases of uplift in the North-East of Denmark after the maximum of the Litorina subsidence. The maximum of the Litorina subsidence did not take place at the same epoch for all parts of Denmark, see V.M. Mikkelsen [25, p. 128] ; for Amager at Copenhagen the maximum transgression was S ö b o r g IV about 2300 B.C. (see: J. Troels-Smith [57, p. 195], Johs. Brøndsted [5, p. 124], and O. Simonsen [45, p. 131] and probably about 1000 years earlier (see Trap: Danmark [55,1, pp. 30, 336]) in the northernmost part of Jutland, where the greatest uplift found by Mrs. Mertz is indicated as 13 m and furthermore situated about 185 km to the north-east of the zero isobase, then this give a total tilting of

$$\frac{13}{185000} \times 206265'' = 14''.50$$

and a mean rate of tilting of

$$\frac{14''.50}{5200 \text{ years}} = 0''.0028 \text{ per year}$$

For Hornbæk on Zealand must be referred to publications by Dr. V. Madsen [22, pp. 86 - 98] and Dr. V. Milthers [28, p. 168], [29, p. 177], and O. Simonsen [45, p. 132] and in similar way is then found a rate of tilt $\frac{11''}{4200} = 0''.0026$, nearly the same as found for Jutland.

Even though these values have a certain uncertainty it is worth comparing them with a value found by the Danish Meteorological Institute by the aid of tidal recording from about 1890 up to now (that is during about 65 years) at ten or nine tidal stations within Denmark. Publications elaborated by state-meteorologist J. Egedal in 1946 [11] and in 1955 [12]. By linear adjustment of annual heights of sea level at each of the tidal stations (indicated in the table below)

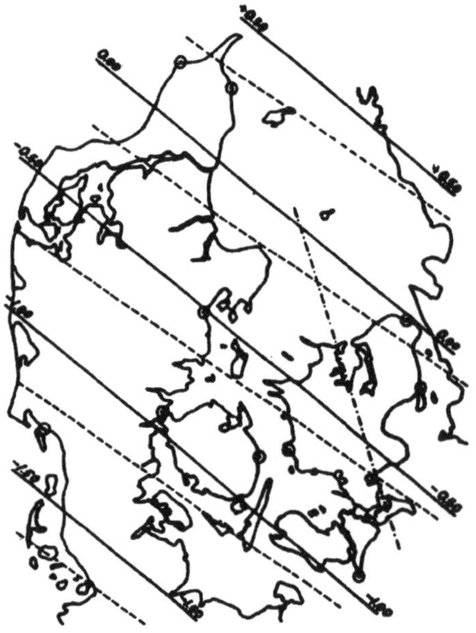

Fig. 2

THE DANISH METEOROLOGICAL INSTITUTE.
STATE-METEOROLOGIST J. EGEDAL.
The apparently yearly upheaval of the earth's crust in relation to Normal Height of the sea level. Units: mm/year

.... Isobases published by Egedal 1946

—— Isobases published by Egedal 1955

⊙ tidal stations

—·— Direction of a relative hinge line found for Zealand by old (1885-1905) and new (1940-1944) levellings by the Danish Geodetic Institute

c_E based on tidal observations up to 1945.0 published 1946			c_E based on tidal observations up to 1955.0 published 1955
København	- 0.23	± 0.27 *) mm/year	- 0.24 mm/year
Hornbæk	+ 0.11	± 0.32 *)	+ 0.02
Korsør	- 0.45	± 0.22 *)	- 0.72
Gedser	- 0.73	± 0.27 *)	- 0.91
Slipshavn	- 0.68	± 0.23 *)	- 0.82
Fredericia	- 0.92 Δ)	± 0.21 *)	- 1.01
Aarhus	- 0.44	± 0.19 *)	- 0.54
Frederikshavn	+ 0.39	± 0.24 *)	
Hirtshals	+ 0.53	± 0.32 *)	+ 0.21
Esbjerg	- 0.98	± 0.33 *)	- 1.29
(land-uplift has the sign +)			

Mr. Egedal finds a coefficient c_E of movement of the land relatively to the sea. Then Mr. Egedal considers the values c_E for all the Danish tidal stations as observations of a tilting of Denmark when supposing that Denmark may be considered as a rigid tilting surface. Egedal makes an adjustment giving the annual tilting of Denmark considered as a whole and Egedal finds astonishing small residuals by this adjustment and he finds an axis of tilt relatively to the sea:

1946 about Nivaa-Nibe
1955 " Hornbæk-Blokhus

and correspondingly may be found for Denmark
(from publication 1946) annual angle of tilt 0"0012 ± 0"00015 **)
(" " 1955) " " " " 0"0013

which quite naturally is smaller than the value found from the Litorina shore-lines. Later we shall compare these results with results obtained by precise levellings within Denmark.

Probably the earth's crust under Denmark is divided by faults into blocks and for that reason the precise levellings may reveal locally conditioned movements with other axes than the axis found in general for Denmark by J. Egedal.

At the General Assembly in Helsinki 1960 Dr. J.R. Rossiter, Director of the Tidal Institute at the University of Liverpool, presented his interesting work [40]

Report on the reduction of sea-level
observations (1940 - 1958) for R.E.U.N.

where R.E.U.N. = Réseau Européen Unifié de Nivellement (having been internationally adjusted 1958 and 1960). Dr. Rossiter's work concerns all tidal stations attached to the United European Levelling Net = U.E.L.N. = R.E.U.N. but in order to use as near as possible a common series of years for recordings at all the tidal stations Dr. Rossiter has been obliged to use only a limited number of years concerning annual heights Z of sea-level, and he has chosen the 19 years 1940 - 1958.

Dr. Rossiter uses besides the tidal observations also meteorological observations within the period of 19 years and calculates meteorological corrections to annual heights of sea-level. Dr. Rossiter uses the meteorological corrected annual heights of sea-level for the period 1940 - 1958 for a simultaneous calculation of secular movement c_R (relatively land to sea) and of the nodal tide. However, from a single period of 19 years at

*) Standard deviations calculated by O. Simonsen [45, p. 106]

Δ) Egedal indicates for Fredericia - 0.97, but a small error has here been corrected [45, p. 106]

**) Standard deviation calculated by O. Simonsen [45, p. 111]

each tidal station, it is not possible with sufficient accuracy to determine the values c_R.

I think this is the reason why Dr. Rossiter as to the Danish tidal stations gets values c_R which do not agree very well with the values c_E found by Mr. Egedal.

The values c_R are numerically greater than c_E so that

$c_R - c_E = - 3.4$ mm/year for Hornbæk

$c_R - c_E = - 3.3$ mm/year for Fredericia

The values c_E found by Mr. Egedal are determined without allowance for meteorological corrections, but in return a still more important circumstance is relevant viz, that the values c_E have been determined for a continuous and long period of about 65 years, which again is in good agreement with the small values of standard deviations for each of the values c_E .

In this connection it is worth mentioning that Ministerialrat Julius Gaye, Hamburg, in 1951 presented a paper [14] concerning changes in heights of sea-level at the Baltic and the North-Sea. Mr. Gaye gives for 20 tidal stations at the Baltic and the North-Sea for the months May, June and October during the latest century curves of mean values for 19 years for the months in question. The largest series of these mean values comprises the interval from 1820/38 to 1930/48.

These curves show clear correlation between neighbouring stations and furthermore they show suddenly great alterations of the tilt of the curve within short series of years.

Ministerialrat Gaye writes [14, p. 4] that these alterations of height of Mean Sea Level within so short a period as 19 years can not be explained as sudden movements of the earth's crust, but must be explained by various climatic causes.

Comparison between old and new levellings

The picture of recent movements within Denmark given by Mr. Egedal is as indicated based on the supposition that the whole Denmark may be considered as a rigid surface and this is as pointed out by Professor Axel Schou [44, p. 55] not quite correct and therefore we get a more detailed picture of recent movements within Denmark by the aid of comparison of old and modern precise levellings. These comparisons shall be given separately for Zealand, Funen, and Falster.

The old precise levelling has been made by the old Danish Geodetic Survey, Den Danske Gradmaaling, 1885 - 1905 and the modern levelling of high precision has been made for the islands mentioned above during 1940 - 1944 by the Danish Geodetic Institute.

For the old levelling net which was levelled along the highroads was placed at nearly each 1800 m a principal bench mark, namely a subsoil mark or a bench mark in an old stable building.

After this introduction I shall give the results from comparisons between the old and the modern levellings for the 3 islands and give information whether possibility exists that these comparisons may to some degree give an impression of the recent secular movements.

Zealand

For this island 225 principal bench marks have been at disposal for the comparison. A zero level MR for the comparison has been defined in the way that for a stable area along the lines Roskilde-Ringsted-Rønnede-Køge-Taastrup the general mean of differences: -(new height - old height)- for 47 principal bench marks has been defined as zero.

In general the differences

$$\text{new height} - \text{old height} = h_{new} - h_{old}$$

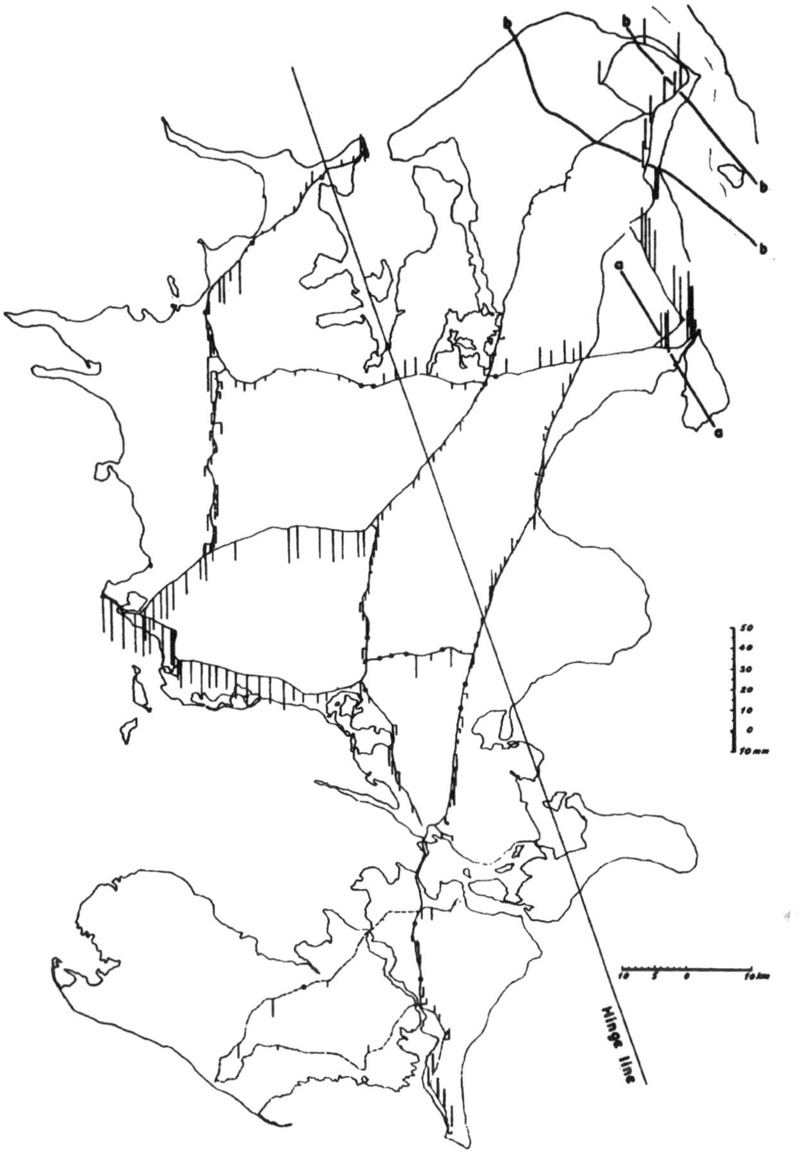

Fig. 3

THE DANISH GEODETIC INSTITUTE. Comparison of old levelling (1885 - 1905) with new levelling (1940 - 1944). Sketch map of differences $h_{new} - h_{old}$ for 225 principal bench marks in Zealand. A relative hinge line has been found by adjustment considering Zealand tilting as a rigid plane. Water crossings 1898 - 1899 Zealand - Falster and levelling 1942 along a bridge (3 km) give the possibility of increasing the synopsis of $h_{new} - h_{old}$ to Falster and Lolland. Only for Zealand $h_{new} - h_{old}$ gives reliable information on secular movement.

GEOLOGICAL SURVEY OF DENMARK AND GEOLOGICAL SOCIETY OF DENMARK.

Fault in the Bed Rock

(a) according to A. Rosenkrantz (1937), (1955-1957)

(b) according to { V. Milthers (1935), Th. Sorgenfrei (1962) }

are greatest in Copenhagen namely + 24 mm,
smallest in Korsør namely - 20 mm.

For the variation 44 mm of the adjusted difference of height from old to new levelling from Copenhagen to Korsør (110 km) has been found [45, p. 119] the mean square error

$$\pm 14 \text{ mm}$$

and therefore a possibility exists for Zealand, that the differences

$$\text{new adjusted height} - \text{old adjusted height} = h_{new} - h_{old}$$

may give an impression of real movements of the bench marks.

Some of my colleagues have made an adjustment [45, pp. 115 - 117], [46] considering Zealand as a rigid plane which was tilting from the first to the second levelling as indicated by the 225 differences $h_{new} - h_{old}$ and as result was found an axis indicated both in fig. 2 and 3 and so we see that the local tilting of Zealand has another orientation than found in general for Denmark by Egedal. But it is worth noticing that the total angle of tilt found by levelling for Zealand [45, pp. 115 - 120] is

$$\theta = 0''060 \pm 0''003$$

and as 44 years have gone between the two levellings this gives an annual rate of tilt

$$\frac{0''060}{44 \text{ years}} = 0''0014 \text{ per year}$$

and the rate of tilt as found by Mr. Egedal for the whole Denmark is

0"0013 (see p. 7).

F u n e n

The director of the Danish Geodetic Institute, Professor Einar Andersen has kindly given the permission to present here some information on the modern levellings of high precision on Funen on which levelling publications not yet have been distributed by the Institute.

On Funen the old precise levellings of 1896 and 1899 have been levelled only in a single line from Middelfart to Knudshoved at Nyborg.

The modern levellings of high precision have been levelled on Funen as ordinary (direct and reverse) along the old levelling line in 1943, but also another, new, levelling line (farther southward) has been levelled in 1944 from Middelfart via Assens and Ringe to Nyborg passing through a levelling network of second order from 1901 - 1904 [15,1].

Along the old levelling line have been found 33 principal bench marks fitted for comparison of the two levellings. For twelve principal bench marks to the East of Odense a nearly constant (or perhaps no) change has taken place from the first to the second levelling, whereas westward to the bench mark GM 933 about 5 km to the East of Odense suddenly an increase of subsidence of about 17 mm taken place up to Odense and going westward from Odense along the levelling line the subsidence increases still more about 18 mm (GM 842).

The levelled distance Odense-Middelfart (GM 842) is about 49 km and therefore the mean square error of the increase 18 mm in subsidence (Odense - Middelfart) is 9,38 mm.

Thus the increase of 18 mm in subsidence from Odense towards West to Middelfart has a certain probability.

As to the new southern line on Funen no subsoil bench marks were constructed for the levellings 1901 - 1904 and therefore only some few bench marks in old buildings may give an impression of possible movement along the southern line. The movements correspond to the movements along the northern line; only the stable part in the East-end of the line is shorter and furthermore the subsidences to the West of Odense are smaller.

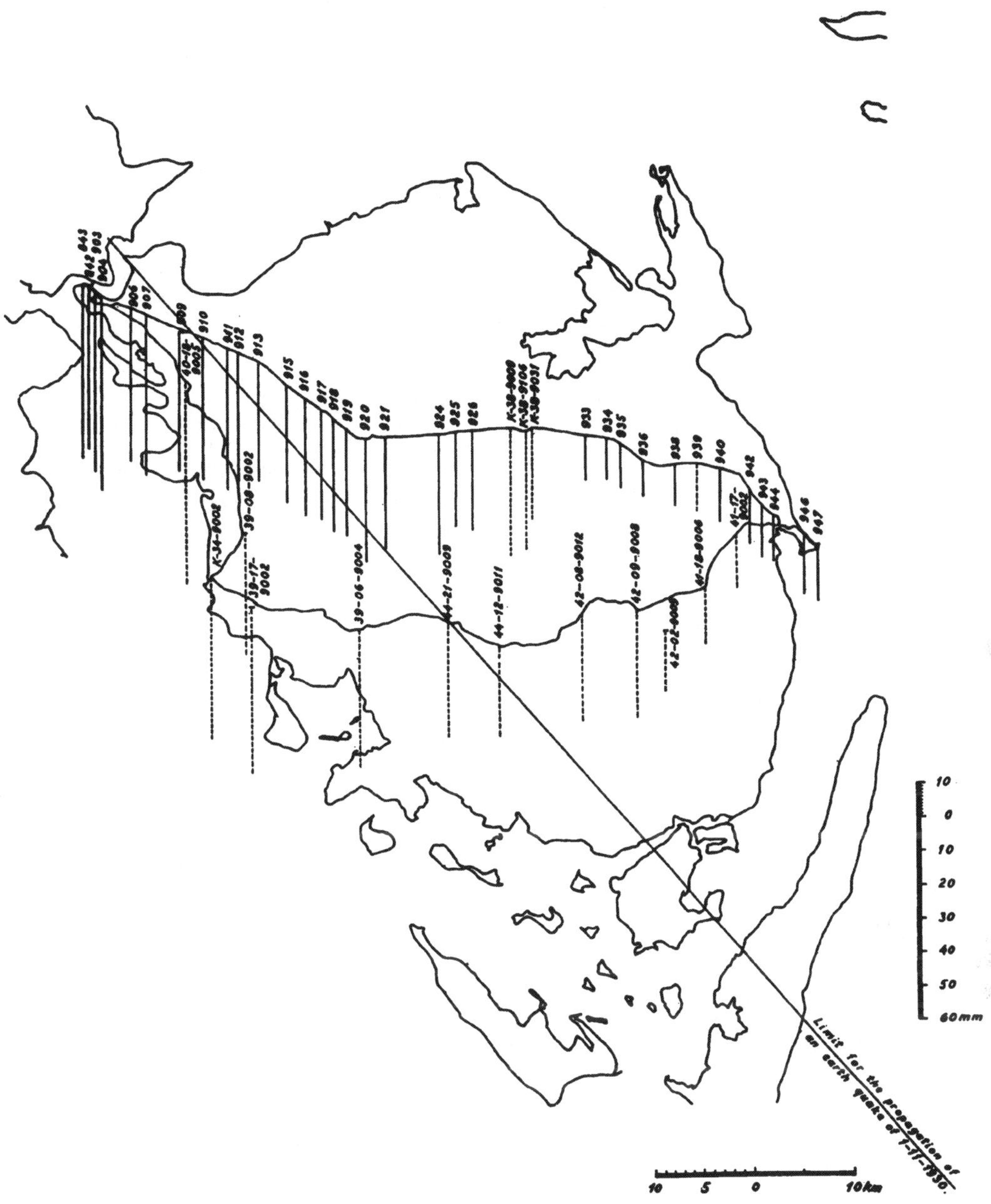

Fig. 4

THE DANISH GEODETIC INSTITUTE.

Funen. Comparison between
old levelling 1896 - 1899 along the northern line
" " net of 1901 - 1904 to the South of this line
and new levelling 1943 - 1944.
Datum for h_{old} given by the old Danish Geodetic Survey.
Datum for h_{new} the same as used for Zealand (MR).

For subsoil bench marks: $h_{new} - h_{old}$ indicated by fully drawn vertical lines
" ordinary " " : $h_{new} - h_{old}$ indicated by dashed vertical lines

The straight line Middelfart - Svendborg indicates limit for the propagation of an earthquake 1/11 1930 with epicentre in the Sound (I. Lehmann 1931)

Falster and Lolland

For Falster and Lolland the differences $h_{new} - h_{old}$ are rather small, of the same order of magnitude as the mean square error, and therefore the formal uplift of Falster relatively to Lolland and of Southern Falster relatively to Northern Falster may be due to observation uncertainty, but these possible displacements seem likely when comparing with the retreat of the ice since the Ice-Age. See figs. 3 and 5.

Comparison between the variation of the differences $h_{new} - h_{old}$ and the retreat of ice border lines during the late glacial period

During the later years many geological researches on the retreat of the ice after the Ice-Age have taken place in Denmark, especially by Dr. V. Milthers [28] [29] [30] and his son Dr. K. Milthers [26] [31] [32].

During the Würm-period the ice covered the northern and eastern part of Jutland and all the Danish islands (see Trap [55,1, pp. 29 - 33])and the ice melted back northwards from Northern Jutland and eastwards or southwards from the remaining part of Denmark. Consequently the ice border lines which will be shown for the Danish islands are of a later stage, namely real late glacial.

Zealand, Falster and Lolland

As to the latest glaciation af Zealand Dr. Milthers's [30] publication is very important especially his glaciomorphological map of Zealand presented here in reduced form in fig. 5.

The Great Belt Glacier has transgressed Zealand to a line that nearly may be indicated by Kulsbjerg in the southernmost part of Zealand and Saltbaek Vig 5 a 8 km NE to Kalundborg and the ice has here transgressed from South against North and melted back in the opposite direction.

For the rest of Zealand the latest stages of ice have transgressed from East as to be seen from the figure and also melted back in the opposite direction. Dr. V. Milthers shows that in an area about 20 km to the north of Copenhagen ice streams from south or south east met the ice streams going east west. Thus the area of Copenhagen has been unloaded for ice relatively late. If we suppose that the magnitude of isostatic movement may depend on in which succession the ice has retreated in various stages in such a way that a possible uplift is greater where the ice has retreated latest then the picture of relative tilting of Zealand found by levelling is in good agreement with the retreat of ice from the northern and eastern Zealand.

For the south-western part of Zealand at Skelskør the values $h_{new} - h_{old}$ show subsidences relatively to the northern parts of West-Zealand, but the ice has (according to V.& K. Milthers) retreated southward from Asnæs and northward and eastward from Røsnæs, and until further we do not know whether Odsherred or the area at Skelskør has at first been exempted from the ice-cover (see K. Milthers [26, p. 136 - 137]). However, in this connection an important fact must be mentioned namely that raised submarine areas (of postglacial age) both in Odsherred and Halsnæs (eastward to Issefjord) clearly indicate that the northern part of Zealand has really risen since postglacial time considerably more than the area at Skelskør (see Trap [55,5, p. 603]) in accordance with the general tilting of Denmark, see pp. 4 - 5 .

For Falster and Lolland the values $h_{new} - h_{old}$ are small and seem to be in accordance with the retreate of the ice, see p. 15.

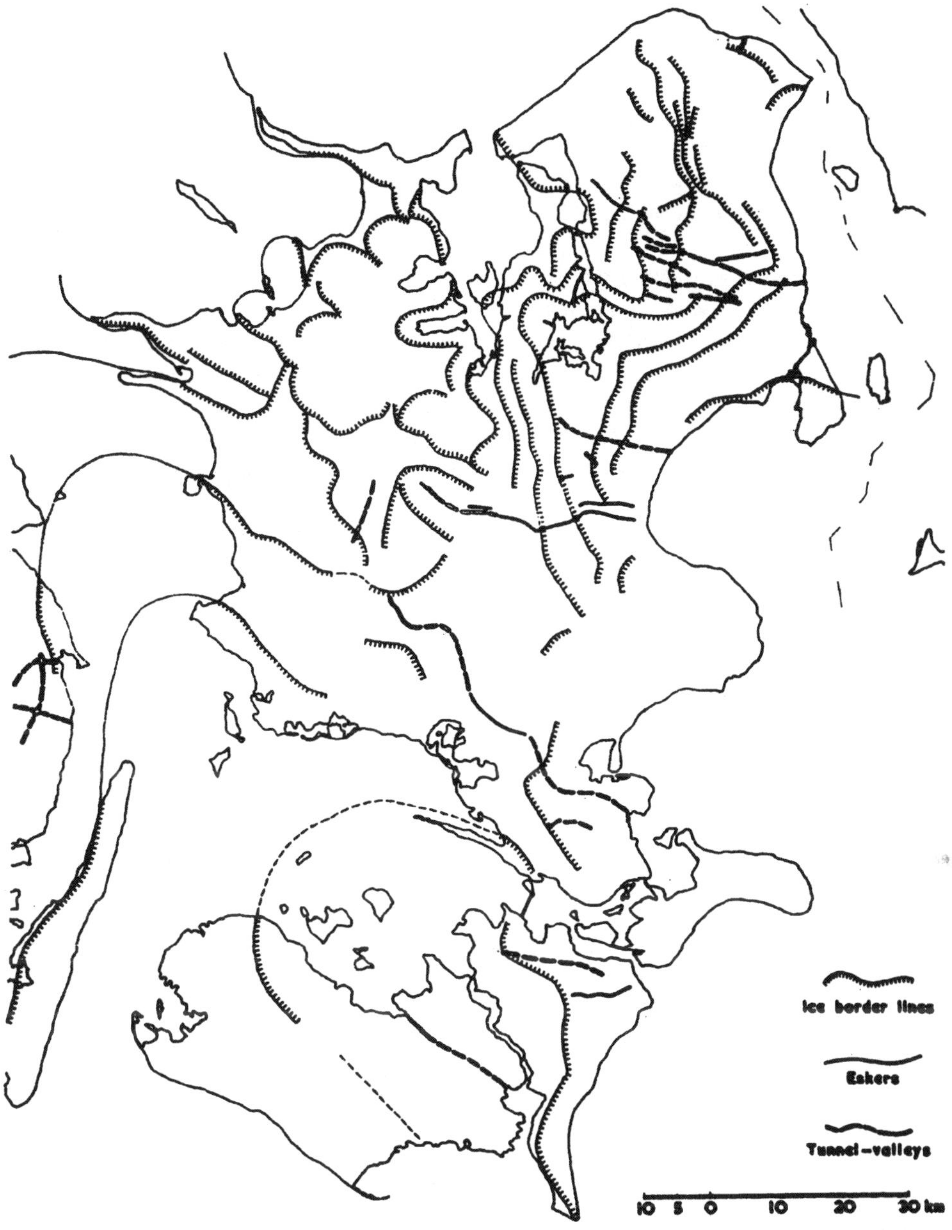

Fig. 5

GEOLOGICAL SURVEY OF DENMARK. V. MILTHERS, (1948). Ice border lines on Zealand, Falster and Lolland. Copy of glaciomorphological map of Denmark without indication of postglacial and lateglacial deposits or other details which have been indicated in colour on the original map.

F u n e n

The very important researches on the glacial landscape of Denmark have been supplemented in 1962 as in this year M. Sc. Per Smed [48] has published the latest results on the glacial landscape of Funen where the problems are more complicated than on Zealand.

During the later years very large and deep excavations have been made at Lindø at Odense Fjerd for construction of big docks. This has given the Danish geologists a very favourable opportunity for further investigations.

In the abstract to his publication on his geomorphological investigations on Funen Mr. Smed writes:

"The most important feature of the large map is that drumlinoids are found in two main directions. In the central region of the island they are stretched NE-SW, and in the coastal parts SE-NW. By means of glacial striae and meltwater deposits etc. it is shown, that this must mean that Funen during the Würm glaciation has been transgressed by at least 2 ice streams, an older in the direction NE-SW, and a younger with direction from the Baltic Sea towards NW. This "Baltic Stream" was divided into two main branches, one of which were moving through the Little Belt depression and the other through the Great Belt depression, while at the same time central Funen was covered by a huge dead ice sheet, remnant of the NE ice stream. On account of the distribution of indicator boulders it is supposed that the two "Belt Ice Streams" have met again in Jutland".

Concerning the Lindø observations Mr. Smed [48, p. 36] gives a possible analysis for the succession

Ice	coming	from	SE	latest phase		
"	"	"	NE	earlier "		
"	"	"	SE	still	earlier	phase
"	"	"	N	"	"	"
"	"	"	E	"	"	"

and for the comparison with recent movements the last phase is the most important and determined with great certainty, whereas the analysis of the elder phases are more complicated.

For an analysis of ice movements on Funen and neighbouring isles Mr. Smed divides the Islands in eight Regions I - VIII. As characteristic features may be stated:
Region I represents a stable centre covered with dead ice round which the glacial movements have taken place.
Region II has practically been without ice when the border line of the Great Belt Glacier was at Stenstrup Gislev and probably Munkebo and the Little Belt Glacier was at Kegleholm Silkeaa.
Region III = Sletten had its latest transgression from SE namely of Baltic ice as an early phase of the Great Belt Glacier, which probably for a long time has stayed at the Munkebo-Gislev line.

Later phases of the Great Belt Glacier are indicated in the regions IV and V and in the regions VI and VII we see how the Little Belt Glacier has retreated against South and South-East through Little Belt.

For Region VIII = Vends Herred, the geomorphological analysis is difficult and it is not yet clear whether the ice transgressed this Region either from NE or from Region III or from Region VII, but Region VIII is that part of Funen from which the ice has retreated at first.

Summarizing these results on the retreat of the ice and comparing with the differences

$$h_{new} - h_{old}$$

found by precise levelling for Funen we see that the stable part from Knudshoved -

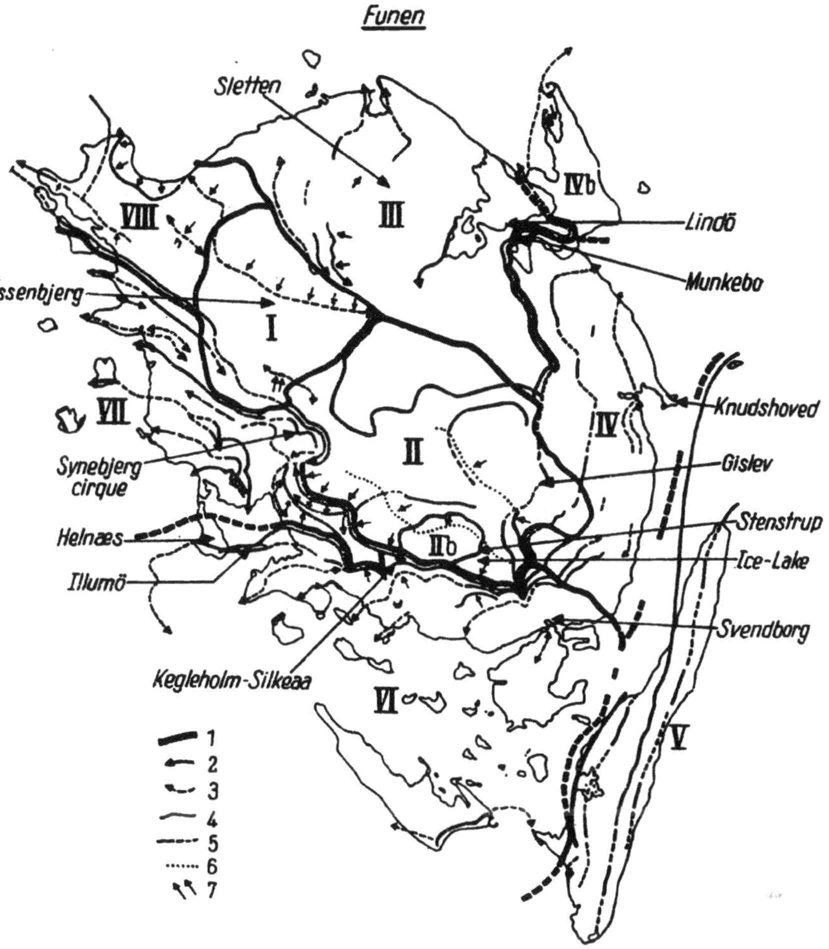

Fig. 6

GEOLOGICAL SOCIETY OF DENMARK. PER SMED (1962). Funen. I - VIII Glacia-morphological Regions,

1. ——— Limit of Region.
2. ——— Ice border line (safe).
3. - - - - Hypothetical ice border line.
4. ═══ Hypothetical line caused by melting of dead-ice.
6. ░░░ Outwash plain.
7. —► Direction of ice-current.

Nyborg to about 5 km before Odense is as far as concerns the eastern half of this line situated in an area where the ice has withdrawn in a late phase and the possible subsidence at Middelfart is situated in an area from which the ice has at first retreated from Funen. Perhaps the dead ice region at Vissenbjerg may give an explanation of why the differences $h_{new} - h_{old}$ decrease rather slowly from Odense to about Vissenbjerg and thereafter decrease quickly at the westernmost part of the levelling line towards the Little Belt.

Comparison between differences $h_{new} - h_{old}$ and Bouguer gravity anomalies

Bouguer gravity anomalies have been published for Zealand and Falster by Professor, Director Einar Andersen [1] and for Funen and Lolland by Mr. Svend Saxov [42] [43].

It is astonishing to see how closely the gravity anomalies of Zealand are correlated with the values of $h_{new} - h_{old}$. Only in the southernmost part and between Sorø and Ringsted agreement is poor. It is uncertain whether isostatic conditions may be valid for Zealand, and generalizations of greater areas of Denmark and the surroundings of Denmark may be erroneous. The gravity anomalies are not in all cases correlatable with the values of $h_{new} - h_{old}$, e.g. not on Falster. When crossing the Sound from Helsingør on Zealand to Hälsingborg in Sweden the gravity anomalies published 1945 by Dr. Nørgaard [33] and confirmed by Swedish-Danish gravity measurements again in 1956 show a sudden increase in the Bouguer anomaly from about zero in Helsingør (Elsinore) to + 20 mGal in Hälsingborg harbour and + 23 mGal in Hälsingborg (Kärnan Grundmur, height 32.8 m).

Observations of Litorina shorelines indicate the possibility that Hälsingborg in Sweden still rises relatively to Helsingør in Denmark in spite of the great positive Bouguer anomalies within Hälsingborg.

Probably tectonic conditions and not isostatic conditions determine possible movements between Helsingør and Hälsingborg as important faults are found here e.g. along the Swedish coast from Kullen to Hälsingborg and to the north of Landskrona and continuing further on in this direction, see Uno Sundelin [54, pp. 7, 15, 39].

For Funen the correlation between Bouguer anomalies and differences $h_{new} - h_{old}$ is less pronounced than for Zealand. Especially for the Region I the gravity anomalies are greater than corresponding to the differences $h_{new} - h_{old}$ and in this connection it is worth mentioning that under Glamsbjærg (situated nearly at the southernmost point of Region I) the rock-formation (here gnejss) is found in a depth of only 833 m below Zero (G.O. Andrup [3, p. 34] and Trap [55.8, p. 9]) and to comparison may be stated that at Ullerslev, between Odense and Nyborg, in a depth of 1037 m [3, p. 34] below the surface only the Perm-formation has been reached.

Other parts of Denmark outside the islands confirm that correlation between $h_{new} - h_{old}$ and Bouguer anomalies may *not* be expected in general.

Some few summarizing remarks on the geological structure of the Bed Rock

Of course also an analysis of the structure of the Bed Rock of the prequaternary deposits might be of great interest for comparison with the values $h_{new} - h_{old}$, but here only some very few summary indications may be given.

Professor Th. Sorgenfrei has published several communications on the Bed Rock of Denmark and here shall be referred to the geological map of Denmark of the formations at the base of the Quaternary deposits published by Professor Sorgenfrei [52] [52a]. In Trap's "Danmark" [55.1, p. 27] Professor Sorgenfrei gives a valuable summary here translated into English:

"When summarizing the structural feature in the whole country, the area of Denmark may be divided into the following regions:

(1) The Fennoscandian Border Zone, to which belong Bornholm, NE-Zealand and probably the northernmost part of Jutland.
(2) A region of subsidence to the west of and parallel with the Fennoscandian Border Zone.
(3) A Salt Dome Province in Northern Jutland.
(4) A region of upheaval on the trend: Møn, SE-Zealand, Northern Lolland, Funen and Southern Jutland (see below).
(5) A region with shallow salt structures in the southernmost part of Jutland.

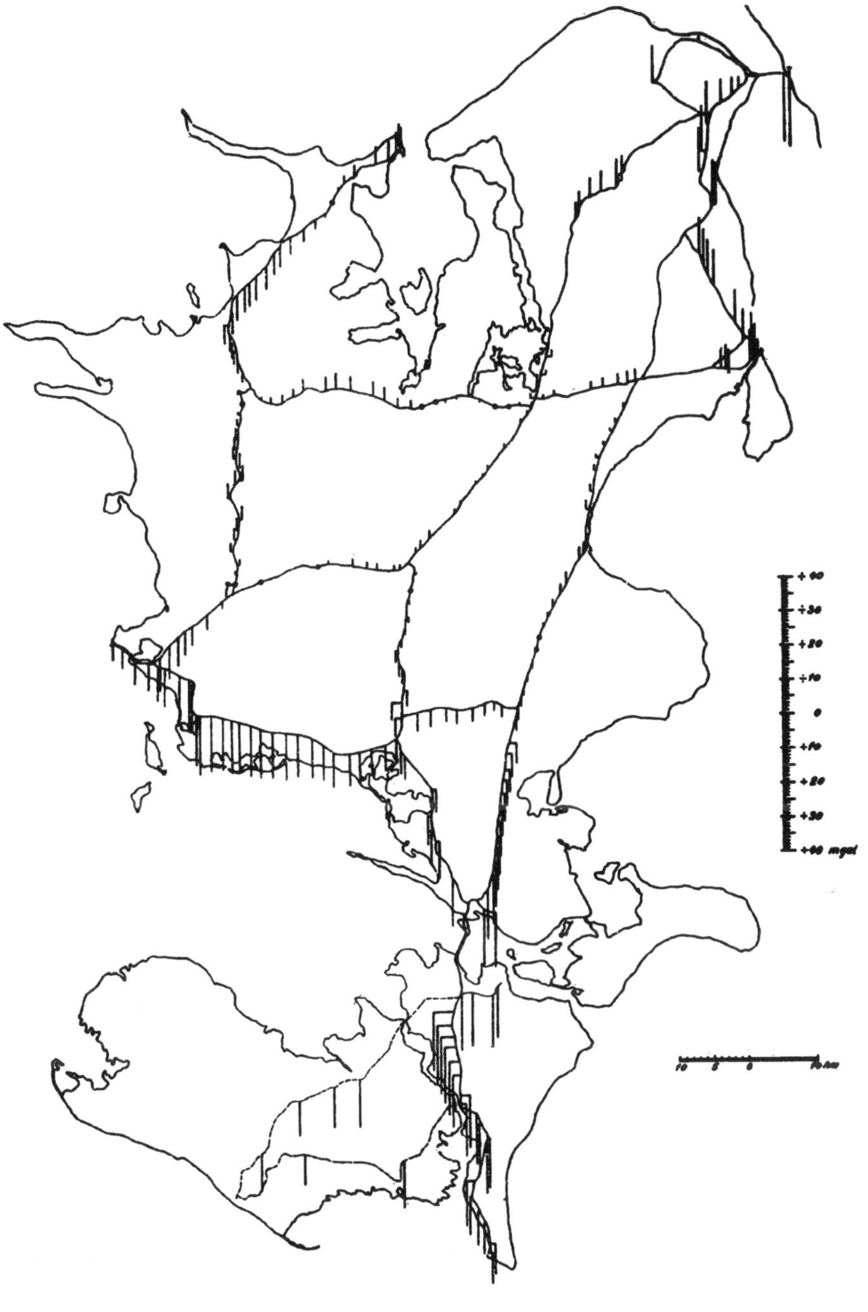

Fig. 7

THE DANISH GEODETIC INSTITUTE. Bouguer gravity anomalies for those bench marks for which $h_{new} - h_{old}$ has been determined, for Zealand and Falster according to Professor EINAR ANDERSEN (1947) and for Lolland according to SVEND SAXOV (1958)

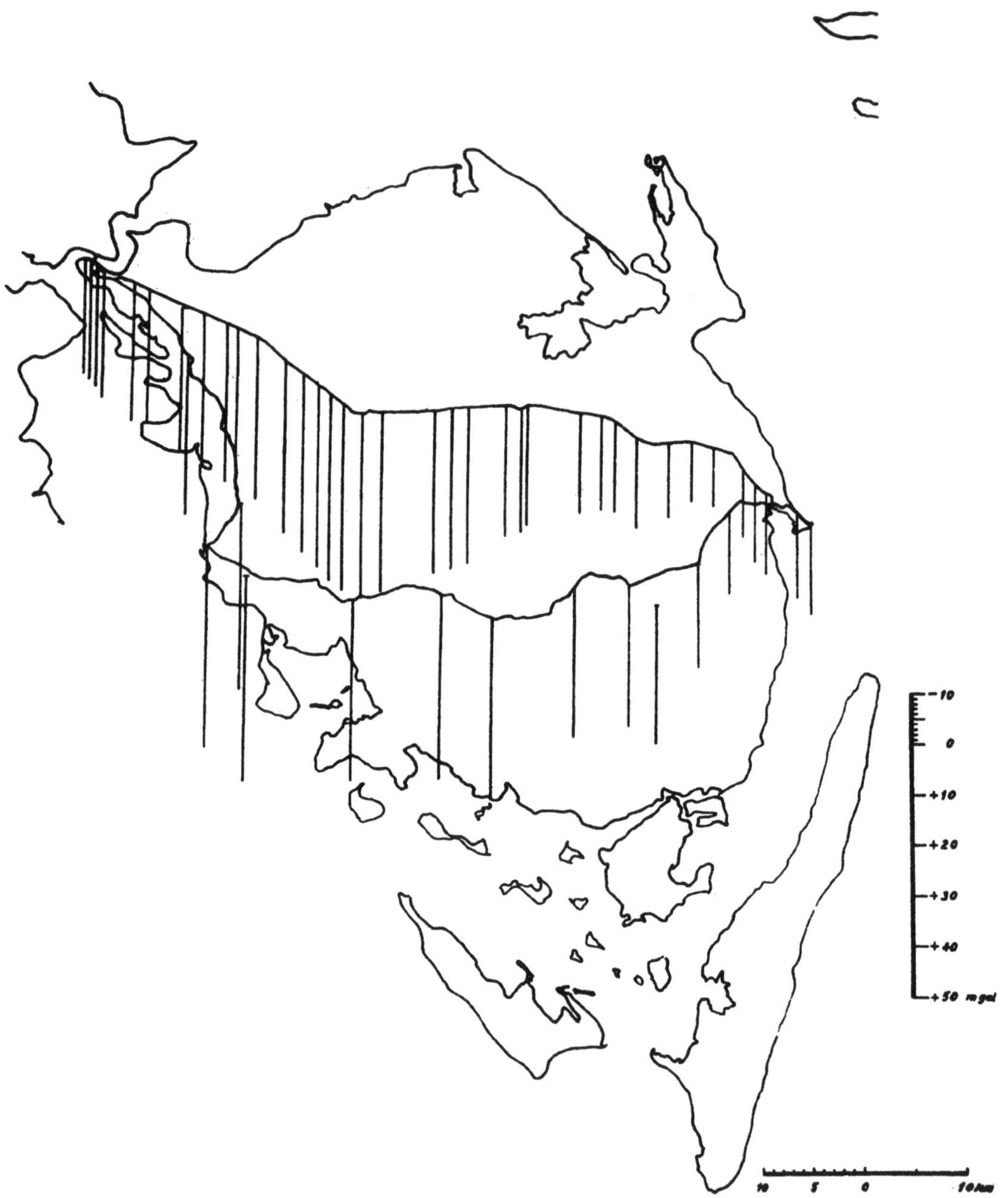

Fig. 8

THE DANISH GEODETIC INSTITUTE. SVEND SAXOV (1955). Funen. Bouguer gravity anomalies for those bench marks for which $h_{new} - h_{old}$ has been determined

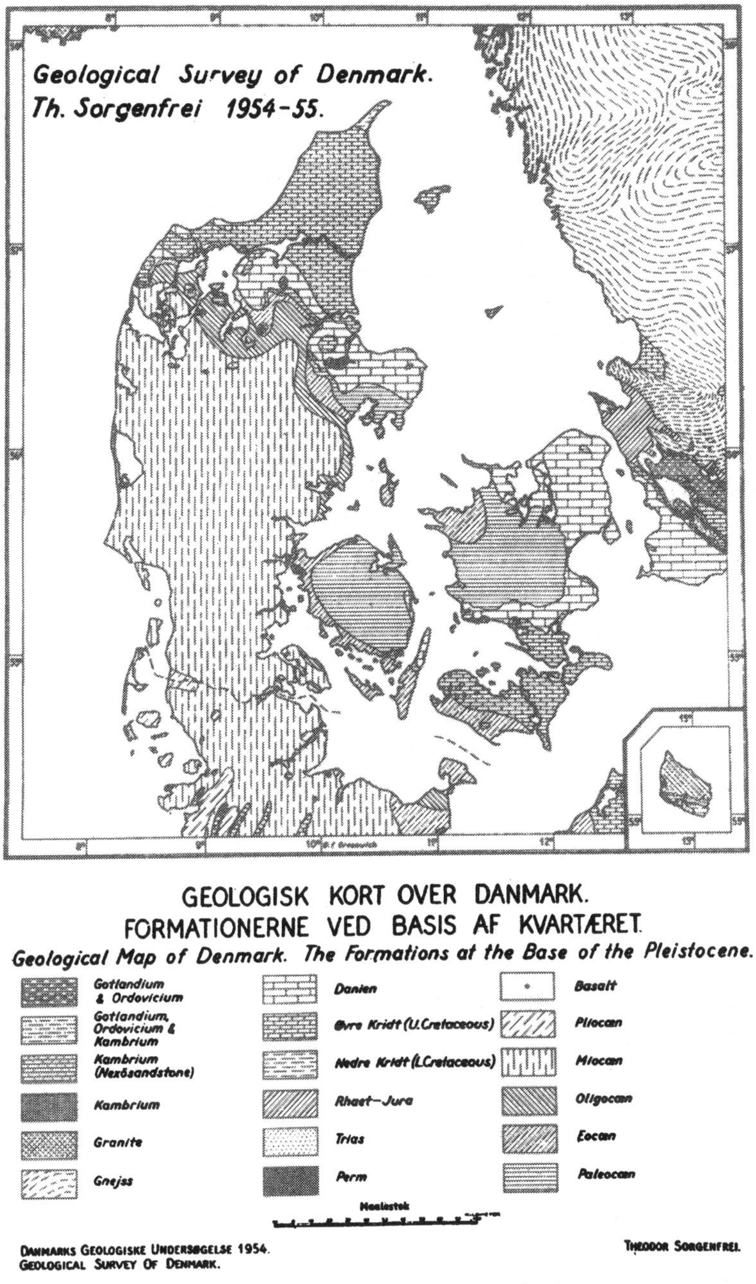

Fig. 9

GEOLOGICAL SURVEY OF DENMARK. Map of the Pre-Pleistocene deposits in Denmark according to TH. SORGENFREI (1954)

Within all these structural regions it is possible to outline a number of local structures and fracture lines".

The region indicated in item (4) has a trend WSW-ESE and Professor Sorgenfrei [50, pp. 492 - 493],[51, p. 170] considers this area as a continuation of the "Tempelburger Achse" and indicates that the tectonic movements that caused this structure of upheaval took place in the Oligocene period.

For Northern Jutland a similar orientation of formation boundaries is found from Aarhus in the direction ESE-WNW nearly parallel to the Forchhammer Line.

Also seismic researches indicate structure lines of this orientation. As an example may be indicated a limit (published by I. Lehmann [20] 1931) for the propagation of an earthquake of 1/11 1930 with epicentre in the Sound, see fig. 4. This limit goes from Svendborg to Middelfart-Strib has been indicated again by Mr. S. Saxov [42, pp. 63 - 64] in 1955 and furthermore he states that the propagation of another earthquake of 23/10 1904 with epicentre to the North of Denmark has on Funen reached as far as to Kauslund (near Middelfart), Odense and Svendborg (see E.G. Harboe [16, pp. 441 - 442]) and these three points are situated at or to the NE of the limiting line Svendborg-Middelfart in mention.

Thus the geological structure also indicates that the possibility exists that Denmark is subject to a general tectonic tilting as indicated by the investigations by Mrs. Ellen Louise Mertz and Mr. J. Egedal, but certainly the country is also divided in local structures which move independently, however, with a rate not deviating considerably in magnitude from the rate for Denmark in general.

As to the local faults only some very few remarks may be given here, see in fig. 3 indication of a fault in the chalk formations from Furesøen to Copenhagen and Amager (A. Rosenkrantz [38] [39] [39a]) and indication of the Graben of the Alnarp-valley at Humlebæk-Esrum-Gilleleje according to V. Milthers [29] and Th. Sorgenfrei [49].

Furthermore (see Trap [55.5, p. 599] a local area of uplift at Slagelse and a local area of subsidence at Sorø as indicated by the situation of the Danian-Paleocene unconformity should be mentioned.

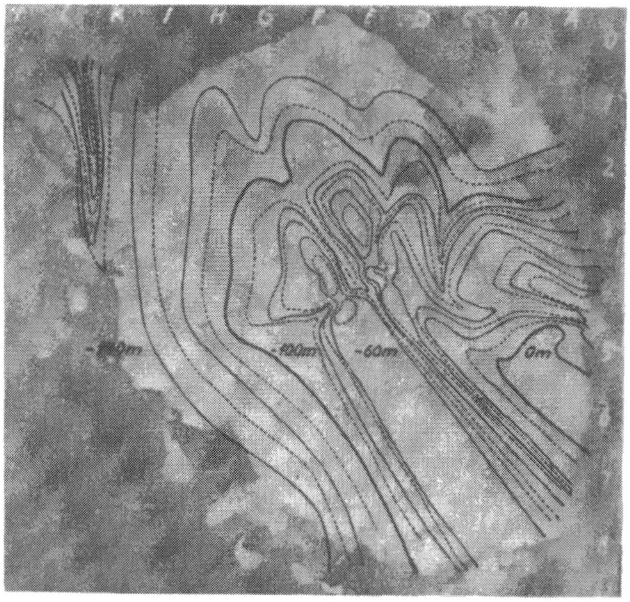

Fig. 10

Funen. Map of the surface of the chalk formations according to G.O. ANDRUP (1960), director of the WATER SUPPLY OF ODENSE

As to Funen Professor Th. Sorgenfrei [53, p. 42] in 1958 in the Danish Society of Civilengineers mentioned characteristic deep valleys in the chalk formations at Odense and in 1960 G.O. Andrup [3, p. 35], director of the Water Supply of Odense, has published a map of the surface of the Danian of Funen which gives the latest information on the structure of Danian at Odense. It may be observed that about 5 km to the North of Ringe the Danian is situated in a height of only 20 m below the sea (see Th. Sorgenfrei [50, p. 491] , [51, p. 170] and Trap [55.9, p. 521]).

Remarks on levelling water crossings between the Danish islands or from them to Jutland or to Sweden

As to levellings of high precision between the Danish islands and from them to Jutland or to Sweden may be stated that the water crossing of

1894 over the Little Belt

and 1898 - 1899 between Zealand and Falster

are both so accurate as they would have been, as if it had been possible during these years to perform spirit levelling such as it has been possible by modern levelling in 1942 and 1943 along bridges.

For the Great Belt and the Sound greater difficulties appeared.

As to the Great Belt must be stated that the old Danish Geodetic Survey published a resulting mean square error across the Belt for trigonometrical levellings 1896 - 1897 of about 5 mm, but a former Director D. la Cour [6, p. 78] of the Danish Meteorological Institute has indicated (1913) that the mean square error ought to be estimated not smaller than ± 16 mm.

The hydrostatic levelling of 1938 under the guidance of Professor Nørlund [34, p. 115] is very accurate and differs only about 0.5 mm from the optical levelling 1896 - 1897 which indicates that the old optical levelling is good but nothing about the mean square error of the old levelling.

If the secular movement across the Great Belt is of the same order of magnitude as the tilting of Zealand it will amount only to 4 mm from 1897 to 1938, so we see that not until new hydrostatic levelling or a new spirit levelling along a bridge may be performed across the Great Belt it will not be possible to find a value for secular movement across the Great Belt with sufficient accuracy.

For the Sound two circumstances are still worse than for the Great Belt. Even though the distance across the Sound is only 4.5 km (whereas the Great Belt is 16 km) the following circumstances must be considered for the Sound:

(1) The refraction during the optical levelling (1896 - 1898) across the Sound was much more unfavourable than across the Great Belt, thanks to the nearly uninterupted sailing of smoking steamers in and out of the Sound with heating effect and irregular variation of the refraction.

(2) The gravity anomalies vary considerably across the Sound from Helsingør to Hälsingborg, namely 23 mGal.

Both Dr. Nørgaard [33] and Professor Rune [41] has estimated the uplift of the geoid caused by this variation in the gravity anomalies and have found different values namely respectively 32 cm and 4.3 cm.

A hydrostatic levelling across the Sound was performed in 1939 by Professor N.E. Nørlund [35].

Also at the Sound we must wait for a new hydrostatic levelling or for spirit levelling along a future bridge for getting detailed information on secular movement across the Sound.

References

[1] Andersen, Einar: Gravity Measurements in Sjælland, Møen, Falster, and Lolland by Means of the Askania-Gravimeter. Geodætisk Instituts Skrifter, 3. Række. Vol. 10. Copenhagen 1947

[2] Andersen, S.A.: Lolland i den sidste istid. Meddelelser fra Dansk Geologisk Forening. Vol. 13, Part 4. Copenhagen 1957. pp. 225 - 235

[3] Andrup, G.O.: Odense-egnens vandforsyning i relation til de geo-hydrologiske forhold i Fyn. Odense 1960

[4] Bertelsen, Ole: See No. 52

[5] Brønsted, Johs.: Stenalderen. Danmarks Oldtid. Bd. 1. Copenhagen 1938

[6] Cour la, D.: Quasinivellement. Publikationer fra Det Danske Meteorologiske Institut. Meddelelser. No. 1. Copenhagen 1913

[7] Den Danske Gradmaaling: Præcisionsnivellement Jylland. Udgivet af Generalmajor V.H.O. Madsen efter General Zachariaes Udkast sammenarbejdet af Oberstløjtnant N.M. Petersen. Publikationer fra Den Danske Gradmaaling, Ny Række. København. No. 3 Copenhagen 1909

[8] Den Danske Gradmaaling: Nivellement over bredere Vanarealer. Udgivet efter afdøde General Zachariaes Manuskript. Med Forord af Generalmajor V.H.O. Madsen. Publikationer fra Den Danske Gradmaaling, Ny Række. København. No. 4 Copenhagen 1909

[9] Den Danske Gradmaaling: Præcisionsnivellement. Fyn, Sjaelland og Falster. Udgivet af Generalmajor V.H.O. Madsen. Bearbejdet af Oberstløjtnant N.M. Petersen. Publikationer fra Den Danske Gradmaaling, Ny Række. København. No. 8 Copenhagen 1911

[10] Doodson, A.T.: Mean Sea Level and Geodesy. Bulletin Géodésique. No. 55, 1960, pp. 69 - 88

[11] Egedal, J.: On the Variations of the Normal Height of the Sea-level round the Danish Coasts. Appendix to Nautical Meteorological Annual. The Danish Meteorological Institute. 1945. Copenhagen 1946

[12] Egedal, J.: On the Variations of the Normal Height of the Sea-level round the Danish Coasts. Appendix to Nautical Meteorological Annual. The Danish Meteorological Institute. København. 1955 Copenhagen 1955

[13,1] Forchhammer, G.: Om den forandrede Vandhøjde ved de danske Kyster. Nordisk Universitets Tidsskr. 2. Aarg. pp. 1 - 23. Copenhagen 1856. Translated to German, see No. 13,2

[13,2] Forchhammer, G.: Über die veränderte Wasserhöhe an den Dänischen Küsten. Von Etatsrath G. Forchhammer. Zeitschrift für allgemeine Erdkunde. Neue Folge. Bd. 1. pp. 473 - 490. Berlin 1856. See No. 13,1

[14] Gaye, Julius: Die Wasserstandsänderungen in der Ostsee und in der Nordsee in den letzten hundert Jahren. "Die Wasserwirtschaft" Sonderheft: Vorträge der gewässerkundlichen Tagung 1951 in Hamburg. Gedruckt in Stuttgart 1952

[15,1] Generalstabens Praecisionsnivellement: Paa Fyn. Udført i Aarene 1901 - 1904. Autograferet Hæfte. Copenhagen

[15,2] Generalstabens Præcisionsnivellement: Paa Lolland-Falster. Udført i Aarene 1910 - 1911. Autograferet Hæfte. Copenhagen

[16] Harboe, E.G.: Jordrystelserne i Danmark ved det skand. jordskælv den 23. Okt. 1904. Meddelelser fra Dansk Geologisk. Forening Vol. 3. Part 16. Copenhagen 1910. pp. 393 - 462

[17] Jensen, Henry: Nivellementer over Store Bælt. Geografisk Tidsskrift. Bd. 48, Copenhagen 1946 - 1947. pp. 140 - 157

[18] Jensen, Henry; Sinding, Erik: On the Difference between the Heights of the Mean Sea-Level at the Self-Recording Tide-Gauges at Korsør and Slipshavn. Geodætisk Institut. Meddelelse. København. No. 20 Copenhagen 1945

[19] Jessen, Axel: Niveauforandringer. Oversigt over Danmarks Geologi. Danmarks Geologiske Undersøgelse 5. Række No. 4 Copenhagen 1928. pp. 149 - 163

[20] Lehmann, I.: Jordskælvet d. 1. Nov. 1930. Naturens Verden, 15. Aarg. Copenhagen 1931. pp. 219 - 235

[21] Madsen, V.H.O.: See Nos. 7, 8, 9

[22] Madsen, Victor: Strandvoldene ved Hornbæk og Alderen af dem. Meddelelser fra Dansk Geologisk Forening. Vol. 10 Part 2, Copenhagen 1942. pp. 83 - 100

[23] Mertz, E.L.: Oversigt over sen- og postglaciale Niveauforandringer i Danmark. Danmarks Geologiske Undersøgelse. 2. Rk. No. 41. Copenhagen 1924

[24] Mertz, E.L.: Nogle betragtninger over niveauforandringer, geofysisk set. Meddelelser fra Dansk Geologisk Forening. Vol. 12. Part 4. Copenhagen 1954. pp. 518 - 520

[25] Mikkelsen, Valdemar M.: Præstø Fjord. The Developement of the Post-Glacial Vegetation and a contribution to the History of the Baltic Sea. (Reprint of: Res Botanicae Danicae. Vol. 13, No. 5) Thesis. Copenhagen 1949

[26] Milthers, K.: Ledeblokke og Landskabsformer i Danmark. Danmarks Geologiske Undersøgelse. 2. Rk. No. 69. Copenhagen 1942

[27] Milthers, K.: See Nos. 31, 32

[28] Milthers, V.: Nordsjællands Geologi. Danmarks Geologiske Undersøgelse. 5. Række No. 3. 1. Edition. Copenhagen 1922

[29] Milthers, V.: Nordsjællands Geologi. Danmarks Geologiske Undersøgelse. 5. Række No. 3. 2. omarbejdede Udgave. Copenhagen 1935

[30] Milthers, V.: The morphology and genesis of the glacial landscape in Denmark. Danmarks Geologiske Undersøgelse. 3. Række No. 28. Copenhagen 1948

[31] Nielsen, A.V.: Dansk Geologisk Forenings ekskursioner til Lindø og NØ - Fyn under ledelse af Keld Milthers. Meddelelser fra Dansk Geologisk Forening. Vol. 14. Part 2, Copenhagen 1959. p. 189

[32] Nielsen, A.V.: Lindø, et af Keld Milthers sidste kvartærgeologiske arbejdsfelter. Meddelelser fra Dansk Geologisk Forening. Vol. 14 Part 4. Copenhagen 1961. pp. 453 - 454

[33] Nørgaard, G.: Tyngdeforhold, Undergrundstruktur og Geoideforløb ved Helsingør og Hälsingborg. Matematisk Tidskrift B. Festskrift til N.E. Nørlund. Copenhagen 1945. pp. 93 - 96

[34] Nørlund, N.E.: Hydrostatisk Nivellement over Store Bælt. Geodætisk Instituts Skrifter, 3. Række. Vol. VI. Copenhagen 1945

[35] Nørlund, N.E.: Hydrostatisk Nivellement over Øresund. Geodætisk Instituts Skrifter, 3. Række. Vol. VIII, Copenhagen 1946

[36] Petersen, N.M.: See Nos. 7, 9

[37] Rosén, P.G.: Precisionsnivellement öfver Öresund. Bihang till K. Svenska Vet. Akad. Handlingar. Bd. 28 Afd. I No. 1. Stockholm 1902

[38] Rosenkrantz, A.: Bemærkninger om det østsjællandske Daniens Stratigrafi og Tektonik. Meddelelser fra Dansk Geologisk Forening. Vol. 9. Part 2. Copenhagen 1937. pp. 199 - 212

[39] Rosenkrantz, A.: Om de strukturelle Forhold i den prækvartære Undergrund i Østsjælland. Meddelelser fra Dansk Geologisk. Forening Bd. 10. Part 2 Copenhagen 1942. pp. 152 - 157

[39a] Rosenkrantz, A.: Kobenhavns klippegrund og trafikproblemer. Salomonsens Leksikon - Tidskrift. 15. Aarg. 1955. Copenhagen 1957 Spalte 653 - 662

[40] Rossiter, J.R.: Report on the Reduction of Sea-Level Observations (1940 - 1958) for R.E.U.N. Permanent Service for Mean Sea Level (Liverpool). Submitted at Helsinki July 1960

[41] Rune, G.A.: Några exempel på tyngdkrafts värdens användning inom Geodesien. Svensk Lantmäteritidskrift, häfte 5. Stockholm 1948

[42] Saxov, Svend: Some Gravity measurements on Fyn. Geodætisk Instituts Skrifter, 3. Række. Tome XXIII. Copenhagen 1955

[43] Saxov, Svend: Gravity on Lolland. Geodætisk Instituts Skrifter, 3. Række. Tome XXVIII. Copenhagen 1958

[44] Sohou, Axel: Det marine Forland. Thesis. Copenhagen 1945

[45] Simonsen, O.: Nivellements-Nul paa Sjælland, Møn og Lolland-Falster med særligt Henblik paa København og Frederiksberg 1845 - 1945. Thesis. Copenhagen 1949

[46] Simonsen, O.: Report on the Secular Movement of Ground Observed in Denmark by Means of Comparing Precise Level Networks of 1885 - 1905 and 1938 - 1948. Bulletin Géodésique, Publication de l'A.I.G. No. 18 Paris 1950. pp. 438 - 441

[47] Sinding, E.: See No. 18

[48] Smed, Per: Studier over den fynske øgruppes glaciale landskabsformer. Meddelelser fra Dansk Geologisk Forening. Vol. 15 Part 1. Copenhagen 1962

[49] Sorgenfrei, Th.: Træk af Alnarp Dalens geol. Opbygning. Meddelelser fra Dansk Geologisk Forening. Vol. 10 Part 5. Copenhagen 1945. pp. 617 - 630

[50] Sorgenfrei, Th.: Nyere undersøgelser over Fyns undergrund. Meddelelser fra Dansk Geologisk Forening. Vol. 11 Part 4. Copenhagen 1949. pp. 490 - 493

[51] Sorgenfrei, Th.: Review of the topography, stratigraphy and structure of the Pre-Pleistocene deposits in southeastern Denmark. Meddelelser fra Dansk Geologisk Forening. Vol. 12 Part 1. Copenhagen 1951. (Danish text)

[52] Sorgenfrei, Th. ; Ole Bertelsen: Geologi og Vandboring. Danmarks Geologiske Undersøgelse. III Række, No. 31. Copenhagen 1954

[52a] Sorgenfrei, Th.: "Denmark" in Lexique Stratigraphique International. Volume I: EUROPE. (Sous la direction du professeur P. Pruvost) Fascicule 2 d: DANEMARK par Th. Sorgenfrei 1955. XX Geologorum Conventus. Mexico 1956

[53] Sorgenfrei, Th.: Geologiske forholds betydning for vandindvindingen. Vandbalance - Foredrag og Diskussioner fra Dansk Ingeniørforenings Kursus i November 1958. Teknisk Forlag. Copenhagen 1959. pp. 22 - 44

[54] Sundelin, Uno.: Hälsingborgstraktens Geologiska Historia, i Bååth: Hälsingborgs Historia. Uppsala 1925. pp. 1 - 71

[55] Trap, J.P.: Trap's Danmark. Femte Udgave

[55,1] Trap, J.P.: Landet og Folket. Vol. I,1 Copenhagen 1958

[55,2] Trap, J.P.: Københavns Amt. Vol. II,3 Copenhagen 1960

[55,3] Trap, J.P.: Frederiksborg Amt. Vol. III Copenhagen 1953

[55,4] Trap, J.P.: Holbæk Amt. Vol. III,2 Copenhagen 1954

[55,5] Trap, J.P.: Sorø Amt. Vol. III,3 Copenhagen 1954

[55,6] Trap, J.P.: Præstø Amt. Vol. IV,1 Copenhagen 1955

[55,7] Trap, J.P.: Maribo Amt. Vol. IV,3 Copenhagen 1955

[55,8] Trap, J.P.: Odense Amt. Vol. V,1 Copenhagen 1956

[55,9] Trap, J.P.: Svendborg Amt. Vol. V,2 Copenhagen 1957

[56] Troels-Smith, J.: Pollenanalytisk Datering af Brabrand-Fundet. Danmarks Geologiske Undersøgelse. 4. Rk. Vol. 2 Part 16. Copenhagen 1937

[57] Troels-Smith, J.: Geologisk Datering af Dyrholm Fundet. Det Kgl. Danske Videnskabernes Selskabs Skrifter. Arkæologisk-kunsthistoriske Skrifter. Vol. 1 No. 1. Copenhagen 1942. pp. 139 - 200

[58] Zachariae, G.: Praecisionsnivellementet over Lillebælt og over Limfjorden. Tillæg til Oversigt over det Kgl. Danske Videnskabernes Selskabs Forhandlinger og dets Medlemmers Arbejder. Udgivet for hvert Aar. København. 1894. pp. 253 - 266. Meddelt i mødet den 16/11 1894

[59] Zachariae, G.: Præcisionsnivellementet over Store Bælt. Oversigt over det Kgl. Danske Videnskabernes Selskabs Forhandlinger og dets Medlemmers Arbejder. Udgivet for hvert Aar. København. 1898 No. 4. pp. 163 - 207. (Presenté à la séance du 11. mars 1898)

[60] Zachariae, G.: Praecisionsnivellementet over Øresund. Oversigt over det Kgl. Danske Videnskabernes Selskabs Forhandlinger og dets Medlemmers Arbejder. Udgivet for hvert Aar. København. 1899 No. 6. pp. 491 - 544. (Presenté à la séance du 19. mai 1899)

[61] Zachariae, G.: See Nos. 7, 8

Discussion

L.-E. Åse (Stockholm):

There is a difference in Denmark compared with Sweden and Finland in the fact that the post-Litorina differences in Denmark seem to indicate a greater difference in tilting than in (at least northern parts of) Sweden and Finland compared with recent movement. Maybe the eustatic changes in post-Litorina times can be the reason of this difference. If one compares a place where the Litorina-transgression (or one of them) is situated at or near the present sea level with a place where there has been a land rise since that time, one can get an error in studying the landtilting from that time compared with the present.

G.W. Lennon (Birkenhead/Cheshire):

I should like to make a short comment upon Dr. Simonsen's remarks in connection with the 19-yearly tide and in particular concerning discrepancies between results by Egedal and Rossiter.

Although we know from theory that the nodal tide is present in the oceans we are not satisfied that we can determine it empirically. If we were able to purify mean sea level observations from all their perturbations then perhaps we could analyse for the nodal tide with some success. Until we are able to do this we shall remain disappointed.

W.A. Magnitsky (Moskau):

What was the thickness of the ice cover in the south-western Denmark?

T. Sorgenfrei (Kopenhagen):

I reply to a question regarding the thickness of the ice in Denmark that it was on a rather small order. Consequently I will not quite second Dr. Simonsen regarding the pure isostatic nature of the recent movements but will emphasize that also other crustal movements were active.

In this connection I point out that the comparison of the gravity anomalies and the uplifts in Denmark made by Dr. Simonsen were very interesting. The anomalies act as structural units but it cannot be predicted whether a unit is positive or negative. Some gravity maxima are positive and others negative and the minima may also be either positive or negative.

The only conclusion is that the anomalies act as units and that one has to be careful regarding prediction of the behaviour of previously defined units within the framework of recent crustal movements.

O. Simonsen (Kopenhagen):

We have heard many interesting papers from different countries on the different methods of work but in spite of the heterogeneity we also find many common features such as inheritance from old structure-lines and interference between tectonic and isostatic movements. So we have the hope that our common work will further the development.

Results of investigations of the latest and recent vertical displacements ot the earth's crust in the Baltic Region and some problems of methodical character

von

V. K. Gudelis (Wilna)

Summary

The study of neotectonic movements (latest and recent) of the Earth's crust is carried out in the East Baltic Area since the middle of the XIX century on the basis of the geologic-geomorphological investigations of the old shorelines of the Baltic Sea, the river and lake terraces etc. The recent movements are investigated as early as 1918 using the data of mareographic observations and precise levellings.

For the territory mentioned the neotectonic maps on the scales 1 : 250.000 and 1 : 5.000.000 as well as sketch-maps of the recent tectonics in Lithuania and Estonia are compiled.

On the ground of the achieved data the following regularities may be stated: 1) the neotectonic movements are connected with the old tectonical structures and follow the tendency of the movements of the latter 2) they are of rhythmical and oscillatory character 3) they take place in forming of new local tectonic structures.

Recent movements are developing on the background of the inheritable tendencies of mobility of the different tectonic structures since the beginning of the Post-glacial time. The latest movements are to be considered as the waves of the Earth's crust of a longer period and the constancy of the movements sign. They are the summary result of the more transitory crustal oscillations of the "recent" type movements. The recent movements are represented by the crustal pulsations of relatively shortduration which take place on the limited areas and are of changeable sign and range of displacement.

It is very probable, that the recent movements are the regulatory ones, which tend to balance the local disturbances in the rate and direction of the general-resulting displacement of the vast crustal blocks. The latest crustal movements occurred on the territory of the East Baltic Area during the Quaternary should be considered as the result of an interference of the strictly-tectonic and glacioisostatic movements with the predominance of the latter ones.

Finally, the author outlines some problems of methodical character concerning the estimation of the sign and rate of the crustal movements, the drawing of the neotectonic maps etc. The program of the neotectonic research-work for the next future in the Baltic Region are discussed too.

Zusammenfassung

Das Studium von neotektonischen Bewegungen (vergangenen und gegenwärtigen) der Erdkruste wurde im Ostbaltischen Gebiet seit der Mitte des 19. Jahrhunderts von den alten Uferlinien der Ostsee, den Fluß- und den Seeterrassen usw. ausgeführt. Die rezenten Bewegungen wurden schon 1918 unter Verwendung der mareographischen Beobachtungen und Präzisionsnivellements untersucht.

Für die erwähnten Gebiete wurden sowohl die neotektonischen Karten mit den Maßstäben 1 : 250 000 und 1 : 5 000 000 als auch die Skizzen der gegenwärtigen tektonischen Bewegungen in Litauen und Estland zusammengestellt.

Auf der Grundlage der gewonnenen Daten können die folgenden Regelmäßigkeiten festgestellt werden:

1) Die neotektonischen Bewegungen sind an die alten tektonischen Strukturen gebunden und folgen der Bewegungstendenz der letzteren.
2) Sie tragen rhytmischen und oszillatorischen Charakter.
3) Sie treten bei der Bildung neuer lokaler Strukturen auf.

Rezente Bewegungen entwickeln sich auf Grund der Bewegungstendenzen in den verschiedenen tektonischen Strukturen seit Beginn der Postglazialzeit. Die jüngsten Bewegungen werden als längerperiodische Wellen der Erdkruste betrachtet bei Konstanthaltung der Bewegungsrichtung. Sie sind das Endresultat der vergänglicheren Krustenoszillationen des "rezenten" Typs der Bewegung. Die rezenten Bewegungen äußern sich in Krustenvariationen von relativ kurzer Zeitdauer, die in begrenzten Gebieten stattfinden und von veränderlicher Richtung und veränderlichem Verschiebungsbereich sind.

Es ist sehr wahrscheinlich, daß die rezenten Bewegungen die Regulatoren sind, welche den lokalen Störungen den Betrag und Richtung der resultierenden Verschiebung der gewaltigen Krustenteile das Gleichgewicht zu halten suchen. Die jüngsten Krustenbewegungen, die auf dem Territorium des ostbaltischen Gebietes während des Quartärs stattfanden, sollten als das Ergebnis einer Überlagerung der rein tektonischen und der glazioisostatischen Bewegungen, mit dem Übergewicht der letzteren, betrachtet werden.

Zum Schluß skizziert der Autor einige Probleme von methodischem Charakter, die die Bestimmung von Richtung und Betrag der Krustenbewegungen, die Zeichnung neotektonischer Karten usw. betreffen. Es wird auch das Programm der neotektonischen Forschungsarbeit in der Baltischen Region für die nächste Zukunft diskutiert.

The history of the neotectonic[x)] (latest movements) investigations of the Baltic Region is about one hundred years old. It was the first investigation of the old Baltic shorelines and later on of lake and river terraces that stimulated the interest in the studies. However, before the first classical works on the history of the Baltic Sea (G. De-Geer, H. Munthe, W. Ramsay, M. Sauramo) had been published the former students were interested not so much in neotectonics as in "eustatics", i.e. fluctuations of the sea level.

F. Schmidt, H. Hausen, W. Ramsay, E. Kraus etc. [8, 9, 17, 19] are the names to mention among the students who laid the foundation of the neotectonic investigations in the Baltic Region. While W. Ramsay first tried to reconstruct the tectonic regime on the basis of the old shoreline analysis in Estonia and Latvia, E. Kraus was the first to use the data on the Quaternary structure and bed-rock topography to interpret the tectonics in the Baltic Region. The absence of detailed geological and geomorphological field of investigations as well as the lack of reliable and all-round data put an obstacle to the progress in studying the Quaternary tectonics of the Baltic Region. This accounts for the fragmentary character, theoretical abstractionism and some errors of methodological character in the former works. They served, however, a start-platform to the subsequent investigations.

---

x) Under neotectonic movements we mean the tectonic deformations of the Earth's crust that took place in Neogene and Quaternary. In their turn they are conventionally subdivided into recent (present) and latest ones.

It was in the last decade that the neotectonic studies of the East Baltic Area became of a larger scale. The main source of information was a detailed study of sea-coast morphology and stratigraphy [5, 9, 10, 19, 24 - 29], of river terraces and river network development [4, 8, 16, 22, 32], of the Quaternary structure [8, 11, 19, 21]. The first summary on the neotectonic activities of the North-Western Russian plain, including the East Baltic Area, was published by J.A. Mescherikov and M.I. Syniagina [15]. In 1956 V. Gudelis made the first attempt to analyse the latest movements, on the basis of the old shorelines studies of the Baltic Sea. This analysis was being specified in the course of time [24, 26]. A new method of interpreting the course of the vertical earth's crust movements, the so-called clinometrical analysis of old shoreline diagrams was suggested [6, 7]. In 1960 the first Regional Conference on Neotectonic Movements in the East Baltic Area was held in Tallin where the Regional Commission (Board) for co-ordinating the investigation works in the Baltic Republics was established.

At present we have at our disposal maps compiled on the latest movements in the East Baltic Area (1 : 2.500.000 and 1 : 5.000.000) as well as schemes of the latest movements in Estonian and Lithuanian S.S.R. and the adjacent regions [3, 12, 28].

The present (recent) tectonics of the Baltic Region were being studied on the basis of precise repeated levellings. Once mareographic (gauge) observations were done to reconstruct the recent tectonic activities in the Baltic Sea coast area [4, 23, 35] . Unfortunately, immediate contacts and co-operation between levelling and gauge efforts have not been ascertained yet. Geophysical methods as well as the method of precise repeated triangulation (for ascertaining horizontal earth's crust displacements) are not applied so far.

The results of the latest and present tectonic studies in the territory under consideration, makes it possible to reveal some regularities and suggest some theoretical views.

The latest movements are characterized as follows:

1) They are related to the old (ancient) tectonic structures and specified by the inheritance of their movement tendencies. The regions of the latest tectonic uplifts, for instance, are generally timed to the positive Paleozoic and Mesozoic structures (the Southern slope of the Baltic Shield, Byelorussian-Lithuanian anticlyse), while regions of the latest subsidences are met, as a rule, in the territory where synclines of that time are common (Polish-Lithuanian syncline).

2) The latest vertical movements (absolute and relative) are distinguished by their oscillatory character and rhythm. In spite of their principal tendency to display tectonic movements of one or another region, the latest movements were of oscillating character, i.e. times of uplift were followed by periods of relative rest or subsidence. The latest movements were deformations of differentiated character both in their sign and velocity but, generally, without tectonic joints (tectoplastic).

Repeated processes of the continental glaciation and deglaciation during the Quaternay (glacial loading and unloading) caused rhythmical character in the display of the latest movements [24, 27].

The Ice Ages correspond to the time of predominating crustal downwarpings while the Interglacials are related to the uplift periods. Thus the Interglacial and the subsequent glacial ages make a uniform rhythm of the latest first order movements. It contains second order rhythms corresponding to Interstages and Stages. The range of individual rhythms as to their amplitude, time and velocity gradient was not equivalent. It was caused by specific conditions of each tectonic rhythm as taken separately. One should bear in mind that at different times, definite structural and dynamic plans of the latest tectonics were being formed. These were depending upon both endogenic and exogenic (glacial control) reasons.

Two phases are observed in each rhythm on the former glaciated territory: a phase of downwarping and a phase of updoming. It is rather convenient to subdivide the phase of updoming into 2 subphases; a) one of total, quick updoming of the earth's crust, rather indifferent in relation to the substratum structures; b) another of differentiated oscillatory movements against the background of inherent development tendencies of the main tectonic structures.

The subphase of total updoming is characterized by spasmodic changes of the gradients in the continuous crustal blocks. These changes, most probably, are caused by deep ruptures in the lithosphere (Gotiglacial and Finiglacial hinge-lines [23]. In this subphase tectonic movements as elastic deformations were formed [18, 25, 27]. The elastic deformations caused by a discharge of stress in the sedimentary cover of the lithosphere, under the influence of glacial unloading were first to come into action. They were distinguished by a high gradient of uplift and moved in gentle-sloped wave - bulge following the receding Ice sheet. Then there followed an interval of a slower uplift of the whole deglaciated area due to plastic deformations taking place in the subcrustal zone (late-Glacial). In the course of the second subphase (post-Glacial) elastic deformations disappear giving way to plastic deformations related to the processes of dedensification (expansion) of the subcrustal matter in situ and to its transflow from the periphery to the centre of glaciation [7, 20].

The continental glaciations (glacioisostatic factor) were a new and powerful process having complicated the course and nature of the tectonic development in the Quaternary of the East Baltic Area. Thus we have every reason to consider the Quaternary type and stage (period) of tectogenesis. The latest movements of the Baltic Region that have taken place in the Quaternary should be considered as complicated, alternating and progressively developing interactions--a total outcome of tectonic proper (oscillatory, epirogenetic) and glacioisostatic movements [14, 24, 25]. The appearance of new local structures and their further development is another characteristic of the latest tectonics of the East Baltic Area. Most of these structures are connected with the folding deformations (flexures, brachy-anticlines, synclines etc.). Disjunctions (faults) seem to play a secondary part and are mostly connected with reviving of old sutures of the tectonic activity. Some neotectonic structures are exceptionally related to the sedimentary cover; others are in close relation to crystalline substratum [8, 11].

The total range of vertical Quaternary movements of the earth's crust in the Baltic Region is estimated 300 to 400 m [27, 28].

The recent movements in the Baltic Region make a natural continuation of the latest movements of the second subphase of the last rhythm (post-Glacial). The recent movements are but a series of cinema stills of the post-Glacial tectonic movements. The recent tectonic represent earth's crust pulsations of comparatively duration developing against the background of inherent tendencies of the movements of separate tectonic structures. These tectonic pulsations, however, take place in a limited territory and differ in its alternating sign and movement rate. It is quite probable that the recent tectonic oscillations are regulatory movements tending to balance local displacements in rate and direction of the total-resulting movement of larger earth's crust blocks [2, 25]. The possibility that the present oscillatory movements consist of pulsations (undulations) of different period and intensity is not excluded. This may lead to interference of movements of different ranges as well.

According to the present levelling data the range of the present vertical oscillations of the earth's crust in the Baltic Region make about 6.5 mm a year [12]. At present this figure should be considered a preliminary one since the network of the repeated highly accurate levelling in the East Baltic Area is open, on the whole, being not connected with the Reference level.

To carry on the further and successful study of the recent tectonic activity of the Baltic Region the following measures should be taken:

1) A repeated levelling of high accuracy should be carried out all over the territory of the East Baltic Area along the main base lines of the levelling network simultaneously.

2) The levelling network of the East Baltic Areas should be connected with that of Finland via the Estonian S.S.R. and Leningrad region [30].

3) Special lines should be arranged for repeated shortspace levelling of high accuracy, for microgravimetrical [13], clinometrical and micromagnetic observations to reveal the nature and character of the present tectonic proper movements.

4) It is necessary to study the degree and character of the influence of geomechanical, geophysical and meteorological factors on the precise establishment of the present tectonic movements (densification and dedensification of grounds, influence of the atmospherical pressure changes, tidal effects in the earth's crust, etc.).

5) A network of gauge observations in close co-operation with levelling works should be arranged.

6) Techniques and methods both of measuring and computing the levelling processes including their standardization should be improved [34].

7) Works should be arranged to study the present horizontal displacements of the earth's crust.

8) A complex and stationary investigation should be carried on to study the supposed elastic tectonic deformations caused on the territories of large storage lakes-reservoirs (water loading) as well as in the areas of concentrated manystoreyed massive constructions [1, 7, 18, 25].

R e f e r e n c e s

[1] Aberberg, G.A.: 1960. Izmenenie vysot punktov nivelirnoj seti goroda Rigi. Neotektoničeskie dviženija v Pribaltike". Tartu

[2] Belousov, V.V.: 1948. Obščaja geotektonika. M.L.

[3] Želnin, G.A.: 1960. Izučenie vertikal'nych dviženij zemnoj kory v Ėstonskoj SSR metodom povtornych nivelirovok. Neotektoničeskie dviženija v Pribaltike. Tartu

[4] Gerasimov, I.P. i Filippov, Ju. V. (red): 1958. Sovremennye vertikal'nye dviženija zemnoj kory na territorii zapadnoj poloviny Evropejskoj časti SSSR. Tr. Cniigaik. vyp. 123, Moskva

[5] Grinberg, Ė.F.: 1957. Pozdnelednikovaja i poslelednikovaja istorija poberež'ja Latvijskoj SSR, Riga

[6] Gudelis, V.K.: 1957. Opredelenija charaktera dviženij zemnoj kory po ėpejrogeničeskim spektram drevneberegovych linij (klinometričeskij analiz). Trudy AN Litovskoj SSR, serija B, I, Vil'njus

[7] Gudelis, V.K.: 1960. O reakcii zemnoj kory na processy gljaciacii i degljaciacii. Mat. soveščanija po neotektoničeskim dviženijam. (On the reaction of the earth's crust to the processes of glaciation and deglaciation), Tartu

[8] Gudelis, V.K.: 1961. Neotektoničeskie dviženija na territorii Pribaltiki v četvertičnom (antropogenovom) periode. Mat. sov. po izyč. četv. perioda. Moskva

[9] Gudelis, V.K.: 1961. Obščie čerty razvitija morskich beregov Vostočnoj Pribaltiki v pozdne- i poslelednikovoe vremja. Tr. In-tuta geologii AN Ėstonskoj SSR. Tallin

[10] Gudelis, V.K.: 1961. Nekotorye voprosy razvitija Baltijskogo morja i neotektoniki golocena. Voprosy golocena. Vil'njus.

[11] Indans, A.P.; Kovalevskij, M.I.; Springis, E.N.: 1960. Neotektoničeskie dviženija na primere Latvijskoj SSR. Mat. sov. po neotektoničeskim dviženijam. Tartu

[12] Lesis, I.P.; Gudelis, V.K.: 1961. Vertikal'nye dviženija zemnoj kory na territorii Jugo-vostočnoj Pribaltiki po dannym točnych nivelirovok. Neotektonika SSSR. Riga

[13] Maazik, V.Ja: 1960. O vozmožnosti primenenija gravimetričeskogo metoda dlja vyjasnenija vertikal'nych dviženij zemnoj kory i ich obuslavlivajuščich pričin. Mat. sov. po neotektoničeskim dviženijam. Tartu

[14] Markov, K.K.: 1960. Gljacioizostatičeskie dviženija zemnoj kory. Mat. sov. po neotektoničeskim dviženijam. Tartu

[15] Meščerjakov, Ju. A.; Sinjagina, M.I.: 1956. Sovremennye dviženija zemnoj kory i metody ich izučenija. Sb. statej dlja XVIII. mežd. geogr. kongressa. M.-L.

[16] Meščerjakov, Ju. A.; Fedorova, R.V.: 1961. O vozraste i proischoždenii terras zapadnoj Dviny. Mat. sov. po izuč. cetv. per. t. P, Moskva

[17] Nikolaev, N.I.; 1949. Novejšaja tektonika SSSR. Moskva

[18] Nikolaev, N.I.: 1960. O prirode sovremennych tektoničeskich dviženij i ich izučenii v rajonach stroitel'stva krupnych gidrotechničeskich sooruženij. Tr. Mosk. geologorazv. In-tuta, t. XXXVIII, Moskva

[19] Orviku, K.K.: 1960. O neotektoničeskich dviženijach v Ėstonskoj SSR na osnove geologičeskich dannych. "Neotektoničeskie dviženija v Pribaltike". Tartu

[20] Subbotin, S.I.: 1957. O pričinach vertikal'nych dviženij zemnoj kory. Moksliniai Pranesimai, t. V, Vilnius

[21] Šul'c,S.S.: 1958. O svjazi rel'efa i novejšej tektoniki so strukturoj severo-zapadnoj i zapadnoj častej Russkoj platformy. Tr. laboratorii aėrometodov. t. VI, Leningrad

[22] Basalykas, A.: 1956. Nemuno upes paleodinamines fazes Alytaus-Kauno atkarpoje ir kai kurie neotektoninio aktyvumo klausimai. Lietuvos TSR MA Darbai, serija B, t. 3, Vilnius

[23] Bikis, J.: 1940. Zemes garozas kustibu ietekme precizas limetnosanas darbos Baltijas juras piekraste. Riga

[24] Gudelis, V.: 1955. Lietuvos TSR Baltijos pajurio geologines raidos velyvajame glaciale ir postglaciale (holocene) pagrindiniai etapai. Vilniaus Valstybinio Universiteto Mokslo Darbai, t. VII, Vilnius

[25] Gudelis, V.: 1955. Lietuvos pajurio neotektoninio aktyvumo klausimu (On the question of the neotectonical activity of the Lithuanian sea-coast). Lietuvos TSR MA darbai, serija B, 3, Vilnius

[26] Gudelis, V.K.: 1956. Lignes anciennes des côtes de la mer Baltique et question des mouvements neoteotoniques sur les territoires de regions Baltiques sovietiques. Resumenos de los trabajos presentados. Mexico

[27] Gudelis, V.: 1958. Kai kurie velyvojo kvartero stratigrafijos ir paleogeografijos klausimai Europoje ir Siaures Amerikoje naujausiais duomenimis. Geografinis metrastis t. 1, Vilnius

[28] Gudelis, V.: 1960. Neoteotonic movements of the territory of the East Baltio Area in the Quaternary period. Colleotanea Acta Geographica Lithuanioa. Vilnius

[29] Gudelis, V.K.: 1961. Latest and recent vertical Earth's crust movements and the morphology of the sea-coast of the East Baltic Area. Association International de Geodesie. Bulletin Geodesique, No. 62, Decembre, Paris

[30] Kaariainen, E.: 1953. On the recent uplift of the Earth's crust in Finland, Fennia, Helsinki

[31] Kraus, E.: 1933. Über tektonische Gegenwartsbewegungen im Ostseegebiet. IV. Hydr. Konferenz der Baltischen Staaten. No. 83, Leningrad

[32] Mescherikov, J.A.: 1961. Recent movements of the Earth's crust in the north-west of the European part of USSR in the light of geological and geomorphological data. Association Inter. de Geodesie. Bulletin Geodesique, No. 62, Paris

[33] Sauramo, M.: 1939. The mode of the land upheaval in Fennoscandia in late-Quaternary time. Fennia, 66, No. 2, Helsinki

[34] Sinyagina, M.I.: 1961. Investigation on recent vertical movements in the Earth's crust in the U.S.S.R. Area. Association Inter. de Geodesie. Bulletin Geodesique, No. 62, Paris

[35] Witting, R.: 1915/1918. Hafsytan, geoidytan och landhojningen utmed Baltiska hafvet och vid Nordsjon. Fennia, 39, Helsingfors

Some results on neotectonic movements of the earth's crust in the Estonian SSR obtained by the method of repeated levellings

von
G. Zhelnin (Tartu)

S u m m a r y

A review is made of the investigations of the vertical movement of the Earth's Crust on the territory of the Estonian S.S.R. based on the method of repeated levellings.

In Estonia levellings were carried out in 1871, 1881, 1897, 1911 and 1933-1940. Because of their insufficient precision and unreliability in marking the above levellings could not be used in our studies with the exception of the 1933-1940 levelling. For this reason the Institute of Physics and Astronomy of the Academy of Sciences of the Estonian S.S.R. began in 1956 to repeat the levelling network along the line of the 1933-1940 levelling.

To this day the repeated levelling network covers the whole territory of the Republic and forms five closed polygons each with a perimeter of 300 or 400 km. The time interval between the levellings varies between 17 and 26 years.

The relative yearly shift between the ends of the lines was calculated for each of the line, bounded by fundamental marks. The networks, consisting of the yearly shifts, was weighted as a free system with respect to the weights.

The absolute shift on the main fundamental mark in Tallin equal to + 2.5 mm per annum served as the starting point.

On the basis of the weighed data a figure was drawn up showing the run of the isobases. In general it confirmed the conclusions which had been made earlier. On the figure the regions where the isobases show deviations have been marked. In a number of cases the geologists have confirmed the possibility of local movements in the regions.

The repeated levelling in the isles Saaremaa and Muhu is being planned.

Another repetition in 1961 of a levelling along a 100 km line has not solved the problem of the reality of short-period pulsations of the Earth's crust. The investigations on this problem will be continued.

Z u s a m m e n f a s s u n g

Es wird von den Untersuchungen der Vertikalbewegungen der Erdkruste auf dem Gebiet der Estnischen SSR, die auf der Methode der Wiederholungsnivellements basieren, berichtet.

In Estland wurden 1871, 1881, 1897, 1911 und 1933-1940 Nivellements durchgeführt. Mit Ausnahme des Nivellements von 1933-1940 konnten die oben genannten Nivellements für die Untersuchungen wegen ihrer ungenügenden Präzision und der Unzuverlässigkeit in der Ausführung nicht verwendet werden. Aus diesem Grunde begann 1956 das Institut für Physik undAstronomie der Akademie der Wissenschaften der Estnischen SSR das Nivellement entlang der Linie des Nivellements von 1933-1940 zu wiederholen.

Heute bedeckt das Netz des Wiederholungsnivellements das gesamte Territorium der

Republik und bildet fünf geschlossene Polygone, jedes mit einem Umfang von 300 - 400 km. Die Zeitspanne zwischen den Nivellements variiert zwischen 17 und 26 Jahren. Die relative jährliche Verschiebung zwischen den Enden der Linien, die durch Fixpunkte festgelegt sind, wurde für jede Linie berechnet. Das aus den jährlichen Verschiebungen bestehende Netz wurde als ein freies System unter Berücksichtigung der Gewichte ausgeglichen.

Die absolute Verschiebung an dem wichtigsten Fixpunkt in Tallin von 2,5 mm /Jahr diente als Ausgangspunkt.

Auf der Grundlage der gewonnenen Daten wurde eine Zeichnung angefertigt, die den Lauf der Isobasen zeigt. Im allgemeinen bestätigt sie die Schlußfolgerungen, die früher gezogen wurden. Auf der Zeichnung sind die Gebiet markiert, wo die Isobasen Abweichungen zeigen. In einer Anzahl von Fällen ist die Möglichkeit von lokalen Bewegungen in den Gebieten durch Geologen bestätigt worden.

Das Wiederholungsnivellement auf den Inseln Saarema und Muhu ist geplant.

Das Problem der Realität von kurzperiodischen Pulsationen der Erdkruste wurde 1961 durch ein Wiederholungsnivellement entlang einer 100 km-Linie nicht gelöst. Die Untersuchungen über dieses Problem werden fortgesetzt.

On the territory of the Estonian S.S.R. levellings were executed in year 1871, 1881, 1897, 1911 and between 1933-and 1940. All these levellings formed, with the exception of the last comparatively short lines along railways. During 1933-1940 the levelling of these lines were partly repeated. Unfortunately by that time greater part of the old levelling marks was lost. Because of their small precision of the older levellings, their unreliable fastening and too small extent in repeating their value for our purposes is limited. The only conclusion that can be drawn from them is that the town-region of Narva is sinking with respect to the town-region of Tapa at the rate of 2.1 mm per year.

However, the error of this value is greater than the result itself.

The more recent levellings in years 1933-1940 offer valuable initial data for making investigations about latest crustal movements on the territory of the Estonian S.S.R.

The following main features of these levellings are of greatest value for our work:

1. The levelling network covered the territory of the Republic nearly uniformly and was tied with the networks of nearby territories.
2. The levelling was performed with high accuracy as is seen from the errors: $\eta = \pm 0.4$ mm, $\lambda = \pm 0.04$ mm.
3. All the knot points were fixed by means of reliable basic-marks.
4. The height-marks (bench-marks) were distributed along the lines in such a number (with distances 1.5 - 2 km) that if one half of the marks were lost the others would completely suffice to carry out the investigations.

On the basis of this estimate of existing levelling networks in the Republic and considering that the first class levelling in the line Narva-Tallin-Parnu undertaken by the Administration of Geodesy and Cartography in 1948 was nearly in its whole extension a repetitional one the Institute of Physics and Astronomy of the Estonian Academy of Sciences commenced in 1956 with the construction of repeated levelling networks. The repeated levellings were performed with the same methods and precision as the 1933-1940 levellings. Every year a line covering 170-200 km was levelled.

The yearly work began and ended at a fundamental mark. All existing height-marks in the lines were included. In the process of work for every line bordered by fundamental marks, table and graph were compiled showing the accumulation of yearly discrepancies between the height differences of neighbouring marks $\frac{\Delta h_2 - \Delta h_1}{t}$ with reference to the first and second levelling.

In compiling these tables and graphs it was supposed that in the period between the first and second levelling the relative rate of vertical displacements of neighbouring marks was constant and that the reliably placed fundamental marks were subject only to tectonic movements.

Further, from the yearly relative uplifts of the endpoints of consecutive lines the rate polygons were constructed and adjusted according to the usual rules.

Till 1960 the repeated levelling net consisted of three closed polygons. In 1961 to these a repeated polygon at the north-east border of the Republic was added and the central polygon was split into two parts. Also a third levelling of the line Lelle-Parnu, conclusions of which caused some doubts, was performed this year.

The more accurate results obtained made it necessary to make some changes in the former schemes of locating the isobases.

At the moment the repeated network consists of five polygons.

The polygons of yearly rates have the following closing errors:

N. of polygon	Perimeter P in km	Closing error in mm/year
I	450	-0.70
II	360	-0.66
III	460	+0.03
IV	310	-0.77
V	345	+0.70

From these data the mean error of yearly land uplift-values per 100 km

$$\mu_v = \pm \sqrt{\frac{1}{n}\left[\frac{t^2}{0.01P}\right]} = \pm\ 0.3 \text{ mm/year}$$

The above net was adjusted with weights as a free net. The weights were calculated from the formula $\frac{1}{P} = \frac{L}{t^2}$ where L means the distance between the knots in km and t the time interval between consecutive levellings in years. The mean value of t is 18 years.

As an initial absolute land uplift value the displacement of the basic marks in Tallin with the value of 2.5 mm per year was accepted. The last value is due to Bikis corrected according to the repeated levellings within the city boundaries of Tallin.

On the following scheme (Fig. 1) the isobases are drawn with the step of 0.2 mm/year.

Their locations were corrected in accordance with the graphs of changes in values of yearly land uplift rates.

On the graph (Fig. 2) the adjusted curve representing the changes in the yearly uplift rates of the Earth's crust along the line was drawn nearer to the upper teeth-points of the broken graph-line. It was settled that under the action of external causes the marks sink as a rule.

Nearly 50 per cent of all left marks carry traces of shocks. A few cases of mark-uplifts always had its explanation (as for example the inclosing of the building or the damaging of the road-tube into which the bench-mark was placed). It can be seen from the present graph that the change in the rate of land uplift does in fact proceed remarkably non-uniformly.

Sudden and comparatively great changes in the land uplift rate occured in a case along a path with a length of 5-10 km, while at neighbouring tracks the change happened to be much smoother. It has an appearance, as if in the regions of sudden changes of the land uplift rate there is a border between blocks lifting with different rates. An analogous apparent discontinuity of lift-rates may be caused, however, from height-errors in the first or second levelling.

The location of these places of sudden changes in certain directions (as may be seen

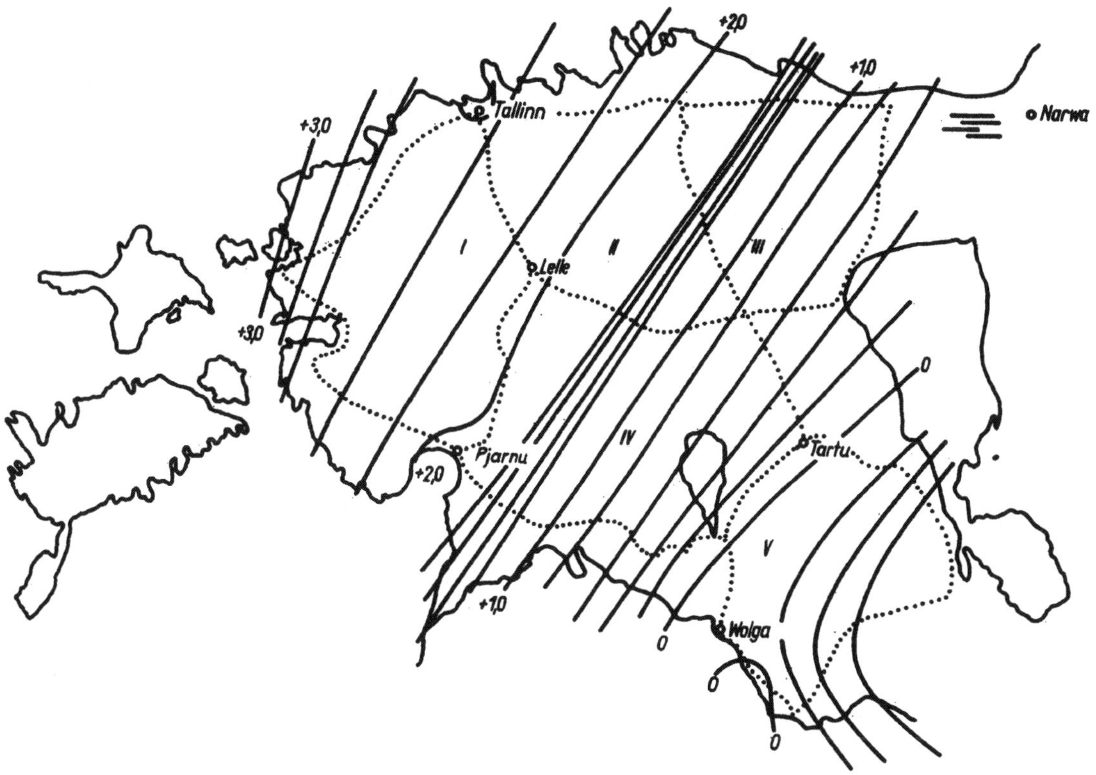

Fig. 1

The location scheme of the isobases on the territory of the Estonian S.S.R.

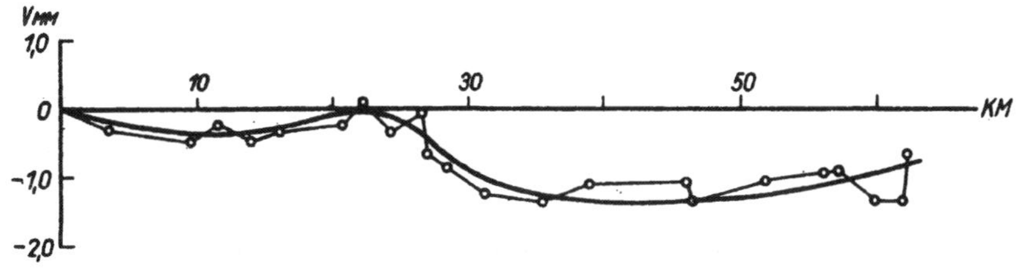

Fig. 2

The graph of variation of yearly land uplift rates of the earth's crust

from the scheme) and, in the first hand, their coincidence with region where zones of tectonic discontinuities are geologically ascertained and the existence of a breakline of post-glacial period is postulated suggest strongly that our former conclusion is the right one.

Inasmuch, as our conclusions concern the processes, occurring at the present time, we have the right of speaking of enduring processes of local movements of the earth's crust on the territory of the republik, of their lasting perhaps beyond our time.

It is seen from the scheme that the block (or blocks) situated to the north-west of the breakline are lifting up with a rate 0.6 - 0.8 mm/year relative to the other block.

The gradient of the change of rate is also different on the blocks. The south-eastern block is inclosing more rapidly than the north-western block. On our scheme also a region of local movements is demarcated at the north-eastern part of the territory of the republic, where the oilstone (kucersite) mines are situated. Here, apparently, the underground mining manifestates itself. In the region of the town Valga the bending through of the earth's crust is observed, the relative rate of sinking amounts to 0.2 mm/year. The earth's crust is sinking also at Tallin. The cause for the last may be seen in the pumping away of the underground water and the evergrowing load upon the ground in connection with great building work.

An analogous situation is observed, however, in a lesser amount, at the town Parnu.

The described scheme cannot be seen upon as final. Its precision does not yet suffice. The error of concluded yearly land uplift rates attains the values 0.2 - 0.3 mm/year. For some lines the third levelling will be necessary.

This in the first hand concerns the lines, repeated levelling of which was finished 12 - 13 years ago.

Research work according to this plan is going on. The range of the levellings will be extended to the islands. From the data that will be obtained by the third levelling it is hoped that changes in rate and character of movements with time can be inferred.

R e f e r e n c e s

1. Zhelnin, G.: Investigation of vertical displacements of the earth's crust on the territory of the Estonian S.S.R. by the method of repeated levellings. Proceedings of the Conference on neotectonic movements. Tartu, 1960
2. Zhelnin, G.: Über die Beständigkeit der Höhenmarken bei Nivellements der Stadt Tallin. Publ. of the Tartu Astron. Obs. XXXIII, No. 3, Tartu, 1958
3. Orviku, K.: Geological data on neotectonic movements of the earth's crust in the Estonian S.S.R. Proceedings of the Conference on neotectonic movements. Tartu, 1960

D i s k u s s i o n

L. A s p l u n d (Stockholm):

We all agree that there are various reasons for vertical movements of the earth's crust to be regarded. In Fennoscandia we have much evidence, that the uplift is caused by the removal of the last ice-cap. It is generally regarded as an effort to reach isostasy, even if the proofs in form of isostatic anomalies actually fail. It is interesting to see that the uplift in this region as far as we know is very regular which seams to exclude heavy movements of tectonic character. This is clearly seen from the well-known map of isobars in Finland. The work which has been carried out in Sweden, and which I found a little too preliminary to be presented here in a formal paper - confirms that the whole area has a very regular uplift. From the paper by Zhelnin we see that the Estonian work fits very nicely in this general picture. It seams therefore astonishing that according to the paper by Gudelis, very close outside this regular region very heavy recent vertical movements (6,5 mm/year) occur.

J. T. K u k k a m ä k i (Helsinki):

As the landuplift values in Finland have been published already in several connections I thought not to write any report on them. However, in order to complete the picture I mention some words about the land uplift in Finland.

The km error of The First Finnish levelling (1892–1910) is ± 1,5 mm and the one of the Second levelling (1934–1955) ± 0,6 mm. The difference of these two levellings has accuracy of ± 6 mm/km and when this is divided with the interved time 44 years one gets an accuracy of ± 0,036 mm/y/km for the land uplift differences in the levelling network of Finland. The absolute values of land uplift are derived from recordings of 12 moneographs at the Finnish coasts the standard error of land uplift in different parts of the network variates ± 0,2 – ± 0,6 mm/y.

It has been interesting to see how much work has been done in order to study the land uplift by the aid of relevellings. Especially we must congratulate our colleagues from the Soviet Union for their numerous investigations.

On the other hand we must pay attention to the fact that the attained results are of very different accuracy and quality.

The areas investigated have d i f f e r e n t g e o l o g i c a l c h a r a c t e r. The a c c u r a c y o f t h e l e v e l l i n g s is different in different areas. Especially one must pay attention to the old levelling as in many cases low accuracy of those makes the results more than questionable.

The reference surface of the land uplift is continuous and common only in very few cases.

It will be of great value if this CRCM will enable to compile a map about the performed determination of the vertical movements of the Earth's crust. To avoid misinterpretation of this map the accuracy of the values should be given in a definite way e.g. in the form of ± mm/y/km.

L. E g y e d (Budapest):

Die Isobathen zeigen in Estland eine Faltungslinie.

J. A. M e s c h e r i k o v (Moskau):

I have no accurate data about the location of epicenters of weak earthquakes in Estonia. I only know that such shocks are registered. The gimballine, as the hinge-lines of Prof. Sauramo, corresponds in Estonia to the margin between large blocks of the earth's crust. Similarly to Denmark (as we heard in an extremely interesting paper by Prof. Simonsen), the position of this hinge-line correlates with the data about the deformation of post-glacial coast lines. Thus we observe a certain inheritance of recent crustal movements.

G. B a r t a (Budapest):

Ich möchte zu den Daten hinzufügen, daß sie vielleicht noch beeinflußt werden durch große Massenbewegungen im Erdinnern.

A. A. I z o t o v (Moskau):

I have an objection to make to the remark of Dr. G. Barta. In the determination of the velocities of the vertical movements of the earth's crust the choice of the reference surface is of no significance. We can use any reference surface for the determination of the velocity of vertical crustal movements. This is especially true when we study the vertical movements in the constant gravitational field of the earth.

Ein Beitrag zur wahrscheinlichen Senkung der deutschen Ostseeküste, abgeleitet aus 100jährigen Wasserstandsbeobachtungen

von

H. Weise (Potsdam)

S u m m a r y

Mean values for a period of fifty years have been calculated from the measurements of the water gauge level made by the station Swinemünde between 1811 and 1943. These mean waters of many years allow to recognize a decrease of the difference in course of time compared with the reference point of the assumed level. If this effect is explained as subsidence of the coast, we get the numerical value of ~0,8 mm per year.

Z u s a m m e n f a s s u n g

Aus den Pegelbeobachtungen der Station Swinemünde von 1811 bis 1943 wurden 50-Jahr-Mittel gebildet. Diese langjährigen Mittelwasser lassen im Laufe der Zeit eine Abnahme der Differenz gegenüber dem Referenzpunkt des Höhensystems erkennen. Deutet man diesen Effekt als Küstensenkung, so erhält man den numerischen Wert von ~ 0,8 mm pro Jahr.

E i n l e i t u n g

Dem Problem der Erdkrustenbewegung wird in vielen Ländern der Erde ein reges Interesse entgegengebracht. Das Studium von Veränderungen der oberen Erdschichten ist nicht nur wichtig in geophysikalischer Hinsicht, sondern auch für grundlegende Messungen der Geodäsie.

Die vorliegende Arbeit soll der Ermittlung der Vertikalkomponente der Erdkrustenbewegung gewidmet sein, die sich entweder aus geodätischen Präzisions-Nivellements oder aus Wasserstandsbeobachtungen der Meeresoberfläche ableiten läßt. Durch wiederholte Nivellements von geologischen Linien lassen sich Veränderungen im Laufe der Jahre mit großer Genauigkeit nachweisen, wobei jedoch ein Bezugspunkt als unveränderlich in seiner Höhenlage vorausgesetzt werden muß. Nehmen weiträumige Gebiete (einschließlich des Referenzpunktes) an einer Schollenbewegung teil, so lassen sich durch nivellitische Meßverfahren keine eindeutigen Nachweise der Veränderungen erbringen.

Die ozeanographische Bearbeitung des genannten Problems geht vom Meeresspiegel aus, der im ideellen Falle als ungestörte Niveaufläche des Schwerepotentials zu betrachten ist. Brauchbare Aussagen ergeben sich jedoch nur aus langjährigen Wasserstandbeobachtungen.

B e o b a c h t u n g s m a t e r i a l

Die vorliegende wissenschaftliche Untersuchung ist methodisch auf die Auswertung von Wasserstandsbeobachtungen begrenzt und räumlich auf eine Station der südlichen Küste der Ostsee, nämlich Swinemünde beschränkt. Swinemünde gehörte bis 1945 zum Pegelnetz des Geodätischen Institutes Potsdam, das aus 9 Stationen an der Ostseeküste bestand.

Für Swinemünde liegt das umfangreichste Beobachtungsmaterial (bis zum Jahre 1811 zurückrechnend) in zuverlässig ausgewerteter Form vor [1] [2] [3] [4] und bildet damit die z.Z. günstigste Voraussetzung für wissenschaftliche Erörterungen. In die rechnerische Auswertung sind folgende Pegelablesungen eingegangen

a) von 1811 bis 1826 beob. Terminwerte für 8^h MEZ
b) von 1827 bis 1870 beob. Terminwerte für 12^h MEZ
c) von 1871 bis 1943 Tagesmittelwerte.

R e c h n e r i s c h e R e d u k t i o n

Von den Komponenten, die den Gleichgewichtszustand des Meeresspiegels stören, läßt sich der Anteil der Gezeiten am sichersten erfassen. Da der Einfluß zudem eine systematische Einwirkung verursacht, kann er nicht vernachlässigt werden. Für die mathematische Behandlung hat man zu unterscheiden zwischen der Mond- und der Sonnentide. Zwar ist die Mondflut mehr als doppelt so groß wie die Sonnenflut [1] , jedoch tritt eine Elimination dieses Effektes bei der Mittelbildung über größere Zeiträume insofern ein, als die in unserer geographischen Breite maßgebliche Hauptmondtide M_2 eine Periode von 12;42h hat. Das bedeutet, daß die M_2-Tide pro Tag etwa 50 Minuten hinter unserer Zeiteinteilung zurückbleibt und im langjährigen Mittel ohne wesentlichen Einfluß auf das Ergebnis ist. Anders verhält es sich mit der Hauptsonnentide S_2 , die eine Periode von genau 12;00h besitzt und folglich den einzelnen Terminwert stets um den gleichen Betrag beeinflußt. Es ist deshalb prinzipiell notwendig, den Effekt der Hauptsonnentide S_2 auf den Wasserstand zu berücksichtigen. Mit den Terminwerten von 1811-1870 ist entsprechend verfahren worden; bei den Tagesmittelwerten konnte eine solche Korrektion aus den vorgenannten Gründen unterbleiben.

Zwar ist das auf diese Weise erhaltene Zahlenmaterial noch nicht homogen, da ein einzelner Terminwert a priori noch keinen repräsentativen Charakter besitzt, doch haben die Untersuchungen von Seibt (1, Seite 34) über einen Zeitraum von 10 Jahren gezeigt, daß zwischen den Jahresmitteln des Wasserstandes abgeleitet aus

a) kontinuierlichen Aufzeichnungen des Pegels und Integration der Kurve und
b) den täglich nur einmal vorgenommenen Ablesungen am Lattenpegel

kein wesentlicher Unterschied besteht. So wurde beispielsweise für den mittleren Wasserstand der Ostsee bei Swinemünde für den Zeitraum von 1870-1879 erhalten nach

a) - 2,412 8m und nach
b) - 2,409 7m

Die auftretende Differenz von ~3 mm kann als unbedeutend bezeichnet werden. Durch diesen Vergleich soll die Wertigkeit der beiden Verfahren zur Bestimmung eines Mittelwasserstandes keineswegs gleichgesetzt werden. Dem Bearbeiter war lediglich daran gelegen, die Grössenordnung der Differenz zwischen den auf diese Weise gewonnenen Werten abzuschätzen und eine Rechtfertigung dafür zu finden, daß in gewissen Zeiträumen nur die zur Verfügung stehenden Terminwerte benutzt wurden. Die Abb. 1 zeigt in graphischer Darstellung die Unterschiede. Man erkennt in diesem Diagramm außerdem deutlich die jährliche Periodizität des mittleren Wasserstandes mit einem Minimum im Frühjahr und einem Maximum im Hochsommer. Auf die Ursachen dieser periodischen Erscheinung soll jedoch nicht näher eingegangen werden.

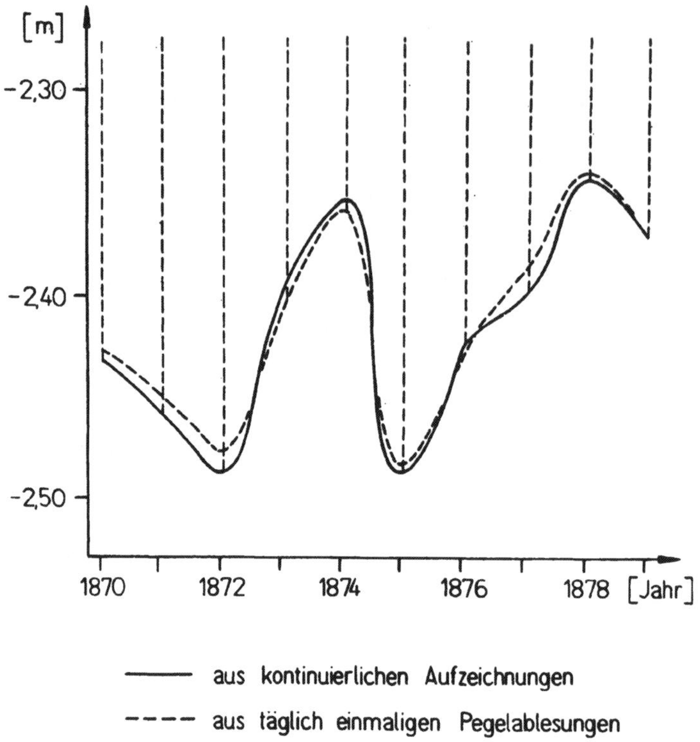

Abb. 1

Mittlerer Wasserstand der Ostsee bei Swinemünde nach Seibt (Monatsmittel aus den Jahren 1870 bis 1879)

Die langjährigen Mittelwasserstände

Es ist unmittelbar einleuchtend, daß e i n Jahresmittel noch keinen, für wissenschaftliche Zwecke geeigneten Bezugspunkt darstellt, sondern man diesen nur aus längeren Beobachtungszeiträumen ableiten kann. Bildet man sogenannte langjährige Mittelwerte M, so darf man annehmen, daß die Wasserstandsschwankungen der einzelnen Jahresmittel nicht mehr ins Gewicht fallen und der Ausdruck M einem Endwert zustrebt, den man dann als Mittelwasser definieren kann. Um die Zufälligkeiten, die den Jahresmitteln noch anhaften, so gering wie möglich zu halten, habe ich das Mittelwasser der Ostsee bei Swinemünde aus jeweils 50 Jahresmitteln abgeleitet. Die so erhaltenen langjährigen Mittelwerte M sind auf die mittlere Epoche bezogen in Abb. 2 graphisch dargestellt. In jedem einzelnen Punkt sind 18 000 Pegelablesungen enthalten. Aus dieser Darstellung ist zu erkennen, daß die ersten 12 Punkte einen alternierenden Charakter haben, von dann ab aber eine steigende Tendenz des Meeresspiegels in bezug auf seinen Referenzpunkt erkennen lassen. Man sieht ferner, daß die Dispersion mit der Zeit geringer wird, was seine Ursache in den fortschreitenden Verbesserungen der Registriereinrichtungen haben dürfte.

Die steigende Tendenz des Mittelswassers ist überraschend, denn bei langjährigen Mittelwerten dürfte man ein Konvergieren der so entstandenen Zahlenreihe erwarten.

Wie ist nun die Veränderung der Differenz zwischen den Werten des Mittelwassers M und dem Referenzpunkt des Höhensystemes zu deuten?

Gutenberg [5] hat bereits vor 20 Jahren die beobachteten Hebungen in Fennoskandia und Nordamerika untersucht und ist zu den Resultaten gekommen, daß diese Erscheinungen

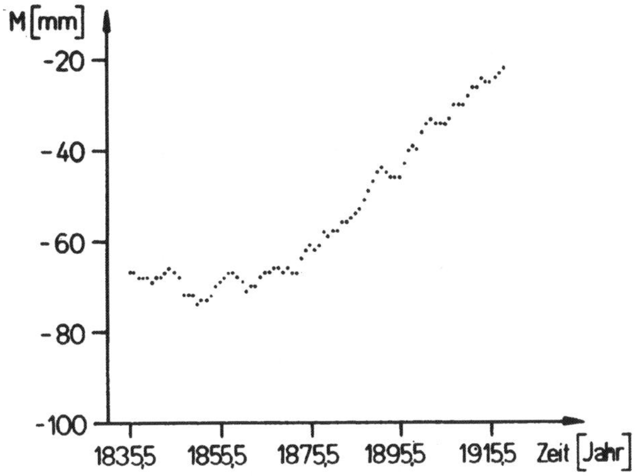

Abb. 2
Mittelwasser der Ostsee bei Swinemünde

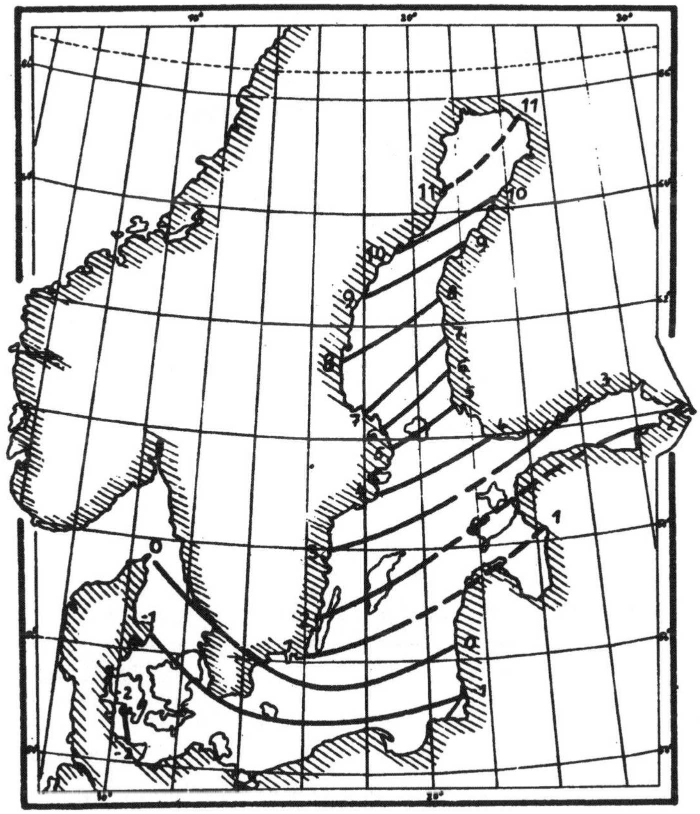

Abb. 3
Mittlere Landhebung nach Witting

in den postglazialen Abschmelzungen des Eises ihre Ursache haben und als Folge der Wiederherstellung des isostatischen Gleichgewichtes gedeutet werden müssen. Als genäherten Zahlenwert für den Anstieg des Meeresspiegels gibt Gutenberg 10 cm pro Jahrhundert an.

Witting [6] veröffentlichte 1943 eine Karte der skandinavischen Länder, die für das Gebiet der Ostsee Isobasen[x)] enthält (siehe Abb. 3). Diese Darstellung läßt erkennen, daß man für das Gebiet der südlichen Ostseeküste eine Senkung von 1 bis 2 mm pro Jahr zu erwarten hat.

Aus dem Beobachtungsmaterial für Swinemünde wurde im Geodätischen Institut Potsdam unter Weglassung der ersten zwölf 50-Jahr-Mittel ein mittlerer Anstieg η nach der Methode der kleinsten Quadrate berechnet.

Als numerischer Wert ergab sich

$$\eta = 0.79 \text{ mm pro Jahr}$$
$$m_\eta = \pm 0.02$$

Dieser Zahlenwert η deckt sich in der Größenordnung mit den Ergebnissen von Gutenberg, jedoch hat dieser den Anstieg ausschließlich als Erhöhung des Meeresspiegels gedeutet, während ich mehr zu den Ansichten Wittings neige und dies als Landsenkung deuten möchte. Möglich wäre auch, daß sich beide Komponenten Landsenkung und Meeresspiegelhebung überlagern und gemeinsam diesen Effekt hervorgerufen haben.

Durch weitere Untersuchungen an anderen Orten der Ostseeküste und Erweiterung des Materials bis in die neueste Zeit, sollen die Aussagen über die vertikale Erdkrustenbewegung vervollständigt werden.

Bei der weiteren Bearbeitung des Themas "Erdkrustenbewegung" ist eine enge Verbindung und eine gute Koordinierung mit den Anliegerstaaten der Ostsee anzustreben.

Durch Vergleich mit geodätischen Meßmethoden könnten die Aussagen der Wasserstandbeobachtungen gefestigt werden.

L i t e r a t u r

[1] Seibt, W.: Das Mittelwasser der Ostsee bei Swinemünde. Veröff. kgl. Geod. Inst. Berlin 1881

[2] Seibt, W.: Das Mittelwasser der Ostsee bei Swinemünde. Veröff. kgl. Geod. Inst. Berlin 1890

[3] Westphal, A.: Das Mittelwasser der Ostsee bei Travemünde, Marienleuchte, Wismar, Warnemünde, Arkona und Swinemünde in den Jahren 1882/1892. Veröff. kgl. Geod. Inst. Berlin 1900

[4] Kühnen, F.: Das Mittelwasser der Ostsee bei Travemünde, Marienleuchte, Wismar, Warnemünde, Arkona, Swinemünde, Pillau, Memel und das Mittelwasser der Nordsee bei Bremerhaven in den Jahren 1898-1910. Veröff. kgl. Geod. Inst. Berlin 1916

[5] Gutenberg, B.: Changes in sea level, postglacial uplift, and the mobility of the Earth's interior. Bull. geol. Soc. Amer., 52 S. 721 - 772, New York, 1941

[6] Witting, R.: Landhöjningen utmed Baltiska Havet under aren 1898 - 1927. Fennia 68, Nr. 1, 1943

---

x) Das sind Linien gleicher Landhebungen

D i s k u s s i o n

G. W. L e n n o n (Birkenhead/Cheshire):

I would like to congratulate Dipl.-Ing. Weise on the availability of a tide gauge record with such a long and reliable history.

I am not too clear, however, as to the method which he uses to eliminate tidal oscillations from the record; am I to understand there the linar and solar tides are removed separately?

H. W ei s e (Potsdam):

Ermittelt man die Differenzen zwischen den beobachteten Pegelwerten für einen bestimmenten Zeitpunkt (Tagesstunde) und den Tagesmitteln, abgeleitet aus kontinuierlichen Aufzeichnungen eines Mareographen für einen längeren Zeitraum, z.B. ein Jahrzehnt, so können aus den so erhaltenen Werten Korrektionen erhalten werden. Setzt man voraus, daß durch das benutzte größere Zeitintervall die Zufälligkeit der bestimmten Differenzen beseitigt sind, so können die korrigierten Terminwerte den aus kontinuierlichen Aufzeichnungen erhaltenen Tagesmitteln gleichgesetzt werden.

J. B ö h m (Prag):

Der Wasserstand eines Meeresspiegels hängt auch von den meteorologischen Verhältnissen ab, besonders von der Richtung und Stärke des Windes. Die gefundenen periodischen Schwankungen oder systematischen Änderungen können teilweise auch durch die langfristigen klimatischen Schwankungen verursacht werden. Wurde dieser Faktor bei den Untersuchungen berücksichtigt?

H. W e i s e (Potsdam):

Windrichtung, Luftdruckverteilung, Temperaturschichtung des Wassers und dgl. mehr haben auf den einzelnen Beobachtungswert großen Einfluß. Für meine Untersuchungen habe ich aber solche Einzelwerte nicht benutzt, sondern sogenannte "langjährige" Mittelwerte verwendet. Der in meiner Abbildung 2 erkennbare Trend der 50-Jahr-Mittel wurde aus einer sehr großen Anzahl von Beobachtungswerten abgeleitet. Wenn man bedenkt, daß jeder der 84 Einzelpunkte im Diagramm (Abb. 2) mindestens 18 000 Pegelablesungen in sich vereinigt, so darf man wohl annehmen, daß alle Zufälligkeiten, zu denen ich auch die klimatischen Einflüsse zählen möchte - kompensiert sind.

L. - E. Å s e (Stockholm):

Es besteht eine gute Übereinstimmung zwischen den Beobachtungen von Weise und denen von Egedal und Fairbridge, wobei die eustatische Hebung während der Zeit seit ungefähr 1880 sich bemerkbar macht.

On the methods of determining the present-day tectonic movements of the seacoast

von

A. S. I o n i n , P. A. K a p l i n und

V. S. M e d v e d e v (Moskau)

S u m m a r y

The present-day tectonic movements of seacoasts are defined by the authors as the vertical movements occurring at the period of the formation of the present-day coastal relief, i.e. the relief of the underwater coastal slope up to the depth of active action of the waves, of present-day coastal accumulative forms and the cliff. This period covers a few thousand years and in general coincides with the historical time. To recognize the character of present-day vertical movements of coasts it is necessary at first to find the processes acting during the period of the formation of coastal relief without interruption and lasting at the present time. These processes are abrasion and accumulation ones.

There are the accumulative beach ridges over sea-level which are coastal forms sensitively reacting on the relative oscillations of sea level and the abrasion terraces are the same ones on the underwater coastal slope.

The repeated instrumental levellings of ridges and investigations on the underwater abrasion terraces by echo-sounding and by diving are the reliable methods of exploring vertical movements of sea coasts. Also careful studying of composition and stratigraphy of nearshore sediments obtained by vibro-piston core sampler are the very trustworthy data about relative changes of sea-level.

The methods of determining the present-day vertical movements of the coasts on the base of geomorphological and lithological data have been used by the authors during the explorations of seacoasts of the Chukotka Sea, the Bering Sea, the Sea of Okhotsk, the Japan Sea, the Black Sea, the White Sea, the Barents Sea and the Baltic Sea.

The material collected in this way made it possible to obtain comparable results, elucidate the general character of present-day relative differentiated movements of the seacoasts of the U.S.S.R. and draw a chart of these movements. As to the direction of the present-day vertical movements, the sea coasts of the U.S.S.R. are divided into a number of regions. Some of them are subsiding, others - rising and finally there are relative stable regions without clear expressed signs of any vertical movements.

On the whole, the rising coasts are prevailing over the territory of the U.S.S.R. It depends on the slowing down of the eustatic rising of the level of the World Ocean and the prevailing of the tectonic factor in the present-day dynamic of seacoasts.

Z u s a m m e n f a s s u n g

Als heutige tektonische Bewegungen der Meeresküsten bezeichnet Vf. die Vertikalbewegungen, die in der Bildungsperiode des heutigen Küstenreliefs auftreten, d.h. des Reliefs des submarinen Küstenhangs bis herauf zur Tiefe der aktiven Welleneinwirkung, der heutigen Küstenakkumulationsbildungen und das Kliff. Diese Periode dauert einige wenige tausend Jahre und fällt im allgemeinen mit der historischen Zeit zusammen. Um den Charakter der rezenten Vertikalbewegungen der Küsten zu erkennen, ist es zuerst notwendig, die Prozesse herauszufinden, die während der Bildungsperiode des Küstenreliefs ohne Unterbrechung wirken und bis heute andauern. Diese Prozesse sind Abrasion und Akkumulationen.

Es gibt die akkumulativen Strandwälle über dem Meeresniveau, das sind Küstenformen, die empfindlich auf die relativen Oszillationen des Meeresniveaus reagieren und die Abrasionsterrassen sind dasselbe auf dem submarinen Küstenhang.

Die wiederholten instrumentellen Nivellements von Rücken sowie Untersuchungen auf den submarinen Abrasionsterrassen durch Echolotung und Tauchen sind die zuverlässigen Methoden der Erkundung vertikaler Bewegungen der Meeresküsten. Auch eine sorgfältige Untersuchung der Zusammensetzung und Stratigraphie der küstennahen Sedimente, die durch "Vibro-piston"-Kerngeräte erhalten wurden, liefern die vertrauenswürdigsten Daten über relative Änderungen des Meeresniveaus.

Die Methoden zur Bestimmung der rezenten Vertikalbewegungen der Küsten auf der Basis geomorphologischer und lithologischer Daten sind von den Autoren während der Erkundung folgender Meeresküsten angewandt: Chukotka-See, Bering-See, Ochotskisches Meer, Japanisches Meer, Schwarzes Meer, Weißes Meer, Bartens-See, Ostsee.

Das auf diese Weise gesammelte Material ermöglichte es, vergleichbare Resultate zu erhalten, den allgemeinen Charakter der rezenten, relativ differenzierten Bewegungen der Meeresküsten der UdSSR zu erhellen und eine Karte dieser Bewegungen zu zeichnen. Entsprechend der Richtung der rezenten Vertikalbewegungen sind die Meeresküsten der UdSSR in eine Anzahl Gebiete geteilt worden. Einige von ihnen senken sich, andere heben sich und schließlich gibt es relativ stabile Gebiete ohne klar ausgeprägte Vorzeichen irgendeiner Vertikalbewegung.

Im ganzen gesehen herrschen die aufsteigenden Küsten auf dem Gebiet der UdSSR vor. Das hängt mit der allmählichen Abnahme der eustatischen Hebung des Weltozeanniveaus und mit dem Vorherrschen des tektonischen Faktors bei der rezenten Dynamik der Meeresküste zusammen.

The seacoasts are situated in different geological structures, undergoing the tectonic movements of different kinds. These movements influence upon coastal processes and are reflected in the coastal relief. Paleogeographical construction including the reconstructions of the ancient coastlines, the correlation of the quaternary marine deposits as well as some problems of modern dynamics of the sea coasts can only be solved correctly if the direction and speed of the coastal tectonic movements in the past and at the present time is found out.

As a rule the present-day tectonic movements of geological structures are inherited from earlier stages of geological history. For this reason the signs of the present-day vertical movements as well as earlier ones there are on the sea coasts. To date there was no clear demarcation between modern and present-day tectonic movements. As a result quite opposite views and opinions on the speed and directions of the tectonic movements for the same regions of coasts were appeared (Budanov and others, 1957, 1960, 1961).

The present-day tectonic movements of sea coasts are defined by the authors as the vertical movements occurring at the period of the formation of the present-day coastal relief, i.e. the relief of the underwater coastal slope up to the depth of active action of the waves, of the present-day coastal accumulative forms and the cliff. This period covers a few thousand years and in general coincides with the historical time.

Repeated levellings is considered as the most reliable and precise method of exploring the tectonic movements of the land. But this method necessitates specially organised work requiring much effort and long intervals between the periods of observation, which is often impossible to do on account of the remoteness and inaccessibility of the regions being explored. That is why the Soviet explorers most widely apply geomorphological and geological methods of which the main one consists in the analysis of the structure of the marine accumulative forms.

To recognize the character ofpresent-day tectonic movements of sea coast it is necessary first of all to find the processes are fixed in coastal relief acting during the period of the formation of coastal relief without interruption and lasting at the present time (Zenkovitch, 1948).

These permanently acting processes are abrasion and accumulation. The accumulative beach ridges over sea-level and abrasion terraces on the underwater coastal slope are the coastal forms sensitively reacting on the relative oscillations of sea level.

With a constant sea level and the sufficient amount of elastic material the absolute heights of the beach ridges complicating the surface of the accumulative forms are approximately the same. If the sea-level was changing while the accumulative forms appeared (on the condition of the homogeneousness of the elastic material and the constancy of the wave regime) the absolute heights of the beach ridges were changing accordingly, from new to older ones (Johnson, 1919, Zenkovitch, 1948). In consequence the slow (1 mm per year) World Ocean's eustatic rising of the level taking place (Gootenberg, 1941) it is possible to say about absolute tectonic rising of the coast the beach ridge's profiles of which show the relative immersion of sea level. A negligible subsidence or stability of individual regions of the coast should be estimated as relative displacements, for they are the algebraic sum of an eustatic rise of the World Ocean level at the lateglacial time and tectonic movements of the earth crust.

The analysis of the profiles of the marine accumulative forms is the most reliable qualitative method of determining the present-day relative movements of the sea coasts. It must be borne in mind not only the tectonic and eustatic factors are responsible for the changings of the heights of the beach ridges but the wave's regime, quantity of debris material and others also are. More often the heights changings of beach ridges of different ages are happened (independently of relative sea level's oscillations) in wave's defenced water areas (bays, gulfs, straits) (Boldyrev, 1960). On this account the levellings of accumulative forms must be done on the open sea coasts.

The studying of the profiles of the underwater slope is the essential supplement to the method of instrumental levellihgs of the coastal accumulative forms.

Abrasion terraces with small gradients well expressed in the bottom relief are the characteristic feature of the morphology of the underwater coastal slope of the rising coasts. Under the same geological structure and wave's regime the width of the underwater abrasion terraces is the less the coastal rising the more intensive. The explanation of this is: when the coast rises, new and new regions of the bottom, which become open to the action of the waves are subject to abrasion. But it must be taken into account that the width of the underwater abrasion terraces depends on resistance of the coastal rocks. Under the intensive emerging of the coast the nearshore parts of the underwater terraces are emerged and the raised benches over sea level are formed. Owing to an extension of the underwater abrasion terraces and raised benches, abrasion on the rising coasts is usually slowed down. "Died cliffs" no longer touched by the waves are often to be found. Wave-out niches raised to a considerable height may be observed on their surface.

On the relative stable coasts the underwater abrasion terraces possess the smallest surface slopes and their most intense working out is taking place until the very surface reaches the utmost width, so that the waves rolling over shallow water are broken, lose their energy and no longer reach the coastal cliffs. The latter are "died" and wide

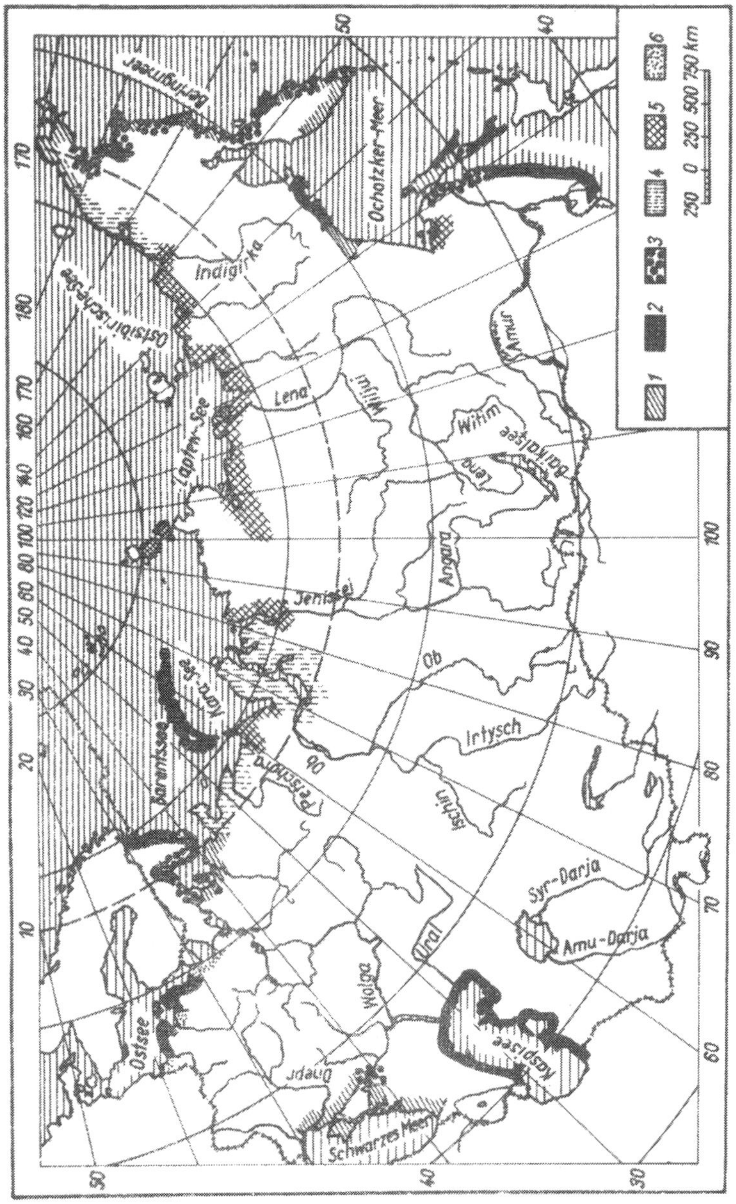

Fig. 1

Map of the recent vertical movements of the sea-coasts of the Soviet Union

Areas on the basis of special investigations of the shore.

1 - areas of recent relative subsidences
2 - areas of recent upheavals
3 - relatively stable coastal areas.

Areas on the basis of data given in literature

4 - areas of recent relative subsidences
5 - areas of recent upheavals
6 - relatively stable coastal areas

beaches or accumulative terraces are formed in their stead. They are made up of material thrown up by the waves from the sea bottom during the further destruction and levelling up of the surface of the underwater abrasion terrace.

There are no abrasion terraces on the underwater slope of the subsiding coasts or they are sank to a considerable depth and buried under cover of the latest sediments. In the near-shore part the bottom is usually a sloping concave abrasion surface with no debris. The "immersed boulder benches" are also characteristic to such coasts (Zenkovitch, 1949). The diving facilitates the finding of the immersed relief of subaerial origin.

It would be taken into account the underwater slope are reconstructed faster than the accumulative forms are originated. Therefore the accumulative forms are the very sensitive to the present tectonic movements of the coasts.

The lithological method, based on the analysis of transgressive and regressive layers of nearshore sediments are used in addition to the above-mentioned ones. For this aim since the 1955 year the vibro-piston core sampler being applied in the Institute of Oceanology, Academy of Sciences of the USSR. It permits to obtain very quickly the five metres columns of nearshore rough sediments, E.N. Nevessky (1961 a, b) used the method of comparative lithological and paleodynamical analysis of nearshore sediments and discovered the ancient accumulative forms buried under the cover of the latest sediments on the bottom of the Black Sea. According to data of fauna analysis the relict accumulative forms of different age mark the ancient coastal lines. The attempts to determine the absolute age of the relict accumulative forms had been made. The knowledge of the absolute age of the accumulative forms allows to determine the speed of the present vertical movements of the coasts.

Naturally the mentioned methods have a lacks. Therefore they would be used successfully if the geographical conditions, geological structure and the peculiarities of separate parts coastal, zone's development would be taken into consideration.

The methods of determining the present-day vertical movements of the coasts on the base of geomorphological and lithological data have been used by the authors during the explorations of seacoasts of the Chukchi Sea, the Bering Sea, the Sea of Okhotsk, the Japan Sea, the Black Sea, the White Sea, the Barents Sea and the Baltic Sea.

The material collected in this way made it possible to obtain comparable results, elucidate the general character of present-day relative differential movements of sea-coasts of the U.S.S.R. and draw a chart of these movements (Fig. 1). As to the direction of the present-day vertical movements the sea coasts of the U.S.S.R. are divided into a number of regions. Some of them are subsiding, other - rising and finally there are relative stable regions without clear expressed signs of any vertical movements.

On the whole, the rising coasts are prevailing over the territory of the U.S.S.R. It depends on the slowing down of the eustatic rising rate of the level of the World Ocean and the prevailing of the tectonic rising within the sea coasts of the U.S.S.R.

R e f e r e n c e s

Boldyrev, V.L.: 1960 (Boldyrev, V.L., Analiz profilja beregovych valov. Izv. Vses. Geograf. ob-va, t. 92)

Budanov, V.I.; Vladimirov, A.T.; Ionin, A.S.; Kaplin, L.A.; Medvedev, V.S.: 1957 (Sovremennye vertikal'nye dviženija beregov dal'nevostočnych morej. Dokl. AN SSSR, t. 116, No. 6)

Budanov, V.I.; Vladimirov, A.T.; Ionin, A.S.; Kaplin, P.A.; Medvedev, V.S.: 1961 (O priznakach sovremennych vertikal'nych dviženij beregov dal'nevostočnych morej. Sb. "Materialy Soveščanija po izučeniju četvertičnogo perioda", t. I, Izd. AN SSSR)

Budanov, V.I.; Ionin, A.S.; Kaplin, P.A.; Medvedev, V.S.: 1960 (Sovremennye vertikal'nye dviženija beregov morej Sovetskogo Sojuza. V kn. "Morskaja geologija". Meždunarodnyj geol. kongress, XXI sessija. Doklady sov. geologov, problema 10. M. Izd. AN SSSR)

Gutenberg, G.: Changes in sea level, postglacial uplift and mobility earth's interior. Bull. Soc. Amer., Vol. 52, No. 5, 1941

Johnson, D.W.: Shore processes and shoreline development. New York, 1919

Nevesskij, E.N.: 1961 (Nekotorye dannye o pozdnelednikovoj évoljucii akvatorii Kapkinitskogo zaliva i nakoplenii v nem donnych osadkov. Trudi In-ta okeanologii AN SSSR, t. 48)

Nevesskij, E.N.: 1961 (O pitmičnosti morskich transgressij. Žurn. "Okeanologija", t. I, vyp. I)

Zenkevič, V.P.: 1948 (Metody opredelenija vertikal'nych dviženij po morfologii beregov. V kn. "Trudy Soveščanija po metodam izučenija i deformacij zemnoj kory". M. Geodezizdat)

Zenkevič, V.P.: 1949 (Glybovyj benč kak pokazatel' pogruženija berega. Trudy IN-ta okeanologii AN SSSR, t. 4)

Zur Methodik der Berechnung des statischen Luftdruckeffektes auf den Wasserstand abgeschlossener Meeresbecken

von
O. M i e h l k e (Warnemünde)

S u m m a r y

The elimination of disturbances of the sea-level is a concern of those persons, too, who are occupied with recent vertical movements of the earth's crust and who start methodically from observations of the water level. In contrast to the studies hitherto made, based upon statistic methods of investigation, and concerning the influence of the atmospheric pressure on the sea-level, the present contributions find a solution of the problem, how to evaluate the static effect of air pressure in secluded basins of the sea. Such a solution postulates the fundamental laws of hydrostatics and easily explains the great variations of the results found with statistical investigations.

Z u s a m m e n f a s s u n g

Die Elimination von Störungen des Meeresspiegels ist ein Anliegen auch für diejenigen Bearbeiter der rezenten vertikalen Erdkrustenbewegungen, die methodisch von Wasserstandsbeobachtungen ausgehen. Der vorliegende Beitrag gibt - im Gegensatz zu den bisherigen, auf statistischen Untersuchungsmethoden basierten Arbeiten über den Einfluß des Luftdruckes auf den Meeresspiegel - eine solche Lösung des Problems der Bestimmung des statischen Luftdruckeffektes für abgeschlossene Meeresbecken, die die Grundgesetze der Hydrostatik postuliert und zwanglos die bei den statistischen Untersuchungen gefundenen großen Streuungen der Resultate erklärt.

Eine derjenigen Methoden, mit denen man versucht, rezente vertikale Erkrustenbewegungen zu studieren, basiert auf der Auswertung von Wasserstandsbeobachtungen. Dabei wird, zunächst idealisiert, vorausgesetzt, daß der Meeresspiegel ein Bezugsniveau bildet, auf das man als Fläche von Festpunkten die Beobachtungen beziehen dürfe. In der Praxis ist nun diese Bezugsfläche in Form des beobachteten Meeresspiegels durch eine Reihe von Faktoren gestört, deformiert, so daß die Voraussetzung in Strenge nicht erfüllt ist. Um dennoch voranzukommen, untersucht man einmal die Störungen dieser Fläche und versucht sie zu eliminieren und zum anderen bearbeitet man langjährige Beobachtungsreihen, um mit Hilfe des Gesetzes der großen Zahl und des Zufalls zu einer "mittleren" Fläche zu gelangen.

Die folgenden Ausführungen sollen einen Beitrag zu den erstgenannten Untersuchungen liefern und sich mit dem seit rund zweihundert Jahren vermuteten, vom Luftdruckfeld verursachten statischen Deformationseffekt der Wasseroberfläche befassen.

In der Vergangenheit haben die uns bekannt gewordenen Autoren beinahe ausnahmelos versucht, mit Hilfe statistischer Auswertemethoden Korrelationen zwischen den Luftdruckänderungen und den damit zu verknüpfenden Wasserstandsänderungen zu finden, wobei man nicht selten Trugschlüssen erlag: Die befriedigende Lösung stand bisher aus, was auch namhafte Autoren, z.B. Doodsen u. Warburg [1] , einräumen. Von dieser befriedigenden Lösung muß man doch mindestens verlangen, daß sie nicht den elementaren Grundgesetzen der Hydrostatik widerspricht. Unter diesem Gesichtspunkt sei folgende Lösung für den Fall eines beliebig geformten, jedoch abgeschlossenen Meeresbeckens vorgetragen: Wir betrachten zunächst zwei Beispiele:

B e i s p i e l A : Vorgegeben sei als Meeresbecken ein langer, schmaler und abgeschlossener Kanal, mit einem durch äquidistante Isobaren angedeuteten Luftdruckfeld an der Wasseroberfläche, wie in Abb. 1 skiziert.

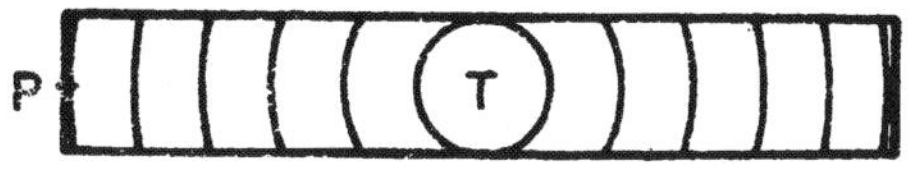

Abb. 1

Grundriß eines langen Kanals mit vorgegebener Luftdruckverteilung an der Oberfläche

Setzt man den Gleichgewichtsfall voraus - was in den folgenden Betrachtungen immer der Fall sein soll - so bilden die Isobaren die Isohypsen der Wasseroberfläche, wobei einer Luftdruckdifferenz vonlMillibar eine Höhendifferenz von etwa 1 Zentimeter hinreichend genau entspricht. Es ergibt sich das Bild 2, das den Kanal im Aufriß zeigt.

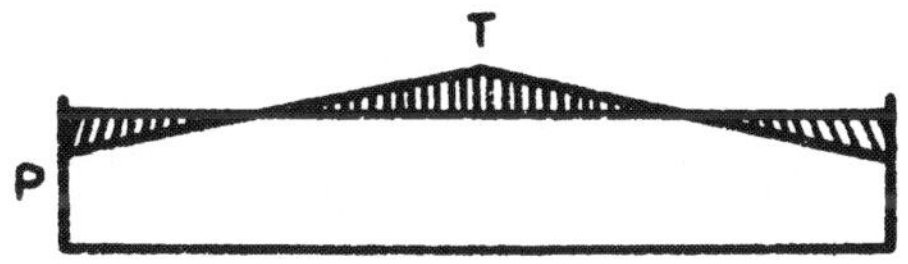

Abb. 2
Aufriß zu Abb. 1

Wünscht man nun eine Wasserstandsbeobachtung am Kanalende im Punkte P so zu korrigieren, daß der Luftdruckeffekt eliminiert wird, so ist leicht einzusehen, daß die halbe Luftdruckdifferenz zwischen dem niedrigsten Druck über der Kanalmitte und dem höchsten Druck im Punkte P, in Millibar gemessen, den numerischen Wert der Korrektur in Zentimetern ergibt. Für einen Punkt, der nicht am Kanalende liegt, gilt der Proportionalitätsfaktor 0,5 nicht mehr, ist jedoch in diesem einfachen Beispiel noch leicht zu ermitteln.

B e i s p i e l B : Die Verhältnisse werden sofort verwickelter, wenn man zu einem Meeresbecken mit dreieckigem Grundriß übergeht, den Abbildung 3 mit einem vorgegebenen Luftdruckfeld darstellen möge.

Hier ist der Proportionalitätsfaktor o,5 auch für den Punkt P nicht mehr angebracht. Die Verhältnisse für die anderen Punkte sind ebenfalls komplizierter geworden, weil der Faktor offensichtlich

1. von der Konfiguration des abgeschlossenen Meeresbeckens abhängt, die in der Praxis sehr verschiedenartig sein kann, wie zum Beispiel die Ostsee, die in diesem Zusammenhang praktisch als abgeschlossen betrachtet wird;

2. abhängig ist von der Lage des Beobachtungsortes an oder auf dem Meere.

Diese Abhängigkeiten muß eine physikalische Lösung des Porblems mindestens berücksichtigen. Sie ist über die sehr durchsichtige Kontinuitätsbedingung zu erzwingen, wenn man davon ausgeht, daß die einmal vorhandene Wassermasse infolge der luftdruckbedingten Deformation der Oberfläche nicht verändert wird, also konstant sein muß. Daraus resul-

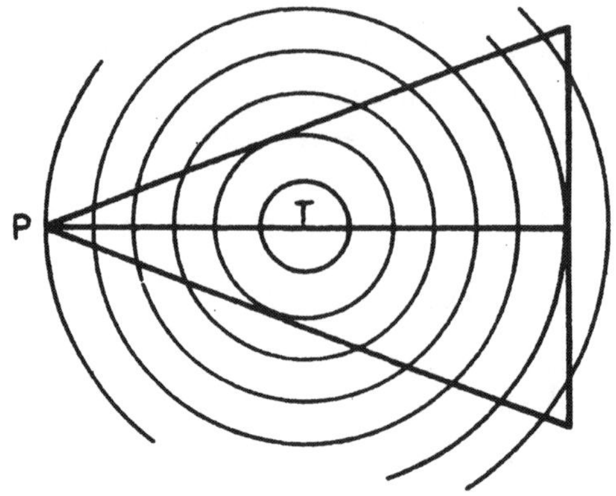

Abb. 3

Grundriß eines abgeschlossenen dreieckigen Meeresgebietes mit vorgegebener Luftdruckverteilung

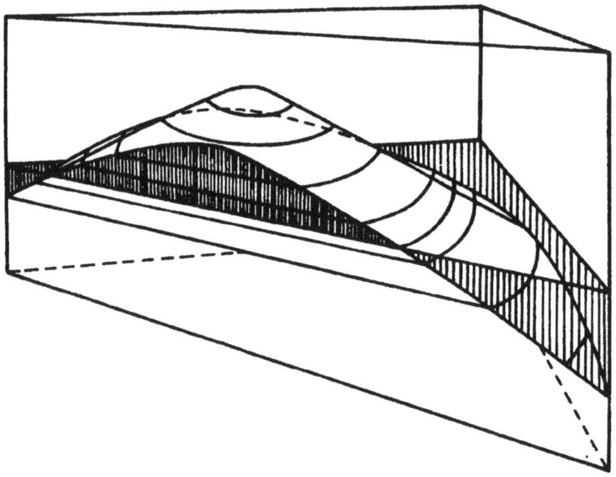

Abb. 4

Perspektivische Ansicht zu Abb. 3

tiert sofort, daß der Wasserspiegel druckbedingt nur so deformiert werden kann, daß die unter den Gebieten höheren Luftdruckes verdrängten Wassermassen gleich jenen sind, die über das ungestörte Niveau unter den Gebieten niedrigeren Luftdruckes emporgedrückt werden. Abbildung 4 möge das andeuten. Jenes Niveau, das diese gleichen Massen trennt, ist dasjenige, das im Falle der Isobarie von der Wasseroberfläche eingenommen wird. Gelingt es, den dem isobarischen Niveau im vorgegebenen Luftdruckfeld zuzuordnenden Luftdruck zu bestimmen, so ist über die bereits genannte statische Grundbeziehung (1 mb ≙ 1 cm WS) für jeden Ort an oder auf dem betrachteten Meer die am beobachteten Wasserstand anzubringene Korrektur bestimmbar, so daß die Luftdruckstörung des Wasserspiegels eliminiert ist.

Nach der Einsicht in die prinzipiellen Erfordernisse ist das Verfahren leicht anzugeben: Man bestimmt schichtweise, entsprechend den Isobarenstufen des Luftdruckfeldes, einmal die unter dem hohen Druck verdrängten und die unter dem niedrigen Druck emporgedrückten Wassermassen und trägt die Werte der Stufensummen in ein Volumen-Druck-Diagramm ein. Ein aus Stetigkeitsgründen glatter Linienzug durch die zusammengehörenden Punkte ergibt zwei Kurven, deren Schnittpunkt den gleichen Verdrängungsvolumina den entsprechenden Luftdruckwert zuordnet. Mit diesem Druckwert ist das isobarische Bezugsniveau festgelegt (Abb. 5).

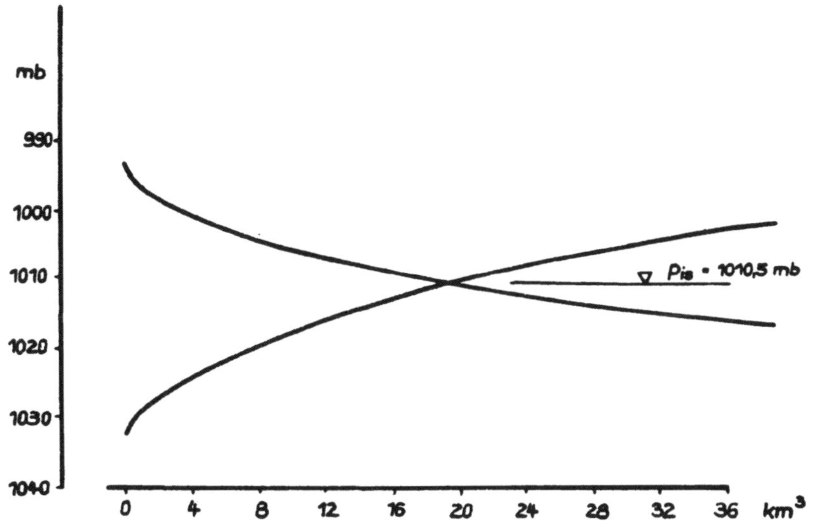

Abb. 5

Graphische Ermittlung des Luftdruckwertes im isobarischen Niveau

Wir wollen das Gesagte an einem schematischen Beispiel erläutern.

Vorgegeben sei ein zylindrisches Meeresbecken der Tiefe H mit dem Radius R. Es herrsche am Anfang überall gleicher Luftdruck, etwa 1000 mb. Dann bildet die Oberfläche eine Niveaufläche.

Über dem Zentrum dieses Beckens entwickle sich eine Zyklone derart, daß der Luftdruck um den enorm hohen Betrag von 5 mb in einer Stunde falle und die Störung sich linear ausbreite, so daß die Isobaren als konzentrische Kreise im Grundriß erscheinen. Nach 2,5 Stunden sei das Druckfeld aufgebaut und bleibe dann konstant, so daß sich ein Achsenschnitt nach Abb. 6 ergibt. Der Radius der Druckstörung ϱ hat dann 500 km erreicht.

Untersucht man den Vorgang nach der angegebenen Methode rechnerisch, was in diesem Falle mittels bestimmter Integrale mit einer variablen Grenze in exakter Form möglich ist, so ergibt sich für drei in radialer Richtung im Abstand von jeweils 200 km aufgestellte Beobachter folgendes:

Beobachter A (bei r = 200 km):
Luftdruckänderung $\Delta p = -7{,}5$ mb
Wasserstandsänderung $\Delta z = +3{,}34$ cm für $R = \varrho$
$\Delta z \to +7{,}5$ cm für $R \to \infty$

Beobachter B (bei r = 400 km):
Luftdruckänderung $\Delta p = -2{,}5$ mb
Wasserstandsänderung $\Delta z = -1{,}66$ cm für $R = \varrho$ (1)
$\Delta z \to +2{,}5$ cm für $R \to \infty$

Beobachter C (bei r = 600 km):
Luftdruckänderung $\Delta p = 0$ mb

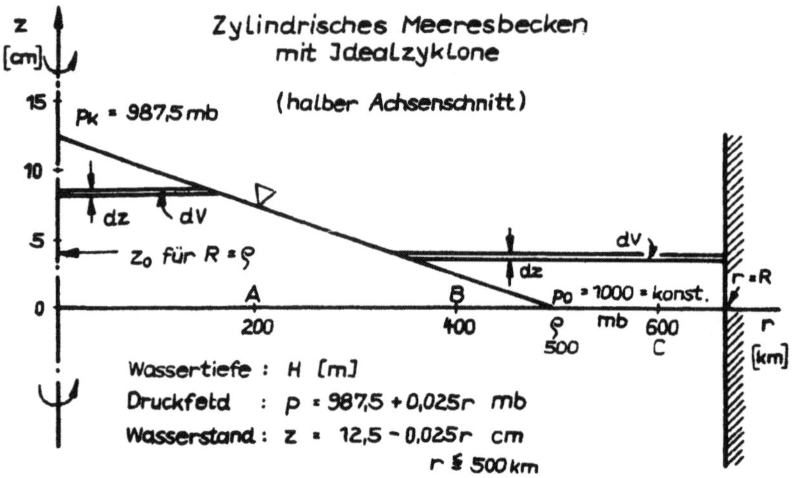

Abb. 6

Halber Achsenschnitt durch ein zylindrisches Meeresbecken mit luftdruckdeformierter Oberfläche

Wasserstandsänderung $\Delta z = -\ 2{,}89$ cm für $R = 600$ km (1)

$\Delta z \rightarrow 0$ cm für $R \rightarrow \infty$

Man stellt in diesem Falle also nicht nur die eingangs entwickelten Behauptungen der Abhängigkeit des Effektes vom Standort und von der Beckenausdehnung fest, sondern erklärt auch völlig zwanglos, daß Wasserstandsänderungen und Luftdruckänderungen das g l e i c h e Vorzeichen haben können.

Weiter läßt sich aus dem Beispiel mit den angegebenen Änderungsgeschwindigkeiten leicht herleiten, daß die zur Herstellung des Gleichgewichtszustandes erforderlichen Strömungsgeschwindigkeiten des Wassers durchaus möglich sind: Die Geschwindigkeiten sind in diesem Falle zeitabhängig. Sie betragen maximal am Ende des schnellen Aufbaus der Störung bei einer Wassertiefe H = 5 m 1,35 kn, bei H= 50 m 0,135 kn und bei H = 100 m 0,068 kn. Solche Geschwindigkeiten sind klein genug, um die Voraussetzung eines Gleichgewichtszustandes zwischen Luftdruckfeld und Wasseroberfläche nahezulegen. Dabei ist noch zu berücksichtigen, daß die Änderungsgeschwindigkeit des Luftdruckes, der ja auf jedes Quadratzentimeter Meeresoberfläche wirkt, für die Abschätzung mit dem enorm hohen Betrag von 5 mb/h angesetzt wurde. Normalerweise sind die beobachteten Druckänderungen der Zeiteinheit wesentlich geringer.

Es verbleiben noch einige Angaben zu der praktischen Arbeitsmethode zu machen: Vorgegeben ist in jedem Falle die Luftdruckverteilung über einem beliebig geformten, abgeschlossenen Meeresgebiet, im allgemeinen in Form einer Arbeitswetterkarte, wie sie im synoptischen Dienst verwendet wird. Die für die Volumenberechnungen nötigen Flächenermittlungen mit einem Planimeter nimmt man praktisch in diesen, meist nicht flächentreuen Karten ohne vorherige Umzeichnung vor. Unser Wetterdienst verwendet Kartenunterdrucke in konformer Kegelprojektion mit Längentreue in 30 und 60 Grad Breite. Entsprechend den breitenabhängigen Maßstäben wurden Umrechnungstabellen errechnet, so daß die Arbeiten hinreichend erleichtert werden. Auch dieses Vorgehen ist auf andere Kartenprojektionen übertragbar.

Fassen wir abschließend zusammen! Die Vorzüge der Methode werden in dem Folgenden gesehen:

1. Es ergibt sich kein Widerspruch zu den Grundgesetzen der Hydrostatik, da diese postuliert werden.

2. Es ist möglich, für ein abgeschlossenes Meeresbecken beliebiger Form den statischen Luftdruckeffekt für ein beliebig vorgegebenes Luftdruckfeld als individuelles Problem zu ermitteln.
3. Die Methode beinhaltet die Möglichkeit der Abschätzung der mit einer durch ein Luftdruckfeld bedingten Deformation der Wasseroberfläche verknüpften Massenverlagerungen.
4. Es konnte gezeigt werden, daß die zur Herstellung des Gleichgewichtes erforderlichen Strömungsgeschwindigkeiten des Wassers klein genug sind, um die Voraussetzung des Gleichgewichtsfalles nahezulegen.
5. Als strömungsauslösende Ursache wird die Bedingung

$$\frac{d}{dt} (p_{is} - P) \neq 0$$

erkannt. Dabei ist p_{is} – der Druckwert des isobarischen Niveaus – nur zeit-, P – der Luftdruckwert des Beobachtungsortes – sowohl zeit- als auch ortsabhängig

und schließlich

6. Im hydrologischen Bereich sind in ihr bei der Anwendung auf die Ostsee prognostische Möglichkeiten dieses Effektes bei Wasserstandsvorhersagen enthalten, was wir hier übergehen können.

Ein Nachteil muß in der Tatsache erblickt werden, daß die Lösung nur auf abgeschlossene oder praktisch als abgeschlossen betrachtbare Meeresgebiete (z.B. die Ostsee) anwendbar ist, weil man hinreichend genau dV = 0 setzen dürfen muß.

L i t e r a t u r

1 Doodson, A.T. u.a.: Admiralty Manual of Tides. London, 1941, Neudruck Juni 1946, spez. S. 239

2 Miehlke, O.: Über die Berechnung des statischen Luftdruckeffektes auf den Wasserstand abgeschlossener Meeresbecken. Diss. Greifswald 1960

Some researches on the velocity of eustatic changes of the world ocean level

von

M. I. S i n i a g i n a (Moskau)

S u m m a r y

The report discusses principles laid by B. Gutenberg as a basis of his determinations of eustatic changes in the level of the World ocean.

Gutenberg's investigations have been repeated on the basis of fuller data on the change of the World ocean level published in the Publications of the International Association of Physical Oceanography.

Conclusions have been arrived at regarding the possibility of a tilt of the European continent from the north southwards with a rate of 0.2 mm per year.

Z u s a m m e n f a s s u n g

Der Bericht behandelt die Prinzipien, die von B. Gutenberg als Grundlage seiner Bestimmungen der eustatischen Änderungen des Weltozeanniveaus aufgestellt worden sind.

Gutenbergs Untersuchungen sind wiederholt worden auf der Grundlage vollständigerer Daten bezüglich der Änderung des Weltozeanniveaus, wie sie in den "Publications of the International Association of Physical Oceanography" veröffentlicht sind.

Die Schlußfolgerungen zeigen, daß die Möglichkeit einer Neigung des europäischen Kontinents von Norden nach Süden mit einem jährlichen Betrag von 0,2 mm besteht.

Lengthy observations have established that the mean sea level does not remain constant in relation to a chosen point of land. On this basis a deduction is made regarding relative movements of the land, considering thus the position of the sea level to be constant.

However, many scientists think that changes of sea level are taking place owing to so-called "eustatic factors".

In his article "Changes in Sea Level, Postglacial Uplift and Mobility of the Earth's Interior" (1941) B. Gutenberg raised many questions relating to the study of recent vertical movements of the crust according to observations over the sea level and gave a quantitative appraisal of the velocity of eustatic changes of the World ocean level. However, until now there has been no discussion of Gutenberg's conclusions in the press, though the data of his researches are widely used in different countries.

The hypotheses laid by Gutenberg as a basis of his study on eustatic changes in the World ocean level can be summarized as follows:

1) the volume of the globe does not change, consequently during a deformation of the Earth's surface there is only a redistribution of the areas of uplift and subsidence; the sum of positive velocity values of recent movements calculated for the territory of the entire globe, should equal the sum of their negative values.

2) one of the most important causes resulting in an eustatic change of the World ocean level is the melting of glaciers in Arctic and Antarctic regions, which can produce an increase of the amount of water in the World ocean. In this case the sum of positive velocity values of the changes of the World ocean level in different points of its coast should be larger than the sum of their negative values (or, what amounts to the same, the sum of negative values of recent movements velocities on the sea coasts should be higher).

A necessary condition for a correct solution of the problem is a determination of the velocity of recent movements for points uniformly distributed throughout the territory of all continents or at least on the coasts of all the seas of the world. For this purpose it is necessary to have a dense network of sea-level posts with observations carried out in a proper way. These posts should be evenly distributed on the coasts of all the seas.

However, at the present time the network of sea-level posts is not evenly distributed on the coasts of seas and oceans; on some coasts there are no sea-level observations at all. That is why Gutenberg resorted to a preliminary solution of the problem, which interested him.

Gutenberg had calculated approximately the average value of sea level changes (in centimeters for 100 years) for 71 postslocated in different parts of the world.

In examining Gutenberg data one can notice that for the majority of coasts included into the research there has been a rise in the sea level, which lead him to a thought about a systematic rise of the entire level of the World ocean. The rate of this rise has been determined by Gutenberg by meaning all the changes of sea level for each coast and proved to be 10 cm in a hundred years (1.1 mm/year).

In 1954 the International Association of Physical Oceanography has published data on the changes of sea level in relation to the land calculated on the basis of available series of sea-level observations at 183 posts located on the coasts of seas and oceans of many countries of the world. However, even a brief description was lacking of the method used for the establishment of the velocity of sea level changes according to the data of oceanographic observations. Consequently, the first thing in carrying out researches was to check the reliability of the values published. For this end the series of sea-level observation points on the coasts of Europe have been processed by the method of least squares.

To obtain the rate of changes in the sea level position for each point error equations have been compiled following a type recommended by V.A. Zenin in his article "Mean sea level and methods of processing results of sea-level observations in a study of recent vertical movements on sea coasts" (Publications of the Central Scientific-Research Institute of Geodesy, Airphotosurvey and Cartography, fasc. 123, Moscow, 1958).

By solving respective normal equations velocity values of the changes of sea levels in Europe have been obtained and an appraisal has been made of the precision of their determination. In comparing velocity values obtained by the method of least squares with the data of the International Association of Physical Oceanography, it has been found that velocities of an overwhelming majority of points differ only within the limits of possible error of computation. Somewhat more exact velocity values have been obtained by deducing them from the series for a longer period of observations (up to 1951). This way, a check of the data provided by the International Association of Physical Oceanography showed that the rates of sea level changes in relation to the land published in its Publications, have been obtained with sufficient strictness and can be used in further researches.

For greater clarity velocities of recent crustal movements (reverse in sign to the velocities of change in the position of sea level) have been plotted on a map of the world, on which the employees of the Institute of Geography of the USSR Academy of

Sciences have outlined the areas of the last glaciation.

In studying Gutenberg's paper the following work has been done:

1) Gutenberg's conclusions on the velocity of eustatic changes in the level of the World ocean has been repeated according to new data obtained from longer series of sea-level observations and processed with greater precision;

2) a study has been effected of eustatic changes in the level of the World ocean using data on the coasts of European seas only, including the coasts which have been under glaciation.

In reviewing the material available we come to the conclusion that during recent years there has been a much greater change in the level of the Atlantic ocean near the coast of the USA. The mean change of level in this ocean on the American coast according to Gutenberg (observation period up to 1939) amounts to + 2.6 mm/y, while according to new data (observation period up to 1951) it is 4.3 mm/y. A similar change in the rate of sea level changes is also recorded in the Gulf of Mexico.

On other coasts used by Gutenberg for the establishment of eustatic changes of the World ocean level no similar rise has been recorded. For these coasts there is a record of both a greater velocity in the change of sea level and its decrease as well.

The mean rate of changes of sea level in relation to the land calculated according to data of 19 coasts (without the Atlantic coast of the USA, the coast of the Gulf of Mexico and the coast of Spain for which there are no new data) amounts according to Gutenberg to 10.4 cm in a century and to 10.7 cm according to new data.

This can lead to a conclusion that the approximation of the establishment of the rate has no effect whatsoever upon the value of eustatic changes of the World ocean level obtained by Gutenberg. Consequently in studying this problem an analysis should be made mainly of the principles, which he laid as a basis for his researches.

Unfortunately, we lack full material on sea level observations in the United States of America and this makes it difficult to form an opinion on the reasons for a break in the diagram showing the progress of the yearly level of the Atlantic ocean. It is possible that this event is associated with tectonic processes in the Earth's crust causing subsidences of the coast.

An absence of data on oceanographic observations along the northern coast of the USA and Canada did not permit to study eustatic changes of the level of the World ocean on all coasts of the globe. This research had to be restricted by the coasts of the seas washing the European continent.

A mean rate of the change of sea level has been obtained for 26 coasts of Europe. The results can be formulated in short:

1) the total length of all sinking coasts (without the coasts of the seas washing the northern territory of the European part of the USSR) according to a very approximate estimate amounts to 12 000 km and the sum of velocity values for them is 17.52 mm/y (the number of coasts 17);

2) total length of all rising coasts by an approximate estimate comes to 10 000 km, and the sum of velocity values for them -- 24.37 mm/y (the number of coasts 9);

3) the mean rate of movements of all sinking coasts of Europe approximately amount to 1 mm/y and corresponds to the value of eustatic changes in the level of the World ocean, computed by Gutenberg;

4) mean rate of recent movements on all rising coasts approximately comes to 3 mm/y.

An estimate of the velocity of a vertical movement of the European continent can be obtained in two ways:

1) inasmuch as the length of sinking coasts in Europe is not very different from the length of rising coasts, it is possible to consider the value of a mean velocity of their uplift calculated for 8 coasts and the value of sinking deduced for 17 coasts as equally reliable. This leads us to a conclusion that the mean rate of vertical movement

of the European continent comes to approximately 1 mm per year;

2) discarding the difference in weights of the data obtained, to take an average of the velocity values of recent movements for all the coasts of Europe. In this case the rate of vertical movement of the territory of Europe approaches zero (+ 0.2 mm a year).

The difference of these two results can be of substantial significance in a study of the causes of the origin of recent deformations of the crust (in this case computations should be repeated with greater precision and on the basis of a more abundant material).

On the basis of the researches made, the impression is created that the European continent is tilted from the north towards the south (its northern part rises with a substantial speed, while the central and southern parts are somewhat sinking).

According to published material a similar slant can be assumed for the continent of Northern America. On other continents sea-level observations are carried out at the present time only over an insignificant part of their coasts, which makes it impossible to draw any conclusions about them.

These researches do not permit, of course, to draw a strict boundary between eustatic changes in the level of the World ocean and between the velocity of recent crustal movements determined according to data of oceanographic observations. And yet they permit to assume that the process of the origin of crustal deformations observed now practically in all countries, is much more complex than it was imagined to be previously and that it needs further study.

Tidal observations as a measure of movements of the earth's crust

von

G. M. Lennon (Liverpool)

Summary

The nature of the problem is posed by a review of the physical forces which interact to determine the level of the sea surface: astronomical tidal forces, meterological perturbations, residual currents over the period of observation due either to gradients caused by fresh water flow or to gradients caused by geostrophic forces and finally effects caused by water density differentials.

The analysis of observations of sea level in order to provide data of geodetic significance falls into two categories:

a) The extraction of secular trends from very long periods of observation as a measure of relative movements of the surface at two or more stations. Here the length of the record serves to average out the shorter period changes of level and the comparison of results at different stations removes eustatic effects.

b) For shorter periods of observation more elaborate techniques are required to eliminate perturbing influences. The 19 yearly tide may place restrictions upon the period of observation presented for analysis. Such problems have recently been considered by the R.E. u.N. in attempts to effect hydrographic levelling.

Requirements for a practical study involve considerations of instrumentation and uniformity of data processing. The recent work of the Permanent Service for Mean Sea Level and its projected studies will be summarised in this light.

Zusammenfassung

Die Besonderheiten des Problems werden deutlich durch eine Betrachtung der physikalischen Kräfte, die aufeinander einwirken, um das Niveau der Meeresoberfläche zu bestimmen: Astronomische Gezeitenkräfte, meteorologische Störungen, Restströme während der Beobachtungsperiode, die entweder auf Gradienten des Frischwasserstromes zurückzuführen sind, oder auf Gradienten, die durch geostrophische Kräfte verursacht werden und schließlich Effekte, die ihre Ursache in Dichteunterschieden des Wassers haben.

Die Analyse von Beobachtungen des Meeresniveaus zur Datengewinnung, die geodätische Bedeutung haben, teilt sich in zwei Kategorien:

a) Die Suche nach säkularen Tendenzen aus sehr langen Beobachtungsperioden, wie die Messung der Relativbewegungen der Oberfläche an zwei oder mehreren Stationen. Hier dient die Aufzeichnung zur Herausmittelung der kurzperiodischen Änderungen des Niveaus, und der Vergleich der Ergebnisse an zwei verschiedenen Stationen beseitigt eustatische Effekte.

b) Für kürzere Beobachtungsperioden werden sorgfältigere Methoden zur Elimination der Störeinflüsse gefordert. Die 19jährige Tide kann Einschränkungen für die Beobachtungsperiode bedeuten, die für die Analyse vorgegeben wurde. Solche Probleme werden gegenwärtig bei Versuchen von der R.B.u.N., die der hydrographischen Nivellierung dienen, bearbeitet.

Eine praktische Untersuchung muß den Forderungen des Instrumentenbaues und der Einheitlichkeit der Datengewinnung Rechnung tragen. Die gegenwärtigen Arbeiten des Permanent Service for Mean Sea Level und seine vorgesehenen Untersuchungen werden unter diesem Gesichtspunkt zusammengefaßt.

In representing the Permanent Service for Mean Sea Level at this Symposium, I wish in the first instance to give a brief review of the problems associated with the use of sea level data in geodetic studies, together with an outline of the techniques which have been used in practical attempts to apply hydrographic principles to this general problem. Secondly I wish to indicate the scope of recent and also projected work of the Permanent Service so that those of our colleagues, not too familiar with our work, may be acquainted with the nature of the data at present available for such studies, with investigations, relevant to the work of this symposium, at present being conducted, and finally with the facilities which the Service is willing to offer to its geodetic colleagues in this connexion.

The nature of the problem

Although the geoid is often linked with a theoretical reference plane defined by the undisturbed surface of the oceanic waters, it is in practice extremely difficult to extract information of geodetic significance from observations of sea level. The level of the sea surface at any one instant of time is the result of the interaction of a great number of physical forces. Among them can be listed:

a) Astronomical tidal forces generating oscillations of level which fall into certain classes or species:
   (I) Shallow water terms in the range, say, 1-8 hours
   (II) Common tidal terms of semidiurnal and diurnal period
   (III) Seasonal variations with periods of 1 year, 6 months, 1 month, 1 fortnight, and finally
   (IV) The Nodal Tide with a period of 18·6 years.
b) Meteorological perturbations of level caused by variations in barometric pressure acting directly upon the sea surface and also indirectly through winds.
c) The effects of variations in water density on the slope of the sea surface.
d) Gradients caused by geostrophic forces applied to coastwise currents or by fresh water flow.
e) Forces linked with the transport of water and energy by sea waves.
f) The effects of changes in volume of oceanic waters - eustasy.

Any value of mean sea level obtained from tide gauge observations, irrespective of the lenght of record used as a basis, necessarily contains a significant contribution from certain if not all of these forces. This being so it is essential that before mean sea level data can be employed in any geodetic investigation, a serious attempt must be made to remove the effects of these perturbations. In this connexion it is significant that although several workers have studied certain aspects of this problem, no one would claim that he has accurately purified mean sea level data so as to define the geodetic reference plane.

Methods of reduction

The process of removal of tidal oscillations is well known, particularly with regard to items (I) and (II), by the application of suitable numerical filters (see publications [1] and [2]). Seasonal variations can be satisfactorily eliminated by a consideration of annual means, but the 19 yearly tide does occasion some concern. Theoretically the 19 yearly tide is present in the oceans and is capable of definition by harmonic analysis. Practical attempts to define this oscillation however, have provided questionable results even when the data used have first been purified of secular variation and meteorological perturbation. The conclusion has been reached that the "noise level" in any set of mean sea level data is relatively too high for the successful determination of the effect. Consideration of this nodal tide has led many investigators to work in units of 19 yrs., but the present climate of opinion questions the need for such restriction in view of the above considerations.

Meteorological perturbations have been investigated by statistical techniques and their contributions effectively reduced from the basic data. In particular, publications [3] and [4] are particularly relevant to the interests of this symposium and to this work further reference will be made later.

The effects of water density variations can be calculated from theory if adequate observational material is available. Such calculations can provide valuable contributions to geodetic studies: see publication [5]. Similarly on the subject of geostrophic gradients, recent work by J. Crease and D.E. Cartwright (publication [5]) in attempts to link the levelling networks of France and England based upon the residual non-tidal current through the Straits of Dover as determined by a submarine cable technique is worthy of note.

Of very considerable interest, and an entirely new concept in this work are the unexpected changes in mean water level which occur when waves approach a coast. It has been found by Longuet-Higgins and Stewart, that small waves and a stepp bottom-slope would involve lower levels on the shore than, say, half a mile offshore, whereas large waves and a shallow bottom-slope may involve a rise of mean level on the shore by as much as 3 feet in storm conditions. These conclusions are in agreement with recent observations by Fairchild and seem to imply that future investigations of mean sea level variations will require the consideration of exposure to wave action and will involve adequate observation of wave conditions. Publication [7] applies.

Finally a measure of the eustatic changes in sea level can only come from a thorough investigation of secular variations in level as determined from long series of tide gauge observations, preferably showing a world-wide coverage. It may be too much to expect that realistic values for this phenomenon will be available for the use of the present generation.

In the face of these most complex problems, it is obvious that great care must be taken in accrediting hydrographic data with the accuracy required by geodesists. Nevertheless, by careful analysis, it has been found possible, in undertaking hydrographic levelling, to produce results of geodetic significance, and experience has shown that prospects for the future in this work can be viewed with optimism.

The role of hydrographic studies in an investigation of movements of the earth's crust

It would appear that there are two distinct ways in which the treatment of hydrographic data can make a contribution to the general study: in the first place as an end in itself in providing a direct measurement of relative movement, and in the second place as a means to an end by its use, in hydrographic levelling, as a technique for the reduction of separate land-based levelling networks to a common reference plane.

In the first of these roles, the study reduces to an investigation of secular variations in level from long series of tide gauge observations. In work of this nature it is often assumed that the variations are linear and that a least-squares fit to the observed data is all that is required to eliminate the shorter period changes of level. Thereafter by comparing resultant trends at a number of stations one can eliminate eustatic changes in level and obtain a measure of relative land movement. In the study of recent movements of the earth's crust, it would be necessary to conduct a more detailed investigation since departures from any linear trend are likely to be quite as important as the overall tendency. It would seem therefore that a systematic attempt would have to be made to purify the series of annual mean levels at each port for all the perturbing influences listed earlier.

Even so, for short period movements of the crust, measured at intervals of one or two years, which appear to be the aim of the present study, it seems unreasonable to expect hydrographic techniques to provide a satisfactory solution. Consequently it may be that the most useful contribution which can be made by sea level studies will be in the field of adjustments to levelling networks and the bridging of sea areas so that separate networks may be reduced to a common datum in order to allow the crustal movements themselves to be studied by survey techniques.

In this respect the work of Cartwright and Crease in the Straits of Dover is a classic example, whereby as a result of a relatively short period of observation it has been estimated that the datum used in France, based upon mean sea level at Marseilles, is 13 cm higher than the British datum based upon mean sea level at Newlyn. The accuracy of this figure is calculated to be within 3 cms, whereas conventional optical or radar methods could not be expected to give an accuracy better than half a metre.

Of equal significance is the work which has been conducted by the Permanent Service for Mean Sea Level in connexion with R.E.U.N. (United European Levelling Network) in attempts, on a larger geographical scale, to purify mean sea level data with particular reference to meteorological perturbations and secular variations. An interesting treatment of the meteorological effects has been devised by Doodson (publication [3]) in the form:

$$Z - \bar{Z} = \sum a_n(B_n - \bar{B}_n)$$

where $Z - \bar{Z}$ is the departure of an annual value of MSL from the overall mean of the period studied at a certain tide gauge station

$B_n - \bar{B}_n$ is the departure of an annual mean barometric pressure from the overall mean for the same period at a selected Meteorological Station

and a_n is a constant determined by a least squares solution.

In publication [4] Rossiter applied these principles to the observed levels at some 49 European tide gauge stations and provided a solution for $Z - \bar{Z}$ in each case in terms of barometric pressure recorded at 3 suitable met. stations. The reduction of scatter in the mean sea level data thereby achieved is remerkable and is indicated in the accompanying diagram. But this was not the ultimate aim of the study. Rossiter was able to proceed from this stage, having defined the response of sea levels to barometric pressures, to calculate the values of MSL appropriate to an isobaric atmosphere. The latter, it is interesting to note, have effectively reduced the anomalies in level, which are apparent along the coast of Europe from the Gulfs of Finland and Bothnia to the Mediterranean, by approximately half. The remaining anomalies, when viewed in the light of Bowden's work on water density (publication [5]) can clearly be reduced much further by similar considerations, and thus provide a valuable and independent check upon conventional methods of precise spirit levelling.

The Work of the Permanent Service for Mean Sea Level

Since the work of the Permanent Service is pertinent to the interests of this study it may be fruitful to outline the scope of its responsibilities at this stage.

A. Collection and publication of MSL data

For many years, the Permanent Service and its precursor, the Mean Sea Level Committee of A.I.O.P., have undertaken the duty of collection of all available MSL data, of extraction of MSL values, where necessary, from tidal records and of publication of this information in the Publication Scientifique Series of A.I.O.P. Data for the period up to and including 1958 have already been published and we are at present engaged upon the task of compiling results for the years 1959 to 1961. The accompanying table indicates the geographical coverage of stations contributing to our most recent publication. Monograph No. 13 of the International Union of Geodesy and Geophysics, 1961, takes the form of a catalogue of all published mean sea level data and is a valuable guide to all workers on the hydrographic problems which confront this meeting. All future MSL data will be stored in the form of punched cards to assist publication and to facilitate subsequent research.

B. Installation and maintenance of tide gauges

In its efforts to increase world coverage and to improve the value of its publications, the Service has been greatly concerned in recent years with the inadequacy of the existing global network of tide gauge stations. Despite repeated efforts to improve the position the Permanent Service is faced by the fundamental difficulty that the initiative, installation coasts and the organisation of maintenance have often to be provided by a local authority with which the Service has no influence. The situation is particularly difficult in countries which are in the process of development, either economically or politically. We feel that the desired increase in the number of properly maintained gauges can only be effected by placing the need for them before the Intergovernmental Oceanographic Commission, and steps have been taken in this direction.

In so far as the interests of this symposium are concerned, the same problems apply and in particular it should be noted that for an adequate estimate of the eustatic change in sea level, the present distribution of gauges is too heavily biassed towards the northern hemisphere. Official support, from this conference, in its efforts in this connexion would be greatly valued by the Permanent Service.

Maintenance in the day to day operation of existing gauges and the treatment of the instrumental records is again a subject of great concern to the Service and in the absence of any direct contact with the local administration of the observing stations the aim has been to give guidance through published notes (e.g. publication [9]).

C. Projected Work of the Permanent Service

The following items, pertinent to the interest of the symposium are already planned to receive attention in the routine programme of the Service:

1) The analysis of the seasonal variations in sea level at selected stations throughout the world has been undertaken and progress is being made with the appropriate calculations.

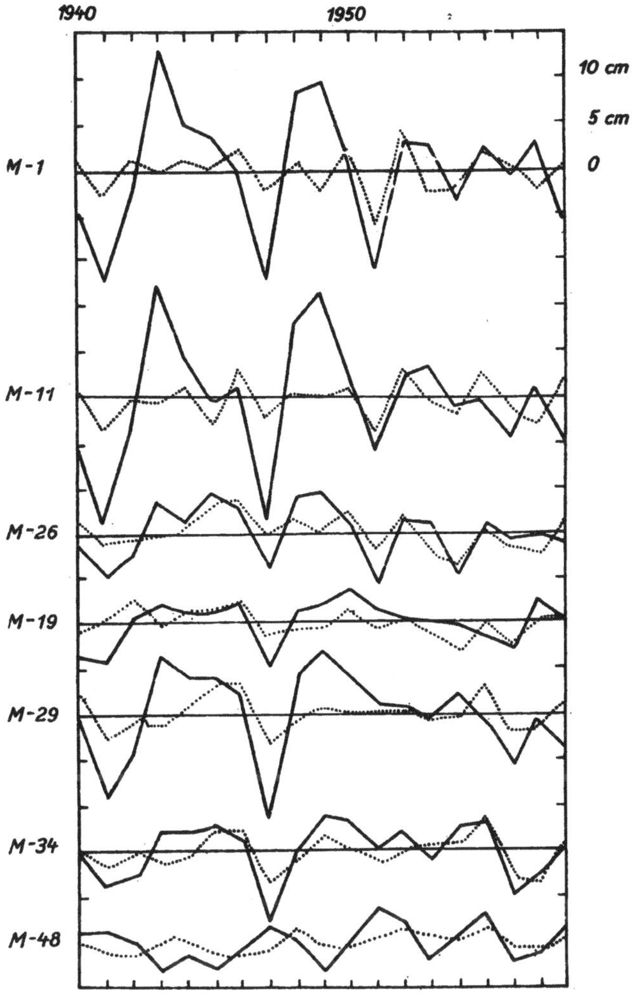

Fig. 1

Annual values of MSL, at selected stations, after removing secular variations.

Crude data: ——
Data corrected for meteorological effects ....

2) It is proposed to conduct a continuing analysis of data, as received, designed to give the most reliable estimates of secular changes in sea level. The results of this study should be invaluable in all investigations involving eustasy. This work will supplement the initial attempt to systematise calculations of secular variations given in publication [10].

3) The investigations outlined earlier, in connexion with the Unification of the European Levelling Network are to continue and in order to promote this work the IAG Special Study Group No. 22 has been formed.

4) It is hoped to publish later this year a comprehensive Mean Sea Level Bibliography.

At many points, the interests of this symposium and that of the Permanent Service converge and it is important to emphasise that the Service is always willing to make available to geodesists, as well as tidal workers, the facilities which it has for the collection and reduction of tide gauge data together with its experience in research on the subject of Mean Sea Level.

S t a t i o n s   c o n t r i b u t i n g   d a t a   f o r   1957/8 p u b l i c a t i o n

Sea or ocean	Number of stations
North Atlantic, Iceland Waters	1
Arctic Ocean	6
Norwegian Sea	27
Baltic Sea and Gulfs of Finland and Bothnia	41
North Sea, English Channel, Irish Sea	34
Atlantic Coasts of Europe	5
Mediterranean and Black Sea	34
Atlantic Coast of Africa	13
West Indian Ocean	4
North Indian Ocean	20
Gulf of Siam and China Sea	12
North-West Pacific Ocean	49
Australasian waters	23
North-East Pacific Ocean	35
Atlantic Coast of South America	14
Caribbean Sea and Gulf of Mexico	21
Atlantic Coast of North America	27
Atlantic Islands	4
Pacific Islands	25
	395

B i b l i o g r a p h y

[1] Note on methods of determining monthly and annual values of mean water level. J.R. Rossiter, International Hydrographic Review, May 1958

[2] A further note on the determination of mean sea level. J. R. Rossiter, International Hydrographic Review, July 1960

[3] Mean sea level and geodesy. A.T. Doodson, Bulletin Géodésique No. 55, 1960

[4] Report on the reduction of sea level observations (1940-58) for R.E.U.N. J.R. Rossiter, Permanent Service for Mean Sea Level, 1960

[5] The effect of water density on the mean slope of the sea surface. K.F. Bowden, Bulletin Géodésique No. 55, 1960

[6] A study of currents in the Straits of Dover. D.E. Cartwright. The Institute of Navigation, 1961

[7] Radiation stress and mass transport in gravity waves with particular application to "Surf Beats". M.S. Longuet-Higgins and R.W. Stewart, Journal of Fluid Mechanics, in press

[8] Catalogue of published mean sea level data (1807-1958), Monograph No. 13, International Union of Geodesy and Geophysics, 1961

[9] Recommendations for operation of tide gauges and reduction of tidal records. L.P. Disney and J.R. Rossiter, Special Publication No. 43, International Hydrographic Review, 1961

[10] Secular variation of sea level. Publication Scientifique No. 13, Association d'Oceanographie Physique, 1954

Diskussion

J. A. Meschcherikov (Moskau):

It is gratifying to know that within this Symposium we have had a number of reports on oceanographic problems and among them very valuable papers of our German colleagues Dr. Miehlke and Weise. At our Symposium I am very glad to see Dr. Lennon, the representative of the Permanent Service on Mean Sea Level. For the solution of our problems the oceanographic (tide-gauge) data are of real importance. An extensive program of the studies of recent movements is being contemplated - the compilation of the summarising maps of the recent movements of continents, compilation of the general world map of recent movements. These tasks cannot be fulfilled without constant application of oceanographic materials. I think we should approve and support the program of researches, planned by Dr. Lennon. It is especially important to continue and extend the systematisation and reduction of the tide-gauge data. The papers, similar to the "Secular variations of the sea level", Publ. Sc. No.13, 1954, are of importance. The generalisation of the tide-gauge data shall allow, for instance, to solve the important problem of the eustatic variations of the ocean level. The researches carried out by Dr. Sinyagina revealed the necessity of checking the data of Prof. Gutenberg. The eustatic rise of the ocean is evidently less than 1 mm/year, as indicated by Gutenberg, and probably not one and the same in different regions of the globe. The tide-gauge data are also important for the solution of the problem of the period of modern oscillating movements. This cannot be determined only on the basis of geodetic data, as these are insufficient at present. At the same time not knowing how the sign and intensity of movements change with time we cannot in fact compile reliable maps of the recent crustal movements.

Crustal movements as revealed by measurements with horizontal pendulums and gravimeter in the course of earthtide observations

von

R. T o m a s c h e k (Breitbrunn/Chiemsee)

Z u s a m m e n f a s s u n g

Nach Besprechung der Deformation der Erde durch die astronomischen Flutkräfte wird gezeigt, daß die von dem Verhalten einer idealen deformierbaren Erde abweichenden charakteristischen regionalen Bewegungen nach 2 verschiedenen Methoden bestimmt werden können. Die erste Methode benutzt die Residualellipsen, die nach Elimination der globalen Deformation einer idealen Erde durch harmonische Analyse gewonnen werden und für Meereseinflüsse korrigiert werden können. Die zweite Methode beruht auf der Beobachtung der Deformationen, die durch großflächige meteorologische Belastungsänderungen eintreten und aus den Gangkurven von Horizontalpendeln und Gravimetern festzustellen sind.

Es wird eine Zusammenstellung der für Europa beobachteten Werte gegeben und der Zusammenhang mit der tektonischen Struktur aufgezeigt, die die beobachteten Vorzugsbewegungen verstehen läßt.

Es wird auf die Notwendigkeit einer internationalen Zusammenarbeit hingewiesen und das Centre International des Marées Terrestres in Brüssel als Kristallisationspunkt vorgeschlagen.

1. I n t r u d u c t i o n

In its cosmic motion the Earth is under the influence of the constantly varying gravitational forces of the Moon and the Sun[1]. These influence not only the solid Earth, but also the oceans. The solid Earth behaves in this case as an elastic, yielding body which follows, as far as we know, these forces without perceptible delay.

The accelerating forces on the surface of the Earth are maximally by the Moon o.14 milligal in vertical and 0.10 mgal in horizontal direction, representing a deflection of the vertical of 22 milliseconds. The effects of the Sun are 0.05 and 0.04 milligals (= 8,5 millisec). The total maximum effects are therefore 0.19 and 0.14 milligals (= 30 msec). The Earth follows these forces by yielding periodically with periods of nearly half a day and nearly one day, but its movements also contain longer periods and even constant deformations.

The equipotential surfaces vary from 78 cm at the time of the New and the Full Moon to 29 cm at the times of Half Moon.

The measurements show that the solid Earth follows these theoretical tides only partially (Fig. 1). At the maximum, the change of hight is about half a meter with a period of about half a day. It is usual to analyse the complicated movements into a sum of sinus oscillations of different groups of periods according to the astronomical

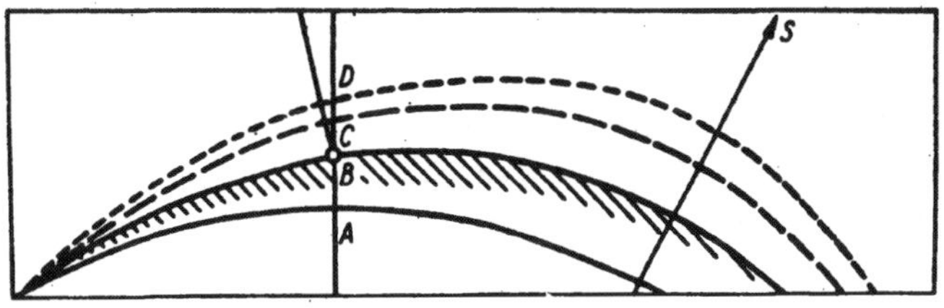

Fig. 1

Elastic deformation of the crust by tidal forces and connection with LOVE's numbers.

rigid earth : A and D: } Surface and equipotential surface
yielding earth: B and C: }

$$W_{el} = W_0 + W_2 + V_2 - g\delta r = W_0 + W_2(1 + k - h)$$

$$V_2 = k\,W_2; \quad g\delta r = h\,W_2$$

$\delta H_{el} = \delta H_{rg}\ (1 + k)$	Deformation of Equipotential Surface
$\delta H_{el} - \delta r = \delta H_{rg}\ (1 + k - h)$	Dislocation of Equ. S. Relative to Observer Observable Equilibrium Tide
$\delta g_{el} = \delta g_{rg}\ (1 + h - \frac{3}{2}k)$	Change of Gravity for observer.

elements contained in the tidal forces. We shall use for our discussion the biggest of these so called partial tides, that is the tide M_2 due to the Moon, having a period of 12,42 hours.

In order to observe this behaviour of the Earth a number of earth tide stations has been established especially during the IGY and mainly two types of instruments have been used, the gravimeters for measurements of the vertical component of the changes of gravity and horizontal pendulums for the horizontal components, that is for the deviations of the vertical, but also registering the tilts of the soil.

2. The results of the tidal measurements

Using these instruments it is possible not only to determine LOVE's characteristic numbers h, k and l (which characterize the elastic properties of the Earth), but also to obtain the deformations of the earth crust regarding its changes in hight and tilt. A great number of observations have been made since the IGY [2].

The observations show proper movements of the earth's crust, different on different places.

I. Firstly one can see that the observed periodic movements differ from the theoretical ones in their phases and amplitudes. There is no doubt that this is primarily due to the loading of the crust by the oceans whose tides contain the same partial tides as the earth tides, but with different relative amplitudes and phases.

It is important, as will be shown below, that these superposed oscillations cannot simply be explained by the elastic deformation of the crust, but that they point towards induced proper regional oscillations of certain parts of the earth crust.

II. There exist other movements of nonperiodical character, Their cause is the loading of large regions by meteorological phenomena, as the change of barometric pressure or rain or snow. Partly they can also be started by movements of fairly distant regions shocked by earthquakes [3]. These induced movements extend in certain cases over nearly half of the surface of the Earth with a duration of several days. Thes nonperiodic movements show in the drift of the records and can be recognized by suitable methods.

Even in these nonperiodic movements characteristic directions can be observed; these are similar to those obtained by the analysis of the periodic movements.

From this the following picture has to be developed: The earth's crust behaves as it would consist of different blocks, like ice-floes covering a tidal sea [4].

3. Methods of determining the regional oscillations

I. Periodical regional movements

The theoretical tilt movements of the horizontal plane under the influence of tidal forces are curves which can be regarded as a superposition of ellipses of the different partial tides mentioned above. The periods are given by the astronomical factors, the amplitudes are calculated for a rigid Earth. These ellipses are the theoretical ellipses T (Fig. 2). Due to the elasticity of the Earth and the subsequent

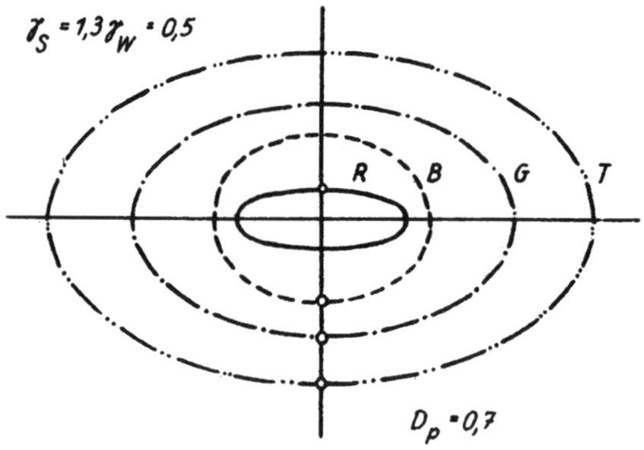

Fig. 2
The different ellipses

deformation another ellipse is formed, the observed ellipse (B), which is obtained by harmonic analysis of the observed data. This ellipse should theoretically be smaller than the observed one by the diminution factor. The ratio between the component observed in the N-S direction to the theoretical one is γ_s , that in the W-E direction γ_w. Generally they are different.

It is very probable that for the minimal deformations by the tidal forces, which are of the order of magnitude of 10^{-8} and which comprise the whole Earth in space, the Earth can be regarded as centrally sysmmetric. That means that we can assume that values of k, h and l for certain types of partial tides can be determined which are valid for the whole Earth. We can therefore calculate a planetary or global Diminutionfactor $D_p = 1 + k - h$ which gives the deformation for the whole Earth regarding the halfdaily periods. The value of D_p should be determined by measurements which comprise the totality of the Earth's globe. Today this is directly only possible for the number k. Its value of k = 0,28 can be obtained with some exactness by the pole movements, discussing Chandler's period. The second number which can be obtained with sufficient accuracy today is the so called gravimetric factor $\delta = 1 + h - \frac{3k}{2}$. The mean value of many measurements over stations which were statistically sufficiently well distributed over the Earth's surface is $\bar{\delta}$ = 1,19 regarding partial tides with a period of half a day [5]. Both equations together give D_p = 0.67.

The mean of all available values of $\bar{\gamma}$ is 0,72 for halfdaily tides. But the distribution of the stations is statistically not quite satisfactory, so it can not be stated how far this mean can be put on a par with D_p. But as it is an independent value it will be taken account of it. As special investigations have shown the difference is not important regarding the questions here under discussion [6].

By multiplying the theoretical ellipsis by D_p a new ellipsis is obtained which shall be called the g l o b a l e l l i p s e which shows the halfdaily tidal movements on an ideal elastic yielding Earth. If this global ellipse G is subtracted from the observed ellipse B another ellipsis is obtained the r e s i d u a l e l l i p s e R which shows the characteristic regional movements as far as they deviate from the theoretically expected tidal movements.

Fig 2 shows the different ellipses a defined above. In this figure it has been assumed that there is no phase difference, but that γ is different; γ_s = 1.3, γ_w = 0.5.

4. E x a m p l e s

Fig. 3 shows the different ellipses for Winsford (Great Britain). As this station is only about 50 km distant from the sea, the maritime influence is nearly as great as the earth tides. The residual ellipse is even larger than the observed one, as there exists a phase difference of 180° between oceanic and terrestrial tides. The maritime influence is a threefold one! The gravitational attraction, the loading which causes an elastic deformation and the influence of the changed potential. These three influences are parallel in their direction. Fig. 4 shows that even after this correction a residual influence remains, which does not coincide with the maritime influences in its direction.

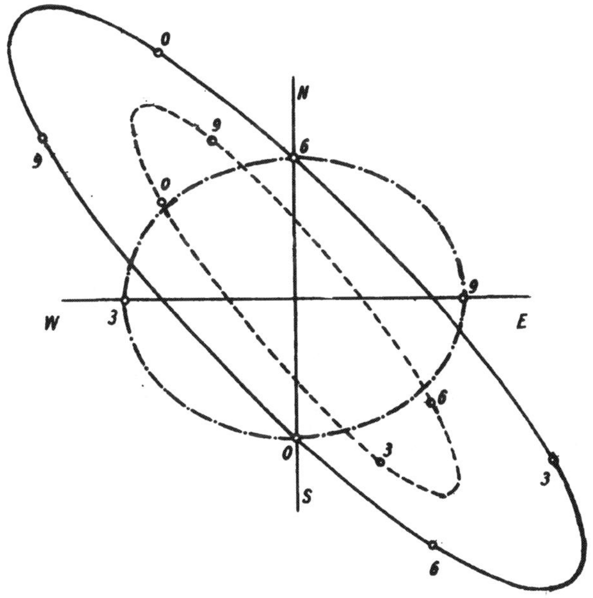

Fig. 3

The ellipses at Winsford (Lat. 53°12'N, Long 2°30'W)

Residual ——
Experimental - - -
Global —·—

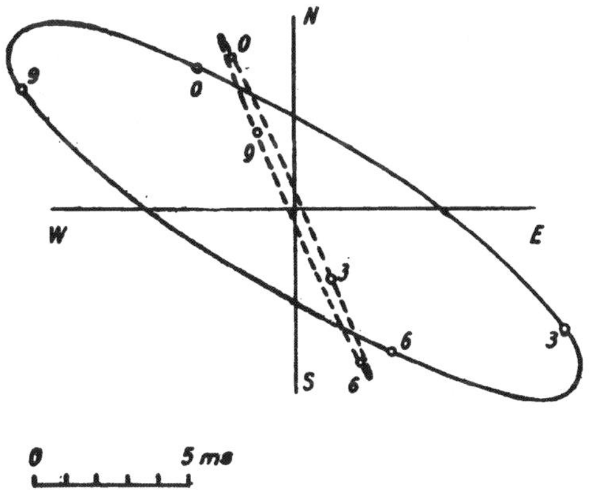

Fig. 4

Corrected residual ellipse at Winsford

Load Tilt ——

Total Tidal Attraction - - -

This corrected residual movement is characteristic for the tectonic proper movements. Another method for studying these tectonic movements will be given later.

Fig. 5 shows the residual ellipses as they have been obtained for the stations of Europe, as far as these have been observed for a sufficient long time. It can easily be seen that there exists a general connection with the tectonics of Europe (See Fig. 6).

Fig. 5
Residual ellipses in Europe

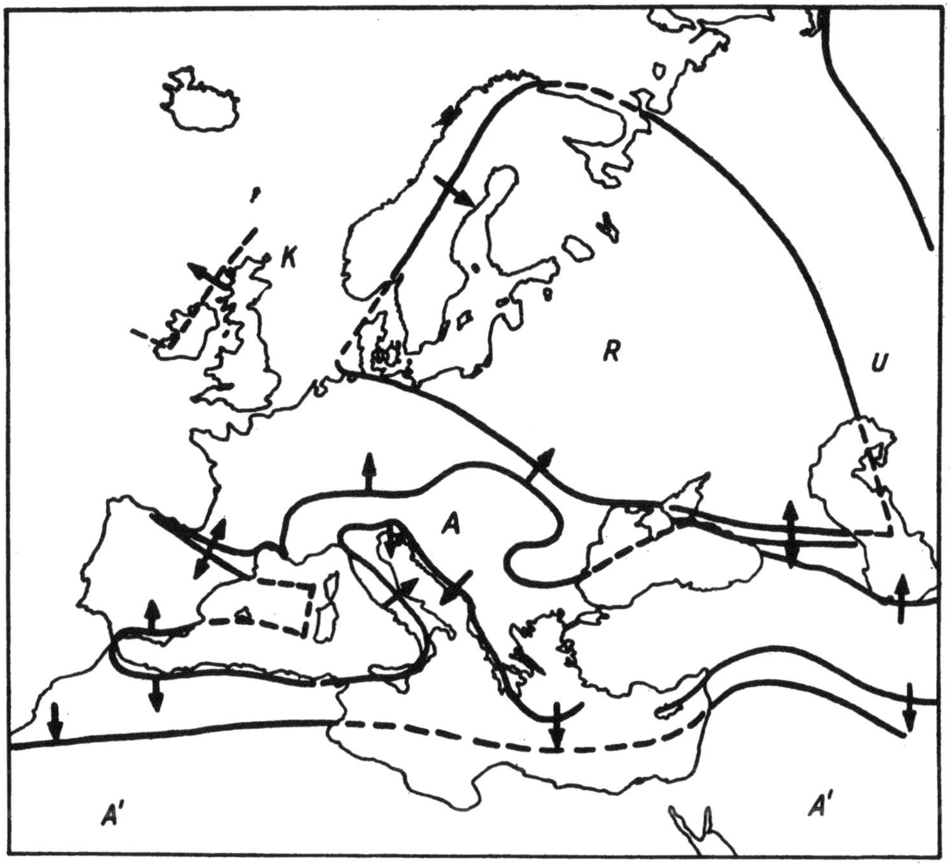

Fig. 6

Tectonics of Europe after KOBER

The two near oceanic stations of Bidston and Winsford show that their movement is mainly determined by the caledonian structure (K), whereas Berchtesgaden shows a marked North-South movement due to the alpine structure (A) which is confirmed by the observations at Trieste and Genoa [7]. Interesting is furthermore the change of the main movement towards the axis NW-SE at the stations near or on the baltic-russian block.

The maritime influence can be easily seen by the diminution of the size of the ellipses when progressing towards the inner of the continent. As we have seen in the former example it is possible to correct for the oceanic influence directly. A provisional correction shows the calculated influence of the main part of the Atlantic according to a procedure which has been first introduced by Nishimura. As can be seen in Fig. 7 this correction is important only in the western part of the continent. It results in tilting the main axes of the ellipses towards North which seems to be in accordance with the more detailed structure of the tectonics [8]

There exist exemptions which have to be discussed in detail. So is the ellipse at Berggiesshübel strikingly larger than the others near this region. The reason seems to be that this station is located near the steep southern slope of the Erzgebirge. The counterpart is the more southerly Station of Březove-Hory, which is shielded by its situation in the heart of the very stable Bohemian massive. Furthermore the stability of baltic-russian shield is quite remarkable, see Stations Tartu and Poltava.

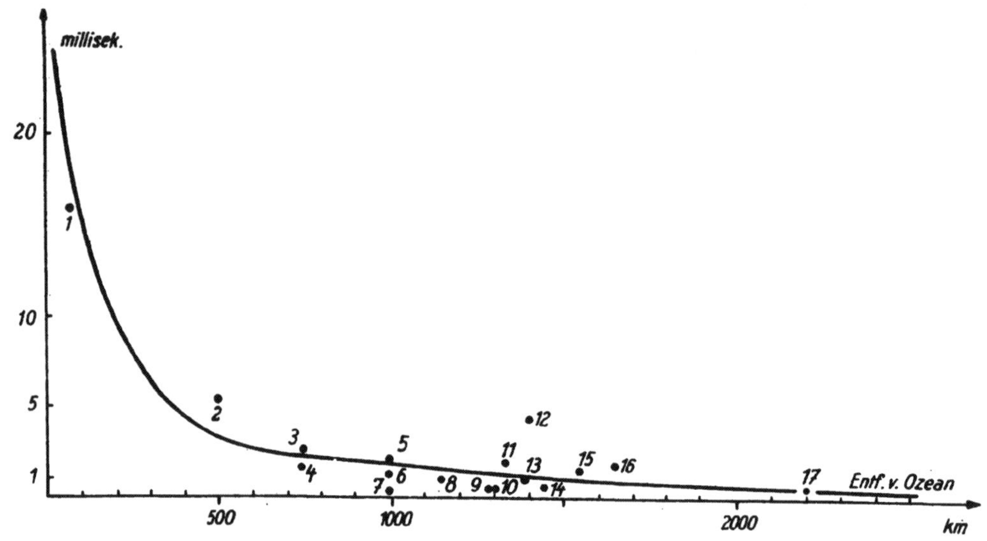

Fig. 7

Atlantic maritime influences in Europe

5. The second method, using nonperiodic movements, mainly due to meteorological loading

There exists another method using the loading by meteorological effects, mainly by changes of the barometric pressure over large regions. There does not yet exist a larger amount of systematically studied material. Fig. 8. shows according to observations of Bossolasco and Cicconi in Genoa, how such an influence shows in the records [9].

The probleme is discussed in Fig. 9 according to observations at Winsford (on the caledonian block). The upper curves show the barometric pressure. L is the local pressure at Winsford, W is the difference between the loading pressure north of Winsford to the Hebrides and the loading pressure South from Winsford to the Channel; the Curve G gives the same loading difference north up to Greenland and south up to Marseille. Figure 9 shows that the tilt of the whole block is really involved in this phenomenon. It is therefore necessary to obtain the acting difference of the barometric pressure by using the isoplethes as could by shown on a special example [4]. In order to obtain the real movements measurements with gravimeters have to be added, as these only give the possibility of determining the vertical displacements. A systematic examination of these cases has not yet been performed in a greater number of cases; this seems to be mainly due to difficulties in the case of evaluating the influence of the local pressure variations on the gravimeters.

In any case the results obtained by the second method confirm the data obtained by the method of residual ellipses and show by this that really movements of extended regions, structurally combined into a tectonic block are the reasons of the observed phenomena. Fig. 10 shows an example of the possible structure of the Earth crust of Europe as Kober has shown from tectonic-geological reasons.

Another proof of the interpretation mentioned above is given by the movement of whole blocks, possibly of the chain of blocks covering whole continents, induced by very distant earthquakes in certain cases. Those movements may last for several days and can be observed on different stations [3]. As Fig. 11 shows the axis of the movement

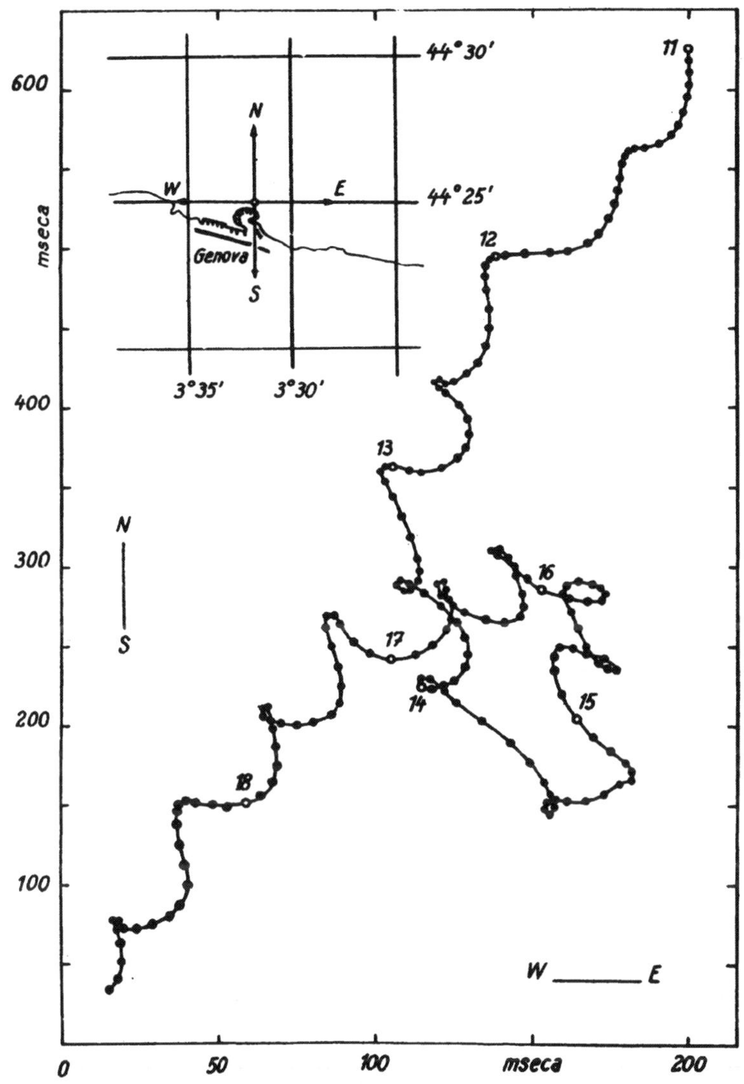

Fig. 8

Influence of barometric loading at (Bossolasco and Cicconi)

coincides also in these cases with the direction determined by the methods described.

Finally it should be considered that only by international organiation of the observations major successes can be obtained and that the International Centre of the Permanent Commission of Earth Tides at Brussels (Director Dr. Paul Melchior) is available as such an organisation.

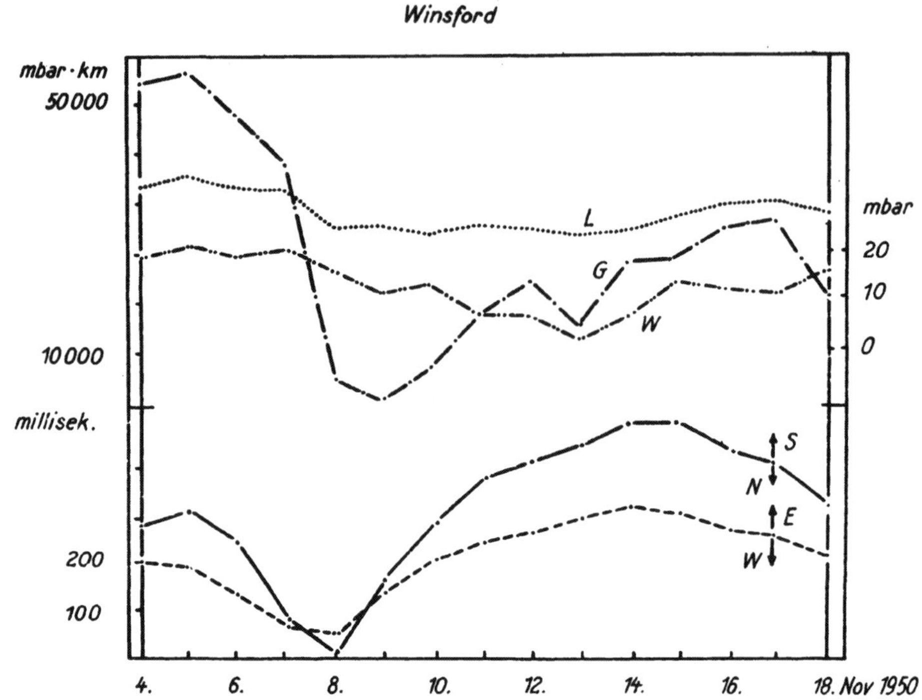

Fig. 9

Barometric pressure distribution (see text) and loading tilt (lower two curves) at Winsford

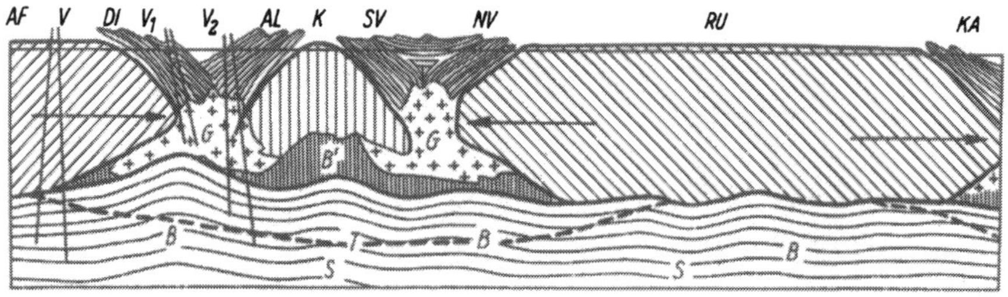

Fig. 10

Example for possible block structure of Europe after Kober

AF	African Mass	NV	North-Variscan Structure
V	Basalt of Foreland	RU	Russian Shield
DI	Dinarides } Alpine Structure	KA	Caledonians
AL	Alps } Alpine Structure	G	Zone of Granitisation. Migma. Sialisation
V_1	Sialic Orogenetic Volcanism "pacific"	B'	Zone of Simatisation
V_2	Mixed Volcanism "atlantic"	B	Sima
K	Alpine-Variscan Foreland	S	Adjusted Zone
SV	South-Variscan Structure	T	Depth of Trough

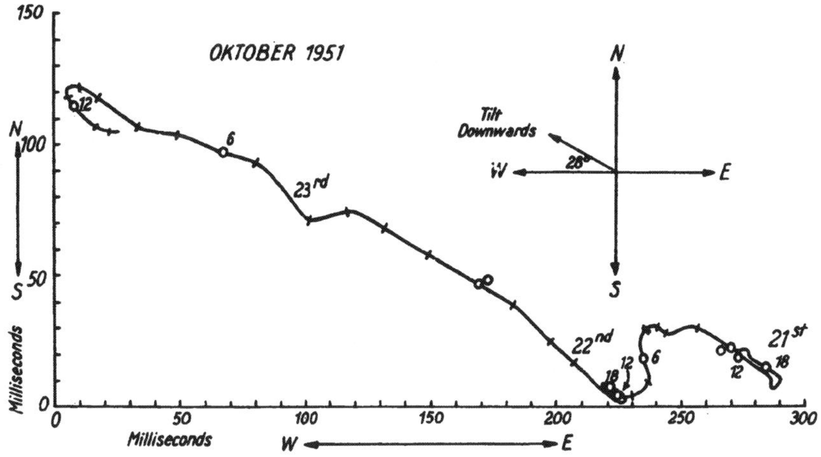

Fig. 11

Tilt at Winsford caused 1951 by earthquake in Formosa (Oct. 21, $21^{h}34$ G.M.T. M = 7)

R e f e r e n c e s

[1] See for example Encycl. Physics Vol. 48, 775 - 845

[2] 3rd Sympos. Int.: Boll. di Geofisica 2, 5 (1960); 4th Sympos. Int.: Commun. Observ. Roy. Belg. 188, Ser. Geoph. 58 (1961). See also (5)

[3] Tomaschek, R.: Nature 176, 24 (1955); Veröff. DGK Reihe A 391, 61 (1961)

[4] Tomaschek, R.: Freiberg. Forschgh. C 80, (1959)

[5] Melchior, P.: Bull. Inform. Mar. terr., Bruxelles (1960)

[6] Tomaschek, R. and Groten, E.: Com. Obs. Roy. Belg. 188, Ser. Geoph. 58, 78 - 93 (1961)

[7] Marussi, A.: Boll. Geod. 19, No. 4, 645 (1960); Zadro, M.: Commun. Observ. Roy. Belg. loc. cit 165 (1960)

[8] Compare Table III in Kober, L.: Tektonische Geologie (1940), Ed. Borntraeger

[9] Bossolasco, M. and Cicconi, G.: Commun. Observ. Roy. Belg. loc. cit. p. 109 (1961)

D i s c u s s i o n

J. D. B o u l a n g e r (Moskau):

Prof. Tomaschek suggests extremely perspective methods of researches of the recent crustal movements. Besides, Prof. Tomaschek introduces a new method which allows to study the inhomogeneities in the crustal structure. The task of our Commission is to work out a recommendation to support these methods and to promote their application in the practice of Geodetic Services and scientific institutions that study the recent crustal movements.

A. Z a t o p e k (Prag):

I was very pleased to hear the very interesting lecture by Prof. Tomaschek. Shortly before my departure from Prague I had a discussion with Dr. Picha. He told me that at present the zero-point deviation as a function of time is studied for the tiltingstation in Příbram, where pendulums are situated in depths of around 1000 m and 1300 m respectively. It was observed that a periodic component is superimposed with a non-periodic one which would correspond to a steady pressure of the bow of Alpes against the Bohemian Mass. This will be systematically studied.

It was also very instructive to see how the orientation of individual ellipses is changing at different stations in connection with the great-tectonical structure.

R. T o m a s c h e k (Breitbrunn):

Die Residualellipsen im Böhmischen Massiv sind, wie schon oben ausgeführt, auffallend klein und es ist daher eine überlagerte kontinuierliche Bewegung günstig festzustellen. Die periodischen Bewegungen infolge des Mitschwingens mit den Erdgezeiten, die aus den harmonischen Analysen vieljähriger Beobachtungen genommen sind, scheinen diejenigen Schwingungen herauszufiltern, die den sehr ausgedehnten Blöcken selbst eigentümlich sind, wie ein projiziertes Bild zeigt.

J. W i t k o w s k i (Poznan):

Interesting conclusions concerning the structure of the crust can be obtained with horizontal pendulums. At the Latitude Station of the Polish Academy of Sciences at Boraniec near Poznan changes of the inclination have been investigated. Some interesting results were obtained. It was found that changes of barometric pressure at considerable distances from the Station are correlated with changes of inclination and one can speak of the existence of a continental block stretching from the Baltic to the Danube. Variations of inclination caused by a passing pressure wave speak in favour of a laminary structure of the surface of the Earth in the region considered.

On secular gravity changes

von

J.D. Boulanger (Moskau)

Data testifying to profound differences between the results of gravity measurement carried out with an interval of serveral years at the same station can be found in numerous literary sources. Sometimes these differences reach such high values that they can hardly be attributed to the measurement errors alone. This enabled some authors to advance the idea that the values of gravity undergo considerable changes in time.

With no calculations to support their opinion, they claimed that secular gravity changes could be caused either by tectonic movements in the Earth's crust or by physical and chemical processes in the subcrustal layer which could bring about transformation of the substance attended by a noticeable change of its density. In their opinion, a change in gravity could likewise be caused by the flow of a substance with a density differing from that of the enveloping substance.

Opinions have also been voiced recently on the changes in the Newtonian attraction constant, on the expanding Earth, etc. All these hypotheses touch in one way or another upon the problem of the constancy of gravity on the Earth's surface. For this reason, it might be useful to revert once more to the question of secular changes in the values of gravity and discuss some of the facts that have been accumulated [10].

The first data on repeated determinations of the values of gravity were published by Abakelia [1], [2]. According to these data, considerable differences have been registered between repeated determinations in the Caucasus in a period slightly over 30 years. For example, in Gudauri the determinations of 1879 and 1910 differed by 101 mgal, in Dusheti by 58 and in Ordjonikidze by 32. Three consecutive determinations of the values of gravity conducted in Baku in 1883, 1902 and 1931 revealed the changes in gravity to a value of 13 and 10 mgal which occurred in the same direction, The work also contained data on big differences recorded between repeated determinations in India.

These large changes in the values of gravity during a comparatively short period of 30-35 years gave rise to certain doubts. With a view to verifying these data, the author of this paper organised in 1935-1936 a special gravimetric expedition to the Caucasus which undertook repeated measurements of gravity at fourteen stations situated in the region of the main Caucasian range (Fig. 1). A special attention was given to the most accurate location of the old stations and determination of their geodetic connection with the new ones.

N.N. Pariisky [3], who was a member of the expedition, thoroughly studied the defects of both old and new determinations and compared the two periods. The results of these investigations are represented in Table 1.

The above data show that the changes in the values of gravity during 25-30 years do not exceed the errors of their determination for eight stations while for four

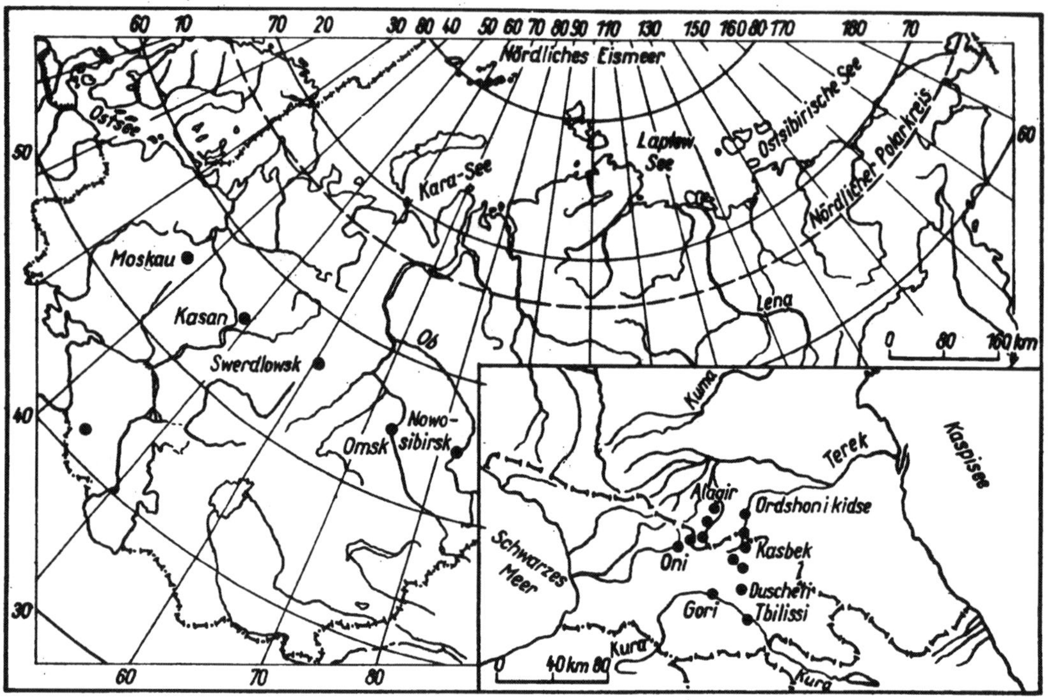

Fig. 1

14 stations for repeated gravity measurements in the region of the main Caucasian range

Table 1

Gravity changes in the Caucasus

Station	Number of years between periods	Gravity change in mgal	Error in mgal
Tbilisi	32	0	± 2.0
Gori	28	+ 4	5.8
Telavi	23	+ 4	4.8
Dusheti	25	+ 2	3.9
Passanauri	25	+ 4	4.0
Gudauri	25	+ 3	4.1
Kazbeghi	25	+ 9	4.6
Lars	24	- 3	4.4
Ordjonikidze	28	+13	5.8
Alagir	25	+ 5	4.0
Sadon	25	+12	4.7
Severny Priyut	25	+ 6	4.2
Glola	25	+ 7	4.2
Oni	25	+ 1	4.2

stations the differences are below the doubled mean error. Two stations showed an increase in gravity by 12 and 13 mgal with an error of ± 5 mgal. As far as these stations are concerned, we are not certain that the old points were identified with sufficient accuracy. Besides, it is difficult to believe that such a large secular change could be observed at one station while at the neighbouring station situated some 30-40 km away no changes at all were recorded.

By 1945 Pariisky had completed to process the gravimetric connections of Tbilisi with stations in the European part of the USSR and directly with Potsdam [4]. His work showed that the values of gravity in Tbilisi in reference to these stations had decreased by 0.4 mgal ±2.0 mgal from 1903-1909 to 1931-1936, i.e., in the course of 25-30 years.

The two works cited above give us good reason to assert that there are no noticeable gravity changes in time, at least in the Caucasus in reference to the plain part of the Soviet Union. Even if they do exist, they are so negligible that they are within the errors of measurement by conventional pendulum instruments.

The numerous determinations of the values of gravity in Washington can serve as an illustration of such secular changes. The gravity acceleration for Washington was determined for the first time in 1875-1876 by Pierce who used the absolute method. His result was: g = 980.098 gal.

After this the Washington station was repeatedly connected with European gravimetric stations. The most important results were obtained by Pierce [5], Putnam, Vening Meinesz [6], Brown [5], Winter [7] and others. A summary of these determinations is shown graphically in Fig. 2

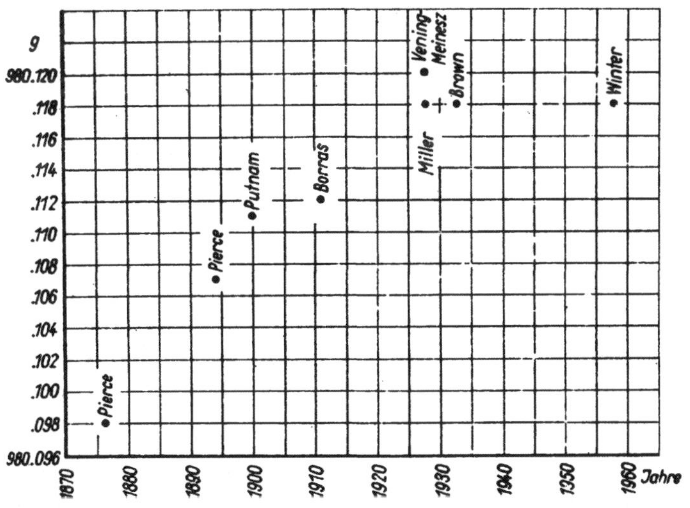

Fig. 2

Results of the repeated connection of gravity determinations in Washington with European stations. After Pierce [5], Putnam, Vening Meinesz [6], Brown [5], Winter [7] and others

If we omit the analysis of the measurement errors we may have an impression that in Washington the gravity was increasing during 50 years in reference to Europe at a rate of 0.4 mgal per year and then, beginning from 1929, this increase ceased. This is an erroneous viewpoint. The analysis of the measurements shows that the determinations carried out prior to 1929 were far from accurate, contain serious errors and cannot therefore be utilised for the study of secular gravity changes. In actual fact, the first period of observations for Washington dates from the early thirties with a mean error of 1-1.5 mgal while the second period commenced about 30 years later with a mean error of less than o.5 mgal. Their good agreement testifies to the fact that here too, just as in the Caucasus, we have no grounds whatever to assume any noticeable secular gravity changes.

In 1950-1954, work was undertaken in the Soviet Union to set up a network of gravimetric reference points. Incidentally, several such points were marked out along the 56th parallel in 1950-1951 between Riga and Klyuchi in Kamchatka. The measurements were done with a set of Nørgaard gravimeters consisting of seven instruments. The gravimeters were transported between the points by aircraft in special thermostats. Each connection between the adjacent stations was determined by at least three independent measurements conducted according to the following scheme: point A-point B-point ZA. The gravimeters were calibrated on an area with a range of 550 mgal whose relative error in the CGS system did not exceed 4×10^{-4}.

Ten years later, in 1961, the measurements were repeated at the same points in Kazan, Sverdlovsk, Omsk and Novosibirsk. Nine ГАЭ-3 gravimeters were used for the purpose. As before, the gravimeters were transported by aircraft in special thermostats. The measurements were conducted according to the former scheme. The gravimeters were not calibrated since numerous experiments had established that the scale factors of these instruments equalled unity and their relative error did not exceed 1×10^{-4} [8] [9].

In both cases we used the same point in the region of Moscow as the initial station. The results of these measurements are shown in Table 2.

Table 2

Gravity measurement in the Moscow-Novosibirsk area

Station	Mean quadratic error of determination		Differences	Mean quadratic error of difference
	1951	1961		1961 - 1962
Moscow	±0.00	±0.00	0.0	±0.00
Kazan	0.18	0.16	0.0	0.24
Sverdlovsk	0.24	0.17	0.0	0.29
Omsk	0.38	0.17	-0.2	0.42
Novosibirsk	0.44	0.18	0.0	0.47

It follows from the data in this table that during the ten years the values of gravity at all the three points situated in a latitudinal direction over more than 3,000 km remained the same within 0.1 mgal and decreased by 0.2 mgal only at one point with an error of ± 0.42 mgal. Besides, we are not quite sure that the old and new stations were identified with the required accuracy.

Thus, the above data show that at present we do not possess trustworthy evidence on secular changes of gravity. If they really exist they are too small to be detected by the measurement methods used for the purpose.

The instruments we have at our disposal have a sensitivity of 0.01-0.02 mgal. The method of numerous group measurements which gave a good account of itself in setting up the network of gravimetric reference points can be effectively used to measure Δg in the system of the given group of gravimeters with an error of 0.01 mgal and even smaller.

When measuring high Δg we face an exceptionally difficult task of calibrating the gravimeters. To day most gravimeters can be calibrated with an accuracy not exceeding $4\text{-}5 \times 10^{-4}$, whereas the guaranteed accuracy of calibration is 1×10^{-3}. Only type ГАЭ gravimeters can be calibrated with an accuracy by one order higher [9]. But these instruments have a large reading error and cannot be employed for such work.

Repeated observations for revealing secular gravity changes should have between them intervals of serveral years and possibly serveral decades. Due to a rapid development of measuring techniques it is inevitable that instruments of other types are used for repeated observations. Therefore, for the results to be comparable they should be obtained strictly within the CGS system. For this purpose, we must evolve such methods of measurement as would make the effect of calibration errors of no essential value.

The following method can be suggested. After marking out a point where gravity changes can be expected for geological and geophysical reasons, choose two control points so that the Δg between them and the main point is close to zero but does not exceed 10 mgal. The control points must be removed by over 500 km from the main point, because, according to the present-day conceptions, the causes that may bring about changes in the values of gravity might be situated 200 to 700 km deep. With this arrangement of the points the change of g at the main point will affect g at the control point by only 0.1 of the maximum value of effect as a result of which 0.9 of the effect value will be detected.

Connections between the main and control points must be established by gravimeters transported by aircraft. It stands to reason that it is impossible in practice to select airfields so that their location satisfies the condition $\Delta g < 10$ mgal. It is expedient, therefore, that connection between the main and control points be effected in a chain-like manner, by means of several links. Some links can be serviced by aircraft, some by helicopters and others by cars. With this arrangement the control points can always be located as far from the main point as possible and the final difference between them reduced close to zero.

If all these measures are observed the control differences Δg can be obtained in the CGS system with a very high accuracy. This method of measurement makes it possible to follow the secular relative changes of the values of gravity with an accuracy of 0.01 mgal, further develop the study of the constancy of the Earth's gravitational field in time and make use of far more effective gravimetric measurements for the investigation of recent crustal movements.

R e f e r e n c e s

[1] Abakelia, M.S.: On the Gravity Changes in Time in Connection with Geotectonic Movements in the Caucasus, Problems of Soviet Geology, 1936, Vol. 6, No. 2, pp 117-122

[2] Abakelia, M.S.: On the Organisation of Gravimetric Observatories in the Transcaucasus, Problems of Soviet Geology, 1936, Vol. 6, No. 5, pp. 452 - 454

[3] Pariisky, N.N.: Investigation of Time Changes of Gravity in the Caucasus, International Geological Congress, XVII session, theses of reports. Moscow-Leningrad, 1937, pp.189-190

[4] Pariisky, N.N.: Gravity Acceleration in Tbilisi, Transactions of Geophysical Institute, No. 7, (134), Moscow-Leningrad, 1949

[5] Brown, E.J.: A Determination of the Relative Values of Gravity at Potsdam and Washington, Coast and Geodetic Survey, Sp. Publ. No. 209, Washington, 1936

[6] Miller, A.H.: Determination of the Relative Values of Gravity at Potsdam, Greenwich, Ottawa and Washington. Publications of the Dominion Observatory, Ottawa, Vol. XI, No. 2, 1931

[7] Winter, P.J.; Valliant, H.P. and Hamilton, A.C.: Pendulum Observations at Ottawa, Ganger, Teddington, Paris, Rome and Bad Harzburg

[8] Kozyakova, K.Y.: Some Results of Calibration of Type CH-3 Gravimeters, Proceedings of the Ac. of Sci. of USSR, Geophysical series, 1956, No. 7

[9] Kozyakova, K.Y.: Results of Calibration of Type ГАЭ-3 Gravimeters, Proceedings of Academy of Sciences of the USSR, Geophysical series, No. 3, 1962

[10] Boulanger, J.D.: On Secular Gravity Changes, Soviet Geology, 1947, No. 25, pp. 3-8

The test of tectonic deformation measurements with the tiltmeter in Kondara (near Dushanbe) 1957-1959

von

A. Ye. Ostrovsky, A. B. Bakhrushin

L. I. Mironova, (Moskau)

Zusammenfassung

Tektonische Deformationen an Brüchen ergeben Neigungen von einigen Zehntel Sekunden im Jahr, das sind 100 mal größere Werte als jene, die auf der Grundlage geologischer Daten bestimmt wurden. Diese großen Differenzen ergeben sich daraus, daß die Neigungsmesser die Deformationen in einem kleinen Gebiet für eine kurze Zeit (von einigen Monaten bis zu einem Jahr) feststellen. Die geologischen und geomorphologischen Methoden geben die Durchschnittsdeformationen während größerer Zeiträume und in großen Gebieten an, so daß die richtungsändernden Deformationen sowohl bezüglich Zeit wie Richtung von Ort zu Ort die allgemeine Neigung verkleinern.

Um die gegenwärtigen Bewegungen mit Neigungsmessern zu studieren, ist es notwendig, Beobachtungen an einer großen Zahl von Punkten auszuführen, weil die tektonischen Deformationen von Ort zu Ort sich sehr ändern.

Kleine Zeiträume, von einigen Monaten bis zu einem Jahr, während denen Deformationen gemessen werden können und eine kontinuierliche Aufzeichnung der Deformationen vorgenommen werden kann, die die genauen zeitlichen Deformationsvariationen wiedergibt, sind die Vorteile der Neigungsmessermethoden von tektonischen Bewegungsmessungen im Vergleich zu den geodätischen Messungen.

During the International Geophysical Year 1957 - 1959 the observations of tidal tilts of the Earth were carried out in Central Asia. Observations were carried out in three districts: in Ashkhabad, Alma-Ata, Dushanbe.

The results of this work were reported at the symposiums on the earth's tides.

Tilts were measured with new instruments with photoelectric recording which had a distant control [10, 11]. The sensitivity of a tiltmeter was "/1000 mm or 0.0001 mm/ ". However, the realisation of such high sensitivity was not always a success, owing to the slow deformations of the ground, on which the tiltmeters were set.

While treating tidal tilts slow deformations were detached from tidal oscillating tilts by means of the method which excludes the drift-curve from observations [12]. This paper deals with the slow monotonous deformations, noticeable on the records of tilts from month and from year to year.

The observations in Alma-Ata as well as in Ashkhabad were carried out at one point, therefore it is difficult to interpret them. As to Kondara (near Dushanbe) tilts were measured at three points of one and the same adit, the comparison of slow tilts is of some interest because these measurements may be taken on any area, though not large.

The adit, in which the tiltmeters were set, was driven for the purpose of studying the

Kondara rupture. It lies along the rupture and discloses it by three cuts (see Fig. 1) which intersect the rupture. Tiltmeters were set at three points: 1,2 and 3. The first pair of instruments was set on the south block, the second on the north one. Both these points were placed in direct closeness from the rupture (2 - 5 meters). The third pair of tiltmeters was set at a distance of 20 meters from the rupture, i.e. the greatest possible removal from it.

Slow deformations, recorded with a tiltmeter, are the results of many causes: 1 - the variation of the Earth's surface temperature; 2 - the variations of atmospheric pressure; 3 - the penetration of water into the ground; 4 - instrumental causes; 5 - the deformations of foundation, on which the instruments are set, and at last, 6 - the deformations of displacement, depending on tectonic procedures.

The geological and geomorphological methods of analysis of vertical movements show that tectonic tilts are extremely little [13] and do not exceed some hundreds of a second a year. However these deformations are mad for large spaces of time -- millions and more years. Besides their estimation is made by taking into consideration large areas. In reality the deformations of the Earth are considerably large areas. In reality the deformations of the Earth are considerably greater than those usually given by geologists.

The measurements of deformations with tiltmeters show that the value of deformations greatly depends on the place of observations.

According to observations in central Asia tilts in some points reached 20 seconds a year, but in other places they range 1 - 3 seconds a year.

While setting the instruments the depth of 20 and more meters the temperature tilts become very little. Thus the daily temperature wave at a depth of 15 meters in Alma-Ata does not exceed 1 - 2 milliseconds of the arc.

The comparison of records of atmospheric pressure with tilts showed, that if the tiltmeters are set on solid rocks -- granites the influence of atmospheric pressure on tilts is insignificant.

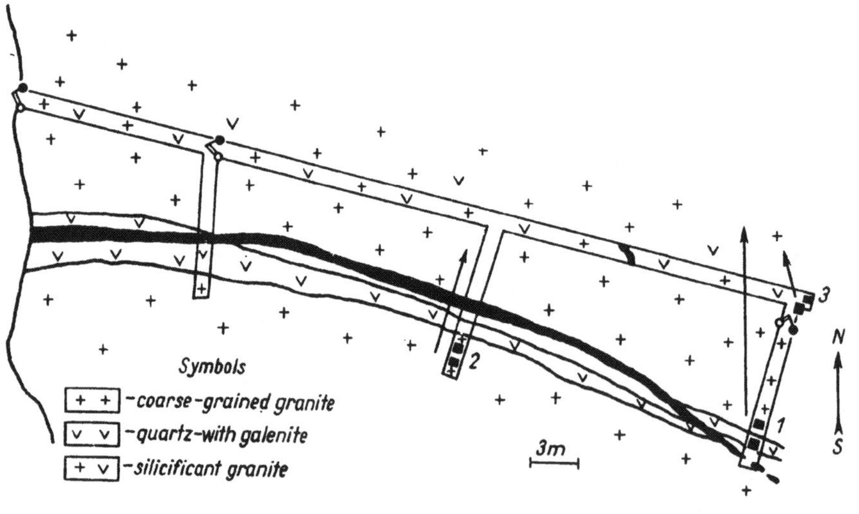

Fig. 1
Tilts in Kondara, pictured as vectors

Instrumental tilts may be greatly decreased by improving the apparata. These fictitious tilts may be considerable instant after setting and decrease quickly afterwards. Besides they may be marked out by the form of the curve. The deformations of the foundation or local deformations may be marked out only by carrying out observations simultaneously at some points. The tilts, connected with the penetration of moisture into the chamber, make great obstacles. To remove them the points of observations must be set at dry places. At the station "Kondara" this last request was ignored and therefore it was very difficult to interpret the tilt curves. However, it was found that it was still possible to determine approximately the tectonic monotonous tilts at three points of the Kondara adit according to the yearly number of observations.

Table I indicates the values of tectonic tilts a year. The tilt for a year in the direction of N-S in the first point reaches 16 seconds of the arc, in the second - 9 seconds and in the third - 3 seconds. Tilts decrease quickly depending on moving off from the rupture. Tilts in the direction of E-W are considerably smaller than in the direction of N-S. The third point is of great importance as to the value of resulting tilts. Here tilt is four times smaller than that in the first point, and three times smaller than that in the second one.

As to the azimuth the tilts in all three points have meridional direction (see tilts, pictured as vectors on Fig. 1). It is easy to explain the deformations of displacement of such direction, if we assume that the north block lowers and the south one rises.

The measurements of tilts do not determine if the south block rises or the north one lowers. One may judge only about the relative movements of one block in comparison with the other.

Deformations damp quickly depending on the increase of distance between the setting of tiltmeters and the rupture.

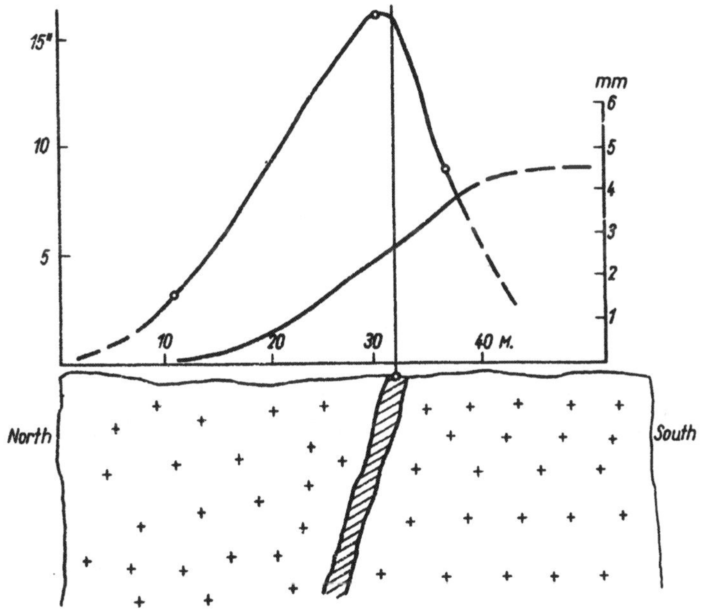

Fig. 2

Curve A: Decrease of tectonic angles of gradient with increase of the distance from the rupture

Curve B: Displacement of one block relative to the other one

Fig. 2 (curve A) shows the decrease of tectonic angles of gradient with increase of the distance from the rupture, and the curve B presents the displacement of one block relative to the other one.

Maximum tilts a year reach 18 seconds, but the displacement of one block relative to the other is about 4 mm a year here.

C o n c l u s i o n

1. Tectonic deformations on the ruptures may give the tilts some tens of seconds of arc a year, i.e. hundreds times greater than those which are determined according to geological data. Such great difference is because tiltmeter fixes deformations on a small area and for a short time (from some months to a year). The geological and geomorphological methods give mean deformations during large spaces of time and on great areas, thus all the sign-varied deformations both in time and passage from place to place decrease the general tilt.

2. To study present movements with a tiltmeter it is necessary to carry out the observations in the great number of points, as the tectonic deformations change greatly from place to place.

3. Short space - from some months to a year during which the deformations may be measured, and the continuous recording of the deformations, which gives the detailed variations of deformations in time, are the advantage of the tiltmeter's method of tectonic movements measurements in comparison with the geodetic measurements.

R e f e r e n c e s

[1] Ostrovskij, A.E.; Bachrušin, A.B.; Mironova, L.I.: Prilivnye naklony zemli po nabljudenijam v Kondare v 1958 g. Sbornik MGG Gravimetričeskie issledovanija, No. I, 1960, AN SSSR

[2] Ostrovskij, A.E.; Choromskij, A.V.; Mironova, L.I.: Rezul'taty nabljudenij prilivnych naklonov zemli v Kounrade v 1957 g. Sbornik MGG. Gravimetričeskie issledovanija, No. I, 1960 g., AN SSSR

[3] Ostrovskij, A.E.; Ionyčev, N.N.; Fandjušina, S.M.: Prilivnye naklony zemli po nabljudenijam v Aščhabade v 1957 - 1958 gg. Sbornik MGG. Gravimetričeskie issledovanija, No. I, 1960 g., AN SSSR

[4] Ostrovskij, A.E.; Matveev, P.S.; Fandjušina, S.M.: Nabljudenija prilivnych naklonov v Poltave v 1958 g. Sbornik No. I MGG. Gravimetričeskie issledovanija, No. I, 1960 g., AN SSSR

[5] Ostrovsky, A.E. and Horomsky, A.V.: Observations of tidal tilts of the earth crust by means of a photoelectric tiltmeter in Kounrad in October - December 1957. Communications de l'observatory Royal de Belgique, No. 142, 1952

[6] Ostrovsky, A.E.; Horomsky, A.V.; Mironova, L.: Results of observations of tidal tilts of the earth's surface by means of a photoelectric tiltmeter in Kounrad in October - December 1957. Bullettino di Geofisica Anno 2, No. 5, Marzo 1960, Trieste

[7] Ostrovsky, A.E.; Matveiev, P.; Fandushina, S.: Observations of tidal tilts of the earth by means of a photoelectric tiltmeter in Poltava, July- October 1958

[8] Ostrovsky, A.E.; Bakhrushin, A.; Mironova, L.: Tidal tilts of the earth according to the observations in 1958 in Kondara (near Stalinabad). Bullettino di geofisica Anno 2, No. 5, Marzo 1960, Trieste

[9] Ostrovsky, A.E.; Jonitchev, H.; Fandushina, S.: Tidal tilts of the Earth according to the observations in Ashkhabad in 1957 and 1958. Bullettino di Geofisica Anno 2, No. 5, Marzo 1960, Trieste

[10] Ostrovskij, A.E.: Sejsmonaklonomer s fotoelektričeskoj registraciej. Bjulleten' Soveta po sejsmologii, No. 6, 1957 g.

[11] Ostrovskij, A.E.: Naklonomer s fotoėlektričeskoj registraciej. Sbornik MGG, No. 2. Gravimetričeskie issledovanija, 1961 g. AN SSSR

[12] Percev, B.P.: Ob učete spolzanija nulja pri nabljudenii uprugich prilivov. Izv. AN SSSR, serija geofiz., No. 4, 1959 g.

[13] Gzovskij, M.V.; Krestnikov, V.N.; Rejsner, G.I.: Geologičeskie metody opredelenija gradienta skorosti tektoničeskich dviženij i nekotorye rezul'taty ich primenenija. Izv. AN SSSR, serija geofiz., No. 8, 1959 g.

Anomalous tilting movement of the ground observed before destructive earthquakes

von

E. Nishimura (Kyoto)

Introduction

Among all the natural disasters the damage caused by destructive earthquake will be the most horrible and fatal because of its unexpected occurrence and inexpressible violence. Seismology tells us in great detail of various phenomena observed at and after the moment of an earthquake occurrence, such as location of the focus, propagation mode of the seismic waves, total energy of the earthquake, and the kindred items. Earthquake

1. Hosokura
2. Izu-Nagaoka
3. Kamioka
4. Ogoya
5. Ŝhima
6. Nagashima
7. Kishû
8. Shionomisaki
9. Yura
10. Şuhara
11. Ōura
12. Akibasan
13. Ide
14. Ōsakayama
15. Kamigamo
16. Ikuno
17. Tottori
18. Aso
19. Makimine
20. Sakurajima

Fig. 1
Location of the observation stations of the Kyoto University

engineering also offers us many considerations concerning the safety factors of building and construction against earthquake. But, we, much to our regret, have little definite knowledge on phenomena occurring just before the earthquake. Finding of the phenomena forerunning earthquake, is considered to be one of the most urgent tasks assigned to every seismologist, because it will greatly serve to investigate the intrinsic nature of the earthquake, namely the accumulation process of earthquake energy, mode of its release, and others on one hand, and to give a warning against destructive earthquake coming on the other, which will considerably reduce disastrous loss, especially of human life.

It is a matter of common knowledge that conspicuous upheaval and subsidence of the ground have often been observed in the epicentral region after occurrence of a destructive earthquake. However, a fact has also been found by repetition of precise levelling that this sort of the ground movements has, in some cases, preceded the earthquake occurrence. In these cases, a continuous observation of the ground movements with some suitable instruments, is considered to be certainly serviceable to the purpose of foretelling earthquake occurrences.

On this point of view, the Kyoto University has made some observations of the ground movements with tiltmeters, extensometers, and other instruments since 1937, and at present about fifty tiltmeters, forty extensometers, ten geomagnetic declinometers, and ten gravity variometers are being in operation at twenty stations in Japan. Location of these observation stations is as shown in Fig. 1. Some peculiar changes of the ground tilt, obtained by the above-mentioned observations and regarded as closely related to the occurrence of recent destructive earthquakes, will briefly be described in the following.

1. Indicational ground tilts within several hours before earthquake occurrence

In this section, some examples will be described on a peculiar change of the ground tilt observed within a period of several hours just before the occurrence of destructive earthquakes.

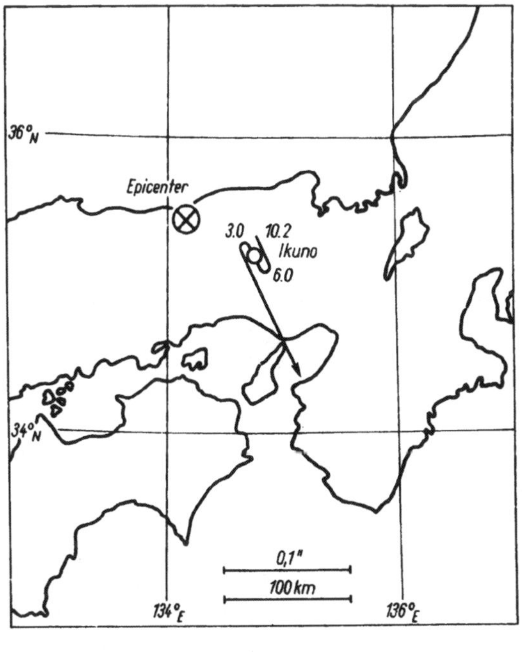

Fig. 2

Tilting movement of the ground observed at Ikuno before occurrence of the Tottori Earthquake on September 10, 1943

a) Case of the Tottori Earthquake

On September 10, 1943, a destructive earthquake occurred in the vicinity of Tottori City where were 7,348 casualities and 13,897 buildings damaged. The position of the epicentre was 134.2°E and 35.5°N, the depth of the focus and seismic magnitude were 10 km and 7.5 in Pasadena Scale respectively. In this case, the nearest observation station to the epicentre was Ikuno, its epicentral distance being 60 km, where a minute but clear characteristic tilting movement of the ground was observed several hours before the occurrence of the earthquake. This peculiar ground tilt is schematically shown in Fig. 2. The direction of arrow line in the figure shows the direction of the downward tilting movement of the ground, its length represents the amount of tilt, and annexed numerals denote the time in hours counted back from the moment of the earthquake occurrence. Although the amount of this forerunning tilting movement is minute, its S-type movement is very peculiar and clearly distinguishable from other usual tilting movements.

b) Case of the Tônankai Earthquake

On December 7, 1944, the great Tônankai Earthquake occurred in the region 20 km off-shore in the open sea of Kumanonada, in consequence of which there were 3,133 casualities and 76,151 buildings damaged, mainly by a tsunami reaching to several metres in height. The position of the epicentre, focal depth, and seismic magnitude were estimated to be 136.2°E and 33.7°N, about 20 km, and 8.0 in Pasadena Scale respectively. In this case, the only observation station near the epicentre was Kamigamo Geophysical Observatory, situated in the northern part of Kyoto City and 160 km distant from the epicentre, also where an extremely minute and characteristic ground tilt of S-type was clearly observed. The schematic diagram of the tilting mode is shown in Fig. 3

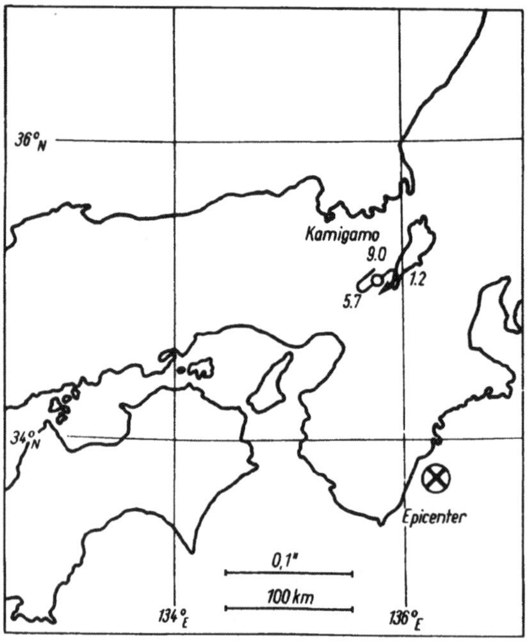

Fig. 3
Tilting movement of the ground observed at Kamigamo before occurrence of the Tônankai Earthquake on December 7, 1944

c) Case of the Nanki Earthquake

On April 26, 1950, the strong Nanki Earthquake occurred in the southern part of Kii Peninsula. The position of the epicentre, focal depth, and seismic magnitude were 135.8°E and 33.9°N, 40 km, and 6.7 in Pasadena Scale respectively. The damage by this earthquake was, however, small by reason of the epicentral region being in an uninhabited area. In this case, the tiltgrams obtained at Tamamizi (at present in suspension), Kamigamo Geophysical Observatory, and Kôchi (at present in suspension), their epicentral distances being 80 km, 120 km, and 200 km respectively, similarly showed the peculiar S-type ground tilts several hours before the earthquake occurrence. The mode of the tilting movements is schematically shown in Fig. 4.

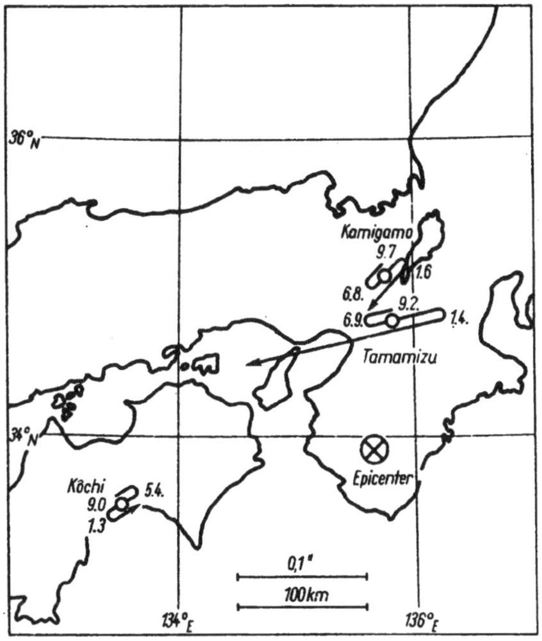

Fig. 4
Tilting movements of the ground observed at Tamamizu, Kamigamo, and Kôchi before occurrence of the Nanki Earthquake on April 26, 1950

Scrutinizing the three cases of the above-mentioned earthquakes, it was established that a great or destructive earthquake had, in some cases, been preceded by a minute and characteristic S-type tilting movement of the ground reaching to 0.1"in angle at a distance of 100 km from the epicentré, in several hours before its occurrence.

2. Indicational ground tilts in several days and several months before earthquake occurrence

Although the above-mentioned ground movements observed several hours before earthquake occurrence are considered to be very suggestive and effective as to the research on the intrinsic nature of the earthquake, it is very difficult in practice to make them serviceable as an early warning of an earthquake occurrence because of their lateness. Concerning the purpose of warning against the occurrence of a destructive earthquake and reducing disastrous loss, it is necessary to find any phenomena, which would forerun the earthquake in sufficient time before its occurrence. But an accurate observation of secular change in the ground tilt or ground strain, available to this purpose, is considerably a difficult task, because of the amount of such a change being generally very minute compared with that of irrelevant disturbance caused by meteorological changes, in addition to, the instrumental defects, the interruption of observation caused by artificial difficulty, and the like.

In spite of these difficulties some cases were recently found, in which slow and peculiar ground tilts and ground strains were observed several months before an earthquake occurrence, and rapid and severe changes appeared several days before. In this section, five examples of such a case will be reported.

a) Case of the Daishôji-oki Earthquake

On March 7, 1952, the strong Daishôji-oki Earthquake occurred, having its epicentre in the region of 40 km distance in the NW-direction from Ogoya observation station. The position of the epicentre, focal depth, and seismic magnitude were 136.2°E and 36.5°N, 20 km, and 6.5 in Pasadena Scale respectively. The mode of the ground tilt observed at Ogoya was as shown in Fig. 5 by a vector diagram, in which the direction and amount of

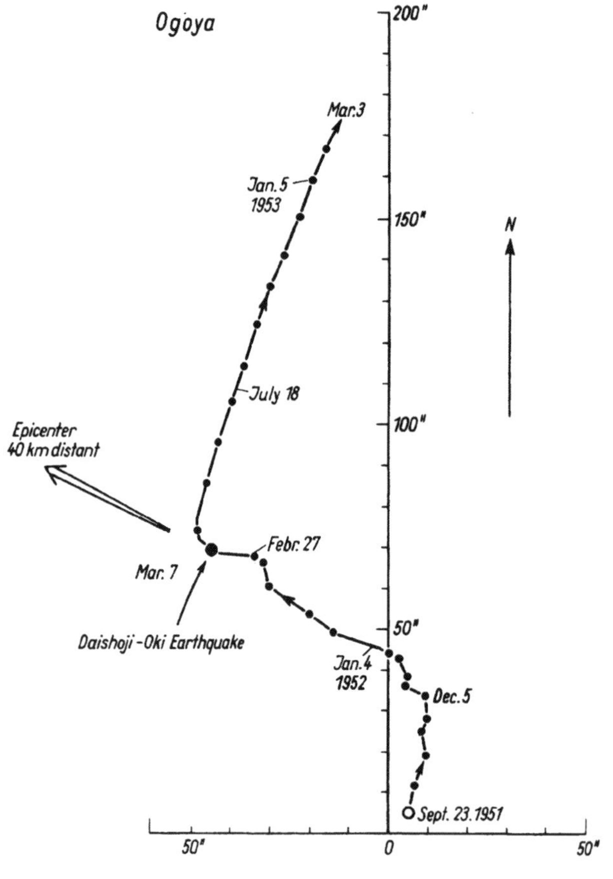

Fig. 5
Vector diagram of the ground tilt observed at Ogoya. Single and double arrows show the time of earthquake occurrence and the direction of epicentre

the ground tilt would be represented by a vector drawn from the origin to a point on the curve, annexed numerals denoting the date of observation, and single and double arrows the time of earthquake occurrence and the direction of epicentre respectively. As seen in the figure, the ground of Ogoya continued its downward tilting in the N-direction from September 23, 1951 till December 5, and since then turned its direction to NW. After February 27, 1952, the westward tilting movement became more and more marked and continued to March 7, namely the day of occurrence of the strong Daishôji-oki Earthquake. Several days after the earthquake, the tilting movement turned its direction to NE, orthogonal to the preceding, and continued it until the present. Namely the ground of Ogoya commenced its downward tilting movement in the direction of the epicentre three months before the earthquake occurrence, increasing its speed several days before, and several days after the occurrence turned its direction orthogonally to the epicentre.

b) Case of the Yoshino Earthquake

On July 18, 1952, the strong Yoshino Earthquake occurred, having its epicentre in the region 55 km, 72 km, and 94 km respectively from Yura, Ide, and Ôsakayama observation stations. The position of the epicentre, focal depth, and seismic magnitude were 135.4°E and 34.1°N, 70 km, and 7.0 in Pasadena Scale respectively. As to the tiltmetric observation at Yura, a vector diagram of the ground tilt is shown in Fig. 6, by means of

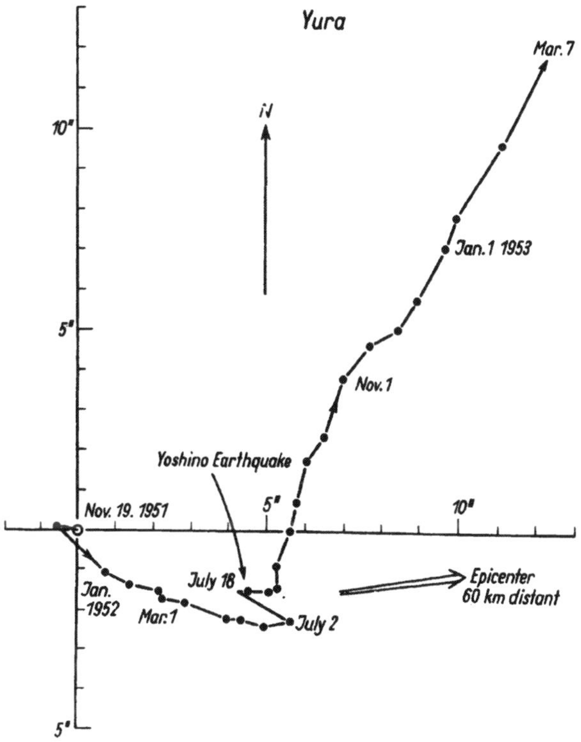

Fig. 6

Vector diagram of the ground tilt observed at Yura. Single and double arrows show the time of earthquake occurrence and the direction of epicentre

the same representation as in Fig. 5. In present case, the downward ground tilt at Yura had been towards the ESE-direction until July 2, 1952, and then turned its direction to WNW to continue till July 16. Two days after the reverse of its direction to E, namely on July 18, the strong Yoshino Earthquake occurred. After the earthquake, the eastward tilting continued till July 30, and then turned its direction to NE, that is to say nearly orthogonal to the preceding. Such a mode of the ground tilt of Yura may be regarded as entirely similar to that of Ogoya in the case of the Daishôji-oki Earthquake.

In addition to above, at Ide and Ôsakayama stations, some remarkable changes in the ground strain were found by extensometric observations in several months before the earthquake occurrence. At Ide, extensometric observations for six directions had been in operation since 1951. On the occasion of this earthquake, some anomalous ground strains, namely rapid expansion and contraction, were observed respectively about four and two months before the earthquake occurrence. At Ôsakayama, gradual expansion of the ground had taken place nearly one year before and rapid expansion three months before the earthquake, and nearly one month after the earthquake, the ground strain changed its behaviour from expansion to contraction.

c) Case of the Ôdaigahara Earthquake

On December 26, 1960, the conspicuous Ôdaigahara Earthquake occurred, its epicentre being in Ôdaigahara Heights of Kii Peninsula. The position of the epicentre, focal depth, and seismic magnitude were 136.2°E and 34.2°N, 60 km, and 6.0 in Pasadena Scale respectively. At that time, there were seven observation stations within 100 km distance around the epicentre, where tiltmetric and extensometric observations were carried out. The mode of the ground tilts observed at those stations was as shown in Fig. 7, in which

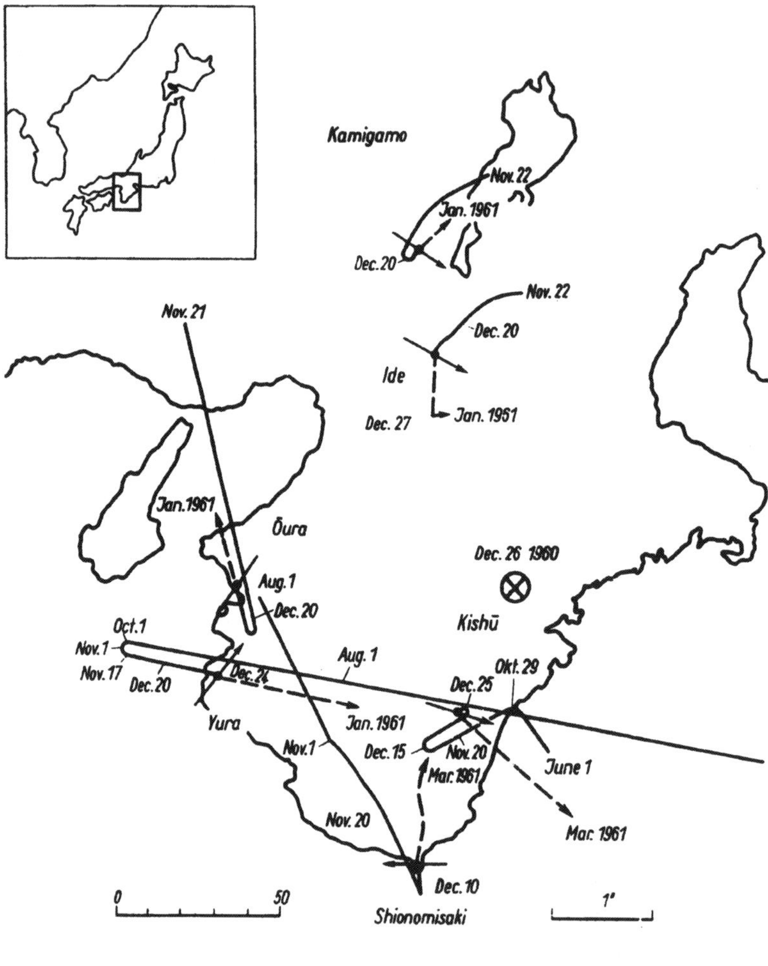

Fig. 7

Abnormal tilting movements of the ground observed before and after occurrence of the Ôdaigahara Earthquake on December 26, 1960. Small arrows show the usual tilting movements at the respective stations

small and straight arrows showed the direction of the usual ground tilts proper to each station, long curved arrows the direction and amount of the abnormal ground tilts deviated from the usual, and annexed numerals the date of observations. Examining the figure in detail, one would observe that the tilting movements of the ground at those stations could be divided into three parts respectively corresponding to three epochs. In the first epoch, at Kishû, Shionomisaki, and Yura stations, situated in the SW-quadrant about the epicentre, where abnormal ground tilts commenced in June of 1960, and continued till the end of October to reach several seconds in angle, and then turned their directions. In this epoch, extensometric observation at Yura also recorded an abnormal expansion of the ground in the N 60°W - S 60° E - direction. In the second epoch, on November 21 or 22, the grounds of Kamigamo, Ide, Ôura, and Akibasan (not expressed in the figure because of being too close to Ôura), situated in the NW-quadrant about the epicentre, commenced their abnormal counter-clockwise tilting movements. Extensometric observations at Ide, Yura, Ôura, and Akibasan also showed abnormal expansion or contraction of the ground since November 10. In the third epoch, namely in the middle of December, the ground tilts at almost every station changed their directions, and several days later the conspicuous Ôdaigahara Earthquake took place. It is difficult to draw a definite conclusion as to whether the above-described ground movements were entirely related with the occurrence of the earthquake or not, because the tilting movements of the ground everywhere used to be largely disturbed by meteorological changes, especially by precipitation and atmospheric pressure change. In the present case, however, based on a fact that the ground of so many stations simultaneously showed similar movements, it might be asserted that the above-mentioned ground movements were closely related to the earthquake occurrence.

d) Case of the Hyûganada Earthquake

On February 27, 1961, the strong Hyûganada Earthquake occurred, accompanying with tsunami, its epicentre being in the region 100 km distant from Makimine observation station, off-shore in the open sea of Hyûga-nada. The position of the epicentre, focal depth, and seismic magnitude were 131.7^{o}E and 31.7^{o}N, 17 km, and 7.2 in Pasadena Scale respectively. For the purpose of investigating earthquakes occur in Hyûga-nada, tiltmetric observations were kept up since 1942 till the present in the underground drift of 165 metres from the ground surface of Makimine, where the amount of the usual ground tilt had been observed as 0.35" in angle per a month in the SSW-direction. The mode of the ground tilt observed at this station is shown in Fig. 8, by the same representation as in Fig. 7 except that a broken arrow showing the direction of the usual ground tilt at Makimine. As seen from the figure, abnormal ground tilt in the opposite direction to the epicentre commenced on February 15 of 1961 to continue till February 23, then reversed its direction, and four days later, namely on February 27, the earthquake occurred, and after that earthquake, the ground movement changed its direction to NNE, namely almost orthogonal to the preceding one.

e) Case of the Kitamino Earthquake

On August 19, 1961, the strong Kitamino Earthquake occurred, its epicentre being in the region 34 km and 60 km respectively distant from Ogoya and Kamioka observation stations. The position of the epicentre, focal depth, and seismic magnitude were 136.8^{o}E and 36.0^{o}N, 40 km, and 7.2 in Pasadena Scale respectively. The amount and direction of the usual ground tilt at Ogoya and Kamioka were 4.0" per a month in ENE, and 3" $\sim$ 4" per a month in NNW, and the mode of abnormal ground tilts at both stations was as shown in Fig. 9, two thick arrows representing the direction and amount of the abnormal ground tilt at respective stations at the moment of the earthquake occurrence. At Ogoya, as seen from the figure, abnormal ground tilt commenced on July 29, 1961, and, after a rolling motion of tilt roughly around the direction of the epicentre, of which double amplitude was estimated as about 1.8" in angle, returned to the initial state on August 24, namely five days after the earthquake. At Kamioka, an abnormal ground tilt commenced on July 25, and after a similar rolling motion of the amplitude of 0.45", returned to the initial state on August 23.

Hitherto, some examples of anomalous and characteristic changes of the ground tilt and ground strain which were considered to precede and be closely related to the occurrence of

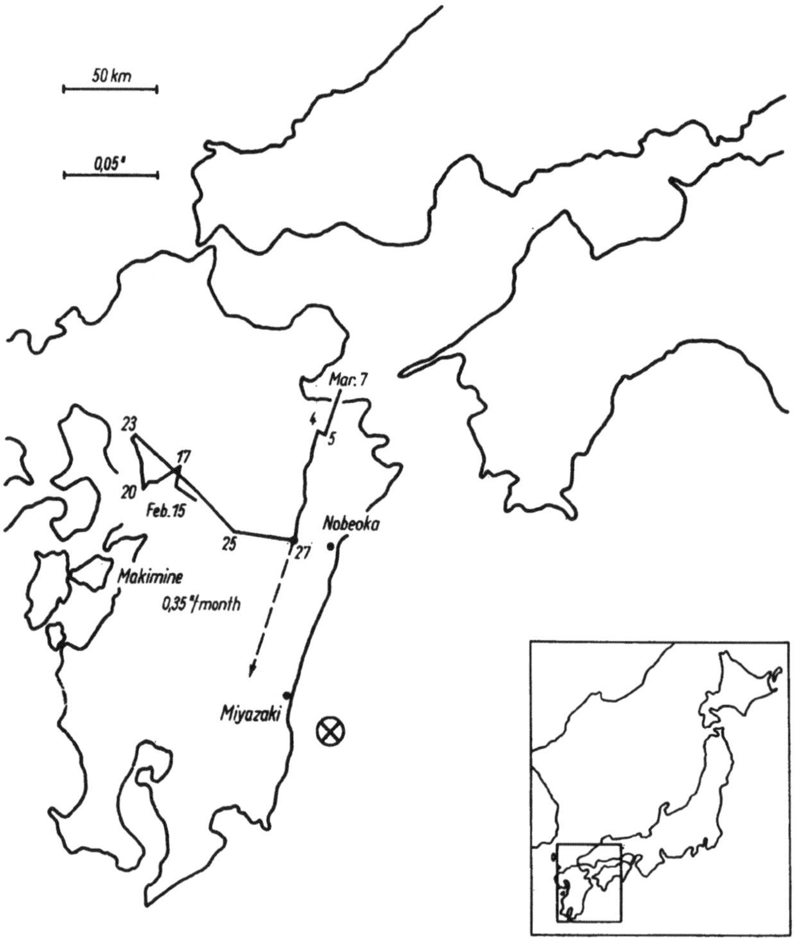

Fig. 8

Abnormal tilting movement of the ground observed at Makimine before and after occurrence of the Hyûganada Earthquake on February 27, 1961. Broken arrow shows the direction of the usual ground tilt

destructive earthquakes, have been reported. Although it is dangerous to derive a definite conclusion from such a few examples, these observational facts are expected to throw some light upon the methods of investigating the essential nature of an earthquake and forecasting its occurrence.

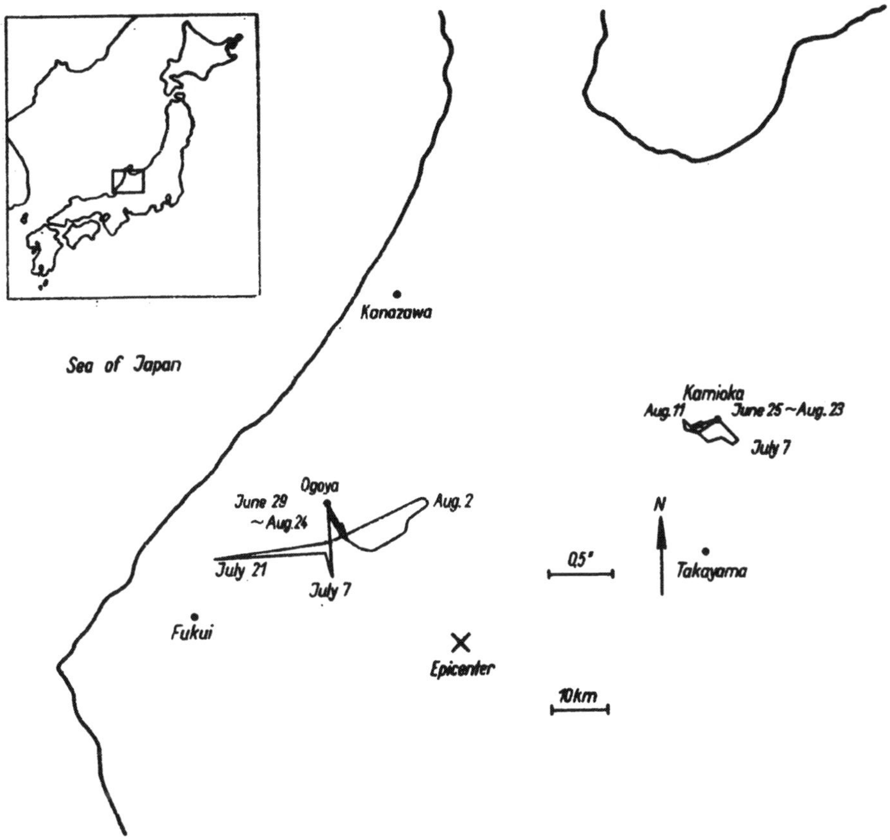

Fig. 9

Abnormal tilting movements of the ground observed at Ogoya and Kamioka before and after occurrence of the Kitamino Earthquake on August 19, 1961. Two thick arrows show the direction and amount of the abnormal ground tilt at the moment of the earthquake occurrence

3. Relation between the secular change of the ground tilt and the occurrence of earthquakes

In this section, on the basis of the observational facts obtained at three stations by tiltmetric observations of long periods, the relation between the secular change of the ground tilt and the occurrence of earthquakes will be described.

a) Case at the Kamigamo Geophysical Observatory

Kamigamo Geophysical Observatory is situated on a hill in the northern suburbs of Kyoto City. Tiltmetric observation at this observatory was commenced early in 1910 by T. Shida with tiltmeters of Rebeur-Paschwitz type, and after several interruptions it was resumed in 1937 by the present author with fused silica tiltmeters of Zöllner suspension type. Since then the observation with these instruments have, without any interruption, been continued until the present time, covering a long period of 25 years.

Annual variation of the ground tilt observed at this observatory was ascertained to be fairly regular and, to a certain degree, corresponding to the variations of atmospheric pressure and temperature. But, by reason of these meteorological variations being regular, the ground tilt represented by the successive annual averages of daily observational data, are considered to indicate the real secular change of the ground tilt, having little or no connection with changes caused by meteorological variations. Secular ground tilts at Kamigamo obtained by such a method, are graphically represented in Fig. 10, in which upward

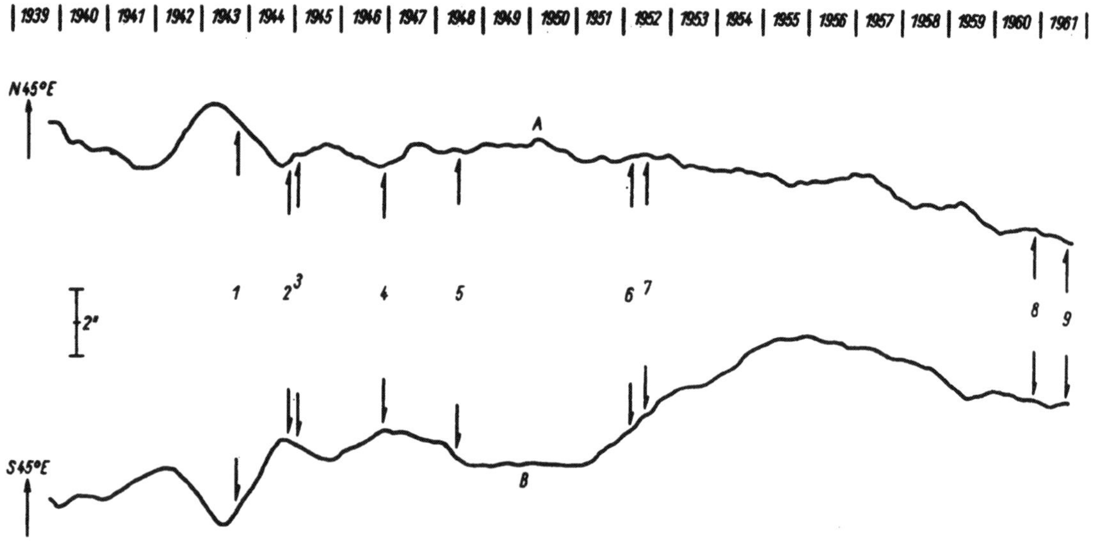

Fig. 10

Relation between the secular ground tilt and the occurrence of the neighbouring destructive earthquakes observed at Kamigamo Geophysical Observatory

movements of A and B curves show the downward ground tilts in the directions of N 45°E and S 45°E respectively.

Nine destructive earthquakes which seismic magnitude exceeded 7 in Pasadena Scale, occurred in the region within a distance of 200 km from Kamigamo during the recent twenty years, the time of occurrence of these nine earthquakes being indicated in the figure by numbers and arrows. Although not so clear in this figure, examining this figure with attention or reforming this into a vector representation, it will become clear that, in the cases of Nos. 1, 2, 4, 5, 6, and 8, the ground of Kamigamo showed some characteristic tilting movements one year or several months before occurrence of respective earthquakes. In view of the amount of the ground tilt being dependent upon the magnitude of earthquake, distance from the epicentre, and mechanism of energy accumulation and others, it is not so strange that such a characteristic ground tilt did not appear in the other cases.

b) Case at the Makimine observation station

Tiltmetric observation at Makimine has been carried out, as already described in the section 2.d , since 1942. Throughout the twenty years, seismic activity in the neighbouring region of Makimine was especially prominent during a period from July of 1960 to March of 1961, namely sixteen rather conspicuous earthquakes successively occurred in the region of Hyûga-nada within a distance of 100 km from Makimine in this period. Relation between the secular ground tilt and seismic activity in Hyûga-nada during twenty months from January of 1960 to August of 1961, is shown in Fig. 11 by the similar representation as in Fig. 10, the additional lower figure represents more minutely the part enclosed by a square of dotted line in the upper one. From this figure, similarly as in the case of Kamigamo, one would find that anomalous and characteristic ground tilts preceded the earthquake occurrence on the occasion of several earthquakes, especially of No. 15 (so-called the Hyûganada Earthquake on February 27) and No. 20.

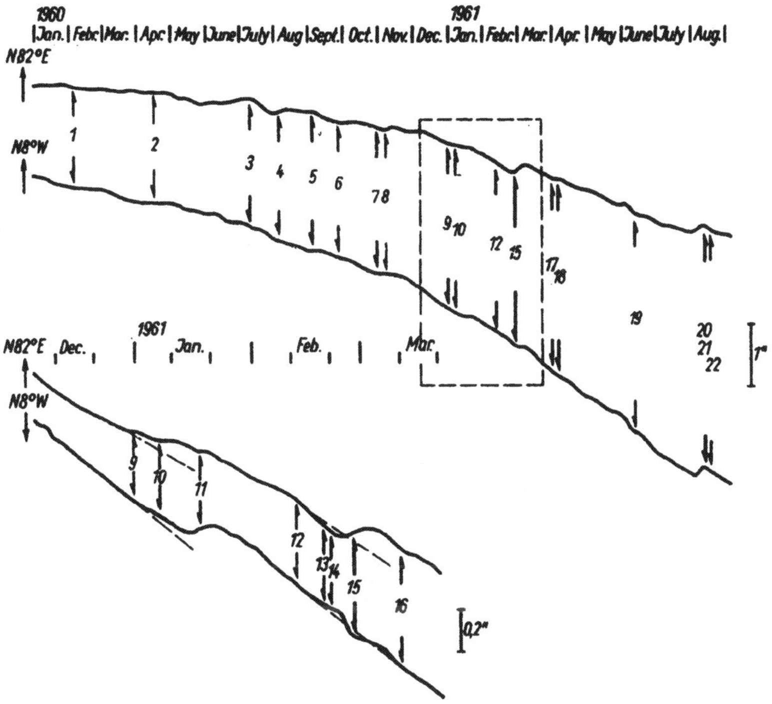

Fig. 11

Relation between the secular ground tilt and the occurrence of neighbouring earthquakes observed at Makimine

c) Case at the Yura observation station

Tiltmetric observation at Yura was commenced in 1951, and six strong or conspicuous earthquakes occurred in the neighbouring region within a distance of 100 km from Yura during ten years from 1951 to 1961. Secular change of the ground tilt at Yura during this period is shown in Fig. 12 by a vector representation. In the figure, six arrows indicate the time of occurrence of those six earthquakes , of which two thick arrows showing so-called Yoshino and Ôdaigahara Earthquakes already described in the section 2. b and c . At the first glance on the figure, one would clearly perceive that the ground tilt at Yura, almost orthogonally, changed its direction with every occurrence of earthquake, especially the ground movements in the cases of the Yoshino and Ôdaigahara Earthquakes were exceedingly peculiar, of which a detailed description was already given in the preciding section.

Only three examples of observational facts on the secular ground tilt were described here. But, in near future, more examples will be reported and some definite conclusion may be derived as to the relation between the secular change of the ground tilt and the occurrence of earthquakes.

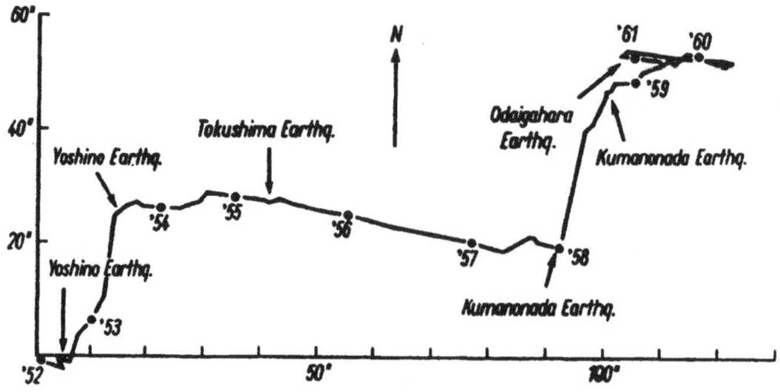

Fig. 12

Vector diagram of the secular ground tilt observed at Yura. Six arrows indicate the time of occurrence of respective earthquakes

Conclusion

There may be different opinions and various interpretations concerning the contents mentioned above. But, the observational facts are indisputable, and it is also difficult to regard the observational facts described above as merely accidental in all cases.

There are various methods considered available for predicting the occurrence of an earthquake; for instance, the repetition of precise levelling, observation of changes in the seismic activity and propagation velocity of seismic waves, investigation of changes in terrestrial magnetism and electricity, and others. But, a continuous observation of the ground movements with tiltmeter, extensometer, gravity variometer, and other instruments will, as one of the most effective method, greatly contribute to advance our knowledge on the essential nature of earthquake and the practical technique of foretelling the earthquakes.

Bibliography

Japanese Contributions to the Study on Recent Movements of the Earth's Crust by means of Continuous and Various Observations

1912

Shida, T.: On the elasticity of the earth and the earth's crust. Part I. Horizontal pendulum observations of the change of plumb line at Kamigamo, Kyoto. Mem. Coll. Sci. and Eng., Kyoto Imp. Univ., Vol. 4, pp. 23 - 174

1922

Sekiguchi, R.: On the tilting of the earth at Jinsen (Chemulpo) due to tidal load. Mem. Imp. Marine Obs. Kobe, Vol. 1, pp. 1 - 17

1927

Ishimoto, M.: Observations sur les variations de l'inclinaison de la surface terrestre (Premier rapport). B.E.R.I., Vol. 2, pp. 1 - 12

1928

Ishimoto, M.: Construction d'un pendule horizontal de quartz et observations sur les variation de l'inclinaison de la surface terrestre. Jap. Journ. Astron. Geophys., Vol. 6, pp. 83 - 118

1929

Takahasi, R.: Tilting motion of the earth crust caused by tidal loading. B.E.R.I., Vol. 6, pp. 85 - 108

Takahasi, R.: Tilting motion of the earth crust caused by secondary undulations of tides in a bay. ibid., Vol. 7, pp. 95 - 102

Ishimoto, M.: Silica tiltmeter. Journ. Seism. Soc. Jap., Ser. 1, Vol. 1, No. 1, pp. 17 - 32. (J)

Takahasi, R.: Tilting motion of the earth crust caused by tidal loading. ibid., Ser. 1, Vol. 1, No. 3, pp. 9 - 26. (J)

Imamura, A.: Inclination of the ground observed before the strong earthquake of July 4, 1929 in Kii District. ibid., Ser. 1, Vol. 1, No. 12, pp. 1 - 5. (J)

1930

Takahasi, R.: Preliminary report on the observation of the tilting of the earth's crust with a pair of water pipes. B.E.R.I., Vol. 8, pp. 143 - 152

Ishimoto, M.: Observations sur les variations de l'inclinaison de la surface terrestre, (Deuxième rapport). Résultats obtenus dans la région épicentrale du tremblement de terre de Sekihara. ibid., Vol. 8, pp. 222 - 236

Takahasi, R.: The eruption of Komagatake. Hokkaido, in 1929. Part IV. Observations of the tilt of the ground accompanying the eruption. ibid., Vol. 8, pp. 290 - 297

Inouye, W., Sugiyama, T.: On earth-tiltings observed at Mt. Tukuba. ibid., Vol. 8, pp. 346 - 363. (J)

Ishimoto, M., Takahasi, R.: Séismes d'Itô et l'observation sur les variation de l'inclinaison de la surface terrestre. ibid., Vol. 8, pp.427 - 458. (J)

B.E.R.I. : Bulletin of the Earthquake Research Institute, Tokyo University
D.P.R.I.B.: Disaster Prevention Research Institute, Kyoto University, Bulletins
B.D.P.R.I.: Bulletin of the Disaster Prevention Research Institute, Kyoto University (in Japanese)
(J) : in Japanese

1932

Inouye, W.: On earth-tiltings observed at Mt. Tukuba. B.E.R.I., Vol. 10, pp. 130 - 144

Takahasi, R.: Tilting motion of the earth's crust observed at Kawana. ibid., Vol. 10, pp. 145 - 170

Takahasi, R.: Tilting motion of the earth's crust observed at Ryozyun (Port Arthur). ibid., Vol. 10, pp. 531 - 559

Takahasi, R.: A note on the tilting motion of the earth's crust observed at Zinsen (Chemulpo). ibid., Vol. 10, pp. 826 - 843

1933

Takahasi, R.: Tilt of the earth's crust observed at the Asama Volcano. B.E.R.I., Vol. 11, pp. 25 - 37

Inouye, W.: Earth-tilts observed at Mt. Tukuba, Japan. ibid., Vol. 11, pp. 693 - 703

Kawase, J.: Observation of changes in inclination at Momoyama, Kyoto. Journ. Seism. Soc. Jap., Ser. 1, Vol. 5, No. 2, pp. 52 - 60. (J)

1934

Takahasi, R.: A new extensometer for measuring crustal deformation. B.E.R.I., Vol. 12, pp. 760 - 775

Kogure, T., Kimura, T., Matsuda, T.: Observation of ground-temperature and ground-tilting. Journ. Seism. Soc. Jap., Ser. 1, Vol. 6, No. 12, pp. 640 - 644. (J)

1935

Minakami, T.: The explosive activities of Volcano Asama in 1935 (Part 1). B.E.R.I., Vol. 13, pp. 629 - 644

1937

Takahasi, R., Minakami, T.: Tilt observations during the recent activities of Volcano Asama. B.E.R.I., Vol. 15, pp. 463 - 491

Yoshiyama, R.: Observation of changes in inclination at Momoyama, Kyoto. Journ. Seism. Soc. Jap., Ser. 1, Vol. 9, No. 2, pp. 56 - 58. (J)

1938

Hagiwara, T.: Observations of changes in the inclination of the earth's surface at Mt. Tukuba. B.E.R.I., Vol. 16, pp. 366 - 371

Minakami, T.: Explosive activities of Volcano Asama and tiltings of the earth's surface. ibid., Vol. 16, pp. 372 - 390

1939

Nishimura, E.: Observation of ground tilting at Aso Volcanological Observatory. Rep. Jap. Soc. Sci. Advancement, Vol. 14, pp. 383 - 384. (J)

Takahasi, R.: Results of continuous observations on the length of a base-line at Komaba, Tokyo. B.E.R.I., Vol. 17, pp. 579 - 589. (J)

1940

Sassa, K.: Ground deformation caused by meteorological changes. Jap. Journ. Astron. Geophys., Vol. 1, pp. 17 - 26. (J)

1941

Hagiwara, T.: Observation of changes in the inclination of the earth's surface at Mt. Tukuba (Second report). B.E.R.I., Vol. 19, pp. 218 - 227

Nishimura, E.: On earth tides. Part II. Particular motion of the earth surface near an active fault. Tikyubutsuri (Geophysics), Kyoto Imp. Univ., Vol. 5, pp. 33 - 50. (J)

1942

Minakami, T.: On volcanic activities and tilting of the earth's surface. B.E.R.I., Vol. 20, pp. 431 - 504

Minakami, T., Matsusita, K., Utibori, S.: Explosive activities of Volcano Kusatu-Sirane during 1938 and 1942 (Part II). ibid., Vol. 20, pp. 505 - 526

1943

Sassa, K.: Ground deformation affected by meteorological variation. Rep. by Hattori Hōkō Kai, No. 10, pp. 158 - 160. (J)

Hagiwara, T.: Movement along the fault that appeared in the Tottori earthquake and change in the inclination of the earth's surface. Journ. Seism. Soc. Jap., Ser. 1, Vol. 15, No. 11, pp. 280 - 284. (J)

1944

Hagiwara, T.: Movement along the fault that appeared in the Tottori earthquake and the change in the inclination of the earth's surface. B.E.R.I., Vol. 22, pp. 66 - 71. (J)

Sassa, K.: Tilting motion of ground observed before and after the destructive Tottori earthquake. Kagaku (Science), Vol. 14, pp. 220 - 221. (J)

1946

Hagiwara, T., Omote, S., Murauchi, S., Akashi, K., Yamada, Z.: The eruption of Mt. Sakura-Zima in 1946. B.E.R.I., Vol. 24, pp. 143 - 159. (J)

1947

Hagiwara, T.: Observation of changes in the inclination of the earth's surface at Mt. Tukuba (Third report). B.E.R.I., Vol. 25, pp. 27 - 32

Obayashi, T., Nagata, T.: Tilting motion of the ground caused by tidal loading at Muroto Peninsula. Rep. Geophys. Inst., Tokyo Univ., No. 25. (J)

Hagiwara, T., Yamada, J.: Observation of change in inclination made with a water-tube tiltmeter at Muroto Peninsula. Adv. Rep. Earthq. Res. Inst., Tokyo Univ., No. 5. (J)

1948

Hagiwara, T., Rikitake, T., Yamada, J.: Observations of the deformation of the earth's surface at Aburatsubo, Miura Peninsula. Part. I. B.E.R.T., Vol. 26, pp. 23 - 27

Sassa, K.: Some problems on the forecasting of the earthquake (I). B.D.P.R.I., No. 1, pp. 13 - 30. (J)

1949

Ozawa, I.: On the strain of earth-crust observed in Osakayama-tunnel. B.D.P.R.I., No. 2, pp. 115 - 121. (J)

Ichinohe, T.: Crustal deformation near Fukui-City after the great earthquake. ibid., No. 2, pp. 131 - 132. (J)

Ozawa, I.: On the secular tilting of the surface crust observed at Kamigamo Geophysical Observatory. ibid., No. 2, pp. 142 - 143. (J)

Hagiwara, T., Rikitake, T., Kasahara, K., Yamada, J.: Observations of the deformation of the earth's surface at Aburatsubo, Miura Peninsula. Part II. B.E.R.I., Vol. 27, pp. 35 - 38

Hagiwara, T., Rikitake, T., Kasahara, K., Yamada, J.: Observations of the deformation of the earth's surface at Aburatsubo, Miura Peninsula. Part III. ibid., Vol. 27, pp. 38 - 44

Hagiwara, T., Rikitake, T., Yamada, J., Kasahara, K.: Observation of change in inclination and of extension of the ground at Kitago, Fukui Prefecture. Adv. Rep. Invest. Fukui Earthq. of 1948, pp. 61 - 63. (J)

Ozawa, I.: On the observations of crustal deformations before and after the Fukui earthquake and it's after-shocks. ibid., pp. 64 - 66. (J)

Nishimura, E., Ichinohe, T., Hosoyama, K.: On the change of the crust and geomagnetism in Fukui District after the great earthquake. ibid., pp. 86 - 87. (J)

Hagiwara, T., Rikitake, T., Yamada, J., Kasahara, K.: Observation of the deformation of the earth's crust at Aburatsubo. Adv. Rep. Earthq. Res. Inst., Tokyo Univ., No. 6. (J)

1950

Hagiwara, T., Rikitake, T.: Observation of the deformation of the earth's surface in the vicinity of the epicentres of the Imaichi earthquake. B.E.R.I., Vol. 28, pp. 435 - 441

Kishinouye, F.: Instrumental recording of a landslide caused by the Imaichi earthquake of Dec. 26, 1949. ibid., Vol. 28, pp. 457 - 459

Nishimura, E.: On earth tides. Trans. Amer. Geophys. Union, Vol. 31. pp. 357 - 376

Ozawa, I.: On the observations of strain of the ground surface using by invar-wire and invar-bar extensometer, and effect of the rain-fall to the change of the ground surface. B.D.P.R.I., No. 3, pp. 69 - 74. (J)

Rikitake, T. (Compiler): Observations with tiltmeters and extensometers. Rep. Special Commit. Study Fukui Earthq., Chap. V, Sec. 3, pp. 124 - 130

Sano, S.: Observation of the earth tilts at Nagoya and Inuyama. Geophys.Abst.Phys. Inst., Nagoya Univ., No. 13, pp. 1 - 8

1951

Hagiwara, T., Kasahara, K., Yamada, J., Saito, S.: Observation of the deformation of the earth's surface at Aburatsubo, Miura Peninsula. Part IV. B.E.R.I., Vol. 29, pp. 455 - 468

Hagiwara, T., Kasahara, K.: Observations of the deformation of the earth's surface at Aburatsubo, Miura Peninsula. Part V. ibid., Vol. 29, pp. 557 - 569

Sassa, K., Nishimura, E.: On phenomena forerunning earthquakes. Trans. Amer. Geophys. Union, Vol. 32, pp. 1 - 6

1952

Hosoyama, K.: Characteristic tilt of the ground that preceded the occurrence of the strong earthquake of Mar. 7, 1952. Journ. Physics of the Earth, Vol. 1, pp. 75 - 81

1953

Nishimura, E.: On some destructive earthquakes observed with the tiltmeter at a great distance. D.P.R.I.B., No. 6, pp. 1 - 15

Hosoyama, K.: On a mercury tiltmeter and its application. ibid., No. 6, pp. 17 - 25

Nishimura, E., Hosoyama, K.: On tilting motion of the ground observed before and after the occurrence of an earthquake. Trans. Amer. Geophys. Union, Vol. 34, pp. 597-599

1954

Nishimura, E., Hosoyama, K., Itō, Y.: Minute tilting movements of the earth's surface, simultaneously observed at several stations in Western Japan. Journ. Geod. Soc. Jap., Vol.1, pp. 49 - 51. (J)

Sassa, K., Nishimura, E.: On phenomena forerunning earthquakes. Publ. Bur. Central Séism. Intern., Sér. A, Trav. Scientif., Fascicule 19, pp. 277 - 285

1955

Ozawa, I.: On the observation of changes of the earth-crust in the time of earthquakes. Journ. Seism. Soc. Jap., Ser. 2, Vol. 8, pp. 45 - 47. (J)

1956

Nishimura, E., Hosoyama, K., Itō, Y.: Micro-tilting motion of the ground. Trans. Amer. Geophys. Union, Vol. 37, pp. 645 - 646

Sassa, K., Nishimura, E.: On phenomena forerunning earthquakes. D.P.R.I.B., No. 13, pp. 1 - 8

Sassa, K.: Some problems on the forecasting of the earthquake. (II). (On the observation of the crustal strain accompanied by the earthquake). D.P.R.I.B., Memorial Issue of the Fifth Anniversary, pp. 3 - 7. (J)

Ozawa, I.: On the observation of crustal deformation at Osaka-yama. ibid., pp. 14 - 19.(J)

Hosoyama, K.: On secular observation of tilting motion of the ground. ibid., pp. 20 - 27. (J)

Takada, M.: Tele-observation of the crustal movement. ibid., pp. 28 -34. (J)

1957

Hosoyama, K.: On the observation of secular phenomena of the tilting motion of the ground. Mem. Coll. Sci., Univ. Kyoto, Ser. A, Vol. 28, pp. 253 - 282

Shimozuru, D., Gotō, K.: Observations of the crustal deformations at Hakamagoshi, Sakurashima Volcano. Bull. Volcanolog. Soc. Japan, Vol. 2, pp. 1 - 5. (J)

Yoshikawa, K.: The observations of the crustal deformations at Sakurajima Volcano. D.P.R.I. Ann., No. 1, pp. 106 - 111. (J)

1958

Nishimura, E.: Some problems on tilting motion of the ground. Comm. Obs. Roy. Belg. N°142, S. Géophys. N°47, pp. 100 - 110

Takada, M.: On the observation of the crustal deformation and meteorological effect on it at Ide Observatory. D.P.R.I.B., No. 23, pp. 1 - 12

1959

Nishimura, E., Kishimoto, Y., Tanaka, Y.: Observed secular tilting motion of the ground which preceded the occurrence of several destructive earthquakes. Tellus, Vol. 11, pp. 130 - 134.

Ozawa, I.: On the observations of crustal deformation at Osakayama. D.P.R.I. Ann., No. 3, pp. 148 - 153. (J)

Takada, M.: On the observing instruments and tele-metrical devices of extensometers and tiltmeters at Ide Observatory. D.P.R.I.B., No. 27, pp. 1 - 27

Takada, M.: On the crustal strain accompanied by a great earthquake. ibid., No. 27, pp. 28 - 46

Tsumura, K.: Earth tidal observation with a water tube tiltmeter. Rep. 3rd Internat. Symp. on Earth Tides, Trieste, pp. 77 - 79

Yoshikawa, K., Kikuchi, S.: On anomalous fluctuation of ground-tilting by oceanic tidal load related with the volcanic activity. ibid., p. 80

Nishimura, E., Tanaka, Y., Tanaka, T.: On a tiltmeter for the observations of secular ground-tilting. ibid. pp. 106 - 107

Ozawa, I.: On an extensometer of new type. ibid., pp. 117 - 118

Takada, M.: On the observing instruments and the tele-metrical devices of extensometers and tiltmeters at Ide Observatory, ibid., pp. 125 - 126

Takada, M.: On the crustal strain accompanied by a great earthquake. ibid., pp. 127 - 128

1960

Ozawa, I.: On the extensometer of which magnifier is a Zöllner suspension type tiltmeter, and the observations of the earth's strains by means of the instruments. Journ. Geod. Soc. Jap., Vol. 6, pp. 1 - 5. (J)

Tsumura, K.: On a high sensitive water tube tiltmeter and its application to the earth tides. ibid., Vol. 6, pp. 85 - 88. (J)

1961

Ozāwa, I.: On the observations of crustal strains due to Chile tide wave. D.P.R.I. Ann., No. 4, pp. 36 - 44. (J)

Tanaka, Y., Tanaka, T.: On the ground tilt and strain caused by the Chile tsunami. ibid., No. 4, pp. 45 - 60. (J)

Itō, Y.: Tilting motion of the ground as related to the volcanic activity of Mt. Aso. D.P.R.I.B., No. 42, pp. 1 - 15

Nishimura, E.: On the continuous observation of the crustal deformation. Journ. Seism. Soc. Jap., Ser. 2, Vol. 14, pp. 260 - 266. (J)

Ozawa, I.: On the comparative observations of the crustal strains by means of a wire-type's extensometer and rod-type's one. Journ. Geod. Soc. Jap., Vol. 7. (J)

1962

Ozawa, I.: On the observations of the crustal movements at some neighbouring points (in Osakayama Observatory). D.P.R.I. Ann., No. 5, pp. 1 - 11. (J)

Nishimura, E., Tanaka, Y., Tanaka, T.: On anomalous crustal deformation observed before some recent earthquakes (Part 1). ibid., No. 5, pp. 28 - 43. (J)

Nishimura, E., Kishimoto, Y., Tanaka, T.: Study on relation between the local shocks and the minute crustal deformation at Wakayama (Part 1). ibid., No. 5, pp. 44 - 56. (J)

Takada, M.: On the observations of crustal deformation at Ide Observatory. ibid., No. 5, pp. 64 - 72. (J)

D i s c u s s i o n

A. Z á t o p e k (Prag):

I should like to ask some questions:
Does there exist or can there be established a correlation between the earthquake magnitude and the value of the corresponding tilting?

E. N i s h i m u r a (Kyoto):

There are only several examples in the long period of 25 years. Consequently we can not discuss in detail the relation between the seismic magnitude and the amount of anomalous grand-tilting. But it is safely said that there is a tendency showing that the large destructive earthquakes also are connected with the large anomalous grand-tilting. I suppose that there is also an azimuthal characteristic in such a case of earthquake with preceding grand-tilting.

A. Z á t o p e k (Prag):

On the examples given there was a monotonous movement (tilting) interrupted by sudden changes partly corresponding to earthquakes. Is it always so that the sense of this monotonous movement is conserved?

E. N i s h i m u r a (Kyoto):

Of the particular vector diagram observed at the Yura observation station, the observed grand-tilting is largely affected by the meteorological change, because of the shallowly seated observation room of 10 m below the ground surface. We only discuss and treat the anomalous grand-tilting which is presumed to be connected with the strong earthquakes which occured in the surrounding area of the station.

W. A. M a g n i t s k y (Moskau):

The tiltmeter stations in the Middle Asia, similar to Japan, observed sudden changes of direction and velocity of tilts before the earthquake occurrence. However, the variations of tilts, which are not followed by an earthquake, are observed as frequently. Therefore, it seems that the problem, tackled by Prof. Nishimura, seems rather complicated. Our colleagues in Japan have better opportunities to study the causes of earthquakes. I would like to wish them successful solution of this important problem in the near future.

E. N i s h i m u r a (Kyoto):

I am very pleased to hear Prof. Magnitsky's suggestion. We also observed sometimes in Japan the same cases as were mentioned by him. I wish to express my gratitude for his kind advice.

Types of crustal movements accompanied with earthquakes

von

S. Miyamura (Tokyo)

Summary

Types of crustal movements accompanied with earthquakes have close relation with types of earthquakes or the seismotectonic zones of the region concerned.

Crustal earthquakes with M ≧ 7 accompany visible crust deformations and those with M ≧ 6 of very shallow foci show geodetic crust movements. Magnitude frequency and maximum magnitude of earthquakes are characteristic for each seismo-tectonic zones to which that group of earthquakes belongs. Crustal movements are one of the characteristics of each seismo-tectonic zones.

Detailed study along this line is only in the beginning and the result reported will be tentative. But, generally speaking, the relations are clear in Japanese area and short summarizing discussion will be given.

Destructive earthquakes of Japan Sea side block tectonic region accompany faultings and/or small dome-ups. Large orogenic earthquakes of the Pacific side of SW Japan accompany continuous pre-seismic downwarping, abrupt co-seismic up-heaval and post-seismic asymptotic partial recovery over the large meizoseismal area along the coast. Main faults have not been found but they might be hidden under the sea.

Other seismic zones of Japan have also their own modes of seismicity and crustal movements.

Comparative study of Japanese seismic zones with special reference to the crustal movements will be presented illustratively.

Zusammenfassung

Die Typen von Krustenbewegungen, die mit Erdbeben verbunden sind, haben enge Beziehungen zu Erdbebentypen oder seismotektonischen Zonen der beteiligten Regionen.

Krustenbewegungen mit M ≧ 7 sind mit sichtbaren Krustendeformationen verbunden, und solche mit M ≧ 6 mit sehr oberflächennahen Herden zeigen geodätische Krustenbewegungen. Magnitude, Frequenz und Maximalmagnitude von Erdbeben sind für jede seismotektonische Zone, zu der diese Erdbebengruppen gehören, charakteristisch. Krustenbewegungen sind eines der Merkmale jeder seismotektonischen Zone.

Genaue Studien auf dieser Grundlage stehen erst am Anfang, und das mitgeteilte Ergebnis wird ein versuchsweises sein. Im allgemeinen sind die Beziehungen im japanischen Gebiet klar, und es wird eine kurze zusammenfassende Erläuterung gegeben.

Zerstörende Erdbeben in der tektonischen Region der Japanischen See sind mit Verwerfungen und/oder kleinen Aufwölbungen verbunden. Ausgedehnte orogenetische Erdbeben der pazifischen Seite von Südwest-Japan sind mit kontinuierlichen vorseismischen Tieffaltungen, plötzlichen co-seismischen Hebungen und nachseismischer asymptotischer, teilweiser Wiederherstellung über das ausgedehnte meizoseismische Gebiet entlang der Küste verbunden. Die Hauptverwerfungen sind nicht gefunden worden, aber sie können sich unter der

See verbergen.

Andere seismische Zonen in Japan haben auch ihre eigenen Seismizitätsmoden und Krustenbewegungen.

Vergleichende Studien der japanischen Seismizitätszonen mit ihrer besonderen Beziehung zu den Krustenbewegungen werden erläutert.

1. Island Arcs Meeting in the Neighbourhood of Japan

An island arc is one of the most characteristic features of the Circum-Pazific belt and several island arcs of different ages meet in the neighbourhood of Japan as shown in Fig. 1.

The Honshu Arc, being the oldest among them, has experienced three main stages in its development and has shifted its axis of orogenic activity to the outer side or the Pacific side.(cf. Fig. 2).

The first stage (V_1 : Inner Honshu Arc) is represented by the Hida gneiss (Hd) and the Sangun schist (Sg), touching each other by the so-called Hida Outer-margin Tectonic Line (1). The former is a so-called high-temperature, low-pressure metamorphic rock and

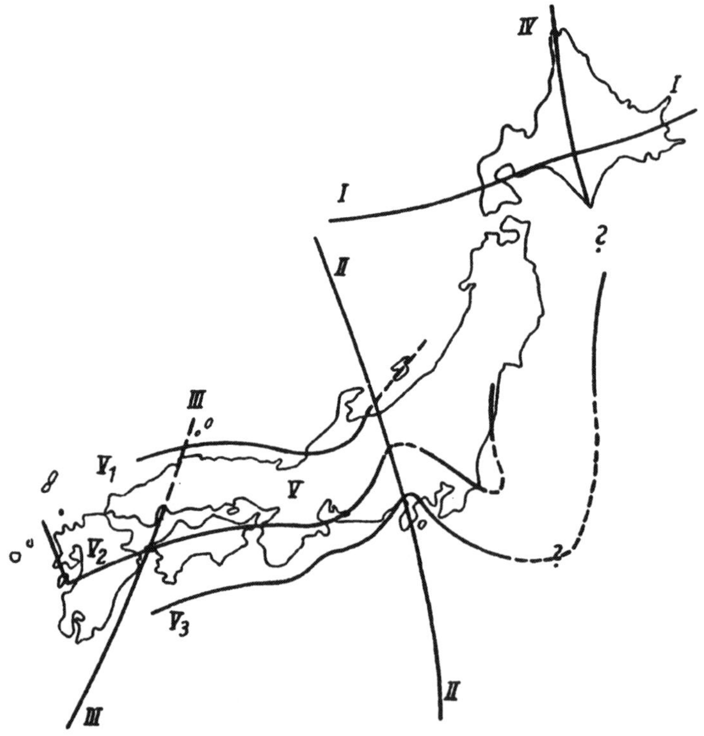

Fig. 1

Island arcs meeting in the Japanese Islands

the latter is a so-called low-temperature, high-pressure metamorphic rock. This first orogenic stage had been accomplished in the late Paleozoic Era as the Akiyoshi Orogeny of T. Kobayashi and it can be regarded now rather as a continental platform than as a part of an island arc, though the continuation with the continent is cut by the recent

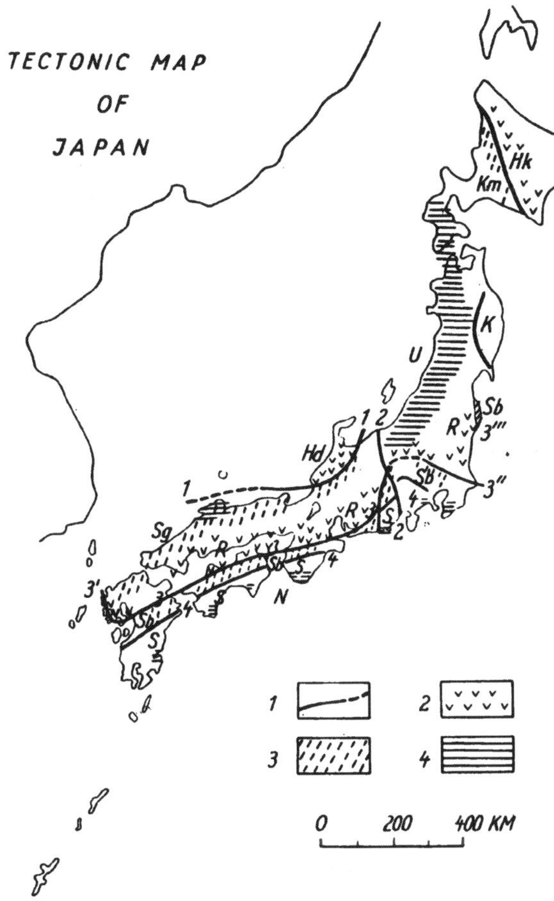

Fig. 2

Tectonic map of Japan

1. Main tectonic lines: 1, Hida Outer-margin Tectonic Line. 2, Fossa-Magna Tectonic Line. 3, Median Tectonic Line. 3', 3", 3"', Extensions of the Median Tectonic Line. 4, Butsuzo-Itogawa Tectonic Line.
2. Basement terrains of low-pressure, high-temperature type metamorphic rocks: Hd, Hida Gneiss Terrain. R. Ryoke Gneiss Terrain. Hk, Hidaka Gneiss Terrain.
3. Basement terrains of high-pressure, low-temperature type metamorphic rocks: Sg, Sangun Schist Terrain. Sb, Sanbagawa Schist Terrain. Km, Kamuikotan Schist Terrain.
4. Other basement terrain: K, Kitakami Paleozoic Terrain
5. Meso-cainozoic terrains: S, Shimanto Mesozoic Terrain. N, Nakamura Tertiary Geosyncline or Outer Honshu Geosyncline. U, Uetsu Neogene Sub-geosyncline

depression of the Japan Sea area. The basement namely, of the Inner Side or Japan Sea Side of Japan, belonging to this oldest Honshu Arc (V_1), shows clear block structure and the recent movements in this area are characterized by this nature.

The second stage (V_2 : Median Honshu Arc) was the Sakawa Orogeny of T. Kobayashi and the axis of the orogeny is exposed along the Median Tectonic Line (3), i.e. just north of the line extends the Ryoke gneiss (R), showing a metamorphic nature similar to the Hida gneiss, and just south along the line is distributed the Sanbagawa schist (Sb), having a metamorphic nature similar to the Sangun schist. This orogenic stage had also been completed in the late Mesozoic Era and it does not indicate any independent activity now. Along this zone we can only notice slight seismic activities and crustal movements at separate places, which might be thought to be a relic of post-orogenic activity.

The third stage (V_3: Outer Honshu Arc) began in the Mesozoic and partly shows us thick, folded sediments as the Shimanto Terrain (S), but out-side of the latter terrain we have a Tertiary Terrain which partly extends now as a geosyncline under the off shore area in the Pacific. The author thinks that this stage is even now in the full swing of its activity, in particular in its most active magmatic intrusion stage and this duly characterizes the nature of the big earthquakes and the crustal movements in this area.

The Honshu Arc has thus been developing since the remote Paleozoic Era, yet it was strongly influenced by the newer arcs crossing it rather perpendicularly, since the beginning of the Tertiary or a somewhat earlier time. Namely, in the west, the Ryukyu Arc (III), in the middle, the Izu-Mariana Arc (II), and in the north, the Kurile Arc (I) meet and cross the Honshu Arc. They have given and are giving even now strong and quite different modes of movements to the Honshu Arc. Accordingly the recent movements of the earth's crust in Japan ought to be studied considering this complex situation of tectonic structure. Especially the west Island, Kyushu, and Central Honshu characterized by the Fossa-Magna Zone (2), need very careful studies. The north island, Hokkaido, where the Kurile Arc, the Honshu Arc and the Hidaka-Sakhalin Arc mentioned below meet also demands careful treatment. But because the nature of the structure is complex, we can undoubtedly expect many interesting and fruitful results from the careful investigations of such regions.

Lastly we must touch upon the existence of the Hidaka-Sakhalin Arc (IV) running en échélon at the northern end of the Honshu Arc, but it is at present seismically rather quiet and from the standpoint of the study on recent movements of the earth's crust it has not yet been well investigated.

2. Seismic Zones of Japan

As is well known, Japan and her vicinity have one of the most densely concentrated epicentres and from the instrumentally determined epicentres it does not seem that any systematic zoning of Japanese earthquakes is shown as formerly pointed out by A. Imamura and others.

However as seismotectonic zoning in and near Japan has been recently proposed by the author because of the fact that the island arcs meeting in the Japanese area have their own seismic activities corresponding to their geo-tectonic ages (S.Miyamura, 1962a, 1962b).

The two youngest arcs, the Kurile and the Izu-Mariana Arcs, are now in the climax of their activities and earthquake foci extend from near the crust to deep into the earth's mantle down to several hundred kilometers. The pattern of such a three-dimensional distribution of earthquake foci in this area is well known and it will not be necessary to give any more words here. This activity is, considering the crustal thickness of NE Japan and in particular that of the off shore area, mainly restricted to the upper mantle, and it indicates that the original motive force of the formation of the island arc or of the orogenic development is due to the energy in the upper mantle.

The magnitude of earthquakes in these zones reaches to the largest size but in comparison with the number of smaller shocks the number of larger earthquakes is restricted. Both seismic zones are often regarded as a continuous Kurile-Mariana Zone, running along the Japan Trench or the Honshu Arc in NE Japan. But the upper mantle seismic activity has a gap off the Sanriku Region (Pacific coast of northern Honshu), and the crustal seismic activity of the Honshu Arc is superposed there, giving the impression of a continuous distribution of epicentres. Thus the author divided the two seismic zones there and assumed that each of them perpendicularly crosses the Honshu Arc.

The Ryukyu Arc has its own seismic activity similar to the former two, but not so active. The trench outside the arc is only clear in the southern part and the penetration of the foci into the mantle is restricted only down to about 300 km. The stage of the island development is between the Kurile or Mariana and the Outer Honshu Arcs. The nature of the magnitude frequency relation of earthquakes is intermediate, too, and the number of larger shocks increases compared to the number of smaller shocks.

This arc crosses the Honshu Arc in Kyushu and in the western part of Honshu and reaches into the Japan Sea. The earthquakes near Hiroshima with depths of several tens of kilometers seem to be the continuation of the Hyuga-nada epicentres (SE off shore of Kyushu) and are considered to belong to the Ryukyu Arc. A. Sugimura bends this zone along the Seto-Inland-Sea, stretching between Hiroshima and Osaka, but the author of this article disagrees with him. Along the Seto-Inland-Sea the writer assumes the deeply penetrated part of seismic activity belonging to the Outer Honshu Arc (V_3).

The most important seismic zone seen from the vast destructiveness of earthquakes to the Japanese Islands is that which belongs to the last-mentioned Outer Honshu Arc. In this zone we have had many first class big earthquakes, among which the recent two in 1944, Dec., 7th, and 1946, Dec., 21st are even now vivid in the memory of the Japanese people. They devastated many towns and cities over the whole Shikoku Island and Central Honshu.

This seismic zone runs along the Pacific coast of SW Japan, but when it crosses the Izu-Mariana Arc in the South Fossa-Magna Region, it is subjected to the influence of the newer arc and bends to the NE and again to the SE parallel to the bending of the Median Tectonic Line as illustrated in Figs. 1 and 2. On the SE trend of this zone we had the Sagami Bay Earthquake or Kanto (Kwanto) Earthquake of 1923, Sept. 1st.

Further, the extension of the zone may be traced as shown in the figures in NE Japan, but it is not so certain at present. The shallow, larger earthquakes off the Boso Peninsula (near Tokyo) and those off the Sanriku Region may be regarded as those belonging to this zone in the author's opinion.

The seismicity of the Outer Honshu Arc Zone is characterized by its intermittent activity of the first class big earthquakes with their after-shock sequences in broader areas, and from the standpoint of magnitude frequency we have here relatively more frequent larger shocks than smaller ones in comparison with the above cited three seismic zones. The author would like to take it as a characteristic of an older stage of the orogenic activity.

The penetration of the seismic activity into the mantle is for this zone weak and does not reach deeper than 100 km. The mantle is found active only in the central part of the Kii Peninsula, Central Japan, (near Nara) (S. Miyamura. 1962a). In the Kanto (near Tokyo) and Seto-Inland-Sea (extending between Osaka and Hiroshima) regions, probably exist such intermediate foci belonging to this zone, but at present it is not clearly established.

Along the Median Tectonic Line we have separate seismic activity of very shallow shocks as mentioned above. Some of them are swarms of tremendous small shocks without any main earthquake, continuing several years (e.g. Swarm activity near Wakayama, Kii Peninsula) (A. Imamura et al., 1932, S. Miyamura, 1959, 1960).

A comprehensive study of this zone as a whole has not yet been made. We call these separate groups of minor seismic activites the Median Honshu Seismic Zone.

The Inner Honshu Arc is as mentioned above a block-tectonic region like a platform and the seismic activity there is suitable for it, i.e. the activity is restricted exclusively to the crust. Though the maximum magnitude of earthquakes in the Inner Honshu Arc Zone does not exceed M = 8 with the exception of the Mino-Owari Earthquake, 1891, M = 8,5, the magnitude frequency relation is quite different from the mantle activity of newer arcs, having much more frequent larger shocks relative to smaller ones.

As those earthquakes are crustal and accordingly rather shallow, they are usually very destructive, although their meizoseismal areas are limited.

Lastly we must mention a special sub-geosyncline, which developed since the Middle Tertiary over Inner Japan especially in NE Japan, namely the Uetsu Geosyncline (U in Fig. 2). It is distributed over the Inner Honshu Arc basement as a veneer with a thickness of a few kilometers. In this region many small folding structures are found to be active even now. (Y. Otuka, 1941) Very shallow shocks of NE Inner Japan can be classified as a special seismic zone named Uetsu Seismic Zone.

Summarizing the above-mentioned description, we get the following seismic zones of Japan; 1) The Kurile Arc Seismic Zone (I), 2) The Izu-Mariana Arc Seismic Zone (II), 3) The Ryukyu Arc Seismic Zone (III), 4) The Outer Honshu Arc Seismic Zone (V_3), 5) The Median Honshu Arc Seismic Zone (V_2) (not in the strict sense of the word "zone"), 6) The Inner Honshu Arc Seismic Zone (V_1) and 7) The Uetsu Seismic Zone (VI), as illustrated in Fig. 3.

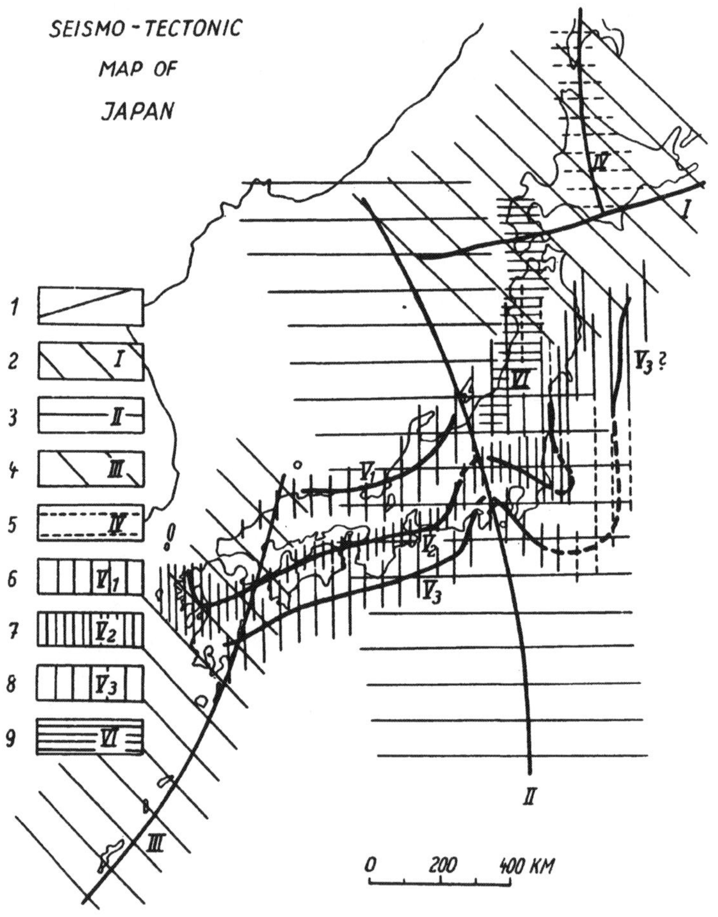

Fig. 3

Seismotectonic map of Japan
1: Axis of the island arc or orogeny; 2: Kurile Arc Seismic Zone; 3: Izu-Mariana Arc Seismic Zone; 4: Ryukyu Arc Seismic Zone; 5: Hidaka-Sakhalin Arc Seismic Zone; 6: Outer Honshu Arc Seismic Zone; 7: Median Honshu Arc Seismic Zone; 8: Inner Honshu Arc Seismic Zone; 9: Uetsu Seismic Zone

3. Crustal Movements Accompanied with Earthquakes

The types of crustal movements accompanied with earthquakes have close relation to the types of earthquakes or to the seismotectonic nature of the earthquakes, namely they depend on the seismic zones to which the earthquakes belong. Crustal earthquakes with $M \geqq 7$, accompany visible deformations of the earth's surface and those with $M \geqq 5$ of very shallow foci show geodetic movements. The big earthquakes of the Outer Honshu Arc Seismic Zone cannot be simply classified as crustal but, of course, they accompany very large crustal movements.

According to the seismic zones mentioned in the previous section the types of crustal movements accompanied with earthquakes in the respective zones will be given below.

Mantle earthquakes of the Izu-Mariana Arc Seismic Zone gave us no information on crustal movements caused by them till now. Even the indication of the tsunami was seldom. The big earthquakes of the Kurile Arc Seismic Zone often cause tsunamis and after certain large earthquakes off Hokkaido, e.g. the Tokachi-oki Earthquake, 1952, there was a conspicuous land deformation as shown in Figs. 4 and 5. But the movements of triangulation points and relevelling results indicate no clear relationship with the earthquake epicentre and also gave no relationship with the movements in recent geological times represented as elevated coastal terraces. Two points are questionable, namely, 1) whether the earthquake is a mantle activity belonging to the Kurile Arc Seismic Zone or is there any other seismic zone of crustal nature along the Kurile Islands as far as off Hokkaido, and 2) whether the detected movements are directly related to the earthquake, for the first measurements were conducted many years before the earthquake took place, whereas the second ones were executed just after the shock.

This type of crustal movements was also obtained by the Sanriku Earthquake in 1933 (cf. Fig. 6), and by one of the Hyuga-nada Earthquakes SE off Kyushu. The latter seems to belong to the Ryukyu Arc Seismic Zone but eventually may not and might be considered to belong to the Outer Honshu Arc Seismic Zone. The former is in the author's opinion

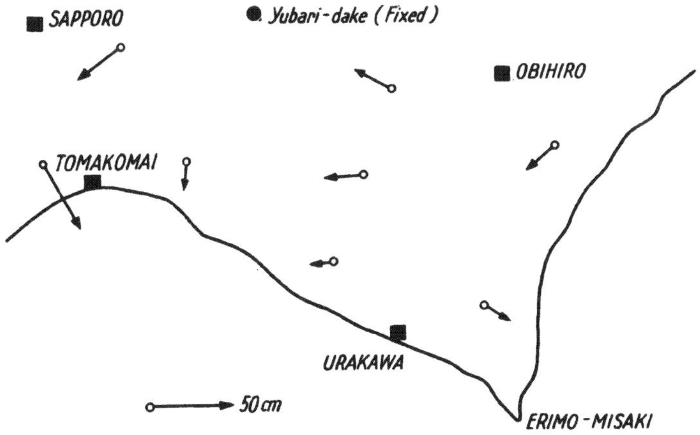

Fig. 4

Horizontal displacements accompanied with the Tokachi-oki Earthquake, 1952. First order triangulation results obtained by the Geographical Survey Institute

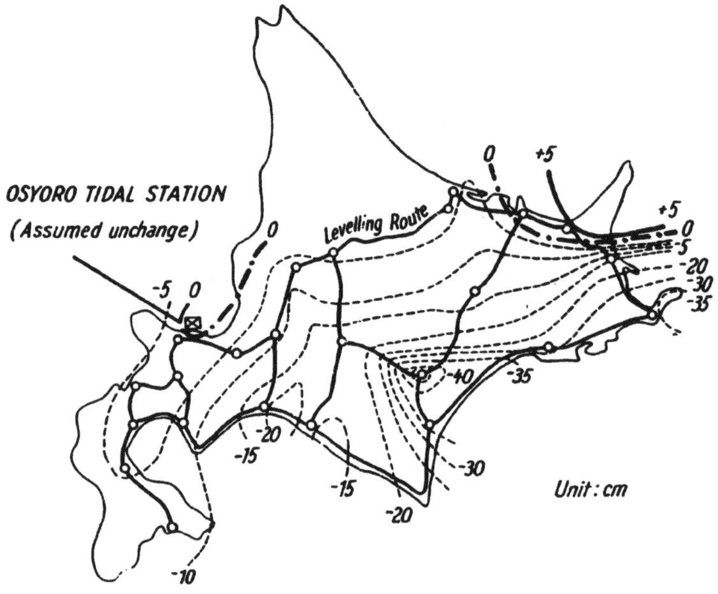

Fig. 5 Vertical displacements accompanied with the Tokachi-oki Earthquake, 1952. First order precise levelling results obtained by the Geographical Survey Institute

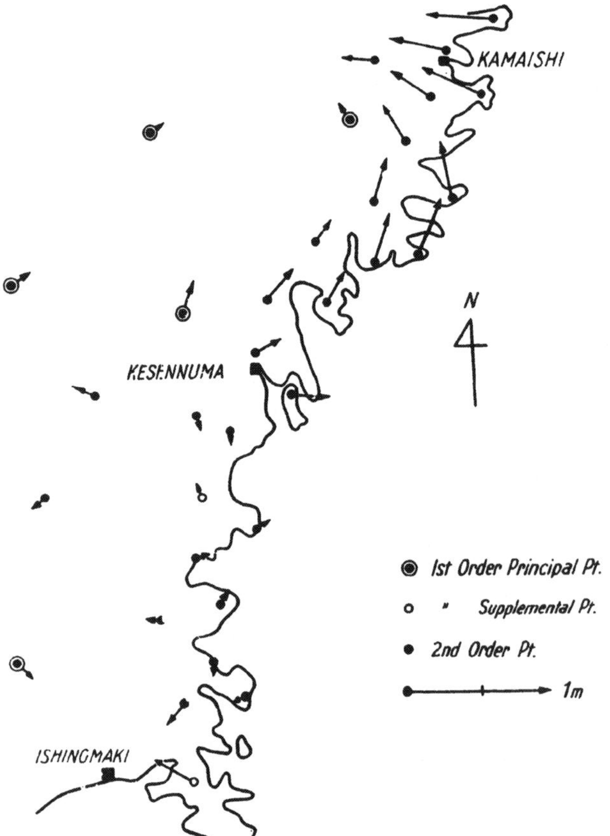

Fig. 6 Horizontal displacements accompanied with the Sanriku-oki Earthquake, 1933. First order triangulation results obtained by the Imperial Military Land Survey

highly probable to be a crustal earthquake belonging to the extension of the Outer Honshu Arc Seismic Zone, but the epicentre was too far from land and the irregular crustal movements cannot be easily explained.

The big earthquakes of the Outer Honshu Arc Seismic Zone show very characteristic types of crustal movements related to them. Before the 1946 earthquake (Nankaido Earthquake) the re-levellings in the Kii Peninsula always showed typical southward dipping, namely the top of the peninsula was sinking, and it was regarded as a secular tendency of the crustal movements showing the warping up of the Japanese Islands. But the height of the coastal terraces there does not coincide with this tendency. It was rather reverse. After the Nankaido Earthquake, the re-levellings were carried out and showed that the reverse movements, namely the upheaval of the top of the peninsula, was very great and after the earthquake it did not fully recover.

The similar situation was also discovered for the big earthquake of 1923 near Tokyo (Kanto Earthquake). The movements at the time of the Kanto Earthquake show clear parallelism with those which remained as coastal terraces (A. Sugimura, 1962).

Horizontal movements of these earthquakes shown in Figs. 7 and 8 indicate large amounts toward the epicentral area. A very interesting explanation of the movements was given by E. Inoue, where the author explained it as the magmatic intrusion of the orogeny now in action in the Outer Honshu Arc Zone (E. Inoue, 1960).

For the small destructive earthquakes in the Median Honshu Arc Seismic Zone only a few cases of crustal movements were measured. The Mikawa Earthquake, 1945, Jan. 13th, showed large upheavals (max. about 1 m) and horizontal displacements (max. about 2 m) over the small epicentral area (Fig. 9). Horizontal shifts were directed toward NE and a clear over-thrusting fault was observed, fringing the upheaval area (H. Tsuya, 1946).

Repeated levelling revealed that the swarm earthquakes of the Wakayama area accompanied the upheavals of the epicentral area during their active periods (A. Imamura et al., 1932, S. Miyamura and A. Okada, 1960). Similar upheavals around the swarm earthquake areas were found in the cases of the Ito Swarm Earthquakes in the Izu Peninsula in 1931-1932, and of the activity of the Volcano Usu in Hokkaido in 1944-1945, which suggests that the very shallow shocks in the Wakayama area may be due to a shallow intrusive volcanism which since the Miocene has been active as a post-orogenic volcanism in SW Japan (e.g. the Kumano acidic rocks in the Kii Peninsula).

Along the Median Tectonic Line we have several similar swarm earthquake areas, but neither seismological nor geodetic investigations have been made there and we have no detailed informations about them.

The Inner Honshu Arc Seismic Zone is one of the most important zones from the view-point of calamity. As mentioned above there occurred many big earthquakes of $M = 6$--8, although not the biggest class of $M \geqq 8$, and as they had the depths of 10 to 40 km they usually gave severe damages. They accompanied distinct crustal deformations of the block-movement type, namely clear seismic faults with large horizontal and vertical displacements. A famous example was given by the movements that accompanied the Mino-Owari Earthquake, 1891. The continuation of this fault line moved later at the time of the Fukui (or Hukui) Earthquake, 1948. The latter fault was hidden under the alluvium of the Fukui Plain and was first detected by the geodetic survey. Photographs taken from the airplane, however, showed a clear échelon trend of fissures over the plain and suggested the existence of the sub-surface fault (N. Nasu, 1949, 1950, 1951).

Another example of well studied crustal movements accompanied with the earthquake of this zone is that of the Tango Earthquake, 1927 (C. Tsubai, 1933, 1939).

Characteristic of the nature of the fault-movements accompanied with these earthquakes were the so-called hinge faults. The horizontal movements were left-hand (sinistral) when the fault trend cut the direction of the Honshu Arc (Mino-Owari fault, Fukui fault, Gomura fault of the Tango Earthquake) and right-hand (dextral) when the fault trend ran

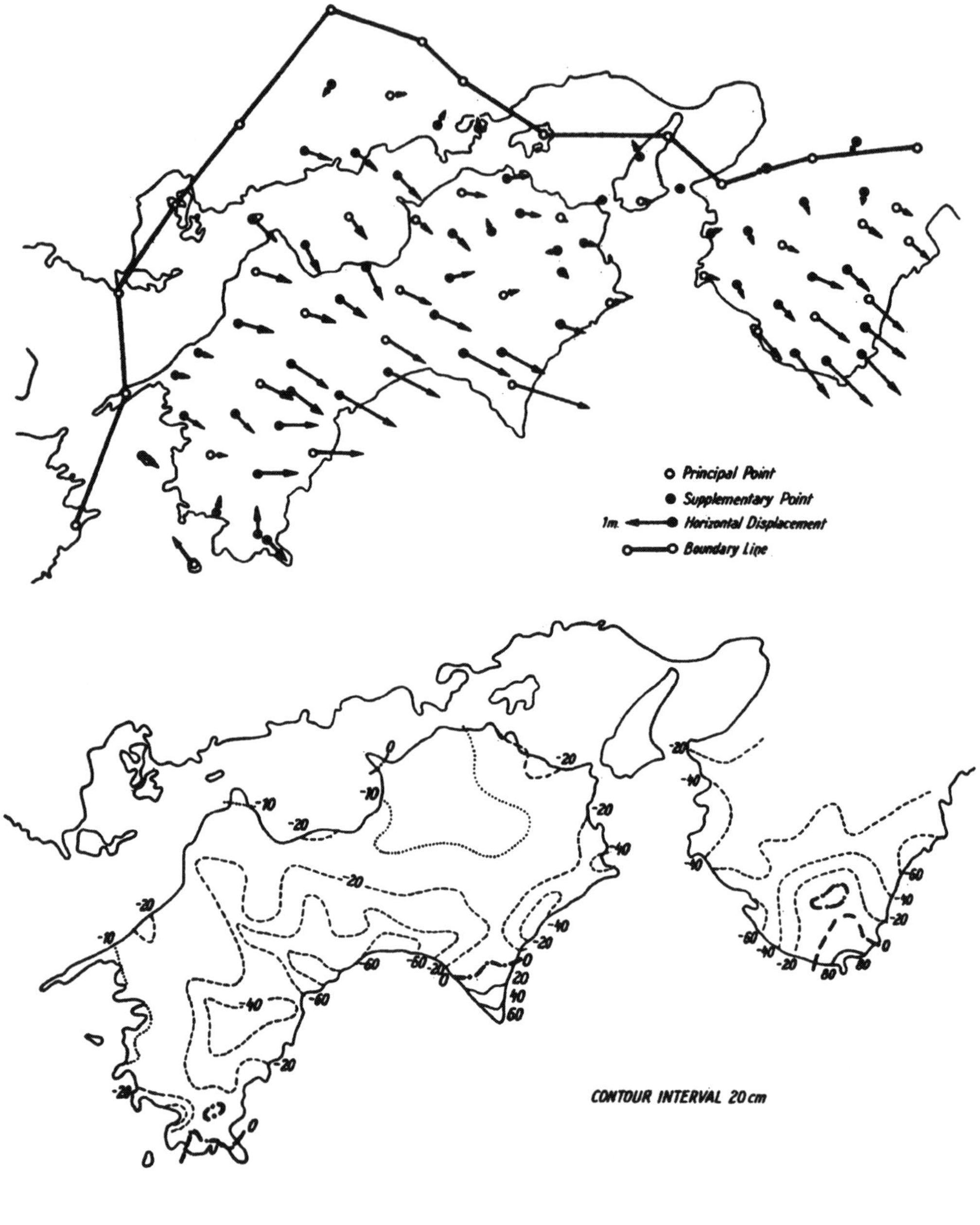

Fig. 7

Horizontal and vertical displacements accompanied with the Nankaido Earthquake, 1946, and partly with the Tokaido (Tonankaido) Earthquake, 1944. First order triangulation and precise levelling results obtained by the Geographical Survey Institute

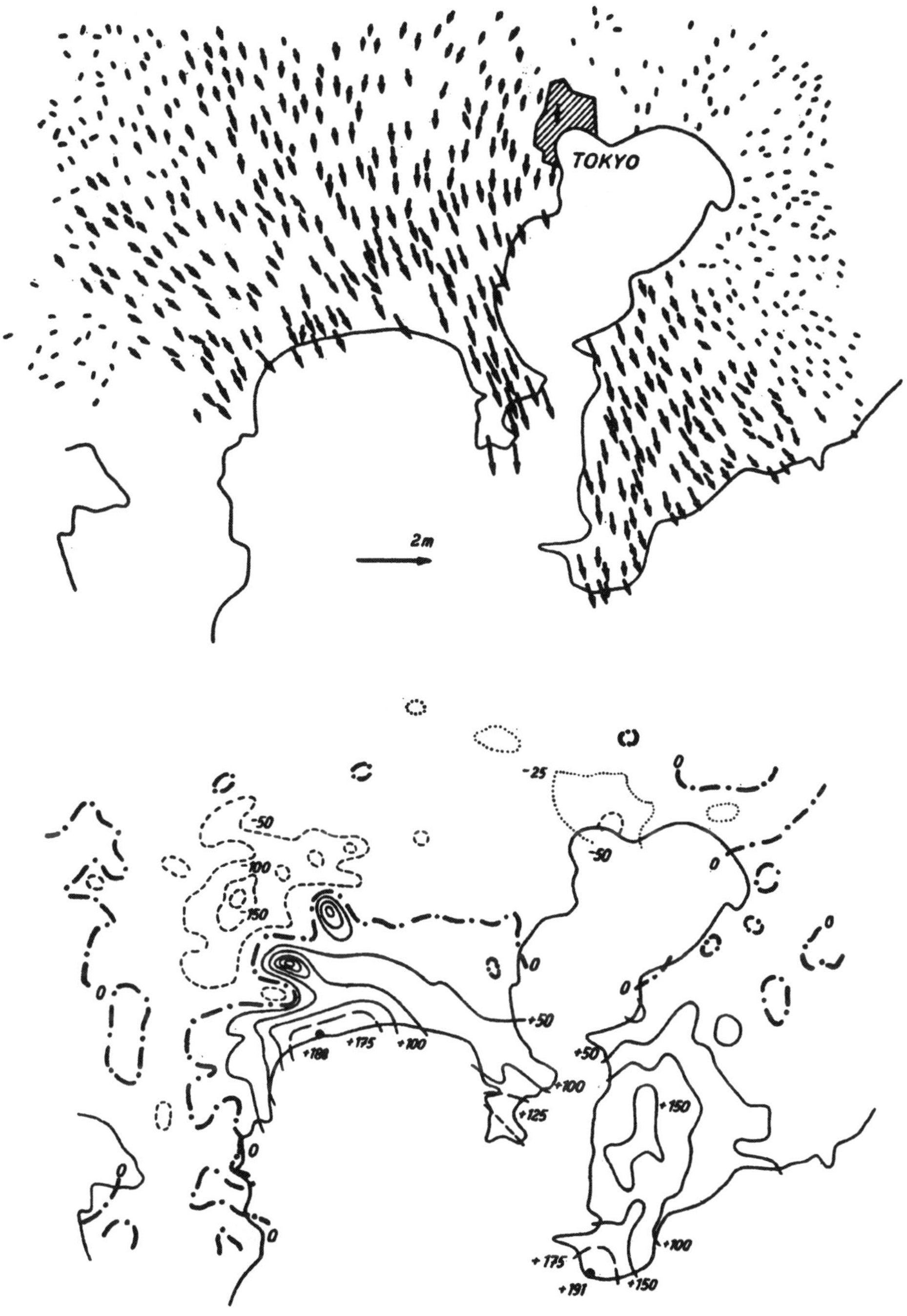

Fig. 8

Horizontal and vertical displacements accompanied with the Kanto (Kwanto) Earthquake, 1923. First and second order triangulation and first order precise levelling results obtained by the Imperial Military Land Survey

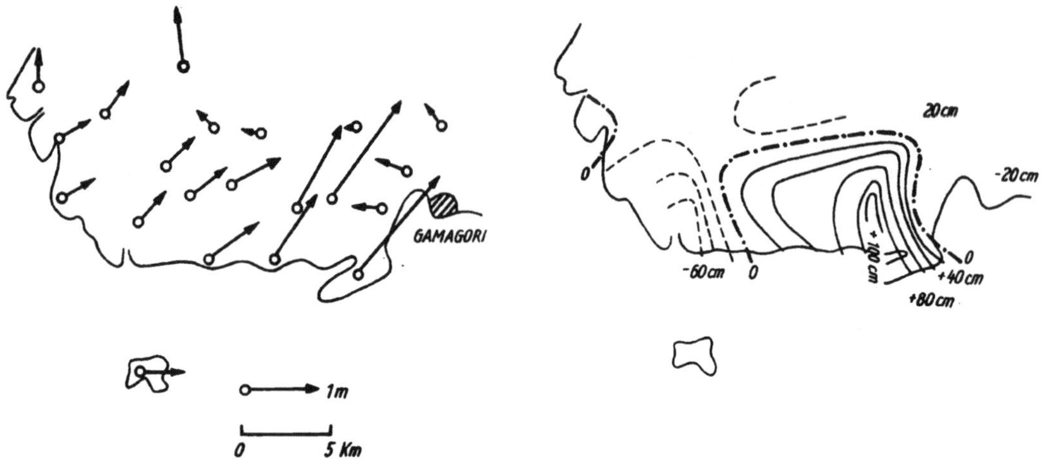

Fig. 9

Horizontal and vertical displacements accompanied with the Mikawa Earthquake, 1945. First order triangulation and precise levelling results obtained by the Geographical Survey Institute

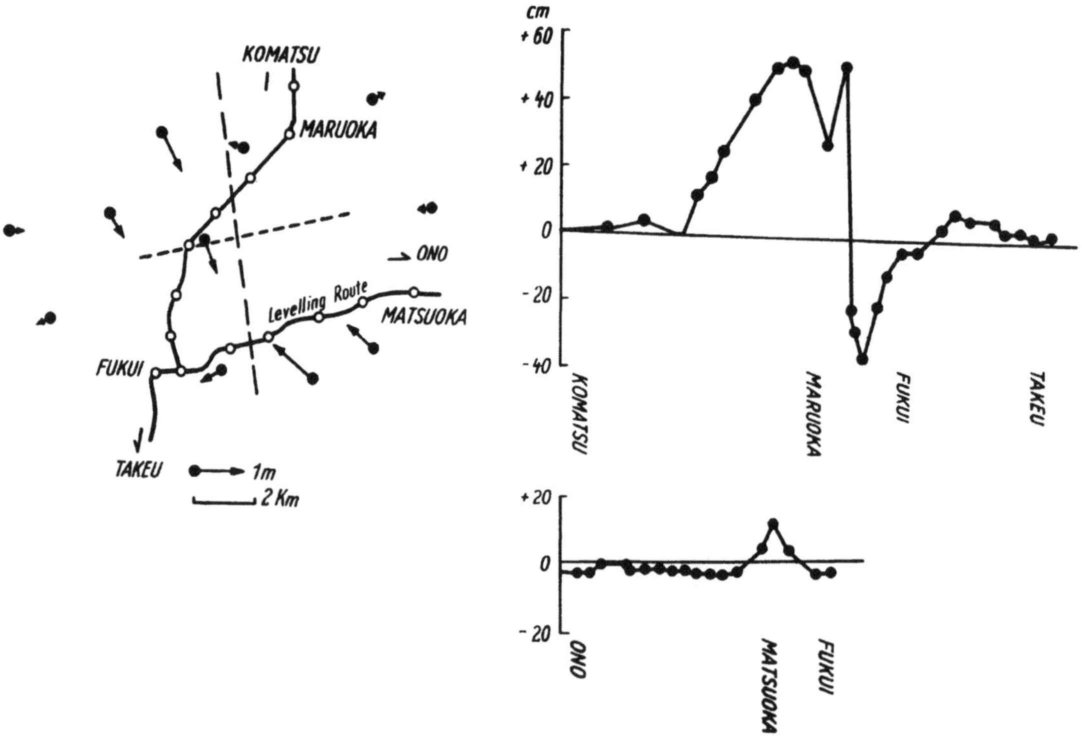

Fig. 10

Horizontal and vertical displacements accompanied with the Fukui (Hukui) Earthquake, 1948. First order triangulation and precise levelling results obtained by the Geographical Survey Institute

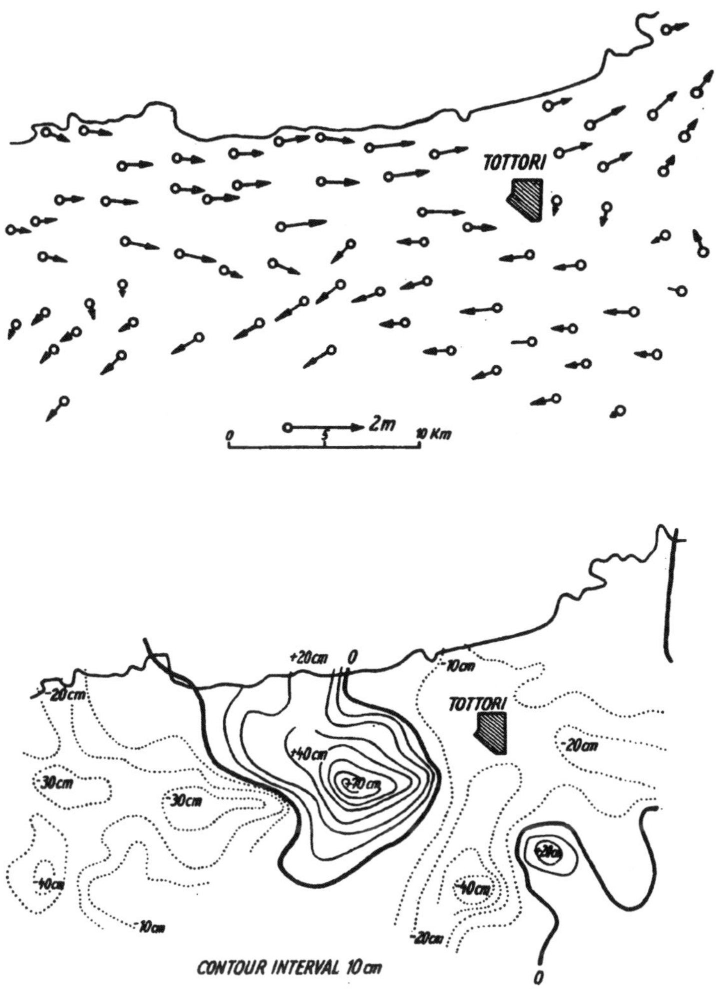

Fig. 11

Horizontal and vertical displacements accompanied with the Tottori Earthquake, 1943. First order triangulation and precise levelling results obtained by the Imperial Military Land Survey

parallel to the Arc (Yamada fault of the Tango Earthquake, Tottori Earthquake faults). (Figs. 10 and 11).

As already noticed (C.F. Richter, 1958), it is interesting that this tendency is also found in the other parts of the Circum-Pacific belt, e.g. in California the San-Andreas Fault which runs along the tectonic trend is right-hand and the Garlock Fault running perpendicularly to it is left-hand. The only difference is that in Japan the fault parallel to the arc trend is less prominent than the perpendicular one and in California it is the reverse. But it must be worth-while to note that the direction cutting the Honshu Arc is parallel just to the trend of the Izu-Mariana Arc.

The amounts of displacements in horizontal and vertical directions were in general of the same order in these earthquake faults of the Inner Honshu Arc Zone. In vertical movements faults are often superposed by the general upheavals of the epicentral regions and when the earthquake is small and the upheaval is not large or not sharp, the fault does not occur and only epicentral doming is observed (e.g. Imaich Earthquake, 1949, Omachi Earthquake, 1918).

In the Japan Sea side of NE Japan there are also very shallow, small shocks, the epicentres of which seem to lie in the veneer composed of the Neogene sediments mentioned above as Uetsu sub-geosyncline. They may be attributed to the activity of the Uetsu sub-geosyncline noted by the existence of active foldings (Y. Otuka, 1941). The crustal movements accompanied with the earthquakes of this zone were measured by re-levellings in several cases. Typical ones were domings in epicentral areas (Sekihara Earthquake, 1927, Nagaoka Earthquake, 1961), but when the space-gradient of the upheaval became large they gave rise to being accompanied with faults (Futatsui or Hutatui Earthquake, 1955) (C. Tsuboi, 1933, 1939, S. Miyamura and Okada, 1956, A. Okada, 1961).

It is difficult to distinguish the shallow earthquakes of NE Inner Japan whether they belong to the basement block movements or to the veneer folding activity. The depth of the focus is the only clue to the solution of this problem but its accuracy is at present not sufficient.

4. Two Types of Crustal Movements before and after the Earthquakes

Crustal movements before the earthquakes, if they have any peculiar behaviours, must be very important for the earthquake predictions. For the big earthquakes of the Outer Honshu Arc Seismic Zone secular southward tiltings of the peninsulae near the epicentres such as the Muroto (Shikohu Is), Kii (Central Honshu), Miura and Boso (Kanto District) Peninsulae were known, but no peculiar change of the mode of movements has ever been found before the following big earthquakes by the re-levellings. In any way large displacements of reverse sense were accompanied with earthquakes. After the earthquake the movements again change their sense and continue to tilt down southward to recover the movements accompanied with earthquakes. But this recovery was always incomplete and left residual upheavals which have been kept as coastal terraces as mentioned above.

Movements of faults before the earthquake have not been measured in Japan but several measurements or surveys were carried out after the earthquakes to detect the movements of the faults. In some cases the movements at the time of earthquakes continued in the same direction and stopped asymptotically (e.g. Tango Earthquake, 1927, Izu Earthquake, 1930) (C. Tsuboi, 1933, 1939), In other cases the faults moved after the earthquake in the opposite direction of the movements at the time of the earthquakes, and also stopped asymptotically (e.g. Tottori Earthquake, 1943) (S. Miyamura, 1944).

A peculiar behaviour of crustal movements before the earthquake, found by geodetic means, is very seldom even in Japan. To obtain the necessary data we must have three surveys before the earthquake at least. Moreover it is necessary that we have enough reason to regard the movements between the first and second surveys as normal or comparatively lasting and those between the second and third as abnormal or rather exceptional.

We have got only a few of such examples in Japan by re-levelling and by other geodetic methods we have none except the areal change of the base line rhombus at Mitaka, near Tokyo, before the Kanto Earthquake, 1923, (C. Tsuboi, 1961) but regrettably this rhombus example does not seem to be strongly convincing. For the Anegawa Earthquake, 1909, which took place in the area of the east coast of the Lake Biwa, a pre-seismic movement was indicated in the re-levelling result (A. Imamura, 1928), but it seems also to be rather doubtful. For the big earthquake of the Outer Honshu Zone A. Imamura was trying to detect the pre-seismic peculiar movements by the repeated re-levellings, but his attempt was hampered by the War. The necessary repeating during the ten years just before the earthquake was not executed. Thus we could not have any information about the pre-seismic behaviour for the Outer Honshu Zone earthquakes as stated above.

Rather hopeful an example was the reverse tendency of general tilt near the epicentral area of the Hutatui (Futatsui) Earthquake, 1955, several years before the shock, detected by re-levellings (S. Miyamura and A. Okada, 1949, 1956). Repeated levellings before the Nagaoka Earthquake, 1961, which occurred just in the epicentral area of the Sekihara Earthquake, 1927, showed a similar reversal a few years before the earthquake (A. Okada, 1961).

By the earth-tide instruments, i.e. tiltmeters, extensometers etc., several examples of peculiar movements preceding the earthquakes were reported (E. Nishimura, 1962).

For the short period prediction the measurements made by the earth-tide instruments will be hopeful and for the long period prediction geodetic re-surveying will be helpful to realize the earthquake predictions together with some other seismological and geophysical methods.

5. Acknowledgement

For the preparation of this manuscript the author owes very much to the colleagues of the "Institut für angewandte Geophysik der Bergakademie Freiberg", and his hearty thanks must be due to them all, especially to Prof. Dr. O. Meisser, Dr. L. Hiersemann, Dr. W. Bachmann and Miss. U. Sauerstein.

References +)

Imamura, A.: 1928. On the topographical changes preceding and following the Anegawa Earthquake of 1909. Proc. Imp. Acad., Japan, 4, 109-111

Imamura, A.; Kodaira, T. and Imamura, H.: 1932. The earthquake swarms of Nagusa and vicinity. Bull. E.R.I., 10, 636-648

Inoue, E.: 1960. Land deformation in Japan. Bull. Geogr. Sur. Inst., 6, 73-134

Miyamura, S.: 1944. Die zwei Verwerfungen beim Tottori-Beben vom 10. Sept. 1943 und ihre Bewegungen nach dem Beben. Bull. E.R.I., 22, 49-59

Miyamura, S.: 1959, 1960. Local earthquakes in the Kii Peninsula, Central Japan. Part. I--IV. Bull. E.R.I., 37, 347-358, 593-608, 609-635; 38, 71-112

Miyamura, S.: 1962a. Seismicity and geotectonics. Zisin (Journ. Seism. Soc., Japan), 15, in press

Miyamura, S.: 1962b. Seismotectonic zoning of Japan. Journ. Phys. of the Earth, in press

Miyamura, S. and Okada, A.: 1949. Vermessung eines Teiles von Nivellierrouten am Fluß Yonesiro. (Zweite Mitteilung) Bull. E.R.I. 27, 141-146

Miyamura, S. and Okada, A.: 1956. Vermessung eines Teiles von Nivellierrouten am Fluß Yonesiro. (Dritte Mitteilung) Höhenveränderung begleitet von einem starken Lokalbeben bei Hutatui, 19. Oktober, 1955, Bull. E.R.I., 34, 373-380

Miyamura, S. and Okada, A.: 1960. Results of levelling resurvey between Wakayama and Gobo, Wakayama Prefecture. Bull. E.R.I., 38, 355-359

+) The references given here are not intended to be complete. Many valuable data and results included in the cited references are re-quoted in the text without mentioning their original sources.

Nasu, N.: 1949. Block movements along the seismic fault (I). (Fukui Fault and others) Bull. E.R.I., 27, 27-34

Nasu, N.: 1950. Levelling and triangulation surveys. Investigation of seismic faults. The Fukui Earthquake of June 28, 1948 (Report of the Special Committee edited by H. Tsuya), 93-123

Nasu, N.: 1951. Block movement along the seismic fault (II). Bull. E.R.I., 29, 283-287

Nishimura, E.: 1962. Anomalous tilting motion of the ground before destructive earthquakes. Symposium on Recent Movements of the Earth's Crust, May, 1962, Leipzig. (Printed in this volume in pp 214

Okada, A.: 1961. Land deformation accompanied with the Nagaoka Earthquake, Feb. 2, 1961. Bull. E.R.I., 39, 537-547

Otuka, Y.: 1941. Active rock folding in Japan. Proc. Imp. Acad. Japan, 17, 518-522

Richter, C.F.: 1958. Elementary Seismology. W.H. Freeman u. Co., San Francisco

Sugimura, A.: 1962. Two examples of recent movements accompanied with earthquakes. Symposium on Recent Movements of the Earth's Crust, May, 1962, Leipzig (Printed in this volume in pp 252

Tsuboi, C.: 1933. Investigations of the deformation of the earth's crust found by precise geodetic means. Jap. Journ. Astr. u. Geophys., 10, 93-248

Tsuboi, C.: 1939. Deformation of the earth's crust as disclosed by geodetic measurements. Ergeb. d. kosm. Phys., 4, 105-168

Tsuboi, C.: 1961. Paper read at the Symposium on Earthquake Occurrence held by the Seism. Soc. of Japan, May, 1961, Tokyo

Tsuya, H.: 1946. The Fukozu Fault. A remarkable fault formed during the Mikawa Earthquake of Jan. 13., 1945. Bull. E.R.I., 26, 59-75

Diskussion

T. Sorgenfrei (Kopenhagen):

I ask for information on the maximum horizontal and vertical displacement.

S. Miyamura (Tokyo):

Horizontal displacement about two metres and vertical about or more than two metres.

T. Sorgenfrei (Kopenhagen):

I ask whether it had been possible to trace single active faults during subsequent disastrous earthquakes.

S. Miyamura (Tokyo):

We have no experience that the same fault moved at the time of two past earthquakes. But the continuation of one past earthquake fault moved at the time of next earthquakes, e.g. the continuation of Neodam-fault of 1891 earthquake moved at the time of Fukui earthquake, 1957. Continuation of earthquake fault is not so long, may be less than 30 ... 50 km.

A. A. I z o t o v (Moskau):

With which methods have the displacements been recorded?

S. M i y a m u r a (Tokyo):

Horizontal displacements were of course determined by triangulation of 1st class and lower classes.

A. A. I z o t o v (Moskau):

Which calculation methods have been used?

S. M i y a m u r a (Tokyo):

One 1st class triangulation point outside the disturbed area and direction from this point to another nearest one point were assumed not to have moved during the two triangulations.

Two examples of recent and pleistocene crustal movements in Japan

von

A. Sugimura (Tokyo)

S u m m a r y

The author describes here two kinds of earth's crust movement, which are principally vertical movement (1) and principally horizontal movement (2).

Upheaval-amount of coastal terrace formed in about 6000 years before the present time in South Kantō (K in Fig. 1), is directly proportional to upheaval-amount of the ground in 1923 earthquake (Figs. 2 and 3), and can be explained as a sum of upheavals by earthquakes.

The Adera dislocation in Central Japan (A in Fig. 1), cut eight river terraces of R. Kiso (Fig. 4), whose formations are approximately known. The movement shows approximately constant direction and constant speed.

R e s u m o

La autoro priskribas tie ĉi du specojn da terkrusta movado, tiuj estas principe vertikala movado (1) kaj principe horizonta movado (2).

Leviĝkvanto de marborda teraso formita en ĉ. 6000 jaroj antau estanto en Suda Kantō (K en Fig. 1), estas rekte proporcia kun leviĝkvanto de tero en 1923 tertremo (Figs. 2 kaj 3), kaj povas esti klarigata kielamaso de leviĝoj en tertremoj.

La Adera dislokiĝo en Centra Japanio (A en Fig. 1) malakordigas ok riverterasojn de Rivero Kiso (Fig. 4), kiuj formaĝoj estas proksimume konataj. La movado montras proksimume konstantan directon (Fig. 5) kaj konstantan rapidecon (Fig. 6).

Z u s a m m e n f a s s u n g

Der Autor beschreibt hier zwei Arten der Erdkrustenbewegung, die in der Hauptsache Vertikalbewegung (1) und Horizontalbewegung (2) sind.

Der Hebungsbetrag der Küstenterrasse, die vor ungefähr 6000 Jahren in Süd-Kantō (K in Bild 1) gebildet wurde, ist direkt proportional der Bodenhebung beim Erdbeben von 1923 (Bilder 2 und 3) und kann als eine Summe von Hebungen bei Erdbeben dargestellt werden.

Die Adera-Verwerfung in Zentraljapan (A in Bild 1) schneidet acht Terrassen des Flusses Kiso (Bild 4), deren Formationen ungefähr bekannt sind. Die Richtung (Bild 5) und die Geschwindigkeit (Bild 6) der Bewegung sind ungefähr konstant.

1. Seismic upheavals in South Kantô

The aerial variation of the altitudes of a coastal terrace in South Kantô was surveyed by A. Sugimura and Y. Naruse (1954; 1955). The age of the terrace is estimated to be about 4,000 B.C. The geographical position is given in Fig. 1. Numbers from 1 to 13 along the linked lines in the figure are the points where the altitudes of the terrace were measured. The result is provided in Fig. 2. The linked lines are taken for the abscissa in the figure. The altitude of the terrace and upheaval during the Kantô earthquake of 1923 are taken for the ordinate on different scales. The curves shown in Fig. 2 mutually resemble in shape. The seismic upheaval is proportional to the altitude of the terrace as shown in Fig. 3 (left), if the lowering of the sea level (6 m) is taken into consideration. It may be expected that the uplift since 6,000 years ago is the result of accumulation of the seismic upheavals.

2. The Adera fault in Central Japan

Nine river terraces are found at Sakasita along the River Kiso in Central Japan. These terraces are separated and offset in height and position across the Adera fault at Sakasita. The geographical position is given in Fig. 1. The vertical component of the faulting is confirmed from the displacement of the terrace surface, and the lateral component from the displacement of the terrace scarp. Recently the writer and his colleague revealed the displacement vectors of the Adera fault by this method (A. Sugimura u. T. Matsuda, in press). The result is provided in Fig. 4. The vertical scale is exaggerated five times on Fig. 5. The numbers of the terraces that are given in the order of increasing age are indicated in the Roman numerals on Fig. 5. It is the present river bed. IX is the highest terrace with the aggradational surface but without the scarp behind and is rather similar to a table. Consequently, the lateral component can not be known for IX.

The ages of most of the terraces can be roughly estimated on the basis of information by many of the Quaternary geologists. Relationship between the age (B.P.) and the net slip is given in Fig. 6, and would lead us to suppose the constant rate of faulting, which averages 50 m / 20,000 years in lateral component.

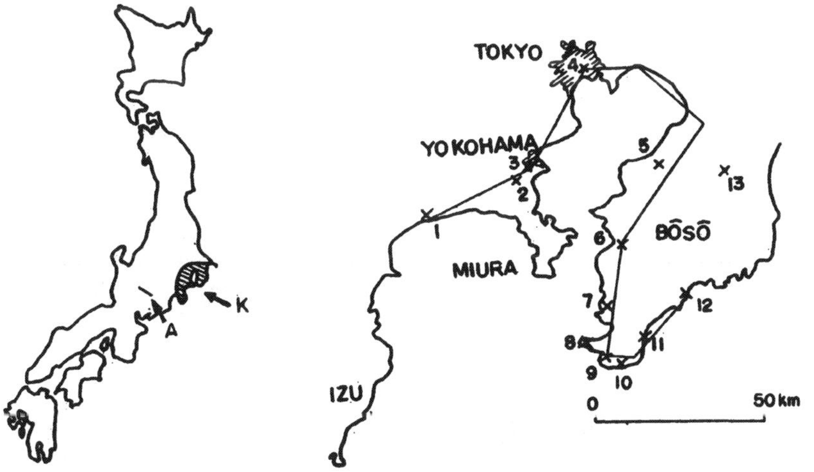

Fig. 1
Location of areas surveyed (left) and location of section in figure 2 (right)
A: the Adera fault.
K: the southern Kantô region
x1 - x13: stations where the altitudes of the Holocene highest terrace were measured in the southern Kantô region

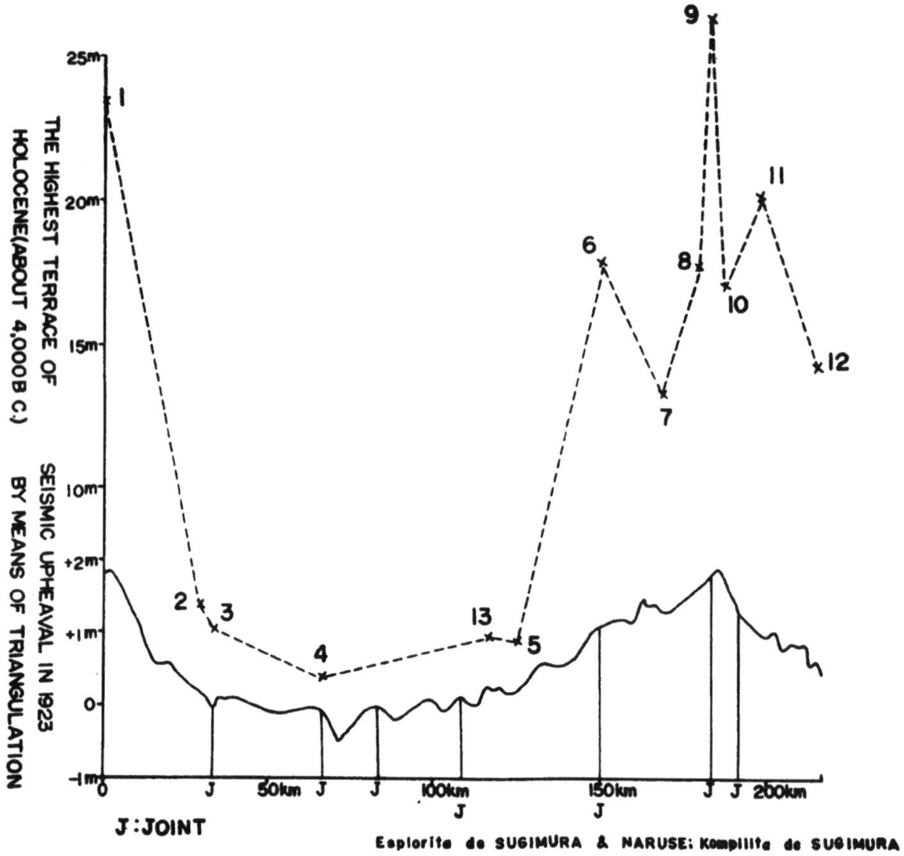

Fig. 2

Spatial variations of the altitude of the Holocene highest terrace (broken line) and of the seismic upheaval in 1923 (solid curve) both along the lines indicated in the right part of figure 1

xl - x 13: same as for fig. 1

J: joints of the lines indicated in the right part of fig. 1
Vertical scale for the seismic upheaval is more exaggerated than that for the terrace

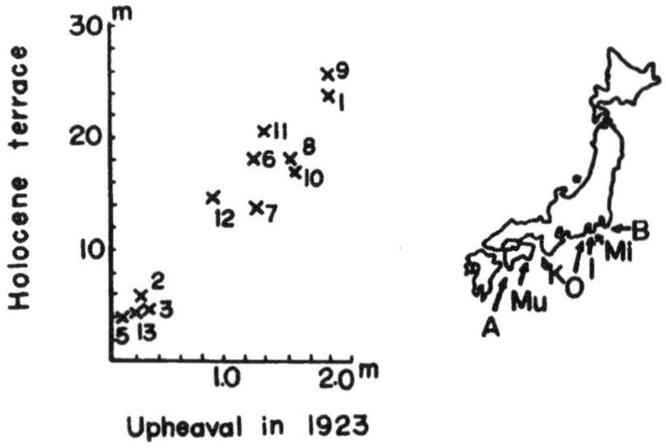

Fig.3

Relationship between y (the altitude of the Holocene highest terrace) and x (the seismic upheaval in 1923) both in meter (left) and distribution of peninsulas showing seismic upheaval (right)

xl - xl3: same as for figs. 1 and 2
A: Asizuri Point;
Mu: Muroto Point;
K: Kii Peninsula;
O: Omaezaki Point;
I: Izu Peninsula;
Mi: Miura Peninsula;
B: Bôsô Peninsula

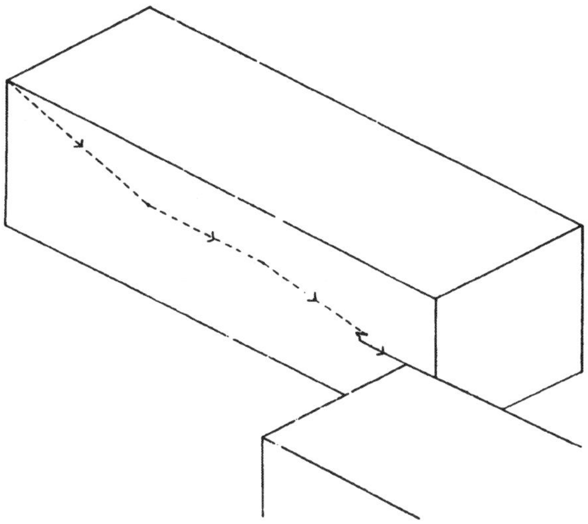

Fig. 4

Diagrammatic representation of displacement vectors
of the Adera fault (view looking north)
Arrow: the direction of the faulting.
Broken line and solid line: the locus of the faulting

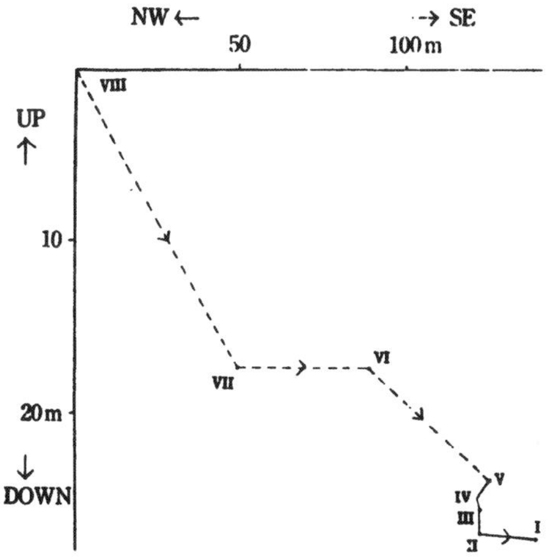

Fig. 5

Displacement vectors of the Adera fault
I - VIII: the number of the river terrace
assigned from the younger to the older.
Vertical scale is exaggerated five times.
The graph shows the relative displacement
of the southwestern side of the fault

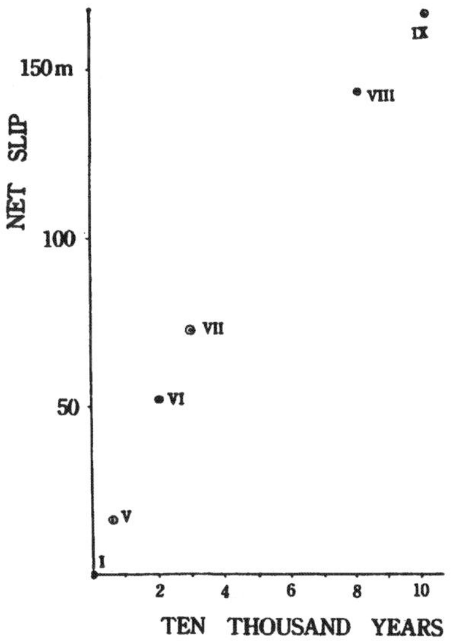

Fig. 6

Net slip of the Adera fault as a
function of absolute geologic time
I, V - IX: same as for fig. 5

R e f e r e n c e s

Sugimura, A.; Matsuda, T.: The Adera fault (in press), Quat. Res., vol. 2 (Japanese with English abstract)

Sugimura, A.; Naruse, Y.(1954; 1955): Changes in sea level, seismic upheavals, and coastal terraces in the southern Kantô region, Japan (I); (II), Japan J. Geol. Geogr., vol. 24, 101-113; vol. 26, 165-176

Rotierende Felder in der Erdkruste

von

N. P a v o n i (Zürich)

S u m m a r y

Recent tectonic processes as well as new research on the focal mechanism of earthquakes show that great importance has to be attached to horizontal displacements along steep fracture planes in the earth's crust. Recent horizontal displacements have been observed in California, Alaska, Japan, Formosa, New Zealand, Sumatra, Baluchistan and Turkey. The following facts and reflections allow us to see these strike-slip movements on a world-wide scale, in spite of their small number:

1.) The earthquake epicenters are concentrated in definite zones stretching over the whole earth's surface.

2.) From a geological point of view the seismic districts coincide with the Cenozoic mountain belts, with certain submarine ridges, and with large fracture systems of the earth's crust.

3.) The zones of active strike-slip displacements may have a length of more than 1000 km.

4.) The direction of displacement and the velocity of horizontal displacement within a definite zone remain the same for a long period.

5.) All of these observations confirm that large deep-seated processes of long duration take place in the earth's crust and mantle.

6.) It appears that in these zones, the Tertiary tectonic features and the present displacements agree amazingly. Therefore it is possible that the correct interpretation of the Cenozoic tectonic features of the earth's crust may show us the interrelation between the widely separated zones of recent horizontal displacement.

The Tertiary tectonic features and recent tectonic phenomena make it possible to distinguish three large rotational fields in the earth's crust (cp. fig. 3), namely the Laurasia Field, the Gondwana Field, and the Pacific Field. These fields rotate very slowly counterclockwise and shift relatively to each other with a velocity of 1m/century. Where these fields touch shear-zones develop with right-handed strike-slip displacements: These are the mobile zones of Cenozoic mountain-building movements and at the same time the belts of earthquakes which are tectonically still active. Due to the relative movements extensive left-handed horizontal strike-slip displacement develop transverely to the right-handed system of horizontal displacements.

Z u s a m m e n f a s s u n g

Rezente tektonische Vorgänge und ebenso die neuen Untersuchungen über den Herdvorgang bei Erdbeben zeigen, daß im heutigen tektonischen Geschehen den horizontalen Verschiebungen an steilstehenden Bruchflächen eine große Bedeutung zukommt. Es wird der Versuch unternommen, diese rezenten Horizontalverschiebungen, wie sie z.B. in Kalifornien, Alaska, Japan, Formosa, Neuseeland, Sumatra, Baluchistan und der Türkei beobachtet wurden, in einen größeren, erdumfassenden Zusammenhang zu stellen.

Folgende Tatsachen und Überlegungen erlauben es, diese Daten, trotz ihrer geringen Zahl, in einen weiteren Rahmen zu stellen:

1.) Die Erdbebenherde konzentrieren sich auf bestimmte, erdumspannende Zonen.

2.) Geologisch gesehen fallen die Erdbebengürtel mit dem känozoischen Gebirgsgürtel, mit gewissen untermeerischen Rücken und großen Bruchsystemen der Erdkruste zusammen.

3.) Die heute aktiven Horizontalverschiebungszonen besitzen zum Teil sehr große Ausdehnungen von bis über 1000 km Länge.

4.) Verschiebungssinn und Verschiebungsgeschwindigkeit innerhalb einer bestimmten Horizontalverscheibungszone bleiben über lange Zeiträume hindurch dieselben.

5.) Diese Beobachtungen sprechen übereinstimmend für große, langandauernde, tiefgreifende Vorgänge in der Erdkruste und im Erdmantel.

6.) Es zeigt sich, daß das tertiäre tektonische Bewegungsbild und die heutige Verschiebung in den bis jetzt bekannten HV-Zonen in ganz erstaunlichem Maße übereinstimmen. Es ist daher möglich, daß die richtige Interpretation des känozoischen Bewegungsbildes der Erdkruste die Zusammenhänge zwischen den weit auseinander liegenden rezenten Horizontalverscheibungszonen zu liefern vermag.

Tertiäres Bewegungsbild und rezente tektonische Vorgänge lassen drei große Rotationsfelder in der Erdkruste (s. Abb. 3), das Laurasiafeld, das Gondwanafeld und das Pazifikfeld, erkennen, die sich sehr langsam im Gegenuhrzeigersinn drehen und mit ca. 1m/Jahrhundert relativ zueinander verschieben. Im Kontaktbereich entstehen Scherzonen mit rechtshändigen Horizontalverschiebungen: Es sind dies die mobilen Zonen känozoischer Gebirgsbildung und zugleich die tektonisch heute noch aktiven Erdbebengürtel. Infolge der relativen Verschiebungen entstehen quer zum rechtshändigen Horizontalverschiebungssystem ausgedehnte linkshändige Horizontalverschiebungen.

1. Einleitung

(1) Die rezenten tektonischen Vorgange und ebenso die Untersuchungen über den Herdvorgang bei Erdbeben (J.H. Hodgson 1957) zeigen, daß den Horizontalverschiebungen im heutigen tektonischen Geschehen große Bedeutung zukommt.

(2) Unter dem Begriff Horizontalverschiebung sei sowohl der Vorgang der relativen gegenseitigen Verschiebung zweier Gesteins- oder Krustenabschnitte in vorwiegend horizontaler Richtung entlang einer steilstehenden Bruchfläche oder Bruchzone als auch die steilstehende Bruchfläche oder Bruchzone selbst, in deren Bereich die Verschiebung erfolgte, verstanden. Synonyme zum Begriff Horizontalverschiebung sind die Bezeichnungen Blattverschiebung, Seitenverschiebung, Transversalverschiebung, Paraphor (H. Cloos, 1936). Die entsprechende französische Bezeichnung ist décrochement, die englischen Bezeichnungen heißen: Heave, shift fault, tear fault, strike-slip fault, wrench fault, transcurrent fault. Der relative Verschiebungssinn ist von oben gesehen rechtshändig (= im Uhrzeigersinn, d.h. relative Verschiebung des rechten Flügels zum Beschauer hin, wenn dieser in Richtung der Verschiebung blickt), oder linkshändig (= im Gegenuhrzeigersinn).

(3) Daß auch vom energiemäßigen Standpunkt aus betrachtet Horizontalverschiebungen ebenso häufig, ja viel häufiger als vertikale Versetzungen, zu erwarten sind, hat H.

Benioff (1958) überzeugend dargelegt. Nach diesem Autor benötigen horizontale Massenbewegungen in der Erdkruste entlang steilstehender Bruchflächen 10-100 mal weniger Energie als entsprechende Massenbewegungen, die der Schwerkraft entgegengesetzt sind.

II. Beobachtungen an heute aktiven Horizontalverschiebungszonen

(4) Rezente, in historischer Zeit gut belegte horizontale Verschiebungen sind aus folgenden Gebieten bekannt geworden: Kalifornien, Alaska, Japan, Formosa, Neuseeland, Sumatra, Baluchistan und Türkei. Der Verfasser war bestrebt, alle Angaben über rezente Horizontalverschiebungen zu sammeln und zu sichten. Da aber die diesbezügliche Literatur außerordentlich zerstreut ist und ihm überdies nur ein Teil dieser Literatur zugänglich ist, ist er außerordentlich dankbar für jede Mitteilung über beobachtete rezente Horizontalverschiebungen. Die uns bekannten Daten haben wir in Abb. 1 schematisch dargestellt.

Neben den an der Erdoberfläche beobachteten relativen Horizontalverschiebungen wurden auch einige durch dicht gescharte Epizentren und Nachbeben gut belegte seismisch aktive Linien eingezeichnet. Theoretische Angaben über den Herdvorgang bei Erdbeben wurden nicht berücksichtigt. Im ganzen gesehen, erscheinen somit die Beobachtungen recht spärlich und unzusammenhängend über die ganze Erdoberfläche zerstreut.

Wenn der Verfasser im folgenden diese relativ spärlichen und weitzerstreuten Anhaltspunkte in einem größeren Zusammenhang zu sehen versucht, so darum, weil es sich dabei um sehr weiträumige großartige Vorgänge in der Erdkruste und im Erdmantel handelt und weil diese Relativverschiebungen nicht beliebig auftreten können, sondern sich notwendigerweise in einen - letzten Endes - erdumfassenden Rahmen einordnen müssen.

(5) Relativ gut erforscht sind die rezenten Horizontalverschiebungen in Japan, in Kalifornien und Neuseeland. Weniger gut bekannt ist die nordanatolische Horizontalverschiebungszone (N. Pavoni, 1962). Die aktiven Horizontalverschiebungszonen besitzen z.T. nachweisbare Längserstreckungen von bis über 1000 km. Detailuntersuchungen zeigen, daß es sich nicht nur um sprunghafte plötzliche Relativverschiebungen, sondern ebenso auch um ein kontinuierliches, gleichmäßiges Kriechen handelt, ohne daß dabei Erdbeben ausgelöst werden. Direkte Messungen und wiederholte Überprüfung geodätischer Vermessungen seit über 50 Jahren (K.V. Steinbrugge u. E.G. Zacher, 1960; D. Tocher, 1960; C.A. Whitten, 1960) ergeben für die ganze San Andreas Fault-Zone Relativverschiebungen von einem bis mehreren Metern pro Jahrhundert. Ganz ähnliche Werte ergaben sich aus der Versetzung jungquartärer Flußterrassen in Neuseeland im Bereich der alpinen Fault-Zone. Verschiebungen gleicher Größenordnung sind auch aus Japan bekannt geworden. Es ergibt sich aus diesen Beobachtungen, daß in den großen Horizontalverschiebungszonen der Erde bei vorsichtiger Schätzung mit einem Meter relativer Verschiebung pro Jahrhundert gerechnet werden kann. Damit ist dem Geologen und Geophysiker ein interessanter Wert, eine Art "Konstante" für gewisse Vorgänge in der Erdkruste in die Hand gegeben.

(6) Das Alter der heute aktiven großen Horizontalverschiebungszonen läßt sich kaum je mit Sicherheit bestimmen. Charakteristisch für die Horizontalverschiebungen ist das Auftreten von Thermalquellen und die starke Bebentätigkeit, ferner die Ablenkung der Flußläufe und die Versetzung von Terrassen und Schuttkegeln. Geomorphologisch sind die Horizontalverschiebungszonen meist durch langgestreckte auffallend geradlinige Talungen parallel zur Längserstreckung der Verschiebungszone ausgezeichnet. Diese Tatsachen deuten daraufhin, daß die großen Horizontalverschiebungszonen ein Alter von Zehntausenden von Jahren besitzen und ihre Aktivität vielfach weit ins Quartär und ins Pliozän hineinreicht.

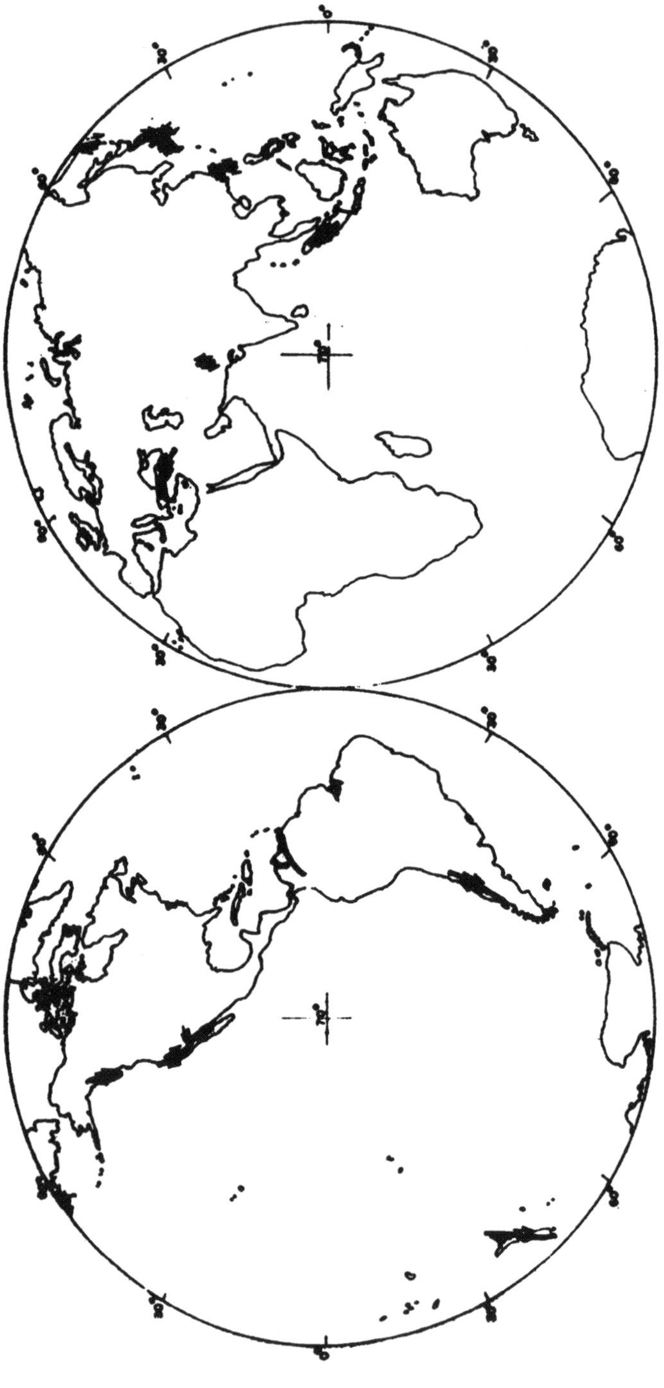

Abb. 1

Heute aktive Horizontalverschiebungen. Pfeile: relativer Verschiebungssinn. Pfeile gestrichelt: Verschiebungssinn auf Grund seismischer Untersuchungen nachgewiesen

(7) Der Verschiebungssinn einer bestimmten Horizontalverschiebung bleibt offenbar über lange Zeiträume und über die ganze Verschiebungszone hinweg derselbe. Dem Verfasser ist kein Beispiel bekannt, wo in historischer Zeit der Verschiebungssinn einer bestimmten Horizontalverschiebung gewechselt hätte. Umgekehrt bestehen zahlreiche Beobachtungen aus historischer Zeit, welche den wiederholten gleichen Sinn der Relativverschiebung bestätigen. Geologische Beobachtungen bekräftigen die historischen Feststellungen und führen sie fort bis in sehr weit zurückliegende Zeiträume.

(8) Wir gelangen anhand der Beobachtungen an den heute aktiven Horizontalverschiebungszonen zu folgenden Aussagen:

a.) Horizontalverschiebungen treten mindestens ebenso häufig auf wie vertikale Versetzungen.

b.) Die aktiven Horizontalverschiebungszonen besitzen große Längserstreckungen von 100 bis über 1 300 km.

c.) Der Verschiebungssinn einer bestimmten Horizontalverschiebungszone bleibt über lange Zeiträume hindurch derselbe.

d.) Die Verschiebungsgeschwindigkeit beträgt in den großen Horizontalverschiebungszonen der Erde bei vorsichtiger Schätzung ca. 1 m/Jahrhundert. Es würde daher von der Gegenwart aus gesehen nicht überraschen, wenn in einzelnen Horizontalverschiebungszonen für den Zeitraum seit Beginn des Tertiärs bis heute Relativverschiebungen von mehreren hundert Kilometern gefunden werden sollten.

III. Horizontalverschiebungen und mesozoisch-tertiäre Tektonik

(9) Überaus vielfältig sind die Beziehungen zwischen Horizontalverschiebung und den daraus entstehenden Bewegungsbildern. Experimente und die konsequente Analyse auf Grund der mechanischen Gesetzmäßigkeiten können hier manche Beziehungen erhellen. Handelt es sich um kompetentes, ungeschichtetes Material, so führt die Horizontalverschiebung an der Erdoberfläche zu einer Zerbrechung der Kruste in einzelne Schollen, die sich in gesetzmäßiger Weise aneinander verschieben. Liegt jedoch über den sich verschiebenden Schollen eine geschichtete Sedimenthaut in inkompetenten Zwischenlagen, so wird diese über den Bewegungsfugen gefaltet (N. Pavoni, 1961). Faltung durch Horizontalverschiebung ist ein Vorgang, der nach der Meinung des Verfassers im tektonischen Geschehen eine sehr bedeutsame Rolle spielt.

(10) Auf Grund der Bewegungsbilder in der Kruste schließt der Geologe auf die Bewegungen, die zur Entstehung dieser Bilder geführt haben. Leider sind die Aussagen keineswegs immer eindeutig. Stets muß man sich bewußt sein, daß es sich um Relativbewegungen handelt. Erschwerend kommt gerade im Falle der Horizontalverschiebungen hinzu, daß es sich dabei um sehr lange andauernde langsame Vorgänge handelt, wobei die Bewegungsbilder durch die Einwirkung der exogenen Vorgänge sehr weitgehend zerstört oder doch stark verwischt werden können.

Vergleicht man in den großen Horizontalverschiebungszonen, z.B. in Kalifornien, auf Neuseeland, in der Türkei, in Baluchistan und auch in Japan den rezenten Verschiebungssinn und die regionale junge Tektonik, so ist man erstaunt zu sehen, *wie gut die heutige Verschiebung und das tertiäre Bewegungsbild übereinstimmen*. Man erhält unbedingt den Eindruck, daß in den meisten Gebieten, wo ein solcher Vergleich möglich ist, heute noch prinzipiell die gleichen Bewegungen und relativen Verschiebungen auftreten, die auch die jungen tektonischen Formen geschaffen haben.

Ausgehend von dieser bemerkenswerten Tatsache, auf die an anderer Stelle im Detail eingegangen werden soll, versuchte der Verfasser das Bewegungsbild der jungen Faltengebirgsgürtel der Erde nach dem Prinzip der Horizontalverschiebungsfaltung (N. Pavoni, 1961) in großen Zügen zu analysieren, um damit möglicherweise die Verbindung zwischen den heute an der Erdoberfläche beobachteten Horizontalverschiebungen zu erhalten. Das stark vereinfachte Resultat dieser Analyse ist in Abb. 2 dargestellt.

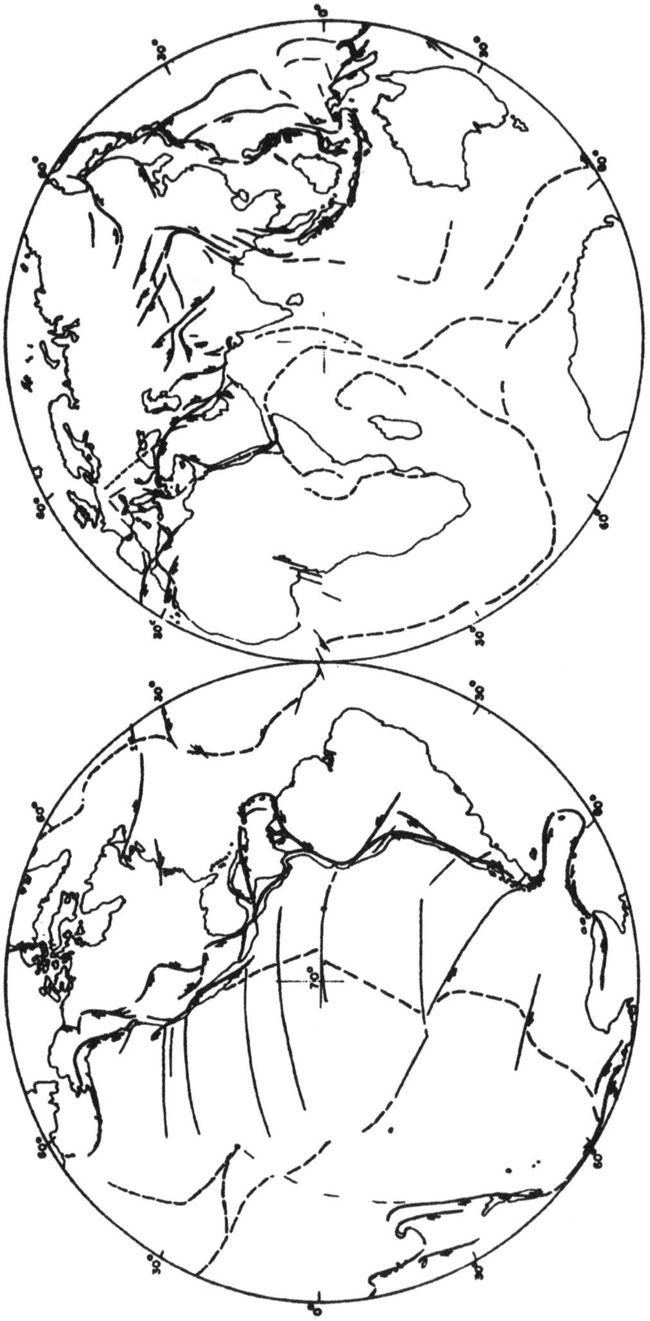

Abb. 2

Analyse des tektonischen Bewegungsbildes im Bereich der jungen Gebirgsgürtel nach dem Prinzip der "Faltung durch Horizontalverschiebung". Kräftige Linien: Horizontalverschiebungen im Untergrund vermutet auf Grund der jungen Faltenstrukturen. Pfeile: relativer Verschiebungssinn. Gestrichelt: Große untermeerische Rücken und Bruchzonen. Dünne Linien: Untermeerische Horizontalverschiebungen vermutet auf Grund der Topographie oder geomagnetischer Messungen

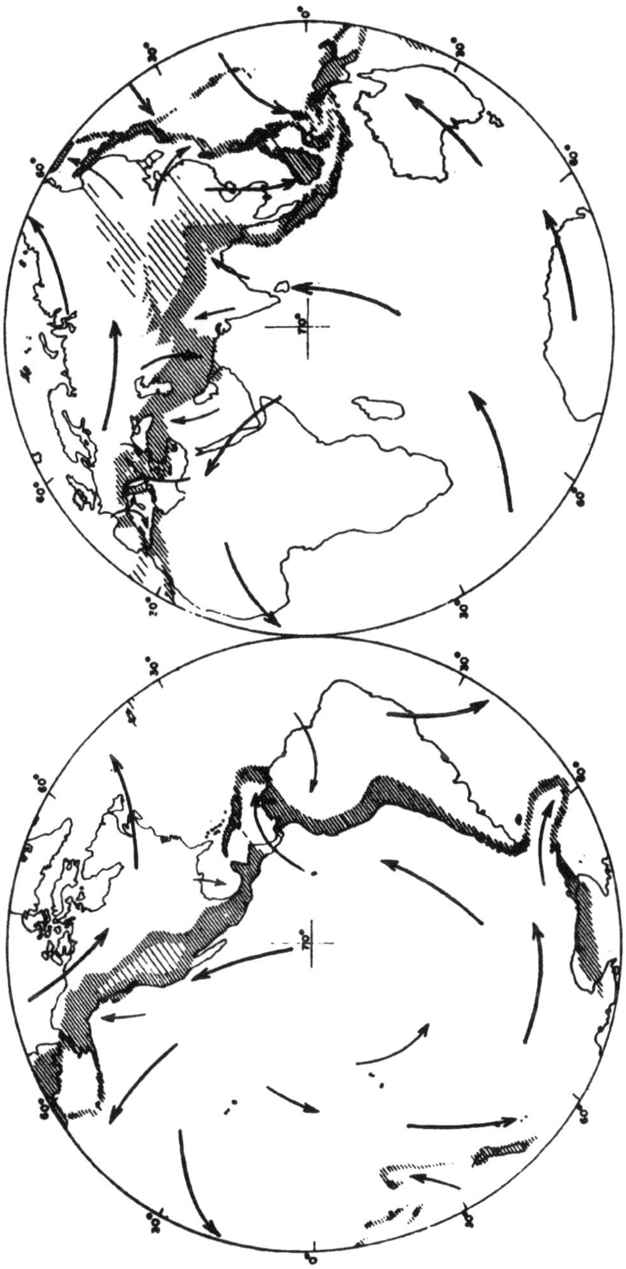

Abb. 3

Die drei Rotationsfelder der Erde. Die Pfeile sollen die generelle Bewegungstendenz im tieferen Untergrund andeuten. Kleinere Pfeile: sekundäre Bewegungen (Wirbelbildung). Eng schraffiert: junge Faltengebirgsgürtel

IV. Rotierende Felder in der Erdkruste?

(11) Ein Vergleich von Abb. 1 und Abb. 2 zeigt eine weitgehende Übereinstimmung und gegenseitige Ergänzung der Verschiebungssysteme. So darf wohl - trotz der zunächst stark hypothetischen Analyse der Faltengebirgsbögen - auf eine einheitliche Ursache für die Entstehung der känozoischen Gebirgsgürtel als auch der rezenten Horizontalverschiebungen geschlossen werden. Andererseits spricht gerade diese Übereinstimmung stark zugunsten einer Faltung durch Horizontalverschiebung in den jungen Faltengebirgen.

(12) Die Verteilung der Epizentren zeigt deutlich, daß die Erdbeben nicht willkürlich zerstreut auftreten, sondern sich auf bestimmte Zonen konzentrieren; es sind dies die heute aktiven tektonischen Zonen. Geologisch gesehen fallen diese Erdbebenzonen mit den jungen Gebirgsgürteln, mit vulkanisch aktiven Gebieten, mit gewissen untermeerischen Rücken und großen Bruchsystemen der Erdkruste zusammen. Ausgehend von tektonisch-stratigraphischen Beobachtungen kommt die Geologie zum Schluß, daß es sich bei den orogenen Zonen um mobile Abschnitte der Erdkruste handelt. Es ist darum nicht überraschend, daß die jungen Gebirgsgürtel mit den tektonisch heute noch aktiven Zonen zusammenfallen. Die Koinzidenz weist sehr eindrücklich darauf hin, daß der Vorgang, der zur Entstehung der känozoischen Faltengebirgsgürtel führte, heute noch nicht abgeschlossen ist.

(13) Anordnung und Verlauf der Erdbebenzonen, ergänzt durch den Verlauf der jungen Gebirgsgürtel, ergeben eine erste Großgliederung der Erdoberfläche in drei Felder, die sich ihrerseits in Teilfelder unterteilen lassen: Das Pazifik-Feld, das Gondwana-Feld und das Laurasia-Feld. Die jungen Gebirgsgürtel liegen im Grenzbereich dieser drei Felder. Die Faltenanalyse nach dem Prinzip der Horizontalverschiebungsfaltung und ebenso die rezenten Horizontalverschiebungen deuten darauf hin, daß es sich dabei um gewaltige Scherzonen in der Erdkruste handelt. Der Verschiebungssinn in Richtung generell parallel zum Gebirgsgürtel ist rechtshändig! Quer zum rechtshändigen System treten linkshändige Horizontalverschiebungen auf, wie das mechanisch zu erwarten ist. Die linkshändigen Verschiebungen besitzen nicht die Geschlossenheit und auch nicht jene Bebentätigkeit wie diejenigen des rechtshändigen Systems. Es scheint also von der heutigen Bebentätigkeit aus gesehen, daß dem rechtshändigen System die primäre Bedeutung zukommt und daß die linkshändigen Verschiebungen eher als eine Folge der rechtshändigen Verschiebungen zu betrachten sind, obwohl natürlich keines der beiden Systeme ohne das andere auskommen kann. Dafür spricht auch die Geologie. Dies würde heißen, daß die drei Felder oder doch wenigstens deren zwei sich langsam im Gegenuhrzeigersinn drehen und sich mit einem bis mehreren Metern pro Jahrhundert relativ zueinander verschieben. In Abb. 3 habe ich die Bewegung der drei Rotationsfelder, wie sie sich aus der Anordnung der Horizontalverschiebungen an der Erdoberfläche ergibt, darzustellen versucht, unter der Annahme, daß alle drei Felder sich drehen.

V. Zur Theorie der horizontalen Rotation

(14) Das Resultat der vorliegenden Analyse mag zunächst überraschen. Bei näherem Zusehen jedoch ergeben sich eine Fülle von Aussagen und neuen Erklärungsmöglichkeiten, gerade auch in bezug auf die Fragen der Gebirgsbildung, die hier höchstens kurz gestreift werden können. Vergessen wir auch nie, daß es sich um ein dreidimensionales Problem handelt. Doch ist schon einiges gewonnen, wenn die großen tektonischen Vorgänge an der Erdoberfläche in einen Zusammenhang gebracht werden können.

Die drei Rotationsfelder sind ungefähr gleich groß. Schematisch dargestellt ergibt

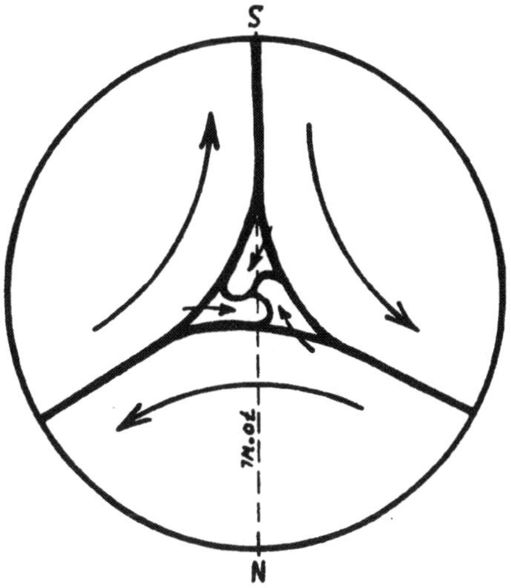

Abb. 4 Schematische Darstellung der drei Rotationsfelder

sich die Anordnung, wie sie in Abb. 4 dargestellt ist. Über die Ursache der Rotation der Felder kann vorläufig nichts ausgesagt werden. Tatsache ist, daß die gegenseitige Verkeilung der Felder in der Kontaktzone die Drehbewegung zu bremsen versucht. Gebirgsbildung wäre gewissenmaßen eine Trägheitsreaktion der Erdkruste, die der Drehung der drei Felder und damit der Ursache für die Drehung entgegenwirkt (N. Pavoni, 1960). Das tektonische Bild an der Erdoberfläche ist ein passives Bewegungsbild, verursacht durch Bewegung in größerer Tiefe. Die Horizontalverschiebungen in der Kruste erscheinen als Komponenten dieser tieferliegenden Bewegung. Aus der Anordnung der Horizontalverschiebungen kann auf die Richtung der Bewegung im tieferen Untergrund geschlossen werden. Wichtig ist, daß die Zeitabfolge berücksichtigt wird. In der vorliegenden Analyse werden nur diejenigen Verschiebungen berücksichtigt, die im Känozoikum aktiv waren.

Am ehesten lassen sich die Bewegungen im Untergrund als S t r ö m u n g e n denken. Daraufhin deuten an der Erdoberfläche die Stauzonen, die Gebirgsbögen und ganz besonders die Inselbögen, die, wie z.B. der Malayische Archipel oder die Fidschi-Inseln, an eigentliche Wirbelbilder erinnern.

(15) D i e T h e o r i e d e r h o r i z o n t a l e n R o t a t i o n, kurz Rotationstheorie, zeigt gegenüber den bisherigen geotektonischen Theorien manche Vorteile, indem sie von einem Minimum an Voraussetzungen ausgeht. So ist es für die Rotationstheorie zum Beispiel nicht von entscheidender Bedeutung, ob die Erde eine Kontraktion oder eine Ausdehnung erfährt. Mit der Kontraktionstheorie wird eine geschlossene feste Kruste angenommen. Im Gegensatz zur Kontraktionstheorie, welche mit einer praktisch unveränderlichen Lage der Kurstenteile rechnet, erlaubt und postuliert die Rotationstheorie weitreichende horizontale Verschiebungen innerhalb der Erdkruste, ein mosaikartiges Spiel sich aneinander vorbeipressender Schollen.

Mit den Kontinentalverschiebungs- und Unterströmungstheorien nimmt die Rotationstheorie weitreichende Verschiebungen in der Erdkruste an, doch ohne das weiträumige Auseinanderdriften von Sialschollen, das zur Öffnung ganzer Ozeanräume geführt haben soll. Das schwierige Raumproblem ist gelöst, indem im Gegensatz zu allen bisherigen Theorien weder ein radiales Auseinanderweichen, noch ein frontales Aufeinanderstoßen, sondern ein seitliches aneinander Vorbeistreichen der Krustenfelder angenommen wird. Bei der Rotationstheorie müssen weder neue Flächen ozeanischen Ausmaßes entstehen, noch müssen bestehende Flächen in großem Ausmaße verschwinden. Die Gesamtoberfläche bleibt im Prinzip erhalten, auch wenn es zu weitreichenden gegenseitigen Krustenverkeilungen kommt. Die Gliederung der Kruste in Kontinentalschollen und Ozeanböden ist nicht von entscheidender Bedeutung, ein Gleiten der Sialschollen im festen Untergrund nicht notwendig. Für große Gebiete der Erdoberfläche ist somit eine Permanenz der Kontinente und Ozeane anzunehmen. Die entscheidenden Veränderungen vollziehen sich in den Scherzonen und in Gebieten starker Verkeilung der Kruste. In der Rotationstheorie ist die Forderung nach isostatischem Ausgleich erfüllt.

(16) Für die Gebirgsbildung ergeben sich neue Aspekte und Erklärungsmöglichkeiten. Von Seiten der Geologen wird im allgemeinen dem Querprofil die hauptsächliche Beachtung geschenkt. Gerade im Querprofil sind Horizontalverschiebungen schwierig zu erkennen. Bewegungsbilder im Querprofil werden fast unwillkürlich auf Bewegungen in Richtung des Profils zurückgeführt. Daher die primäre Bedeutung, welche die bisherigen Gebirgsbildungstheorien dem frontalen Aufeinanderstoßen zweier Krustenabschnitte beimessen.

Es zeigt sich, daß Gebirgsbildung durch Horizontalverschiebungen nicht nur alle Bewegungsbilder des frontalen Zusammenstoßens - denn es kommt auch hier, allerdings sekundär, zu frontalen Zusammenstößen quer zum Gebirgsstreichen- sondern auch zahlreiche weitere Phänomene der Geosynklinalbildung zwanglos und einfach zu erklären vermag. Innerhalb einer Scherzone kommt es zur Entstehung langgestreckter Scherspäne (Abb. 5a) Diese

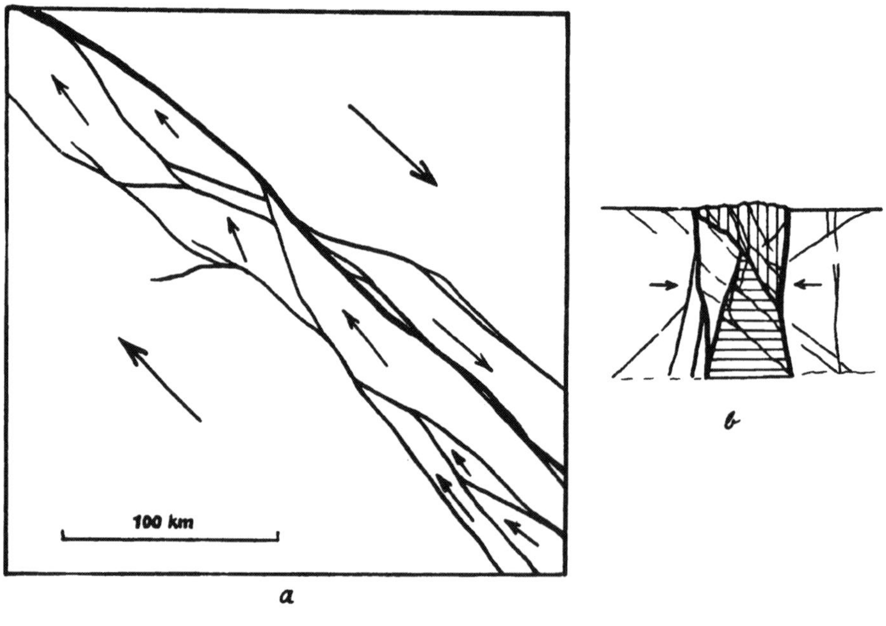

Abb. 5

a) Zerscherung der Kruste im Bereich zweier aneinander vorbeistreichender Felder in linsen- und keilförmige Einzelschollen (Scherspäne). Grundriß

b) Scherzone und Scherspäne im Querschnitt. Vertikalschraffierte Scholle wird herausgepreßt, horizontalschraffierte Scholle hinabgepreßt. Dick gezeichnete Linien: große Horizontalverschiebungen und Mylonitzonen

sind im Querschnitt linsenförmig oder keilförmig zugespitzt (Abb. 5b). Je nach dem werden diese Scherspäne im Verlaufe der Zusammenpressung und Verschiebung herausgepreßt oder hinabgepreßt. Diese Scherspäne entsprechen den langgestreckten Schwellenzonen, eng benachbart den tief absinkenden Trögen, wie sie im Bereich der Geosynklinalen so charakteristisch sind. Kräftige Reliefunterschiede werden so geschaffen, welche ihrerseits Anlaß geben zu untermeerischen Rutschungen, Abgleitungen und turbidity currents. Es ist sehr wohl zu verstehen, daß sich vulkanische Erscheinungen, vor allem ophiolithische Intrusionen und Ergüsse im Bereiche dieser mobilen Horizontalverschiebungszonen ganz besonders häufig vorfinden und den Verschiebungsvorgang begünstigen.

(17) Es sei an dieser Stelle noch kurz auf die Entstehung der Gebirgsbögen und Gebirgsschlingen hingewiesen, die ebenfalls durch ein primäres aneinander Vorbeigleiten benachbarter Krustenfelder erklärt werden kann. Diese Bogenbildung läßt sich als Wirbelbildung deuten. Sie führt zur gegenseitigen Verkeilung der Rotationsfelder in deren Grenzbereich. Das primäre aneinander Vorbeigleiten führt sekundär zu Bewegungen quer zur primären Bewegung (Abb. 6).

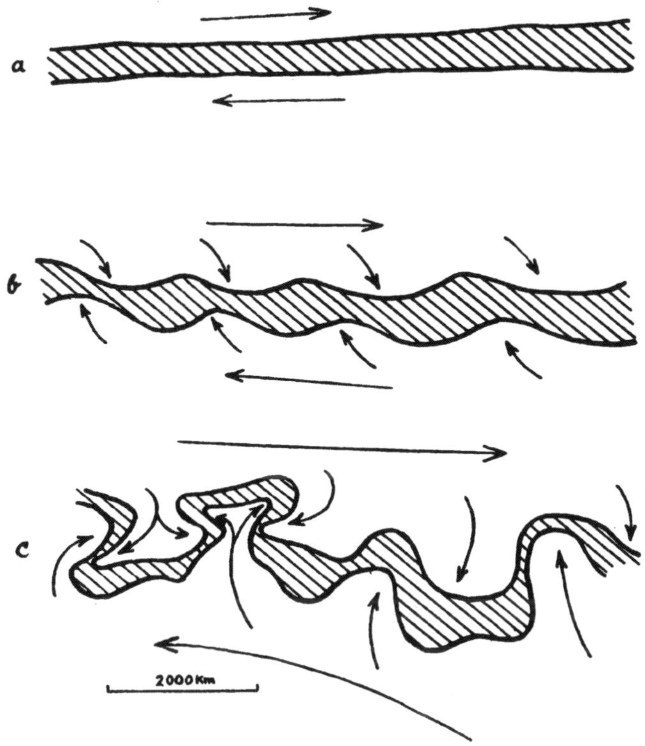

Abb. 6

Deformation einer primären Scherzone (a) bis zur vollständigen gegenseitigen Verkeilung der beiden Krustenfelder (c). Grundriß

VI. Nachwort

(18) Rezente Tektonik und die Analyse der känozoischen Gebirgsgürtel nach dem Prinzip der Horizontalverschiebungsfaltung sprechen übereinstimmend für das Vorhandensein von großen Rotationsfeldern in der Erdkruste. Der erste, der unseres Wissens von einer Rotation des N Pazifiks, ebenfalls im Gegenuhrzeigersinn, gesprochen hat, war S. Fujiwhara (1925, 1927). Er hat auch zahlreiche Experimente zur Abklärung tektonischer Zusammenhänge ausgeführt. 1957 haben P.St. Amand und H. Benioff, ausgehend von tektonischen und seismischen Untersuchungen im Randbereich des N Pazifiks auf eine Drehbewegung des Pazifiks im Gegenuhrzeigersinn geschlossen. Ganz unabhängig davon gelangte der Verfasser vor wenigen Jahren auf Grund eigener Untersuchungen, vor allem auf Grund tektonischer und geomorphologischer Betrachtungen zur oben dargelegten Hypothese dreier Rotationsfelder in der Erdkru-

ste. Er hat darüber 1960 kurz berichtet. Die weiteren Untersuchungen haben ihn in seiner Auffassung bestärkt, daß die Rotationstheorie eine fruchtbare Arbeitshypothese darstellt. Sie steht in guter Übereinstimmung mit den geologischen und geophysikalischen Beobachtungen. Sie läßt sich auch mit den palaeomagnetischen Resultaten (R.R. Doel u. A. Cox, 1961), auch wenn diese mit allem Vorbehalt zur Kenntnis genommen seien, in Einklang bringen.

(19) Fast unübersehbar groß ist die Aufgabe, die sich hier weiterhin stellt. Insbesondere gilt es, weitere Tatsachen über rezente Horizontalverschiebungen zu sammeln, die regionale junge Tektonik und die Zusammenhänge zwischen junger Tektonik und heutiger Bewegung in den heute aktiven Horizontalverschiebungszonen im Detail wie auch in großen Zügen möglichst genau darzustellen und abzuklären. Ergänzende Angaben und kritische Mitteilungen werden mit Dank entgegengenommen werden.

Es ist mir ein Bedürfnis, auch an dieser Stelle Herrn Prof.Dr. F. Gassmann, Leiter des Institutes für Geophysik der ETH in Zürich, für seine vielen wertvollen Hinweise und sein großes Verständnis, welches er meiner Arbeit entgegenbrachte, ganz herzlich zu danken.

L i t e r a t u r

Benioff, H.: Circum-Pacific tectonics. Publ. of the Dominion Observatory, Ottawa (1957) XX, No. 2, 395-402

Bucher, W.H.: The deformation of the earth's crust. Princeton Univ. Press, Princeton (1933), 518 p.

Carey, S.W.: Continental drift, a Symposium Univ. Tasmania. Geol. Dept., Hobart (1958), Tasmania, 363 p.

Cloos, H.: Einführung in die Geologie, ein Lehrbuch der Inneren Dynamik. Gebr. Borntraeger, Berlin (1936) 508 p.

Coulomb, J.: La constitution physique de la terre. Albin Michel, Paris (1952) 284 p.

Doell. R.R. u. Cox, A.: Paleomagnetism. Advances in Geophysics (1961) 8, 221 - 313

De Sitter, L.U.: Structural geology. McGraw-Hill, New York (1956) 4th impression, 552 p.

Fujiwhara, S.: Torsional form on the face of the earth. Jap. Journ. Astr. u. Geophysics (1925) III/1, 103 - 114

Fujiwhara, S.: On the échelon structure of Japanese volcanic ranges and its significance from the vertical point of view. Gerlands Beitr. Geoph. (1927) XVI, 1 - 14

Fujiwhara, S.; Tsujimura, T. u. Kusamitsu, S.: On the Earth-Vortex, échelon faults and allied phenomena. Erg. Kosm. Physik (1933) II, 303 - 360

Gutenberg, B. u. Richter, C.F.: Seismicity of the earth and associated phenomena. Rev. ed., Princeton Univ. Press, Princeton (1954) 312 p.

Hodgson, J.H. Editor: The mechanics of faulting, with special reference to the fault-plane work (a Symposium). Publ. Dominion Obs., Ottawa (1957) XX/2, 418 p.

Kraus, E.: Vergleichende Baugeschichte der Gebirge. Akademie-Verlag (1951) Berlin, 588 p.

Lee, J.S.: The Geology of China. Th. Murby u. Co., London (1939) 528 p.

Menard, H.W.: The East Pacific Rise. Science (1960) 132, 1737-1746

Pavoni, N.: Das Bewegungsbild der Erdkruste. Vierteljahrsschr. Naturf. Ges. Zürich (1960) 105, 184-196

Pavoni, N.: Faltung durch Horizontalverschiebung. Eclogae geol. Helv. (1961) 54/2,515-534

Pavoni, N.: Die Nordanatolische Horizontalverschiebung. Geol. Rdsch. (1962) 51/1, 16 p.

Reed, R.D. u. Hollister, J.S.: Structural evolution of Southern California. Bull. Am. Petr. Geol. (1936) 20, 1529-1721

Richter, C.F.: Elementary seismology. W.H. Freeman,,San Francisco (1958) 767 p.

Richter, G.: Ein Fiederflexurensystem am Rand eines spanischen Miozänbeckens. Nachr. Ges. d. Wiss. Göttingen, Math.-Phys. Kl. (1931), 80-96

Riedel, W.: Mechanik geologischer Bruchterscheinungen. Centralbl. f. Min. Abt.B., (1929) 354-368

St. Amand, P.: Geological and geophysical synthesis of the tectonics of portions of British Columbia, the Yukon Territory and Alaska. Bull. Geol. Soc. Am., (1957) 68, 1343 - 1370

St. Amand, P.: Circum-Pacific orogeny. Publ. Dominion Obs., Ottawa (1957) XX/2, 403-411

Schofield, J.C.: Some theoretical structures associated with transcurrent faulting applied to the Alpine Fault. New Zealand Journ. Geol. Geoph. (1960) 3, Aug. 60, 461 - 466

Steinbrugge, K.V. u. Zacher, E.G.: Fault Creep and Property Damage. Bull. Seism. Soc. Am. (1960) 50,No. 3, 389 - 415

Suggate, R.P.: The Interpretation of Progressive Fault Displacements of Flights of Terraces. New Zealand Journ. of Geol. Geophys. 3 (1960) No. 3, 364-374

Tocher, D.: Creep rate and Related Measurements at Vineyard, California. Bull. Seism. Soc. Am. 52 (1960) No. 3, 389-415

Umbgrove, J.H.F.: The pulse of the earth. 2.ed. M. Nijhoff. The Hague (1947) 357 p.

Wellmann, H.W.: New Zealands Quarternary Tectonics. Geol. Rundsch. 43 (1955), 248-257

Whitten, C.A.: Horizontal Movement in the Earth Crust. Journ. Geoph. Research, 65 (1960) No. 9, Sept. 1960, 2859 - 2844

Whitten, C.A. u. Claire, C.N.: Analysis of Geodetic Measurements along the San Andreas Fault. Bull. Seism. Soc. Am. 50 (1960) No. 3, 389-415

D i s k u s s i o n

Z a t o p e k , A . (Prag):

Benioff war der erste, der die horizontalen Bewegungen an den großen Dislokationen am Rand der Kontinente systematisch studierte. Er hat diese auch mit den Kriechbewegungen der Kontinentalschollen gegenüber dem Ozean erklärt. Die pazifischen Kontinentalränder bewegen sich danach im Uhrzeigersinn. Es ist nur eine scheinbare Ausnahme in der Umgebung der Kermadec-Inseln und des Tongagrabens, wo die Linie doppelt erscheint. Daran kann man aber die Grenze des australischen Kontinents erkennen. Die wirklichen Bewegungen entlang der pazifischen Küste entstehen durch eine Kombination von den Kriechbewegungen am Rand der Kontinente und werden sicherlich auch durch die Spannung beeinflußt, die sich aus der Lage der Erdbebenherde ergibt, die mit wachsender Herdtiefe immer mehr unter den Kontinent einfallen.

P a v o n i , N . (Zürich):

Die Grenzzone zwischen zwei benachbarten Rotationsfeldern setzt sich offenbar innerhalb der Kruste im allgemeinen nicht vertikal in die Tiefe fort. Dies geschieht nur dort, wo beiderseitig gleichartiges Krustenmaterial aneinanderstößt. Da, wo spezifisch leichteres Krustenmaterial an die spezifisch schwerere ozeaniche Kruste stößt, wird die Kontaktzone schräg unter den Kontinent einfallen. Da es aber im Grenzbereich zweier aneinander vorbeistreichender Felder zu Verkeilungen, zu einer Wirbelbildung kommt, wie zum Beispiel beim Fidschi-Inselbogen, ist die beobachtete "abweichende" Verteilung der Bebenherde z.B. im Gebiet der Salomon-Inseln durchaus zu verstehen. Tatsächlich erhält man den Eindruck, daß die Kruste im Bereich von Australien einen kräftigen Vorstoß ins Pazifik-Feld ausführt und sich dabei an ihrem Außenrande gegen den Pazifik hin unter "Wirbelablösung" aufsplittert.

W i t k o w s k i , J . (Poznan):

Können die im oberen Mantel möglichen machtigen Strömungen als abklingende Reste einer z.Z. des flüssigen Stadiums unserer Erde herrschenden Zonenumdrehung aufgefaßt werden?

P a v o n i , N . (Zürich):

Über die Ursachen der Rotation der Felder möchte ich mich nicht näher äußern, da sie mir unbekannt sind. Man kann sich verschiedene Mechanismen vorstellen. Tatsächlich erinnern gerade im Gebiet des Pazifik die Bewegungsbilder der Kruste, so wie wir sie an der Oberfläche erkennen können, unbedingt an Strömungsbilder. Ich denke z.B. an den gewaltigen Bogen des Malaischen Archipels, an den Fidschi-Inselbogen, an den Scotia-Inselbogen und an den Antillen-Bogen. Im Bereich der kontinentalen Kruste kommen entsprechende Strömungsbilder nur sehr unvollständig zum Ausdruck. Wenn der Nachweis geführt werden kann, daß es sich bei den Bewegungsbildern an der Erdoberfläche teilweise um Strömungsbilder handelt, so wäre dies ein Nachweis für Strömungen im Erdmantel.

D u m i t r a s h k o , N. W. (Moskau):

What role do the vertical crustal movements play in the tectonic processes?

P a v o n i , N . (Zürich):

Vertikale und horizontale Bewegungen spielen natürlich beide die ihnen zukommende bedeutsame Rolle bei den geotektonischen Prozessen. Ich möchte nochmals darauf hinweisen, daß relative Massenverlagerungen innerhalb der Erdkruste in horizontaler Richtung entlang steilstehender Verschiebungsflächen um ein vielfaches weniger Energie benötigen als entsprechende Massenbewegungen entgegen der Schwerkraft.

Junge Erdkrustenbewegungen und die Erdbebentätigkeit auf dem Gebiet der Tschechoslowakei

von

A. Zátopek (Prag)

Summary

All seismically active districts of Czechoslovakia, primarily those of Carpathian Mountains and those in which slow or sudden earthwork is to be expected as a result of human activity, are to be investigated not only seismically but also by means of other methods especially the geodetic method. First of all, levellings of high precision have repeatedly been made with great success in the district of greatest seismoactivity, i.e. in the Komárno basin in Southern Slovakia, where the existance of extremely weak seismic shicks could be proved too, by highly sensitive seismometers. In the future this micro-activity will be observed permanently in a specially equipped seismic station at Šrobárovo. Another seismic district with great activity is the line of the river Waag in Western Slovakia. Microactivity could not be proved there owing to the disturbances of the soil caused by industrial plants which do not allow the necessary accuracy of measurement. The third seismic district with very shallow hypocentres situated in Eastern Slovakia appears to be a very interesting object even for the future geodetic and complex geophysical tests.

In the much quieter Bohemian massif which forms the western part of Czechoslovakian territory, a detailed geodetic investigation should be made. Much depends in this respect on the "Egerland" (region of the river Eger), which forms a single seismic district with frequent earthquakes together with the adjoining "Vogtland", as well Bohemia (the district from Trutnov to Náchod), where the most energetic earthquakes of Bohemia took place.

The zones of higher seismic activity statistically discovered in the study of the propagation of the quakes of the Eastern Alps through Bohemian territory, which indicate the outlines of the large block structure of the Bohemian massif, should also be investigated geodetically in the future.

In the author's opinion further dynamic information could be gathered in the districts mentioned above from the final errors of the measuring of the canvas of surveys covering longer spaces of time.

Zusammenfassung

Alle seismisch aktiven Gebiete der Tschechoslowakei, in erster Linie die karpatischen und diejenigen, bei denen infolge der menschlichen Tätigkeit langsame oder plötzliche Bodenbewegungen erwartet werden, sind nicht nur seismisch, sondern auch mittels anderer Methoden, insbesondere der geodätischen, untersucht worden. Zunächst wurden im Gebiet der lebhaftesten Seismoaktivität, d.h. im Komárno-Becken in der Südslowakei, wiederholte Nivellements hoher Genauigkeit erfolgreich eingesetzt, wobei auch die Existenz von äußerst schwachen Erdstößen mit hochempfindlichen Seismometern nachgewiesen wurde. Für die Zukunft wird diese Mikrotätigkeit dauernd an einer speziell ausgerüsteten seismischen Station in Šrobárovo verfolgt werden. Ein weiteres seismisches Gebiet mit lebhafter Tätigkeit ist die Waaglinie in der Westslowakei. Hier konnte die Mikrotätigkeit infolge der durch Industrie verursachten Bodenunruhe, die die notwendige Meßgenauigkeit nicht zuläßt, nicht nachgewiesen werden. Das dritte, in der Ostslowakei gelegene Erdbebengebiet mit sehr flachen Herden erscheint als ein recht interessantes Objekt auch für die künftige geodätische und komplexe geophysikalische Untersuchung.

In der viel ruhigeren Böhmischen Masse, die den Westteil des Staatsgebiets aufbaut, sollte eine detaillierte geodätische Untersuchung durchgeführt werden. Es kam dabei besonders auf das Egerland an, das mit dem angrenzenden Vogtland ein einziges Schwarmbebengebiet bildet, ebenso auf Nordostböhmen (Gegend von Trutnov bis Náchod), wo die energiereichsten Erdbeben der Böhmischen Masse stattgefunden haben.

Auch die beim Studium der Ausbreitung von ostalpinen Beben durch die Böhmische Masse statistisch erschlossenen Zonen höherer seismischer Beweglichkeit, die die Umrisse der Großblockstruktur der Böhmischen Masse andeuten, sollten in der Zukunft einer geodätischen Untersuchung unterzogen werden.

Weitere dynamische Auskünfte könnte man nach der Ansicht des Verfassers in den angeführten Gebieten aus den Abschlußfehlern der sich über längere Zeitspannen erstreckenden Netzmessungen gewinnen.

E i n l e i t u n g

In der Tschechoslowakei wird eine systematische und komplexe Erforschung der jungen, geologisch ziemlich gut bekannten Erdkrustenbewegungen auch mittels geophysikalischer und geodätischer Methoden vorgesehen. In einigen Gebieten befinden sich derartige Arbeiten bereits im Gange. Vor allem sollen, nicht nur aus wissenschaftlichem Interesse, sondern auch aus volkswirtschaftlichen Gründen die seismisch tätigen Gebiete untersucht werden. Ebenso soll die Dynamik derjenigen Gebiete, wo infolge der menschlichen Tätigkeit langsame oder plötzliche Bodenbewegungen zu erwarten sind, wie z.B. in Bergbaugebieten oder in der Umgebung von großen Talsperren, dauernd verfolgt werden.

Das Ziel meiner Mitteilung besteht in einer objektiven Erschließung der gegenwärtig beweglichen Gebiete auf Grund einer auf modernen seismologischen Arbeitsmethoden beruhenden quantitativen Klassifikation ihrer Bedeutung. Diese Angaben sind gewissermaßen für die Reihenfolge und Methodik der künftigen Untersuchungen maßgebend. Da sich manche von diesen Gebieten über die Staatsgrenzen hinaus ausbreiten, werden die künftigen Forschungen eine engere Zusammenarbeit der Spezialisten der in Betracht kommenden Länder erfordern. Außerdem möchte ich auf einige Möglichkeiten, weitere dynamische Auskünfte zu gewinnen, hinweisen.

G e o l o g i s c h e s

Geologisch gehört das tschechoslowakische Staatsgebiet zu zwei Gebirgssystemen. Böhmen und der nordwestliche Teil von Mähren gehören zu der sogenannten Böhmischen Masse (weiter nur BM); der südöstliche Teil von Mähren und die Slowakei bilden den Westteil des Karpatensystems. Die beiden Systeme berühren sich längs der sogenannten Mährischen Pforte (etwa die Linie Ostrava-Znojmo). Die geologische Entwicklung und der tektonische Aufbau beider Systeme sind wesentlich verschieden. Während die BM ein altes, im Wesen kristallinisches Massiv mit einer Zone intensiver varistischer Tektogenese darstellt, haben sich die Karpaten zu der heutigen Form erst im Tertiär entwickelt und diese Entwicklung geht noch weiter vor sich. Die BM hat im Nordwesten ihre natürliche Fortsetzung in den Formationen von Sachsen und Thüringen; von diesen wird sie durch ein System von steilen, in große Tiefen hinabreichenden Randbrüchen der Richtung NO/SW geteilt. Weitere dazu parallele Bruchzonen verlaufen auf dem Gebiet von Böhmen. In der Richtung NW/SO verlaufen korrespondierende Systeme von Randbrüchen, die die BM gegen Nordost und Südwest abgrenzen. Auch hier existieren im Innern der BM parallele Bruchsysteme. Wichtige Störungszonen verlaufen in meridionaler Richtung in West- und Mittelböhmen und über Brno in Mähren. Obwohl die genannten postherzynischen Bruchsysteme der BM während der alpinen Faltung im Tertiär teilweise neu aktiviert wurden, repräsentiert die gegenwärtige BM ein relativ sehr ausgeglichenes System von Schollen, deren Kopplungen so fest sind, daß ein Spannungsausgleich durch Erdbeben größeren Maßstabs nur sehr selten und zwar ausschließlich in Randbruchzonen vorkommt.

Im Gegensatz dazu sind die Brüche des Karpatensystems, also auch der Westkarpaten, in der Gegenwart noch immer aktiv geblieben. Die Bewegungen an den in den Richtungen NO/SW und NW/SO orientierten Bruchsystemen, besonders längs der Waaglinie und im Komárno-Becken (Südslowakei), ebenso wie die Randbrüche der Kleinen Karpaten, an denen erhebliche Vertikalverschiebungen stattgefunden haben, werden sogar in den jüngsten Sedimentschichten beobachtet. Als Ganzes erscheinen die Westkarpaten als ein gegenüber der Böhmischen Masse sinkendes Gebiet. Die Kopplungen einzelner Schollen in den Karpaten sind in der Regel relativ sehr los. Dementsprechend ist der Spannungsausgleich durch zahlreiche, meistens energiearme Erdbeben üblich; in tieferen Partien kann jedoch im Laufe längerer Zeitabschnitte verhältnismäßig große Energie aufgespeichert werden, die sich dann in Form eines starken Bebens auflösen kann.

S e i s m i z i t ä t

Das Gebiet der Tschechoslowakei im Ganzen genommen wird gewöhnlich als seismisch relativ ruhig angesehen. Tatsächlich ist die Erdbebentätigkeit hier, was die Anzahl der wahrgenommenen Erdbeben und die ausgelöste seismische Energie anbelangt, im Vergleich mit den seismischen Gebieten in Südeuropa, schwach und im Verhältnis mit den Gebieten hoher Seismizität, wie z.B. Japan, belanglos. Trotzdem sind hier im Laufe der Geschichte starke Erdbeben vorgekommen, die zahlreiche Gebäude zertrümmert haben. Auch menschliche Opfer sind zu beklagen. Wie schon angedeutet, wurden bei diesen Erdbeben oft auch angrenzende Teile von Deutschland, Österreich, Ungarn und Polen in Mitleidenschaft gezogen. Umgekehrt wurde in vielen Fällen von den in diesen Ländern gelegenen Herden die seismische Energie in das tschechoslowakische Gebiet eingestrahlt, so daß die geologisch-geophysikalischen Zusammenhänge nur in gemeinsamer Zusammenarbeit erfolgreich untersucht werden können.

Wie schon oben erwähnt, ist die seismische Tätigkeit in der Böhmischen Masse viel schwächer als in den Karpaten. Die Untersuchungen der Erdbebenmagnitude haben ermöglicht, die Größenordnung der ausgelösten Herdenergie zu bestimmen. Die energetische Bilanz der autochthonen Erdbeben der BM läßt sich für die letzten 100 Jahre in Einheiten der Größenordnung 10^{19} Erg ausdrücken, dagegen ergibt sich die in den außerhalb der BM gelegenen Herden freigelöste Energie, wo die Wirkungen in der BM beobachtet werden, etwa 100 mal größer.

Ganz umgekehrt sind die Verhältnisse in den Westkarpaten. Die in den autochthonen Herden in den letzten 100 Jahren freigelöste Energie beträgt mehr als 10^{21} Erg; dagegen der von den äußeren Herden stammende Energiebetrag ist in Einheiten von 10^{20} Erg ausdrückbar.

Die größten Erdbeben der BM besaßen eine Energie von der Größenordnung 10^{19} Erg. Die stärksten Stöße in der Slowakei ergeben Energien, die 10^{21} Erg erreichen und den Energien entsprechen, die im anliegenden Pannonischen Becken bei den dortigen größten Erdbeben ausgelöst werden, was aber der Größenordnung nach bloß ein Tausendstel der Energie der starken südeuropäischen Beben ausmacht und einige Hunderttausendstel von der Energie der großen japanischen Beben repräsentiert. Es sei noch bemerkt, daß der verhältnismäßig grossen Anzahl von Kleinbeben nur ein kleiner Bruchteil der Gesamtenergie zukommt, so daß die Energiebilanz ausschließlich durch die relativ größeren Beben bestimmt wird.

Von den in der BM gelegenen Herden liegen einige im Böhmerwald; ein wichtiges Herdgebiet, das mit dem anliegenden Vogtland ein einziges Schwarmbebengebiet bildet, befindet sich im westlichen Winkel von Böhmen und längs des Erzgebirges; in den Sudeten liegen die energiereichsten Herde der BM am Fuße des Riesengebirges; noch weiter nach Osten findet man Erdbebenherde im Gebiet von Großen Gesenken. Meistens sind die hier wahrgenommenen Erdbeben schwach, die Epizentralintensitäten betragen in der Regel 3° - 4° der M.-C.-S.-Skala; höhere Intensitäten kommen einmal in einigen Jahrzehnten vor. Die stärksten Erschütterungen haben in der BM kaum 7° überschritten. Die Magnituden der stärksten Stöße ergeben sich zu etwa 5 Magnitudeneinheiten. Außer den eben genannten Herdgebieten befinden sich vereinzelt Epizentren von Kleinbeben an der Böhmisch-Mährischen Höhe und im Innern von Böhmen. Diesen letzten Stößen kommt seismisch keine Bedeutung zu, sie beweisen jedoch, daß auch die in den mittleren Partien des Böhmischen Blocks befindlichen Schollen noch nicht zur Ruhe gekommen sind. Als gemeinsame Ursache der Erdbebentätigkeit in der BM wird meistens der gegen Norden gerichtete Druck des Alpenbogens betrachtet.

Die Ergebnisse von Neigungsbeobachtungen an der tiefen Gezeitenstation in Příbram, wo der systematische Gang der Ruhelage von Pendeln und gewisse sprunghafte (nicht mit den Gebirgsschlägen zusammenfallende) Bewegungen auf den tektonischen Ursprung hinweisen, scheinen diese Ansicht zu unterstützen. Um der Erkenntnis dieser Erscheinungen und ihrem Zusammenhang mit den künstlich hervorgerufenen Bewegungen näher zu kommen, werden in den Bergbaugebieten spezielle geodätische Messungen über und unter Tage angestellt und in den Bergwerken von Příbram und Kladno auch seismische Untersuchungen eingesetzt. An neuen großen Talsperren sollen nach ihrer Füllung durch wiederholte Feinnivellements die nachfolgenden Ausgleichsbewegungen verfolgt werden.

Eine bemerkenswerte Rolle spielen in der BM die Erdbeben, die von den ostalpinischen Herden kommend sich nach Norden hin mit anomal schwacher Dämpfung ausbreiten, wobei sie in die Karpaten nur wenig hineindringen und auch in den Alpen selbst schnell abklingen. Für diese Beben konnten in der BM bevorzugte Zonen höherer Beweglichkeit nachgewiesen werden, die häufig (nicht immer) mit den an der Oberfläche sichtbaren tektonischen Störungszonen zusammenfallen und die Hauptzüge der tieferen Blockstruktur der BM anzudeuten scheinen.

In den Westkarpaten konzentriert sich die seismische Tätigkeit auf das Innere des Karpatenbogens. Die äußere Flyschzone erscheint als ein erdbebenarmer Gürtel. Die schon erwähnten losen Kopplungen in dem karpatischen Schollenmosaik kommen durch zahlreiche Kleinbeben mit den größtenteils in den obersten 5 km gelegenen Herden zum Ausdruck; aber auch bei den größten Seismen des Westkarpatengebietes finden wir Herdtiefen, die 10 km nur wenig überschreiten, während eine Herdtiefe von rund 25 km bei den größeren Beben der BM ganz üblich ist. Auch dieser Unterschied beweist die höhere Mobilität der Karpaten.

Wollen wir nun den Gebieten der stärksten westkarpatischen Erdbebentätigkeit nähere Aufmerksamkeit widmen!

1. Die seismische Linie Žilina Bratislava zieht sich längs der schon erwähnten Bruchzone, längs der mittleren Waag und an den beiden Seiten der Kleinen Karpaten entlang und enthält drei aktive Zonen. Die nördlichste mit dem größten Erdbeben bei Žilina (1858), dessen Epizentralintensität mit 9° angegeben wird, wobei sich die Magnitude zu etwa 5 und die Energie zu 10^{20} Erg ergeben. Die mittlere Zone wurde durch zahlreiche und starke Beben in der Gegend von Dobrá Voda (in der ungarischen Literatur Jókö) bekannt; das stärkste Erdbeben vom Jahre 1906 besaß eine Epizentralintensität von 9° (oder etwas mehr) M.-C.-S.; für die Magnitude ergibt sich etwa 6 und die Energie etwa 10^{21} Erg. Die südliche Zone befindet sich am südwestlichen Ende der Linie und enthält Herde mit Maximalintensität 7° M.-C.-S. Die Energie ist wesentlich kleiner als bei den vorhergehenden Fällen. Dieselbe Intensität gebührt den stärksten ostalpinen Stößen, die bei Bratislava in diese Zone vordringen.

2. An einer parallelen, etwas östlicher gelegenen Epizentrenlinie befindet sich das Becken von Komárno, das erdbebenreichste und auch energiestärkste Gebiet der Tschecho-

slowakei, das sich auch am ungarischen Donauufer weiter nach SW verfolgen läßt. Das stärkste hiesige Erdbeben vom Jahre 1763 mit der Epizentralintensität beinahe 10° stellt die mächtigste in den Westkarpaten bekannte seismische Erscheinung dar; die Magnitude ergibt sich zu 6,2 Magnitudeneinheiten und die Energie zu mehr als 10^{21} Erg. Das Becken von Komárno als Ganzes sinkt, aber einzelne Schollen besitzen individuelle Geschwindigkeiten. Das wurde nicht nur geologisch, sondern auch nach älteren Höhenmessungen festgestellt und in den letzten Jahren durch die hier eingesetzten wiederholten Präzisionsnivellements, von denen in einer anderen Mitteilung berichtet wird, bewiesen. Interessant ist zu bemerken, daß in diesem Gebiet mittels hochempfindlicher Seismometer die Existenz von äußerst schwachen, mit üblichen Instrumenten unfaßbaren Seismen (die z.B. in Zentralasien festgestellt wurden) bewiesen wurde. Analoge Messungen haben im Gebiet von Žilina wegen Industriebodenunruhe nicht zum Ziel geführt.

3. Ein drittes erdbebenreiches Gebiet mit Maximalintensitäten von 7° befindet sich in der Ostslowakei. Diese Erdbeben zeichnen sich durch einen sehr kleinen Bereich des makroseismischen Feldes, also durch relativ sehr kleine bei hoher Epizentralintensität ausgelöste Energien, und durch ganz kleine Herdtiefen aus, was durch die tektonische Zerstückelung dieser Gegend bedingt wird. Die seismische Tätigkeit in dieser Zone hängt schon mit der regen seismischen Aktivität der Karpatoukraine die einen ähnlichen Charakter aufweist, zusammen.

Die seismischen Gegenden in der Umgebung der Hohen Tatra, bzw. in der Mittelslowakei und im südlichen Grenzgebiet der Slowakei sind trotz bedeutender Anzahl von verspürten Erdstößen (selten über 5°) von sekundärer Bedeutung. Es ist noch zu bemerken, daß in der Süd- und Mittelslowakei die Ausbreitung der aus dem Gebiet von Salgótarján in Nordungarn kommenden Erdbeben weitgehend durch die oberflächennahe Tektonik beeinflußt wird und auf die Lage der energieleitenden Brüche hinweist, wie an dem nordungarischen Erdbeben vom Jahre 1956 gezeigt werden konnte.

Schlußfolgerungen und Perspektive

Aus dem vorliegenden Gesamtbild der aktuellen seismischen Beweglichkeit des tschechoslowakischen Staatsgebiets ergeben sich folgende Schlüsse:

1. Für die weiteren Untersuchungen würde es sich empfehlen, den Einsatz der im Komárno-Becken erfolgreich angewandten wiederholten Präzisionsnivellements der Reihe nach auf die seismische Zone Bratislava-Žilina und das Erdbebengebiet der Ostslowakei zu erweitern und ähnliche Messungen auch in der BM, besonders im seismischen Gebiet von Nordostböhmen und im Egerer Becken unter Zuhilfenahme von geophysikalischen Methoden und schon vorhandenen Ergebnissen durchzuführen.

2. Es ist erwünscht, diese Aufgaben möglichst komplex zu lösen und in den Grenzgebieten die Zusammenarbeit der Fachleute von betreffenden Ländern zu organisieren und zu unterstützen.

3. Die statistisch festgestellten Zonen höherer seismischer Beweglichkeit sollten einer geodätischen Untersuchung unterzogen werden.

Nach der Ansicht des Verfassers besteht die Möglichkeit, neue wertvolle dynamische Auskünfte auch mittels einer Analyse von Schlußfehlern der über längere Zeitspannen erstreckten Netzmessungen zu gewinnen.

Diskussion

Z. Kowalczyk (Kraków):

Mit welcher Genauigkeit kann man die Koordinaten der Bebenherde bestimmen?

A. Zátopek (Prag):

Die Genauigkeit hängt von vielen Umständen ab. Wenn es sich, wie es bei historischen Beben üblich ist, um makroseismische Methoden handelt, kann das Ergebnis ziemlich genau (± einige Kilometer) sein; es wird aber sehr viel auf die Herdtiefe ankommen und auch von der Neigung der Diskontinuitäts-(Bruch-)fläche abhängen. Für große Beben hat der Herd manchmal große Dimensionen, so daß er einen beträchtlichen Raum einnehmen kann. Die in der Geodäsie üblichen Genauigkeiten kommen hier wenig in Betracht.

Z. Kowalczyk (Kraków):

In welcher Weise - mit welcher Beziehung stellen Sie in der Tschechoslowakei die Bebenenergie fest?

A. Zátopek (Prag):

In der ČSSR wurde eine ausgedehnte, die Erdbebenmagnituden betreffende Untersuchung durchgeführt, so daß die Magnituden für die Wellen PH,PV, PPH, SH, L und M in gegebenen Fällen für die Epizentraldistanzen von 1° bis 160° für normale Beben bestimmt werden können. Aus diesen wurde die Energie berechnet nach den neuen, kritisch gewerteten Formeln und auch unabhängig davon auf direktem Wege unter gewissen physikalischen Voraussetzungen.

Z. Kowalczyk (Kraków):

Die Hauptstörungen, welche Prof. Zátopek gezeigt hat, stimmen mit den Richtungen von Sprüngen überein, die wir in Polen in Gruben feststellen.

N. W. Dumitrashko (Moskau):

In the Soviet Union the zones of the earthquake epicenters are connected with the zones of the contrasting modern tectonic movements and of contrasting relief. Is this correlation observed in ČSSR?

A. Zátopek (Prag):

Die Böhmische Masse besitzt eine Erdbebentätigkeit, die als Folge des nordwärts gerichteten Alpendruckes erklärt wird, also gewissermaßen benimmt sich die Böhmische Masse, die steigt, passiv. Die Karpaten dagegen, die im ganzen ein Senkungsgebiet darstellen, sind verhältnismäßig sehr aktiv. Ich würde noch auf die Ostslowakei aufmerksam machen, wo die Verhältnisse denen in der Karpatoukraine ähneln. Dort ist die tektonische Zerstückelung so groß und die Schollenkopplungen so lose, daß das Erdbeben oft praktisch auf einem einzigen, manchmal auch morphologisch beschränkten Block begrenzt bleibt.

L. Egyed (Budapest):

1. Die Donau fließt entlang einer Verwerfungsfläche. Das bestätigen auch die Erdbeben.

2. Das Erdbebengebiet zwischen Slowakei und Böhmen liegt entlang einem Riesengraben, der auch jetzt sinkt.

Über die Epizentrumberechnung am Erdellipsoid und Rückschlüsse auf die Fortpflanzungsgeschwindigkeiten

von

A. Tárczy-Hornoch (Sopron)

Summary

The seismic districts often are districts of intense crustal movements, too. Therefore the preciser determination of the foci and epicentres is also important for the movements of the earth's crust. The inclusion of the distant stations for the purpose of increasing the accuracy requires a compensation on the terrestrial ellipsoid. Appropriate equations are set up. After the determination of the epicentre, the velocities of propagation in the different directions can be determined. The variations from the average velocity of propagation may allow to draw certain conclusions regarding the structure, resp. extension of the foci.

Zusammenfassung

Die Erdbebengebiete sind oft auch Gebiete der stärkeren Erdkrustenbewegungen. Die genauere Festlegung der Herde und Epizentren ist daher auch für die Erdkrustenbewegungen von Bedeutung. Durch Einbeziehung auch der weit liegenden Stationen zur Erhöhung der Genauigkeit ist eine Ausgleichung auf dem Erdellipsoid erforderlich. Aufstellung der entsprechenden Vermittlungsgleichungen. Nach Ermittlung des Epizentrums können die Fortpflanzungsgeschwindigkeiten in den einzelnen Richtungen bestimmt werden. Die Abweichungen von der durchschnittlichen Fortpflanzungsgeschwindigkeit in den einzelnen Richtungen können gewisse Rückschlüsse auf den Aufbau bzw. die Ausdehnung des Herdes geben.

Wir haben in einer früheren Studie [1] die Berechnung von Erdbebenherden bis auf die kugelförmig angenommene Erde behandelt, dort am Schluß auf die Berechnung am Erdellipsoid hingewiesen und bemerkt, daß die bestentsprechende Kugel mit dem Radius $R = \sqrt{MN}$ meist entspricht, wobei M den Meridian- und N den Querkrümmungshalbmesser bedeutet.

Will man aber das Epizentrum möglichst genau berechnen, so erfordert dies die Einbeziehung möglichst vieler und so auch weit liegender Stationen. In diesem Falle, besonders bei einer angestrebten erhöhten Genauigkeit und bei Verwendung der Oberflächenwellen kann die Berechnung auf dem Erdellipsoid schon erforderlich sein und zwar 1.) weil hier die zugelassenen Abweichungen geringer sind und 2.) weil bei den vielen und verschieden weit liegenden Stationen man mit einer einzigen bestentsprechenden Kugel nicht auskommt und die Rechnung auf verschiedenen Kugeln umständlich wird.

Wir wollen zunächst sehen, was für Unterschiede zwischen den Berechnungen auf der bestentsprechenden Kugel und am Erdellipsoid entstehen können. Zu diesem Zwecke weisen wir auf die in der Fußnote auf S. 48 in [1] angegebenen Zahlenwerte hin, wonach die Entfernung Kiruna - Uppsala auf der bestentsprechenden Kugel 897,80 km, wogegen die ellipsoidische Entfernung 900.08 km beträgt. Der Unterschied von 2,28 km ist schon merklich und bei höheren Genauigkeitsforderungen nicht vernachlässigbar. Noch weniger wird dies bei noch größeren Entfernungen, da die Abweichungen stark zunehmen und man dabei noch an verschiedenen bestentsprechenden Kugeln rechnen muß.

Es ist natürlich nicht neu, daß man bei den Epizentrenberechnungen die ellipsoidische Erdgestalt berücksichtigt. So hat schon Wiechert in [2] selbst bei Entfernungen von nur einigen hundert km - allerdings nicht ausgesprochen zur Herdbestimmung - ellipsoidische Korrektionen vorgeschlagen, während u.a. bei Bullen diese in [3; S. 172] bereits in Verbindung mit der Herdbestimmung erscheinen. Wir wollen hier einen anderen Weg wählen, welcher einer jeden gewünschten Rechengenauigkeit entsprechen kann.

Unseren folgenden Berechnungen ist ein einheitliches Erdellipsoid zugrundegelegt. Es ist wichtig, daß die besten Werte des Erdellipsoids in Rechnung gestellt werden, teils weil dadurch die Rechenarbeit nicht wächst, teils weil so die durch die ungenauen Werte verursachten Fehler herabgesetzt werden.

Es seien im folgenden die ellipsoidischen geographischen Breiten φ_i und Längenunterschiede λ_i der einzelnen Stationen i gegeben und deren Einlaufzeitpunkte t_i beobachtet worden. Die gesuchten Koordinaten des Epizentrums seien φ und λ, für die durch Näherungsrechnungen auf der Kugel (vgl. [1; Gl. 20]) Näherungswerte φ_0 und λ_0 bekannt und dazu die plausibelsten Zuschläge $\delta\varphi$ und $\delta\lambda$ zu berechnen sind.

Wir berechnen zunächst mit den gegebenen φ_i , λ_i die geographischen Koordinaten der Stationen und mit den Koordinaten φ_0, λ_0 des Näherungswertes des Epizentrums die Längen und Azimute der geodätischen Linien, d.h. der Linien der kürzesten Verbindung am Erdellipsoid zwischen den einzelnen Stationen und dem Näherungspunkte. Diese Entfernungen seien $s_{i,0}$, während die Azimute in i mit $\alpha_{i,0}$, jene im Näherungspunkt nach i mit $\alpha_{0,i}$ bezeichnet werden mögen.

Das zu behandelnde Problem, aus den gegebenen geographischen Koordinaten zweier Punkte die ellipsoidische Entfernung zu berechnen, ist eine Aufgabe der höheren Geodäsie und wird dort als die zweite geodätische Hauptaufgabe genannt. Die einschlägige Fachliteratur enthält eine ganze Reihe von Lösungen, wobei allerdings zu bemerken ist, daß die noch vor 20 - 30 Jahren angegebenen Lösungen bis kaum über 200 km Entfernungen für unsere Zwecke nicht ausreichen. Nun haben aber die Großraumtriangulierungen, Navigationsaufgaben, die Verbindung der Kontinente miteinander usw. in neuerer Zeit mehrere Lösungen hervorgebracht, die eine Berechnung der ellipsoidischen Länge bei 1000 km auf 1 - 2 mm gestatten und auch die Berechnungen bis 20 000 km mit nur mäßig größeren Fehlern von etwa 1 - 2 cm erlauben. Es soll hier nur kurz u.a. auf die Lösungen von Levallois und Dupuy [4], Bodenmüller [5], sowie Cimbalnik [6] hingewiesen werden, die meist auch die zur rascheren Berechnung dienenden Tabellen enthalten.

Für unseren vorliegenden Zweck sind die vorher angegebenen Genauigkeiten zu groß. Wir streben etwa eine Genauigkeit von nur 1 - 10 m (die untere Grenze nur in besonderen Fällen) an, und dementsprechend können die angegebenen Reihen früher abgebrochen und die Koeffizienten mit geringerer Stellenzahl angegeben werden. Auch soll auf die Erfordernisse für die Berechnungen mit elektronischen Rechenmaschinen mehr Rücksicht genommen werden, für die u.a. die trigonometrischen Funktionen und Reihenentwicklungen weniger günstig sind. Wir haben diese Erfordernisse mehr berücksichtigt und Beziehungen benützt, welche die zulässigen Vernachlässigungen von etwa 1 - 10 m ausnützen. Unser Mitarbeiter G.Szádeczky-Kardoss hat die entsprechende Anpassung der dazu sehr geeigneten Levallois-Dupuyschen Beziehungen vorgenommen. Sie sollen als Anhang angeführt werden. Die mit $\llcorner$ abgegrenzten Glieder können bei einer Genauigkeitsforderung von nur 10 m entfallen.

Um die Vermittlungsgleichungen für die Ausgleichung aufzustellen, sind noch die Zusammenhänge erforderlich, die zwischen den Veränderungen von φ_0 und λ_0 um $\delta\varphi$ und $\delta\lambda$, sowie der dadurch verursachten Veränderung δs von $s_{i,0}$ bestehen. Hierzu können wir die von Helmert [7; S. 281, Gl. (8)] aufgestellte Beziehung benützen, die im Falle, wo nur φ_0 und λ_0 Zuschläge in Sekunden erhalten, folgend geschrieben werden kann (Die Vorzeichen sind hier anders, da wir die Azimute vom Norden rechnen und die geographischen Längen nach Osten fortlaufend bis 360° zählen. Dies gilt auch für Punkte auf der südlichen Halbkugel, falls die entsprechenden φ-Werte negativ eingesetzt werden):

$$\delta s_i = - a \frac{1 - e^2}{\rho W_0^3} \cos \alpha_{0,i} \, \delta \varphi - \frac{a}{\rho W_0} \cos \varphi_0 \sin \alpha_{0,i} \, \delta \lambda \qquad (1)$$

Dabei bedeuten: ρ = Umwandlungsfaktor = 206 265
a = die große Halbachse des Erdellipsoids
b = die kleine Halbachse des Erdellipsoids

$$e^2 = \frac{a^2 - b^2}{a^2} = \text{das Quadrat der I. Exzentrizität}$$

$$W_0 = \sqrt{1 - e^2 \sin^2 \varphi_0}$$

Man erhält mithin bei Zuordnung der Verbesserung w_i zum beobachteten Einlaufzeitpunkt t_i sowie bei der unbekannten Herdzeit t und v durchschnittlicher Fortpflanzungsgeschwindigkeit der Vermittlungsgleichung:

$$s_i = s_{i,0} + \delta s_i = s_{i,0} - a \frac{1 - e^2}{\rho W_0^3} \cos \alpha_{0,i} \, \delta \varphi - \frac{a}{\rho W_0} \cos \varphi_0 \sin \alpha_{0,i} \, \delta \lambda =$$

$$= (t_i + w_i - t) \, v \qquad (2)$$

Auf der rechten Seite der Gl. erscheint der zum Zeitraum $(t_i + w_i - t)$ entsprechende Wellenweg. Daraus ergibt sich die Verbesserungsgleichung:

$$w_i = - a \frac{1 - e^2}{\rho v W_0^3} \cos \alpha_{0,i} \, \delta \varphi - \frac{a}{\rho v W_0} \cos \varphi_0 \sin \alpha_{0,i} \, \delta \lambda + t + \left(\frac{s_{i,0}}{v} - t_i\right)$$
.
.
.
$$(3)$$

Man hat hier 3 Unbekannte: $\delta \varphi$, $\delta \lambda$ und t, die aus den Verbesserungsgleichungen (3) in bekannter Weise (vgl. [8; S. 125 - 127]) berechnet werden können. Man wird auch für t vorteilhaft einen Näherungswert t_0 einführen und nur dessen Zuschlag Δt aus der Ausgleichung berechnen, weil man dadurch kleinere Zahlenwerte für die Unbekannte zu ermitteln hat.

Es wird oft - besonders bei vielen Stationsbeobachtungen - ratsam erscheinen, den Wert der durchschnittlichen Fortpflanzungsgeschwindigkeit v selbst durch die Ausgleichung zu bestimmen. Man sucht in diesem Falle zum Näherungswert v_0 den Zuschlag Δv. Durch Einführung von $v_0 + \Delta v$ an Stelle von v in Gl. (2) erhält man durch Entwicklung und Vernachlässigung der kleinen Glieder zweiter Ordnung die Verbesserungsgleichung:

$$w_i = \Delta t - a \frac{1 - e^2}{\rho v_0 W_0^3} \cos \alpha_{0,i} \, \delta \varphi - \frac{a}{\rho v_0 W_0} \cos \varphi_0 \sin \alpha_{0,i} \, \delta \lambda -$$

$$- \frac{(t_i - t_0)}{v_0} \Delta v + \left\{ \frac{s_{i,0}}{v_0} - (t_i - t_0) \right\} \qquad (4)$$

Es soll hier bemerkt werden, daß in jenem Falle, wo bei der Berechnung der Zuschläge die ellipsoidische Erdform durch eine Kugel mit dem Halbmesser R ersetzt werden kann, e^2 = Null und $W_0 = 1$ zu setzen ist und man aus Gl. (4)

$$v_i = \Delta t - \frac{R}{\rho v_0} \cos \alpha_{0,i}\, \delta \varphi - \frac{R}{\rho v_0} \cos \varphi_0 \sin \alpha_{0,i}\, \delta \lambda - \frac{t_i - t_0}{v_0} \Delta v + \left\{ \frac{s_{i,0}}{v_0} - (t_i - t_0) \right\} \qquad (4a)$$

erhält. Diese Gl. ähnelt[1] – mit Ausnahme des anders ausgedrückten reinen Gliedes – der für die Kugelform aufgestellten Gl. (22a) in [1; S. 59], falls im Sinne der Beziehung (24) auch die Fortpflanzungsgeschwindigkeit als Unbekannte eingeführt wird. Sie ist dennoch wesentlich allgemeiner, da hier im reinen Glied – im Gegensatz zum reinen Gliede in [1; Gl. (22a)] – die der ellipsoidischen Entfernung $s_{i,0}$ entsprechende Zeit erscheint und daher die Berechnung nach Gl. (4a) der Berechnung am Ellipsoid entspricht, vorausgesetzt, daß in den Koeffizienten für die Zuschlagsglieder $\delta\varphi$ und $\delta\lambda$ die Beziehungen

$$a \frac{1 - e^2}{W_0^3} \triangleq R$$

und (5)

$$\frac{a}{W_0} \triangleq R$$

gesetzt werden können. Weil dies eine Vereinfachung der Rechnung bedeutet, sollen die Grenzen dieser Zulässigkeit abgeschätzt werden.

Der Wert in Gl. (4)

$$\frac{1 - e^2}{W_0^3} = \frac{1 - e^2}{(1 - e^2 \sin^2 \varphi_0)^{\frac{3}{2}}} \qquad (6)$$

weicht von eins dann am stärksten ab, wenn $\sin^2 \varphi = 0$ ist[2], d.h. am Äquator, da in diesem Falle der Quotient $1 - e^2 = 1 - 0{,}006\,693\,4 = 0{,}993\,306\,6$ am kleinsten wird. Der fragliche Faktor wäre dann in Metern:

$$a \frac{1 - e^2}{W_0^3} = 6\,378\,245 \cdot 0{,}993\,306\,6 = 6\,335\,552{,}9 \text{ m}$$

Wenn wir an Stelle dieses Wertes den von Gutenberg angegebenen Wert 6 371 200 [9; S. 7] für R verwenden, so gibt dies im Faktor einen Fehler von rund 0,5 %, d.h. der errechnete Zuschlag $\delta\varphi$ in Metern wird um 0,5 % seines Betrages unsicher. Nimmt man die noch zulässige Unsicherheit in 10 m an, so bedeutet dies, daß der Näherungswert von φ_0 auf etwa 1' ≙ 2 km genau sein muß, wenn wir die Zuschläge auch im ungünstigen Falle nach Gl. (4a) berechnen wollen. Für λ_0 sind etwas größere Abweichungen zulässig, auf die wir hier nicht eingehen wollen. Für größere Zuschläge bis etwa 30 km können wir für R den Wert $\sqrt{MN}$ des Näherungspunktes nehmen.

Im übrigen sind auch die Berechnungen mit Gl. (4) nicht so umständlich, falls wir entsprechende Tabellen für die Koeffizienten zusammenstellen, aus welchen die Werte als Funktionen von φ_0 entnommen werden können.

Die dem berechneten Näherungswerte relativ nahe liegenden Stationen läßt man bei der strengen Berechnung des Epizentrums aus, oder man führt entsprechende Korrektionen für die Herdtiefe ein.

---

1) Bezüglich der Vorzeichen siehe Bemerkung auf Seite 279. In [1] wurde ein System mit nach Westen zunehmender Zählung der geographischen Längen gewählt, wie bei Helmert in [7; S. 281, Gl. (8)] und Jeffreys (The Earth usw. Cambridge, 1959, S. 64).

2) Dies kann auch durch Bildung der Extremwerte der Funktion (6) nach φ mathematisch nachgewiesen werden.

Die Genauigkeit der Bestimmung kann bei der gleichen Anzahl von Stationen dadurch erhöht werden, daß in den Stationen auch die Einlaufazimute α_i' beobachtet werden.

Die Lösungen für die zweite geodätische Hauptaufgabe liefern - wie erwähnt - auch die für den Näherungswert φ_0, λ_0 entsprechende Azimute $\alpha_{i,0}$. Andererseits sind auch die Änderungen $\delta\alpha_i$ dieses Azimutes bei Änderung von φ_0 und λ_0 um $\delta\varphi$ und $\delta\lambda$ bereits bekannt (vgl. u.a. [10; S. 181, Gl. 32.12]). Es ist demnach

$$\delta\alpha_i = \frac{\rho a}{s_{i,0}} \cdot \frac{1-e^2}{W_0^3} \sin\alpha_{0,i}\,\delta\varphi - \frac{\rho a}{s_{i,0}\,W_i} \cos\varphi_0 \cos\alpha_{0,i}\,\delta\lambda \qquad (7)$$

Es wird mithin die Vermittlungsgleichung hier:

$$\alpha_i = \alpha_i' + v_{\alpha_i} = \alpha_{i,0} + \frac{\rho a}{s_{i,0}} \cdot \frac{1-e^2}{W_0^3} \sin\alpha_{0,i}\,\delta\varphi - \frac{\rho a}{s_{i,0}\,W_i} \cos\varphi_0 \cos\alpha_{0,i}\,\delta\lambda$$

und daraus die Verbesserungsgleichung:

$$v_{\alpha_i} = \frac{\rho a}{s_{i,0}} \cdot \frac{1-e^2}{W_0^3} \sin\alpha_{0,i}\,\delta\varphi - \frac{\rho a}{s_{i,0}\,W_i} \cos\varphi_0 \cos\alpha_{0,i}\,\delta\lambda + (\alpha_{i,0} - \alpha_i') \qquad (8)$$

Hat man nur Azimutbeobachtungen, so gleicht man nur diese nach [8; S. 125 - 128] aus. Sind aber auch die Einlaufzeiten beobachtet worden, so muß man die Gl.(4a) und (8) gemeinsam ausgleichen. Da entsteht zunächst das Problem, daß in den Gl. (4a) die Verbesserungen Zeitverbesserungen, in Gl. (8) dagegen Winkelverbesserungen sind. Die Ausgleichung kann aber dennoch streng richtig in einem erfolgen, wenn nur die Gewichte umgekehrt proportional den Quadraten der entsprechenden mittleren Fehler angenommen werden, wie darauf schon Gauß hingewiesen hatte. Die mit den entsprechenden Gewichten multiplizierten Verbesserungsquadrate und somit auch deren Summe werden so dimensionslos, und es können daher auch Beobachtungen in verschiedenen Maßeinheiten leicht gemeinsam ausgeglichen werden.

Sollten die berechneten Zuschläge größer ausfallen, so sind die ursprünglichen Näherungswerte damit zu verbessern und die Ausgleichung mit diesen besseren Näherungswerten zu wiederholen. Um diese zweite Rechnung zu ersparen, ist es wichtig, gute Näherungswerte in die Ausgleichung einzuführen.

Hat man eine entsprechend große Anzahl von überschüssigen Beobachtungen, darunter auch mehrere mit Azimutbeobachtungen, so erhält man aus der Ausgleichung einen relativ guten Wert für die durchschnittliche Fortpflanzungsgeschwindigkeit der zur Berechnung gewählten Welle. Sind nun einzelne Zeitverbesserungen v bedeutend größer als die zu erwartenden Beobachtungsfehler der Einlaufzeitpunkte, so kann dies oft als Zeichen einer von der durchschnittlichen abweichenden Fortpflanzungsgeschwindigkeit angesehen werden. Eine gute Näherung für den plausibelsten Wert der Fortpflanzungsgeschwindigkeit in den einzelnen Richtungen kann dann verschieden angegeben werden (vgl. [11; S. 150 - 151]). Ein etwas anderer Weg ist der folgende: Die Fortpflanzungsgeschwindigkeit nach i kann auch folgend ermittelt werden:

$$v_i = \frac{t_i - t}{s_i} \doteq \frac{t_i - t}{s_{i,0}}, \qquad (9)$$

wobei aus den mittleren Fehlern von t_i und t auch jener von v_i angegeben werden kann. Dabei ist aber zu beachten, daß t und t_i voneinander nicht unabhängig sind, da in der Berechnung von t auch t_i mitwirkte. Es müßten daher, streng genommen, die Beziehungen (16) - (19) in [8; S. 113] angewendet werden. Einen zulässigen Näherungswert für den mittleren Fehler können wir aber etwa nach dem Fehlerfortpflanzungsgesetz für unabhängige Werte nach der Beziehung:

$$\mu_{v,i} = \frac{\sqrt{\mu_{t,i}^2 + \mu_t^2}}{s_{i,0}} \tag{10}$$

angeben. Aus den so erhaltenen Werten v_i und $\mu_{v,i}$ sowie aus den Abweichungen in den beobachteten Azimuten kann man gewisse Rückschlüsse auf die Eigenschaft des zurückgelegten Weges ziehen und unter Berücksichtigung der weiteren Parameter petrographische und tektonische Folgerungen erhalten, auf die wir hier nicht weiter eingehen wollen.

Dagegen soll noch kurz darauf hingewiesen werden, daß die größeren Zeitverbesserungen auch durch die räumliche, oder ebene Ausdehnung des Herdes verursacht werden können. Letztere kann z.B. von der Richtung der Stationen vom Herd abhängige halbperiodenförmige Veränderungen in den Zeitverbesserungen verursachen, die erkannt werden können. Das aufgeworfene Problem muß daher komplex untersucht werden, und die unter Gl. (9) angegebene Beziehung gilt richtiger nur dann, wenn der Herd mit guter Näherung punktförmig angenommen werden kann, wie etwa bei den Sprengungen.

Die Folgerungen können erweitert werden, falls wir verschiedene Gattungen der Wellen in die Berechnung einbeziehen. Das ist natürlich möglich, nur sind für diese dann getrennte durchschnittliche Fortpflanzungsgeschwindigkeiten als Unbekannte in die Ausgleichung einzuführen, falls wir aus bestimmten Gründen diese als Unbekannte annehmen.

Die Ermittlung der Herdtiefe kann dann aus den näher gelegenen Stationen u.a. so erfolgen, wie dies in den Gl. (25) - (26) in [1; S. 60] angegeben wurde.

Die genauere Bestimmung der Erdbebenzentren kann zur genaueren Lokalisierung der Dislokationsflächen führen, deren Kenntnis andererseits bei der Planung von Höhennetzen und Höhenmessungen für die Festlegung von Erdkrustenbewegungen von Vorteil ist. Für die Forschung der Erdkrustenbewegung sind daher genauere Lokalisierungen der Erdbebenzentren empfehlenswert. Dazu sollen diese Ausführungen rechentechnisch beitragen.

Die Anregung zu dieser Untersuchung verdanke ich meinem hochverehrten Freund O. Meißer.

Anhang

Zu unserer Berechnung scheint es - wie bereits erwähnt - am vorteilhaftesten, das Verfahren von Levallois und Dupuy [4] anzupassen, da bei dieser Methode relativ wenig Winkelfunktionen aufgeschlagen werden müssen und die aus den Tabellen zu entnehmenden Hilfsgrößen weniger veränderlich, d.h. empfindlich auf die Ungenauigkeit der Näherungswerte sind, weshalb diese Methode rasch konvergiert. Die sonst vorteilhaften Methoden mit den Normalschnitten scheiden hier aus, da bei den in Frage stehenden Entfernungen die Unterschiede zwischen Normalschnitt und geodätischer Linie die gewählten zulässigen Grenzen überschreiten.

Wir führen in Ergänzung noch die Abplattung

$$\alpha = \frac{a - b}{a}$$

und das Quadrat der zweiten numerischen Exzentrizität

$$e'^2 = \frac{a^2 - b^2}{b^2}$$

ein, die für ein angenommenes Ellipsoid konstant sind.

Die Berechnungen sind - Entfernungen von 20 000 km vorausgesetzt,- bei einer Rechenschärfe von 1 m in der Regel achtstellig, bei 10 m siebenstellig vorzunehmen. Das Rückaufschlagen der Winkelfunktionen muß bei 1 m auf 0,01", bei 10 m auf 0,1" erfolgen. Es genügt überall die lineare Interpolation.

Der Berechnungsvorgang ist bei den gegebenen Koordinaten φ_i, λ_i der Station i und φ_0, λ_0 des Näherungswertes des Epizentrums der folgende:

1.) Berechnung der reduzierten geographischen Breiten ψ_0 und ψ_i, sowie deren erforderlichen Funktionen

$$\operatorname{ctg} \psi_0 = \frac{a}{b} \operatorname{ctg} \varphi_0 \qquad \operatorname{ctg} \psi_i = \frac{a}{b} \operatorname{ctg} \varphi_i$$

$$\sin\psi_0 \ \sin\psi_i = \dots . \qquad \cos\psi_0 \ \cos\psi_i = \dots .$$

2.) Der erste Näherungswert $\psi_m^{(1)}$ der auf die Besselsche Hilfskugel bezogenen maximalen reduzierten Breite ψ_m der geodätischen Linie, sowie die ersten Näherungswerte $\omega_0^{(1)}$ und $\omega_i^{(1)}$ der sphärischen Bogenlängen ω_0 und ω_i können aus einem entsprechenden Graphikon, wie ein solches in [5; S. 26] und [12; S. 1002] zu finden ist, auf 4 Stellen entnommen werden. Der Unterschied $\Delta\omega^{(1)}$ der in erster Näherung erhaltenen Bogenlängen $\omega_i^{(1)}$ und $\omega_0^{(1)}$ also:

$$\Delta\omega^{(1)} = \omega_i^{(1)} - \omega_0^{(1)}$$

kann zur Berechnung der ersten Näherung $\Delta\lambda_k^{(1)}$ des auf die Besselsche Hilfs k u g e l bezogenen Längenunterschiedes $\Delta\lambda_k$ verwendet werden. (Der Längenuntersohied $\Delta\lambda$ bezieht sich auf das Ellipsoid). Also:

$$\Delta\lambda_k^{(1)} = \Delta\lambda + \cos\psi_m^{(1)} \ (\alpha\,\Delta\omega^{(1)}) = \Delta\lambda + \varepsilon^{(1)}$$

3.) Berechnung der Werte der zweiten Näherung (mit Ausnahme von $\Delta\lambda_k^{(2)}$).

Nach der Formel der sphärischen Trigonometrie wird der zweite Näherungswert $\Delta\omega^{(2)}$ der Differenz der Bogenlängen:

$$\cos\Delta\omega^{(2)} = \sin\psi_0 \ \sin\psi_i + \cos\psi_0 \ \cos\psi_i \ \cos\ \Delta\lambda_k^{(1)}$$

(6stellig)

Weiter der zweite Näherungswert $\psi_m^{(2)}$ der reduzierten maximalen Breite:

$$\cos\psi_m^{(2)} = \frac{\cos\psi_0 \ \cos\psi_i \ \sin\ \lambda_k^{(1)}}{\sin\Delta\,\omega^{(2)}}$$

(8- bzw. 7stellig)

sowie die zweiten Näherungen der Bogenlängen:

$$\sin\omega_0^{(2)} = \frac{\sin\psi_0}{\sin\psi_m^{(2)}} \qquad \sin\omega_i^{(2)} = \frac{\sin\psi_i}{\sin\psi_m^{(2)}}$$

(8- bzw. 7stellig)

Daraus berechnen wir den dritten Näherungswert der Differenz der Bogenlängen:

$$\Delta\omega^{(3)} = \omega_i^{(2)} - \omega_0^{(2)}$$

Weiter entnehmen wir aus Tabellen (z.B. aus [4; Tabelle III]) die zu $\omega_0^{(2)}$ und $\omega_i^{(2)}$ gehörigen Werte $B_2 W_2$ und bilden deren Differenz $B_2\,\Delta\,W_2$.
Diese sind für unsere Zwecke in Altsekunden umzurechnen.

Mit den beiden Werten $\Delta\omega^{(3)}$ und $B_2\,\Delta\,W_2$ rechnen wir den zweiten Näherungswert $\Delta\lambda_k^{(2)}$ des Längenunterschiedes auf der Hilfskugel auf 0,01" bzw. 0,1" genau:

$$\Delta\lambda_k^{(2)} = \Delta\lambda + \cos\psi_m^{(2)} \left(\alpha\Delta\omega^{(3)} - \sin^2\psi_m^{(2)} \; B_2\Delta W_2 \right) = \Delta\lambda + \varepsilon^{(2)}$$

4.) Berechnung des vierten Näherungswertes $\Delta\omega^{(4)}$ der Differenz der Bogenlängen:

$$\cos\Delta\omega^{(4)} = \sin\psi_0 \sin\psi_i + \cos\psi_0 \cos\psi_i \cos \Delta\lambda_k^{(2)}$$

(8- bzw. 7stellig)

Aus den Tabellen entnehmen wir die zu $\omega_0^{(2)}$ und $\omega_i^{(2)}$ gehörigen Werte A_2 W_2 bzw. A_4 W_4 und bilden deren Differenzen $A_2\Delta W_2$, $A_4\Delta W_4$

Weiter berechnen wir die Hilfsgröße

$$u^2 = 10^2 \; e'^2 \; \sin^2\psi_m^{(2)}$$

5.) Dann wird die Länge $s_{i,0}$ der geodätischen Linie:

$$s_{i,0} = \frac{b}{\rho} \left(\Delta\omega^{(4)} + u^2 A_2 \Delta W_2 - \underline{u^4} A_4 \Delta W_4 \right).$$

Bei einer Rechenschärfe auf 10 m kann in dieser letzten Gl. das letzte Glied $- u^4 A_4 \Delta W_4$ weggelassen werden. Ähnliches ist auch bei 1 m Rechenschärfe bei merklich geringeren Entfernungen als 20 000 km zulässig, in welchem Falle sich auch die Anzahl der erforderlichen Näherungswerte und Stellenzahl verringern kann. Dies kann schon im Laufe der Rechnung wahrgenommen werden.

Als Zahlenbeispiel soll ein in der Fachliteratur schon wiederholt behandeltes Beispiel mit einer 15 000 km langen geodätischen Linie gewählt werden. Weil diese Beispiele sich auf das Hayford-Ellipsoid beziehen, wollen auch wir (um die Vergleiche in den Vereinfachungen leichter anstellen zu können) dieses wählen, obwohl die Berechnungen auf dem Krassowskischen Ellipsoid genauer sind. Die entsprechenden Tabellen für das Krassowskische Ellipsoid sind noch zu berechnen. Unsere Berechnung sollen hier in Altgraden durchgeführt werden.

Es seien:

$$\varphi_0 = +\,50^\circ\; 00'\; 00{,}00'' \qquad \lambda_0 = +\;10^\circ\; 00'\; 00{,}00''$$
$$\varphi_i = -\,62^\circ\; 57'\; 03{,}20'' \qquad \lambda_i = +\,105^\circ\; 05'\; 38{,}30''$$

und das Verhältnis a/b = 1,003 378 38

1.) Berechnung der reduzierten geographischen Breiten und deren erforderlichen Funktionen:

$\operatorname{ctg}\psi_0$	$= +\ 0{,}841\ 934\ 43$	$\operatorname{ctg}\psi_i$	$= -\ 0{,}512\ 330\ 63$
ψ_0	$= +\ 49^\circ\ 54'\ 17{,}35''$	ψ_i	$= -\ 62^\circ\ 52'\ 21{,}17''$
$\sin\psi_0$	$= +\ 0{,}764\ 975\ 58$	$\sin\psi_i$	$= -\ 0{,}889\ 994\ 44$
$\cos\psi_0$	$= +\ 0{,}644\ 059\ 28$	$\cos\psi_i$	$= +\ 0{,}455\ 971\ 39$
$\sin\psi_0\ \cos\psi_i$	$= -\ 0{,}680\ 824\ 01$	$\cos\psi_0\ \cos\psi_i$	$= +\ 0{,}293\ 672\ 60$

2.) Die erste Näherung aus dem Graphikon:

$\psi_m^{(1)} = 65{,}5^\circ$

$$\omega_i^{(1)} = +\ 168{,}2^\circ$$
$$\omega_0^{(1)} = +\ \ 33{,}1^\circ$$
$$\Delta\omega^{(1)} = +\ 135{,}1^\circ = 486\ 360''$$

Damit rechnen sich:

$\cos \psi_m^{(1)} = 0,4147$

$\alpha = 0,003367$

$\Delta\lambda = 95^{\circ}\ 05'\ 38,30''$

$\varepsilon^{(1)} = +\ \ 11'\ 18,96''$

$\Delta\lambda_k^{(1)} = 95^{\circ}\ 16'\ 57,26''$

$\sin \Delta\lambda_k^{(1)} = +0,995\ 753$

$\cos \Delta\lambda_k^{(1)} = -0,092\ 068$

3.) **Zweite Näherung:**

$\cos \Delta\omega^{(2)} = -\ 0,707\ 862$

$\Delta\omega^{(2)} = 135^{\circ}\ 03'\ 40,7''$

$\sin \Delta\omega^{(2)} = +\ 0,706\ 349\ 79$

$\sin \omega_i^{(2)} = +\ 0,977\ 715\ 93$

$\sin \omega_0^{(2)} = -\ 0,840\ 374\ 70$

$\cos \psi_m^{(2)} = +\ 0,413\ 995$

$\psi_m^{(2)} = 65^{\circ}\ 32'\ 38,25''$

$\sin \psi_m^{(2)} = +\ 0,910\ 279\ 16$

$\omega_i^{(2)} = +\ 77^{\circ}\ 52'\ 53,82''$

$\omega_0^{(2)} = -\ 57^{\circ}\ 10'\ 46,95''$

$\Delta\omega^{(3)} = 135^{\circ}\ 03'\ 40,77'' = 486\ 220,8\ ''$

Aus Tabellen:

	$\omega_i^{(2)}$	$\omega_0^{(2)}$	
$B_2 W_2 =$	$+\ 0,6769''$	$-\ 0,3182''$	$B_2 \Delta W_2 = +\ 0,9951''$

$\alpha = 0,003\ 367\ 003$ $\sin^2 \psi_m^{(2)} = 0,828\ 608$

$\alpha\Delta\omega^{(3)} = +\ 1637,11''$

$-\ \sin^2\psi_m^{(2)} B_2 \Delta W_2 = -\ 0,82''$

$\varepsilon^{(2)}/\cos \psi_m^{(2)} = +\ 1636,29''$

$\Delta\lambda = 95^{\circ}\ 05'\ 38,30''$

$\varepsilon^{(2)} = +\ \ 11'\ 17,42''$

$\Delta\lambda_k^{(2)} = 95^{\circ}\ 16'\ 55,72''$

4.) **Letzte Näherung:**

$\cos \Delta\lambda_k^{(2)} = 0,092\ 060\ 28$

$\cos \Delta\omega^{(4)} = -0,707\ 859\ 59$

$\Delta\omega^{(4)} = 135^{\circ}\ 03'\ 39,71''$

$u^2 = 0,560\ 816$

$u^4 = 0,314\ 514$

Aus Tabellen:

	$\omega_i^{(2)}$	$\omega_0^{(2)}$	
$A_2 W_2$	$+\ 595,089''$	$-\ 279,739''$	$A_2 \Delta W_2 = +\ 874,828''$
$A_4 W_4$	$+\ 0,989''$	$-\ 0,317''$	$A_4 \Delta W_4 = +\ 1,306''$

$\Delta\omega^{(4)} = 135^{\circ}\ 03'\ 39,71''$

$+\ u^2 A_2 \Delta W_2 = +\ \ 8'\ 10,61''$

$-\ u^4 A_4 \Delta W_4 = -\ \ 0,41''$

$135^{\circ}\ 11'\ 49,91'' = 486\ 709,91'' = 2,359\ 636\ 23$

Mit b = 6 356 911,9 m wird daraus

$s_{i,0} = 14\ 999\ 999,6$ m

Die Rechenvernachlässigung beträgt nur 0,4 m, und die Berechnung kann mit einer gewöhnlichen Tischmaschine bei einiger Übung dennoch in etwa 40 Minuten bewältigt werden. Es würde im vorliegenden Falle auch eine Berechnung mit weniger Stellenzahlen genügen, was die Rechenarbeit noch weiter herabsetzt.

Literatur

[1] Zur Herdbestimmung von Erdbeben.
Geofisica pura e applicata, Bd. 49 (1961) II , S. 43 - 60

[2] K. E. Bullen: An Introduction to the Theory of Seismology.
Cambridge, Second Edition (1953)

[3] E. Wiechert: Entfernungsberechnungen von Orten auf der Erde bei kleineren Abständen.
Zeitschr. f. Geophysik, Bd. I (1924 - 1925), S. 177 - 182

[4] J.J. Levallois; M. Dupuy: Note sur le calcul des grandes géodesiques.
Publications techniques de l'Institut Géographique National, Paris (1952)

[5] H. Bodemüller: Die geodätischen Linien des Rotationsellipsoids und die Lösung der geodätischen Hauptaufgaben für große Strecken unter Berücksichtigung der Bessel-Helmertschen Lösungsmethode.
Deutsche Geod. Kommission, Reihe B, No. 13 (1954)

[6] M. Cimbalnik: Lösung der geodätischen Hauptaufgaben für große Entfernungen.
Studia Geophysica et geodaetica, 1957, S. 10 - 45

[7] Helmert: Die mathematischen und physikalischen Theorien der höheren Geodäsie.
Bd. I, Leipzig (1880)

[8] W. Grossmann: Grundzüge der Ausgleichungsrechnung, II. erweiterte Auflage.
Verlag Springer (1961)

[9] B. Gutenberg: Physics of the Earth's Interior.
Acad. Press, New York London (1959)

[10] F. Baeschlin: Lehrbuch der Geodäsie
Orell Verlag, Zürich (1948)

[11] Zur Ortung seismischer Herde in Bergbaugebieten.
Gerlands Beiträge zur Geophysik, 1961, S. 137 - 151

[12] Jordan - Eggert - Kneissl: Handbuch der Vermessungskunde, Bd. IV.
Metzlersche Verlagsbuchhandlung, Stuttgart (1959)

Diskussion

Ch. Teupser (Jena):

Wie schon vorher von Prof. Zátopek bemerkt, ist ein Erdbebenherd sicher bedeutend größer als 10 m. Ein bis zwei km Genauigkeit dürften in der Seismologie sowieso die untere Genauigkeitsgrenze sein. Die Zugrundelegung einer Kugel wird also ausreichend sein. In Kürze wird von Ullmann und Maaz (W. Ullmann und R. Maaz: Theoretische Untersuchung über die Ortung von Erdbeben. Veröff. des Inst. für Bodendynamik und Erdbebenforschung, Jena, Heft 72) ein Verfahren veröffentlicht werden, das es gestattet, auf der Kugel ohne Näherungsrechnungen Erdbebenherde zu orten.

A. Zátopek (Prag):

Was die Ausmaße der natürlichen Beben anbelangt, hat Dr. Teupser vollkommen recht. Die Ausführungen von Prof. Tárczy-Hornoch hätten aber Sinn, wenn es sich um die Feststellung der Koordinaten einer Explosion handeln würde, wo der Herd praktisch punktförmig ist. Auch wären hier die Unterlagen für den Fall einer geheim durchgeführten unterirdischen Kernexplosion. Offen bleibt die Frage der mittleren Geschwindigkeit. Ändert sich diese sehr stark, besonders bei kleinen Epizentralentfernungen, wo die Wellen nicht tief in die Erde hineindringen, so können die Wellengeschwindigkeiten wirklich sehr stark schwanken je nach den Eigenschaften der oberflächennahen Schichten der Erdkruste.

A. Tárczy-Hornoch (Sopron):

Mit Rücksicht auf die Kürze der zur Verfügung stehenden Zeit möchte ich nur auf die Berechnung am Ellipsoid antworten.

In meinem, an die Teilnehmer verteilten Vortragstext habe ich sofort am Anfang zahlenmäßig nachgewiesen, daß der Unterschied selbst zwischen der ellipsoidischen und jener der im fraglichen Bereich bestentsprechenden Kugel schon bei 900 km Entfernung 2,3 km erreichen kann und so größer ist, als die von Herrn Teupser als entsprechend erachtete Genauigkeit von 2 km. Der Unterschied zwischen beiden kann mit der Zunahme der Entfernung mehr als mit der ersten Potenz anwachsen und so bei 10 000 km noch ganz bedeutend größere Werte annehmen. Will man also entsprechend unserer, im Vortrag betonten Zielsetzung auch die weitliegenden Stationen in die Berechnung einbeziehen, so ist die Berechnung am Ellipsoid sehr angebracht, zumal der dadurch ausgeschaltete Fehler genau und sicher erfaßt und folglich der entstehende Gesamtfehler sicher herabgesetzt werden kann.

Wir haben unsere Berechnung für eine Rechenschärfe von 1 - 10 m aufgestellt. Vorteilhaft wird ein Zehntel der anzustrebenden Genauigkeit als Rechenschärfe gewählt. Unsere Rechenschärfe ist aber keinesfalls bindend. Will jemand nur eine Genauigkeit oder eine Rechenschärfe von 1 - 2 km, so kann man einfach noch weitere Glieder der Reihen weglassen, bzw. die Berechnung weiter vereinfachen. Nicht aber umgekehrt: Die für 1 - 2 km Rechenschärfe angegebenen Beziehungen können nicht ohne weiteres in genauere verwandelt werden. Deshalb wurde unserer Untersuchung die angegebene Rechenschärfe zugrunde gelegt.

Über Bewegungstendenzen in der Erdrinde. Ein Beitrag zur Seismotektonik

von

E. K r a u s (München)

S u m m a r y

The attempt is made to point out through examples some of the most conspicuous motion phenomena of the earth's crust already investigated in detail in geology. First of all the present large and often violent actions of the earth round about the Pacific Ocean are examined from the geological point of view. In comparison with them the geosynclinal-orogenic processes, which had to be explained subsequently, take place on a smaller scale with regard to the time and the space of their formation, though they produce the orogenic movements from the combination of which our continents have originated. These processes of formation of the continents are only discussed inasmuch as their present peculiar features get intelligible from them.

In selecting the examples the leading principle has been the interest of geophysics, especially that in those motion phenomena which are active still nowadays or which are expected at least to be going on.

The present state can only be understood in its character, if we regard it connected with the development of these phenomena and with the stage of their development.

Till now the modern conception of the planetary motion phenomena could be formed only to some extent by the help of geophysical methods. In the majority of cases our geological ideas were exclusively due to the observations at the earth's surface and in boreholes. These ideas can and should be examined and completed by the methods of geophysics, the more because our experience and our reflections force us to conclude that the subcrustal displacements of masses are to be looked upon as the chief agents of the endogenous tendencies. This account is given from the point of view that recent geological results or hypotheses should be known if geophysical experiments and completions towards depth are to be a valuable help.

Thus some experiments made up to now are briefly discussed such as the geological interpretation of an earthquake, studies of the seismotectonics of the Tyrolean Mountains, investigations with the aid of extensive measurements in the area of the Baltic Sea, and an unsuccessful attempt to measure the dephts of the pits in the northern parts of the Alps and to evaluate them geologically.
I should be very glad if my arguments could be conducive to a very close co-operation between geophysics and geology as far as possible with the universal collaboration of the specialists of all countries.

Zusammenfassung

Es wurde versucht, einige der auffallendsten und geologisch bereits näher untersuchten Bewegungstendenzen in der Erdrinde an Beispielen aufzuzeigen. Zunächst wurden die oft sehr gewaltsamen und großzügigen Gegenwarts-Aktionen der Erde rings um den Pazifischen Ozean geologisch betrachtet. Ihnen gegenüber vollziehen sich die geosynklinal-orogenen Vorgänge, die anschließend zu erläutern waren, in nach Zeit und Raum ihrer Gestaltung kleinerem Maßstab; obwohl gerade sie die Orogene produzieren, aus deren Kombination unsere Kontinente aufgebaut sind. Diese Bildungstendenzen der Kontinente wurden insoweit besprochen als daraus ihre heutige Eigenart verständlich wird.

Bei Auswahl der Beispiele war leitend das Interesse der Geophysik, besonders an jenen Bewegungstendenzen, die heute noch wirken, oder von denen man dies wenigstens erwarten kann. Erst gesehen im Zusammenhang ihrer Entwicklung und des Stadiums derselben wird ja der Gegenwarts-Zustand wesensmäßig verstehbar.

Die moderne Auffassung der planetaren Bewegungstendenzen konnte bisher nur teilweise schon unter Auswertung geophysikalischer Mithilfe gewonnen werden. Weit überwiegend boten nur die Beobachtungen an der Erdoberfläche und in Bohrungen die Grundlagen für unsere geologischen Vorstellungen. Diese können und sollten mit den Methoden der Geophysik geprüft und ergänzt werden. Dies umsomehr, als unsere Erfahrungen und Überlegungen immer mehr zu dem Schluß zwingen, daß es die subkrustalen Massenverlagerungen sind, welche als Hauptakteure unserer endogenen Tendenzen zu gelten haben. Dieser Bericht wurde in dem Gedanken abgefaßt, daß aber die modernen geologischen Ergebnisse oder Hypothesen bekannt sein sollten, wenn die geophysikalische Erprobung und Ergänzung nach der Tiefe hin zur wertvollen Hilfe werden soll.

Es wurden daher einige bisherige Versuche kurz besprochen, so die geologische Bearbeitung eines Erdbebens, Studien über die Seismotektonik der Tiroler Alpen, Untersuchungen unter Zuhilfenahme ausgedehnter Messungen im Ostsee-Umkreis und ein nicht geglückter Versuch, die nordalpine Narbentiefe auszumessen und geologisch zu werten.

Ich würde mich freuen, wenn meine Ausführungen beitragen könnten zu einer recht engen Zusammenarbeit zwischen Geophysik und Geologie unter möglichst allseitiger Teilnahme der Spezialisten in allen Ländern.

Einleitung

Es erscheint klar, daß die gegenwärtigen Bewegungen der Erdrinde, welche die Geophysik zusammen mit den anderen Zustands-Problemen erforscht, dem Momentbild eines ungeheueren Filmes gleichen, des Films der Erdentwicklung.

Weil alles "Gegenwärtige" also "Rezente" geworden ist, so wird sein Wesen allein verstehbar aus seinem Werden, aus seiner Entwicklung. Dieser Grundsatz leitet ja auch beispielsweise die experimentelle Methode. Ortslage und Zeit des Auftretens, Intensität, Eigenart und Bedeutung der krustalen Gegenwartsbewegungen gewinnen eine mögliche und notwendige Beleuchtung dann, wenn wir von deren Tendenzen erfahren. Diese ergeben sich aus der Erdgeschichte. Sie stellte ja u.a. das Nacheinander der Erdkrusten-Bewegungen und damit oft oder meist auch ursächliche Zusammenhänge fest.

Eine riesige, kaum mehr überschaubare Summe von Beobachtungen auf allen Kontinenten erlaubt es uns heute bereits baugeschichtliche Entwicklungslinien abzuleiten, auch solche allgemeiner Art. Und nur von forschungsmäßig Fernerstehenden, besonders von allein kleinanalytisch tätigen Kollegen kann man mitunter noch hören, wir seien nicht so weit, und eine Deutung der geotektonischen Entwicklung, welche krustale Bewegungs-Tendenzen aufzeigen könnte, müsse erst langsam begonnen werden.

Nachfolgend sei versucht, an Beispielen einige Haupttendenzen der Erdrinden-Entwicklung aufzuzeigen, um so rezent beobachtete oder aus naher Vergangenheit erschlossene Bewegungen der Erdkruste als Teilakte desjenigen Entwicklungs-Ablaufes sichtbar zu machen, dem sie angehören. Dabei wollen wir uns nicht mit den vulkangeschichtlichen sondern allein mit den tektonischen oder seismotektonischen Erscheinungen beschäftigen. Nach einer Beleuchtung der Beobachtungen über dem randpazifischen, als dem weitaus tätigsten Erdrinde-Labor seien die gegenwärtigen geosynklinalen, also orokinetischen Bewegungs-Tendenzen, dann die epikontinentalen Rezent-Bewegungen in Übersichten untersucht. Sie führen abschließend zu Gedanken über die Platzschaffung in der Tiefe.

I Die rezenten Tendenzen rings um den Pazifischen Ozean

Lange ist bekannt, daß die stärksten und häufigsten Bewegungen der Erdrinde heute geknüpft sind an den Umkreis des Stillen Ozeans. Alle Übersichtskarten der vulkanischen und seismischen Gegenwartstätigkeit zeigen dies. In systemvoller Tiefenanordnung sehen wir am Ostrande Asiens und in Indonesien die Lage der Hypozentren. Als gewaltig überfüttert mit pazifisch differenzierten Magmen erkennen wir die Westrandzone der Neuen, der amerikanischen Welt. Diese Schmelzflüsse sind, neben arealbasaltischen, typisch für den Gesamtumkreis des Pazifiks. Daher wurde diese Randzone mit dem Namen des darin besonders verbreiteten Gesteins als "Andesitring" bezeichnet.

Neben diesen gemeinsamen Schmelzfluß-Aufstiegen kennzeichnen diesen Ring noch andere Eigenschaften. Die Hypozentren der Welt-, Groß- und Kleinbeben erschüttern ihn häufig. Eine überall verbreitete Rand-Zertrümmerung ist gemeinsam. Eine fest zusammenhängende Westrandzone von tektonischen Gräben begrenzt den Süd- und Nordamerikanischen Kontinent. Eine riesige "Vulkanogene Zone" der Abspaltung kupiert den Ostrand Asiens und begleitet ihn von den Alēuten im hohen Norden und von der Tschuktschen-Halbinsel bis weit nach Süden. Die so auffallenden Inselgirlanden und die kontinuierliche Folge der Tiefseegräben oder -rinnen vermehrt diese Gemeinsamkeiten gleichfalls von den Alēuten bis zu dem Tonga-Kermadek-Rinnenzug im Süden. Solche randlichen Tiefseerinnen, die offensichtlich jüngeren Datums sind, begleiten auch den Westen Südamerikas. Hebungen und Senkungen in den küstennahen Randgebieten der Kontinente und des Ozeans sind nicht selten, und randliche Gebirge umgürten den Ozean.

Schon mindestens während des Paläozoikums und durch das ganze Meso- und Neozoikum kann man eine solche randpazifische Aktivität ableiten.

1951 begründete ich, daß eine ganze Reihe unterscheidbarer Geosynklinalen die Ränder des offenbar uralten Pazifischen Ozeans begleiten. Auf Kosten seines randlichen Untergrundes sind die benachbarten Kontinente während geosynklinaler Gebirgsbildung in den Ozean vorgewachsen. Über 1000 km weit drangen pazifische Schmelzflüsse in den Untergrund von Nord- und Südamerika vor und sind, z.B. als mesozoische Porphyritformationen, dort aufgestiegen. Noch weiter drang das pazifische Magma unter mitgeschleppten und dabei zertrümmerten Orogenbögen zwischen den beiden amerikanischen Teilkontinenten gegen Osten vor. Es bildeten sich die N o r d a n t i l l e n b ö g e n ; ebenso wie sich zwischen dem ostwärts vorgebogenen Südamerika und dem gleichfalls östlich vorgebogenen Antarktischen Kontinent der S ü d a n t i l l e n b o g e n vorschob.

Völlig abweichend ist die Struktur am pazifischen Westrand, also a n d e n O s t k ü s t e n v o n A s i e n u n d A u s t r a l i e n . Hier gibt es auch Inselbögen mit zertrümmerten Orogenstrukturen, wie sie die besonders gut untersuchten japanischen Inseln erkennen lassen. Aber diese Inselbögen kehren dem Pazifik nicht den Rücken, sondern sie bauten sich in diesen konvex hinein. Auch kennen wir heute näher die von den russischen Geologen als "Vulkanogene Zone" benannte, riesige Abriß-Zone, die den ganzen nördlichen Rand Asiens begleitet als rund 200 km breites Band. Sie schnitt die Gebirgsstrukturen Asiens teils quer ab, teils nützte sie deren Schwächezonen aus. Diese riesige, gleichartige Dehnungszone hat aufklaffende Spalten, aus denen überwiegend basische bis mittelbasische Schmelzflüsse in gewaltigen Mengen emporgequollen sind. Sie haben vorwiegend senones und alttertiäres Alter (E. Kraus, 1959).

Die Inselgirlanden selbst zeigen eine mesozoische und tertiärpleistozäne Schollenstruktur. Auch sie wurde von älteren und jüngeren Magmen in Menge durchsetzt. Noch heute ist ja der Reichtum an tätigen Vulkanen und an Erdbeben besonders von Japan bekannt.

Gegen Süden dürften die Randmeere, das Japanische und Gelbe Meer, das Ost- und Südchinesische Meer gleichfalls durch ein erhebliches A b r ü c k e n r a n d l i c h e r T e i l e A s i e n s g e g e n O s t e n entstanden sein. So kam der bekannte japanische Geologe Professor Dr. T. Kobayashi 1933 nach sehr gründlichen Vergleichen zu der Auffassung, daß Teile der Japanbögen einst nordwestlicher lagen, nämlich an Korea und SW am Sichota Alin Asiens. J a p a n s e i v o n A s i e n a b g e d r i f t e t g e g e n S O u n d h i n t e r i h m v e r s a n k d a s b r e i t e J a p a n m e e r .

Bekanntlich steigen nun die von Hypozentren der Erdbeben besetzten Bewegungsflächen aus rund 600 km Tiefe unter dem Küstengebiet Asiens ungefähr gegen Osten an bis zu den Flachbebenherden unter den Tiefseerinnen. Das gleiche wiederholt sich in Indonesien. Die internen Herdverschiebungen großer Erdbeben wurden von L. Hiersemann 1958 geprüft. Es ergaben sich dabei neben Abschiebungen mitunter Überschiebungen, insbesondere für die über 300 km tiefen Hypozentren im Bereiche der japanischen Inseln bezeichnenderweise B l a t t v e r s c h i e b u n g e n i n R i c h t u n g W N W - O S O .

Aus der ostasiatischen Randstruktur entnahm ich 1959, daß diese durch Hypozentren aktivierten Bewegungsflächen d i e o b e r e G r e n z z o n e n f l ä c h e e i n e s a u s d e r T i e f e d e s a s i a t i s c h e n K o n t i n e n t e s g e g e n O s t e n b i s S ü d o s t e n a u f s t e i g e n d e n , s i m a t i s c h e n G e o p l a s m a s t r o m e s b i l d e n . Auf dem ozeanwärts weiterrückenden Unterstrom wurden östlich Randteile abgerissen von Asien und auf dem Stromrücken in durchschnittlich ostsüdöstlicher Richtung unter Zertrümmerung als Inselbögen fortgeschleppt.

Solche überaus großzügige Tendenzen dürften es sein, welche die Unzahl der Vulkane, der Erdbeben und der Hebungen und Senkungen der Erdrinde om Ostrande Asiens verursachten und sich so als heute noch tätig erweisen. Aus den teils aus 600 km Hypozentren-Tiefe gegen Süden nach einer Tiefseerinne aufsteigenden, W-O-streichenden, teils S-N-streichenden Hypozentrenflächen Indonesiens dürfte hier auf ähnliche simatische Unterströme mit ähnlichen Verschleppungsfolgen geschlossen werden können.

Dieselbe unterströmende Tendenz in allgemein OSO-Richtung scheinen uns die ganz überwiegend in dieser Richtung längsgestreckten Untiefen- und Inselaufragungen Polynesiens im Inneren des Pazifiks anzudeuten.

Dieselbe Unterstromtendenz kann - trotz der Mannigfalt ihrer Teilerscheinungen - wohl am Ostrand unseres Urozeans die pazifischen Schmelzflüsse 1000 km unter die amerikanischen Kontinente hinein verlagert und die riesigen Plutone samt der ungewöhnlichen Kofferfaltung dort veranlaßt haben. Nur sie dürfte - zwischen den Kontinentalplatten weniger gehemmt - die pazifischen Magmenströme mit den verschleppten Orogentrümmern über ihnen bis zu den Nord- und Südantillenbögen so weit in das magmatisch fremde atlantische Gebiet vorgetrieben haben.

Ein allgemeiner simatischer OSO-Unterstrom, der außerdem mindestens im NW-Pazifik und wohl auch an der kalifornischen Küste eine Richtungs-Komponente im Gegensinne des Uhrzeigers besaß, dürfte also die Gesamterscheinungen des Andesitringes beherrschen. Zu ihnen gehört aber wohl überall die Tendenz der geosynklinalen Orogenbildung. Deren Produkte dürften durch Abdriftung in den Inselbögen des Westens und in jenen der Nord- und Südantillen zertrümmert worden sein. Ostdrift wird durch ihre magmatische Überlastung besonders in den Anden zur Bildung ungewöhnlicher Orogene geführt haben.

II Geosynklinale Gegenwarts - Tendenzen

1) Die Alpen im ganzen

Geosynklinalen, aus denen bekanntlich die Orogene erwachsen, zeigen -soweit bekannt- selten mehr als 1500 km Länge und vielleicht 600 km Breite. Ihre Entwicklungsgeschichte ist begrenzt. Denn wir unterscheiden dem Alter nach archäische, alt- und jungalgonkische, alt- und jungpaläozoische, mesozoische und alpidische Geosynklinalen. Zwar enthalten deren Orogene oft ältere in einer vorangegangenen Gebirgsbildungs-Ära teilkonsolidierte Schollen. Doch konnten solche noch miteingebaut werden während der jüngeren Hauptorokinese, die durch ihre Merkmale der Konsolidation bis heute für die Altersfestsetzung maßgebend wird. Und diese Altersspanne überschreitet die Dauer einer Ära, soweit bekannt, nicht.

Aus diesen Gründen ist ersichtlich, daß räumlich und zeitlich das Phänomen der "Orokinese". also der geosynklinalen Faltengebirgs- oder "Orogen"-Bildung, durchweg beschränkter ist als das Phänomen der eben besprochenen Andesitringbildung durch die Aktion einer planetar verbreiteten und durch mehrere Erdalter wirksam gewesenen Sima-Unterströmung. Daher haben wir auch festgestellt, daß in den riesigen zirkumpazifischen Andesitring eine größere Anzahl von Geosynklinalen eingebaut sind, welche verschiedenen Bildungsären angehört haben und sich offenbar noch heute bilden.

Die Hauptentwicklungszeit der Orogene vollzieht sich submarin in Randregionen nicht aller Ozeane, nur der "Urozeane" Poseidon, Tethys und des noch immer tätigen Pazifischen Ozeans. Dazu gehören also nicht die "Neuozeane" Atlantik, Indik und Arktik. Seit Beginn des Mesozoikums gab es in letzteren keine Orokinese. Über die gegenwärtig sich vollziehende submarin-geosynklinale Gebirgsbildung wissen wir geologisch und geophysikalisch noch sehr wenig. Sie vollzieht sich offenbar heute noch im Andesitring und im Malayischen Archipel.

Umso mehr wurden die jetzigen Festländer untersucht hinsichtlich der offenbaren Begleiterscheinungen, den Erdbeben, der Schwere, des Magnetismus und natürlich der geologischen Erscheinungen selbst. Man hat einen recht verschieden tiefen Orogen-Tiefgang gefunden bei den alten und bei den jungen Gebirgen. Man kennt deren verschieden ausgeprägte Unterschwere.

Erdgeschichtlich gesehen dürfte der Grund für diese Unterschiede zum Teil wohl der Umstand sein, daß sich unsere jüngeren Gebirge, deren Gebirgsrelief noch nicht abgetragen ist, in einem verschiedenen Entwicklungsstadium ihres Orokinetischen Zyklus befinden.

Wie schon gesagt, vollzieht sich die Hauptzeit der geosynklinalen Orokinese submarin. Nur kleinere Gebirgsteile des werdenden Orogens tauchen mitunter Insel-bildend aus dem Meere empor, um sich dann wieder zu senken. Nach diesem ersten Hauptstadium der "Tief- oder Hyporokinese" steigt schließlich der schon fast fertig eingeengte und bei der fortschreitenden Absenkung mit dickem Sediment und von unten mit sialischen Differentiaten durchsetzte Orogenkörper allmählich und in Etappen höher und höher über den Meeresspiegel. Dieses "Hoch- oder Epiorokinetische" Stadium, das Spätstadium also, ist aus der Menge der Abtragungs-Schuttmassen ersichtlich, welche im humiden Klima die Flüsse aus dem immer höher steigenden Gebirgsrelief in die Senken und in das Vorland hinaustransportieren. Dieses Sediment ist die "Molasse". Es wird deshalb normalerweise sehr mächtig, weil sich das Vorland des aufsteigenden Gebirgskörpers gleichzeitig senkt. Wo dies - vielleicht wegen Schwächezonen parallel zum Gebirgsstreichen - besonders tief absinkt, da wird auch das Molasse-Sediment besonders mächtig und bildet breite Mulden.

Nicht allein die regelmäßig eintretende E n d - H e b u n g ist leicht verständlich, weil die langdauernde Geosynklinalabsenkung vorher die spec. leichtere Geosynklinalfüllung tief hinab in die spec. schwerere Salsima- (Basalt) bzw. Sima-Unterlage geraten ließ. Beim Nachlassen dieser Absenkung muß daher der Sialkörper aus isostatischen Gründen im Spätstadium des orokinetischen Zyklus zum Gebirgsrelief emporsteigen. Ebenso notwendig erscheint uns ein A b s i n k e n d e s V o r l a n d e s z u r " V o r t i e f e " w ä h r e n d d e s A u f s t i e g e s d e s O r o g e n s . D e n n d e s s e n U n t e r l a g e m u ß s i c h j a g l e i c h z e i t i g f ü l l e n , u n d d a s F ü l l m a t e r i a l s t r ö m t a u s d e m V o r l a n d h e r a n , s o d a ß s i c h d i e s e s g l e i c h z e i t i g a b s e n k t .

Auch der v e r s c h i e d e n e T i e f g a n g d e s O r o g e n s erklärt sich zwanglos je nach dem Zyklus-Stadium, in dem sich das betreffende Orogen oder der Orogenteil jeweils befindet. Der größte Tiefgang dürfte erreicht sein während der letzten großen Absenkungszeit. Das ist noch submarin; und am tiefsten sinkt der mediane, generalstreichende Raum unter dem orogenen Scheitel, also über der Längszone ab, welche am meisten nach abwärts verliert. D a s i s t d i e " N a r b e " oder die "Verschluckungszone", ein von O. Ampferer gebrauchter Ausdruck. Erlahmt nun die geosynklinale Absenk-Tendenz, so steigt der Orogenkörper und sein Tiefgang schwindet. Am stärksten schwindet er während der stärksten Abtragung des Orogens, die wir aus der Mächtigkeit des gleichzeitig in das am stärksten versinkende Vorland eingefüllten Molasseschuttes ablesen können.

D i e s e s S t a d i u m i s t b e i d e n A l p e n b e r e i t s ü b e r s c h r i t t e n . Denn deren Hauptmolasse-Schutt ging zuerst, im Rupel, oder Mitteloligozän, als Schlamm in die wegen der starken Absenkung zunächst flachmarin gewordene Vortiefe, dann als Flußgeröll und Flußschlick nach Zufüllung des Meeresarmes über das Festland. Dies war im Oligozän, und nach kurzer Rückkehr des Vortiefenmeeres und dessen Zufüllung in miozäner Zeit wieder auf dem Vortiefen-Festland. Anders als in der Po-Vortiefe wurde in der schweiz-süddeutsch-österreichischen Vortiefe während des Pliozäns und Pleistozäns dann nur noch sehr wenig Sediment aus dem Gebirge antransportiert (E. Kraus, 1951 a).

Man wollte nun seismisch und gravimetrisch prüfen, ob eine stark abgesunkene Narbe existiert, was sich aus der Neigung, mit der die Oberkante des Vorland-Kristallins unter das Gebirge und in diese tiefliegende Narbenzone hinabtaucht, ersehen lassen sollte.

Die seismischen Reflexions-Messungen der Erdölgeologie in Südbayern und in Österreich zeigten nun für eine kurze Strecke südlich von Bad Tölz (H. Reich 1957) nur eine flache Kristallin-Oberkante, in Oberösterreich (E. Braumüller, F. Aberer, V. Jenisch in R. Janoschek 1961) sinkt diese mit flacher Neigung südwärts bis zum Alpenrand ab. Man hat südlich von Bad Tölz die Messungen noch nicht bis zu der nahe am Südrand der Nordkalkalpen anzunehmenden Narbenzone ausgeführt, konnte also über deren Tieflage nichts aussagen. Man hat sich außerdem noch nicht bekannt gemacht mit den geologisch erkannten Bewegungstendenzen, also auch nicht damit, daß in diesem Alpenteil infolge der s c h o n o l i g o z ä n - m i o z ä n v o r s i c h g e g a n g e n e n i s o s t a t i s c h e n H e b u n g d e s O r o g e n s n a t ü r l i c h a u c h d e s s e n N a r b e n z o n e h e u t e n i c h t m e h r d i e t i e f e L a g e w i e v o r h e r gegen den Abschluß der hyporokinetischen Hauptzeit haben kann. Es empfiehlt sich vor der Durchführung von an sich sehr wünschenswerten geophysikalischen Meßkontrollen Kenntnis zu nehmen von dem Stand der gegenwärtigen baugeschichtlichen Ergebnisse und von dem im Gebirge erreichten Entwicklungs-Stadium.

Wie sich diesbezüglich die tektonisch etwas jüngeren Westalpen und deren Beziehung zur Vortiefe im Westen verhalten und in welchem Entwicklungsstadium bzw. in welchem isostatischen Zustand zur Zeit sich der Süd- und der Ostrand der Alpen befindet, können neue geophysikalische Messungen, die im Einklang mit den modernen Auffassungen angesetzt sind, feststellen.

2) Die Eigenart der geosynklinalen Narbenzonen

Seit 1930 und in den zitierten Werken wurden von mir die Narbenzonen, also die bei der geosynklinalen Absenkung besonders bevorzugten Mediangebiete der Orogene, mit ihrer steilen Fächerstellung und Divergenz untersucht. Auch sie seien hier kurz besprochen, weil deren Strukturgeschichte kinetisch in typischen Gebieten noch nachklingt. Ich kennzeichnete besonders die "N a r b e n s c h n i t t e" . Sie sind in der Nähe dieser Narbenzonen immer wieder zu erkennen. Denn sie haben sich erwiesen als sehr tiefeinschneidende H a u p t s c h w ä c h e f l ä c h e n , an denen die Orogene häufig median zerfallen. Auch insofern, als sie mit ihrer Zertrümmerung der Erosion geringsten Widerstand leisten und diese auf ihnen L ä n g s t ä l e r auszubilden vermag.

In der masseerfüllten, subgeosynklinalen Tiefe ist kein Platz für die tatsächlich erwiesene geosynklinale Absenkung. Platz kann nur durch säkulares Abwandern tiefliegender "Geoplasma"-Massen von Kubikkilometer Umfang geboten werden, die weiter absinken und dann seitlich abströmen; anderen nach, die irgendwo wieder aufsteigen in Dehnungsgebieten und dort den Massenüberschuß abliefern gegen die Wasser- oder Lufthülle. So wird nicht allein die geosynklinale Absenkung über dem Narbengebiet möglich sondern außerdem eine Einengung. Denn der im Narbenbereich entstandene Unterdruck durch Absenkung veranlaßt die beiderseitige Nachbarschaft gegen dieses Unterdruckgebiet heranzurücken und dem Abwärtsstrom gleichfalls zu folgen. Über dieser Nachbarschaft, den "Flanken", lastet aber das hangende Gestein. Es wird durch Reibungskupplung narbenwärts mitgeschleppt; von beiden Seiten also. Daher stauen sich diese Hangendmassen zu Falten, Schuppen, Abscherungsdecken. Das heißt: s o g e s c h i e h t d i e o r o g e n e E i n e n g u n g . Diese erreicht allmählich ihr Maximum, so daß weiterhin neue Staustrukturen nur beiderseits w e i t e r a u ß e n entstehen können. Das geht so auswärts weiter und ist die kinetische Ursache für die A u s w ä r t s - M i g r a t i o n , die wir in allen typischen Orogenstrukturen ebenso beobachten wie die gleichfalls nach außen gerichtete Vergenz. Diese muß ja infolge der Unterschiebung narbenwärts unten entstehen.

Nun sind die Gesteinsmassen einer Geosynklinale nie in beiden Flanken gleich kompetent. Daher wird bald die eine, bald die andere Flanke mit schmiegsamerem Gestein gegen den Raum der Absenknarbe herankommen und leichter dem Abwärtsstrom folgen können als die andere Flanke. D i e F o l g e m u ß d a s A u f r e i ß e n s e h r t i e f a b w ä r t s r e i c h e n d e r B e w e g u n g s f l ä c h e n s e i n . E s s i n d u n s e r e " N a r b e n s c h n i t t e " .

Diese Vorstellung entstand nicht am Schreibtisch, sondern bei der Beobachtung eben solcher steilstehender, oft schwer verruschelter Störungsflächen, die sich immer in Narben- also auch Scheitelnähe halten und ungefähr dem Generalstreichen der orogenen Medianzone folgen.

Es kann gar nicht ausbleiben, daß solche nach oben ausstreichenden steilen und lang fortstreichenden Trümmerzonen von der Erosion ausgegraben werden, so daß sich lange Gebirgs-Längstäler, durch flache Talwasserscheiden miteinander verbunden, herausbilden.

Diese weder durch Faltung noch Decken- oder Schuppenbewegung verständlichen Längsstrukturen wurden früher ihrer Steilstellung und Zertrümmerung wegen als "Wurzeln" angesehen, aus denen bei einem mechanisch freilich unvorstellbarem Ausquetschvorgang die Schweizer Decken 50, ja 100 km weit nach NW herausgedrückt worden sein sollten.

In den Alpen handelt es sich da um die "racines externes" von M. Lugeon, um den insubrischen W-C-Talzug des Veltlins, der NO vom Adamello durch die Judikarien-Querstörung nach NNO verworfen, im Pustertal-Gailtal bis zur Donati-Linie weiterzieht. D i e s e m s ü d a l p i n e n N a r b e n z u g s t e h t e i n n o r d -

alpiner gegenüber. Er bildet im Westalpenbogen den steilen Fächerbau des Brianconnais, zieht nach NO unter dem Rand der Bernharddecke an den Südrand der Are-Gotthardt-Masse, weiter bis an den Rhein, um nun etwas nördlicher durch das Klostertal-Arlberg-Unterinntal als Narbenschnitt schräg auszuscheren und von der Pinzgautalung dem oberen Ennstal-Pongau-Paltental weiter nach Osten fortzuziehen.

Von den sehr zahlreichen Beispielen dieser Art durchmißt das großartigste von NW kommend den Karakorum, wendet sich mit dem oberen Industal über eine Talwasserscheide in das Sangpo-Brahmaputra-Tal nach Osten. Das sind rund 2000 km Talungen! Dieser Narbenzug mit Narbenschnitt trennt ungefähr den südvergenten Himalaja von dem NO-vergenten Karakorum und dem nordvergenten Transhimalaja Tibets.

Ein anderer, fast geradliniger Narbenschnitt zieht mit seinen durch wenige Talwasserscheiden verbundenen Längstälern in Nordanatolien aus dem Marmarameer-Bruch im Westen über die "paphlagonische Narbe" von E. Nowak durch das Kelkittal in das Erzincan-Tal und über Erzerum nach Osten weiter. Hier ist die Talung 1100 km lang nachzuweisen. Daß die Narbenschnitt-Tendenz noch heute nicht zu Ende ist , hat der Erdbebenreichtum mit dem katastrophalen Beben von Erzincan und jenem von Erzerum bewiesen: Es äußert sich so noch die ererbte Labilität der schon primärgeosynklinal geschwächten Narbenschnitte.

Diese erweisen sich als steilstehende Hauptschwäche-Flächen alter und junger Gebirge. Denn längs diesen zertrümmerten Flächen haben sich von jeher dann, wenn sich kräftigere Dehnungen in der Erdkruste auswirkten, die Orogenstrukturen, also ganz ausgedehnte Gebirgsflanken median getrennt. Die eine Flanke ist noch heute als Gebirge sichtbar; die andere, entgegengesetzte vergente aber ist ganz oder teilweise entweder in einen Ozean abgesunken oder heute fast eingedeckt von jungen Schichten.

So ist die atlantische Flanke der altpaläozoischen Namaiden, die in Südafrika mit Ostvergenz der atlantischen Küste folgen, unter dem Meer verschwunden. So liegt die Ostflanke der norwegischen Caledoniden mit Ostvergenz oben, ist die westvergente Westflanke teilweise im Atlantik versunken und liegt erst wieder in Schottland über dem Meeresspiegel. So fehlt den NW-vergenten Appalachen des westlichen Nordamerikas der abgesunkene SO-Flügel, dessen Kennzeichen aber kürzlich im Atlantik erkannt worden sein sollen. So ragt nur noch die westvergente Flanke des Ural als Gebirge vollständig hoch. Der Hauptteil der Ostflanke liegt in Westsibirien unter Tertiär begraben.

Unter diesen Umständen ist es kein Wunder, wenn man früher die Faltengebirge als einseitig zusammengeschobene Strukturen ansah.

3) Geosynklinale Spät-Einengungen

Bekannt ist die allgemeine Erscheinung der Orogene, daß sich im Verlaufe ihrer Strukturentwicklung der Hauptschauplatz ihres Zusammenschubes von dem medianen Innenstreifen (also der Narbenzone unten, der Scheitelzone oben) in den beiden Seitenflanken allmählich gegen außen verlagert. Die Einengung "migriert" mit der Richtung der gegen außen gerichteten Vergenz, also in der Überfaltungs-, Überschuppungs- oder Überschiebungsrichtung. Allmählich beginnt zugleich der schon eingeengte Innenteil des Orogens sich aus dem Meeresniveau isostatisch höher und höher zu heben. Von diesen Aufwölbungs-Stadien, die mit Flexur- und Bruchbildung verknüpft sind, sprachen wir oben.

Kann man einmal in alpidischen Orogenen jüngere Horizontalbewegungen beobachten, so liegen sie in diesem Stadium genähert dem Außenrand in der Vortiefe. Ein solches Spätstadium ist an der Erdoberfläche der süddeutsch-österreichischen Alpenvortiefe überschritten und nicht mehr sichtbar. Denn aquitane bis miozäne Sedimente liegen hier in der Molasse bereits diskordant und selbst waagerecht über dem gegen Süden unterschobenen Untergrund. Daß der bei der Gebirgsbildung aktive Untergrund voraquitant noch horizontal beansprucht wurde, das haben 2 Erdölbohrungen erwiesen.

Die alpenrandnahe Bohrung Perwang 1 in Oberösterreich durchsank nämlich zwischen 1,6 und 2,6 km Tiefe Teilschuppen bestehend aus Oberkreide, Eozän und mitteloligozäner Molasse. Und der mit lauter Dehnungsanzeichen (Abschiebungen) versehene Unterbau der Vortiefe, nämlich die Südfortsetzung der Böhmischen Masse mit germanischer Sedimentfazies bedeckt, unterschob dieses Schuppenpaket, wobei es doch wohl durch den Unterdruck in der Nordalpinen Narbentiefe herangezogen wurde (R. Janoschek 1961). In Südbayern haben die zahlreichen Erdölbohrungen im Norden der breiten subalpinen Molassemulden, unter der bei Nurnau ebenfalls jüngste Verschuppungen erbohrt wurden, die aus den flachen Abschiebungen ersichtliche Dehnung des voraquitanen Unterbaues bestätigt.

Es erscheint nicht mehr möglich, die durch exakt beobachtete Bohrungen bestätigten modernen Anschauungen von Unterschiebungen mit Dehnungs-Strukturen, von der Zunahme horizontaler Beanspruchungen nach unten, wo folglich die aktiven Bewegungsangriffe zu Hause sind, von subkrustalen Bewegungstendenzen ganz allgemein nicht zu beachten. Auch hier regt wieder ein Beispiel die enge Zusammenarbeit der wertvollen erdphysikalischen Forschung an mit den modernen baugeschichtlichen Bemühungen.

Bei der Deutung seismischer Ergebnisse kommt es auch sehr an auf die Beachtung des auch innerhalb desselben Orogens keineswegs völlig gleichen Entwicklungszustandes der beiden Flanken. Oft ist die eine "proximal", das heißt gegen den stabilen Rand des benachbarten Kontinents gewachsen. Die andere ist dann "distal", nämlich ohne jene Beengung hinaus gewachsen in den anders gearteten Ozeanboden (E. Kraus 1959). Proximale Flanken reifen meist früher.

Ebenso müssen die Sonderumstände im Falle des "Fortbaues" eines Gebirges berücksichtigt werden, in dem sich eine im Generalstreichen seitlich weiterwachsende Geosynklinale verlängert. Das ist somit keine Migration, die nur senkrecht zum Gebirgsstreichen vor sich geht. Am Ostrand der Alpen ist heute ein ab Burdigal begonnener Niederbruch der ganzen Fortsetzung zu sehen. Über dem versunkenen Unterbau hat sich eine recht mächtige jungtertiäre Schichtenfolge sedimentiert. Darin scheinen bisher nur Bruchstrukturen und Radialbewegungen beobachtet zu sein. Man wird aber hier auch an die Möglichkeit eines einengenden Fortbaues denken müssen.

4) Seismotektonische Zusammenhänge

Wenn man absieht von der vulkanischen und plutonischen Bodenaktivität, sowie von den durch vergleichende Höhenmessung, erdmagnetische Messung und durch Schweremessung erreichbaren Zustands-Feststellungen der Erdkruste, so bleibt als wichtige Methode der geophysischen Erdforschung die seismische zur Prüfung der rezent-tektonischen Erdrindentendenz übrig. Man hat die Kenntnis der Zusammenhänge zwischen tektonischer Struktur und Erdbeben als "Seismotektonik" bezeichnet. Ein regional begrenztes Beispiel sei hier gegeben.

Am 8. Oktober 1930 gab es mit seinem Epizentrum bei Namlos (nahe Berwang, Lechtaler Alpen, Tirol), also in einem engeren Arbeitsfeld der Allgäuer und Lechtaler Alpen, ein Erdbeben der Schütterungs-Stärke bis 7-8 der Mercalli-Sieberg-Skala. Das Schütterungsgebiet reichte vom Rheintalgraben makroseismisch bis zur Böhmischen Masse und südwärts bis zum Inntal, also viel weiter in W-O als quer zum alpinen Generalstreichen gegen S. Ich konnte von den Erdbebenstationen und von privater Seite rund 3000 Erdbebenbeobachtungen, damals in Riga, als Kenner der betreffenden Geologie seismotektonisch auswerten. Es geschah unter möglichster Beachtung der jeweils vorliegenden Untergrundsart, also sowohl der Mächtigkeit des lockeren Pleistozäns und der Felsnähe wie auch der bekannten Unterbau-Strukturen in Süddeutschland und in den Alpen. Es ergaben sich klare Beziehungen zwischen der Schütterstärke und der gegenwärtigen Untergrundstruktur, deren Brüche nur teilweise schon so weit verkittet waren, daß sie bei Fortpflanzung der elastischen Stoßwellen keine erkennbare Rolle mehr spielten. Kartenmäßig konnten diese Verhältnisse, dazu die beobachteten Vor- und Nachbeben sowie die Verteilung der beobachteten seismischen Geräusche nach ihrer Art festgehalten werden. Natürlich wurde auch das mikroseismische Material auf die Bodenreaktion im Hypozentrum untersucht, über dem auch ein Höhleneinsturz, vorbereitet durch Gipsauslaugung in den Raibler Schichten, geschah. Die Erdbebenstationen in Zürich, Wien und Göttingen hatten Seismogramme geliefert.

Der Gedanke bei Erstellung der Karten war die Absicht, eine Vergleichsgrundlage zu schaffen für entsprechende seismotektonische Beobachtungen an späteren Erdbeben, um so etwas zu erfahren über diagenetische Untergrunds-Veränderungen im Laufe der Zeit (E. Kraus 1932).

Im Gefolge dieser Studien haben sich aus einer tektonischen Diskussion und Wertung von schon früher bearbeiteten Erdbeben bemerkenswerte Abhängigkeiten der elastischen Wellen-Ausbreitung von dem tektonischen Bau der ganzen westlichen Ostalpen erarbeiten lassen (E. Kraus 1932). Das Ergebnis, ergänzt durch Beobachtungen im Karpaten-Bereich, ließ den damals vermuteten, heute für die Karpaten und die geotektonischen Zusammenhänge im Pyrenäen-Apennin-Bereich etwas abzuändernden Verlauf der nord- und südalpinen Scheitel- bzw. Narbenzonen noch in der rezenten Krusten-Regsamkeit gut erkennen. Die damals angenommene Ostfortsetzung der südalpinen Narbenzone in Ungarn wurde seitdem durch die Bohrungen annähernd bestätigt.

Mit diesen seismotektonischen Erscheinungen traten wir bereits etwas heraus aus dem Raum der rezent-geosynklinalen Erdrinden-Bewegungen.

III. Epikontinentale Rezent-Bewegungen

Auch unsere Kontinente sind in junger Zeit und in der Gegenwart in Bewegung. Freilich erweisen sie sich gegenüber den Nachwirkungen alpidischer Geosynklinalen oder gar gegenüber den Bewegungstendenzen im Umkreis des Pazifischen Ozeans als ungleich ruhiger. Von den noch immer nicht exakt erwiesenen kontinentalen Horizontalbewegungen sei zunächst abgesehen. Solche sind einstweilen nur geologisch aus paläoklimatischen, paläomagnetischen, faziellen und baugeschichtlichen Gründen abzuleiten.

1) Hebungen und Senkungen

Noch immer sind Radialbewegungen an tektonischen Störungen, sind Pult-Kippungen sowie undative Auf- und Abwölbungen im Gange. Bekannt ist die zuletzt besonders von M. Sauramo genau analysierte nacheiszeitliche Heraushebung des in uralter Zeit schon geosynklinal konsolidierten Orogen-Unterbaues, der in Schweden an die 300 m aufgewölbt wurde. Aber auch im hohen Norden Finnlands und auf Inseln, auf der kanadischen Halbinsel Labrador (über 3000 m) oder in Teilen Asiens (Sibiria, Anabar u.a.) wurden algonkische oder paläozoische Rumpfebenen in jüngerer Zeit teilweise einige tausend Meter hoch gehoben, wie wir dies heute messen können. Das Zentralmassiv der Sahara (Ahaggar u.a. 3000 m), der schon paläozoisch konsolidierte Hohe Atlas (bis heute 4000 m) sind auffallende Beispiele. Wenn sich, ähnlich wie der Antiatlas (über 2000 m) die mitteleuropäischen Horste z.B. oligozän zu Mittelgebirgshöhen heraushoben, so hat man auch hier keine unmittelbare Nachwirkungen der jungpaläozoischen Geosynklinalen vor sich, in denen sie einst konsolidiert wurden, sondern j ü n g e r e V e r b i e g u n g e n d e s f e r t i g e n K o n t i n e n t s. Die Sinkbecken dürften in ähnlicher Maße abgesunken sein.

Sehr fallen die schon vor der kambrischen Meerestransgression bis hinab zum katazonalen Stockwerk abgetragenen Rumpfmassive und kontinentale Stücke von solchen auf. So der Baltische Schild Fennoskandias, Laurentia, Brasilia, Hochafrika, Vorderindien oder Zentralaustralia. Wenn nämlich deren Beharrung im kontinental-konsolidierten Ruhezustand auf einer bedeutenden Mächtigkeit ihrer Sialplatten beruht, so muß diese im Maße ihres starken Abtragungs-Verlustes oben durch starke Sial-Ankristallisation unten wieder ergänzt worden sein und ergänzt werden. Diese Überlegung berührt wesentliche isostatische Gedanken. Sie berührt außerdem die Annahme, daß nicht allein unter absinkenden Geosynklinalsohlen eine ausgeprägte, in diesem Falle pazifische Differentiation der Geoplasmaunterlage stattfindet. Auch eine subkontinentale, hier jedoch zu atlantischen Produkten sialischer Art führende Differentiation - offenbar des basaltischen Salsimas (E. Kraus 1928) - muß unter nichtgeosynklinalen Zuständen in subkontinentaler Ruhe tätig sein.

S e n k u n g e n . Wie schon angedeutet, entspricht offenbar den lange Zeit nach ihrer orokinetischen Konsolidation herausgehobenen Kontinentalgebieten eine Abwärtsbewegung anderer kontinentaler Erdrindenteile. Diese Senkungen dürften insgesamt von gleichem Ausmaße gewesen sein wie die Heraushebungen. Und es ist keinesfalls anzunehmen, daß es überall nur eine einzige späte Heraushebungs- und Absenkungsphase gegeben hat.

Einfacher als die vergangenen Phasen der Emporwölbungen an Anzeichen wechselnder Hebungs- und Abtragungsstadien erkannt werden können, ist es, die jeweiligen Zeiten und Ausmaße der Senkungen aus den Schichtprofilen der Sinkbecken-Füllungen abzuleiten. Freilich sind terrestrische Füll-Fazies meist schwerer zu datieren, sofern man nicht eine absolute Zeitbestimmung mitheranziehen kann.

Wiederum handelt es sich bei diesen Bemühungen um die Erkennung von Bewegungstendenzen der Erdrinde und darum, wie weit diese noch das säkulare Ab- und -Auf des gegenwärtigen kontinentalen Zustandes beherrschen. Es ist ein Verdienst besonders der russischen Kollegen aus der Untersuchung der versunkenen Becken-Füllungen durch Bohrungen und geophysikalische Messungen hinsichtlich der kontinentalen "Atmungs"-Bewegung der Erde interessante Ergebnisse angebahnt zu haben. Es liegen ja sogar erste Isohypsen-Darstellungen solcher heute gefüllter Beckenböden vor.

Erstaunlich intensiv ist die Zerklüftung des alten, kontinentalen Kristallin-Untergrundes wie z.B. jene des Baltischen Schildes. Die Klüfte schließen sich zwar im allgemeinen gegen unten. Doch legen auch sie Zeugnis ab von den vielfältigen Verbiegungen unseres Unterbaukristallins. Die Fortpflanzung der elastischen Oberflächen-Wellen dürfte sich hier unter besonderen Bedingungen vollziehen. In erdbebenreichen Kalkstein- und Dolomitplatten wird man sich sogar fragen, ob die zahllosen Klüfte dort nicht offengehalten werden durch die mechanische Wirkung der Erdbebenwellen; ob sie vielleicht sogar vertieft und vermehrt werden.

2) Zyklische Kontinental-Bewegung

Bei seiner zusammenfassenden Untersuchung der nachalgonkischen Bewegungsgeschichte Europas (Band II, 1124 ff.) konnte S. von Bubnoff 1935 ziemlich regelmäßige kontinentale Zyklen der Absenkung und der Emporwölbung erkennen. Dabei dreht die Total-Undation als zugleich die Bau-Richtung anordnende Macht aus einer N-S-Richtung die Küstenrichtungen als Anzeichen der Reliefausrichtung jeweils über eine NW-SO-Ausrichtung in eine O-W-Richtung. Eine ganze Reihe solcher Drehungsphasen im Gegensinne des Uhrzeigers konnte so abgeleitet werden - wenn auch manche Teilbewegungen sich diesem Gesamtschema nicht fügten.

Dieses wurde zunächst abgeleitet aus dem europäischen Russland. Ich erwähne dieses Ergebnis nicht nur weil ich glaube, die Leistung des europäischen "Cyklons" (E. Kraus 1951 b, S. 485) in SW-Europa wiedergefunden zu haben. Es dürfte auch möglich werden, diese Stadien der Hoch- und Tieflage sehr ausgedehnter Oberflächen durch systematische, exakte Messungen in langen Jahren festzustellen und so jenen Zyklon subkrustaler Säkular-Bewegungen nachzuweisen. Denn als bewirkenden Antrieb wird man sich eine subkrustale Massenströmung im Gegensinn des Uhrzeigers vorstellen können. Derartiges ist bekanntlich schon für den Untergrund des nordwestlichen Pazifischen Ozeans von S. Fujiwhara 1925, 1927 und K. Motizuki 1940 (E. Kraus 1951 b, S. 458) angenommen worden.

3) Rezente Bewegungen im Ostsee-Umkreis

Praktisch große Bedeutung haben Höhenmessungen und deren Vergleich mit denjenigen Messungen, die längs der Meeresküste gewonnen wurden. Auch hier lassen erkannte Tendenzen der Vergangenheit Schlüsse auf die nähere Zukunft zu. Ausgedehnte Deichbauten mußten die menschliche Kultur beispielsweise in den Niederlanden oder an der norddeutschen Küste schützen vor Einbrüchen des Meeres nicht allein wegen der wechselvollen, bald hohen, bald tiefen Meeresspiegel sondern auch infolge der Tendenz zu Landsenkungen.

Gelegentlich der IV. Hydrologischen Konferenz der Baltischen Staaten in Leningrad (September 1933) wurden die für den Ostsee-Umkreis wesentlichen Teilfragen erörtert und die Referate publiziert. Als Nr. 83 derselben konnte ich 1933 die "Tektonischen Gegenwartsbewegungen im Ostseegebiet" näher darlegen. Zur Abbildung kamen dabei:

Die säkularen Küstenbewegungen,

die mittlere Landhebung 1898/1912 in cm je Jahr nach R. Witting 1918

der Landhöhenwechsel an der schwedischen Küste 1825/1927 in cm je Jahr nach Folke Bergsten 1930

die Isobasen der Litorinaküste

ein Übersichtsversuch der Transgressions-Isobasen der Tapes/Litorinazeit nach V. Tanner 1929

die Isobasen des höchsten spätglazialen Meerestrandes in Nordeuropa nach A. Gavelin

die Schwerekarte und magnetischen Werte, die quartärbewegten Störungen und die Gebiete rezenter Hebungen und Senkungen

die jungen Hebungs- und Senkungszonen, sowie die Eisrandlagen im Ostbaltikum nach E. Kraus 1928

Die geologischen Zusammenhänge wurden geschildert und Deutungsversuche gegeben. Es stellten sich überaus wechselvolle Teilbewegungen heraus, aus denen die großen, so einheitlich aussehenden Bewegungstendenzen sich zusammensetzen. Dies wurde zurückgeführt auf verwickelte Einzelvorgänge innerhalb des subkrustal strömenden Geoplasmas, bzw. auf dessen Zustandsänderungen.

Die seismische Tätigkeit erwies sich nicht mehr so sehr gebunden an die postglaziale Langkuppel-Aufwölbung Skandinaviens als an die heutigen, großen Reliefsenken des atlantischen Uferrandes und des Bottnischen Meerbusens. Dabei dürfte die wechselnde Mächtigkeit der lockeren Quartärdecke eine große Rolle spielen. Die Reliefbedeutung der pleistozänen Eisbedeckung wurde untersucht, um tektonische Einflüsse herauszufinden. Die beschränkte Mitwirkung der Eis-Belastung und Entlastung, der "Eisisostasie" also, ergab sich gegenüber der Leistung der normal-endogenen Hebungs- und Senkungstendenzen des kontinentalen Untergrundes.

4) Die großen Randabbrüche und Randgräben der Kontinente

Bei der Untersuchung der ozeanischen Kontinentalränder hat sich gezeigt, daß diese zumeist mit einem ziemlich steilen Abfall gegen den Ozeanboden enden. Die Randschärfe ist meist noch betont durch die von O. Jessen 1948 im Zusammenhang untersuchten Randschwellen der Kontinente. Die bei Bildung solcher Schwellen wirksam gewesenen und offenbar noch tätigen Faktoren, also Tendenzen, können hier nicht erörtert werden. Jedenfalls haben sich noch in junger Zeit in der Grenzzone der Sial-Platten über und neben der basischen Unterlage des Ozeanuntergrundes bedeutende Massenverlagerungen abgespielt.

Wir hörten bereits von der zusammenhängenden Grabenzone am Westrand des süd- und nordamerikanischen Kontinents. Sie zieht vom Antarktischen Kontinent nach Alaska, also vom südlichen zum nördliche Polarkreis. Dieser planetare randpazifische Grabeneinbruch greift aber nicht weit vor in die amerikanischen Kontinente und steht in Beziehung zu den Tiefseerinnen, die ihm parallel laufen.

Noch weitere Gründe veranlassen uns die Genese dieses ostpazifischen Grabenzuges abzutrennen von den Gräben und Niederbrüchen der Kontinente gegen den Atlatischen, Indischen und Arktischen Ozean.

Beiderseits umrahmen den Atlantik solche Einbruchsgräben. Nähern wir uns ihnen vom Osten, so treffen wir bald, Ozean-parallel, den Grabenzug der "Mittelmeer-Mjösenseezone" von H. Stille. Gegen sie bricht die Iberische Halbinsel südostwärts ab. Der Abbruch setzt sich fort mit dem Rhonetal-Graben, der zugleich Vortiefen-Bedeutung für den Westalpenbogen besitzt. Über den Saône-Graben nördlich Lyon zunächst gegen N fortsetzend wird die Grabenzone dann 120 km gegen O in den Rheintalgraben versetzt. Dessen NNO fortsetzende Spalten begrenzen weiterhin in der Wetterau das Rheinische Schiefergebirge gegen Osten und ziehen, sich wechselvoll zersplitternd, nördlich in den Leinetalgraben. Die "rheinische" Linienfortsetzung wird im norddeutschen Untergrund gekennzeichnet durch Basaltaufstiege, durch die rheinische Anordnung der Salzhorst-Aufragungen sowie durch den mächtig gefüllten Buntsandsteintrog Kassel - Hannover - Aller. SO Hamburg und in Schleswig-Holstein kehrt die rheinische Anordnung der Salzstöcke wieder. Über Großen Belt und Kattegat schneidet dann unsere rheinische Störungszone nordwärts tief in den Umriß Skandinaviens ein. Die Kerbe wird durch den Oslograben gebildet, der weiter in den Mjösenseegraben zieht. Weil in gleicher NNO-Richtung nun die sehr scharfe tektonische Grenze des Baltischen Schildes gegen die norwegischen Caledoniden liegt, kann man annehmen, daß auch diese altpaläozoische Grenzlinie noch die Fortsetzung unserer der Atlantikküste ungefähr gleichlaufenden Graben- und Spaltenlinie bedeutet.

Geologisch hat unsere rheinische Zerspaltung wahllos alte und junge, meist quer liegende Strukturen zerhackt. So das Rheinische Schiefergebirge oder die Kristallinmassen des Französischen Zentralplateaus und des Baltischen Schildes. Die häufigen tertiär-vulkanischen Aufstiege von basaltischen und anderen Schmelzflüssen der rheinischen Gräben längs der Trennzone beweisen wohl, daß die hier lang hinziehenden Spalten tief in den Untergrund hinabschneiden.

Weniger weit verfolgbar, aber gleichfalls zur atlantischen Küste parallel und unbekümmert die alten Strukturen des Französischen Zentralplateaus durchpflügend sehen wir westliche Gräben des Loire-Oberlaufes, den breiteren Graben der Limagne mit dem Allier-Tale, sowie - im W das ausgedehnte Vulkangebiet des Cantal und des Mont Dore abschliessend - den scharfen NNO-Schnitt des "Kohlenkanals" mit seiner Südfortsetzung bis ins obere Aveyrontal. Von diesem 300 km NNO, also wieder "rheinisch" durchschneidenden Spaltenzug zweigen 2 ähnliche gegen NNW ab. Der Morvan-Horst im N wurde gegen W und O isoliert und beiderseits im Mesozoikum begleitet von gleichfalls südnördlichen Spalten. Auch wo die Trierer Grabenbucht in das Rheinische Schiefergebirge einsank, gibt es wieder eine große Zahl rheinisch gerichteter Spalten, wie sie auch Südteil und Südfortsetzung der Vogesen zerhackt haben.

In Großbritannien hat der grabenartige Einbruch der Irischen See und des Minchkanals Irland abgetrennt von England/Schottland. Ein scharfer Blattverschiebungs-Schnitt trennte - wieder in NNO-Richtung - die Grampian Mts. von den NW-Highlands in Schottland. Mit ähnlicher Hauptrichtung brach das kaledonische Gebirge Norwegens westwärts zum Atlantischen Ozean nieder. Sowohl Spitzbergen wie Grönland wurden durch randliche Abbrüche beschnitten gegen den Atlantischen Ozean.

Gleiches kann nun gesagt werden von der afrikanischen, der nord- und südamerikanischen Küste des Atlantik. Daher hat man ja den "A t l a n t i s c h e n K ü s t e n t y p u s" als quer abschneidenden Typ definiert. Er beherrscht in gleicher Weise die Küsten des Indischen wie die des Arktischen Ozeans. Die Bruchlinie des unteren St. Lorenzstromes und jene des triadischen Newark-Grabens in Nordamerika - sie mögen als weitere Beispiele f ü r d a s w e i t e b e i d e r s e i t i g e V o r g r e i f e n d e r a t l a n t i s c h e n Z e r t r ü m m e r u n g i n e i n e n N a c h b a r k o n t i n e n t gelten.

Wir stehen hier vor einer m i t d e m M e s o z o i k u m e i n s e t z e n d e n E i n b r u c h s t e n d e n z, welche die von den atlantischen Einbruchsküsten umgrenzten "Neuozeane" erst bildete und so das heutige Erdantlitz schuf.

In der Zeit des geologischen Fixismus-Gedankens galt es als selbstverständlich, daß an der Stelle der heutigen Neuozeane oder doch von ausgedehnten Brücken-Kontinenten lauter Festländer gelegen hätten. Diese seien eben im Laufe der nachpaläozoischen Zeit eingebrochen. Unlösbar war freilich in diesem Falle die Wasserbilanz und die Frage, wie da unten P l a t z geschaffen worden war. Es mehren sich jetzt die Anzeichen dafür, daß es sich in diesen Fällen um Folgen einer planetar verbreiteten K o n t i n e n t a l v e r s c h i e b u n g handeln muß.

5) Die Platzfrage in der Tiefe

Man hat sich früher wenig überlegt, w a s m a n d e r m a s s e n e r f ü l l t e n T i e f e z u m u t e t e bei der Feststellung der ausgedehnten Hebung und Senkung kontinentaler Gebiete, bei der Bildung ihrer ausgedehnten Gräben, der geosynklinalen Absenkungen und Hochgebirgs-Aufstiege - und nun gar bei dem Gedanken, es könnte der überwiegend mesozoische Einbruch der drei Weltozeane, welche das heutige Erdbild beherrschen, in seine unverlagerte Tiefe hinab möglich gewesen sein. W i r s i n d h e u t e g e z w u n g e n z u d e r V o r s t e l l u n g v o n z w a r s e h r l a n g s a m e n a b e r ü b e r a u s v o l u m i n ö s e n G e o p l a s m a - V e r l a g e r u n g e n u n t e r d e r E r d r i n d e. Diese rücken, platzbietend, nach unten und seitlich ab vom Untergrund großer Senkungsgebiete. Sie steigen dort wieder empor, wo die Oberfläche sich mächtig hochwölbt und dabei gedehnt wird, sowie durchdrungen wird von gewaltigen Schmelzflußmassen.

Werden diese U n t e r s t r ö m u n g s - T e n d e n z e n unter einem Kontinent zu einer überwiegend ähnlichen Bewegungsrichtung "gleichgeschaltet", so schleppen sie diesen Kontinent auf ihrem Rücken mit sich fort. Trennen sich solche Groß-Unterströme, so können sie sialische Kontinentalplatten auseinander reißen. Die geringe Reißfestigkeit der Gesteine ist ja bekannt. D a s i s t d e r h e u t i g e G r u n d g e d a n k e d e r K o n t i n e n t a l v e r s c h i e b u n g.

Wir haben auch die sehr langsam und sehr langdauernden Regungen der Kontinente hier erwähnt, weil sie ohne Zweifel nicht zufällig kurz vor der Zeit, in der Menschen begonnen haben, erdgeschichtlich zu denken, beendet wurden. Die Kontinente verlagern sich dank ihrer subkrustalen Tendenzen noch jetzt hebend und senkend, und wohl auch bald dahin, bald dorthin. So setzen sich ebenso die Bruch- und Grabenbildungen noch heute durch.

In einem nicht allzu umfangreichen Bericht können sie bezüglich ihrer heutigen Ortslage und Tendenz - soweit dies überhaupt schon bekannt ist - näher nicht geschildert werden. Es wird immer der jeweils zuständige Fachgeologe zu Rate zu ziehen sein.

L i t e r a t u r

Bubnoff, S. von: Geologie von Europa Bd. II. Verlag Gebr. Bornträger, Berlin 1935

Fujiwhara, S.: Torsional form on the face of the earth. Japan Journal Astron. and Geophysics 3, 103, 1925

Fujiwhara, S.: On the earth-vortex, echelon faults and allied phenomena. Ergebn. der kosm. Physik 2, 303 - 360, 1934

Fujiwhara, S.; Takayama, T.: On the mechanism of the great Sagami Bay earthquake. Res. Inst. Univ. Tokyo 6, 1929

Hiersemann, L.: Die Erkenntnisse über den Herdvorgang bei großen Erdbeben und ihre Bedeutung für die Geotektonik. Z.d. Bergakademie Freiberg/Sa., 10, Nr. 5/6, 343-350, 1958

Janoschek, R.: Über den Stand der Aufschlußarbeiten in der Molassezone Oberösterreichs, Erdöl-Zeitschr. Verlag Urban Wien-Hamburg, Heft 5, Mai 1961

Janoschek, R.: Überblick über den Aufbau der Neogengebiete Österreichs. Mitt. Geol.Ges. Wien 52, 149 - 158 (1959), Wien 1960

Jessen, O.: Die Randschwellen der Kontinente. Peterm. Geogr. Mitt. Ergänzungsh. 241, Gotha 1948

Kobayashi, T.: The Sakawa orogenic cycle and its bearing on the origin of the Japanese Islands. J.Fac. Sc. Imp. Univers. Tokyo Sect. II, Vol. V/7, 230 - 578 p. 1941

Kraus, E.: Der Orogene Zyklus und seine Stadien. Centralbl. f. Miner. etc. Jg. 1927, B, Nr. 6, 216 - 233

Kraus, E.: Der geomechanische Typus der Mittelrheinischen Masse und der Orogene Zyklus C. Rend. Congrès géol. Intern. Madrid 1926, III, 1031 - 1076, Madrid 1929

Kraus, E.: Das Wachstum der Kontinente nach der Zyklus-Theorie. Geol. Rundschau 19, 353 - 493, 1928

Kraus, E.: Seismotektonik der Tiroler Alpen. Gerl. Beitr. Geophysik 30, 1/2, 96 - 135, 1931

Kraus, E.: Die Bewegung des Erdbebens am 8. Oktober 1930 im süddeutschen Bau. 50 Ber.d. Naturw. Ver.f. Schwaben u. Neuburg, 93 S., Augsburg 1932

Kraus, E.: Über tektonische Gegenwartsbewegungen im Ostseegebiet. IV. Hydrol. Konf. d. Balt. Staaten, Leningrad Sept. 1933. Report Nr. 83, 49 S., 8 Abb., Taf. Verlag Hydrol. Staats-Inst. des Volkskomm. f. schwere Industrie UdSSR 1933

Kraus, E.: Der alpine Bauplan, 352 S. Gebr. Bornträger, Berlin 1936

Kraus, E.: Zur Baugeschichte des Alpenvorlandes. Z. Deutsch. Geol.Ges. 102, 216 - 226 (1950), Hannover 1951

Kraus, E.: Die Baugeschichte der Alpen, 2 Bände, Akademie-Verlag Berlin 1951 a

Kraus, E.: Vergleichende Baugeschichte der Gebirge. Akademie-Verlag Berlin 1951 b

Kraus, E.: Über die jüngsten Bewegungstendenzen der Alpen. Geol. Rundschau 44, 77 - 107, 1955

Kraus, E.: Zur Kenntnis der Orogene Anatoliens. Berg- u. Hütt. Monatsh. Montan. Hochschule Leoben 101, 25 f. 1956

Kraus, E.: Über die Strukturen der orokinetischen Narbe. R.v. Klebelsberg-Festschrift, Geol. Ges. Wien, Mitt. 48, Wien 1957

Kraus, E.: 50 Jahre Unterströmungstheorie. Denkschr. f. S.v. Bubnoff, Z. Geologie 7, 261 - 283, 1958

Kraus, E.: Die Entwicklungsgeschichte der Kontinente und Ozeane. 285 S. Akademie-Verlag Berlin 1959

Motizuki, K.: The origin of the Sitito-Mariana. ArcGeogr. Rev. Japan 16. 219 - 332, 1940

Reich, H.: Unser Beitrag zum Geophys. Jahr. Z. Heidelberger Portländer, Zementw. Heidelb. A.G. 5 S., 1957

Reich, H.: Die geol. Ergebnisse seismischer Registrierungen großer Sprengungen in Deutschland. Geof. pura e appl. Milano 40, 41 - 48, 1958/II

Secular crustal movements: some results and tasks of researches

von

J. A. M e s c h e r i k o w (Moskau)

S u m m a r y

The paper reviews the main data on secular crustal movements obtained in different countries through application of geodetic, oceanographic and geologo-geomorphological methods. New valuable evidence on secular movements of the Earth's crust has been recently obtained in the USSR, Poland, Netherlands, Finland, Great Britain, Italy, Bulgaria, Hungary, Czechoslovakia, USA, Canada and Japana.

The factual material obtained allows certain assumptions to be made on the regularities of manifestation and the nature of secular movements. Secular crustal movement has been recorded everywhere. Secular upheavals and subsidences occur interchangeably and have the appearance of "waves" whose length is larger on the platforms and shorter in the orogenic regions. The predominant rates of secular movements amounts to several millimetres per year. An assumption has been made that the period of recent oscillatory movements is in the order of centuries and millennia. The tectonic nature of secular movements is noted; the glacio-isostatic factor is almost completely absent in the contemporary epoch. The secular (slow) movements are connected with the fast (seismic) movements and affect the course of many exogenic processes; the study of the essence of these interrelations is an important objective of research. The problem of extensive horizontal continental drifts requires thorough investigation. In regions where the horizontal movements have been thoroughly studied their intimate connection with vertical movements has been established. The horizontal and vertical movements should obviosly be regarded as separate components of complex spatial deformations of crustal blocks.

Z u s a m m e n f a s s u n g

Die Arbeit berichtet über die Hauptdaten säkularer Krustenbewegungen, die in verschiedenen Ländern durch Anwendung geodätischer, ozeanographischer und geologisch-geomorphologischer Methoden erhalten wurden. Neue wertvolle Beweise über säkulare Bewegungen der Erdkruste sind gegenwärtig in der UdSSR, Polen, den Niederlanden, Finnland, Großbritannien, Italien, Bulgarien, Ungarn, USA, Czechoslowakei, Kanada und Japan erhalten worden.

Das Tatsachenmaterial erlaubt bestimmte Annahmen über die Regelmäßigkeiten, die Erscheinung und die Natur säkularer Bewegungen. Säkulare Krustenbewegungen sind überall aufgezeichnet worden. Säkulare Hebungen und Senkungen treten wechselweise auf und haben die Erscheinungsform von "Wälen", deren Länge auf den Plattformen größer und in den Orogengebieten kürzer ist. Die vorherrschenden Beträge säkularer Bewegungen liegen bei mehreren Millimetern pro Jahr. Es ist die Annahme gemacht worden, daß die Periode der rezenten Schwingungsbewegungen in der Größenordnung von Jahrhunderten und Jahrtausenden liegt. Die tektonische Natur der säkularen Bewegungen wird besonders hervorgehoben; der glazioisostatische Faktor fehlt fast vollständig in der gegenwärtigen Epoche. Die säkularen (langsamen) Bewegungen sind mit den schnellen (seismischen) Bewegungen vergesellschaftet und beeinflussen den Verlauf vieler exogener Prozesse. Das Studium des Wesens dieser Beziehungen ist ein bedeutsames Forschungsobjekt. Das Problem ausgedehnter, horizontaler Kontinentalverschiebungen erfordert eine gründliche Untersuchung. In Gebieten, wo Horizontalbewegungen gründlich studiert worden sind, ist ihr enger Zusammenhang mit Vertikalbewegungen festgestellt worden. Die horizontalen und vertikalen Bewegungen könnten offensichtlich als getrennte Komponenten komplexer räumlicher Deformationen von Krustenblöcken aufgefaßt werden.

The existence of secular crustal movements is not a novel discovery: it was recognised in remote past by such prominent researchers as Celcius, Lomonosov, Lyell, Darwin, Suess, Karpinsky, Obruchev, Sauramo and Gutenberg who displayed a considerable interest in this phenomenon.

However, large-scale investigations of crustal movements were launched only recently when the improvement of research methods and accumulation of new data, especially the data provided by repeated levellings in the control network of various countries, threw a new light on the problem of secular movements of the Earth's crust. Today this problem by right ranks among the most vital theoretical and practical problems of the science of Earth and continues to enjoy an ever increasing attention.

Despite the successes achieved in various countries, our knowledge of slow movements of the Earth's crust remains, generally speaking, very scanty. Nevertheless, in summing up the first results of the study of neotectonics, certain tentative opinions can be put forward.

1. Changes of the sign and intensity of recent movements in space

In the light of the new data we have no ground to believe that secular movements are characteristic only of separate, especially mobile sections of the Earth's crust whereas the other sections are stable. Recent movements manifest themselves wherever their study is undertaken. As in geological past, uplifts and subsidences of the Earth's crust in the contemporary epoch have an interchangeable character. To all appearance, the sections of recent uplift and subsidence in platform and orogenic regions differ in size. For exemple, the profile (Fig. 1) of the Russian platform gives waves of recent movements some 600-800 km long; the profile of Japanese Islands exhibits waves extending for 180-250 km. In many districts the zones of upheaval pass into the zones of subsidence very sharphly; the transition zones are noted for their increased gradients of movement and have (in plan) rectilinear contours or the form of arcs of large radius. In all probability, this feature reflects a deep-seated block structure of the Earth's crust.

The predominant rates of recent vertical movements come to several millimetres per year. Intensities of several centimetres per year are less frequent. In some mainly orogenic districts, the intensities of the Earth's surface displacements reach several decimetres per year. It can be assumed, however, that the recent tectonics of the orogenic regions are characterised not so much by the higher intensity of movement as by a greater differentiation of movements and increased gradients. Comparison of the profiles (see Fig. 1) shows incidentally that the gradients of the movement rates on the Japanese Islands are by one order higher than on the Russian platform. These problems can be solved solely along the lines of compiling summary maps of recent movements of large territories.

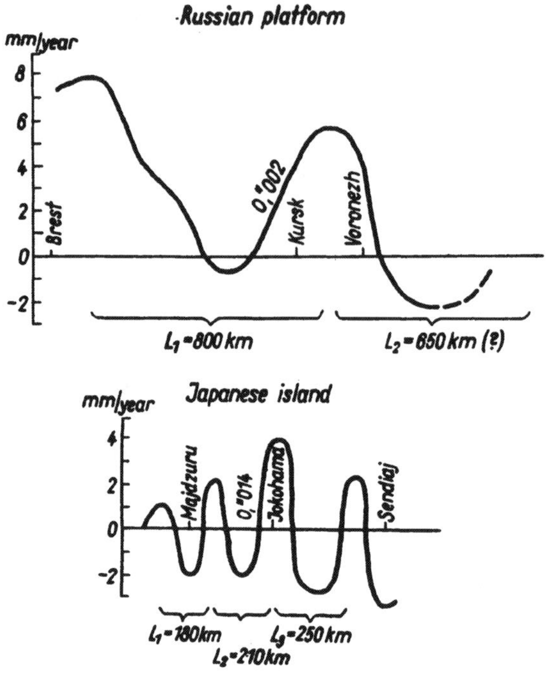

Fig. 1

Curves of intensity of recent vertical movements of the Russian platform and Japanese islands (mm/year)

2. Changes of the sign and intensity of recent movements with time

Secular crustal movements are complicated apparently by short-period variations of various origin. Except for the Earth's tides, the knowledge of these variations is clearly inadequate and we have therefore little to say about their periods, amplitudes and their nature. An opinion has been voiced that short-period variations of the Earth's crust (with a period of several years and an amplitude of several centimetres) almost completely conceal the effect of secular movements when the time interval between the levellings is not long enough. The opinion was that an interval of 80-100 years was needed between levellings to detect secular movements. We cannot agree with this. Analysis of the materials provided by precise levellings shows that when the interval between two successive levellings is very brief (some 8-10 years) the rates of movement become indeed excessive and the general picture of the movements frequently fails to agree with the geological-geomorphological concepts. At the same time experience goes to show that if the interval between the levellings is increased to 20-25 years and over we shall be able to record stable, apparently secular tendencies of movements. The principal proofs to corroborate this point are as follows: 1) mutual coordination of the results on various lines of levelling conducted at different periods but forming a network of repeated levelling; 2) good agreement between the geodetic results and the geological-geomorphological data on secular movements.

It can be assumed that an interval of 20-25 years between the levellings is sufficient for the secular movement to compensate for or entirely eliminate the effect of the short-period variations and the errors of levelling. It is interesting to note that a minimum of twenty years is also necessary to estimate the mean average level of the sea. It is possible that this coincidence is not fortuitous and really reflects the affinity between the nature of short-period variations of the Earth's crust and the surface of the Earth's water envelope. It should be stressed, however, that these conclusions need additional substantiation by the direct study of the short-period movements of the Earth's crust.

The study of the course of secular movements is another difficult problem. Simple computations show that the intensities of secular movements cannot by applied to lengthy geological periods. Thus, if the upheaval of the Middle Russian zone of elevation had retained its intensity (about 5 mm/year) only during the Quarternary period (0.5-1 million years), this plain would have given rise to mountains over 2-3 thousand metres high. Therefore, secular movements should in the course of time change not only their intensity but also their sign.

The problem of the period of secular movements of the Earth's crust cannot be solved so far without drawing upon geological, geomorphological and archaeological data since the duration of secular movements is clearly outside the terms of precise geodetic and tide-gauge observations. According to some data, it can be assumed that the period of secular movements covers not less, but rather more, than 600-700 years. Archaeological evidence indicates that the full period of the Earth's crust variation (uplift-subsidence-new uplift) in the district of Pozzuoli (Italy) is estimated at approximately 2,000 years and extends from the 2nd century B.C. to the 18th century A.D. According to our geomorphological investigations, the distribution of regions of recent upheaval and subsidence in the north-west of the Russian plain had formed about 6,000-7,000 years ago.

3. The nature of secular movements

Different approach can be taken to elucidate the nature of secular movements. Various geophysical data - gravimetric, geomagnetic, geothermic - as well as geological and geomorphological materials - can be undoubtedly used with great advantage for the purpose.

The author has made an attempt to appraise objectively the relationship between the secular movements and the structure on the basis of quantitative indices. With this aim in view, a map of secular movements of the western half of the European part of U.S.S.R. was superimposed on a tectonic map (Fig. 2). The areas of the sections have been measured where the recent uplifts of subsidences corresponded to the positive blocks or negative blocks of the basement as well the sections where the recent movements and the basement structure were unlike in their sign. We have found that the geostructural elements corresponded to the sign of the recent movements approximately on two-thirds (69%) of the area (see the Table below).

We have also found out that the positive blocks of the basement are distinguished by a more stable course of secular movements: recent upheaval extends to 78% of their area. The regions of negative blocks proved less stable in respect to the sign of recent movements: subsidences were registered only on 57% of their area. Geomorphological data have shown that the sections of the basement nearest to the surface have proved to be most stable in the sign of movement (uplift) in recent time. The regions of negative blocks and anticlinal structural elements experienced numerous changes of the movements sign.

The high factor of correlation between the recent movements and the geological structure leaves no doubt as to the fact that the source of secular movements is concealed in the same endogenetic processes which have caused the formation of the elements of the crustal structure.

Fig. 2

Effect of geological structure on recent movements of the western part of the Russian platform
1 - positive (+) and negative (-) blocks of crystalline and folded basement; 2 - recent uplifts (+) and subsidences (-); 3 - sections where structure and movements are alike in their sign; 4 - sections where structures and movements are unlike in their sign

Areas of negative and positive tectonic blocks and the sections of recent upheaval and subsindece in the Western Part of the Russian Platform

Geostructure	Movement	Area %	Total area %
a) For the entire studied territory:			
Positive blocks (+)	Uplifts (+)	44	69
Negative blocks (-)	Subsidences (-)	25	
Positive blocks (+)	Subsidences (-)	12	31
Negative blocks (-)	Uplifts (+)	19	
		100	100
b) For positive tectonic blocks:			
	Uplifts (+)	78	
	Subsidences (-)	22	
		100	
c) For negative blocks:			
	Subsidences (-)	57	
	Uplifts (+)	43	
		100	

However, the coincidence between the movements and the structure is never complete. It cannot be complete from a theoretical point of view if we take into account the high rates of secular movements which considerably exceed the mean rates of tectonic movements during geologically lengthy periods of time. There is not the least doubt that in the course of time secular movements of each section of the Earth's crust must change not only their intensity but also their sign. It can be further assumed that for the positive blocks of the Russian platform basements such changes, during which the disagreement between the structure and the movement comes to the fore, must be of a relatively short duration taking only 100-78 = 22% of the time required for the variation cycle. For negative blocks the period of such disagreement comes to about 100 - 57 = 43% of the cycle.

Although these values are rather arbitrary we can assume that the curves of the variation movements are always asymmetrical in profile (in time), the ascending component being predominant within the limits of structural positive blocks and the descending component of the vertical movement - within the limits of negative blocks.

This thesis can be confirmed by certain archaeological evidence. Let us come back to the example of Pozzuoli where, as has been pointed out, the full period of variation movements comprised approximately 2,000 years. There, the stage of upheaval which began in the 16th century continued only up to the close of the 18th century after which a new subsidence took place which also goes on at present. We see that the stage of upheaval which needed about 200-300 years took only 10-15% of the time of the variation cycle, a fact which is in good agreement with the tendency towards subsidence typical of the given section of the Mediterranean geosyncline zone.

In our investigations the role of glacioisostasy can hardly be isolated from the study of the nature of secular movements. The sum total of facts at our disposal does not allow us to relegate the distribution of the section of recent upheaval and subsidence of the Earth's crust to the glacioisostatic factors alone. It would be a mistake, however, to ignore the glacioisostatic factor altogether. Apparently, during the first half of the Holocene (14,000-,7000 years ago), after the ice caps have melted away, the crustal movements in Fennoscandia and the adjoining regions were entirely determined by the glacioisostatic processes. However, in the second half of the Holocene (7,000 years up till now) the glacioisostatic equalisation apparently ended and the leadership was again taken by the tectonic processes proper. Today they determine the principle features of the crustal movement.

4. Slow movements and seismic shocks are two different forms of manifestation of the endogenetic forces. There is an obvious interconnection between them which is not yet clear. Comparison of the map of the recent movements of the Russian plain with the data on the distribution of the epicentres of slight earthquakes testifies to the fact that all epicentres have a tendency to be associated with the zones of increased gradients of movements on the boundary between the uplifts and subsidences, closer however to the upheaval sections. But, not all the zones of higher gradients have epicentres. For example, the recent Krivoi Rog uplift is no less intensive that the middle Russian, but, as distinct from the latter, it is not associated with earthquakes. Complex connections between seismic phenomena and secular movements of unlike sign have been discovered also on other regions for which the corresponding data are available. An important task of further research lies in elucidating the essence of the interrelation between the slow and fast movements. Its solution will make it possible to use the data on slow displacements to prognosticate seismic shocks. From this point of view, the research into slow recent movements in seismic regions must be stimulated in every way. At the same time we must remember that very young faults of flexures in the stage of formation are very frequently seismogenic. They cannot always be

detected by purely geological methods but a great effect can be derived from geomorphological observations, rerunning precise levellings and tide-gauge observations.

5. Effect of secular movements on exogenetic processes

Slow upheavals or subsidences of crustal blocks exercise a noticeable effect on landscape-making and the dynamics of its development. To give an objective appraisal of the effect of recent movements on river erosion the map of recent movements was compared with the map of the main types of river-beds of the European part of the USSR. Within each section of recent subsidence and recent upheaval we computed the length of unmeandrine (down-cutting) river-beds and determined their percentage in relation to the total length of the river-beds in the given contour. Measurements were taken separately for the region of the Valdai (Würm) glaciation and for the extraglacial region. As the map (Fig. 3) shows the regime of rivers in the investigated territory reflects first and

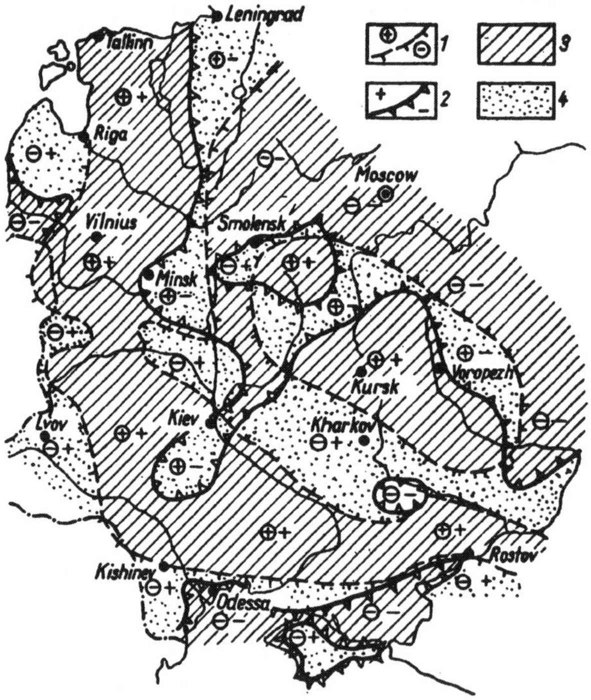

Fig. 3

Effect of recent tectonic movements of the Russian platform on river erosion.
1 - isometric lines of the rates of recent movements (0; +5 mm/year); 2 - boundary of the last glaciation; 3 - length of un-meandrine rivers (in per cent of the total length of the rivers in the given Contour); 4 - regions where the length on un-meandrine rivers is 50% and over

foremost the effect of the paleogeographical factor: as distinct from the extraglacial region, the glacial region is distinguished by a general increase in river entrenchment. However, within each of these two paleogeographical zones the local intensification or weakening of river erosion undoubtedly reflects the effect of non-uniform secular movements. Thus, in the glacial region the maximum length of the un-meandrine rivers

(81%) was registered in the region of the mid-stream of Western Dvina within the contour where the intensity of uplift exceeded 5 mm/year. In the region of subsidence, in the lower reaches of the Nieman, the length of the down-cutting rivers dropped to 67%. In the extraglacial region the maximum indices of the length of the down-cutting (un-meandrine) rivers were obtained on the sections of upheaval with an intensity higher than 5 mm/year - middle Russian (84%), Krivoi Rog (60%) and Lvov (50%) sections.

Many geographers, geologists, hydraulic engineers and melioration experts still underestimate the effect of secular movements on river erosion, the regime of underground waters, the bogging-up processes, peat formation, gully erosion and slope processes. It is necessary to give a greater impetus to systematic study of the recent exogenetic processes which are the indicators of secular movements. Comparison of maps and aerophotographs of various years can be especially effective in this respect. It can be said without exaggeration that the analysis of landscape changes on the basis of the materials of repeated mapping provides geomorphologists and geographers with a means to study secular movements which is not less effective than the method of repeated levelling.

6. Horizontal movements of the Earth's crust

The horizontal crustal movements have been studied to a far less extent than the vertical movements. The most interesting and complete materials on horizontal movements were amassed by Japanese scientists. Comparison of the data on horizontal displacements and vertical movements in Japan allowed them to surmise the existence of connection between these two kinds of movements. Thus, in the centre of Kyushu island (Fig. 4),

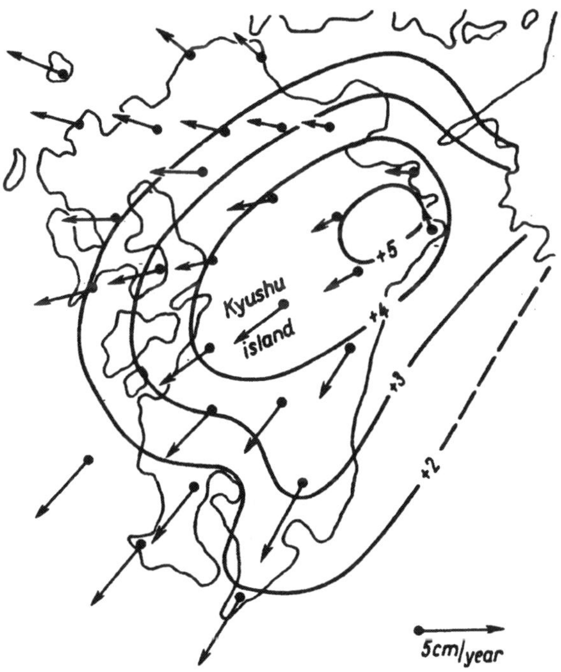

Fig. 4

Ratio between vertical and horizontal crustal movements on Island Kyushu
1 - Isometric lines of the rates of vertical movement (mm/year); 2 - vectors of the rates of horizontal movements (for a scale a vector corresponding to a rate of 5 cm/year is given)

when the vertical movement is most intensive the horizontal displacements are relatively small. Upheaval diminishes towards the periphery of the island but the horizontal movements increase in intensity. The general configuration of the vectors shows that here an enormous rising block of the Earth's crust tends to spread out. Other crustal blocks sink and are compressed. It can be concluded that the subdivision of the displacements of the Earth's crust into vertical and horizontal is rather arbitrary. In actual fact these movements are but the s e p a r a t e c o m p o n e n t s of c o m p l e x t h r e e - d i m e n s i o n a l d e f o r m a t i o n s o f c r u s t a l b l o c k s .

If we adhere to this viewpoint on horizontal and vertical movements we shall arrive at the conclusion that the rates of recent horizontal movements as well as the rates of vertical movements cannot be extrapolated over geologically lengthy periods of time. Like the vertical component of movement, the horizontal component should to all appearance change in the course of time not only its intensity but also the sign (direction) so that the total effect of movement, on a geological time scale, must essentially decrease. We are far from negating entirely horizontal displacements of large crustal blocks on a planetary scale. We only believe that, at first, reliable data must be obtained on such movements with the help of the latest astronomical and geodetic methods. This field of investigation of recent movements, more than any other field, calls for a close cooperation among scientists of different countries and continents.

D i s k u s s i o n

R o b b i n s , A.R. (Oxford):

Prof. Mescherikov in his interesting paper, also in his valuable paper which was distributed to delegates before this Symposium, has quoted recent evidence of crustal movement from England and I would like to comment on this. In his first paper Prof. Mescherikov reproduces a map from Colonel Edge's paper given at Toronto, which has contours ranging from 0 in the South to 0,5 ft (15 cm) in the North of Great Britain. Prof. Mescherikov interprets these as contours of crustal movements but in England we say that we do not know if they are or not. We are only certain that they are differences between the second and third geodetic levellings. Experiments with unequal illumination of levelling staves have shown that systematic error can occur in levelling along North-South-Lines, when the North staff is always in shadow. The experiments are unfortunately not yet conclusive, but they do show that the difference of 15 cm between the two levellings could be attributable entirely to this source of systematic error. It is then of the utmost importance, that the interpretation of all scientific measurements be made with great care and criticism.

If I may continue, I would like to add that this is why I consider this Commission and this Symposium to be so important. Scientists from many different disciplines gather together and study the results each has obtained from entirely different points of view. When these results agree, we may feel more sure that we are approaching the truth.

E g y e d , L. (Budapest):

1.) The small deformations of the crust can be treated as deformations of a swimming slab having at their edges torsional moments. This can be described by a differential equation of the form:

$$\frac{d^4 y}{dx^4} + \lambda y = 0$$

This can have also a periodical solution with a wave length - as I remember - of 600-800 km; in accordance with the date of Mescherikov.

2.) The horizontal movements shown for the northern part of Japan can be considered as an effect of all-directional tensional forces.

M e s c h e r i k o v , J. A. (Moskau):

I thank Dr. Robbins and Prof. Egyed for their remarks and comments to my report. It seems to me that the results of levelling in England are in concord with the geological and geomorphological data about the crustal movements in that country, and also with the oceanographic (tide-gauge) observations. I have in mind, for instance, the works of Prof. Linton, Dr. Valentin and others. This correlation, as Dr. Robbins noted in a general form, allows to conclude, that the data of levelling might be used to understand the crustal movements. As to the differences in the length of the waves in the platform and orogenic regions, these are, probably, connected not with the thickness of the earth's crust, but rather with its other properties, especially with its solidity.

Mitteilung über eine vorläufige Untersuchung über die neuen vertikalen Bewegungen der Erdkruste in Bulgarien

von

W. K. Hristov, J. Gâlâbov (Sofia)

In dem militär-topographischen Dienst in Bulgarien ist eine vorläufige Untersuchung über die vertikalen Bewegungen der Erdkruste in Bulgarien, Oberfläche 110 000 km^2, durch das Vergleichen zweier Nivellements ausgeführt worden. Zwischen den zwei Nivellements ist an verschiedenen Orten ein Zeitintervall von 11 bis 35 Jahren verflossen. Der gemeinsame Teil umfaßt auf dem ganzen Land acht geschlossene Nivellementsschleifen erster Ordnung mit einer Gesamtlänge von 3344 km und 585 km Linien zweiter Ordnung auf einem Gebiet in Südbulgarien.

Zur Ableitung der Geschwindigkeiten der Bewegungen der Erdkruste sind die unmittelbar gemessenen Höhenunterschiede $h_{AB(1)}$ und $h_{AB(2)}$ zwischen je zwei Bolzen A und B im ersten (1) und zweiten (2) Nivellement benutzt worden, wobei im Zeitintervall t_{AB} die Geschwindigkeit als konstant angenommen wurde. Für die relative Geschwindigkeit c_{AB} des zweiten Bolzens B bezüglich des ersten Bolzens A haben wir folglich

$$c_{AB} = \frac{h_{AB(2)} - h_{AB(1)}}{t_{AB}}$$

unter der Voraussetzung, daß die zwei Höhenunterschiede $h_{AB(1)}$ und $h_{AB(2)}$ fehlerlos sind, was natürlich unmöglich ist.

Der letzte Umstand verlangt eine Ausgleichung der Geschwindigkeiten c_{AB}, c_{BC}, c_{CD} unter der Bedingung, daß für eine geschlossene Nivellementsschleife erster Ordnung die Summe der Geschwindigkeiten Null sei. Da $h_{AB(1)}$ und $h_{AB(2)}$ unabhängige Größen sind, so ist der mittlere Fehler der nichtausgeglichenen Größe c_{AB} proportional der Quadratwurzel aus der Länge L_{AB} und selbstverständlich umgekehrt proportional dem Zeitintervall t_{AB}. Für das Gewicht P_{AB} der nichtausgeglichenen Größe c_{AB} ist entsprechend

$$P_{AB} = \frac{50}{L_{AB}} \left(\frac{t_{AB}}{20}\right)^2$$

gesetzt worden, da man Gewicht Eins für L = 50 km und t = 20 Jahren festgelegt hat.

Die Ausgleichung hat einen mittleren Fehler der Gewichtseinheit von 1.11 mm pro Jahr gezeigt.

Nachdem man die ausgeglichenen Werte c_{AB}, c_{BC}, c_{CD} ... erhalten hat, weiterhin durch aneinander folgendes Hinzufügen, ausgehend von dem Pegel in Burgas, hat man die relativen jährlichen Geschwindigkeiten der gemeinsamen Bolzen der beiden Nivellements bezüglich Burgas erhalten. Das gibt uns die Möglichkeit, die Isolinien der relativen jährlichen Geschwindigkeiten, wenn auch nicht ganz sicher, aufzustellen, wobei diese Geschwindigkeiten später durch eine additive Konstante korrigiert werden müssen, um in absolute Geschwindigkeiten übergeführt zu werden.

Da die Isolinien der jährlichen Geschwindigkeiten nicht genug definiert gezeichnet werden können, so sind auf dem Diagramm - samt den Nivellementslinien - die Gebiete des Hebens in Millimeter pro Jahr: 0 - 1 mm, 1 - 2 mm, 2 - 3 mm, 3 - 4 mm, 4 - 5 mm und des Senkens 0 - 1 mm pro Jahr bezüglich Burgas angegeben worden.

Es wäre interessant, die jährlichen Geschwindigkeiten auf eine absolute Skala zu reduzieren und mit den absoluten jährlichen Geschwindigkeiten in den anderen Ländern zu vergleichen, woraus bestimmtere Schlüsse gezogen werden können.

Aus der vorgelegten Karte ersieht man, daß sich in Bulgarien ein Gebiet auszeichnet, das sich im Heben befindet und das sich von Südwesten nach Nordosten erstreckt und zwar mit einer steigernden Tendenz. An dieses diagonale Gebiet grenzen zwei andere Gebiete, in Nordwest - und Südostbulgarien mit relativ schwacher Bewegung. Die in Frage kommenden vertikalen Bewegungen können geologisch-geomorphologisch auf folgende Weise interpretiert werden:

Die Bewegungen im Diagonalgebiet

Der südwestliche Teil dieses Gebietes mit relativer Erhebung bezüglich Burgas 1 - 2 mm pro Jahr, umfaßt die inneren Balkaniden, Rila-Rhodopen-Massive (die Ostrhodopen ausgeschlossen), sowie die Ebenen und Becken in diesem Gebiet. Die letzteren charakterisieren sich mit mächtigen pliozänen Ablagerungen (bis zu einigen hundert Metern) und holozänen alluvialen Ablagerungen (bis zu einigen zehn Metern), was besagt, daß diese Ebenen und Becken während des Pliozäns und Holozäns gesunken sind und gegenwärtig samt dem ganzen Gebirgsgebiete sich erheben, d.h., die gegenwärtigen Bewegungen zeigen einen inversen Charakter gegenüber denjenigen im Pliozän und Holozän.

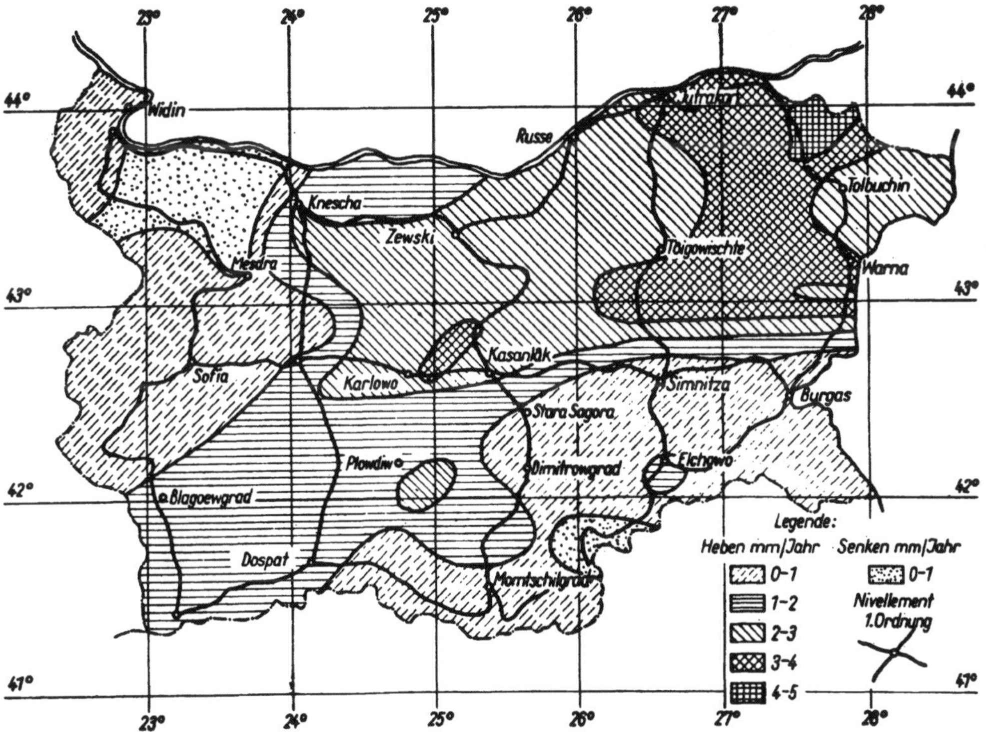

Abb. 1

Rezente Vertikalbewegungen der Erdkruste in Bulgarien. Ausgeführt von: S. Tzenew, A. Schumkow, J. Wassilew

Der mittlere Teil des Diagonalgebietes mit relativer Erhebung 2 - 3 mm pro Jahr umfaßt den westlichen Abhang der Aufwölbung in Nordostbulgarien und den mittleren Teil der geosynklinalen Struktur von Stara-Planina. Endlich der nordöstliche Teil des gleichen Gebietes mit einer Erhebung über 3 mm pro Jahr umfaßt sowohl die höchsten Teile dieser Aufwölbung, wie auch den östlichen Teil der geosynklinalen Struktur von Stara-Planina. Dieser letztere Teil des Diagonalgebietes zeigt die ausgesprochenste Tendenz zum Heben.

Die Teile von Bulgarien, die nordwestlich und südöstlich von dem Diagonalgebiet liegen, zeichnen sich durch eine relativ kleine Höhenbewegung aus. Hierzu gehören die Depression bei Lom westlich vom Iskarfluß und das Niederthraktische Becken. Auf diesem Becken liegen mächtige pliozäne und holozäne alluviale Ablagerungen. Längs der Donau ist die Dicke der holozänen alluvialen Ablagerungen rund 25 m, also liegt deren Grundfläche unterhalb des Meeresniveaus. Die gegenwärtigen Bewegungen in diesen Teilen haben das gleiche Vorzeichen wie die Bewegungen in Pliozän und Holozän.

Folglich behalten in Bulgarien, allgemein gesagt, die gegenwärtigen vertikalen Bewegungen mit wenigen Ausnahmen das gleiche Vorzeichen wie im Neozän.

Diskussion

J. A. Mescherikov (Moskau):

I would like to note the interesting and completely new materials presented by our Bulgarian colleagues. From the geological point of view of special interest is the fact that within Bulgaria the recent deformations of the crust are discovered to be transversal to the structural elements of the crust. The main structural elements of Bulgaria have latitudinal direction, and among the recent risings and subsidings the "waves" of meridional direction are distinguished. The meridional "waves" of modern movements are also discovered on the Russian platform and in the Caucasus. Probably, this is a wide regularity inherent to the modern stage of the tectogenesis. I would like to wish success to our Bulgarian colleagues in their studies of the recent movements. According to the details of the materials the Bulgarian data, evidently, could be compared only with the data of the Japanese colleagues. The special value of these data is that they characterise the recent movements of mountainous territories where precise relevellings are seldom done.

Ein Versuch der Analyse von Längsprofilen der Flüsse zwecks Untersuchung der rezenten Erdkrustenbewegungen

von

L. E. Setunskaja (Moskau)

Summary

A series of natural factors exercises an influence on the form and the character of the longitudinal profiles of rivers. Among them the crustal movements are of the utmost importance. As a rule the longitudinal profiles of rivers undergo deformations in the sections which are characterised by a relative tectonic activity; the height of the fall increases, differences of height appear in the river beds.

The map of the river gradients compiled for the area of the Central and Southern Ukraine shows that sections with deformations of the longitudinal profiles of rivers form groups in certain tracts (cones of abnormal gradients). These zones are coordinated locally to places with increased gradients of recent movements of the earth's crust. In particular they run through the marginal districts of the intense recent elevations of the earth's crust (which were determined by means of methods of repeated levelling operations) in the area of the so-called Kriwoj Rog-Elevation, as if they would surround it.

The comparison of the values of the real river gradients with the theoretical ones (calculated empirically) shows that the zones with abnormal river gradients also incline to the slopes of the elevations, whereas the vault of the elevation is an area of relatively small gradients.

Zusammenfassung

Auf die Form und den Charakter von Längsprofilen der Flüsse übt eine Reihe von natürlichen Faktoren einen Einfluß aus. Unter ihnen sind die Erdkrustenbewegungen von wesentlichster Bedeutung. In der Regel erfahren die Flußlängsprofile auf den Abschnitten, die durch eine relative tektonische Aktivität gekennzeichnet sind, Deformationen - die Größe des Gefälls nimmt zu, es tauchen Höhenunterschiede im Flußbett auf.

Die Karte der Flußgefälle, zusammengestellt für den Raum von Zentral- und Südukraine, zeigte, daß Abschnitte mit Deformationen der Flußlängsprofile sich in bestimmten Streifen gruppieren (Zonen der abnormen Gefälle). Diese Zonen sind örtlich auf Stellen mit erhöhten Gradienten von rezenten Erdkrustenbewegungen abgestimmt. Insbesondere verlaufen sie durch die Randbezirke der intensiven rezenten Erhebung der Erdkruste (welche mit Hilfe von Verfahren der wiederholten Nivellierung ermittelt wurden) im Raum der sogenannten Kriwoj Rog-Erhebung, als wenn sie sie umsäumen würden.

Ein Vergleich der Größen der tatsächlichen Flußgefälle mit den theoretischen (empirisch berechneten) zeigte, daß die Zonen mit übernormalen Flußgefällen auch zu den Abhängen der Erhebungen neigen, während das Gewölbe der Erhebung ein Raum von relativ geringen Gefällen ist.

Als einer der Untersuchungswege von rezenten tektonischen Bewegungen der Erdrinde kann die Analyse von Längsprofilen der Flüsse dienen. Unter den Faktoren, welche die Form und den Charakter des Längsprofils beeinflussen, haben die tektonischen Bewegungen eine besonders große Bedeutung. Viele Forscher weisen auf eine gesetzmäßige Verbindung zwischen der Längsprofilform und den Erdkrustenbewegungen für verschiedene Gebiete der Sowjetunion hin (I.P. Gerassimov, 1954, M.S. Koshurina, 1956, J.A. Mescherikov, 1960 u.a.).

Von uns wurden Versuchsarbeiten ausgeführt, die den Zweck verfolgten, die Längsprofile der Flüsse für die Gebiete der Zentral- und Südukraine zu analysieren. In tektonischer Hinsicht umfaßt der untersuchte Raum den zentralen Teil des Ukrainischen Kristallmassivs und Teile der Dnepr-Donez- und der Schwarzmeerküsten-Vertiefungen, die das Massiv vom Norden und Süden umsäumen. Die Wahl eben dieses Raumes ist dadurch bedingt, daß für ihn Angaben über die rezenten Erdkrustenbewegungen vorliegen. Mit Hilfe solcher Verfahren, wie das wiederholte Nivellieren, wurde hier eine intensive Erhebung der Erdkruste mit einer maximalen Geschwindigkeit von +10 mm/Jahr (Kriwoj Rog-Erhebung) ermittelt. Die Erhebung hat keinen krassen Ausdruck im Relief bekommen, folglich braucht der Einfluß des letzteren nicht den Einfluß der tektonischen Bewegungen auf die Form und den Charakter der Längsprofile von Flüssen zu tarnen.

Es existieren verschiedene Verfahren zur Analyse der Längsprofile von Flüssen. Wir haben das von J.A. Mescherikov (1960) vorgeschlagene Verfahren ausgewertet. Das Wesen des Verfahrens besteht in der Zusammenstellung von Karten der Flußgefälle, einer Aussonderung der Strecken mit erhöhten abnormen Gefällen und ihrer nachfolgenden Interpretierung. Indem man davon ausging, daß für Flüsse ebener Territorien eine schwach konkave Form des Längsprofils mit einer allmählichen Verringerung der Neigungen in der Richtung zur Mündung hin kennzeichnend ist, wurden die Fälle der Vergrößerung der Neigungen flußabwärts als abnorm betrachtet und die entsprechenden Abschnitte wurden auf der Karte ausgesondert (Abb. 1). Die Flußgefälle wurden nach den topographischen Karten auf

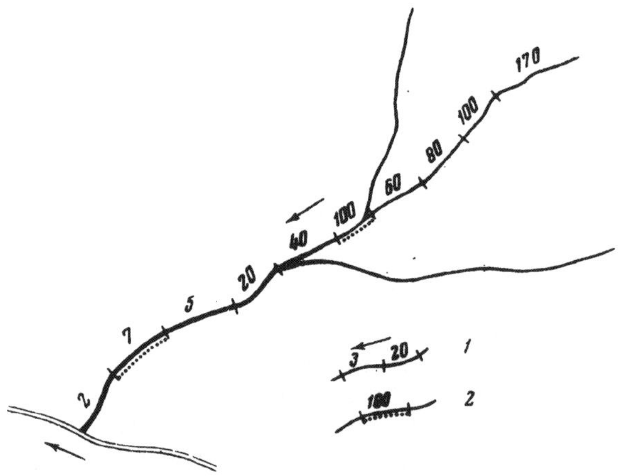

Abb. 1

Schematischer Plan eines ebenen Flusses.

1 – Größe der Gefälle des Längsprofils (cm/km)

2 – abnorme Abschnitte des Längsprofils

Grund der Höhenabzeichen des Wasserspiegels berechnet. Die Berechnung erfolgte nach der Formel

$$i = \frac{h}{l}, \quad \text{wo}$$

h - die Differenz zwischen zwei benachbarten Höhenabzeichen des Wasserspiegels und
l - die Entfernung zwischen ihnen (gemessen mit Hilfe eines Mikrometer-Meßgeräts mit einer Schenkelöffnung von 1 - 2 mm) ist.

Die auf den Karten nach den einzelnen Flüssen ausgesonderten abnormen Abschnitte wurden in Zonen vereint (Abb. 2). Im untersuchten Raum verlaufen zwei Hauptzonen. Die

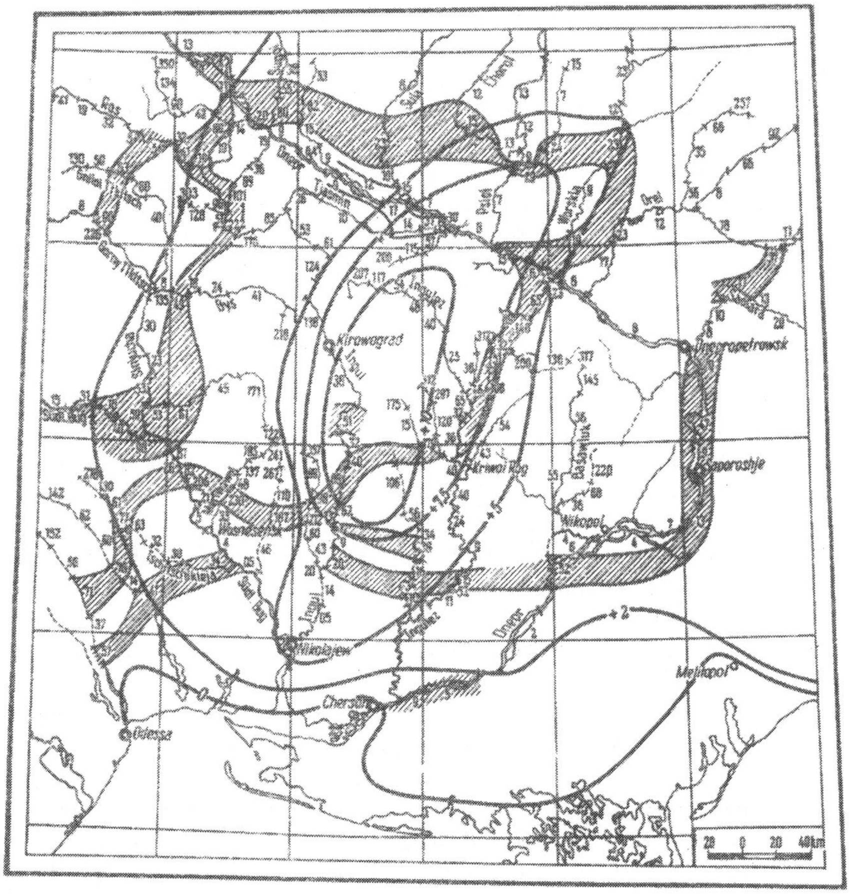

Abb. 2

Karte der Flußabschnitte mit abnormen Gefällen

1 - Größen der Gefälle (cm/km)
2 - Zonen der abnormen Gefälle
3 - Isolinien der Geschwindigkeiten von rezenten Erdkrustenbewegungen (mm/Jahr)

erste von ihnen stellt einen Ring dar, welcher hauptsächlich die Kriwoj-Rog-Erhebung einsäumt und teilweise sein Gewölbe schneidet. Die zweite hat die Form eines Halbringes, welcher auf einzelnen Abschnitten parallel der ersten Zone verläuft.

Das Vorhandensein von Abschnitten mit abnormen Flußgefällen kann durch eine Reihe von Ursachen bedingt sein: durch die örtlichen Bedingungen des Mesoreliefs, durch die Ungleichartigkeit der lithologisch-petrographischen Zusammensetzung des Flußbettes, durch die Besonderheiten des hydrologischen Flußregimes, durch die rezenten Erdkrustenbewegungen.

Die Analyse der Karte, der Vergleich ihrer mit den hypsometrischen und geologischen, mit den geologisch-geomorphologischen Querschnitten und mit anderem Material gestatten, die Natur von einzelnen Abschnitten der abnormen Flußgefälle festzustellen. Insbesondere wurde ein Zusammenhang zwischen den Deformationen der Längsprofile und den geologisch-strukturellen Bedingungen, wie auch mit der Lithologie des Gesteins festgestellt. So sind zum Beispiel die Anomaliender Gefälle für eine Reihe von Flüssen durch den Ausgang von Kristallgestein in die Täler bedingt (Südlicher Bug abwärts von der Stadt Wosnessensk, Tjasmin im Mündungsteil u.a.). Es sind abnorme Gefälle an der Grenze zwischen der Ukrainischen Kristalltafel und der Schwarzmeerküstenvertiefung längs der Flüsse Ingul, Gromokleja, Gniloi Elanez und einiger anderer fixiert worden. Die Ursache der Anomalien kann nicht völlig durch die geologisch-strukturellen Bedingungen, wie auch nicht durch die Zusammensetzung des das Flußbett bildenden Gesteins erklärt werden. Die Deformationen der Längsprofile von Flüssen finden auch in den Grenzen des Verbreitungsfeldes von gleichartigem Gestein und sogar von sedimentärem Gestein statt. Die Anomalien voll und ganz auf Rechnung des hydrologischen Faktors zu setzen, darf man auch nicht. Durch ein und dieselben Zonen von abnormen Gefällen sind Flüsse mit verschiedenen abfließenden Wassermengen und einem unterschiedlichen hydrologischen Regime umfaßt (Dnepr, Ingulez, Südl. Bug, Tschitschikleja).

Die Übereinstimmung der Ringzone von abnormen Gefällen mit dem Umriß der Kriwoj-Rog-Erhebung zeugt, unserer Ansicht nach, von der bedeutenden Rolle der rezenten Erdkrustenbewegungen. Diese Zone fällt hauptsächlich auf die Peripherieteile der Erhebung, als ob sie sie umsäumen würde, und erfaßt bloß teilweise sein Gewölbe, was sichtbar damit verbunden ist, daß die abnormen Gefälle zu den Stellen mit erhöhten Bewegungsgradienten der Erdkruste neigen. Gleichzeitig wird auch eine gewisse Nichtübereinstimmung in der Zonenlage beobachtet: die zentrale Ringlinie fällt nicht mit der Achse der maximalen Erhebung zusammen, sondern erweist sich ein wenig westlich verschoben. Im ganzen jedoch zeigte die Karte der Flußabschnitte mit abnormen Gefällen eine ziemlich gute Übereinstimmung mit den Angaben der wiederholten Nivellierung.

Bei der Untersuchung von tektonischen Bewegungen mit Hilfe der Analyse von Flußlängsprofilen stellt ein bestimmtes Interesse die Aussonderung von Zonen mit relativ großen und relativ geringen Gefällen dar. Auf die Größe der Gefälle übt jedoch, wie es schon vermerkt wurde, ein Komplex von Naturfaktoren einen Einfluß aus. Einer der Hauptfaktoren ist die Oberflächengestaltung. Wir haben versucht, den Einfluß des Flußbeckenreliefs auf die Form seines Flußlängenprofils auszuschließen und ein "normales Profil" des Flusses zu ermitteln, welches sich bei den gegebenen morphometrischen Daten des Beckens gestalten müßte. Der Vergleich solch eines Profils mit dem realen gestattet es, Abschnitte mit erhöhten und verringerten Gefällen, die nicht mit dem Flußbeckenrelief verbunden sind, zu finden. Zur Berechnung eines "normalen Profils" wurde die Formel von P.W. Iwanow ausgewertet, die die Gleichung einer Parabel n-ten Grades ist,

$$h = H \left(\frac{l}{L}\right)^n$$

welche (Parabel) durch den höchsten Punkt des Längsprofils verläuft, ihren Gipfel in seinem tiefsten Punkt hat, und deren Achse senkrecht nach oben durch den tiefsten Punkt gerichtet ist. Eine Parabel n-ten Grades ist eine theoretische Kurve, die ihren Quali-

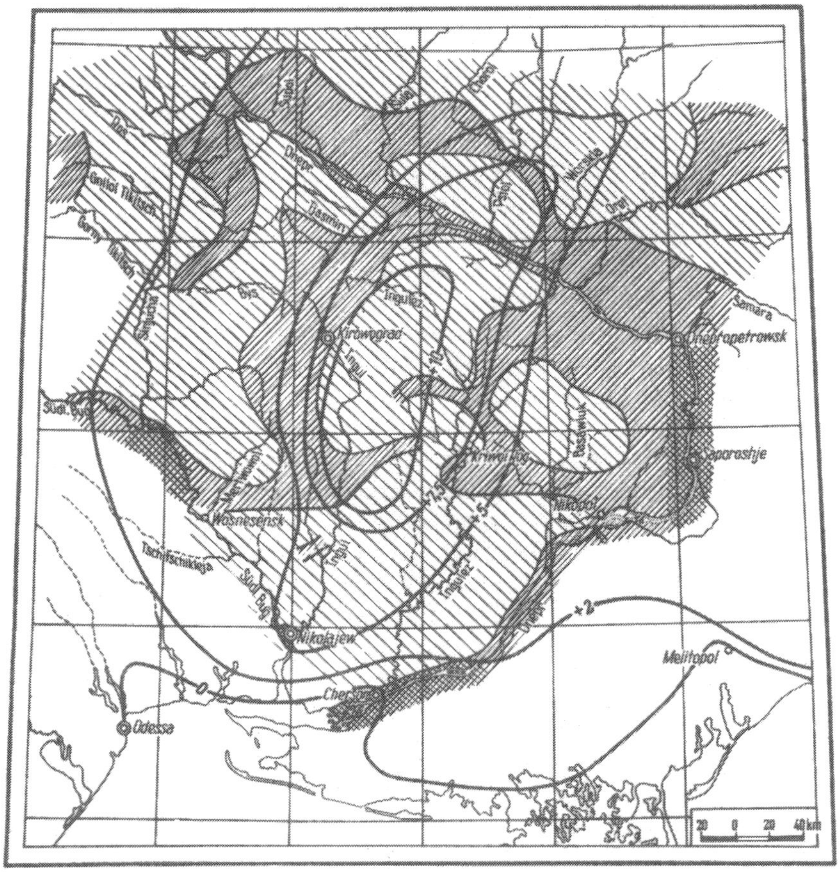

Abb. 3

Karte der Zonen mit verminderten und erhöhten Flußgefällen

1 - Zonen der erhöhten Flußgefälle (101 - 500 %)
2 - Zonen der erhöhten Flußgefälle (mehr als 500 %)
3 - Zonen der verminderten Flußgefälle (weniger als 100 %)
4 - Isolinien der Geschwindigkeiten von rezenten Erdkrustenbewegungen (mm/Jahr)

täten nach sich stark der Form von Flußlängsprofilen nähert. Die Glieder der Gleichung haben folgende Bedeutung: h = der Höhenüberschuß über dem Tiefpunkt des Profils, d.h. über der Mündung; H = der Höhenunterschied zwischen Quelle und Mündung; l = die Entfernung von der Mündung bis zum Punkt mit dem Höhenüberschuß h; L = die Gesamtlänge des Flusses; n = Index der Profilform, der dem Verhältnis $\frac{Q_b}{Q_n}$ gleicht, wo Q_b der über der Profillinie liegende Teil der Fläche des Rechtecks mit der Höhe H und der Basis L ist, in welches das Profil eingezeichnet ist. Q_n ist also entsprechend die Fläche, die unter der Profillinie liegt. Die Parabel n-ten Grades hat bei der angewiesenen Lage ihres Gipfels und ihrer Achse in ihrem Gipfel eine Neigung, die Null gleich ist; solch eine Neigung wird bloß an den Mündungen der Flüsse beobachtet, die in Seen oder Meere einmünden. Zur Berechnung der Profile von Nebenflüssen wurde die von P.W. Iwanow vorgeschlagene Gleichung einer abgeschrägten Parabel n-ten Grades verwendet

$$h = H_1 \left(\frac{l}{L}\right) + H_2 \left(\frac{l}{L}\right) \text{, wo } H_1 = L \times J_0$$

(J_0 = die Neigung im Flußmündungsteil; es wurde angenommen, daß sie der Neigung des Hauptflusses an der Mündungsstelle des Nebenflusses gleicht).

$$H_2 = H - H_1$$

Der Index der Längsprofilform n ist laut Angaben von P. W. Iwanow für Flüsse der Russischen Ebene im Durchschnitt gleich zwei. Eben diese Größe wurde von uns bei der Berechnung eines "normalen" Flußprofils eingesetzt. Nach diesen Formeln wurden die theoretischen Flußgefälle für die Abschnitte berechnet, für welche laut der topographischen Karten die tatsächlichen Gefälle bestimmt wurden. Danach wurde folgerichtig für jeden Fluß, Abschnitt nach Abschnitt, ein Vergleich der tatsächlichen und der nach den Formeln berechneten (theoretischen) Gefälle vorgenommen. Die theoretischen Gefälle wurden als 100 % angenommen, und, von ihnen ausgehend, wurden die tatsächlichen berechnet. Im Ergebnis wurden Bezirke mit einem Flußgefälle ausgesondert, welches das theoretische Gefälle übertrifft, und Bezirke mit einem Flußgefälle unterhalb des theoretischen (Abb. 3). Indem man nun diese beiden Karten miteinander vergleicht, kann man ersehen, daß die Lage der Zonen mit erhöhtem Gefälle in den allgemeinen Zügen eine Ähnlichkeit mit der Lage der abnormen Gefälle aufweist. Es tritt auch ein die Kriwoj Rog-Erhebung umreißender Ring zu Tage. Folglich neigen die Bezirke mit einem erhöhten Gefälle zu dem Randteilen der Erhebung. Das Gewölbe der Erhebung ist größtenteils ein Raum mit vermindertem Gefälle. Das Ermittlungsverfahren von Räumen mit erhöhtem und vermindertem Flußgefälle kann als zusätzliches bei der Untersuchung von rezenten Erdkrustenbewegungen herangezogen werden. Um dieses Verfahren endgültig verwendungsreif zu machen, muß noch eine Reihe von Forschungsarbeiten vorgenommen werden. Außer der Formel von P.W. Iwanow existiert eine Reihe anderer Formeln zur Berechnung des "normalen" Flußlängsprofils. Welche von diesen Formeln die vollkommenste und für die angewiesenen Ziele die passendste sei, ist eine Frage, die eine weitere Untersuchung erfordert.

L i t e r a t u r

Gerasimov, I.P.: Primenenie geomorfologičeskich metodov pri sejsmotektoničeskich issledovanijach. Tr. Geofiz. in-ta No. 25 (152). M. 1954

Ivanov, P.V.: Metod količestvennoj charakteristiki formy prodol'nogo profilja reki. Izv. Vses. geogr. ob-va, 1951, t. 83, vyp. 6

Kožurina, M.S.: Opyt issledovanija prodol'nych profilej nekotorych rek. Prikarpat'ja dlja celej tektoničeskogo analiza. Izv. Vses. geogr. ob-va 1956. t. 88, vyp. I

Meščerjakov, Ju. A. i Fil'kin, V.A.: Opyt sostavlenija karty padenij rek i ee tektoničeskoj interpretacii. V kn. "Voprosy primenenija kartografičeskich metodov pri geografičeskich issledovanijach". M., 1960

Sovremennye vertikal'nye dviženija zemnoj kory na territorii zapadnoj poloviny Evropejskoj časti SSSR. Pod obšč. red. I.P. Gerasimova i Ju. V. Filippova. Trudy Centr. naučn. - ussled. in-ta geodezii aeros'emki i kartogr. vyp. 123 M., 1958

D i s k u s s i o n

O. M i e h l k e (Warnemünde):

Wenn ich recht verstanden habe, wurden bei diesen Untersuchungen die Wasserspiegelgefälle - nicht die Sohlengefälle - der Flüsse untersucht. Nun hängt das Spiegelgefälle nicht zuletzt auch von der Abflußmenge und von der Rauhigkeit des Flußbettes ab. Das sind zwei Faktoren, die mehr und weniger stark wechseln können. Daraus resultiert meine Frage: "Wie wurden Veränderungen in den Abflußmengen - auf Grund verschiedener Niederschlagsmengen in dem Einzugsgebiet des Flusses - und Änderungen in den Rauhigkeiten des Flußbettes - auf Grund von Änderungen in der Bodenstruktur durch Auswaschungen und Sedimentationen - analysiert und berücksichtigt? Wurden beispielsweise Siebanalysen des Bodenmaterials des Flußbettes in den einzelnen Flußabschnitten gemacht und ihre zeitlichen Änderungen bestimmt und verfolgt?"

The study of recent tectonic movements of the earth's crust in the territory of the USSR

von

I. P. Gerasimov (Moskau)

The study of recent tectonic movements of the Earth's crust by means of geodetic, oceanographic, geophysical and geomorphological methods is one of the most urgent problems of today. This problem is remarkable in that it provides vast possibilities for the tempting idea to apply more or less objective, direct and quantitative methods of research to a field of theoretical knowledge where qualitative opinions, indirect approach and mainly subjective ideas prevail to this day. This possibility essentially broadens the scope of the classical actualistic approach to the study of important geological phenomena.

Besides, we hope to utilise this research to set up a reliable basis for a first-hand prognostication of seismic catastrophes or insignificant slow displacements of the Earth's crust which cause misalignment of capital buildings, gradual shift of the coastal outlines of lakes and seas, dislocation of river-beds and progressive bogging-up of localities or, conversely, decrease in the level of underground waters and other similar phenomena.

This broad variety of theoretical and practical prospects prompts us to develop this research in every way. At the same time, we must be very cautious in evaluating the results obtained from the work already accomplished. So far, these results are too much of a general character to be of any practical value. Nevertheless, this must not discourage us since the problem under consideration is very complicated both in its scientific content and the methodical features. All this is forcefully borne out by the experience of corresponding investigations conducted in the territory of the U.S.S.R.

Russian scientific literature provides abundant, although not infrequently contradictory, evidence indicating to considerable recent vertical displacements of separate sections of the Earth's crust. These tectonic displacements occurred both in the pre-historic and historic time. In addition to the numerous seismic dislocations of this kind observed in Crimea and the Caucasus, in the Tien-Shands, the Pamirs, and in the mountains of Southern Siberia, an important place in a series of such phenomena is held by the archaeological and historical displacements of the coastal outlines of the Black, Caspian and Aral seas, Lake Baikal and by certain other similar manifestations. The combination of these data has long ago enabled Soviet scientists to tackle the problem of recent tectonic movements of the Earth's crust in the territory of the U.S.S.R. although these data were clearly inadequate for the elaboration of appropriate scientific methods. In recent years, the activity of a small group of researchers who displayed a considerable interest in the development of scientific research has radically changed the situation.

An important landmark in the study of recent tectonic crustal movements in the territory of the U.S.S.R. was the publication in 1958 of the first monograph on this subject entitled "Recent Vertical Crustal Movements in the Territory of the Western Half of the European Part of the U.S.S.R." (Transactions of the Central Scientific Research Institute for Geodesy, Air Surveying and Cartography, issue 123). This work which was the result of the joint effort of a group of authors (A.V. Zhigavo, V.A. Zenin, L.G. Kamanin, J.A. Mescherikov, M.I. Sinyagina, Y.V. Filippov; edited by Y.V. Filippov and the author of this paper) outlined the results of extensive special research conducted with a view to systematising the materials of the levellings, tide-gauge observations and field engineering and geological inspection of the principal bench-marks and tide-gauges and the general geomorphological study of the repeated levelling areas.

This work resulted in the compilation of a catalogue illustrating the rates of recent vertical movements of the Earth's crust computed on the basis of 2,500 points of repeated levellings conducted from 1883 to 1953 on 122 levelling lines with a total length of 35,000 km as well as in a summary map showing the rates of recent tectonic movements compiled on the basis of the catalogue materials with the use of geomorphological data. In addition to the rates of recent tectonic movements of many individual points, this map shows two isometric lines of general upheavals and subsidences in conformity with the following scale:

U p l i f t s in mm/year: 1) over 10, 2) from 7.5 to 10, 3) from 5 to 7.5, 4) from 2 to 5, 5) from 0 to 2.

S u b s i d e n c e s in mm/year: 6) from 0 to 2, 7) over 2.

Below (see Fig. 1) we give a reduced copy of the 1958 map. This simple and illustrative map hardly needs, we hope, any special explanations.

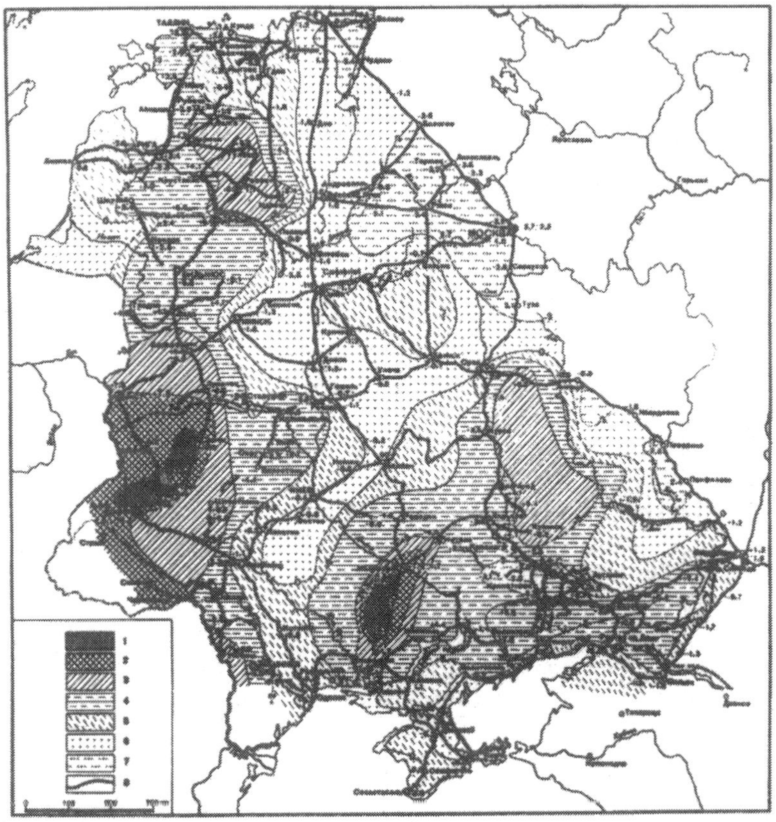

Fig. 1

Map of the rates of recent tectonic movements of the western half of the European part of the USSR
Uplifts in mm/year: 1) over 10, 2) from 7.5 to 10, 3) from 5 to 7.5, 4) from 2 to 5, 5) from 0 to 2
Subsidences in mm/year: 6) from 0 to 2, 7) over 2, 8) repeated levelling lines

The monograph published in 1958 was very important from a methodical point of view for the subsequent development of research into recent tectonic movements. First of all, it demonstrated an obvious advantage of the complex method of research which permits mutual control and correction of the data provided by various methods (geodetic, oceanographic and geomorphological). A good illustration of this point is provided by the study of the degree of the trustworthiness of the elevation marks of individual datum points which underwent accidental local displacements. Bench-marks of this type usually give on precise levelling curves sharp bends which disturb the smooth flow of the curve of the entire profile (see Fig. 2). Field inspection of such bench-marks (and tide-gauges) invariably testifies to a local subsidence, slide or any other similar disturbance.

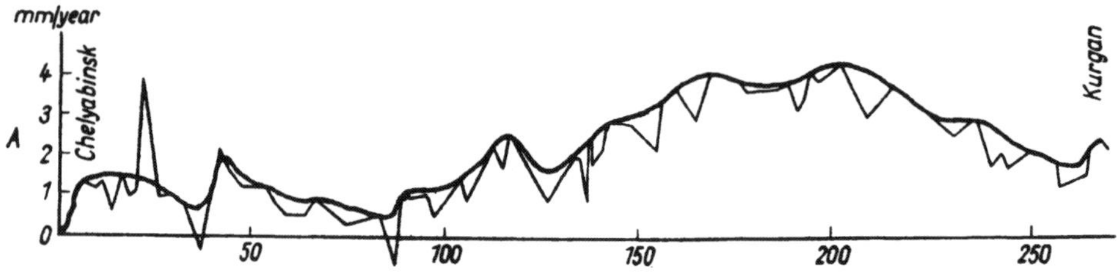

Fig. 2

Graph of the rates of recent tectonic movements in Chelyabinsk-Kurgan area.
1) Curve of the rates of recent tectonic movements plotted according to geodetic data.
2) Curve of the rates of recent tectonic movements corrected (aligned) according to the data of the field inspection of the bench-marks.

It should be noted, however, that alongside this positive evidence of the value of the complex approach, the monograph also reveals an insufficient versatility of the methods used to study recent tectonic movements. In this connection, an intensive research has been carried out in the methodical sphere in recent years which has yielded certain positive results.

An obvious advantage of the method of network investigations (in closed-type levelling grounds) has been established for geodetic research over the study of only separate levelling profiles. The above work has shown that the analysis of the network of the levelling grounds can be used to verify the data by the disagreements in the system of levelling grounds and to determine the most probable final indices on the basis of numerous excessive determinations. Of course, the method of closed-type levelling grounds cannot unfortunately be used for the computation of the rates of tectonic movements in every case because we often lack the appropriate initial data.

Three years after the map of recent tectonic movements was published it was verified by means of geodetic methods. We have pointed out elsewhere that this map was compiled with the help of the lines of repeated levelling conducted before 1953. But in the course of 1953 - 1959, a number of new repeated levelling lines were laid in the west of the European part of the U.S.S.R. totalling about 7,000 km. V. A. M a t s k o v a made use of this new material to publish in 1961 her work entitles "Verification of the Map of the Rates of Recent Vertical Crustal Movements in the Territory of the Western Half of the European Part of the U.S.S.R." It should be mentioned at this juncture that the new research conducted in 1959 - 1960 made it possible not only to supplement and verify the materials of 1958 but also essentially increase the accuracy of the respective data by excluding the most doubtful data and reducing the gaps in some of the levelling grounds. According to V.A. Matskova, the possible errors in the determination of the absolute intensity of vertical movements at more than 3,500 points equalled $\pm$ 0.7 to $\pm$ 1.1 mm/year instead of the former ± 1.5 to ± 2.3 mm/year.

But, the main result yielded by this verification was that the pattern of movements obtained formerly was confirmed in all its aspects. According to the new geodetic data "the general character of the distribution of the rates of recent tectonic movements -- the subsidence of the north-eastern part, the uplift of the western and the south-western parts and a relatively stable position of the central part in the territory under study -- remained the same as on the previous map. The rates on the new, more accurate map varied from - 6.3 mm/year to + 14.6 mm/year." (V.A. Matskova, 1961, p. 65), i.e., in principle they were the same as before. From this point of view, of special interest was the comparison of the second levelling data which enabled us to elucidate (in part) the problem of the degree of the constancy of the tendency of the recent tectonic movement development. Some of the data borrowed from Matskova's work are represented in the graph below (Fig. 3 shows the Ternopol-Kursk profile as established by three levellings with the time interval from 15 to 25 years).

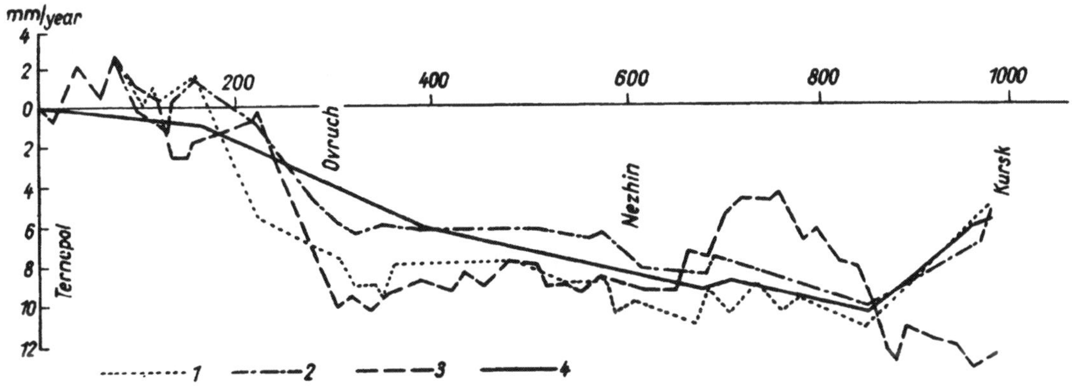

Fig. 3

Relative intensities of recent tectonic movements in Ternopol-Kursk area obtained from the comparison of the results of levellings conducted in various years.

1) by I and II levellings (interval 15 years);
2) by I and III levellings (interval 25 years);
3) by II and III levellings (interval 8 years);
4) by I and III levellings (equalised rates of movements).

Conferences on the study of recent tectonic movements in the territory of the U.S.S.R. held regularly throughout the recent years testify to the growing interest in this problem. In the course of the last five years three such conferences were arranged in the U.S.S.R. (in 1958, 1959 and 1961). These conferences held in Moscow were organised on the initiative of the Geographical Institute of the Academy of Sciences of the U.S.S.R. and other research bodies. The sessions of the first conference were attended only by 50 specialists and eight papers were presented and discussed. At the second conference which took place in the following year the number of participants increased to 70 and the number of papers to 22. The third conference arranged two years later was attended by 300 scientists and fifty-five papers were read at it. Although these figures are primarily of a purely quantitative character, they nevertheless convincingly demonstrate the ever-growing interest shown for this problem in the U.S.S.R.

The analysis of the work of these conferences from the view-point of their scientific results has a much greater importance for the present communication. However, it is hardly possible and necessary to give a detailed account of these problems in this paper but a brief summary review may be of some interest.

The interdepartamental conferences have revealed a rapid growth in the regional investigations conducted in the territory of the U.S.S.R. for the purpose of studying recent tectonic movements. In addition to the geodetic work done to verify and supplement the monograph and the first summary map of the west of the European part of the U.S.S.R. with fresh data, all the conferences dealt with the results of the complex research carried out in the same sphere during the last few years in a number of districts of the western section of the European part of the U.S.S.R. (in the Volga area, in the North Caucasus, in the Urals area and the Baltic area); the first profile investigations in the Western Siberia and Kazakhstan, the first geomorphological research into the problem of recent tectonic movements around the Caucasus; the programme, plans and preliminary considerations on the conductance of respective research in Siberia (Lake Baikal), the Ukraine, in the mountainous districts of the Central Asia, etc. Thus, the study of recent tectonic movements in the territory of the U.S.S.R. has clearly acquired in latest year a broad regional scope which is so much needed for the comparison of territories with a different morphological structure.

The conferences have likewise demonstrated an obvious and rapid development of prospecting methodical work, especially in the field of geomorphology. From this point of view, special attention was focussed at all the three conferences on how to utilise the methods of study and mapping of the thickness and the facies of flood-plain alluvial deposits as the indicators of the neotectonic and recent tectonic movements; on regular investigations of the nature of longitudinal profiles of the present river-beds and a series of river terraces of a local or cyclic importance, with the use of field and mapping methods;on the study of tectonic distortions of lake and sea terraces, on the analysis of the connection of the development of active gullies with districts of recent tectonic uplift and subsidence (this investigation for the Volga area upland gave a negative result); on a comparative study of the old and new surface contour maps to determine the rearrangements in the drainage system which were caused by recent tectonic movements, etc., etc.

It goes without saying that the search for new geomorphological methods for the study of recent tectonic movements can in no way be considered complete. Quite the reverse, despite a number of undoubted methodical successes (for example, as regards the facies and the thickness of alluvium, distortions of the present river beds, the study of terrace spectra, etc.) only the first steps have been taken in this field. What we need is a further intensive and active search and daring efforts, especially in the field of quantitative objective methods for the study of various specific features of the present relief and their connection with the recent tectonic movements.

A great importance was attached at the conferences to the geophysical (gravimetric and other) methods used for the study of recent tectonic movements. Already at the second conference held in 1959 we discussed the first experience in comparing the curves of the rates of recent tectonic movements with the curves of the anomalies of the gravity force which allowed us to mark out, but not yet explain, at least four types of corresponding relationships (direct, inverse, non-gradient and indefinite). At the third conference the main attention was drawn to the possibility of using the dip-gauge method for the study of tectonic movements. To all appearance, this method allows tectonic displacements to be studied in incomparably shorter periods (several months) as compared with all other methods.

The last conference also concentrated on the utilisation of the triangulation method for the study of recent horizontal displacement of crustal sections. The problem of recent horizontal drift of the continents was also taken up for discussion. It became clear, however, that the accuracy of the conventional present-day astronomical longitudinal determinations proved, as before, insufficient.

Much attention was focussed at the conferences on the necessity to set up for the study of recent tectonic movements special grounds where periodic geodetic, geophysical and geomorphological observations and research could be made according to a definite programme. The first such ground was created in the area of Lake Baskunchak. Here, in 1951, on the initiative of the Geographical Institute of the Academy of Sciences of the U.S.S.R. and the Moscow Institute of Engineers of Geodesy, Air Surveying and Cartography the first geodetic and geomorphological ground was set up for the study of recent tectonic movements. Immediately after this, we performed the first levelling which was repeated seven years later (in 1953). The map attached to the work of A. Pevnev which will be presented at this conference illustrates the results of these investigations. The map shows that the salt-dome upland of Mt. Bolshoye Bogdo and the adjoining depression of Lake Baskunchak experience a permanent vertical displacement which in the region of the lake is represented by subsidence of 5 - 12 mm (i.e., by 1 mm/yr) and on the upland by an uplift of 34 - 58 mm (0.6 mm/year).

It stands to reason that the study of recent tectonic movements can now in no way be limited to only one levelling area of a Baskunchak type. For this reason, the conference of 1961 passed a decision to establish in the shortest time possible a new experimental geodetic, geophysical and geomorphological levelling area in Crimea and to recommend to consider the question on the creation of a network of similar grounds in other, most interesting and important districts (particularly, in seismically active districts such as the Caucasus, the area of Lake Baikal, the Central Asia, etc.).

The general review of these investigations poses the question as to what scientific conclusions on the nature of recent tectonic movements in the territory of the U.S.S.R. can be at present regarded as more or less proved on the basis of the investigations carried out and what is as yet not quite clear, disputable and unknown in this complicated problem. As we have pointed out above, it is far from easy to answer this questions in view of a great complexity of the scientific content of this problem and an inadequate elaboration of the methods of its study. In its most general form, the following answer can be tentatively suggested:

1. The investigations of recent tectonic movements carried out in the territory of the U.S.S.R. show that these movements occur e v e r y w h e r e and c o n t i n u a l l y. There is every ground to believe that, figuratively speaking, our hard and seemingly stable Earth's surface appears to be in constant movement; it always and everywhere experiences differentiated undulating variations resembling long secular inhalations in one sections and exhalations in the others.

2. The absolute intensities of recent tectonic movements in various sections of the Earth's crust are undoubtedly different; their mean annual values come to whole millimetres and it is only on some areas with a catastrophic rhythm that they reach annually tens of millimetres. It is extremely important to achieve a coincidence in principle of the general order of these values provided by different methods (geodetic, oceanographic, historical, archaeological, etc.) and for various countries (primarily, for a number of European countries, Japan, the U.S.A., etc.). This will be a sufficient proof of their trustworthiness.

3. We observe a definite, although far from complete (i.e., with many deviations), connection of the signs of recent tectonic movements and their distribution in space, firstly, with large present-day orographic elements of dry-land by the uplift-upland and subsidence-lowland pattern and, secondly, with the principal geological structures (especially neotectonic) by the uplift-anticline of horst and subsidence-syncline or fault trough pattern. This gives us good reason to assume an inheritance over many years of definite tendencies of the geo-tectonic development of separate crustal sections which by no means precludes the necessity to acknowledge the periodic or fluctuating (in time) character of tectonic movements.

4. For all that, it is the coincidence of these two most general regularities in the dynamics of recent tectonic movements (their cyclic character and inheritance) that makes the problem vague and needing further study. Its further elaboration should obviously proceed along the lines of elucidating the following two points:

a) the rhythm and cyclic character of recent tectonic crustal movements;

b) spatial localisation of recent tectonic movements of different sign.

5) As before, we are not as yet certain as to the concrete causes of recent tectonic movements of the Earth's crust, i.e., the geophysical and geochemical essence of the deep-going processes in the Earth's crust of which these movements are the external outcome. Scanty knowledge of this most important aspect of the problem makes it impossible to give theoretically a scientifically-substatiated characterisation of different genetic types of recent tectonic movements, i.e., their classification.

It is quite clear that the way towards the solution of this most cardinal problem should be sought in an all round development of complex methods of research and the accumulation of more factual material than is available at present.

References

Recent vertical crustal movements in the territory of the western half of the European part of the USSR, edited by I.P. Gerasimov and Y.V. Filippov. Transactions of the Central Scientific Research Institute for Geodesy, Air Surveying and Cartography. Issue 123, Moscow, 1958

Recent tectonic movements of the earth's crust and the methods of their study. Academy of Sciences of the U.S.S.R., Moscow, 1961

L.E. Setunskaya: "Conference on the Problems of Recent Tectonic Movements of the Earth's Crust." Proceedings of the Academy of Sciences of the USSR, Geographical series, Nr. 3, 1958

L.E. Setunskaya: "The Second Interdepartmental Conference on the Problem of Recent Tectonic Movements in the Territory of the U.S.S.R." Proceedings of the Academy of the U.S.S.R., Geographical series, Nr. 3, 1959

L.E. Setunskaya, E.A. Finko: "The Third Interdepartmental Conference on the Study of Recent Tectonic Movements of the Earth's Crust." Proceedings of the Academy of Sciences of the U.S.S.R., Geographical series, Nr. 2, 1962

Study of recent crustal movements in the region of repeated precise levellings in the Ternopol-Kursk area

von

S. K. Gorelov, Z. I. Martynova
V. A. Matskova (Moskau)

Summary

The article reviews the results of integrated research into the geodetic, cartographic and geologo-geomorphological data obtained from triple precise levelling in the Ternopol-Kursk area.

A comparative analysis of the precise levellings, the changes in the drainage gradient (during the last 15-20 years) and the geological structure of young (Holocene - Upper Pleistocene) terraces shows that these data are in good agreement. The discrepancies in these data are attributed mainly to the changes in the sign of tectonic movements during the Holocene-Upper-Pleistocene period.

The recent tectonic movements are not always directly and intimately connected with the ancient geological structures, which fact can be accounted for by the complexity and irreversibility of the process of the tectonic development of the territory.

Zusammenfassung

Der Artikel behandelt die Ergebnisse der vervollständigten Untersuchungen bezüglich der geodätischen, kartographischen und geologisch-geomorphologischen Daten, die von einem dreifachen Präzisionsnivellement im Ternopol-Kursker Gebiet erhalten wurden.

Eine vergleichende Analyse der Präzisionsnivellements, der Änderungen im Drainage-Gradienten (während der letzten 15-20 Jahre) und der geologischen Struktur der jungen (Holozän- Oberes Pleistozän) Terrassen zeigen, daß die Werte gut übereinstimmen. Die Unterschiede in diesen Daten sind hauptsächlich durch die Änderung in den Bewegungsrichtungen während der Holozän- Obere Pleistozän-Periode bedingt.

Die gegenwärtigen tektonischen Bewegungen sind nicht immer direkt und zwangsläufig mit den alten geologischen Strukturen verknüpft, was durch die Komplexität und Irreversibilität des Prozesses der tektonischen Entwicklung der Territorien erklärt werden kann.

1. Tasks and Methods of Research

The area we undertook to investigate presents a considerable interest from the point of view of recent tectonic movements in a platform district. The importance of the study is enhanced due to the triple precise levelling which allows these data to be utilised for determining the sign and gradients of the rates of recent crustal movements.

The materials provided by repeated levelling were supplemented with the data of a comparative analysis of large-scale accurate maps of various years and also with the data of the study of the geological-geomorphological features of the territory. This predetermined a complex approach to the problem and essentially helped towards a successful solution of the task.

Levellings in the Ternopol-Kursk area were conducted in 1933, 1948 and 1956. The field geological inspection showed that most bench-marks were in favourable geological and geomorphological conditions and well preserved.

Only individual bench-marks were displaced under the action of external (exogenetic) factors. On the graph of the curves of the rates of recent movements (Fig. 1) such

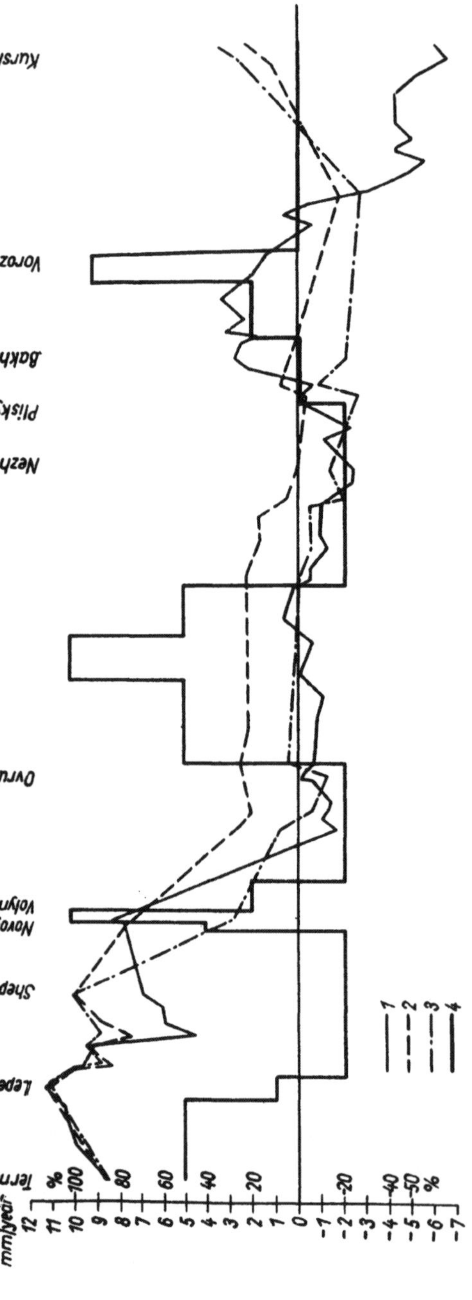

Fig. 1

Graphs of the rates of recent vertical movements and the changes in the river gradients from 1942 to 1956 in the triple levelling area Ternopol-Kursk

1 - levelling of 1948 and 1956 (interval 8 years);

2 - ditto, of 1933 and 1956 (interval 23 years);

3 - ditto, of 1933 and 1948 (interval 15 years);

4 - graph of the changes in the river gradient (during the period of 1942 - 1956 in %%)

sections exhibit extremely sharp changes (peaks) of the curves. However, as can be seen in Fig. 1, these changes in no way affect the characteristic smooth shape of the curves.

The curves of the rates of recent movements (Fig. 1) were computed in reference to the bench-marks in Ternopol and equalised for the system of the levelling area in the west of the European part of the U.S.S.R. (curves 1, 2 and 3 on the graph).

Analysis of the maps consisted mainly in studying the changes of river gradients, since, these data which typify the trend and values of the changes in the present-day river-bed erosion are excellently comparable with the data provided by repeated levellings. They are also comparable in time as was shown by 1942 and 1956 edition maps used for the analysis.

This work was a certain methodological novelty. The changes in the drainage system gradients were computed on large-scale maps by the formula

$$N = \frac{K_2 - K_1}{K_1};$$

where

$$K_1 = \frac{i_1}{l_1}; \quad K_2 = \frac{i_2}{l_2}$$

N -- increase or decrease in river drainage gradients on a definite section of the river in %% to 1942;

K_1 -- gradient of the section on 1942 edition maps;

K_2 -- gradient of the section on 1956 edition maps;

i_1 -- difference in waterline on a definite section of the river on 1942 edition maps;

i_2 -- difference in waterline on a definite section of the river on 1956 edition maps;

l_1 -- length of the river section on 1942 edition maps;

l_2 -- length of the river section on 1956 edition maps.

These investigations helped to compile a map showing the changes in the river drainage gradients in the region for the last fourteen years (Fig. 2). These materials also served as the basis for compiling a graph of the changes in the drainage system gradients (curve 4 in Fig. 1).

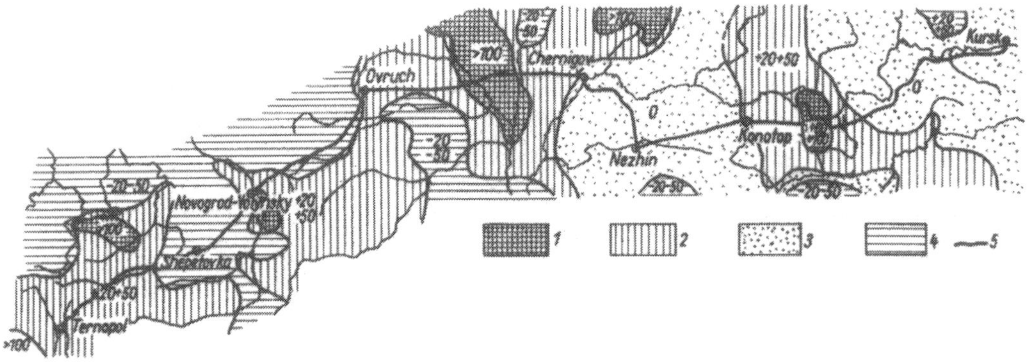

Fig. 2

Sketch map of the changes in the drainage system gradient during the period of 1942–1956.

1 - districts of increased river gradient by 100 % and more;
2 - districts of increased river gradient by 20 - 50%;
3 - districts of increased river gradient of stable position of river gradient;
4 - districts of increased river gradient of decrease in river gradient by 20 - 50%;
5 - direction of repeated levellings

The geological-geomorphological research consisted in studying flood-plain terraces and low terraces above the flood plain which date in this district from Holocene and Upper Pleistocene (Bondarchuk, 1949; Zamory, 1954, et al.). We studied the relative heights of the terraces, the specific features of the morphology of their surface and the composition and mode of occurrence of alluvial deposits. These materials were utilised for compiling sketch maps of alluvium lithofacies on the flood-plain terraces and low terraces above the flood plain (Fig. 3 and 4) which characterise the main trends in the development of the drainage system in Holocene and Upper Pleistocene.

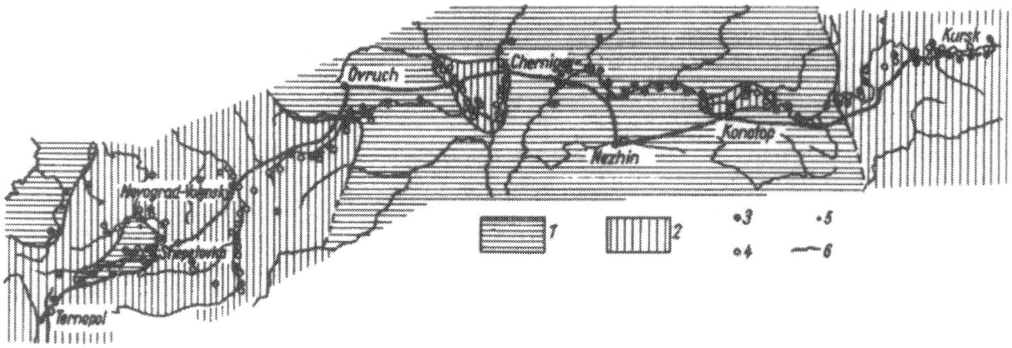

Fig. 3

Sketch map of facies of flood-plain alluvium

1 - districts of flood-plain facies of alluvium,
2 - ditto, of river-bed facies of alluvium,
3 - sections of flood-plain deposits,
4 - ditto, of river-bed deposits,
5 - ditto, of bayou deposits,
6 - direction of repeated levellings

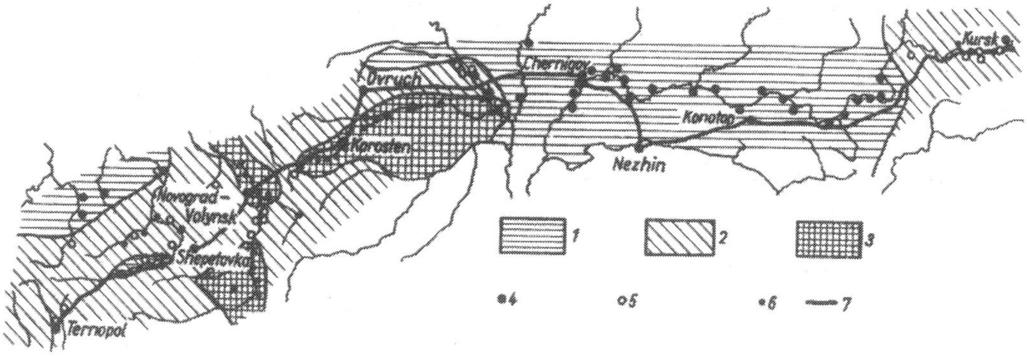

Fig. 4

Sketch map of the facies of alluvium of the first terrace above the flood plain.

1 - districts of flood-plain facies of alluvium;

2 - ditto, of river-bed facies;

3 - districts of denudation and compound terraces;

4 - sections of flood-plain deposits;

5 - ditto, of river-bed deposits;

6 - ditto, of compound terraces;

7 - direction of repeated levellings

2. Principal Results of the Research

The data of the triple levellings indicated to a general recent distortion of the region in the direction from Ternopol to Kursk. The amount of this distortion between the levellings of 1933 and 1956 was estimated at 10-12 mm/year.

Against the background of this distortion, which confirms the general laws governing recent crustal movements, there occur individual upheavals and subsidences of the Earth's crust. The maximum rates of recent movements (upheavals) in individual regions reach +12 mm/year. For local uplifts the following sections can be pointed out: Ternopol-Novograd Volynsky, Bakhmach-Vorozhba (mainly by the data of the 1948-1956 levellings), Kursk area (by the data of levellings of 1948-1933 and 1933-1956). Recent subsidences were recorded in an extensive area between Novograd-Volynsky and Makhmach, on the section between Isyaslavl and Lepesovka. Most sections of subsidence are noted for their relative character (in reference to the bench-mark in Ternopol). Comparatively small territories near Nezhin-Bakhmach and Vorozhba-Dyakonovo were registered as sections of absolute subsidence of the Earth's crust.

Changes in sign detected in comparing the levellings of different years constitute a characteristic feature of recent movements. These changes are especially prominent in Kursk area (according to the levellings of 1948 and 1933 and 1933-1956) and near Shepetovka (see Fig. 1).

Analysis of the factual material shows that the changes in the sign of recent movements (except for Kursk area) cannot be attributed to the effect of external conditions (subsidence of soil, erosion of surface, etc.) and apparently reflect the tectonic features of the territory formation. As we shall see further on, this is also borne out by the comparative analysis of repeated levellings and the results of the study of Holocene flood-plain terraces.

It is difficult on the basis of the data considered here to estimate the full period during which the movement sign changed. If we proceed from the comparison of the repeated levellings and the results of the study of Holocene terraces (compare Figs. 1 and 3) this period can be estimated at several thousand years (see also Gorelov, 1961).

A comparative analysis of the changes in the drainage system gradients and repeated levellings shows, on the whole, a good agreement of these data. This will be especially clear if we compare the curves of the changes in the rates of recent movements and the changes in the drainage system gradients (see.Fig. 1).

Figure 1 shows that the connection between these data is proved by the fact that the regions of recent upheaval (according to the material of repeated levellings) correspond to the sections of an intensive increase in the river gradients during the last 15-20 years by 20-50%, 100% and more (basins of the River Pripyat, Bakhmach-Vorozhba area, etc.) while the regions of recent subsidence correspond to the sections of decrease in the drainage system gradients, not over 50% (areas of Nezhin-Bakmach, Novograd Volynsky-Ovruch, Lepesovka-Shepetovka, etc.).

A good agreement has been registered on the whole also between the data of repeated levellings, the changes in the drainage system gradients and the results of the study of flood-plain terraces. Here, the direct connection is corroborated by the fact that, as a rule, in the districts of recent upheaval of the Earth's crust we observe the development of stepped flood plains with a good draining surface and a predominance of weakly graded river-bed deposits in alluvium; regions of recent subsidence have for the most part uniform level flood plains with bogged-up surfaces, extensive development of peat bogs and aleurite clayey composition of the alluvial deposits (compare Figs. 1 and 3).

On the basis of experience gained in other regions of the U.S.S.R. (see Coll. "Recent Tectonic Crustal Movements and Methods of Their Study", 1961), the features of flood-plain alluvium pointed out above can be interpreted as manifestations of a young tectonic upheaval and subsidence of the territory.

Mention should also be made of certain disagreement between these data. For example, the recent subsidence of the Earth's crust in Kursk area (according to 1948-1956 levellings) was registered in the regions with a predominant development of coarse-grained "river-bed" lithofacies of flood-palin alluvium (see Figs. 1 and 3) which indicates a tendency towards the uplift of the territory in Holocene.

The analysis of the factual material makes it possible to reveal two main causes explaining the disagreement between the materials provided by repeated levellings and the river-bed study: a) changes in the sign of movement over the period of Holocene (with an all-round account taken of the non-tectonic factors which cause the displacement of bench-marks; Shepetovka area, etc.); b) disagreements due to the scanty data provided by repeated levellings during the period from 1933 to 1948 (Kursk area).

As applied to the geological-geomorphological features of the first terrace above the flood plain the data of repeated levellings and the changes in the river gradients show a less close relationship which can be logically explained by the fact that these data are not comparable in time and by the fluctuating nature of the neotectonic movements of the Earth's crust (Nikolayev, 1949). Good agreement between geodetic and mapping data and the results of the study of the first terraces above the flood plains which characterise, respectively, the recent and upper Pleistocene periods of the tectonic development of the territory was obtained for the following districts: Ternopol-Lepesovka, Shepetovka-Novograd Volynsky, Nezhin-Bakhmach, Vorozhba-Dyakonovo (compare Figs. 1 and 4).

The ancient geological structures in the levelling area are expressed only in a most general form in recent and Holocene-Upper Pleistocene movements of the Earth's crust. These data reflect only some of the principal structural elements of the district - the Ukrainian crystalline massive, Dnieper-Donets depression, Voronezh anticline, etc.

Conclusions

1. The study of the materials of triple precise levellings and the mapping data of the Ternopol-Kursk area show, generally, similar tendencies of the recent tectonic developments of separate regions.

2. Inherited nature of recent crustal movements has been noted in relation to the movements during the Holocene and Upper Pleistocene.

3. A comparative analysis of geodetic, mapping and geologo-geomorphological data allows the conclusion to be made on the changes in the sign of movements of separate regions during Holocene and Upper Pleistocene.

References

Bondarchuk, V.G.: Geomorphology of the Ukrainian SSR, Kiev, 1949

Goryelov, S.K.: Young and Recent Tectonic Movements of the South-East of the Russion Plain, Coll. "Recent Tectonic Crustal Movements and Methods of Their Study", 1961.

Zamory, P.K.: Quaternary Deposits in the Ukrainian SSR, Kiev, 1954

Coll. "Recent Tectonic Crustal Movements and Methods of Their Study", Moscow, 1961

Nikolayev, N.I.: Neotectonics of the USSR Moscow-Leningrad, publication of the Academy of the Academy of Sciences of the USSR.

Recent vertical movements of the earth's crust in the Volga-Ural region and methods of their study

von

A. P. Rozhdestvensky, Y. E. Zhurenko (Ufa)

Summary

The purpose of researches undertaken by the authors of this paper was: 1) to give a geological interpretation of geodetic data on the velocities of recent tectonic movements along the line of relevelling, which crosses the Volga-Ural area from the west eastwards following the Bezenchuk-Ufa-Cheliabinsk railway line and 2) to work out methods of a qualitative areal study of recent tectonic movements for territories with an extremely scattered network of precision levelling.

Main conclusions:

1. An analysis of geodetic data along the line mentioned at the stretch Bezenchuk-Ufa-Urzhumka permitted to establish a general development of sufficiently intense recent vertical movements of the Earth's crust within the eastern part of the Russian platform, in the Sub-Uralian depression and the folded area of Southern Urals. All these fundamental geostructural elements of the Volga-Ural area are involved into a general recent tectonic uplift as single consolidated segments of the lithosphere. At the same time there is a definite differentiation of recent movements in individual large morphostructures. Maximum velocities of recent movements on a platform come to +5 mm/year and in Southern Urals -- about +7 mm/year.

2. The velocity chart of recent vertical movements along the line of relevelling indicates a sufficiently full accord with the topographic profile of recent relief and the morphology of the structure of relief-forming rocks (morphostructures) crossed by the levelling line. Especially distinctly such relations are seen on positive and negative morphostructures of the 1 order (Shkapovo-Romashkino arch, Kama-Belaia depression, transition to the folded area of Southern Urals near the Ashinsk overthrust etc.). A similar dependence is established also for latest tectonic movements embracing the Neogene and Quaternary periods.

3. Relations between the curve of recent movements and the character of structural forms of the lower horizons of the sedimentary mantle and the relief of the crystalline basement makes us to suggest that these movements are determined not only by the shoves of the basement but also by some other causes. Among the latter an important role belongs, apparently, to the transformation of tangential stresses originating in the sedimentary mantle in the process of a subsidence of big segments of the lithosphere, as well as to salt tectonics.

4. Quite reliable methods of a qualitative evaluation of recent vertical movements are 1) study of changes in the sinuosity of rivers during a definite period of time according to heterochronous cartographic material and 2) an analysis of the grades of longitudinal river profiles. The application of these methods in an area adjoining the line relevelling from the south and north on the stretch Abdulino-Ufa-Asha and the comparison of the obtained results with relevelling data permitted to compile a

schematic map of recent tectonic movements of the territory studied. The map outlines areas: a) of the most intense recent uplifts, b) of less intense uplifts, c) of local clearly expressed subsidences and d) of poorly defined relative subsidences.

This method can be recommended for a qualitative study of recent tectonic movements on territories not covered by lines of precise levelling.

Zusammenfassung

Mit den von den Autoren unternommenen Forschungen wurden folgende Zwecke verfolgt: 1) Die geodätischen Daten über die Geschwindigkeiten rezenter tektonischer Bewegungen entlang einer Linie von Wiederholungsnivellements geologisch zu interpretieren. Diese Linie kreuzt das Wolga-Ural-Gebiet von Westen nach Osten und folgt der Eisenbahnlinie Besentschuk-Ufa-Cheliabinsk. 2) Methoden auszuarbeiten für ein qualitatives Flächenstudium rezenter tektonischer Bewegungen in Gebieten mit einem extrem dichten Netz von Präzisionsnivellements.

Hauptschlußfolgerungen:

1. Eine Analyse der geodätischen Daten entlang der erwähnten Linie an der Strecke Besentschuk-Ufa-Urshumka erlaubte, die allgemeine Entwicklung ausreichend intensiver rezenter Vertikalbewegungen der Erdkruste innerhalb des örtlichen Teils der russischen Plattform, in der suburalischen Senke und in dem gefalteten Gebiet des Südurals festzulegen. Alle diese fundamentalen geostrukturellen Elemente des Wolga-Ural-Gebietes sind einbezogen in eine allgemeine rezente tektonische Hebung als einzelne verfestigte Abschnitte der Lithosphäre. Zur selben Zeit gibt es eine bestimmte Differenzierung der rezenten Bewegungen in individuelle Morphostrukturen. Maximalgeschwindigkeiten rezenter Bewegungen auf einer Plattform gehen bis zu +5mm/Jahr und im Südural ungefähr +7 mm/Jahr.

2. Die Geschwindigkeitskarte rezenter Vertikalbewegungen entlang der Linie von Wiederholungsnivellements zeigt eine ausreichend gute Übereinstimmung mit dem topographischen Profil des rezenten Reliefs und der Morphologie der Struktur reliefbildender Gesteine (Morphostrukturen), die von der Nivellementslinie gekreuzt werden. Besonders deutlich sind solche Beziehungen auf positiven und negativen Morphostrukturen 1. Ordnung zu sehen (Shkapovo-Romashkino-Bogen, Kama-Belaja-Senke, Übergang zu dem gefalteten Gebiet des Südurals in der Nähe der Ashinsk-Überschiebung usw.). Eine ähnliche Abhängigkeit wird auch für die spätesten tektonischen Bewegungen, die neogene und quartäre Perioden umfassen, festgestellt.

3. Beziehungen zwischen der Kurve der rezenten Bewegungen und dem Charakter der Strukturformen der unteren Horizonte des Sedimentmantels sowie dem Relief des kristallinen Untergrundes lassen uns annehmen, daß diese Bewegungen nicht nur durch den Schub des Untergrundes, sondern auch durch einige andere Ursachen bestimmt werden. Unter den letzteren kommt wahrscheinlich der Umwandlung der Tangentialspannungen, die in dem Sedimentmantel bei dem Prozeß des Absenkens großer Abschnitte der Lithosphäre entstehen, wie auch der Salztektonik eine bedeutende Rolle zu.

4. Ganz brauchbare Methoden einer qualitativen Bestimmung rezenter Vertikalbewegungen sind 1) Die Untersuchung der Änderung des sinusförmigen Verlaufs von Flüssen während einer bestimmten Zeitperiode entsprechend dem kartographischen Material aus verschiedenen Zeiten. 2) Eine Analyse der Stufen longitudinaler Flußprofile. Die Anwendung dieser Methoden auf ein Gebiet, das an die Linie der Wiederholungsnivellements vom Süden und Norden der Strecke Abdulino-Ufa-Ascha angrenzt, und der Vergleich der erhaltenen Resultate mit den Daten des Wiederholungsnivellements erlaubten, eine Karte der rezenten tektonischen Bewegungen für das untersuchte Gebiet zusammenzustellen.

Die Karte zeigt Gebiete: a) der intensivsten rezenten Hebungen, b) weniger intensiver Hebungen, c) örtlich deutlich sichtbarer Senkungen und d) nur wenig bestimmter relativer Senkungen.

Diese Methode kann empfohlen werden für eine qualitative Untersuchung rezenter tektonischer Bewegungen auf Gebieten, die noch nicht mit Präzisionsnivellementslinien überzogen sind.

There is one line of high-precision relevelling in the Volga-Ural region, which is suitable for the study of recent vertical tectonic movements. It crosses the central part of this area from the west eastwards and goes along the Bezenchuk-Ufa-Cheliabinsk railway line. First levellings here were done in 1887, 1904-1911 and the next ones in 1927-1943.

The purpose of researches by the authors of this paper was to: 1) give a geological interpretation of geodetic data on the recent movements in the lithosphere along the mentioned line of relevelling and 2) to work out a method of a qualitative areal evaluation of recent tectonic movements in the territories of an extremely scattered network of precise levelling.

1. To solve the first task researches have been effected on the mentioned levelling line along the stretch Bezenchuk-Ufa-Urzhumka. Geological-engineering conditions for the fixation of beacons available here have been investigated, as well as their degree of preservation. The suitability of the line for a study of recent movements has also been checked.

This line crosses heterogeneous structural elements -- the eastern part of the Russian platform, the Hercynian Sub-Uralian depression and the adjoining western part of the folded area of Southern Urals. A considerable distance between the geodetic point: (an average of 15-25 km) permits to use this line for a study of recent vertical movements mainly within the largest tectonic structures of relief-forming rocks of *morphostructures* of the first order (I.P. Gerasimov, 1946, 1959; Yu. A. Mescherikov, 1960; A.P. Rozhdestvensky and Yu. E. Zhurenko, 1961).

In the platform part of the territory, where the relief has been formed on a sheeted basement, the levelling line crosses (Fig. 1) the following morphostructures of the first order: Zhiguli arch (south-eastern extremity), Shkapovo-Romashkino arch, Stavropol depression and Kama-Belaia depression (A.P. Rozhdestvensky and Yu.E. Zhurenko, 1961). Geomorphologically positive structures find their expression in relief protrusions (Zhiguli mountains, Bugulma-Belebeevo elevation), while the negative structures - in large relief subsidences (Stavropol and Kama-Belaia depressions).

In the Kama-Belaia depression the relief-forming substratum is of a very complicated structure. Its western part is formed on a sheeted platform basement, whereas the eastern part - on a folded-block basement of the Hercynian marginal depression.

In the east the depression adjoins the folded mountain area of Southern Urals. At the stretch, where this boundary is crossed by the relevelling line, it is represented by a big disjunctive deformation of a NW trend - the Ashinsk overthrust. A number of structural complexes are distinguished within the adjoining part of Southern Urals: Karatau, Bakalo-Satka, Zlatoust (A.I. Olli and V.A. Romanov, 1959). On the whole they can be regarded as folded positive morphostructures of the first order.

Fig. 2, upper chart (1) represents a curve of mean yearly velocities of recent vertical movements. Generally speaking, this curve rather well corresponds to the topographic profile of the landscape and to the structure of relief-forming rocks. In the western part of the route (between Bezenchuk and Sarai-Gir) there are negative values of the mean yearly velocities of recent movements as compared with the "zero" bench mark in Bezenchuk. Further east, beginning with the Bugulma-Belebei elevation and throughout the rest of the route (up to Ur humka) the movement velocities are positive everywhere.

These data indicate a differentiated character of recent movements manifestations depending upon the types of morphostructures. Positive morphostructures are characterized

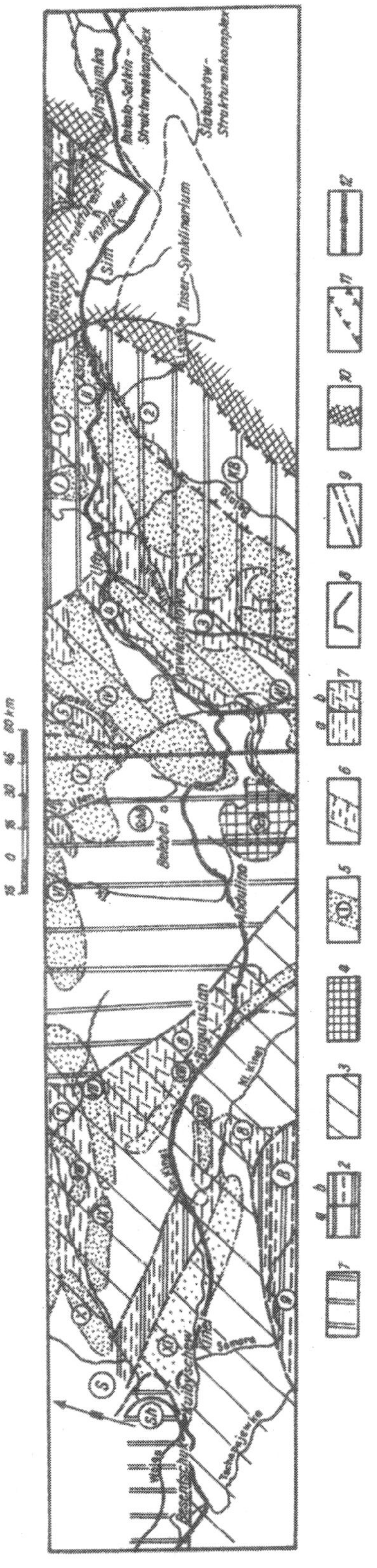

Fig. 1

Scheme of morphostructures in the central part of the Volga-Ural region crossed by the relevelling line.

1 - positive morphostructures of the 1 order (arches): A - Shkapovo-Romashkino and B - Zhiguli arches; 2 - negative morphostructures of the 1 order (depressions): a) concordant (upright): C - Kama-Belaia, D - Stavropol depression; b) inverse (discordant, inverted): E - Buzuluk depression; 3 - flanks of morphostructures.

Positive morphostructures of the 2 order: 4 - tectonic "plateaus" complicating the arches: F - Shkapovo; 5 - tectonic uplifts complicating the arches, flanks and depressions: I - Staro-Petrovsk, II - Riazan-Okhlebnino, III - Fedorovka-Sterlibashevo, IV - Saraevo-Aslikul, V - Kandry-Chekmagush, VI - Bavly-Tiumazy, VII - Baituganovo, VIII - Sernovodsk-Shugurovo, IX - Orlovka, X - Rakovka, XI - Zhiguli, XII - Malaia Kinel, XIII - Bolshaia Kinel; 6 - big anticlinal folds of the Yuriuzan-Ai depression.

Negative morphostructures of the 2 order: 7 - tectonic depressions (synclines, throughs), complicating depressions, flanks, arches: a) concordant (upright) and b) inverse (discordant, inverted): 1) Saldybash-Ufa, 2) Sima, 3) Urshak, 4) Dioma, 5) Chermasan, 6) Bugurusian, 7) Kliavlino, 8) Malaia Kinel, 9) Samara; 8 - tectonic faults; 9 - boundaries of morphostructures; 10 - folded region of Southern Ural; 11 - boundaries of the Hercynian Sub-Uralian Marginal depression; 12 - points of relevelling.

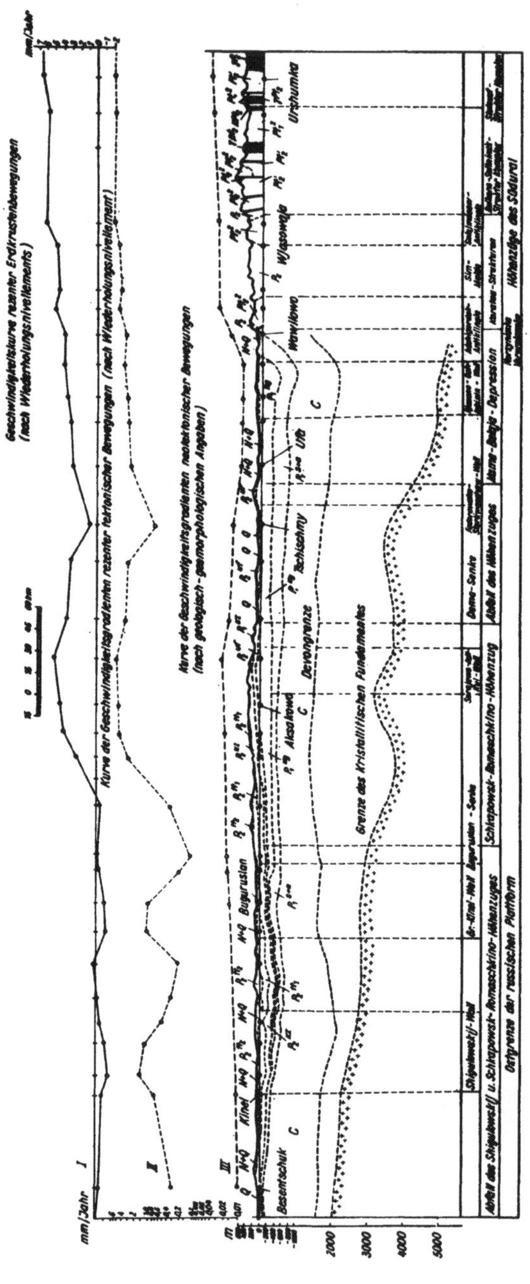

Fig. 2

Velocity chart of recent and latest crustal movements and schematic section along the relevelling line Besentschuk-Ufa-Urshumka by relative recent uplifts, whereas negative - relative subsidences. Very interesting is the established recent activity of certain large disjunctive deformations (the Ashinsk upthrust, for instance).

The middle chart (II, Fig. 2) shows that the highest velocity gradients of recent movements are characteristic for the folded regions of Southern Urals and lower ones - for the area of the Sub-Uralian depression and the eastern part of the Russian platform. On the platform there are, however, individual stretches with higher velocity gradients, which found its reflection in the corresponding peaks of the curve on diagram II. Such stretches, as a rule, correspond to positive morphostructures of the 1 and 2 orders (Shkapovo-Romashkino arch, Zhiguli and Bolshaia Kinel iplifts, etc.) and the subsided parts - to negative morphostructures (Dioma depression, etc.). A similar differentiation of velocity gradients of recent movements depending upon the character of morphostructures can be noticed also in the folded area of Southern Urals (Sulaiman anticline, Sima trough).

Chart III (Fig. 2) shows that a similar picture of gradient changes is outlined also for summary velocities of the Neogene-Quaternary crustal movements. We also find here a direct dependence between these gradients and large morphostructures (Shkapovo-Romashkino arch and corresponding maximum gradients; a sharp jump of latest movements is recorded on the stretch of the Ashinsk overthrust, etc.).

A comparison of curves on Charts II and III indicates their similarity, which shows a definite inheritance of recent movements from neotectonic movements.

The abovegiven data outline the existence of a direct connection between specific features in the manifestations of recent tectonic movements, the relief of the Earth's crust, the main morphostructures and the latest tectonics. At the same time, within the platform structures of deeper (Sub-Kungurian) horizons of the sedimentary mantle, just as the relief of the crystalline basement, do not display ubiquitously a direct connection with the character of the curve of recent movements. Apparently, recent movements are determined n o t o n l y by the movements of the basement. There is a possibility that along with basement shoves, other causes of recent vertical movements can be transformations of tangential stresses, originating within the sedimentary mantle, when there are downwarps of large lithosphere segments, as well as manifestations of salt tectonics.

2. To solve the second task certain methods have been used, which permit to evaluate qualitatively recent tectonic activity of a definite territory covered by an extremely scattered network of relevelling. These methods included: 1) study of changes in the sinuosity of river beds during a definite comparable period of time and 2) analysis of grades in longitudinal river profiles. Studies took place on an area adjoining from the south and north the relevelling line on the stretch Abdulino-Ufa-Asha.

S t u d y o f r i v e r b e d s i n u o s i t y . This method is based on the recognition of the concept that depending upon the sign of recent tectonic movements river sinuosity diminishes on the territory of ascending movements and increases on the stretches of descending movements. Changes in sinuosity are established by a comparison of heterochronous cartographic material. By the precision of time interval dating this method comes close to the geodetic method, but different from the latter provides a qualitative and not a quantitative picture (A.P. Rozhdestvensky and I.K. Ziniakhina, 1959; A.P. Rozhdestvensky and Yu.E. Zhurenko, 1961).

For comparison hachure maps of Bashkiria have been used compiled in 1835-1850 and maps published in 1942. The ratio of sinuosity for the same stretches of beds has been determined separately by each map. The ratio of sinuosity coefficients of the first and second periods designated by us by the letter K, determines the change character in the sinuosity of river beds during nearly one hundred years. This way a starting point has been arrived at for the compilation of a schematic map of river sinuosity for the territory under study (Fig. 3).

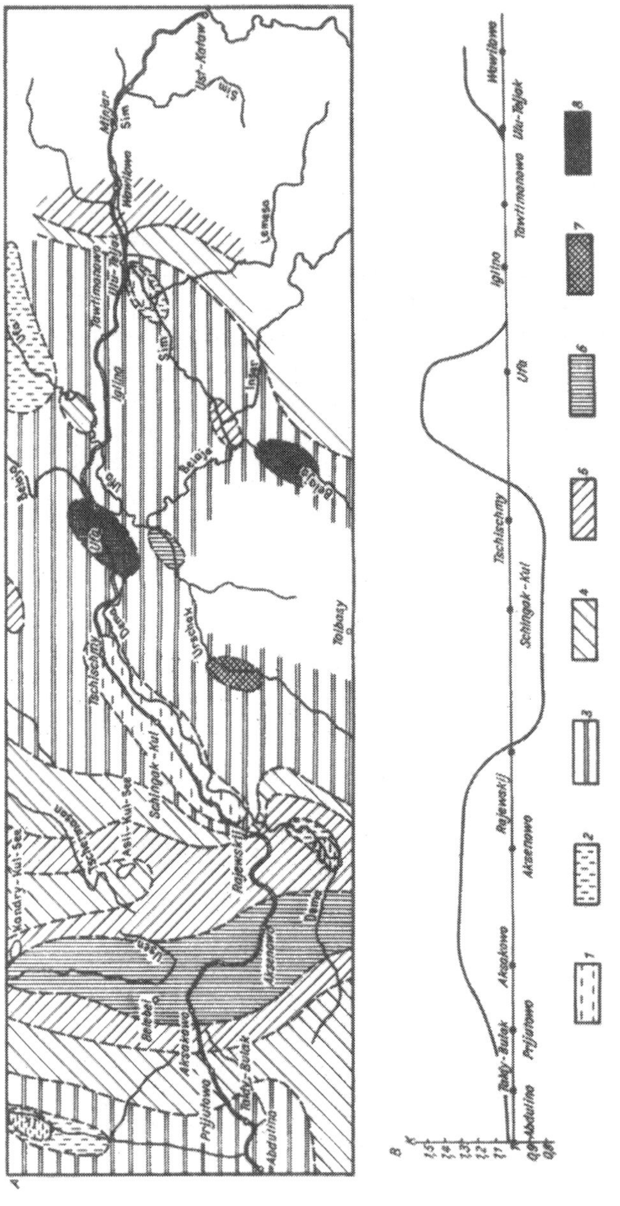

Fig. 3

Sinuosity changes of river beds in Western Bashkiria

a - schematic map of sinuosity changes of river beds; 1 - areas characterized by an increase in the sinuosity of river beds, K = 0.8; 2 - ditto, K = 0.9; 3 - areas characterized by an absence of sinuosity changes of river beds, K = 1.0; 4 - areas characterized by a lesser sinuosity of river beds, K = 1.1; 5 - ditto, K = 1.2; 6 - ditto, K = 1.3; 7 - ditto, K = 1.4; 8 - ditto, K = 1.5 and more; b - curve showing changes of sinuosity coefficients of river beds (along the line of relevelling)

In comparing these data with a morphostructural scheme for the structure of this area, it became possible to establish a general regularity: stretches characterized by smaller or an absence of sinuosity changes during the comparable time interval, correspond in general rather precisely to large positive morphostructures (Shkapovo-Romashkino arch) and to smaller positive structures, which complicate them (like, for instance, the Saraevo-Aslikul, Sterlisbashevo-Fedorovka uplifts, etc.). An increase in sinuosity characteristic for negative morphostructures (Dioma trough, Kama-Belaia depression, etc.).

G r a d e a n a l y s i s o f l o n g i t u d i n a l r i v e r p r o f i l e s has been made according to a method based on a strict estimate of the river order (according to R. Horton, 1945), Yu.E. Zhurenko, 1959, 1960). It was agreed to consider as rivers of the first order those, which have no tributaries (elementary streams). Two rivers of the first order in confluing form a river of the second order and two rivers of the second order at their confluence form a river of the third order, etc. After the order of the rivers has been determined, their grades on different stretches have been established. On the basis of such computations a summary table of average grades of river beds has been compiled for all the orders of rivers (Fig. 4). For such a table arithmetic mean gradients of longitudinal river profiles have been determined for each order separately. To obtain objective data for a grade comparison on rivers of different orders by definite intervals, a coefficient of proportion is deduced. This has been achieved by dividing the figure for the arithmetic mean of river gradient of the first, second and subsequent orders by the arithmetic mean of the river grade of the highest order. The calculated coefficients of proportion were then multiplied by an arbitrarily chosen scale (in our example a scale equal to a grade of 1:100,000 is chosen). As result, for each river order original smallest grade intervals of their beds have been determined. The subsequent compilation of the table is based on the principle of a hypsometric scale.

According to the table data on the grades of individual river stretches were plotted on a map with a segregation of areas, which could be interpreted as regions with commensurate (same type) river bed grades (Fig. 4-a).

Because of a computation of averages for individual stretches of the river beds, the data obtained are more or less free from the influence exerted upon the character of deformation of the longitudinal river profile of both lithological factor and of the turbulent stream movement. For this reason big unevennesses and breaks in the longitudinal river profiles indicated on the maps, are taken by us as a reflection of recent tectonic movements.

The distribution of abnormal grade stretches is generally in good accord with definite structure types of relief-forming rocks. Very distinctly stands out the zone of higher grades in the area of the Shkapovo-Romashkino arch; as a whole the Kama-Belaia depression is characterized by lower grades; there is a sharp jump towards an increase at the transition of this depression into positive morphostructures of the folded area of Southern Urals.

A comparison of data on changes in sinuosity and grades of river beds with a diagram of recent vertical movements (Figs. 2, 3 and 4) shows that areas with the greatest grades and least sinuosity correspond to territories of the most intense recent tectonic uplifts and, on the contrary, areas with least grades and a maximum sinuosity of rivers correspond to territories of recent subsidences.

As a result of data summarization on grades and sinuosity of rivers a schematic map of recent tectonic movements of the territory of Western Bashkiria has been compiled (Fig. 5). A spatial distribution of outlined areas shows that most intense recent uplifts characterize large positive morphostructures. Relative subsidences are characteristic for big negative morphostructures. The same dependence can be recorded also for the majority of morphostructures of lower orders (uplifts, troughs, etc.). However, in a number of

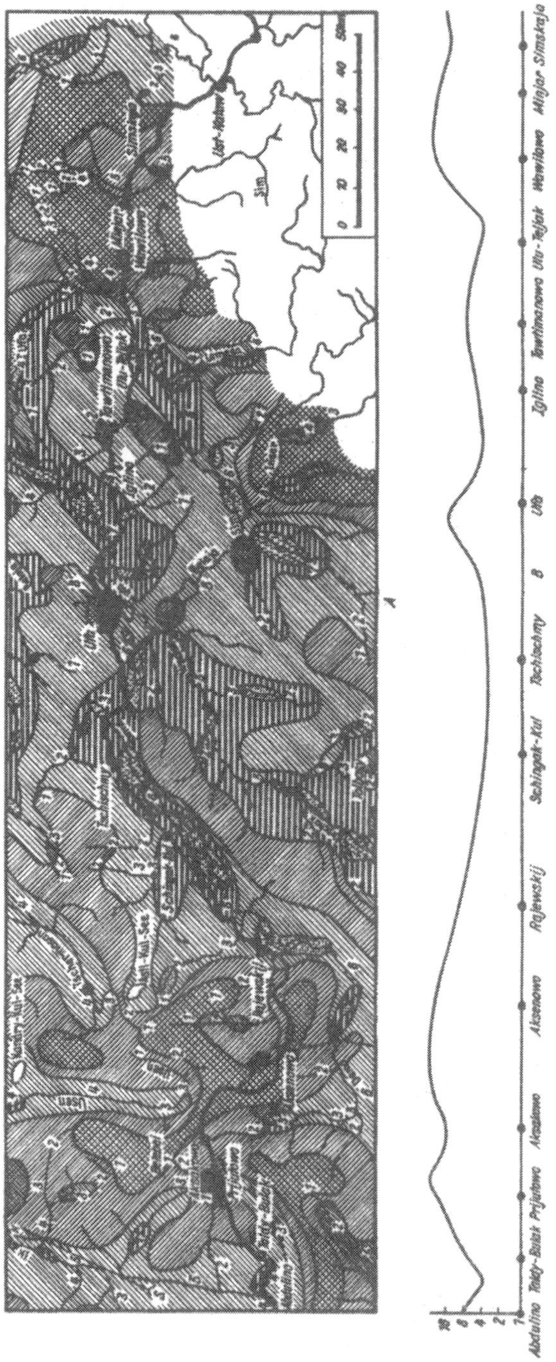

Neigungsintervalle der Flußläufe cm/km		Flußfolge I	II	III	IV	V	VI	VII	VIII
	1	0-100	0-65	0-40	0-20	0-9	0-5	0-3	0-1
	2	100-200	65-130	40-80	20-40	9-18	5-10	3-6	1-2
	3	200-400	130-260	80-160	40-80	18-36	10-20	6-12	2-4
	4	400-800	260-520	160-320	80-160	36-72	20-40	12-24	4-8
	5	800-1600	520-1040	320-640	160-320	72-144	40-80	24-48	8-16
	6	1600 und mehr	1040 und mehr	640 und mehr	320 und mehr	144 und mehr	80 und mehr	48 und mehr	16 und mehr

C

Fig. 4

Grades of river beds in Western Bashkiria

A - schematic map of grades in river beds; B - curve of relative grade values along the line of relevelling (ratio of actual grade values in the given region and the least grades of the rivers of the same order for the entire territory studied); C - table showing mean grades of river beds

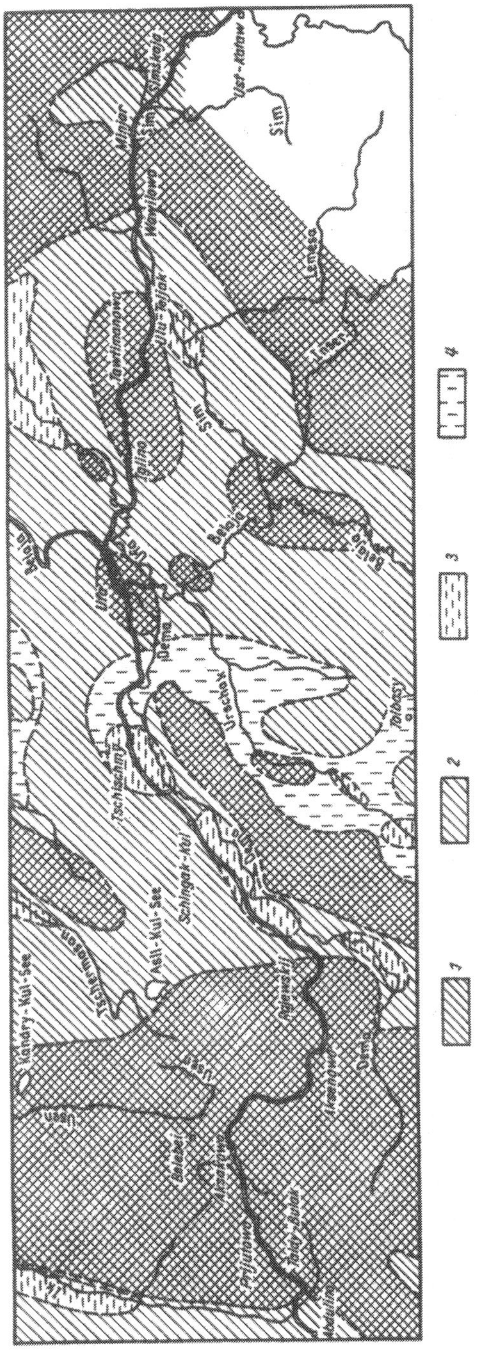

Fig. 5 Schematic map of recent tectonic movements.

1 - areas of most intense uplifts; 2 - areas of less intense uplifts (relatively stable); 3 - areas of distinctly expressed local subsidences; 4 - areas of poorly expressed relative subsidences

cases it is possible to establish that recent uplifts involve some negative structures (for instance, within the Kama-Belaia depression, etc.).

A comparison of the data obtained with the information on Holocene and earlier (Pleistocene and Neogene) movements of the Earth's crust shows that in many cases these movements coincide by their direction and place of manifestation. This permits to speak of an inheritance in the development of structural elements during the neotectonic stage.

Geomorphological interpretation of the data provided by repeated levelling in the low-mountain areas of Kazakhstan

(Arys - Alma Ata - Semipalatinsk)

von

E. A. Finko und E. Y. Rantsman (Moskau)

Summary

The paper outlines a geomorphological interpretation of the data provided by the precise repeated levelling in the Arys-Semipalatinsk area.

The area runs through nine large morphostructural elements which are the sections of an Epihercynic platform changed in various ways by the action of recent tectonics.

Two types of relationships, direct and inverse, are established between the rates of recent movements and the relief, with the direct relationship being predominant in the area.

A detailed research has been made into the inverse relationships in the zone where Kazakhstan shield joins Ob-Irtysh depression and in Chu-Ili mountains.

The geomorphological analysis has shown that in case of reverse relationships the recent movements inherit the tendencies of tectonic development of the last stage of the Quaternary period which have not left their imprint on the general features of the relief.

It follows from this that the data of the recent movements allow us to determine areas with an appearing rearrangement of the structural plan.

In conclusion we point out that in the given profile no essential changes have been registered in the intensities of recent movements for organic or platform regions. The difference lies only in the degree of the differentiation of the movements which is much more larger in the orogens.

Zusammenfassung

Die Arbeit gibt eine geomorphologische Deutung der Daten, die durch ein Präzisionswiederholungsnivellement im Gebiet von Arys-Semipalatinsk erhalten wurden.

Das Gebiet verläuft durch 9 große morphostrukturelle Elemente, die die Abschnitte einer epiherzynischen Plattform sind, welche auf verschiedene Weise durch die Wirkung rezenter Tektonik verändert wurde.

Zwei Arten von Beziehungen, direkte und umgekehrte, wurden zwischen dem Maß der rezenten Bewegungen und dem Relief festgestellt, wobei der direkte Zusammenhang in dem Gebiet vorherrschend ist.

Es ist eine detaillierte Untersuchung der umgekehrten Zusammenhänge in der Zone unternommen worden, wo der kasachstanische Schild mit der Ob-Irtysh-Senke zusammentrifft, und in den Chu-Ili-Bergen.

Die geomorphologische Analyse hat gezeigt, daß im Falle umgekehrter Zusammenhänge die rezenten Bewegungen die Tendenzen der tektonischen Entwicklung des letzten Quartärstadiums, das seinen Einfluß auf die allgemeinen Merkmale des Reliefs noch nicht verloren hat, geerbt haben.

Daraus folgt, daß die Daten der rezenten Bewegungen uns erlauben, Gebiete mit einer scheinbaren Neuanordnung des Strukturplanes zu bestimmen.

Zum Abschluß können wir feststellen, daß in dem gegebenen Profil keine wesentlichen Änderungen der Stärke der rezenten Bewegungen in organischen oder Plattformgebieten registriert worden sind. Der Unterschied liegt nur im Grad der Differenzierung der Bewegungen, die in den Orogenen viel größer sind.

Since 1959 the Geographical Institute of the Academy of Sciences of the U.S.S.R. has been engaged in studying the recent tectonic movements of the Earth's crust in the Eastern and Southern Kazakhstan in the repeated levelling area Arys - Semipalatinsk.

This area runs through nine large morphostructural regions which are the sections of the Epihercynic platform changed in a various degree by recent tectonics. They are: I - Syr-Darya depression, II - area where Karatau range joins Talass Alatau range, III - Chu-Talass depression, IV - Kendyktas range where it joins Chu-Ili mountains, V - Ili depression, VI - Jungar Alatau, VII - depression at Lake Balkhash, VIII - Kazakhstan shield, IX - Ob - Irtysh depression at the joint with Altai.

The engineering and geological inspection of the area has showed that nearly half of the bench-marks have not been affected by unfavourable exogenetic factors and are displaced due to tectonic causes. For this reason the area can be used with advantage for the study of recent tectonic displacements of the Earth's crust in this profile.

The graph of the rates of recent vertical movements (Fig. 1A) gives an idea of the nature of recent displacements of the Earth's crust. In so far as this line is not reduced to the tide gauge datum in Kronstadt the intensities indicated in the graph show not the absolute annual displacement of the points but only the relative displacement, the bench-mark at Arys being taken as the arbitrary datum. The maximum amplitude of intensities reaches 12 mm/year.

Direct comparison of this graph with the hypsometric profile of the repeated levelling line (Fig. 1B) revealed two types of relationships between the orography and the rates of recent movements. These are a d i r e c t relationship at which the maximum rates of recent movements occur on most elevated sections and the minimum rates on the lowest sections and an i n v e r s e relationship when the highest rates are registered in low-lying sections of the hypsometric profile and the slowest rates are observed in orographically elevated territories. The field geomorphological investigations have shown that these types of relationship are by no means accidental. Let us illustrate this by examples.

The direct type predominates along the line.

A d i r e c t r e l a t i o n s h i p is recorded in the area of Syr-Darya depression which at present undergoes a certain subsidence and in the adjoining mountainous district of Bolshoi Karatau at the junction with Talass Alatau which now experiences a relative uplift.

The manifestations of the recent relative subsidence of Syr-Darya depression are expressed most clearly in its western part (west of Arys Station). The tendency towards subsidence is seen here in the morphology of the valley and the composition River Arys alluvium:

1) down the valley the number of terraces decreases; a high flood plain and a second terrace taper out west of Arys Station; 2) the height of the first terrace above the flood plain decreases twofold as compared to its height in the mid-stream (from 4 to 2 metres); 3) in its low reaches River Arys noticeably meanders; the river-bed gradients equal only 0.0003; 4) a clayey fraction predominates in the composition of the deposits of the present river-bed, flood plain and the first terrace.

The geological and structural materials testify to the fact that this region of a prolonged and stable subsidence has accumulated thick friable series (up to 2,000 metres according to geophysical data) and caused the formation of a lowland plain.

According to the data of repeated levelling, this region is distinguished by the smallest relative rates of recent movements along the entire area (Fig. 1A).

As we move further to the east and approach the point where Karatau range joins Talass Alatau range the rates of recent movement increase and reach their maximum where the line intersects the south-east extremity of Bolshoi Karatau range. Absolute height of the locality also increases in this direction.

The manifestations of upheaval in the area of the present maximum rates can be seen in the morphology of the valley and the composition of alluvium in the upper reaches of River Arys: up the valley the number of terraces increases due to the entrenchment and the conversion of the high flood plain into the first terrace above the flood plain; 2) the height of the first terrace above the flood plain increases almost twofold as compared to its height in the mid-stream (from 4 to 8 metres), and morphologically the first terrace becomes the second terrace; 3) in the upper reaches of the river its gradients sharply increase -- up to 0.0135; 4) shingle is found in large amounts in the composition of the deposits of the present river-bed, flood plain and terraces.

The area of the recent maximum upheaval (Chokpak siding) is the watershed of the rivers of Syr-Darya and Chu-Talass depressions. Valleys and hills of medium height are extensively developed here. The plains are crisscrossed by many narrow and deep gullies (up to 10 metres). From a structural point of view this territory lies directly on the continuation of the axial uplift of Boroldai range (south-eastern extremity of Bolshoi Karatau). The smooth relief of gently sloping hills and flat plains testifies to the fact that this relative intensity of uplift (12 mm/year) could have arisen here only in recent time.

Within the area under examination a direct relationship with similar geomorphological manifestations of young upheavals and subsidences has been registered where Karatau joins Chu-Talass depression, Jungar Alatau joins Ili depression and the depression at Lake Balkhash and Kazakhstan shield joins the depression at Lake Balkhash.

E x a m p l e s o f i n v e r s e r e l a t i o n s h i p s are observed in Chu-Ili mountains and in the area where the Kazakhstan shield joins Ob-Irtysh depression.

F o r C h u - I l i m o u n t a i n s which is the most elevated region (area of siding No. 83) the graph shows a section of the lowest relative rates of recent movements and, conversely, the regions of relative subsidence (area of Espe Station) are distinguished by the maximum rates of movement.

This disagreement can be explained by the different structural position of these areas and by the different young tendencies of their tectonic development.

In Chu-Ili Mountain the line of repeated levelling intersects two blocks-- one rising and the other sinking (Fig. 1).

The r i s i n g b l o c k is intersected by the repeated levelling line in its north-western part (Espe area). In the Lower Quaternary period this section of the block sank thereby accumulating sheared conglomerate. As time went on, the region of the accumulation of sheared conglomerate was unraised and dismembered, the uplift spreading in the north-western direction. Today, it is at its maximum in the extreme north-western part of the block -- in the area of Espe Station which is borne out by the morphology of the River Chokpar valley. If the total depth of the river entrenchment into the sheared conglomerate decreases in the north-western direction (down the valley) from 30 metres in the area of Dalai-Kainar siding to 8 metres near Espe Station, the recent entrenchment is far more active in the lower part of the valley (area of Espe Station) where the river-bed cuts directly into the sheared conglomerate which also forms the flood plain socle.

It should be stressed that the recent relative uplift of the extreme north-western part of the block has not received hypsometric expression and the area of Espe Station lies at an altitude of about 500 metres whereas the maximum heights of the rising block

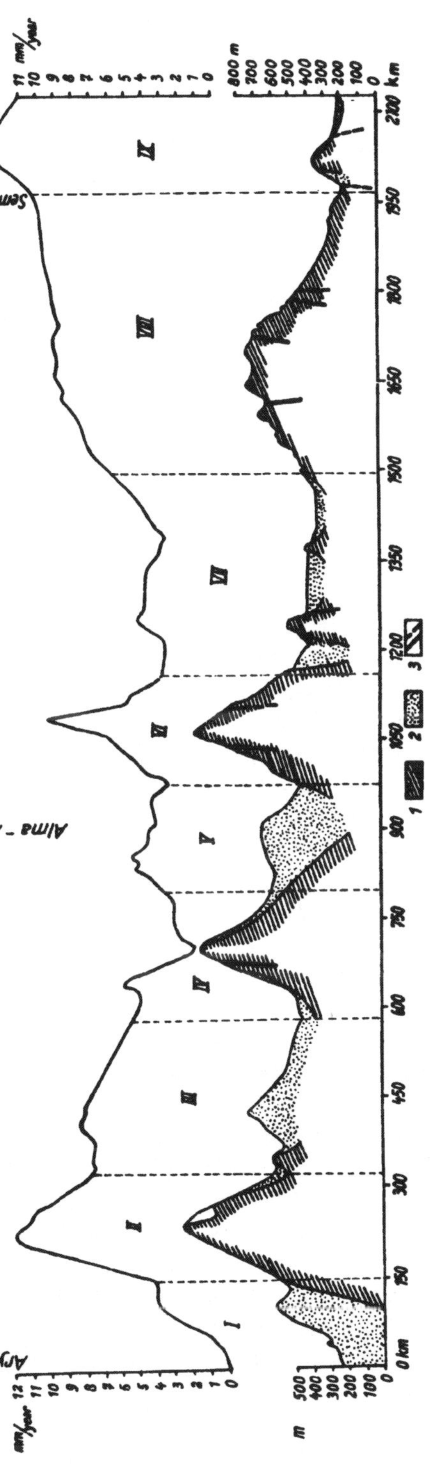

Fig. 1

Repeated levelling line Arys-Alma Ata-Semipalatinsk.

A. Rates of recent tectonic movements

1. Curve plotted according to geodetic data and corrected in the field inspection of bench-marks.

B. Hypsometric profile with large morphostructural elements.

1. Pre-Paleozoic and Paleozoic rocks
2. Mesocenozoic deposits
3. Lines of tectonic disturbances
   a) established,
   b) supposed

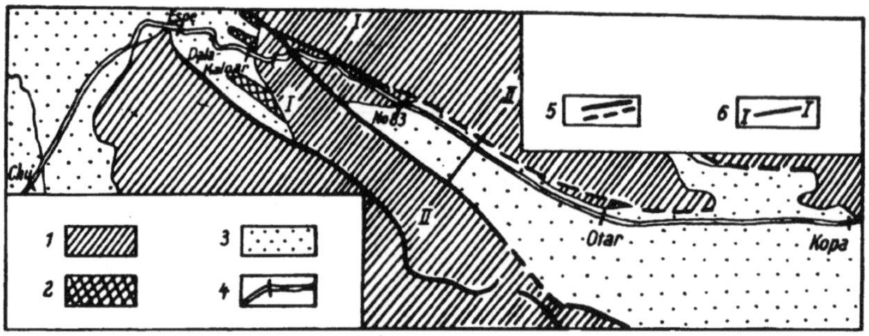

Fig. 2

Schematic structural plan of Cope fault trough area

1. Outcrops of ancient rocks (Pt, Pz)
2. Tertiary deposits
3. Quaternary deposits
4. Repeated levelling lines
5. Dislocation with a break in continuity displacing the Quaternary deposits
   a) observed, b) supposed
6. Lines of profiles

within the profile exceed 700 metres.

In the sinking block, known as the Cope fault trough, the process of subsidence gradually took place during the Quaternary period in the north-western direction. This found its reflection in the morphology of the Cope fault trough and in the distribution of friable deposits in it; morphologically, the south-eastern part of the Cope fault trough is represented by an aggradation plain (area of Otar Station) and the north-western part by a weakly dismembered peneplain with very weak dismembered friable deposits (Fig. 3). The total amplitude of the fault trough subsidence relative to the blocks in which it is confined is less in the north-western part than in the south-eastern section (see. Fig. 3). The recent subsidence of the north-western part of the fault trough which occupies hypsometrically the most elevated, watershed position (siding No. 83) found an indirect expression in the high level of underground waters which caused an extensive development here of moisture-loving plants (sedge, chii and reed rise onto the watersheds) and an abundance of springs.

In the same way as the uplift in the area of Espe, the subsidence of the north-western part of the fault trough are young processes which are not yet reflected in the general features of the relief but which nevertheless manifest themselves in a number of other ways pointed out above.

Let us now discuss an example of inverse relationships -- the area where Kazakhstan shield joins Ob-Irtysh depression. From an orographical point of view the following sections can be pointed out here: the slope of Kazakhstan highland, decreasing from 700 to 200 metres as it runs down to Irtysh valley and Belyagach upland with H = 360 metres. The distribution of the rates in the graph does not correspond to the relief. The minimum rates are registered in the most elevated part of the shield which is about 700 metres high; the maximum rates are recorded in Irtysh valley near Semipalatinsk and in the area of Belyagach upland (area around Semipalatinsk). This discrepancy in the orography and recent displacement is due to the peculiarities of young tectonic movements in this area revealed by geomorphological methods.

In the area around Semipalatinsk there occur young active upheavals connected with Altai structures which have caused an essential rearrangement of Ob-Irtysh watershed and the drainage system (Fig. 4).

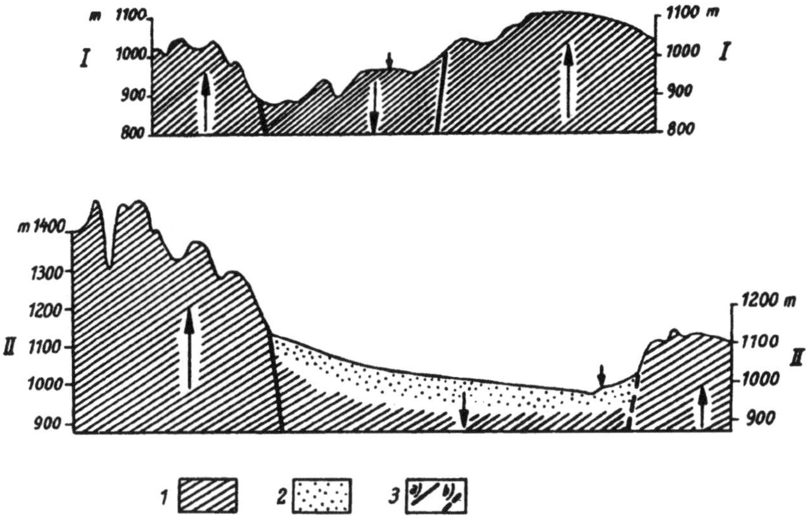

Fig. 3

Schematic geological profiles of Cope fault trough area

1. Outcrops of ancient rocks (Pt, Pz)
2. Cenozoic deposits
3. Dislocation with a break in continuity displacing the Quaternary deposits
   a) observed, b) supposed

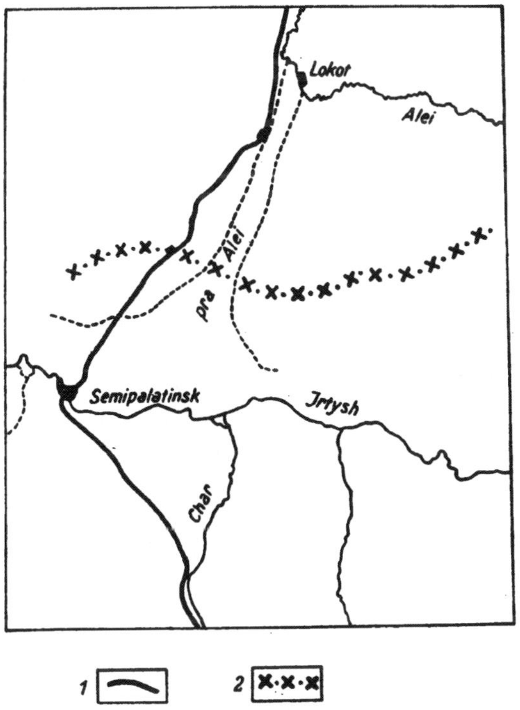

Fig. 4

Diagram of the drainage system rearrangement in the area around Semipalatinsk

1. Repeated levelling lines
2. Watershed formed by the Upper Pleistocene and Holocene movements

The valley of former River Alei reached Irtysh in the area of Semipalatinsk. Today this is a dead section of the valley. Its dewatering is due to the Upper Pleistocene and Holocene movements in the area of the present watershed. The young age of the latter is established from the analysis of the river terraces of the present Alei and Irtysh.

Two sections can be pointed out in Alei valley (near the village of Lokot): the latitudinal (young) and the meridional (ancient). The latitudinal section is characterised by a number of young loamy terraces (flood plain, high flood plain, flood terrace with the maximum heights of up to 5.5 metres). The ancient section of the valley is distinguished by a number of high (up to 18 metres) sandy terraces. A common feature in the appearance of the valley today is the level of the high flood plain which accounts for a recent rearrangement of the watershed and the drainage system. In the area of Lokot there takes place an active recent entrenchment of the Alei river-bed, the formation of a young flood plain and the conversion of the high flood plain into the first terrace.

The manifestations of young movements in Irtysh valley are as follows: a) an active recent entrenchment of Irtysh, the formation of the first terrace at the expense of the high flood plain, merging of islands and their attachment to the bank attended by the development of low flood plain pockets, b) extensive development of socle flood plains and low terraces, c) predominance of shingle deposits in the composition of the present alluvium of the river-bed and the flood plain, d) insufficient thickness of present alluvium revealed by drilling in Irtysh river-bed.

To summarise, the recent movements in the area around Semipalatinsk inherit the tendencies of the tectonic development of the Upper Pleistocene and Holocene which left their imprint on the morphological structure of this region.

In the area of the northern slope of Kazakhstan highland, more complicated relationships exist between the recent and ancient tectonics. The morphostructure of Kazakhstan highland reflects the summary course of recent movements which date from the Lower Oligocene. According to the data provided by the geomorphological analysis, the recent movements reflect the tendencies of the Quaternary period when the northern slope of the highland was drawn into the Altai uplift. There are the following signs of the Quaternary uplifts in this area: a) very insignificant thickness of the quaternary deposits (within the first dozens of metres), b) the presence of socle flood plains at a number of reach rivers, c) shingle composition of alluvium of the river-bed, the flood plain and the first terrace of River Char which intersects the northern slope of the shield parallel to the second levelling line.

The almost intact sections of peneplain with a Mesozoic crust of weathering and the absence of displacements of the low terraces in the valley of River Char indicate to the general non-differentiated character of the Quaternary uplifts.

The above indicated discrepancies between the hypsometry and the intensities of recent movements in the profile under examination, as shown by the data of the geomorphological analysis, are due therefore to the fact that the recent movements inherit the tendencies of the tectonic development typical of only very last stages of the Quaternary period. These tendencies, however, fail to correspond to the summary course of the recent movements. Hence, the study of recent movements makes it possible to bring to light the appearing rearrangement of the structural plan which has as yet not found its expression in the general features of the relief.

The task of further geomorphological investigations in the interpretation of repeated levelling data is to increase the number of indices which could be used for the evaluation of the nature of the movements of the youngest stages of the Quaternary period.

It should be pointed out in conclusion that, as the geodetic data show, there is no essential difference in the intensities of recent movements in the orogenic and platform regions. However, it would be premature to apply this conclusion to all orogenic regions since the quantitative data at our disposal can be used only for the peripheral zones of the orogens. For all that, we can be quite certain as to the differences in the degree of the differentiation of the movements which is considerably larger in the orogens.

Tectonique actuell du Caucase

von

N. V. Doumitraschko, D. A. Lilienberg (Moskau)

Resumé

La présente communication envisage les généralités concernant les mouvements récents de l'écorce terrestre au Caucase. Les auteurs ont fait l'analyse des particularités de la méthode complexe de l'étude de la tectonique récente dans les pays montagneux. Pour les types principaux des mouvements récents, pour les régions où ils se font voir, pour leur succession ou la non-coïncidence avec des structures néotectoniques les auteurs présentent une caractéristique spéciale.

L'étude des mouvements tectoniques récents présente un intérêt spécial dans les régions des géosynclinaux et en particulier, dans les jeunes pays montagneux (type - Caucase), car les déplacements récents de la croûte terrestre se manifestent ici avec plus d'intensité, de différenciation et contrast.

Jusqu'au milieu du XX-ème siècle l'étude des mouvements récents du Caucase se faisait d'une manière disparate et épisodique. Leur investigation systématique commença dès 1950 par l'Institut de Géographie de l'Académie des Sciences de l'URSS en coordination avec l'Institut Central de Recherches Scientifique de Géodésie, de Photographie aérienne et de Cartographie. L'Administration Hydrographique et l'Institut de la Physique de la Terre de l'Académie des Sciences de l'URSS.

L'information obtenue en résultat de ces investigations permet de donner une caractéristique générale des mouvements tectoniques récents du Caucase et du pays Précaucasien de leur amplitude, de la particularité des manifestations en temps et en espace, de la succession ou de la non-coïncidence avec les structures plus anciennes.

L'étude de la tectonique actuelle du Caucase fut exécutée à l'aide des méthodes géodésiques, océanographiques et géologiques-géomorphologiques. Néanmoins, l'application des deux premières méthodes représente encore pour le Caucase un certain nombre de difficultés, car les nivellements réitératifs passent le long des voies ferrées, qui ceignent le Caucase selon la périphérie - dans les régions des avant-pays et des plaines au pied des montagnés, et ne recoupent point ses parties centrales (hormis la chaîne de montagnes Souram). C'est pourquoi les tracés existants des nivellements réitératifs ne présentent encore point la possibilité de juger sur les amplitudes réelles des mouvements dans les confins des systèmes de montagnes propres.

Une grande importance devant l'utilisation des données concernant les nivellements réitératifs doit être apportée au déplacement des signes géodésiques provoqué par l'influence exogène et antropogène: entassement des sous-sols; effondrements; variations du niveau des eaux de sous-sol; processus de crioturbations; salinification; déflation; pompage de pétrole et de gaz; pression sur le sous-sol des grandes constructions; des moyens des transport; modifications de la pression atmosphérique; variations des niveaux des bassins et des mers, etc. L'investigation exécutée par D.A. Lilienberg sur la stabilité des fondations des signes geodésiques le long des tracés du Caucase Oriental (Ordjonikidze-Makhatch-Kala-Bakou-Tbilissi, presqu'île d'Apschéron et les avant pays Adjinoours, le Caucase Mineur) a révélé que presque tous les bâtiments ici ont subi à un degré quelconque des dégats par les charges d'effondrement et les charges dynamiques des moyens de transport. Des conclusions analogiques ont été faites par N.V. Doumitraschko et S.L. Kouschev pour une série de signes dans la zone des sédiments loessiformes du Caucase Occidental et du Pays Central Précaucasien, ainsi que dans la région de Poti. De règle, il faut exclure complètement de l'analyse des données de nivellements les signes posés sur les passages des ponts, sur les bâtiments des réservoirs d'eau, etc. A.V. Jivago (Jivago et d'autres, 1958) a reconnu mal fondés les témoignages d'une série de footstocks du littoral de la Mer Noire, et selon des observations de D.A. Lilienbarg - certains postes de mesurements des niveaux du littoral Caucasien de la Mer Caspienne.

Comme les investigations exécutées ont-elles montré les résultats des nivellements réitératifs et des mesurements des niveaux peuvent être utilisée pour l'evaluation des mouvements récents seulement concernant les plus grandes structures morphologiques du Caucase. L'insécurité de maints signes forcent à les utiliser avec une grande précaution pour l'analyse des mouvements des structures particulières. Il faut aussi avoir en vue que la précision de la transmission du signe absolu à partir de zéro du footstock de Cronstadt est commensurable avec les valeurs des vitesses des mouvements récents ($\pm$ 2 mm par an) à la périphérie du Caucase (Siniaguina et Orlenko, 1959). Voici pourquoi l'attention principale est portée à l'analyse des mouvements relatifs des grandes structures morphologiques.

Les méthodes géologiques-géomorphologiques présentent la possibilité de révéler la caractéristique des influences exogènes et antropogènes sur les constructions, dans lesquelles les signes géodésiques sont situés et ceux du mesurement des niveaux, mais ils sont loin de pouvoir être appliqués pour établir le caractère des déformations tectoniques de la croûte terrestre au cours du temps comparativement court entre les nivellements réitératifs (près de 20-50 années, et en certains cas encore moins), car les manifestations des processus exogènes dans les régions montagneuses et dans les avant-pays sont si intenses qu'elles masquent à un degré considérable l'effet des actions tectoniques, exprimées par des variations d'intensité très minimes (millimètres et parties de millimètres par an). C'est pourquoi le rôle des méthodes géologiques-géomorphologiques d'investigation sur la tectonique récente dans les pays montagneux consiste considérablement à établir qualitativements les tendances des mouvements tectoniques récents et holocènes et leur succession des structures neotectoniques et quaternaires. Les objectifs principaux des observations géologiques-géomorphologiques sont: les rives actuelles, les terrasses marines et lacustrines, les lits actuels des fleuves, les plaines d'inondation et les terrasses, leur caractère, les profils longitudinaux et latitudinaux des vallées, de même que les particularités de la structure des flancs des faîtes de montagnes (en tenant compte de l'influence du climat), l'étude de la composition et de l'épaisseur des sédiments alluviaux, proluviaux et déluviaux, etc.).(Doumitraschko et Lilienberg, 1954). Particulièrement signicative est l'étude du déplacement du lit du Kouban vers l'ouest là, où il coule le long du flanc du plateau de Stavropol (au hord de Nevinnomyssk), englobé d'élèvements, ainsi que de la migration du lit du Koura dans la région de la plaine de Gori vers le sud (dû à l'élèvement du Caucase Majeur).

L'analyse complexe des données géologiques-géomorphologiques pour les parties montagneuses du Caucase permet d'établir les régions d'élevements suivantes: 1-2) régions des élèvements en voute actuelles et holocènes du Caucase Majeur et Mineur, compliquées par des mouvements différentiels en bloc et plicatifs, succédés en traits généraux des structures néotectoniques du miocène, pliocène et antropogène; 3) régions des mouvements récents et holocènes différentiels, contrasts et fortement contrastes, succédés de même des structures néogène-antropogènes néotectoniques. Ils sont situés à la périphérie des régions d'élèvements en voûte du Caucase Majeur et Mineur, de préférence le long de leurs flancs méridionaux et des régions adjointes d'abaissements. Aux dernières se rapportent; 4) les avant-fosses du Kouban et du Terek dans le pays Précaucasien et 5) ceux de Rion et du Koura-Arax dans la Transcaucasie, ainsi que les dépressions des Mers Noire et Caspienne. Les ensellements succédés aussi des plus grandes régions d'abaissements néogènes-antropogènes, qui, au cours de leur développement, sont entraînés graduellement dans les élèvements en voûtes voisines du Caucase Majeur et Mineur. En relation à cela la surface des dépressions au cours de l'antropogène et, particulièrement, de l'holocène et de l'étape actuelle, diminue et leur structure se complique, dû au développement dans leurs confins à la limite des régions d'élèvements adjointes des structures plicatives. Ainsi, une partie des dépressions a subi "une transformation des structures tectoniques", mais dans leurs parties considérables au cours de l'holocène et de l'étape actuelle continue l'abaissement. Dans les régions des élèvements prédomine la succession de mouvements ascendants. Toute la partie montagneuse du Caucase Majeur et de ses avant-pays, ainsi que le faîte de Souram subissent un élèvement intense, tandis que les parties centrales des pays-bas du Koura et du Rion - des abaissements actuelles. Sur le littoral Caspien du Caucase, l'aval de la rivière Kourssartchai au sud de l'embouchure de la rivière Samour s'abaisse d'une manière intense, et avec moins d'intensité - la bande étroite entre les terrains en aval du Terek et près de la ville Makhatch-Kala. D'ici jusqu'à la ville Bakou prédominent des abaissements et des élèvements, dont le maximum se rapporte à l'extrémité de la structure en voûte du Caucase Sud-Est. Sur le littoral de la Mer Noire au Caucase a lieu un tableau compliqué de mouvements tectoniques différenciés - élèvements et abaissements; les premières se rapportent aux parties du littoral fondées des roches consolidés, les deuxièmes - aux deltas alluvionnaires des grands fleuves et du pays-bas du Rion. Le caractère des mouvements tectoniques atuels sur le littoral de la Mer Noire au Caucase est masqué par la transgression actuelle de la Mer Noire, qui a commencé pendant l'holocène.

Au pays Précaucasien dominent de faibles élèvements et des abaissements actuelles, ayant leur maximum dans les confins de l'avant-fosse du Terek et de la partie occidentale de celle du Kouban. En comparaison avec ces régions, le plateau de Stavropol est une aire des élèvements relatives; ceci est confirmé par l'encorbellement du littoral gauche du Kouban dans la région, adjointe au plateau, par la structure en socles des plaines d'inondation et par la composition et l'épaisseur de leur alluvion dans les confins du plateau de Stavropol.

La plus forte intensité des mouvements tectoniques actuels se rapporte aux régions des grands abaissements, particulièrement de ceux du Koura, Rion et Terek, et atteint 4-6 mm par an; dans d'autres régions la vitesse des abaissements ne surpasse ordinairement point 1-2 mm par an. Les vitesses des élèvements actuelles atteignent 1,5 - 2, plus rarement - 3 mm par an; ceci s'explique par la distribution des tracés des nivellements réitératifs, en général à la périphérie du Caucase Majeur, par les tendances successives des mouvements tectoniques actuels, car l'amplitude des élèvements néogènes-antropogènes et holocènes au Caucase Majeur est presque par deux fois moindre (jusqu'à 5 km et plus), que les abaissements dans les dépressions pour le même laps de temps (plus de 10 km).

Outre les lents mouvements tectoniques - élèvements et abaissements - caractéristiques pour le Caucase sont des rapides mouvements tectoniques actuels, qui conditionnent sa séismicité.

La comparaison des mouvements tectoniques rapides - tremblements de terre - avec les structures morphologiques du Caucase révèle une claire coïncidence entre les régions séismiques principales et les zones des mouvements différentiels tectoniques contrastes. De règle, les phénomènes séismiques sont révélés d'une manière particulièrement vive le long des limites des structures morphologiques à diverses directions de mouvements les plus récents et, particulièrement, de mouvements quaternaires - des régions d'élèvements et d'abaissement, accompagnées souvent par des fractures, fréquemment fractures de fond. Les phénomènes séismiques sont en particulier bien exprimés dans les régions d'intersection des structures de direction générale Caucasienne avec l'élèvement méridionale transversale néotectonique, située sur la ligne Stavropol-Elbruz - massif Dzirul - massif cristallin Arzakan en Arménie.

Il existe aussi d'autres fractures moins nettement labourées de "direction anticaucasienne". Ainsi par exemple, l'élèvement méridionale Kurdamiro-Saatlin du fondement de la dépréssion du Koura, en s'intersectant avec les structures sublatitufinales des montagnes du Caucase Sud-Est, conditionne la grande activité séismique de la région Schemakhine (Doumitraschko, Lilienberg, 1954; Doumitraschko et al., 1961; Lilienberg, 1961; Kirillova et al., 1960). Les fractures jouent un grand rôle dans la formation de la dépression de la Mer Caspienne et sont accompagnées par des tremblements de terre (Agabekov et Akhmedbeili, 1956; Lilienberg, 1959) et par des phénomènes actuels de volcanisme à boue.

Les confins des régions avec des mouvements tectoniques à divers signes (+) coïncident avec la distribution des anomalies concernant la pesanteur et les différences dans la constitution du champ de gravité, ce qui indique une liaison entre la tectonique actuelle et les structures de fond du Caucase. Les plus actives en relation séismique sont les zones de mouvements contrastes, le long desquelles au cours du quaternaire (y compris le quaternaire supérieur) avait lieu une reconstruction intense des structures et du relief. En guise d'exemple de cela peut servir le flanc sud du Caucase Majeur, sur lequel "se justapose" la dépression d'Alazano-Agritchai (Lilienberg, 1959).

Schema des mouvements téctoniques récents du Caucase

Légende

I. Mouvements tectoniques holocènes d'après les données géologiques et géomorphologiques.
   Elèvements: 1 - Les plus intensifs, 2 - intensifs, 3 - moyens, 4 - faibles, 5 - différentiels, avec tendance d'élèvement
   Abaissements: 6 - intensifs, 7 - moyens, 8 - faibles.

II. Mouvements Contrastes (succédé du pliocène): 9 - mouvements contrastes intensifs, 10 - mouvements contrastes.

III. Vitesse des mouvements téctoniques récents (mm/ans) d'après les données instrumentales (selon M. I. Siniagina et W. A. Matzkova)
   Elèvements: II - plus de +4, 12 de +4 à +2, 13 - de +2 à -0.
   Abaissements: 14 de 0 à -2, 15 - de -2 à -5, 16 - plus de -5.

IV. Les données séismiques: 17 - épicentres des grands séismes.

V. Elèvements néotéctoniques: 18 - isobases des élèvements (2000, 3000, ...) et des abaissements (-2000, -3000, ...) en m, 19 - les plus grandes fractures actives, 20 - flexures néotectoniques, 21 - les plus grands volcans, 22 - laccolithes, 23 - salses.

VI. Amplitudes Sommaires des mouvements pléistocènes (en somme)
   Elèvements (en m): 24 - plus de 400, 25 - de 400 à - 200, 26 - de 200 à -100, 27 - moins 100.
   Abaissements (en m): 28 - moins 200, 29 - de 200 à -500, 30 - plus de 500.

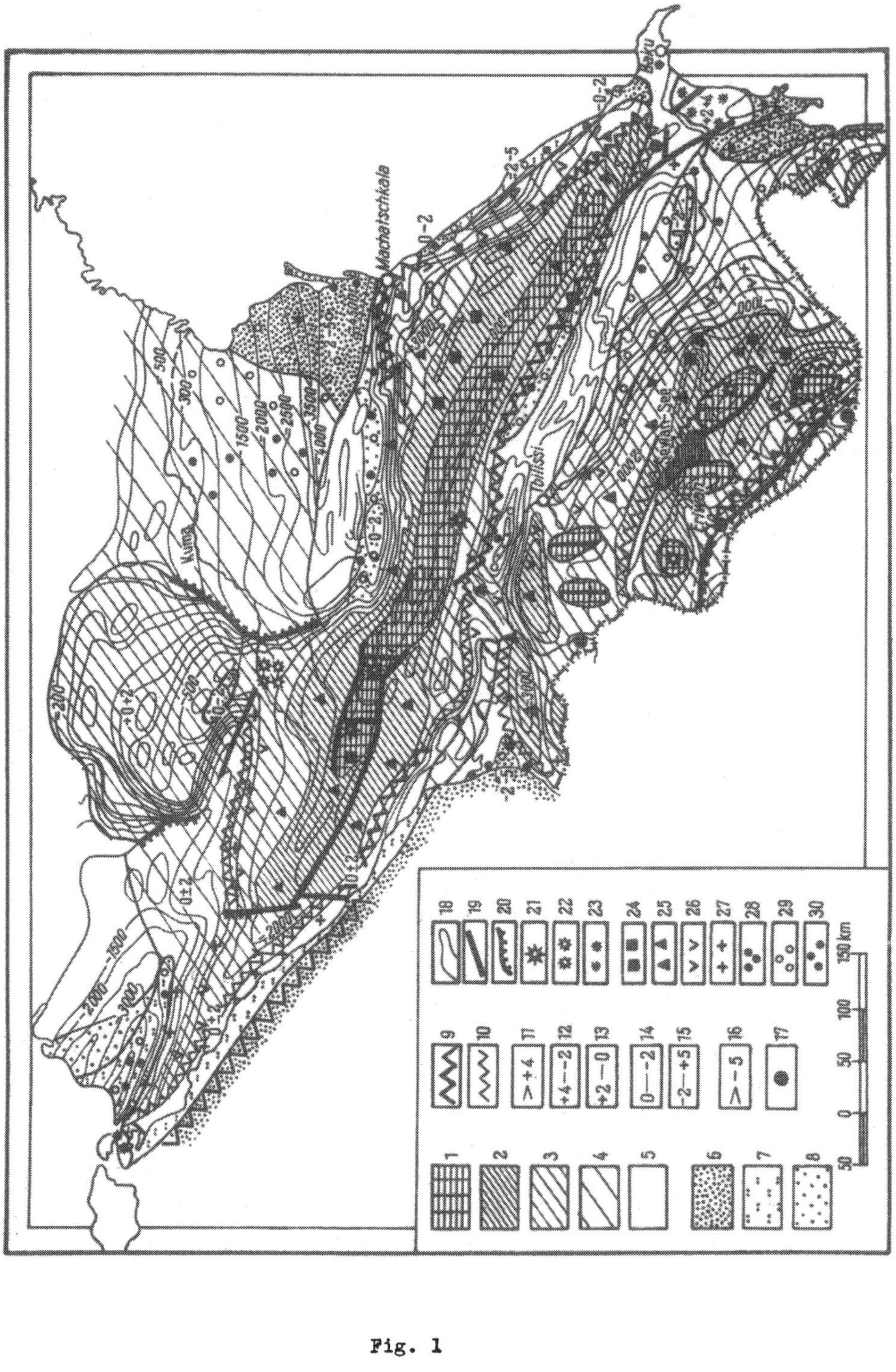

Fig. 1

Literature

Agabekov, M.G.; Akhmedbeili, F.C.: Sur l'étude de la néotectonique en Azerbaidjan. "Izv. Ac. Nauk Azerb. SSR, série géolog.-géograph.", 1957, No. 7

Bregman, R.G.: Altitudes géodésiques en relation au problème du niveau de la Mer Caspienne, "Météorol. i Gydrologia", 1947, No. 1

Doumitraschko, N.V.; Lilienberg, D.A.: Application des méthodes géomorphologiques dans les investigations séismotectoniques. "Travaux de l'Institut de Géophysique", vol. 25 (152). Problèmes du pronostic des tremblements de terre, Ed. Ac. Sci. URSS, M. 1954

Doumitraschko, N.V.; Milanovski, E.C.; Khain, V.E.: Tectonique récente du Caucase. Tectonique récente. Ac. Sci. URSS, M., 1961

Kirillova, I.V.; Lustikh, E.N.; Rastvorova, V.A.; Sorski, A.A.; Khain, V.E.: Analyse du développement géotectonique et de la séismicité du Caucase. Ed. Ac. Sci. URSS, M., 1960

Lilienberg, D.A.: Particularités du relief et mouvements tectoniques récents de la géologie et néotectonique du Daghestan. "Matér. de la Conférence de l'Union Soviétique sur l'étude de la période quaternaire", vol. 2, ed. Acad. Sci. URSS, 1961

Rikhter, V.G.: Sur l'évaluation de la méthode de nivellements réitératifs au cours de l'étude des mouvements tectoniques actuels. "Bull. MOIP", division géol., 1957, vol. 32, No. 2

Siniaguina, M.I.: Mouvements verticaux actuels du littoral de la Mer Caspienne. "Géodésia et Kartographia", 1959, No. 8

Mouvements verticaux modernes de la croûte terrestre sur le territoire de la moitié occidentale de la partie Européenne de l'URSS. "Travaux de l'Institut Central de recherches scientifiques de Géodésie, de levés aériens et de cartographie", cah. 123, M., 1958

Schlepnev, N.I.: Variations épeirogéniques dans la Transcaucasie et les niveaux des mers Noire et Caspienne. "Izv. Ak. Nauk Azerb. SSR", 1947, No. 2

Diskussion

I. P. Destombes (Paris):

Fait remarquer que dans quelques bassins hydromineraux francais qu'il a pu étudier avec quelque détail, les directions des failles thermales, vivantes en quelque sarte, sont soit fortement obliques soit nettement orthogonales vis à vis des directions tertiaires.

La séismicité actuelle du Caucase en direction méridienne sur laquelle les auteurs de la communication, ainsi que Monsieur Egyed intervenu à leur suite ont alteré l'attention, est intéressante à souligner.

L. Egyed (Budapest):

Die Bewegungen, die von den Autoren beobachtet wurden, stimmen gut mit den Untersuchungen von Herrn Pavoni überein.

On the inheritance of the contemporary movements of the earth's crust on the territory of the Ukrainian SSR from older movements

von

I. L. Sokolovsky (Kiew)

Summary

The neotectonic and contemporary movements of the Earth's crust embrace the entire territory of the Southwestern part of the Russian platform and the adjoining mountain structures of the Carpathians and the Crimea. The neotectonic (Neogene-Quaternary) movements are variable in time and space in respect to direction of motion and intensity, but are, as a whole, unitypical within the limits of separate structural elements, although formed in most cases at earlier stages of geological history, but still continuing their development. The contemporary movements directly continue the neotectonic movements: in this sense, they are inherited, and reflect the present stage of the tectonic development of the Earth's crust.

Zusammenfassung

Die neotektonischen und derzeitigen Bewegungen der Erdkruste schließen das gesamte Territorium des südwestlichen Teiles der Russischen Hochebene und der angrenzenden Bergstrukturen der Karpaten und der Krim ein. Die neotektonischen Bewegungen variieren bezüglich Bewegungsrichtung und Intensität mit der Zeit und dem Ort. Sie sind aber als ein Ganzes, Einheitliches innerhalb der Grenzen der trennenden Strukturelemente aufzufassen, obgleich sie in den meisten Fällen in früheren geologischen Zeitepochen einsetzten und noch heute ihre Entwicklung fortsetzen. Die gegenwärtigen Bewegungen setzen direkt die neotektonischen Bewegungen fort: in diesem Sinne sind sie gewissermaßen erblich und spiegeln die gegenwärtige Stufe der tektonischen Erdkruste wider.

The beginning of the latest neotectonic stage of development of the structure on the territory of the Ukraine occurred at the end of the Neogene and coincides with the completion of the Alpine orogenic movements, as a result of which the mountain structures of the Carpathian and Crimean regions became attached to the Russian platform.

The neotectonic movements are manifested in the change of the coastlines of marine basins, the high position of marine and continental deposits of a certain age and their corresponding morphological levels.

After the regression of the Kharkov Sea (Early and Middle Oligocene), the greater part of the territory of the Ukrainian SSR became dry land, which was covered by sea basins in isolated areas only and for brief periods. The Neogene transgressions on the platform part of the territory embraces mainly the Black Sea area depression, the Volyn-Podolian platform, the Forecarpathian depressions, they occupied a relatively smaller area than the Paleogene transgressions. During the Aquitanian-Helvetian sea basins existed on the left-bank part of the Black Sea area and invaded the southeastern part of the Dnieper-Donets depression (basin of the Orel and Samara rivers), the Forecarpathian sag, the south-western slope of the Russian Platform and the Transcarpathian depressions. During the Tortonian (Tarkhan and Chokrak times) the sea of the Black Sea area reached only to Zaporoshye, but expanded considerably in the part of the platform adjoining the Fore-carpathian sag; during Konka times the sea covered the northern and western part of the Bessarabian platform; it expanded still more during the Early and Middle Sarmatian, but during the Upper Sarmatian the sea remained only in the Black Sea area and Mukachev depressions, where it was preserved up to the Pontian, and during the Cimmerian and Kuyalnik it occupied only a small part of the Sivash area. The Quaternary marine basins occupied small areas only, and underwent local fluctuating movements.

A quantitative analysis of the summary amplitudes of the neotectonic movements during the Neogene-Quaternary period, carried out with allowance made for the contemporary altitude of the roof of the Paleogene marine deposits, the facial peculiarities of these deposits and the altitude of the corresponding geomorphological levels has led to the conclusion that in the platform part of the territory of the Ukrainian SSR the region of maximum uplifts with a summary amplitude of over 300 m is located in the middle of the Volyn-Podolian platform, to the west of the city of Ternopol; furthermore, this area underwent inversion of movement during the Sarmatian and began to rise only after the regression of the Middle Sarmatian Sea (the summary amplitude is defined as the difference between the altitudes of the given areas at the beginning and at the end of the stage in question).

The rest of the Volyn-Podolian platform underwent uplift with a summary amplitude of 100 - 300 m (in the south, 200 - 300 m; in the north, when there was no inversion of movements during the Sarmatian and the uplifts developed during a longer time, 100 - 200 m).

The Ukrainian Crystalline Shield also underwent maximum uplifts with a summary amplitude of 200 - 300 m in its western part (the Vinnitsa district), where the uplift began only after the Middle Sarmatian. In the eastern direction the summary amplitudes of the uplifts decreased to 100 m east of Kirovograd, to 50 m in the Krivoy Rog district, diminished still more in the district of Zaporozhye, and again increased to 200 - 300 m in the northern part of the Azov massif. In the Zaporozhye district subsidences took place up to the Middle Sarmatian, and in this area uplifting movements continued to lag behind after the inversion; whereas the western part of the shield and the middle of the Volyn-Podolian platform, having undergone inversion, were uplifted with maximum intensity.

The northern part of the Bessarabian platform underwent, during the neotectonic stage, uplifts having a summary amplitude which decreased from north to south from 100 - 200 m to zero; whereas the southern part subsided with a summary amplitude of up to -- 200 m. The Black Sea area depression subsided with a summary amplitude of up to -- 300 m (in the Sivash Area); the Dobrudzha sag, up to -- 500 m; the Dnieper - Donets depression on the whole uplifted with a summary amplitude of 50 - 100 m, although some areas underwent movements varying in amplitude and sign; the Donets basin uplifted with a summary amplitude of up to 300 m. Complicated tectonic movements occurred in the geosynclinal regions — the Carpathians were uplifted by degrees with a summary amplitude of up to 1500 m; intense subsidences prevailed in the Forecarpathian foredeep up to the Middle Sarmatian, which led to the accumulation of a thick Lower Miocene; later the subsidences were followed by uplifts; which gave a summary amplitude of movements of — 2000 m. The Upper Tisen and Chop-Mukachev depressions subsided intensely during the Miocene; at the close of the Miocene and beginning of the Pliocene inversion of movements occurred in the Upper Tisen and northern Chop-Mukachev depressions, whereas the main part of the Chop-Mukachev depression continued to subside. Along the joint of the subsiding and uplifting parts of the Transcarpathian depressions there occurred volcanic eruptions and outbursts, which formed the effusive-tuff stratum of the Vygorlat-Huta volcanic ridge. The summary amplitude of the subsidence of the Transcarpathian depression considerably exceeds

1000 m; the amplitude of the Upper Miocene -- Quaternary uplifts in the Upper Tisen depression and in the northern part of the Chop-Mukachev depression is 500 m.

Mountainous Crimea uplifted intensely, but not uniformly, with an amplitude of up to 1500 m throughout the neotectonic stage. The Alma depression subsided with a summary amplitude of about 200 m; the Indol depression, with an amplitude of over 500 m. The intense sagging of the Kerch peninsula at the close of the Negene was followed by differentiated uplifts, which resulted in an extremely complicated picture of the distribution of positive and negative summary amplitudes.

An analysis of the summary amplitudes of the neotectonic movements for a shorter interval - from the end of the Middle Sarmatian - conducted with the aim of studying the latest movements of the Earth's crust during the formation of the relief on a considerable part of the territory of the Ukrainian SSR showed that in the regions undergoing inversion of movements during the neotectonic stage (central part of the Volyn-Podolian platform, Forecarpathian and Transcarpathian sags, part of the Kerch peninsular, etc.) the summary amplitude of the neotectonic movements is greater than for the entire Neogene-Quaternary period. The isobase + 300 m contours the entire central part of the Volyn-Podolian platform, the western part of the Ukrainian Crystalline Shield, passes to the south of Kishinov on the Bessarabian platform; the isobase + 200 m embraces the central part of the shield, the Volynian elevation, the southern part of the Bessarabian platform, the most elevated part of the Azov area massif and the main part of the Donets Basin (only the axial part being uplifted with a summary amplitude of over 300 m); uplifts with a summary amplitude of 100 - 200 m embrace the northern part of the Volyn-Podolian platform, the northern and eastern parts of the Ukrainian crystalline shield, the Dnieper-Donets depression (except for its central part which underwent a slower uplift with an amplitude of less than 50 - 100 m), the northern part of the Black Sea area depression and the Tarkhankut uplift; uplifts with a summary amplitude of 0 - 100 m were established for the northern part of the Black Sea area depression. Districts with negative summary uplift amplitudes are located in the axial part of the Black Sea area depression and in the eastern part of the Dobrudzha sag and the subsided Dobrudzha slope.

At present - as can be seen by considering the results of rerun levellings, performed and processed by workers of the Central Research Institute for Geodesy, Aerial Surveying and Cartography and the Institute of Geography of the Academy of Sciences of the Ukrainian SSR and the results of geologo-geomorphological investigations - the movements of the Earth's crust on the described territory attain considerable intensity.

The most intense uplifts (over 100 mm per year) are those of the Carpathian region, the Styr-Goryn interfluve fairly accurately coinciding with the contour of occurrence of Riphean effusive rocks, a small part of the Bessarabian plateau south of Faleshty and the district of the Ingul-Ingulets anticlinorium. A rate of uplift from 7.5 to 10 mm per year distinguishes practically the whole of the Volyn-Podolian platform (to the west of the line Zaleshki-Terebovlya-Slavuta-Korets), the entire zone of the Krivoy Rog structures, the Shepetovka-Novograd-Volynsky district and a narrow strip surrounding the maximum uplift south of Faleshty. Uplift at a rate of 5 to 7.5 mm per year was noted for the northwestern part of the Ukrainian crystalline shield, the northern part of the Black Sea area depression and part of the southwestern slope of the Voronezh massif in the enterfluve of the Oskol and Northern Donets. The rest of the described territory is distinguished by uplifts at a rate of 2 - 5 mm per year; the only exception being strips of about 100 km in width crossing the crystalline shield between the Dniester valley in the vicinity of Mogilyov-Podolsky and the Dnieper valley in the vicinity of Kiev, strips which stretch along the subsidence zone in the northern part of the Dnieper-Donets depression (in the vicinity of Chernigov there is a local area of uplifts with a rate of over 2 mm per year) in the North Donets sag, in the left-bank area of the Dniester and in the Sivash part of the Black Sea area (except for the local Perekop elevation where uplifts at a rate of 0 - 2 mm per year were established).

The subsidence zones occupy a relatively small area. They belong to the northern part of the Dnieper-Donets depression, the Dobrudzha sag and the subsided slope of Dobrudzha, the edge of the Black Sea area depression, the Sivash part of the North Crimean sag and the Indol depression.

The author has considered only three maps which characterise the development of neotectonic movements during various intervals of time. This is, naturally, insufficient for clarifying the detailed history of the neotectonic movements both for the territory as a whole as well as for separate regions. However, there are grounds for formulating certain conclusions based on the data that have been presented:

1. Neotectonic movements embrace the entire territory of the southwestern part of the Russian platform and the adjoining mountain structures of the Carpathians and the Crimea. They are manifested in various parts with varying direction and intensity. Change in the direction of the movements has been recorded on the Volyn-Podolian platform, in the Fore-carpathian foredeep, in the Upper Tisen depression, in the Kerch folding zone, in the northern part of the Dnieper-Donets depression and in a number of other localities; there is no doubt that more detailed studies will establish a wider occurrence of inversion of neotectonic movements. The ratio of the movements have also changed with time, which is confirmed by a number of geomorphological features, most of all river terraces, indicating alternations of periods of activation of movement and relative quiescence, and the comparison of data on contemporary movements, the rate of which attains 10 and more mm per year with summary amplitudes indicating a considerably slower average rate of movements -- of the order of several thousandths or, at most, hundredths of a millimetre per year.

2. During the neotectonic stage movements of the Earth's crust have occurred under concrete conditions in earlier formed structures, which continue their development during Neogene-Quaternary times; they are, therefore, unitypical within these structures.

3. Contemporary movements of the Earth's crust continue directly the neotectonic movements. In this respect they are inherited and reflect the contemporary stage of the tectonic development of the Earth's crust.

Diskussion

A. Zátopek (Prag):

Mich würde speziell interessieren, ob auf dem Gebiet der Karpatenukraine die Hebungen und Senkungen etwas detaillierter verfolgt wurden, da ich dieses Gebiet auch selbst studierte.

I. L. Sokolovsky (Kiew):

Sie wurden studiert, Einzelheiten sind in meiner Arbeit angegeben.

Latest and present tectonic movements in the central part of the Kola Peninsula

von

M. K. Grahve and V. J. Evzerov (Moskau)

Summary

The data on vertical movements in the central areas of the Kola Peninsula are mostly concerned in the Khibini tundra. As for the Lovozero massif, there are only general conceptions of the possibility for local upheaval.

On the sides of the Lovozero Lake the authors have stated developments of Late- and Post-Glacial lake sediments and terraces corresponding to the lake level of 162 m. The similar sediments and relief forms on the eastern slope of the Lovozero tundra are elevated up to 172 m. The difference in the shore-line positions indicates to Post-Glacial mountain upheavals for about 10 metres. The upheaval is confirmed by analyses of the shore morphology and seismic survey data. The Seidozero depression, that is in the centre of the Lovozero tundra, is of the tectonic origin and its features indicate to relative subsidence. Therefore, the upheaval of the Lovozero massif is uneven and of block character.

The southern and northern sides of the Lovozero Lake have features of submersion while the eastern shore is probably neutral. The geological and magnetometric data indicate to the presence of meridional fractures directed to the lake from the South. By all probability, the subsidence of the Lovozero basin is going on precisely along these fractures. Thus, the latest tectonic movements in the central part of the Kola Peninsula are characterized by differentiation both in amplitude and, probably, in sign.

The Baltic shield and its northeastern part - the Kola Peninsula - have been studied for many years as an area of vertical movements of the earth's crust. The epeirogenetic upheaval of this territory, recorded by geological features, is more exactly pronounced by the inclination of Late- and Post-Glacial marine terraces. W. Ramsay (1898) was the first to compile an isobase map of the Post-Glacial upheaval in the Kola Peninsula, and in this map it is obvious that the rate of the movement of the earth's crust was the highest in the southwestern part of the territory.

A.A. Polkanov (1937) stated that "the total amplitude of the upheaval in the Kola Peninsula is probably equal to some hundred metres" (p. 75), and the uplift in the peripheral parts of the Peninsula is considerably slower than in the centre; for the last three thousand years the average rate of the upheaval equals to 0.25 - 0.5 cm per year at the sea-coast, but as A.A. Polkanov states it is 10 times higher in the central part though he notes that this value is exaggerated. The latest investigations by M.A. Lavrova (1939, 1960) have proved the common epeirogenetic scheme in the Kola Peninsula and made it possible to specify the isobase map.

Against the common regular upheaval of the peninsula there may be distinguished some areas of disharmonious movements along tectonic lines. In analysing the epeirogenetic spectrum V. Tanner (1930) stated young tectonic disturbances of Late-Glacial sea levels in the Ribachiy Peninsula. A.A. Polkanov (1937) states that the given disharmony of movements can take place at the margin between the Kildin Island and the continent. There are known some other examples of young movements along the breaks in peripheral parts of the Kola Peninsula.

As to the information of the upheaval character in the central parts of the peninsula they mostly refer to the Khibini massif. The probability of differentiated Late-Glacial upheavals in the Khibini more energetic relatively to the adjacent territory was noticed by W. Ramsay (1898, Fig. 8). Further, the features of young movements different by sign were stated by G.D. Richter along the sides of the Imandra Lake (1934). As Richter states the relative Post-Glacial upheaval of the Khibini, recorded in the terrace levels on the eastern bank of the Imandra Lake, is about 40 - 45 m. These conceptions have been improved by new data. A.D. Armand (1960) has shown that the terraces along the eastern side of the Imandra Lake are the relics of local lakes impounded with stagnant ice. They are located at various levels and are not correlated. Therefore they can't be used as the base for a quantitative characteristic of vertical movements of the Khibini massif. Besides the present upheaval of the Khibini can be proved not only by the morphology of the Khibini side of the Imandra Lake given by G.D. Richter in detail, but also by the structures of the eastern mountain slope adjacent to the Umbozero Lake. Here, so as on the western slope of the Khibini, the upheaval features are reflected in the coastal spots adjacent to the mountains.

By the level of the old dry bed of glacial stream over the present level of the Umbozero Lake, A.D. Armand (1959) stated the differential upheaval of the Khibini massif for the Post-Glacial period to be about 1.5 - 2.0 m or of order 0.3 mm per year. The present tectonic movements in the Khibini are proved by the observations of the seismic station of "Apatiti". In the mountain massif and in the meridional fault westward coinciding with the Imandra Lake there have been recorded several earthquake epicentres up to 4 units. In addition there may be distinguished other two zones of weak earthquakes connected with the areas, epicentres of which couldn't be exactly recorded (Panasenko, 1959).

As to the Lovozero massif there have been only general conceptions of a possible local upheaval and they are based on the resemblance of its structure with that of the Khibini (Ramsay, 1898; Lavrova, 1939, 1960 etc.). A.D. Armand (1959) noticed that in the part adjacent to the Lovozero tundra the eastern side of the Umbozero Lake is distinctly characterized by features of a relative upheaval. The investigations carried out by the authors in the area of the Lovozero tundra since 1959 have broadened the understanding of latest and present tectonic movements in the central part of the Kola Peninsula.

The Lovozero tundra, as well as the Khibini, is a Paleozoic intrusion of nepheline-syenite injected into Archean gneisses. It forms a horseshoe-shaped massif in the central part of which there is an eastward open depression with the Seidozero Lake. In the East and West the mountains are limited with meridionally elongated lakes of Lovozero and Umbozero, respectively.The main relief feature of the Lovozero massif is in its plateau, flat character. Probably, it is the relic of an old peneplain elevating over the surrounding plain for about 600 - 800 m. The massif is limited by steep slopes and is divided by a net of deep vallays; it carries some evident marks of glaciation in forms of corries, troughs and moraines (Abb. 1).

One of the youngest components of the stratigraphic column of the region is the stratum of Late- and Post-Glacial lake deposits which is developed along the banks of Lovozero Lake and is called as the Lovozero stratum (Grahve and others, 1962). It comprises the flat lake plain and has been traced along the eastern slope of the Lovozero tundra. In the examined cross-sections of the stratum there may be noticed the prevalence of sands different in grain size with horizontal, criss-cross or wave-like bedding; horizons of well rounded pebbles and boulders may be also found there.

The stratum in the whole is characterized by a gradual increase of the material

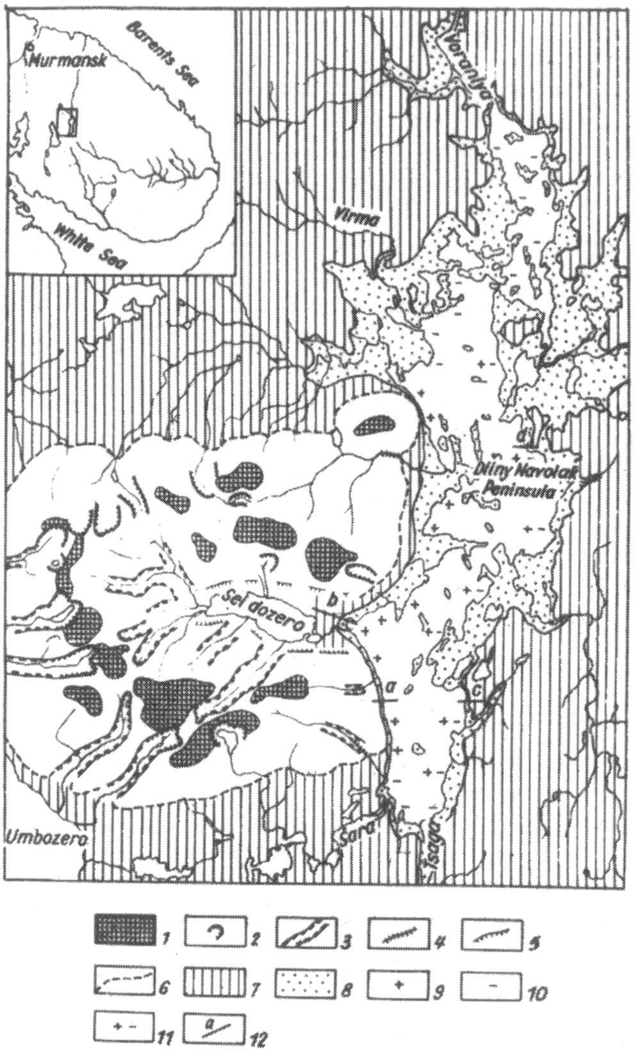

Abb. 1

Lovozero Mountain Massif

1 - Plateaus; 2 - Corries; 3 - trough valleys; 4 - entrenchments; 5 - slopes of the Seidozero kettlehole; 6 - boundaries of the mountain massif; 7 - glacial accumulation plain; 8 - lake plain; 9 - uplifting banks; 10 depression banks; 11 - banks mostly of neutral position; 12 - profile lines

grains upward the cross-section. The sand and pebble occurences in the limits of the lake plain, the character of their bedding and the regular change of the granulometric composition prove the sedimentation of these deposits in a basin the depth of which gradually decreased in the course of time. The stratigraphic setting of the Lovozero sands is characterized by their general unconformable superposition over the moraine and by fluvioglacial deposits of the last mantle glaciation. It is also characterized by the pollen composition. The upper part of the Lovozero stratum is mostly rich in pollen of pine (Pinus up to 26 pct.) and birch (Betula, 20 pct.); the percentage of alder pollen (Alnus) is lower (14 pct.). This corresponds to the first period of the Post-Glacial development of vegetation in the Kola Peninsula (Lavrova, 1960). The above-mentioned indicates to the

Late- and Post-Glacial age of the stratum. Another more exact treatment is still unavailable.

On the western and eastern sides of the Lovozero Lake, and also on the islands, the Lovozero stratum traced around the lake as a continuous development is slightly different in the granulometric composition. Coarse- and middle-grained sands are prevailable on the western side. On the eastern side fine-grained sands and sandy loam are found in amounts equal to the above-mentioned. The tops of nearly all cross-sections are composed by pebble. The study of the petrographical composition of the elastic material indicates that its main part was transported into the Lovozero stratum in the result of erosion of the mantle moraine. The amount of erratic pebble is not higher than 48 - 56 pct. even down the slopes of the Lovozero tundra, where the homeland rocks should be prevailable.

The absolute levels of superface lake deposits round the whole perimeter of the lake, except the foot of the Lovozero tundra, are in the limits of 162 m, i.e. 8 - 9 m higher than the present lake level x). This level also comprises rock benches and aggradation terraces xx) developed on the slopes of morainic and fluvio-glacial hills, and on tł slopes of bedrock elevations along the eastern side of the Lovozero Lake (Abb. 2) and also on the most high islands. Besides terraces there may be also found sandy-pebble beach barriers and zones of boulder accumulations probably washed out of the moraine. No moraine lacking the features of rewashing has been found on the lake plain down the level of 162 m.

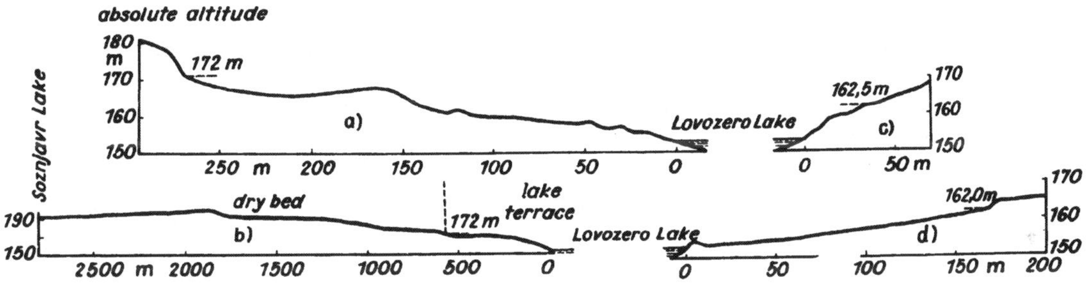

Abb. 2

Profiles across the banks of the Lovozero Lake

a - western side down the slope of the Lovozero tundra

b - dry bed in the interlake space between the Seidozero and the Lovozero lakes

c - eastern side down the slope of the morainic hill (after A.D. Armand)

d - eastern side at the foot of the base rock elevation

The profile settings see in Abb. 1

---

x) Seasonal variations are in the limits of absolute levels 153 and 154 m

xx) The altitude position of the terraces was determined by protractor

The common character of the relief and absolute levels of the lake plain, the identity of deposits composing its cover, the presence of terraces and beach barriers corresponding to the same level and the regular change of the lithological composition of sediments in the vertical direction - all these generally prove the fact that in the area of the Lovozero Lake in the Late- and Post-Glacial period there was a common lake basin of more larger area.

Nowadays the lake drainage is only northward through the Voronya River running in a wide depression open to the Barents Sea. In the past there probably was an obstacle in this depression and it resulted in a higher level of the ancient Lovozero. No tectonic features, which could form such a dam, have been stated. It is most evident that this was due to masses of dead ice, the existence of which in the North, in the middle part of the Voronya River, is stated (N.N. Armand, 1960) for the period of mantle ice degradation. Rich water supply to this basin was due to melting of mountain ice and remains of mantle ice.

On the eastern slope of the Lovozero tundra the lake deposits of the above-mentioned type are advanced up to the absolute levels of 172 - 173 m. Nearly at the same level there is a series of pebble beach barriers elevating regularly from the bank to the mountains (Abb. 2a). We state that the anomalously elevated hypsometric positions of lake deposits and the related to them relief forms at the foot of the mountains justify a relative (and absolute) upheaval of the Lovozero massif in the Post-Glacial time in comparison with the analogous developments in other parts of the coast (Abb. 2c, d). In accordance with the above-mentioned data the upheaval amplitude supposed to be of order 19 m or about 1.2 mm per year in average.

The character of the Lovozero banks adjacent to the eastern slope of the mountain massif indicates to a uplifting direction of their development. Its features have been stated by analyses of airborne photos and by geomorphological observations, i.e. by means used by G.D. Richter in the area of the Imandra Lake (Richter, 1934) and by A.D. Armand in the area of the Umbozero Lake (Armand, 1959).

The banks of the Lovozero Lake are rectilinear as a rule. There may be found small lakes separated from the main water area with plain swamped spots. The straightforwardness of the banks is distributed in one place by the cape and peninsula of Dliny Navolok. They are formed by small hills, formerly islands, as they are connected with the continent by a plain belt dried probably not so long ago. In the western bank there are bays drainaged in their tops (adjacent to the mountains) and rich in silt. As Richter states this may be also considered as a feature of regressive movement of coast line. The character of the Lovozero banks removed from the massif is rather different and shows that the land sinking was predominant over the water level. This will be discussed in detail later on.

The present upheaval of the massif may be proved by the morphology of mountain valleys on the eastern slope of the Lovozero tundra. In the upper and middle courses the valleys mostly show well-preserved trough profiles. In the lower courses the trough bed is cut with the present stream channel. Therefore the valley resembles a narrow steep-walled canyon cutting into the trough and outcropping bedrocks. This indicates to intense erosion in the Post-Glacial period. As far as the western slope is oncerned the geomorphological features of positive movements have been stated by A.D. Armand (1959).

Due to the discussed questions geophysical studies are also of interest. In the Lovozero massif the seismic survey has located some seismic reflecting plane spots. They may be traced in the zone of contact of the intrusion with wall rocks at the depth of several kilometres and dip from the massif at the angle of about 12 - 45° (as G.N. Shablinsky says). The geophysicists engaged in these investigations consider the mentioned spots to be the traces of past and present tectonic movements. These movements change physical features of separate rock localities and therefore form a physical boundary causing reflections.

Thus, all these data show a rather intense Post-Glacial upheaval of the mountain massif in the whole. Against the background of the common upheaval some particular areas are probably subjected to relative movements of reverse sign. In this respect the area of the Seidozero Lake may be used as an example. As it has been mentioned above, the latter is in a deep kettlehole nearly in the middle of the Lovozero massif and is separated from the Lovozero Lake by an interlake space formed with thick masses of Quaternary deposits. As vertical electric soundings (V.E.S.) show, the maximum of soft sediment thickness is about 170 m (Grahve and others, 1962). The basement of the kettlehole under the Seidozero Lake is fixed at the absolute level of 90 m. Taking into account that the lake altitude is at the level of 189 m and its depth does not exceed 12 m it would be clear how high is the thickness of soft rocks filling up the Seidozero kettlehole.

The morphological characteristic feature of this kettlehole is in a straight line shape of its northern and southern slopes (Abb. 1). Including their altitude the total depth of the Seidozero depression is about 460 - 520 m and of order 2.5 km wide at the upper level. The interpretation of seismic and partly V.E.S. data makes it possible to suppose that under the northern slope of the kettlehole in the interlake area there is a sublatitudinal zone of intense crushing probably corresponding to the tectonic line buried under Quaternary deposits.

The lower position of the base rock bed, which is scarcely conformable to the possible base level of erosion in the Lovozero basin, the outline and depth of the kettlehole, the thickness of accumulated materials and geophysical informations - all these prove its tectonic origin. Now the relative continuous depression of the kettlehole is also not excluded. These phenomena may be evident from some morphological features of the bank zone of the Seidozero Lake. Along the northern bank there are relict lakes separated from the main water basin; in the western part of the lake there is a large creek being filled up with drift of the Elmaraiok River. This creek is expected to be separated by a subaqueous dam connecting a chain of small sandy islands. The following fact is also of great interest. At present the drainage of the Seidozero Lake runs to the Lovozero Lake through the Seidjavriok River in the southern part of the interlake area. In the northern part there is a dry bed starting from the Soznjavr Lake (on the coast of the Seidozero Lake) and entering the Lovozero interlake plain at absolute levels 172 - 173 m. By the position and character of the dry bed crossing the moraine it is evident that in the Late- and Post-Glacial time, in the period of the high level of the Lovozero Lake, there was a drainage from the Seidosero Lake. In the longitudinal profile of this bed there may be traced a bend, from which the dip is in counter-directions (Abb. 2b). As the form and width of the valley are constant through the whole stretch, the origin of the mentioned bend can't be explained by a sort of beheading. So, it may be supposed that this bend is originated either due to the interlake area uplifting or, most probably, to the continuous dipping of the Seidozero kettlehole. It is evident from the abovementioned that the common upheaval of the Lovozero massif goes on irregularly and may be of block movement type.

The lifting up western part of the coast line of the Lovozero Lake has been characterized above. The southern and northern sides and also a part of the western coast removed from the massif are characterized by features of lake advance. In the South the mouths of comparatively large rivers such as the Sara and Tsaga end in wide-open estuaries. The delta of the Tsaga River drifting sand in great amounts is completely submerged. Its surface is complicated with a series of about 10 - 12 subaqueous bars visible in airborne photographs. The most part of the coast is of abrasive character. One of the features of the dipping coastal zone is in the well-preserved glacial scars covering the rounded outcrops of base rocks at the water line westward from the Tsaga mouth. G.D. Richter (1934) supposes that the presence of glacial scars indicated to a recent release of rocks from glacial accumulations covering their surface.

Attention should be paid to a developed ruggedness of the coast generally plain and low lying on the northern side of the lake. There is a great number of little peninsulas, islands and bays here. They are all surrounded with a wide belt of sandy shoaliness. In airborne photos there may be seen subaqueous small rounded depressions (relics of lakes in the old coastal belt) and subaqueous river channels entering the lake (e.g. the Virma River). Besides there are seen submerged bars, beach barriers and small sandy islands. Thus, the northern and southern parts of the coast are characterized by evident features of the lake transgression.

The eastern bank is less definite in this respect. No pronounced features either of upheaval or of depression have been found there. There may be located abrasive as well as accumulative spots on the coast; the lake terrace corresponding to the Post-Glacial upper level of the Lovozero Lake is not subjected to any noticeable altitude variations and everywhere is at the level of about 162 m. Nearly along the whole coast there may be traced a single beach barrier perfectly depicted in the relief and this may be considered, as G.D. Richter states, as a feature of the neutral or weakly uplifting coast.

As some geologists have observed (from personal announcement of M.T. Kozlov) from the South a series of young or fresh meridional fractures in base rocks up to the Lovozero Lake, and along the eastern coast, the magnetometric survey has traced a regional tectonic line of submeridional orientation. In comparing these geological data with the above given geomorphological one can suppose that the latter justify a relative depression of the Lovozero Lake kettlehole along the fractures.

The given material shows that against the background of the present and latest regional upheaval of the Kola Peninsula in its Central Part there may be fixed a differentiation of latest tectonic movements as in amplitude and possibly in sign. The relatively rapid upheaval of the Lovozero massif, that is going on irregularly in its separate parts, is followed by a depression of the Lovozero Lake kettlehole. Further geomorphological data together with those of geological and geophysical prospectings will give a more tight correlation of these movements with the tectonic features of the basement and insure their quantiative characteristics.

R e f e r e n c e s

Armand, A.D.: "Short Geomorphological Characteristics of the Umbozero Lake Banks". Proceedings of the Karelian and Kola Filials of the Academy of Sciences of the USSR. No. 3, 1959

Armand, A.D.: "History of the Relief Formation and Quaternary Deposits of the Khibini Tundra". Articles on the Problems of Geomorphology and Geology of the Sedimentary Cover in the Kola Peninsula, v. 1, Kola Filial of the Academy of Sciences of the USSR, Apatiti, 1950

Armand, N.N.: "On the Last Glaciation in the North-Eastern Part of the Kola Peninsula". Articles on the Problems of Geomorphology and Geology of Sedimentary Cover in the Kola Peninsula", v. 1, Kola Filial of the Academy of Sciences of the USSR, Apatiti, 1960

Grahve, M.K.; Evzerov, V.J.; Likhachev, A.S.; Spitsin, A.N.: New Data on Soft Deposits and Relief Formation in the Seidozero Region (Lovozero tundra) - Articles on the Problems of Geomorphology and Geology of the Sedimentary Cover in the Kola Peninsula, v. 2, Kola Filial of the Academy of Sciences of the USSR, 1962 (in print)

Lavrova, M.A.: Quaternary Geology of the Kola Peninsula. The Academy of Sciences of the USSR, Moscow - Leningrad, 1960

Lavrova, M.A.: On the Problem of Epeirogenic Variations of the Kola Peninsula in the Quaternary Period. Works of the Soviet Section'INQUA, v. IV, 1939

Panasenko, G.D.: Earthquakes in the North-Eastern Part of the Baltic Shield. Articles on Seismic and Glaciological Studies during the International Geophysical Year, No. 2, Ac. Sci. of USSR, 1959

Polkanov, A.A.: History of Quaternary Geology in the North-Western Part of the Kola Peninsula. Works of the Soviet Section INQUA, v. III, 1937

Richter, G.D.: Physiogeographic History of the Imandra Lake and its Basin. Geographo-Economical Research Institute of the Leningrad State University, v. V, Gostekhizdat, Leningrad, 1934

Ramsay, W.: Über die Geologische Entwicklung der Halbinsel Kola in der Quartärzeit. Fennia, No. 1, 1898

Tanner, V.: Studier över kvartärsistement i Fennoskandias nordliga delar, IV. Om nivaförändringarna och grundragen av dem Geografiska utvecklingen efter istiden i ishavsfinland samt om homotaxin av Fennoskandias kvartära marina avlayringar. Fennia, LIII, No. 1, 1930

D i s c u s s i o n

T. S o r g e n f r e i (Kopenhagen):

I ask if the author was quite certain regarding the contemporaniety of the 162 m and the 172 m levels. I also want information on the potential occurrence of marine deposits in the area.

Furthermore I ask whether stable levelling points had been established in the area.

I finally point out that the interpretation of the observations made was a rather difficult task which, however, now had been initiated.

I express the hope that the author and his collaborators would succeed to proceed their work and improve the amount of evidence on dating and the crustal movements.

M. K. G r a h v e (Moskau):

These levels are of course contemporary. They correlate with the one series of sediments observed continuously along the perimeter of the Lovozero lake. At all the observational points the profile was of one type, moreover, the mineralogical analysis confirm its unity. No remnants of fauna have been discovered in these sediments, in different points the single type of the pollen and spores content has been determined as well as the diatomic flora. The pollen diagrams were being drafted, but they cannot be presented in the final form as the study is not finished yet. These deposits are not marine sediments. Hipsometric characteristics of the surface indicate that the basin could be connected with the sea only in its northern part along the valley of the river Voronia. As the report shows the valley was, evidently, partly blocked by dead ice which caused the high level of the water. The absence of marine fauna and the presence of the diatomic confirm the lake origin of the basin. I have no data on relevelling, as these studies were not conducted in the region.

Über die Bedeutung der Neotektonik in der morphologischen Entwicklung der polnischen Küste in der Nachlitorinazeit

von

B. Rosa (Toruń)

S u m m a r y

Two geomorphological research methods of recent movements of the earth's crust are dealt with in this lecture. They have been applied in that region of the coast of the Baltic Sea which belongs to Poland. The first of these methods is based on the investigation of the position of the buried plains of denudation and their contact with the cliffs. The second, however, is connected with the research of the internal structure of the old beach-sand-barriers dating from the litorina period.

The results of the research of the neotectonics of the coast hitherto obtained are treated in brief.

The tectonic factor of the post-litorina period was the chief or leading factor of the morphological development of the sea-shore. The eustatic factor had only a comparatively slight influence on the changes of the shore lines at that time.

Z u s a m m e n f a s s u n g

Der Vortrag behandelt zwei geomorphologische Forschungsmethoden der neuesten Bewegungen der Erdkruste, die im Bereich der polnischen Ostseeküste angewendet wurden. Die erste von diesen Methoden beruht auf der Untersuchung der Lage der verdeckten Abrasionsflächen und ihrem Kontakt mit den Fliffs. Die zweite steht dagegen in Verbindung mit der Erforschung der inneren Struktur der alten Strandwälle aus der Litorinazeit.

Die bisherigen Forschungsergebnisse über die Neotektonik der Küste werden kurz behandelt.

Der tektonische Faktor in der Nachlitorinazeit war der Haupt- oder der führende Faktor der morphologischen Entwicklung des Meeresstrandes. Der eustatische Faktor in dieser Zeit beeinflußte die Strandlinienveränderungen nur verhältnismäßig wenig.

E i n l e i t u n g

Die Nachlitorinazeit, die etwa die letzten 4000 Jahre umfaßt, bedeutet im Gebiet der südlichen Ostseeküste eine Periode der allgemeinen Verminderung des großen postglazialen Transgressionsvorganges der Ostsee. Schon im Anfang dieser Zeitperiode entstand eine neue dynamische Situation an der Küste, deren Hauptmerkmale die verhältnismäßig geringen eustatischen Schwankungen des Meeresspiegels waren [13]. In dieser neuen dynamischen Situation kam es plötzlich zur wesentlichen Umgestaltung der allgemeinen Küstendynamik. Die Folge dieser Umgestaltung war vor allem eine Anpassung der früheren, aus der Litorinazeit stammenden Strandlinie an die neuen, veränderten Verhältnisse der Küstendynamik. Ähnliche Veränderungen traten ebenfalls in anderen Küstengebieten auf [1, 5]. Dies alles weist auf eine große Schwächung des Einflusses des eustatischen Faktors auf die morphologische Entwicklung der Küste in der Nachlitorinazeit hin.

In dieser neuen dynamischen Situation erlangte der tektonische Faktor eine besondere

Bedeutung, der in den regional veränderten und langsamen Bewegungen des ganzen Küstengebietes zur Auswirkung kam. Die neotektonische Bewegung wurde jetzt einer der Hauptfaktoren der morphologischen Entwicklung der Küste.

Der tektonische Faktor hatte zweifellos einen bedeutenden Einfluß auf die Verschiebungen der Strandlinie auch in früheren Zeiten gehabt, wie z.B. während der Ancylus-, Mastogloia- oder Litorinatransgressionsphasen. Damals verspätete bzw. beschleunigte er die Verschiebungen der Strandlinie in südlicher Richtung. Damals spielte jedoch der eustatische Faktor in dieser Hinsicht eine führende Rolle [6]. Nun wissen wir aber noch nicht genau, welcher gegenseitige Zusammenhang zwischen diesen beiden eustatischen und tektonischen Faktoren damals bestanden hat. Von dem Gebiet der südlichen Ostseeküste, die jetzt hauptsächlich am Meeresboden gelegen ist, kennen wir leider bisher noch nicht den früheren Verlauf der ältesten Strandlinien des Meeres. Diesbezügliche Forschungen sind bei uns erst im Anfangsstadium. Dagegen können wir viel mehr über die Rolle dieser beiden Faktoren, insbesondere über die Rolle des neotektonischen Faktors in der Nachlitorinazeit sagen. Die Bedeutung dieses Faktors in der nachlitorinazeitlichen Entwicklung der Küste kann man durch den Vergleich des Verlaufs der letzten litorinazeitlichen Strandlinie mit der gegenwärtigen Strandlinie der südlichen Ostsee nachweisen. Im Gebiet der polnischen Küste ist die litorinazeitliche Strandlinie durch die alten Strandformen und Ablagerungen, die am Ende der Litorinatransgression entstanden sind, angedeutet [2, 13].

F o r s c h u n g s m e t h o d e n

Die höchste und gleichzeitig die jüngste Strandlinie des Litorinameeres zeigt jetzt im Gebiet der polnischen Ostseeküste folgende, nur teilweise bewahrte, alte Strandformen auf: tote Kliffe, alte Strandwälle sowie alleralteste Vordünen des Litorinastrandes. Tote Kliffe zeigen sie in Abrasionsabschnitten der Küste. Jedoch gehören diese Formen zur Seltenheit, da sie sich ebenfalls auch nach der Litorinazeit vorwiegend weiterentwickelten. Alte Strandwälle zeigen den maximalen Bereich des Litorinameeres in den Akkumulationsabschnitten der Küste, wo die Wanderung der Strandlinie während der Litorinatransgression verhältnismäßig schnell und sprungweise verlief. Die ältesten Vordünen zeigen ihn dagegen dort, wo die Strandlinienwanderung langsam und kontinuierlich war.

Auf der Zeichnung 1 ist mit schwarzer Linie der maximale Bereich des Litorinameeres angedeutet. Er zeichnet sich durch eine größere Anzahl von sanften Buchten und Landvorsprüngen gegenüber der jetzigen, besser ausgeglichenen Strandlinie aus.

Die möglichst genaue Bezeichnung der maximalen Wasserspiegelhöhe des Litorinameeres für die einzelnen, weit voneinander entfernten Strandabschnitte wurde in morphologischer Hinsicht ein grundsätzliches Problem für die Entdeckung und Feststellung der neotektonischen Bewegungen der Erdkruste an der südlichen Ostseeküste.

Die in dieser Hinsicht durchgeführten Geländeforschungen erlaubten vor allem festzustellen, daß in einigen Küstenteilen das maximale Niveau des Litorinameeres höher, in anderen dagegen niedriger als heute gewesen war. Diese scheinbar in Widerspruch stehende Feststellung, die endgültig und folgerichtig noch nachzuweisen wäre, bildete gleichzeitig die Entdeckung der neotektonischen Bewegungen im Gebiet der ganzen erforschten Küste.

Die alten Strandformen und Ablagerungen, die den Litorinameeresspiegel unter NN zeigen, senkten sich nämlich während der Nachlitorinazeit, und jetzt weisen sie darauf hin, daß das ganze Küstengebiet einer langsamen Senkung unterlag im Gegensatz zu anderen, wo sich eine relative Hebung des Landes bemerkbar machte und die alten Strandformen den maximalen Litorinameeresspiegel über NN zeigen.

Auf der Zeichnung 1 zeigt die schwarze Linie die heutige Lage der litorinazeitlichen Strandlinie im Verhältnis zum jetzigen Meeresniveau an. Die heutige Lage dieser Linie beweist, daß sich das Küstengebiet der Bucht von Gdańsk während der Nachlitorinazeit senkte, der mittlere Teil der Küste Polens dagegen einer relativen Hebung unterlag, schließlich der westliche Teil der Küste mit Ausnahme der Dźwirzyno-Gegend eine Senkungsbewegung er-

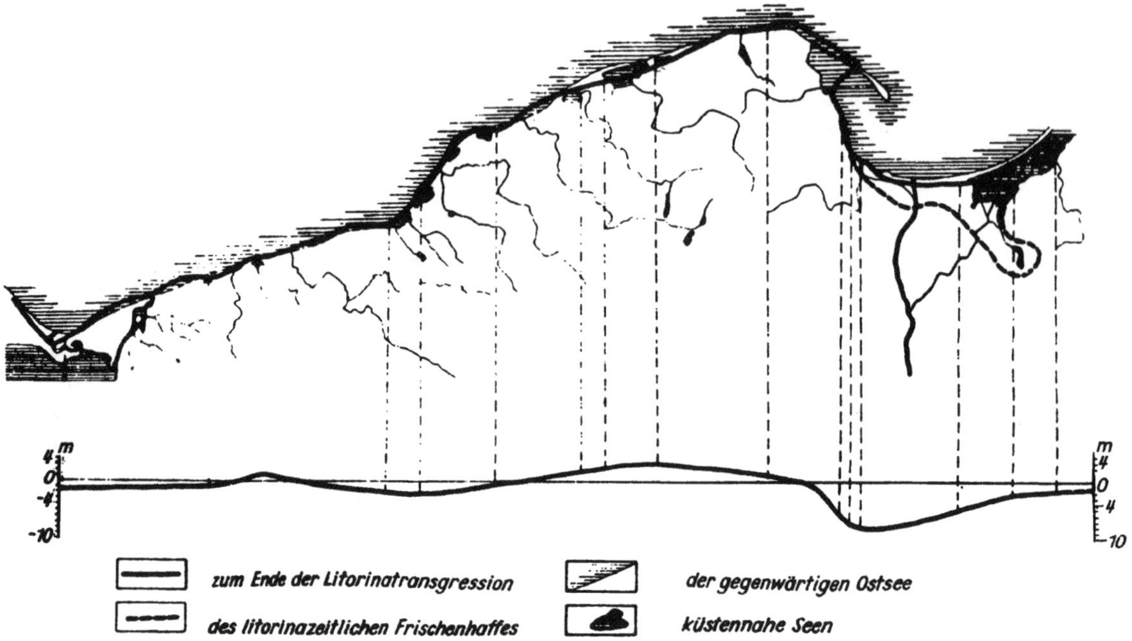

Abb. 1
Die Strandlinie der Litorinatransgression

neut aufwies.

Obiges beweist, daß alle geomorphologischen Methoden, die wir zum Zweck der Feststellung der Höhe des alten Meeresspiegels anwenden, gleichzeitig der Erforschung der neueren tektonischen Bewegungen der Küstengebiete dienen können.

Im Gebiet der südlichen Ostseeküste zeigten sich folgende zwei geomorphologischen Methoden hierzu besonders geeignet.

1. die Methode, die auf der Erforschung der begrabenen Abrasionsfläche und ihrer Kontakte mit den oft teilweise begrabenen Kliffüßen, alten Strandwällen und ältesten Vordünen beruht und

2. die Methode der Strukturanalysen des inneren Baues des proximalen Teiles der alten Strandwälle.

Diese beiden Methoden möchte ich am Beispiel der Łebagegend darstellen:

Ad 1. Auf der Zeichnung 2 wird ein geologischer Querschnitt durch die küstennahe Niederung in der Umgebung von Łeba schematisch dargestellt. Die pleistozänen Bildungen repräsentieren hier die kiesigen und grobsandigen Schichten der glazifluvialen Akkumulation. Von der Meeresseite her sind sie durch die begrabene Abrasionsfläche geschnitten. In einzelnen Bohrungen zeigen Meeresgerölle, kleine Steinchen und Kies diese Abrasionsfläche als das steinige, 30 bis 50 cm mächtige Basispflaster an.

Im Untergrund der Łeba-Nehrung und des Sarbsko-Sees ist diese Abrasionsfläche durch eine bis zu 3 m mächtige Sandschicht bedeckt [2]. Diese letzte Schicht, vom Meeressand aufgebaut, entstand am Boden der alten aktiven Zone des Strandes und enthält jetzt reiche Meeresfossilien mit Scrobicularia piperata u.a. Die Fossilien in der sandigen Schicht lassen also das litorinazeitliche Alter der Ostseetransgression in der Umgebung von Łeba bestimmen, dessen Spur vor allem in der begrabenen Abrasionsfläche festgelegt worden ist.

Mit Hilfe von Bohrungen wurde weiter nachgewiesen, daß die begrabene Abrasionsfläche mit einem alten Strandwall, der an der südlichen Seite des Sarbsko-Sees liegt, in Kontakt

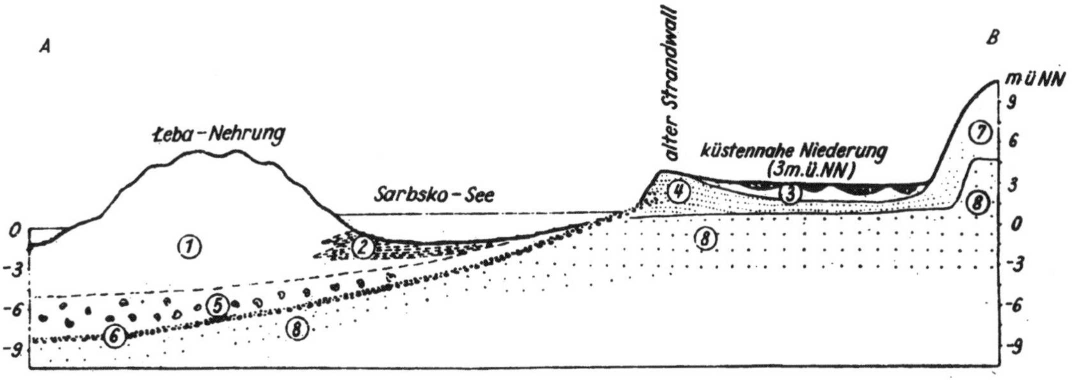

Abb. 2

Geologischer Querschnitt durch die küstennahe Niederung östlich von Łeba (Schema)

1. Sand
2. limnische Absätze
3. Flugsand und Moorerde
4. Sand, Kies, Steinchen und Gerölle
5. Meeressand mit Cardium, Macoma und Scrobicularia
6. Steine, Steinchen und Kies-Transgressionskontakt
7. Sand der spätglazialen Dünen
8. glazifluviale Ablagerungen

ist. Der Kontakt der Fläche mit dem Strandwall liegt jetzt in einer Höhe von etwa 1 m über NN. Deshalb ist es klar, daß der Spiegel des Litorinameeres in der Gegend von Łeba am Ende der Transgression mindestens um 1 m höher im Gegensatz zum heutigen Meeresniveau gelegen war.

Mit Hilfe dieser Methode wurde z.B. der höchste Litorinameeresspiegel östlich von Sopot in einer Tiefe von etwa -7 bis -6 m bestimmt, in Suchacz dagegen der höchste litorinazeitliche Spiegel des Frischen Haffs in einer Tiefe von etwa 2 m unter NN. In den beiden letzten Beispielen haben wir es mit dem Kontakt der begrabenen Abrasionsflächen und der teilweise auch begrabenen alten Kliffüße zu tun.

Die dargestellte Methode erlaubt jedoch den alten Meeresspiegel nicht genau zu bestimmen. Deshalb wurde sie durch eine zweite Methode ergänzt.

Ad 2. Der innere Bau des proximalen Teiles des alten, neben dem Sarbsko-See gelegenen Strandwalles ist auf der Zeichnung 3 dargestellt. Der proximale Teil des Walles enthält 23 verschiedene Schichten und Schichtchen. Tief gelegene Schichten, von 23 bis 5, sind entweder ausschließlich von Sand oder ausschließlich von Kies und Geröll aufgebaut. Die Geröllschichtchen des Walles sind nur 6 bis 10 cm mächtig und liegen fast horizontal. Sie reichen nicht so weit ins Innere des Walles (1,5 bis 3,5 m). Die oberen Schichten des Walles, von 4 bis 1 einschließlich, sind vom Meeressand und Kies aufgebaut und enthalten das grobe Material, Gerölle und Steinchen, räumlich ungeordnet untereinander.

Die nur vom Meeressand gebildeten und auf der beigefügten Zeichnung mit den Zahlen 23, 21, 19, 16, 15, 14, 13, 11, 8 und 6 bezeichneten Schichten entstanden unter dem Wasser am Boden der alten aktiven Zone des Strandes unterhalb der Brandungskante. Ihre Entstehung stand in Verbindung mit der Wanderung der Sandbänke, die sich am Meeresboden der alten aktiven Zone des Strandes des Litorinameeres entwickelten. Ähnliche, jetzt am Meeresboden wandernde Sandbänke hatte K.W. Geib [4] an der pommerschen Küste zum ersten Mal beschrieben. Die Neigungsrichtungen der Lawinen inmitten der auf der beigefügten Zeichnung dargestellten Sandschichten weisen weiter darauf hin, daß die alten Sandbänke am Boden des Litorinastrandes überwiegend nach Osten wanderten, also unter dem Einfluß der Winde, die hauptsächlich aus westlichen Sektoren wehten.

Die kleinen Schichtchen, die aber fast ausschließlich aus Meeresgeröllen aufgebaut sind und auf der Zeichnung mit den Zahlen 18, 9 und 5 bezeichnet wurden, entstanden auch

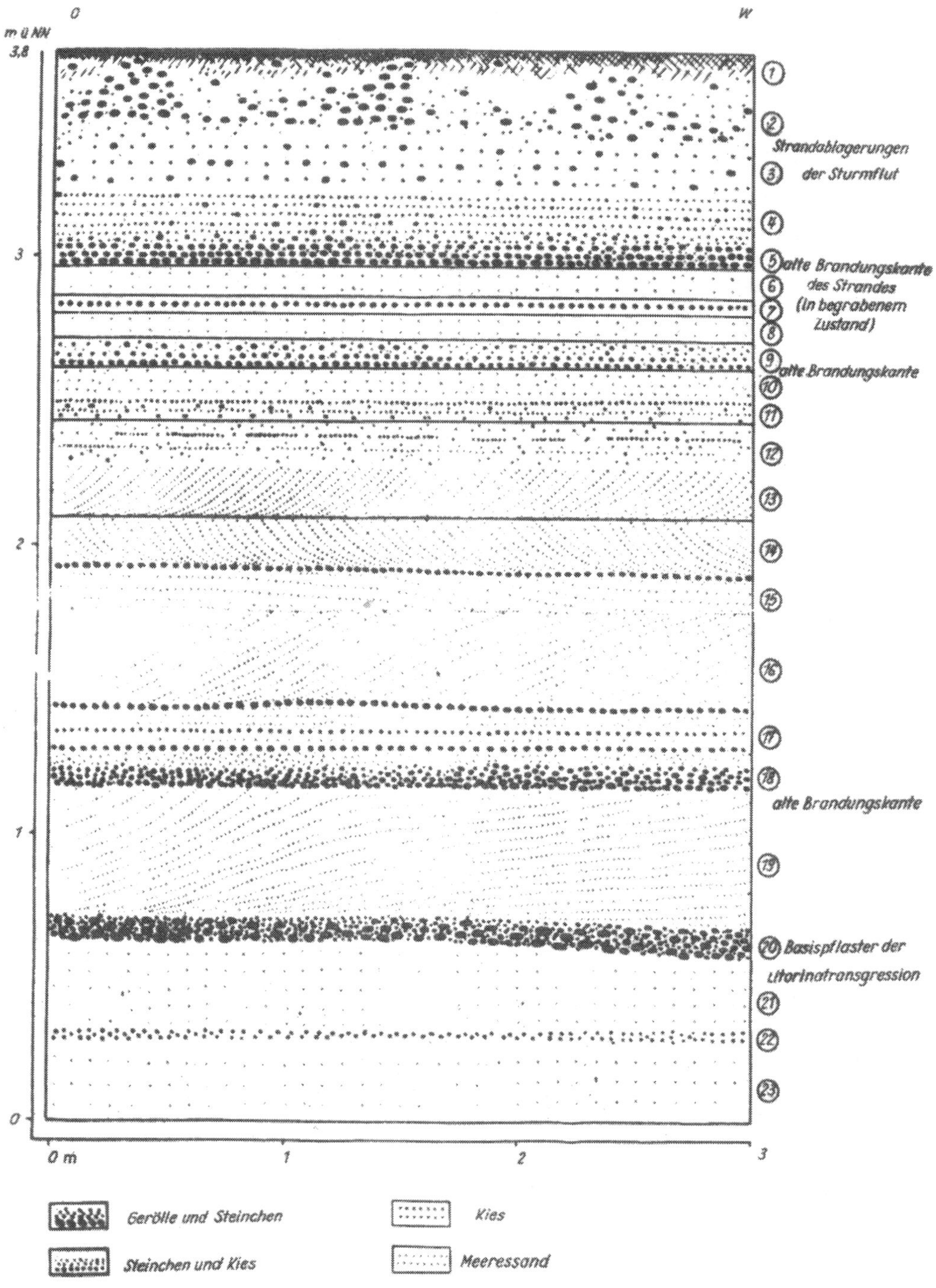

Abb. 3

Alter Strandwall von Nowecin (Neuhof)
Die innere Struktur des Walles

in der aktiven Zone des Litorinastrandes. In begrabenem Zustande repräsentieren sie jetzt alte litorinazeitliche Brandungskanten, die in der Zeit der niedrigen Wasserstände des Litorinameeres in der Brandungszone des Strandes entstanden sind. Die Entstehung der steinigen Schichtchen stand in Verbindung mit dem Akkumulationsvorgang von Sand, Kies und Steinchen auf der Leeseite des Strandwalles und zugleich auch mit dem langsamen Vorrücken der alten Strandkante und der ganzen Brandungszone auf dem sandigen Meeresboden. Aus diesem Grund sind diese Schichtchen von unten ganz deutlich und scharf von den darunterliegenden Sandschichten abgegrenzt. Nach oben dagegen gehen sie ohne scharfe Grenze in kiesige und sandige Schichten über. Diese undeutliche Grenze zwischen den ineinander übergehenden steinreichen Schichtchen und sandigen Schichten der alten Strandwälle hat eine große Bedeutung für die möglichst genaue Bestimmung der ehemaligen Meeresspiegelhöhe. In unserem Beispiel zeigen sie ganz genau die mittleren Höhen während der Zeiten der niedrigen Stände des Litorinameeres an. Die Genauigkeit der Bestimmung beträgt ± 5 cm.

In der Gegend von Łeba erlaubte die dargestellte Methode festzustellen, daß die maximale Wasserspiegelhöhe des Litorinameeres um 3,10 m ± 5 cm über NN gelegen war. Dies weist gleichzeitig darauf hin, daß sich das Land in der Umgebung von Łeba um den gleichen Wert ungefähr in den letzten 4000 Jahren erhob. Es muß hierzu gesagt werden, daß wir es hier mit einer relativen Bewegung der Erdkruste in bezug auf andere Teile der Küste zu tun haben. Der mittlere Geschwindigkeitswert dieser Bewegung betrug also 3/4 mm/Jahr.

Die dargestellte Methode findet aber nur in diesen Fällen Anwendung, wo die frühere Küstenlinie alte Strandwälle angibt. Diese Methode spielt also eine ergänzende Rolle im Verhältnis zur ersten Hauptmethode.

Die erste Hauptmethode besitzt einen weiteren Anwendungsbereich.In den Fällen, in denen die begrabene Abrasionsfläche sich mit den toten Kliffen berührt, erlaubt sie die ehemalige Meeresspiegelhöhe oder die Spiegelhöhe des Haffs sowie der küstennahen Seen mit einer Genauigkeit von etwa 0,5 m zu bestimmen. Die größte Schwierigkeit entsteht in dem Falle, wo die Abrasionsfläche sich oft nicht ganz deutlich mit der ältesten Vordüne des Strandes verbindet, z.B. im Untergrunde der Nehrung des Jamno-Sees. Die genaue Bestimmung des ehemaligen Meeresniveaus ist dort ohne eingehende spezielle Forschungen kaum möglich.

Meiner Meinung nach können beide dargestellte Methoden einen weiten Anwendungskreis in den Forschungen über alte, jetzt am Meeresboden gelegene Strandlinien der Ostsee finden, insbesondere in den Teilen des Bodens, wo die Verschiebung der ehemaligen Strandlinie des Meeres sprungweise und in Gestalt von Strandkatastrophen vor sich ging.

Ich will noch hinzufügen, daß in unserem Küstengebiet auch andere geomorphologische und geologische Methoden der Forschung der neotektonischen Bewegungen der Erdkruste ihre Anwendung finden. Sie sind schon allgemein aus der Fachliteratur bekannt [9, 16]. Beispielsweise gebe ich hier die Methode der eingehenden stratigraphischen Analyse der Ablagerungen, die die Strandterrassen bilden, an, weiter die Methode des Höhenunterschiedes zwischen den ältesten und jüngsten Vor- und Walldünen der Küste, die Methode der Strukturanalyse des inneren Baues und der Lage der Oberfläche der alten, teilweise auch begrabenen oder von Torf bedeckten Rücklaufdelten und andere.

Bei uns führt man auch Geländeforschungen über die Struktur der Alluvionen am Boden der Flußtäler durch. Die letztgenannten Forschungen, die kompliziertere und weniger sichere Ergebnisse bezüglich der neotektonischen Bewegungen des Gebietes geben, bestätigen im allgemeinen die Resultate der Forschung an der Küste.

Forschungsergebnisse

Die Zeichnung 1 gibt die bisherigen Forschungsergebnisse über die Feststellung der maximalen Wasserspiegelhöhe des Litorinameeres im Gebiet der polnischen Küste wieder. Sie stellt gleichzeitig die neotektonischen Tendenzen des ganzen erforschten Küstengebietes dar. Die weiteren Forschungen, die wir in bezug auf eine möglichst genaue Altersbestimmung der begrabenen litorinazeitlichen Abrasionsflächen noch weiterführen müssen, werden zweifellos im einzelnen dieses Bild verändern, jedoch - meiner Meinung nach - unter Beibehaltung seines allgemeinen Charakters.

Die festgestellten neotektonischen Tendenzen des Gebietes finden einerseits ihre Bestätigung im geologischen Bau des pommerschen Landes, anderseits gestatten sie ebenfalls, viele charakteristische Merkmale der gegenwärtigen morphologischen Entwicklung der Küste zu erklären.

Im Gebiet der Bucht von Gdańsk und der Weichseldeltaniederung befindet sich der kristalline Sockel des Urgebirges in einer Tiefe von 2500 m bis 3500 m [10, 15]. In diesem Gebiet haben wir es jetzt mit einer neotektonischen Senkungstendenz des Festlandes zu tun. Im mittleren Teil der Küste, insbesondere in der Umgebung von Łeba, wo der präkambrische Untergrund sich nur in einer Tiefe von etwa 1500 m befindet [15], wurde eine relative Hebungstendenz des Landes festgestellt. Zwischen diesen beiden Gebieten kann man, von neotektonischen Tendenzen des Landes ausgehend, eine tiefgelegene Verwerfungszone längs der westlichen Küste der Bucht von Gdańsk vermuten.

Die zweite, gegenwärtig tektonisch aktive Zone in Pommern bildet das kujawisch-pommersche Paraantiklinorium, das die permisch mesozoische Schichtfolge umfaßt [7, 11, 12]. Im Binnenland haben wir jetzt gewisse Hinweise geomorphologischer Art, die zwar nur vermuten lassen, daß die im Bereich dieses Antiklinoriums, insbesondere die in seinem nord-östlichen Teil gelegenen Antiklinen, einer relativen Hebung unterliegen. Dies geht hauptsächlich aus den Forschungen über die Struktur der Alluvionen der pommerschen Flußtäler hervor. Auf der Zeichnung 1 dagegen ist das erwähnte Merkmal des Antiklinoriums in Gestalt einer Verbiegung der litorinazeitlichen Strandlinie in der Gegend von Dźwirzyno/Deep am Kampersee ersichtlich. Der Untergrund dieses Ortes, und zwar unter den pleistozänen Ablagerungen, weist antiklinale Struktur von Juraschichten auf [15].

Die neotektonischen Tendenzen des Festlandes treten besonders deutlich in der gegenwärtigen morphologischen Entwicklung der Küste hervor. Im mittleren Teil der Küste erfolgte in der Nachlitorinazeit eine geringe, aber allgemeine Regression der Ostsee. In Verbindung mit dieser Regression formte sich hier eine der längsten Nehrungen der pommerschen Küste (85 km). Auf dieser so langen Akkumulationsstrecke der Küste hält sich gegenwärtig entschieden das Gleichgewicht der Dynamik der aktiven Zone des Strandes.

Am Strande der Łebanehrung überwiegen sogar die gegenwärtigen Küstenaufbauvorgänge und der eolische Transport des sandigen Materials. Dagegen am Ostseeboden, 3 bis 5 km von der Łebanehrung entfernt, unterliegt der Meeresgrund dem Einfluß des jetzigen Abrasionsvorganges. Diese Erscheinungen finden offensichtlich ihre Erklärung in der Tatsache der geringen Erhebung des Gebietes.

Im westlichen Teil der Küste in Verbindung mit der allgemeinen Senkung des Landes erfolgte, abgesehen von dem Fall der Dźwirzyno-Gegend, eine geringe Transgression der Ostsee in der Nachlitorinazeit. Diese Transgression dauert gegenwärtig weiterhin an. Dieselbe Erscheinung können wir auch an der westlichen Küste der Bucht von Gdańsk feststellen.

Da wir die gegenwärtige geomorphologische Entwicklung der Küste ganz deutlich in Verbindung mit den neotektonischen Bewegungen stellen können, kann man also feststellen, daß der neotektonische Faktor im Verhältnis zum eustatischen eine führende Rolle in dieser Entwicklung spielt. Dies betrifft auch die ganze Nachlitorinazeit.

Literatur

[1] Bołdyrjew, W.L.; Dołotow, Ju.S.: 1961, Otmiranije bjeregowych akkumulatiwnych form kak jawlenija, charakterizu - juszczeje samjedlenje posleledníkowoj transgressii. Ensv Teaduste Akadeemia Geoloogia Instituudi Uurimused, Bd. VIII, s. 21 - 31. Tallinn

[2] Bülow, K. v.: 1933, Ein neuer Fund von Litorinaablagerungen Dohrniana, Bd. 12, Stettin

[3] Galon, R.: 1961, Morphology of the Noteć-Warta (or Toruń - Eberswalde) Ice Marginal Streamway. Warsaw

[4] Geib, K.W.: 1944, Meeresgeologische Untersuchungen im Bereich der ostpommerschen Küste von Stolpmünde bis zur ehemaligen deutsch-polnischen Grenze. Geolog. d. Meere u. Binnengewässer, Bd. 7, Berlin

[5] Gudelis, W.K.: 1961. Neotektoniczeskije dwiżenia na teritorii Pribałtiki w czetwjertioznom (antropogenowom) periodje. Materiały wsesojuznogo sowjeszczanija po izuczeniju czetwjerticznogo perioda, Bd. II, Moskwa

[6] Hurtig, Th.: 1955. Zur Frage der postdiluvialen Überflutung der südwestlichen Ostseeküste. P.M. Jg. 99, Gotha

[7] Jaskowiak, M.: 1961. Budowa geologiczna synklinorium szczecińskiego. Przeglad Geolog., Nr. 8, Warszawa

[8] Kliewe, H.: 1960. Die Insel Usedom. Berlin

[9] Leotjew, O.K.: 1955. Geomorfołogija morskich bjeregow i dna. Moskwa

[10] Pazdro, Z.: 1959. Budowa geologiczna regionu gdańskiego. Rocznik Pol. Tow. Geolog., Bd. XXIX, H. 4, Kraków

[11] Pawłowski, St.: 1955. Przyczynek do poznania ruchów pionowych skorupy ziemskiej w Polsce. Przeglad Geolog. H. 4, Warszawa

[12] Pozaryski, Wł.: 1957. Południowo-zachodnia krawedź Fenno-Sarmacji. Kwartalnik Geolog., Bd. 1, H. 3 - 4, Warszawa

[13] Rosa, B.: 1959. Die postglaziale Transgression an der polnischen Küste. Geogr. Berichte, H. 10/11, Berlin

[14] Sauramo, M.: 1958. Die Geschichte der Ostsee. Helsinki

[15] Sokołowski, St.; Znosko, J.: 1959. Projekt mapy tektonicznej Polski ... Kwartal. Geolog., Bd. 3, H. 1, Warszawa

[16] Zenkowicz, W.P.: 1946. Dinamika i morfołogija morskich bjeregow. Moskwa-Leningrad

D i s k u s s i o n

T. S o r g e n f r e i (Hellerup):

Die Festlegung der rezenten Trans- und Regressionen auf Grund geomorphologischer Kennzeichen im Küstenbereich ist sehr schwierig, da die verschiedenen endogenen Vorgänge der Erosion und Sedimentation die Kenntnisse der Auswirkung der einzelnen Faktoren der Krustenbewegung sehr erschweren.

Ich bitte deshalb um Auskünfte über die Datierungen der Ablagerungen.

Die Untersuchung im Sedimentationsbereich außerhalb der Küste ist sehr wichtig, da man hier bessere Möglichkeit einer vollständigen Registrierung hat als im Strandbereich. In Dänemark hat man mehrere postglaziale Trans- und Regressionen entdeckt. Man darf deshalb damit rechnen, daß das postglaziale Bewegungsbild sich nicht so ganz einfach abgespielt hat. Ich wünsche den polnischen Forschern guten Erfolg bei den begonnenen Untersuchungen.

J. E. M o j s k i (Warschau):

Zu den Bemerkungen von Prof. Sorgenfrei möchte ich nur hinzufügen, daß in den letzten Jahren bei den Untersuchungen der postglazialen Sedimente der polnischen Küste ganz neue Methoden angewendet wurden, vor allem Pollenanalyse und C_{14}-Daten.

L.-E. Å s e (Stockholm):

Ich möchte hinzufügen, daß die Datierung der Litorina-Transgression (oder die Transgressionen) sehr notwendig ist, vielleicht auch mit C_{14}-Datierungen, so daß man sicher sein kann, daß die Strandlinie, die man studiert, wirklich überall synchron ist. Dazu ist auch zu sagen, daß der Windstau oft für die Lage der Standlinien sehr wichtig ist.

J. E. M o j s k i (Warschau):

Leider kann ich im einzelnen auf die Frage von Åse nicht antworten.

Structure tectonique et mouvements récents de la croûte terrestre en Bulgarie

von

D. J a r a n o f f (Sofia)

S u m m a r y

The comparison of the neotectonic and of the recent movements with the gravitational field allows to distinguish the two following types of principal structures:

1. Regions with a crust which is subject to extensive transformations still today. These regions are neither compensated isostatically nor is it possible to discover tectonic effects in them resulting from adjacent regions. Here we are dealing with large tectonic regions subject to changes of their own, which are very active in the depth but don't affect the surface all of a sudden. In Bulgaria it is "la Plate-forme Moesienne", which is not entirely comparable with the Russian plateau in this regard.

2. Regions in which apparent superficial transformations of the structure take place under the tectonic influence of an adjacent region, but without structural changes in depth at the present moment, regions in which isostasy can appear on its own accord. This is the case with the Alpine orogeny in Bulgarian territory, inclusive of the massif of the Rhodopes as a medium mass in this orogeny.

Z u s a m m e n f a s s u n g

Der Vergleich der neotektonischen und rezenten Bewegungen mit dem Schwerefeld führt dazu, die zwei folgenden Typen von Hauptstrukturen zu unterscheiden:

1. Krustengebiete, die heute noch weitgehenden Veränderungen unterworfen sind. Diese Gebiete sind weder isostatisch kompensiert, noch sind tektonische Einflüsse aus den Nachbarregionen möglich. Es handelt sich hier um die großen tektonischen Gebiete mit ihren eigenen Veränderungen, die in der Tiefe sehr rege vor sich gehen, die aber nicht plötzlich bis zur Oberfläche dringen. In Bulgarien ist es "la Plateforme Moesienne", ein Plateau, das in dieser Hinsicht nicht ganz mit dem russischen Plateau vergleichbar ist.

2. Gebiete, in denen scheinbar eine Umwandlung an der Oberfläche der Struktur unter dem tektonischen Einfluß eines benachbarten Gebietes, aber ohne Veränderungen der Tiefenstruktur im gegenwärtigen Augenblick vor sich geht, wo sich die Isostasie frei einstellen kann. Das ist der Fall bei der alpinen Orogenese im bulgarischen Gebiet, einschließlich des Rhodopenmassivs als einer Mittelmasse in diesem Orogen.

L'object de cette communication sera l'étude, sur l'exemple du petit territoire de la Bulgarie, des questions concernantes

1° La constance des tendances tectoniques,

2° Le mode de transformation, par endroits, du plan structural ancien

et

3° Les possibilités offertes par l'étude comparée des mouvements néotectoniques et récents et du champ de gravitation pour pouvoir expliquer les causes de la constance des transformations du plan structural.

1. La constance des tendances tectonique de la Plate-forme Moesienne

Si l'on veut présenter une région typique à constance des tendances tectoniques on devra s'occuper de la partie bulgare de la Plate-forme Moesienne. Elle est une plate-forme épihercynienne dans le nord de la Bulgarie tandis qu'au nord, en Roumanie, elle est plus ancienne. Après les plissements et les intrusions pendant le Carbonifère inférieur ont eu lieu de grandes fracturations - l'étage structural du Carbonifère supérieur et du Permien a été déposé, avec d'énormes masses volcaniques. Ensuite cet espace a été soumis à de très larges ondulations à petite amplitude, dont l'emplacement n'a pas du tout changé jusqu'à nos jours - comme c'est le cas avec le bombement dans le nord-est de la Bulgarie et le bombement de Vidin-Craiova - ou change ait constamment dans une direction donnée comme c'est le cas avec la fosse périphérique.

Le bombement de Vidin-Craiova, constaté d'abord par la mesure du champ de gravitation, prouvé par nombre de forages profonds, est caractérisé par la trop petite épaisseur du Jurassique inférieur et moyen, du Crétacé inférieur, l'absence du Paléogène et la toute petite épaisseur du Néogène. Ce bombement a peu à peu perdu sa tendance d'exhaussement, surtout pendant les époques géologiques récentes, étant englobé dans la fosse périphérique, et continue à s'affaisser.

La fosse périphérique de la Plate-forme (l'avant-fosse, la "Vorlandsenke") a reçu la physionomie bien définie d'une fosse surtout pendant le Tithonien et le Valanginien, avec un axe immédiatement au nord de la crête du Balkan Central. Un léger déplacement vers le nord a eu lieu pendant le Valanginien et le Hauterivien; après cela la fosse apparaît durant l'Aptien avec un maximum de subsidence sur la ligne Roman - Loveč - Târnovo. Pendant l'Albien la subsidence est maximum sur l'axe Krivodol - Svištov; même pendant l'Albien a eu lieu un déplacement très évident toujours vers le nord: d'après l'étude micropaléontologique des carottes des forages profonds (paléontologiste P. Jovčeva) l'Albien est représenté dans la partie méridionale de la fosse par l'Albien inférieur qui est absent dans la partie septentrionale de la fosse, tandis que l'Albien supérieur y est présent qui manque dans la partie méridionale de la même dépression. Une nouvelle migration a eu lieu pendant le Crétacé supérieur, l'axe de la fosse de cette époque s'étend de Dâlgodelci jusqu'à Nikopol. Pendant le Paléogène l'axe est au nord de l'anticlinorium du Balkan Occidental, pendant le Néogène inférieur il est absent - c'est l'époque du soulèvement global de tout le territoire de la Bulgarie - pour s'installer au Néogène supérieur sur l'emplacement de la Dépression de Lom.

A notre avis cette migration, très typique pour toutes les avant-fosses en général, est une preuve assez convaincante que dans le cas de cette dépression néogène nous avons une néotectonique qui est nouvelle seulement au point de vue de son âge mais pas du point de vue du caractère de l'activité. Elle est seulement une nouvelle expression des anciennes tendances, c'est-à-dire cette partie de la Plate-forme se comporte d'une manière très conservative.

Les plus anciennes parties de la fosse, affectées par plusieurs mouvements pas très intenses mais très compliqués qui ont donné naissance à quelques grandes flexures (à la surface, mais qui reflètent, paraît-il, l'existence de failles profondes), à des plis plats et des décrochements, ne bougent pas, paraît-il, pendant le Néogène et le Quaternaire.

Dans le domaine de la Plate-forme le bombement d'une évolution quasi permanente c'est le bombement situé dans le nord-est de la Bulgarie. Ici on trouve le socle hercynien plissé (Dévonien moyen et supérieur) seulement à 900 mètres de la surface. Le socle est recouvert par le Dogger transgressif, sur celui-ci reposent les dépôts carbonates du Jurassique supérieur et du Valanginien ainsi qu'une mince couverture du Crétacé inférieur. Ensuite a eu lieu de nouveau un exhaussement continuel et qui continue, d'après le nivellement répété, jusqu'à nos jours.

En général la Plate-forme Moesienne montre une constance des tendances tectoniques qui se reflètent jusqu'aujourd'hui dans les mouvements récents.

2. Mouvements néotectoniques et récents dans le tectorogène alpin du territoire bulgare

Le tectorogène alpin consiste dans le territoire bulgare dans les unités tectoniques suivantes:

Le Massif des Rhodopes proprement dit et les fragments de celui-ci qui n'étaient pas complètement engloutis par les nouvelles zones tectoniques qui se sont developpées sur l'emplacement de celui-ci;

La zone triasique-jurassique périrhodopéenne consolidée le Crétacé inférieur.

La zone des ruptures du Crétacé supérieur et du Paléocène;

L'anticlinorium du Balkan Occidental;

La zone paléogène périrhodopéenne.

Les premières ruptures en liaison avec les mouvements néotectoniques ont eu lieu au commencement du Sarmatien dans les environs de Kjustendil et de Sofia. Dans les environs de Kjustendil la reconstruction de l'ancien socle cristallin a eu lieu deux fois - pendant la formation des deux zones périrhodopéennes, dans les environs de Sofia - même trois fois - pendant la formation des zones nommées ainsi que pendant la formation de la zone des ruptures du Crétacé supérieur et du Paléocène. Évidemment les premiers accès des mouvements néotectoniques, encore faibles, ont pu se produire seulement sur les compartiments détruits du Massif.

Mais les assauts décisifs des mouvements néotectoniques égéens pendant le Pliocène supérieur et le Villafranchien ont pu vaincre même la rigidité du Massif des Rhodopes proprement dit et nous voyons qu' à cette époque la compartimentation par fracturation de tout le tectorogène alpin dans le territoire bulgare, c'est-à-dire le Balkan et toute la Bulgarie méridionale, est complète.

La fracturation s'est produite surtout le long d'une directrice préférée qui est simplement la prolongation des grandes cassures de la mer Égée - les fractures NW-SE. Une de ces fractures - de Struma - est marquée dans le territoire bulgare et jougoslave par des épanchements dacitiques d'âge pliocène moyen et par une très grande activité séismique.

Une autre directrice préférée est celle WSW-ENE, parallèle au grand fossé de la mer Egée septentrionale qui conditionne la côte abrupte au sud du Mont Athos, de l'île de Samothrace et la forme du Golfe Saronique. Les failles à cette directrice sont généralement plus courtes et situées entre deux failles majeures de directrice NW-SE.

Les failles à ces deux directrices sont généralement nouveaux-nées et nous pouvons les indiquer comme failles égéennes témoignant la progression de la zone tectonique égéenne vers le Nord. Une limite a été mise à cette progression par la Plate-forme.

L'activation de la croûte terrestre dans la Bulgarie méridionale sous l'influence égéenne a eu comme conséquence la renaissance des failles endémiques et spécialement d'une grande partie des ruptures de la zone du Crétacé supérieur et du Paléocène et des failles qui longent les bombements très anciens du socle cristallin du Massif des Rhodopes.

En ce qui concerne l'amplitude des failles, elle varie d'un mètre et demi pour celles qui se sont produites en 1928 à l'est de Plovdiv lors du dernier grand tremblement de terre à 3000 mètres sur la bordure du grand horst de Pirin.

3. Néotectonique, tectonique récente et champ de gravitation

Les failles pliocènes, quaternaires et récentes à grande amplitude dans la Bulgarie méridionale sont étroitement liées avec la structure interne de la croûte terrestre. Elles correspondent aux plus forts gradients du champ de gravitation et les horst les plus élevés - généralement (dans le Massif des Rhodopes - sans exception) aux minima de ce champ. D'autre part dans le Balkan et dans la Bulgarie méridionale les minima correspondent sans exception aux affleurements granitiques. Dans le véritable Massif des Rhodopes (à l'est de la Struma) ainsi que dans la zone des ruptures du Crétacé supérieur et du Paléocène les efforts des mouvements néotectoniques et récents égéens étaient strictement contrôlés par les conditions géologiques locales: l'activation néotectonique du tectorogène alpin en Bulgarie par l'immixtion de l'influence tectonique égéenne a eu comme conséquence la plaine mise en action des tendances isostatiques. Sur la Plate-forme, au contraire, les maxima du champ de gravitation correspondent aux régions à tendances constantes à surélévation et les minima - aux régions à tendances constantes à subsidence. Ici, dans la Plate-forme, nous paraît-il, l'isostasie n'a pas du tout pu se produire à cause d'une transformation profonde de la Plate-forme dans la voie d'une basaltification comme c'était le cas avec l'aire occupée par les parties profondes de la mer Noire, où la croûte basaltique vient à la surface, d'après les études récentes de Neprotchnov.

Conclusion

La comparaison des mouvements néotectoniques et récents avec le champ de gravitation permet du séparer les deux types mégastructuraux suivants:

1° Régions à une croûte en voie de transformation profonde jusqu'à l'heure actuelle; dans ces régions ne sont possibles ni les phénomènes d'isostasie, ni les immixtions des influences tectoniques des régions voisines. Ce sont les grandes régions tectoniques à leur propre vie qui est très active en profondeur mais qui ne se traduit pas brutalement à la surface. En Bulgarie c'est la Plate-forme Moesienne qui à ce point de vue n'est pas tout à fait comparable à la Plate-forme Russe.

2° Régions à transformation apparente - superficielle - du plan structural sous l'influence tectonique d'une autre région voisine, mais sans changement de la structure profonde à l'heure actuelle, où l'isostasie a libre jeu. C'est le cas avec le tectorogène alpin dans le territoire bulgare, y compris le Massif des Rhodopes comme une masse médiane dans le tectorogène.

Diskussion

A. Zátopek (Prag)

Il a été parlé sur les grands tremblements de long de la nouvelle ligne active dans la région de Plovdiv. Est-ce qu'on observe à présent des élévations et dépressions relativement à cette rupture?

Existe-t-il quelques données quantitatives sur ces mouvements provenant des dernières années?

J'ai demandé parce que les données de ce genre sont très importantes pour la construction de la carte séismotectonique de l'Europe. Les zones considérées représentent le maximum sur le continent européen.

D. Jaranoff (Sofia)

Les régions qui se sont affaissées pendant le tremblement de terre de Plovdiv en

1928 ont continué à s'affaisser ensuite, mais nous n'avons pas de données plus concrètes sur le mode de cet affaissement. Une station séismique, qui sera installée à Dimitrovgrad devra nous donner des données plus concrètes.

A. P. Rozhdestvensky (Ufa):

I would like to know, whether horizontal displacements were observed during strong earthquakes in Bulgaria.

D. Jaranoff (Sofia):

Nous n'avons pas de données sur les mouvements horizontaux, mais ils sont à supposer.

N. V. Dumitrashko (Moskau):

The paper by Prof. Jaranoff has many interesting aspects. I think it is necessary to note in a resolution that the study of the recent and modern tectonics of the mountainous countries is desirable. These researches should be conducted jointly by the geodesists and geomorphologists with geologists and specialists in tectonics.

Ausmaß der holozänen Krustenbewegungen in Ungarn

von

M. P é c s i (Budapest)

S u m m a r y

In the large and small Hungarian lowlands one has succeeded in determining the extent of the holocene and pleistocene movements of the crust by means of the thickness of the quaternary sediments deposited in the basins, by the investigation of the deformation of the outer terraces of the mountains and basins as well as of the smaller outer fault troughs.

The total thickness of the pleistocene layers in the larger and smaller separable basins of the lowland amounts to 100 - 600 m, in the small lowland to 50 - 250 m. The accumulations of the holocene depositions are considerable, especially in the outer fault troughs of the lowlands and amount to 10 - 20 m.

In the outer regions of the plains the terraces of the alluvial cones of the Danube suddenly go down into the depth to pass then into a normal stratigraphic bedding along a band of inflection.

The zone of rupture of the terraces on the border of the lowland coincides in historical times, too, with the districts of increased seismicity (e.g. the seismic district to the south of Budapest and on the eastern border of the small lowland).

Z u s a m m e n f a s s u n g

In der Großen und in der Kleinen Ungarischen Tiefebene ist es gelungen, das Maß der holozänen und pleistozänen Krustenbewegungen mit Hilfe der Mächtigkeit der in den Becken aufgeschütteten quartären Sedimente durch die Untersuchung der Deformation der Randterrassen, der Gebirge und Becken sowie der kleineren Randsenken zu bestimmen.

Die Gesamtmächtigkeit der pleistozänen Schichten in den voneinander abtrennbaren größeren und kleineren Becken der Tiefebene beträgt 100 - 600 m, in der Kleinen Tiefebene 50 - 250 m. Die Anhäufungen der Ablagerungen aus dem Holozän sind besonders in den Randsenken der Tiefebenen beträchtlich und betragen 10 - 20 m.

In den Randgebieten der Ebenen sinken die Schuttkegelterrassen der Donau plötzlich in die Tiefe, um sodann entlang eines Inflexionsstreifens in eine normale stratigraphische Schichtung überzugehen.

Die Zone des Abbruchs der Terrassen am Rande der Tiefebene fällt auch in historischen Zeiten mit den Streifen erhöhter Seismizität zusammen (z.B. die Erdbebenzone südlich von Budapest und am Ostrande der Kleinen Tiefebene).

Mit der reliefbedingenden Rolle der jungen Krustenbewegungen haben sich im letzten Jahrzehnt Aufsätze in zunehmender Anzahl beschäftigt. Jeder der Forscher war dabei bestrebt, Daten zu fixieren und zu zahlenmäßigen Erkenntnissen und Auswertungen zu gelangen. Auch der vorliegende Aufsatz hat sich eine ähnliche Aufgabe zum Ziel gesetzt und versucht, die oberflächenformende Rolle der mit geomorphologischen Methoden erkennbaren jungen Krustenbewegungen aufzuzeigen.

Uns ist, auf die Ergebnisse der terrassenmorphologischen Forschungen stützend, die Möglichkeit geboten, die zeitlichen Rythmen und das Ausmaß der pleistozänen und holozänen Krustenbewegungen auch in absoluten Zahlenwerten mit annähernder Genauigkeit anzugeben.

Über die reliefformende Rolle der jungen Krustenbewegungen in Ungarn wurden im Laufe der entwicklungsgeschichtlichen Untersuchungen der Täler mit Hilfe geomorphologischer Methoden von mehreren Geographen Angaben geliefert (A. Kéz 1937; E. Szádeczky-Kardoss 1938; B. Bulla 1956; M. Pécsi 1958; Ádám-Marosi-Szilárd 1959; Z. Borsy 1961). Wir können dabei diesbezüglich auch den karstmorphologischen Forschungen mehrere wertvolle Daten verdanken (Z.P. Szabó 1957; S. Láng 1955).

1. Der Zusammenhang zwischen der Deformierung der Terrassenniveaus und den jungen Krustenbewegungen

a) In den früheren terrassenmorphologischen Studien wurden die Terrassensysteme der ungarischen Flüsse so aufgefaßt, als ob die relativen Höhen der Terrassen in der ganzen Länge des Flusses, mit wenigen Ausnahmen, annähernd gleichbleibend wären. Erst E. Szádeczky-Kardoss hat erkannt, (1938) daß im Donauabschnitt der Kleinen Tiefebene ("Kisalföld") die Terrassen, die Kleine Tiefebene erreichend, der Reihe nach niedriger werden, bis sie endlich im Becken selbst, in normale Schichtung übergehen. Bezüglich des Donauabschnittes der Großen Tiefebene ("Nagyalföld") ist B. Bulla ähnlicher Meinung, mit Ausnahme der Terrasse No. II./[a] des Pleistozänendes.

b) Der Verfasser dieses Aufsatzes hat im Laufe seiner jahrzehntelangen Untersuchungen im Donautal erfahren können, daß die relative Höhe der Donauterrassen nicht allein am Rande des Mittelgebirges, in der Zone des Austritts auf die Tiefebenen, Schwankungen aufweist, sondern auch innerhalb des Mittelgebirges. Die in den Höhenwerten der älteren Terrassen sich erweisenden ganz bedeutenden Schwankungen wurden erst im Laufe der Kartierung der Terrassen in der Umgebung von Budapest bekannt (M. Pécsi 1956).

Was auch immer unsere Vorstellung über die Ursachen der Entstehung der Donauterrassen sei oder der Terrassen überhaupt, sollen die Terrassen ihre Entstehung klimatischen oder tektonischen Ursachen oder den beiden gemeinsam verdanken, so ist es außer Zweifel, daß man in der gegenwärtigen Lage der Donauterrassen lokale, tektonisch bedingte Höhenveränderungen nachweisen kann. Die Donauterrassen haben also nach ihrer Ausbildung ihre ursprüngliche Lage nicht starr bewahrt. Hierauf konnte auch aus geologischen (Vadász 1953; Noszky d. Ae. 1935), aus tektonischen (F. Pávai Vájna 1938), geomorphologischen (S. Láng 1955), geomechanischen (E.R. Schmidt 1951), geophysikalischen (V. Scheffer 1949, L. Egyed 1957) und geodätischen (Bendefy 1955) Angaben geschlossen werden. Demgemäß befanden sich das Mittelgebirge und sein Vorraum auch noch während des Pleistozäns und des Holozäns in Bewegung . Diese Bewegungen waren jedoch innerhalb des Mittelgebirgs-Abschnittes der Donau von verschiedener Intensität.

Nach dem oben Erwähnten waren wir gezwungen, in den terrassenmorphologischen Forschungen solche Methoden und analytische Untersuchungen einzuführen, mit deren Hilfe wir die Parallelisierung der in verschiedene Höhen gehobenen jungen Terrassenflächen mit entsprechender Genauigkeit durchführen können (M. Pécsi und Frau Pécsi 1960).

2. Die Folgerungen, die in bezug auf die Zeit und das Ausmaß der jungen Krustenbewegungen aus der Lage der Terrassen gezogen werden können

Wir haben die Kartierung der Terrassen des ungarischen Donauabschnittes mit Anwendung von sehr eingehenden und vielseitigen analytischen Untersuchungen durchgeführt. Als Ergebnis der Parallelisierung sind die Lageverhältnisse der Donauterrassen, die Zusammenhänge der Terrassenhorizonte und Beckensedimente durch die beigefügte Abbildung 1 dargestellt.

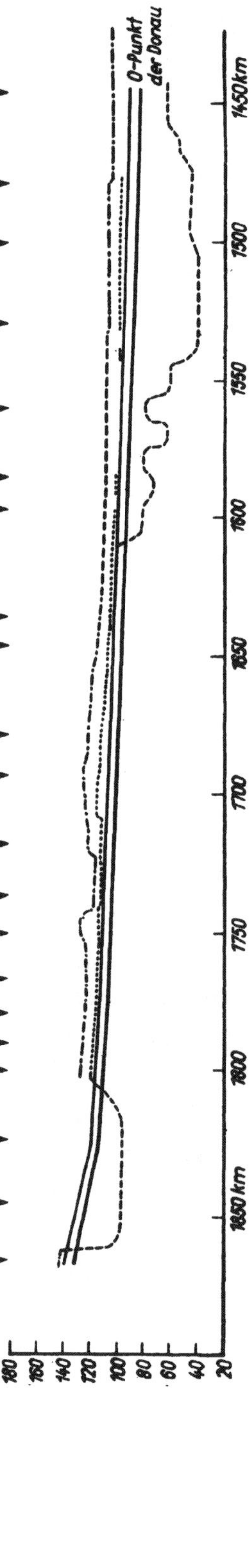

Fig. 1

Die Höhenlage der Donauterrassen in Ungarn
Von Dr. Márton Pécsi

Die niedrigeren Terrassen

- Kurve des O-Punktes der Donau
- Terrasse I bzw. höherer Horizont des Überschwemmungsgebietes
- Terrasse IIa aus dem Ende des Jungpleistozäns
- Terrasse IIb aus dem Anfang des Jungpleistozäns

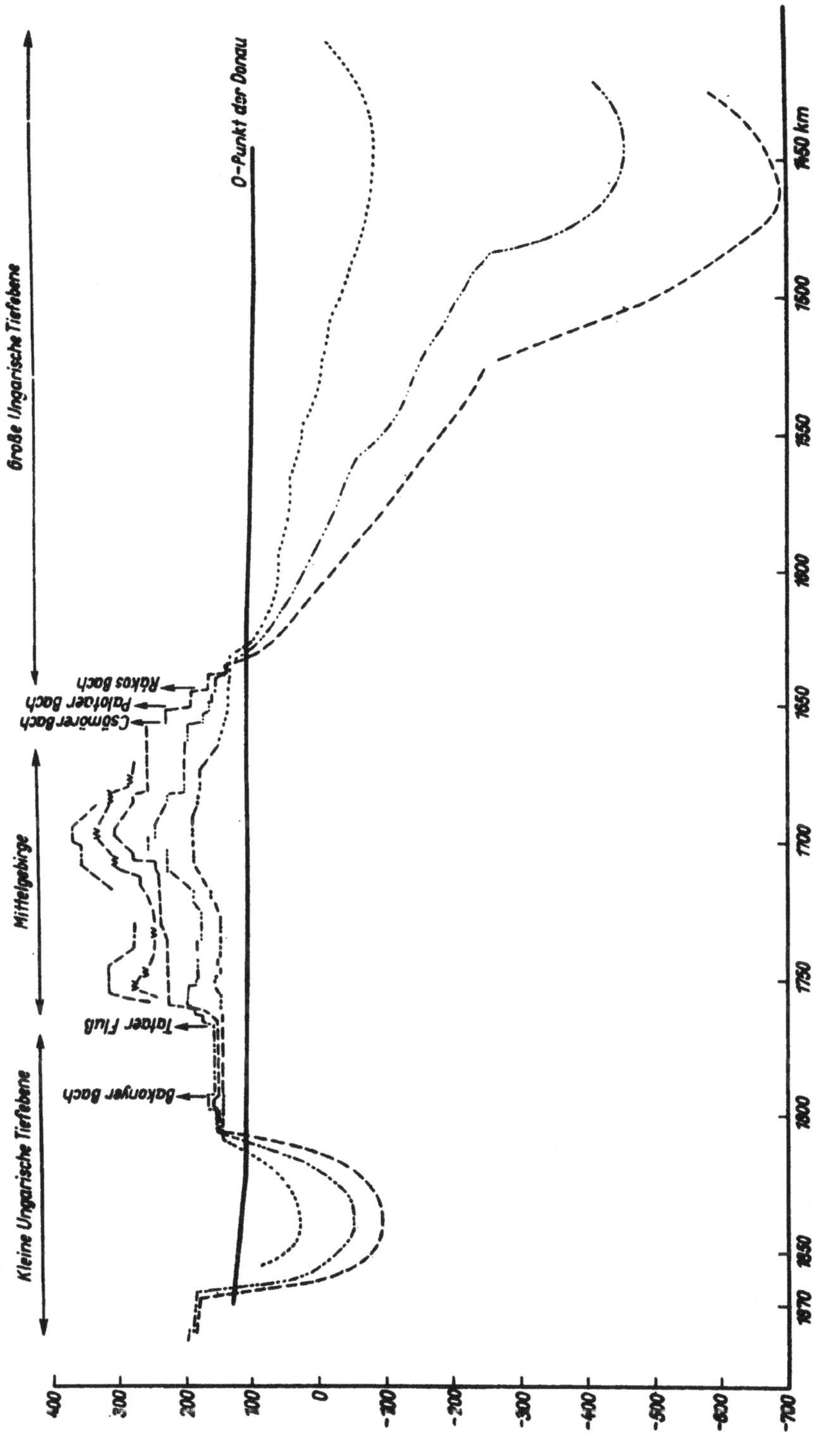

Fig. 2

Die Höhenlage der Donauterrassen in Ungarn. Von Dr. Márton Pécsi. Die höheren Terrassen

Kurve des 0-Punktes der Donau

Terrasse III aus dem Mittelpleistozän

Terrasse IV aus dem älteren Pleistozän

Terrasse V aus dem Altpleistozän

Terrasse VI aus dem Ende des Pliozäns oder Anfang des Pleistozäns

Terrasse VII aus dem Ende des Pliozäns (?)

Da die ältesten Donauterrassen Ungarns, beziehungsweise die fluviatilen Schotterablagerungen des Donaubeckens nicht älter als Spätoberpliozän sind, müßten die an der Abbildung 1 veranschaulichten beträchtlichen Dislokationen im Laufe des Quartärs vor sich gegangen sein. Die ältesten Terrassen innerhalb des Mittelgebirgs-Abschnittes erhoben sich, - gemäß der gegenwärtigen Gefällekurve,- bis 200 - 250 m. In der Mitte der Kleinen Tiefebene sind die diesen Terrassen entsprechenden Ablagerungen um 200 - 250 m abgesunken. Diese Senkung beträgt in der Großen Tiefebene in der Mitte des Gebietes zwischen Donau und Theiss 150 - 200 m. Weiter südlich erreicht das Ausmaß der Senkung der den ältesten Donauterrassen entsprechenden Ablagerungen bereits 250 - 450 m. Die größte quartäre Senkung der Großen Tiefebene erreicht in der Mitte des jenseits der Theiss gelegenen Gebietes selbst eine Tiefe von 1000 m (J. Sümeghy 1954; M. Pécsi 1959; E. Vadász 1960).

An den Terrassen des Donautales, in den Becken, an der Lage der fluviatilen Ablagerungen und an ihren genetischen Zusammenhängen lassen sich Orte, Ausmaße annähernd auch die Zeitmomente der jungen Krustenbewegungen ablesen. Überdies ist auch die Tatsache ablesbar, daß manche alten Bruchsysteme im Quartär, ja sogar bis heute immer wieder auflebten, und gerade entlang diesen Systemen sind auch die Auslösungen der Erdbeben zu erwarten.

Aus der Lage der Donauterrassen und aus ihrem Längsprofil kann die Folgerung gezogen werden, daß die ausgedehnteste rezente Senkung auf dem Gebiete der Kleinen Tiefebene liegt. In dem sogenannten Györer (Raaber) Becken der Kleinen Tiefebene konnte infolge der bis in die Gegenwart reichenden periodischen Senkung keine einzige Terrasse entstehen. Hier ist ein fast 100 km langer und 40 - 60 km breiter mächtiger Schuttkegel entstanden. Dieser jungen, auch in der Postglazialzeit andauernden Senkung gehört auch das Fertö-Hanság Becken an, an dessen in geringem Maße abgesunkenen Rande Kryoturbationserscheinungen des Würm unter den rezenten Bildungen in geringerer oder größerer Tiefe beobachtet werden können.

Der Einfluß der rezenten Krustenbewegungen im Donauabschnitt südlich von Budapest spiegelt gut die holozänen Richtungsänderungen des Flußbettes der Donau wider. Damals hat sich die Senke von Südbudapest-Soroksár-Ócsa herausgebildet. In dieser Senke floß ein Donauarm und erst durch spätere Bewegungen bedingt, schlug der Soroksárer Donauarm seine gegenwärtige Richtung ein.

Diese Bruchrichtungen waren an den Oberflächenspalten, die während der Erdbeben im Jahre 1956 in der südlichen Umgebung von Budapest entstanden sind, gut abmeßbar.

In der Postglazialzeit ist die Donausenke von Kalocsa-Baja mit einem Ruck weiter gesunken. In der Jetztzeit hat sich die Südbácskaer Terrasse der Donau und der Theiss herausgebildet.

Auch jenseits der Theiss sind im Holozän voneinander abgesonderte Senken entstanden (J. Cholnoky 1936; M. Pécsi 1959; J. Sümeghy 1942). Die Theiss ist noch am Ende des Pleistozäns südlich von der heutigen Nyirség nach den Körösflüssen ausgerichtet, gegen die Mitte der Großen Tiefebene ausgelaufen. Während der postglazialen Bewegungsphase ist die Ebene von Szatmár-Bereg eingesunken, infolgedessen schlug die Theiss ihren Weg nach Norden ein, und mit Umgehung der Nyirség meißelte sie für sich an dem Nordrande der Tiefebene ein neues Bett aus. Der Theissabschnitt der Jászáger-Becken stellt ebenfalls eine holozäne Senke dar. Die in der Jászság abgeteuften Bohrungen haben unter den mächtigen holozänen Flußablagerungen mit pleistozänem Solifluctionsmaterial vermischten, in bräunlichen Lehm eingebetteten Gehängeschutt und Schottermaterial zutage gebracht.

Die Grabensenken im südlichen Vorraum des transdanubischen Mittelgebirges sind ebenfalls auf junge Krustenbewegungen zurückzuführen. Obwohl das Einsinken der Gräben des Plattensees und des Velenceer Sees bereits am Ende des Pleistozäns begann, (B. Bulla 1943; Ádám-Marosi-Szilárd 1959) kann ihre weitere Senkung indessen auch zu Beginn des Holozäns nachgewiesen werden (Ádám, Sümeghy, Zólyomi). Unter dem Einfluß der frühholozänen Beckensenkung ist die in der Postglazialzeit ausgemeißelte Terrasse des die westliche Beckenabteilung des Velenceer Sees durchquerenden Wasserlaufes obsequent geworden.

Nach unseren Beobachtungen sind die größten Schwankungen in der Höhenlage der gleichaltrigen Terrassen meistens entlang der Richtung der großtektonischen Linien zu finden.

An den Rändern der Tiefebenen verniedrigen die Terrassen ganz plötzlich und gehen dann entlang einem Inflexionsstreifen in normale Beckenausfüllungen über. Dies geschieht in der Weise, daß das Ausmaß der Erniedrigung der höchsten Terrassen am größten ist, während im Beckeninneren die den höchsten Terrassen entsprechenden Ablagerungen in die größte Tiefe gesunken sind. Die Zone der Randabbrüche der Tiefebenen fällt mit dem Streifen überein, der auch in geschichtlicher Zeit als am meisten erdbebengefährdetes Gebiet gilt. Die Möglichkeit der Auslösung der Erdbeben liegt dort am nächsten, wo im Verlauf der gleichartigen Terrassen die größten Brüche und Sprunghöhen zu beobachten sind (M.V. Gzovskij 1957).

Auf dem Gebiete Ungarns können drei bedeutendere Phasen der jungquartären Krustenbewegungen festgestellt werden (M. Pécsi 1958).
a) Die präglaziale-frühpleistozäne Phase, und b) die Spätmindel - Mindel-Riss interglaziale Phase sind die bedeutendsten, und daran schließen sich noch als dritte c) die letzten glazialen-postglazialen Bewegungen an. Die Auswirkung der letzteren dauert bis heute an. Ihre reliefbildende Wirkung ist bedeutend geringer (10 - 20 m) als die der beiden ersterwähnten Phasen.

L i t e r a t u r

Ádám, L.: A Velencei-tó és a Zámolyi-medence kialakulása. - Die Ausgestaltung des Velenceer Sees und des Zámolyer Beckens. Földrajzi Közlemények, 1955

Ádám-Marosi-Szilárd: A Mezöföld természeti földrajza. - Physische Geographie des Mezöföld. 1959. Budapest

Bendefy, L.: Szintváltozások Dunátul térségében korszerü, szabatos szintezések alapján. Müszaki Tudományok Oszt. Közleménye. XX. 1956

Borsy, Z.: A Nyirség természeti földrajza. - Physische Geographie der Nyirség. Budapest. 1961

Bulla, B.: A Magyar medence pliocén és pleistocén teraszai. - Pliozäne und pleistozäne Terrassen des ungarischen Beckens. Földrajzi Közlemények. 1941

Bulla, B.: Über die Ausbildung und das Alter des Balaton-Sees. Földrajzi Közlemények 1943. idegennyelvü kiadása

Bulla, B.: Folyóteraszproblémák. - Flußterrassenprobleme. Földrajzi Közlemények. 1956

Egyed, L.: A kéregmozgások okai és a magyarországi kéregmozgások. - Die Ursachen der Krustenbewegungen und die Krustenbewegungen in Ungarn. Geofizikai Közlöny. 1957

Gzovskij, M.V.: Tektonofizičeskoe obosnovanie geologi českich kriteriev sejsmičnosti. II. Izv. Akad. nauk SSSR. Serija geofiz. 1957. No. 3

Kéz, A.: Flußterrassen im Ungarischen Becken. Peterm. Mitt. 1937

Kéz, A.: Ujabb teraszmegfigyelések a Zala mentén. _ Neuere Terrassenbeobachtungen neben dem Zalafluß. Földrajzi Közlemények. 1943

Láng, S.: A Mátra és a Börzsöny természeti földrajza. - Physische Geographie der Mátra und Börzsöny-Gebirge. Budapest. 1955

Noszky, J. sen.: Adatok a Visegrádi Dunaszoros teraszképződményeinek geológiai ismeretéhez. - Beiträge zur geologischen Kenntnis der Terrassenbildungen der Visegrader Donauenge. Földt. Intézet Evi Jel. 1933-35. évről.

Pávai-Vajna, F.: Az 1938. évi Budapest-környéki kiegészitö felvételi jelentésem. - Ergänzender geologischer Aufnahmebericht des Jahres 1938 aus der Umgebung von Budapest. Földtani Intézet Evi Jel. 1936-38. évről

Pécsi, M.: Ujabb völgyfejlödéstörténeti és morfológiai adatok a Duna-völgy Pozsony (Bratislava) Budapest közötti szakaszáról. - Neuere talentwicklungsgeschichtliche und morphologische Angaben vom Donautal zwischen Bratislava und Budapest. Földrajzi Ertesitö 1956

Pécsi, M.: Das Ausmaß der quartären tektonischen Bewegungen im ungarischen Abschnitt des Donautales. Peterm. Mitt. 1958

Pécsi, M.: A magyarországi Duna-völgy kialakulása és felszinalaktana. - Entwicklung und Morphologie des Donautales in Ungarn. Budapest 1959. Földrajzi Monográfiák III

Pécsi, M. - Pécsi-Donáth, E.: Méthodes de recherche d'histoire del'évolution des vallées et des terrasses. Annales Univ. Sc. Budapestiensis. Sectio Geol. 1960. III

Pécsi, M.: Der Schuttkegel der Donau auf der Großen Ungarischen Tiefebene. Annales Univ. Sc. Budapestiensis Sectio Geol. 1960. III

Scheffer, V.: Adatok a Kárpát-medencék regionális geofizikájához. - Angaben zur regionalen Geophysik des Karpatenbeckens. Geofizikai Közl. 1959

Schmidt, E.R.: Die Bruchstruktur der ungarischen Zwischenmassen. - A magyar közbensö tömeg töréses szerkezete. Debreceni Szemle 1931

Sümeghy, J.: A Velencei-tó kialakulása. - Die Entwicklung des Velenceer Sees. Földt. Int. Evi Jel. 1944. évröl

Sümeghy, J.: Magyarország talajvizviszonyai. - Grundwasserverhältnisse Ungarns. Budapest 1954

Szabó, P.Z.: A fiatal kéregmozgások geomorfológiai és népgazdasági jelentösége Déldunántulon. - The geomorphological and economical significance of young crustal movements in Southern Transdanubia. Dunántuli Tud. Gyüjtemény. 1955. 4.

Szabó, P.Z.: Délkelet-Dunántul felszinfejlödési kérdései. - Die genetischen Probleme des Formenbildes im SO-Dunántul. Földr. Ert. 1957

Szádeczky-Kardoss, E.: Geologie der Rumpfungarnländischen Kleinen Tiefebene. Sopron 1938

Vadász, E.: Geologie von Ungarn. - Magyarország földtana. Budapest 1960

D i s k u s s i o n

M. K. G r a h v e (Moskau):

What methods were used in the identification of terraces with different altitudes?

M. P é c s i (Budapest):

Vor allem war die genaue Kenntnis der gegenseitigen Höhenbeziehungen der Terrassen im ganzen ungarischen Abschnitt des Donautales notwendig. Weiterhin wurde mit dem Barometer die präzise Messung der Terrassenhöhen durchgeführt; es folgte an zahlreichen Terrassenaufschlüssen die genaue Untersuchung der Erscheinungen der Solifluktion, des Verwitterungszustandes des Schotters, die genaue Klärung der Schichtenfolge, anschließend die gründliche mineralogisch-petrographische Untersuchung des Schottermaterials, die schwermineralogische Untersuchung des Sandes, die Feststellung des Abrollungspfades des Schotters von Terrassen verschiedener Höhe und die Auswertung der paläontologischen und archäologischen Funde.

Quartäre Tektonik am Ostabfall der Vogesen: Der Stand der Forschung

von

H. Vogt (Straßburg)

Der Einbruch des Oberrheingrabens, der bereits im Trias leicht vorgezeichnet war [10] , wird im Oligozän deutlich. Diesen Krustenbewegungen entsprechen korrelative Geröllablagerungen [15]. Die Morphologie des Ostabfalls der Vogesen zeichnet sich durch einen sehr ausgeprägten, als Bruchstufe erkannten Steilabfall aus, der die niedrigeren Formen der aus mesozoischem und eozänem Kalkstein und Mergel bestehenden und durch Verwerfungen durchzogenen Vorhügel überragt. Zwischen diese Hügel schieben sich die quartären Schuttkegel der aus den Vogesen fließenden Flüsse und Bäche ein. Anderswo, z.B. in dem Gebiet um Zabern, bringt ein Abwechseln von Horsten und Bruchsenken den Vogesensandstein vor dem Steilabfall wieder zum Ausstreichen.

Der Einfluß der quartären Tektonik auf die Ausbildung dieser Formengemeinschaft wurde von verschiedenen Autoren hervorgehoben. Wurden die Randverwerfungen wiederbelebt? Welches ist der Anteil der quartären Tektonik am Einfallen oder an der Heraushebung? Gibt es im Quartär differentielle Krustenbewegungen?

H. Baulig [12] hat gleichmäßige Höhenniveaus festgestellt, die nach den letzten differentiellen, auf jeden Fall prä-quartären, Krustenbewegungen zustandegekommen sind. Die Vogesentäler mündeten sämtlich um 300 m Höhe über der Ebene. Der Höhenunterschied kann entweder durch eine gleichmäßige Heraushebung der Vogesen ohne differentielle Krustenbewegungen und Verbiegungen oder durch eustatische Veränderungen erklärt werden. Dieser zweiten Erklärung gibt H. Baulig den Vorzug. Der randliche Steilabfall der Vogesen wäre also eine tertiäre und exhumierte Bruchstufe.

Doch hat sich die Erkenntnis quartärer Verformungen z.T. bereits vor H. Bauligs Veröffentlichungen auch anderen Forschern aufgedrängt.

A. Briquet [4] hat zwischen der Fecht und der Breusch das starke Gefälle der Aufliegefläche der älteren Schuttkegel herausgestellt (30 - 70 ‰), und er kann es nur durch eine Schrägstellung erklären. Diese Geschiebeablagerungen tauchen unter die jüngeren unter. Die Schrägstellung findet sich bei allen Oberflächenformen des Westrandes der Ebene. Briquet setzt diese gekippten Ablagerungen den schweizerischen Deckenschottern gleich. J. Vonfelt [16] schreibt sie dem älteren Quartär zu.

Diese Bemerkungen werden durch das Auffinden von Verwerfungen ergänzt. H. Klähn [8] hat eine solche zwischen dem Sandlöß und den quartären Schottern am Bollenberg, SW Ruffach, beschrieben. W. Wagner [17]schreibt eine Verwerfung zwischen dem älteren Löß und den Rheinablagerungen in Achenheim dem Riss-Würm-Interglazial zu.

Die quartären differentiellen Bewegungen in der Rheinebene, auf die schon Gutzwiller [7] und A. Briquet [3, 5] aufmerksam machten, wurden von N. Theobald [12] systematisiert. Er beschreibt z.B. die Schrägstellung des Sundgaues gegen N, während die Gegend zwischen Lörrach und dem Isteiner Klotz stabil zu bleiben scheint, auch die Einsenkung der Würmterrasse in der Mitte des Grabens, in der Höhe von Marokolsheim. Die morphologische Auswirkung dieser Senkung wird durch die Ablagerung der post-glazialen

Sedimente des Ried bekundet.

Eine Synthese der quartären Tektonik der Horstabfälle wurde von O. Wittmann [18] vorgelegt. Er erkennt drei Phasen für den gesamten Oberrheingraben: Eine wallachische Phase im Günz-Mindel-Interglazial, eine ortenauische Phase vom prä-Riss bis nach Riss I und eine brisgovische Phase im Würm. Der Autor verfügt über keine Beispiele aus der Randzone des westlichen Horstes südlich der Hardt. Diese Phasengliederung scheint verfrüht, und der Autor trägt den klimatischen Einflüssen auf die Ablagerungen der Schuttfächer nicht genügend Rechnung. Ähnliches gilt für W. Wagner [17]: Aus dem Abkappen altquartären Sandes durch Risslöß in Achenheim kann nicht auf eine Heraushebung im Mindel-Riss-Interglazial geschlossen werden.

Seit 1948 hat J. Tricart die Frage des Abfalls der Vogesen mittels der modernen geomorphologischen Methoden, an deren Ausarbeitung er großen Anteil hat, wieder aufgegriffen.

Das morphologische Bild zeigt, daß der Steilabfall der Vogesen oft zweigeteilt ist in einen unteren, massiven, geradlinigen, schartungslosen Teil und einen oberen, zerschnittenen Teil. Diese Morphologie ist besonders südlich Zabern ausgeprägt [14]. Der obere geschartete Teil besteht aus widerstandsfähigem Konglomerat, der untere, geradlinige, aus schlecht verfestigtem Sandstein. Der Unterschied in der Widerstandsfähigkeit zu den an seinem Fuße befindlichen pliozänen Ablagerungen ist gering. Es kann sich nur um eine rezente Bruchstufe handeln.

Viel Tatsachenmaterial, aus dem sich heute bereits eine Gesamtschau über die quartäre Tektonik am Ostabfall der Vogesen herausschält, wurde von J. Tricart und J. Vonfelt zusammengetragen.

Die älteren Ablagerungen der zwischen Ruffach und Schlettstadt aus den Vogesen fließenden Gewässer sind ausnahmslos als sich kreuzende Terrassen angeordnet. Talaufwärts sind sie stufenförmig angelegt, in der Ebene sind sie übereinander gelagert. Die Kreuzungslinie befindet sich meist ungefähr an dem Kontakt zwischen den Vorhügeln und der Ebene [16]. Die Vorhügel wurden also über die Ebene herausgehoben. Verschiedene Indizien zeigen, daß dies oft auf abrupte Weise geschah, da die Ablagerungen des älteren Quartärs sich meist in verhältnismäßig großer Tiefe unter der Ebene befinden (z.B. 84 m östl. Ostheim).

Betrachtungen über die petrographische Zusammensetzung der altquartären Ablagerungen ergeben im Quartär eine relative Absenkung der Vorhügel um 200 m in bezug auf das Gebirge. Die Zone der Vorhügel ist also ein Scharnier, das starken tektonischen Einflüssen unterworfen war.

J. Vonfelt [16] hat schöne Verwerfungen aus dem Mindel-Riss-Interglazial aufgezeigt. Es handelt sich um eine Verbiegungszone, innerhalb derer sämtliche Bruchstufen ursprünglich sind. Die zunehmende Höhe talaufwärts der Rißablagerungen über den Würmablagerungen im Weilertal zeigt, daß sich das Gebirge im Quartär um mindestens 350 m bei Reichenweier herausgehoben hat. Die heutige Höhe der Vogesen ist also hauptsächlich der quartären Tektonik zuzuschreiben. Der gesamte, im Quartär entstandene Höhenunterschied beträgt mindestens 400 m zwischen Ruffach und Schlettstadt, also 400 m/1000 Jahre, was den Zahlen Quirings [11] für die anderen rheinischen Gebirge gleichkommt. Die quartäre Heraushebung ist der oligozänen gegenüber originell, da sie die Zone der Vorhügel, die im Oligozän durch Ablagerungen überdeckt war, einschließt. Die Heraushebung neigt dazu, sich nach der Ebene hin fortzusetzen.

Diese Krustenbewegungen sind nicht einfach, sondern von zahlreichen Unregelmäßigkeiten begleitet.

In dem Gebiet des Schuttfächers des Gießen hat J. Tricart [13] mehrere, die Rißablagerungen störende Verwerfungen bewiesen, die etliche Fächer unterschiedlicher Bewegungen abgrenzen.

In dem Gebiet des Schuttfächers der Zorn hat G. Pacquetet [9] tertiäre Verwerfungen wiederbelebende Schollenbewegungen festgestellt. Einer endpliozänen, ziemlich anhaltenden Heraushebung folgt eine ruhige Periode mit Ausnahme einer relativen Einsenkung zwischen Zabern und Dettwiler. Im Mindel-Riss-Interglazial wird der ganze Schuttfächer gegen E schräg gestellt, und sekundäre Brüche kommen in ihm vor. J. Tricart [14] hat eine der Randverwerfung parallele und ebenfalls nach E schauende Verwerfung in den Mindelablagerungen beobachtet. Sie ist prä-Riss. In dem talabwärtigen Teil des Schuttfächers zwischen Stephansfeld und Hoerdt scheint eine gelinde Schrägstellung nach E die Ablagerungen der Neuen Tundrazeit zu erfassen. Soll man daraus schließen, daß die Heraushebung des Horstes nach E nach der Mitte des Grabens fortzuschreiten neigt?

L i t e r a t u r

[1] Baulig, H.: Questions de morphologie vosgienne et rhénane. Ann. Géogr. 1922, S. 132 - 154

[2] Baulig, H.: Quelques problèmes de morphologie vosgienne. In.: Dubois, G., Baulig, H., Rothe, E.: L'Alsace géologique, géographique et géophysique. (Bibl. Jean Macé n° 1) Strasbourg o.D.

[3] Briquet, A.: Alluvions anciennes et mouvements du sol dans la plaine d'Alsace. C.-R. Somm. Soc. Géol. France 1923, S. 17 - 18, 1925, S. 86 - 87

[4] Briquet, A.: Les alluvions quaternaires de la plaine d'Alsace entre la Fecht et la Bruche. Bull. Serv. Carte Géol. d'Als. et de Lorraine 1928, S. 55 - 76

[5] Briquet, A.: La terrasse à berge haute du Rhin moyen. Bull. Serv. Carte Géol. d'Als. et de Lorraine 1928, S. 263 - 268

[6] Briquet, A.: Le Quaternaire de l'Alsace. Bull. Soc. Géol. France 1930, S. 977 - 1014

[7] Gutzwiller, A.: Die Gliederung der diluvialen Schotter in der Umgebung von Basel. Verh. der naturforsch. Gesellsch. in Basel 1912, S. 57 - 75

[8] Klähn, H.: Die Geologie der Umgebung von Colmar. Ein Beitrag zur Geologie zwischen Lauch und Fecht. Bull. Soc. Hist. Nat. Colmar 1914 - 1915, 1916 - 1917

[9] Pacquetet, C.: Le cône alluvial de la Zorn. Manuskript, 1962

[10] Perriaux, J.: Contribution à la géologie des Vosges gréseuses. Mém. Serv. Carte Géol. d'Als. et de Lorraine n° 18, Strasbourg, 1961

[11] Quiring, H.: Eine Quartärisobasenkarte Westdeutschlands. Peterm. Mittlg. 1928, S. 144 sqq.

[12] Theobald, N.: Carte de la base des formations alluviales dans le Sud du Fossé rhénan. Mém. Serv. Carte Géol. d'Als. et de Lorraine n° 9, Strasbourg 1948

[13] Tricart, J.: Le cône de déjection du Gießen, près de Sélestat. Feuille Colmar au 1/80 000 Bull. Serv. Carte Géol. France 1955

[14] Tricart, J.: Tectonique quaternaire aux environs de Saverne. Bull. Serv. Carte Géol. d'Als. et de Lorraine 1956, S. 85 - 92

[15] Tricart, J.; Vonfelt, J.: La signification paléogéographiques des conglomérats oligocènes de la bordure vosgienne. Bull. Serv. Carte Géol. d'Als. et de Lorraine 1955/1, S. 105 - 118

[16] Vonfelt, J.: La bordure vosgienne entre Sélestat et Rouffach. Rev. Géomorph. Dyn. 1955/1, S. 7 - 33

[17] Wagner, W.: Die Diluvialaufschlüsse von Achenheim-Hangenbieten bei Straßburg 1939 Oberrhein. geol. Abhdlg. 1940

[18] Wittmann, O.: Tektonik und Klima und deren Phasen im oberrheinischen Diluvium. Zeitschr. d. Deutsch. geol. Gesellsch. 1953/5/6

D i s k u s s i o n

J.F. G e l l e r t (Potsdam)

Bezieht sich die Angabe einer Vertikalbewegung der Vogesen im Quartär in 400 m auf eine Heraushebung gegenüber dem Niveau der Oberrheinebene, oder ist das eine relative Bewegung zwischen dem Spannungsfeld des Oberrheingrabens und dem Hebungsgebiet der Vogesen.

H. V o g t (Strasbourg)

Bei dem Höhenunterschied von mindestens 400 m handelt es sich um die Gesamtbewegung.

Le rapprochement de l'Europe et de l'Afrique depuis le miocène supérieur

von

G. Dubourdieu (Paris)

N'ayant pas la possibilité de me rendre à Leipzig, il me semble opportun de confier à M. Destombes la teneur de mes recherches récentes dans le domaine de la Méditerranée occidentale. Car ces travaux conduisent à envisager, à partir d'observations mécaniques, un resserrement très tardif de la Méditerranée ou, si l'on préfère, un rapprochement de l'Afrique et de l'Europe. Ils rejoignent ainsi les préoccupations du Congrès sur les déplacements continentaux.

I. Les cisaillements profonds de l'Afrique du Nord et leur caractère cumulatif

Mes premières constatations ont été faites aux confins algéro-tunisiens. Des séries de fossés d'effondrement, encore actifs, sont décrochées selon des axes de cisaillement orientés du Nord-Est au Sud-Ouest. Les dislocations affectent le bâti profond et il est parfois difficile de les discerner en surface, là où affleurent de puissantes formations crétacées. Mais, pourtant, les tracés des accidents internes sont jalonnés par des anomalies tectoniques dans la couverture, à savoir des petits charriages d'un type particulier ou, encore, des injections horizontales de Trias plastique.

Les décrochements internes de l'Est maghrébin se sont produits après le dépôt des sédiments marins du Miocène et leurs amplitudes atteignent des valeurs de 10 à 25 km. D'autre part, ils sont parallèles entre eux et ils ont tous fonctionné dans le même sens relatif. Par rapport aux compartiments orientaux, les compartiments occidentaux se sont déplacés vers le Sud-Ouest.

Or cette tectonique du socle n'intéresse pas seulement les confins algéro-tunisiens, mais toute l'Afrique du Nord. Les nombreuses cartes publiées par les géologues français depuis la guerre montrent que, de la Tunisie au Maroc, tout le Maghreb est cisaillé par des accidents parallèles et de même sens. Et ces accidents décalent plus ou moins, en surface, toutes les lignes directrices de l'architecture géologique. Les fronts des diverses unités charriées au Miocène sont ainsi décrochés, qu'il s'agisse des nappes sud-telliennes, des chevauchements du métamorphique kabyle, des grands charriages en provenance de l'ancien domaine méditerranéen. Les axes des plis de fond sont aussi perturbés et, notamment, celui du Haut-Atlas marocain dans le région au Sud de Marrakech.

Les travaux cités en références donnent toutes les précisions voulues sur les diverses dislocations et leurs caractères de détail.

Mais un fait considérable doit être souligné. Le parallélisme et le sens constant des décrochements de l'Afrique du Nord oblige à additionner tous les déplacements unitaires. Par rapport au Maroc occidental supposé fixe, la région de Tunis s'est avancée de 350 ou 400 km vers le Nord-Est depuis le Miocène supérieur. D'autre part, l'obliquité des accidents par rapport à la côte méditerranéenne implique un allongement de 200 km environ de la distance de Tunis à Tanger. L'Afrique du Nord s'est déformée tout entière comme un jeu de cartes qu'on pousse sur une table.

II. Le rétrécissement de la Méditerranée depuis le Miocène sup.

Le phénomène précédent a nécessairement, à cause de son ampleur, des conséquences lointaines. Il oblige à concevoir, en particulier, un rétrécissement de la Méditerranée. Et voici pourquoi.

Les côtes méditerranéennes d'Espagne et de France ne portent pas les marques suffisantes d'une déformation comparable, par son intensité, à celle de l'Afrique. Si donc le site de Nice ne s'est pas déplacé, par rapport à Gibraltar, d'une quantité analogue à celle qui mesure le déplacement de Tunis par rapport à Tanger, c'est que, de toute nécessité, Tunis s'est rapproché de Nice.

Mais alors ce rétrécissement de la Méditerranée devrait avoir perturbé le pont géologique constitué par la Sicile et l'Italie. Or tout donne à penser que les choses se sont produites ainsi. La comparaison des directions structurales de la Sicile et de la Tunisie évoque une rotation récente de l'île, rotation de 30 ou 40° dans le sens des aiguilles d'une montre. Et, d'autre part, les vicissitudes tectoniques de la côte est de l'Adriatique s'harmonisent avec un déplacement de l'Italie méridionale en direction du Nord-Est. Je pense, en particulier, à la grande inflexion des chaînes helléno-dinariques au niveau du vieil accident de Scutari-Pec.

D'autres faits sont relatés dans les documents cités. J'indique simplement que l'obligation de concevoir un rétrécissement "postalpin" de la Méditerranée, obligation liée à la mécanique du Maghreb, semble fournir une explication à divers problèmes dynamiques soulevés en Sicile, en Italie ou même plus à l'Est.

III. Conclusions

Les données précédentes - qui exigent un rapprochement tardif des continents africain et européen - impliquent que le domaine méditerranéen ait eu, par rapport aux continents de bordure, une résistance mécanique relativement faible. L'obstacle créé par la soudure continentale de Gibraltar-Tanger déclanche, par réaction, le cisaillement en chaîne du Maghreb. Et, plus à l'Est, l'arc sicilo-italien se ploie pour absorber la poussée des blocs continentaux.

Ces vues impliquent à la fois une différence fondamentale des propriétés physiques entre les bâtis continentaux et la croûte sous-marine - et de très vastes déplacements géologiques.

Je crois d'ailleurs que le mouvement relatif entre l'Afrique et l'Europe n'absorbe qu'une partie des forces en oeuvre. D'autres phénomènes suggèrent que l'Europe s'est elle-même déplacée vers le Nord ou le Nord-Ouest au moment où l'Afrique se rapprochait de nos régions. Je pense ici à la disjonction très probable du bloc corso-sarde, bloc qui pouvait être soudé jadis au bâti européen.

Les migrations continentales réelles seraient donc énormes. Et, d'autre part, la répartition géographique des séismes actuels - notamment en Afrique du Nord - pourrait indiquer que ces migrations se poursuivent de nos jours.

P u b l i c a t i o n s

[1] Etude géologique de la région de l'Ouenza (Confins algéro-tunisiens). Publications du Service de la Carte Géologique de l'Algérie (Nouvelle série), bulletin n° 10, Alger 1956

[2] Esquisse géologique du Djebel Mesloula (Algérie orientale). Publications du Service de la Carte Géologique de l'Algérie (Nouvelle série), bulletin n° 21, Alger 1959

[3] La déformation tectonique de l'Afrique du Nord. Comptes Rendus de l'Académie des Sciences, t. 249, p. 2799, Paris 1959

[4] Tectonique et tremblements de terre en Algérie du Nord. Comptes Rendus de l'Académie des Sciences, t. 250, p. 884, Paris 1960

[5] Vues tectoniques sur la région d'Alger. Comptes Rendus de l'Académie des Sciences, t. 251, p. 2736, Paris 1960

[6] Réactions wegenériennes en Afrique du Nord? Comptes Rendus de l'Académie des Sciences, t. 251, p. 2998, Paris 1960

[7] Dynamique wegenérienne de l'Afrique du Nord. Livre à la mémoire de Paul Fallot, Paris (sous presse)

[8] Le déplacement de l'Europe occidentale. Comptes Rendus de l'Académie des Sciences, t. 254, p. 510, 1962

[9] Déplacement et déformation du pourtour de la Méditerranée occidentale depuis la fin du Miocène. Comptes Rendus de l'Académie des Sciences, t. 254, p. 2029, Paris 1962

Le Service International de l'Heure et le déplacement des continents

von

A. Stoyko (Paris)

Depuis 1923 le Bureau International de l'Heure centralise les résultats de comparaison des services horaires d'Amérique, d'Asie et d'Europe. Cette documentation importante permet de déterminer la variation des longitudes entre ces continents.

Les valeurs des longitudes observées dépendent du mouvement du pôle terrestre. Ainsi, avant de faire la comparaison des longitudes de divers stations, il est nécessaire de tenir compte de l'influence du mouvement du pôle instantané d'après les résultats de Service International des Latitudes (S.I.L.).

En discutant les résultats des services horaires pour la période de 35 années (de 1925 à 1960) nous avons trouvé une diminution progressive des différences des longitudes entre l'Amérique du Nord et l'Europe. Cette diminution annuelle des différences des longitudes est égale à $0^s{,}0012$. Elle correspond à une diminution apparente de distance de 39 cm en un an, ce qui donne plus de 10 m pour la période étudiée ou, plutôt, elle correspond à la déviation relative des verticales de l'ordre de $0''{,}5$. Le réchauffement des régions polaires nord constaté dernièrement, ainsi que le rehaussement du niveau moyen de l'Océan Atlantique Nord enregistré par les marégraphes des ports, peuvent expliquer cette déviation de la verticale dans la direction Est-Ouest.

Jusqu'à l'année 1935 on remarque une augmentation des différences des longitudes entre l'Europe et le Japon. La valeur de cette augmentation est incertaine, mais à partir de l'année 1935 on constate une diminution régulière des différences des longitudes.

On remarque pour l'Océan Pacifique à partir de l'année 1935 une augmentation des différences des longitudes entre les côtes Ouest et Est; sa valeur est d'environ $0^s{,}0017$ par an.

Du fait que la variation calculée des différences des longitudes dépend du déplacement du pôle instantané, on peut, inversement, trouver le mouvement du pôle instantané d'après les résultats de la détermination des longitudes.

Nous avons calculé la variation séculaire du mouvement du pôle instantané pour la période 1925-1960 d'après les résultats du Service International des Latitudes (S.I.L.). De plus, nous avons calculé le mouvement du pôle instantané en utilisant les résultats de variations des longitudes pour la même periode. Le tableau ci-dessous contient les résultats correspondants:

Méthode	Variation annualle	Direction
S.I.L.	$0''{,}0026$	61° W
Longitudes	0,0044	101° E
Long. sans Tokyo	0,0210	62° E

On compte les angles à partir du Sud.

On voit d'après ce tableau que la direction du mouvement séculaire du pôle instantané déterminé d'après les résultats des services horaires est presque opposée à la direction du mouvement du pôle déterminé d'après le S.I.L.: au lieu de 101° E dans le premier cas, on trouve 61° W dans le second.

L'utilisation des coordonnées du pôle déterminées par le S.I.L. augmente la variation des longitudes au lieu de la diminuer. Cela montre l'existence de vraies variation des longitudes qui peuvent provenir ou de déplacement des continents ou de la déviation des verticales dans le sens Est-Ouest.

Par conséquent, la détermination du mouvement séculaire du pole instantané d'après les observations de l'heure et de la latitude, mélangées ensemble, présente un certain danger.

Comme le poids avec lequel les observations de la latitude entrent dans la détermination du mouvement du pôle instantané est plus grand que celui des observations de l'heure, on trouve en combinant les deux résultats du tableau ci-dessus que le mouvement séculaire du pôle est nul.

La révision des longitudes conventionnelles nous a montré que la majorité des longitudes des observatoires participant dans le Service International de l'Heure exige des corrections négatives. Ainsi, par exemple, toutes les stations de l'Amérique du Nord et du Sud exigent les corrections négatives; cela confirme la conclusion précédente concernant la diminution des longitudes d'Amérique par rapport à celles de l'Europe.

D i s k u s s i o n

L. E g y e d (Budapest):

Je voudrais faire une remarque et une recommendation.

1. Les observations des longitudes ne sont pas capables de déterminer la dérive des continents, parce qu'ils ne tiennent pas compte de la possibilité de la variation du rayon de la terre.

2. Pour la détermination de la dérive du continent je propose de faire de très précises mesures concernant la distance des deux côtés des forces océaniques, p.e. en Islande ou en Afrique équatoriale, parce qu'on peut attendre seulement en ces places les plus grands mouvements horizontaux resultant d'une dérive des continents.

A.R. R o b b i n s (Oxford):

I would like to comment on these two important papers. I was very pleased to see that they both emphasized the difficulty of the interpretation of astronomical measurements in terms of earth movement, and I would like to add the following example. The Observatories of Herstmonceux (the now "Greenwich" Observatory of England), of Washington and of Richmond in USA, make time observations with Photo Zenith Tubes whose internal accuracy is 1 or 2 milliseconds. Yet the observed difference of longitude between Herstmonceux and Washington, also between Washington and Richmond and between Richmond and Herstmonceux, varies through about 20 milliseconds during a period of about a year. This is one order greater than the error of the observations. It is believed that this variation can only be explained by errors in star places, in particular by insufficiently accurate knowledge of their proper motions - this possibility being also mentioned by Dr. Stoyko.

It may well take about 50 years to determine proper motions sufficiently accurately to give star places correct to 1 millisecond. Even our error of this size equals 0".015, or about 1/2 metre cosine latitude in distance on the ground - that is to say 25 to 30 cms in these latitudes Consequently continental drift must be very large before it can be detected by astronomical observations.

I would like to emphasize one further point, also stressed in both these papers. The astronomical latitude and longitude of a point determines only one thing without doubt - namely direction of the vertical (i.e. of gravity), in space at this point. Consequently variations of latitude and of longitude to not necessarily signify horizontal earth movements. What happens when the land tilts? Does the direction of the vertical not tilt also? And, if it is possible for continents to drift, is it not also possible for masses within the earth to drift? This could also alter the direction of gravity.

Thus, as I said in the contact of levelling, it is of the utmost importance that the results of all geodetic measurements be interpreted with the greatest care and criticism before they may be held to signify crustal movement.

L. C a h i e r r e (Paris):

Je pense que l'utilisation des satellites artificiels permettra d'avoir des possibilités beaucoup plus grandes pour l'étude de cette question.

Quant à la première remarque, Mme A. Stoyko a pris soin de dire qu'il fallait être très prudent dans l'exploitation des résultats des observations, et que ceux-ci pouvaient correspondre, non pas forcement à une différence de distance des continents, mais à une variation de forme de la terre.

La variation des latitudes et leur relation avec le mouvement de l'écorce terrestre

von

N. Stoyko (Paris)

Le problème de la détermination du mouvement horizontal de l'écorce terrestre d'après les observations de latitude est lié avec le problème de la détermination de position du pôle instantané. Pour déterminer les coordonnées x et y du pôle instantané on utilise, dans le cas du Service International des Latitudes (S.I.L.), la formule suivante:

$$\Delta\varphi_i = x \cos \lambda_i + y \sin \lambda_i + z, \qquad (1)$$

où $\Delta\varphi_i$ est la différence entre la latitude observée et la latitude initiale adoptée de la station, dont la longitude conventionnelle est λ_i et z est le terme de Kimura qu'on suppose le même pour toutes les stations du S.I.L. Si, par exemple, dans le S.I.L. participent trois stations seulement il est impossible de séparer le déplacement du pôle instantané du mouvement éventuel des stations de latitude. Donc, il est nécessaire que le nombre des stations utilisées soit sensiblement plus grand que trois.

Ainsi, il est difficile de déterminer d'après les résultats du S.I.L. les mouvements possibles des stations de latitude: on n'a que quatre stations qui participent d'une façon quasi continue à ce service: Carloforte, Gaithersburg, Mizusawa et Ukiah.

On trouve pour la période de 60 années les variations apparentes suivantes des latitudes des stations du S.I.L. par rapport au pôle moyen de l'époque: Carloforte + 0ʺ045, Gaithersburg + 0ʺ200, Mizusawa - 0ʺ185 et Ukiah + 0ʺ160. Les coordonnées du pôle instantané ont varié pendant la même période de + 0ʺ078 (x) et + 0ʺ163 (y). D'après ces données, on ne peut pas déterminer les variations des latitudes par rapport à ce pôle mobile et, par conséquent, le mouvement horizontal de l'écorce terrestre. Les écarts qu'on trouve dans les variations des latitudes présentent, plus tôt, une allure accidentelle.

Nous avons recalculé les coordonnées du pôle instantané en utilisant les observations astronomiques du S.I.L. de 1900,0 à 1961,0 et en adoptant les mêmes latitudes initiales pour toute la période et pour chaque station. Nous avons utilisé toujours la formule (1). On ne trouve pas de sauts brusques dans les valeurs moyennes annuelles des coordonnées du pôle instantané calculées par la méthode d'Orlov. On peut tracer une ligne droite pour toute la periode pour chaque coordonnée. On a des oscillations des valeurs des coordonnées d'un côte et de l'autre de la ligne droite correspondante avec l'écart maximal ne dépassant pas 0ʺ05 et les écarts moyens égaux respectivement ± 0ʺ018 pour la coordonnée x et à ± 0ʺ019 pour y. Pour la période 1955-1960 l'écart moyen pour la coordonnée y est égal à 0ʺ011 seulement.

Ainsi, on voit qu'on ne peut pas déceler de changements brusques importants dans le mouvement séculaire du pôle instantané et les écarts moyens ne dépassent pas les limites des erreurs accidentelles. On peut remarquer seulement l'existence d'une période de

25-26 ans avec une faible amplitude.

En 1955, d'après la décision de l'Assemblée Générale de l'Union Astronomique Internationale, on a creé le Service International Rapide des Latitudes (S.I.R.). Dans ce service entrent les services independants de latitude en plus de quelques stations du S.I.L. Le nombre total des services participants au S.I.R. atteint actuellement 25.

Dans ce service entrent des stations qui se trouvent presque à la même longitude et pour lesquelles l'influence du mouvement du pôle est pratiquement la même. Par conséquent, les différences de leurs latitudes ne dépendent pas du mouvement du pôle instantané. Ces stations permettent d'étudier le mouvement de l' écorce terrestre dans la direction du méridien.

Nous avons dans ce cas, par exemple, les 3 stations d'Amérique du Nord qui utilisent les lunettes zénithales photographiques (P.Z.T.). Ce sont Ottawa, Washington et Richmond. Nous donnons dans le tableau ci-dessous les valeurs moyennes annuelles de leurs latitude par raport au S.I.R.

	Ottawa	Washington	Richmond
1956	37"369	17"275	47"221
1957	320	212	223
1958	311	208	226
1959	362	177	181
1960	372	171	196
1961	406	105	196
Var.ann.	+ 0"011	- 0"029	- 0"008

On remarque tout de suite que la latitude de Washington a une tendance nette à diminuer progressivement, tandis que celle d'Ottawa à augmenter. Par conséquent, l'influence de ces observatoires sur le mouvement du pôle est de sens opposé. La différence des latitudes entre Ottawa et Washington augmente de 0"040 par an ce qui correspond à 123 cm. en distance, si cette variation provient exclusivement du mouvement relatif de l' écorce.

Mais comme cette valeur de la variation annuelle est assez grande d'autres causes doivent également exister: variation des erreurs systématiques instrumentales, erreurs dans les mouvements propres des étoiles utilisées, déviation des verticales, etc.

Nous donnons aussi, comme exemples, les variations des différences des latitudes entre Poulkovo (Pu) et Poltava (Pl), ainsi qu'entre Mizusawa (Mz) et Tokyo (To) qui se trouvent respectivement sur les mêmes méridiens.

	1954	1955	1956	1957	1958	1959	Var.ann.
Pu-Pl	+ 0"024	+ 12	+ 21	- 33	+ 13	- 36	- 0"010
Mz-To	+ 0"055	+ 24	- 12	- 21	- 37	- 9	- 0"015

On remarque d'après ce tableau une diminution des latitudes des stations Nord par rapport à celle du Sud. Cela est confirmé également par la variation de la différence des latitudes entre Washington et Richmond. Les intervalles trop courts ne permettent pas de tirer des conclusions sûres concernant le mouvement de l'écorce terrestre.

L'accroissement du nombre des services de latitude, l'amélioration des catalogues utilisés pour les observations astronomiques et, surtout, l'amélioration des mouvements propres des étoiles, permettront d'améliorer notre connaissance sur les mouvements horizontaux de l'écorce terrestre.

Pour obtenir une plus grande précision, il est nécessaire d'utiliser pour la détermination des latitudes des instruments qui utilisent des étoiles du catalogue fondamental (FK4), dont les mouvements propres sont bien connus (Astrolabe Danjon, par exemple).

Continental drift in the light of modern astronomical data and prospects of further researches in this field

von

K. P. S c h e g l o v (Taschkent)

Modern astronomic methods began to be applied in a study of continental drift very soon after the appearance in 1912 of A. Wegener's paper "The origin of continents and oceans" [1] . These methods include radiotelegraphic determinations of geographic longitudes for points on different continents.

It should be mentioned that old longitudinal determinations used by A. Wegener and some of his adherents proved to be unsuitable for the detection of continental movements owing to their insufficient precision [2].

1) International Longitude Operations.

In 1919 on the initiative of the Paris Longitude Service, the first International Longitude Operation has been started, which included highly precise determinations of geographical longitudes of a great number of points on different continents. It was thought that a repetition of a similar investigation several years later will permit to establish whether longitudes were subjected to secular changes caused by a shift of the continents.

These studies were carried out in 1926. Forty two observatories were participating. Special attention was paid to the most accurate determination of the longitudes for the following points: Alger (Africa), Zi Ka Wei (Eastern Asia) and San Diego (North America) forming a closed polygon around the globe. As result of the investigations longitudes for a number of points have been determined with a maximum accuracy possible at that time. The data obtained were encouraging. Longitude differences between the angular points of the main polygon were accompanied by errors not exceeding $\pm 0^s.006$ and the error of closure amounted to $\pm 0^s.007$ [3].

A confidence prevailed that a repetition of such an investigation after a period of several years will solve the problem of continental shift. The second International Longitude Operation has taken place in 1933.

In 1952 N. Stoyko and P. Dubois have published the results of two international investigations [4] . In summarizing observations of several stations, the authors compute mean longitude changes during 7 years for the following continents:

Eurasia and Africa: E.A.
North America: N.A.
Oceania, Australia and Japan: O.J.

In computing the differences in the continents' longitude changes during the same period, they get:

$$OJ - EA = - 0^s.0063 \pm 0^s.0170$$
$$EA - NA = + 0.0136 \pm 0.0151$$
$$NA - OJ = - 0.0073 \pm 0.0196$$

The sign " - " corresponds to a move away, " + " - to an approach of the continents.

Though the errors of the given differences indicate the nonreality of the obtained displacements, the authors hope to obtain more positive data as result of the third International Longitude Operation during 1957 - 1959. It is too early yet to say whether their prognostics will be correct. The extremely extensive material accumulated as result of these researches consisting of 4000 series of astronomical observations and 1.000.000 receptions of radio time signals is being worked on at the present time by the Bureau International de l'Heure. The conclusions will be published.

Special mention should be made of longitude determinations in 1927 and 1933 by modern methods for the point Kornok in the south of Greenland. As Prof. Nörlund said in his Darwin lecture of 1937 [5] in both cases virtually indentical results have been obtained. A. Wegener's idea of Greenland's drift at a rate of 9 to 32 m a year has not been confirmed.

2) Movements of continents and the results of continuous measurements of longitude differences between continents.

Thanks to daily receptions of a great number of exact radio time signals, a possibility appeared in a number of observatories to control the mutual stability of longitudes. Beginning with 1920 this work is done by the bureau International de l'Heure in Paris. By using continuous determinations of the moments of time signals throughout the year it becomes possible to diminish quite substantially the random and systematic errors and to obtain annual corrections for the preliminary longitudes of the observatories.

In one of his papers N. Stoyko [6] on the basis of observation results during 20 years comes to the conclusion that in the interval between 1920 and 1929 North America moved away from Europe at a rate of 36 cm a year and Japan from Europe - at a rate of 171 cm a year. For the time interval between 1930 to 1939 he established an approach of America by 7 cm and a drift away of Japan by 13 cm a year. This way, says the author, during this time interval there has been no secular term of longitude discovered.

In 1955 A. Stoyko used in the solution of this problem data collected by the Bureau International de l'Heure. She formed a triangle with the following apexes: Northern America (Ottawa, Washington), Europe (Greenwich, Paris) and Asia (Tokyo, Zi Ka Wei). For other dates she gives differences of mean longitudes (in $0^s.001$):

Continents	1926	1933	1940	1947	1953
America - Europe	+ 12.7	+ 16.2	- 3.7	- 10.6	- 14.6
Europe - Asia	- 7.8	+ 1.9	- 3.5	+ 4.2	+ 5.4
Asia - America	- 4.9	- 18.1	+ 7.2	+ 6.2	+ 9.2

These figures do not confirm the movements of continents - such is the conclusion of the author of the research [7].

Actually, the mean error in longitude differences for two continents during the period from 1926 to 1940 amounts to ± 9.3 cm and for the period from 1940 to 1953 it is ± 5.3 cm. These errors indicate a progress in the precision of longitude difference determinations but deprive the changes of their real meaning.

In 1959 A. Stoyko comes back to this problem. She analyses observations for the period from 1927 to 1957 for two American observatories (Ottawa and Washington) and six European (Greenwich, Hamburg, Neuchâtel, Paris, Potsdam, Uccle). The author comes to the conclusion that during this period there has been a rapproachment of America and Europe at a rate of 39 cm a year [8].

Below are figures for the changes in longitude differences ΔL with their mean square error:

	W-E	O-E	AN-E	AN-P	W-P
ΔL	$-0^s.00126$ ± 19	$-0^s.00105$ ± 16	$-0^s.00116$ ± 12	$-0^s.00133$ ± 17	$-0^s.00143$ ± 19

(W here is Washington, O - Ottawa, P - Paris, AN - North America, E - Europe).

Conclusions of two papers by the same author and based on approximately the same material proved to be contradictory.

The results of A. Stoyko's last paper are disputed by W. Markowitz [9]. Pointing out that A. Stoyko did not take into consideration the proper motions of stars observed with the Washington zenith tube and taking them into account by connecting them with the data of the most accurate modern star catalogues, W. Markowitz comes to the conclusion that during the interval from 1934 to 1955 there have been no changes in longitude differences between Washington and European observatories.

3) On the accuracy of longitude determinations.

The results of International Longitude Operations have shown that the true accuracy of longitude determinations is much below the mean square errors attributed to them. The accuracy of longitude determination does, indeed, not depend so much upon occasional errors, which can be reduced to infinitesimal small values by using extensive observation material, as upon errors of systematics, from which it is extremely difficult to get rid of.

In a number of papers, which came out after 1933, it has been shown that astronomical time determinations contain such systematic errors, which can not be eliminated from observations on the basis of our present state of knowledge [10]. Especially dangerous proved to be instrumental errors, which change with time [11]. Important are also personal errors of observers, errors of star catalogues and errors caused by the environment, in which the observations are taking place [12].

To eliminate systematic errors lengthy - at least a year-long - observations should be used. Such studies are taking place in the Bureau International de l'Heure, where every year corrections are introduced to conventional longitudes of observatories, that participate in the researches on the problem of time. It should be remembered, however, that these corrections indexed K^1 represent the entire sum of systematic errors not excluded during the year of observations. These errors can be reduced by using as datum lines for comparison longitude values of determinations for many years. But in this case the research will last several decades and during such a period, as experience has shown, observations lose their homogeneity. It should be also born in mind that there is not always a justification for the use of longitude observations at different observatories for a determination of the mean longitude of a continent. While in well equipped and experienced observatories K^1 values are of a low variance, in others it is quite substantial. According to data of the last ten years we have obtained mean square errors of the annual correction of longitude for Washington $\pm 0^s.0062$, for Paris $\pm 0^s.0064$, for Greenwich $\pm 0^s.0092$, for Neuchâtel $\pm 0^s.0198$.

To introduce into treatment rough data which soever would be their weight amounts to a deterioration of definitely better data. However, we are far from an idea of excluding longitude determinations from our means of solving the problem of the movement of continents.

In certain cases longitude determinations have already made it possible to diminish substantially the values of continental drift prognosticated by Wegener. The question still remains open: is the rate of their displacement within the limits of errors of longitude determinations and what are these values?

Returning to the paper previously quoted [4], we see that changes in longitude differences between North America and Eurasia according to the results of International Longitude Operations of 1926 and 1933 are accompanied by an average error of $\pm 0^s.0151$, consequently, the error in the value of a yearly displacement of continents should be accepted as $\pm 0^{s}.0022$, which for latitude 43° (an average between Washington and Paris) will correspond to 75 cm. Inasmuch as the dispersion of measured values can come to a tripled mean square error, a displacement of the continents at a rate below 2 m a year should not be looked for on the basis of longitude studies of 1926 and 1933. Because of an improvement in the accuracy of longitude determinations during the last 25 years N. Stoyko and P. Dubois think it possible as result of the third Longitude Operation of 1957 - 1959 to obtain a mean error of an annual displacement of America in respect to Europe amounting to $\pm 0^{s}.0004$, which corresponds to 12 cm.

If one agrees with their rather optimistic, from our point of view, computations, it is possible to expect that the hypothetic movement of continents can be detected if it exceeds 30 cm a year. D. Yu. Belotzerkovsky investigated the problem of accuracy in longitudes obtained as result of a yearly cycle of observations in seven observatories working along a plan of seventeen-year series of observations [13]. The mean error of such a longitude came to $\pm 0^{s}.011$. Thus an error in the change of longitude difference for two periods of observations should come to $\pm 0^{s}.022$ or in a linear measure to 7.5 m. This makes it clear that to establish a movement of continents with a precision up to decimeters, the interval between redeterminations of longitudes should be extended to several decades.

Prospects

A further use of the longitude method in the study of continent movement is related to a higher accuracy in time determinations from astronomical observations. The progress in instruments and methods of observations noticeable during recent years permits to hope that the error in the determination of longitude differences can be brought down to $\pm 0^{s}.003$, which for latitude 45° amounts to 1 m in linear measures. Under such conditions one can expect to detect a displacement of continents by one meter if the periods of redetermination will be separated by an interval of 4 - 5 years. To discover a continent displacement in decimeters and centimeters these periods should be separated by correspondingly longer time intervals. It would be also advisable to use latitude observations in the detection of continental drifts. Here we will also need a perfection of the theory of accounting for periodic and nonpolar latitude changes, the absence of which will not provide the necessary accuracy of research.

Finally, it is necessary to look for new methods of making a link between the continents. Promising in this respect seems the use of synchronous observations of artificial satellites. However, these problems are not sufficiently studied yet both theoretically and practically.

References

[1] Wegener, A.: Proischoždenie materikov i okeanov. M.-L. 1935

[2] Ščeglov, V.P.: Mirovye opredelenija dolgot kak metod izučenija dviženija materikov. Trudy soveščanija po metodam izučenija dviženij i deformacij zemnoj kory. M. 1948

[3] Lambert, A.: Revision des longitudes mondiales. Toulouse, 1928

[4] Stoyko, N. et Dubois, P.: La deuxième opèration internationale des longitudes octobre-novembre 1933, Paris 1952

[5] Nörlund, N.E.: Astronomical longitude and azimuth determinations. Monthly notices, v. 97, No 7, 1937

[6] Stoyko, N.: Sur des variations séculaires et périodiques des longitudes. Comptes rendus des séances de l'Académie de Sciences. t. 214. Paris, 1942

[7] Stoyko, A.: La collaboration internationale dans le domaine de l'heure. Le bureau international del'heure et la détermination des longitudes mondiales par T.S.F. Suisse Horlogère, N.I., 1955

[8] Stoyko, A.: Sur la variation séculaire des longitudes entre l'Amérique du Nord et l'Europe. Comptes rendus des séances de l'Académie des Sciences, t. 248, Paris, 1959

[9] Markowitz, W.: Latitude and longitude, and the secular motion of the pole. Methodes and techniques in Geophysics. London, 1960

[10] Zverev, M.S.: Trudy Gosudarstvennogo astronomičeskogo instituta im. Šternberga, t. XVIII, vyp. 2, M. 1949

[11] Ščeglov, V.P.: Opyt issledovanija nekotorych sistematičeskich ošibok pri opredelenijach vremeni perenosnymi passažnymi instrumentami. A.Ž. t. XXVII, M. 1950

[12] Pil'nik, G.P.: Sezonnye ošibki v nabljudenijach služby vremeni. A.Ž. t. XXVIII, M. 1961

[13] Belocerkovskij, D.Ju.: Sostojanie voprosa o dolgotach služb vremeni. Trudy 14 astrometričeskoj konferencii SSSR, L. 1960

Discussion

J. Witkowski (Poznań):

Die Ausführungen von Prof. Scheglov setzen die Frage von der Bestimmung der Länge und Breite ins richtige Licht.

Da die Instrumente, die dazu gebraucht werden, noch nicht die nötige Genauigkeit haben, und die Sternkataloge verbesserungsbedürftig sind, kann augenblicklich noch keine endgültige Antwort auf die Frage der Verschiebung der Kontinente gegeben werden.

Die zweite von Prof. Egyed gestellte Frage (Bemerkungen zu den vorgelesenen Berichten von N. und A. Stoyko) findet eine vorläufige Antwort in den Beobachtungen von Sternbedeckungen durch den Mond. Es wurde nämlich festgestellt, daß Beobachtungen, die in Japan und auf dem Asiatischen Kontinent gemacht wurden, auf eine anormale Verschiebung der Vertikale hindeuten, deren wahrscheinliche Ursache das dazwischenliegende Diskontinuitätsgebiet ist.

A. A. Izotov (Moskau):

We can only regret that at this Symposium we do not have unquestionable data necessary for the solution of the problem of continental drift. Therefore, it seems to me, that one of the tasks of such Symposia should be the consideration of the ways and methods of obtaining the necessary data. I think it would be important to discuss the organisation of new determinations of latitudes and longitudes. This includes the problems of the choice and location of stations, of equipment and methods of highly precise astronomical observations. The Soviet delegation had the honour to submit certain suggestions along that line, which are given in the report by Prof. Boulanger and Dr. Mescherikov. I want to again draw your attention to these proposals which are directed to create a basis for the solution of the problem.

A. V. K o n d r a s h k o v (Moskau):

Besides purely astronomical observations the artificial earth satellites with radioelectronic equipment for the measurement of distances could be used for the study of the continental drift. The observations with satellites provide data of the same and even greater accuracy than the astronomical determinations. The resolutions of the Symposium should note the possibility and necessity of application of the satellite data in the study of that problem.

The importance of determining the relativ horizontal movements in continental and island-arc areas

von

F. A. Vening Meinesz (Amersfoort)

It is clear that the subject of relative horizontal movements at the earth's surface is of the greatest importance for geodesy; the obviously affect the triangulations and the mapping. But they are also highly important for the geophysics of the solid earth because they provide most valuable information about the processes going on in the earth. This paper does not touch on another important subject, the relative movements of continents as a whole.

As it is well knwon many earthquakes are caused by shifts along a faultplane through the crust of the two crustal blocks on boths sides with regard to each other. In most cases this relative shift has a horizontal component. After the earthquake in Sumatra of 1892 the director of the Triangulation Service of the East Indian Archipelago, J.J.A. Müller, remeasured the triangulation-net and found that two stations at a mutual distance of about 27 km on a line roughly parallel to the length axis of the Island of Sumatra were over 1.23 m and 1.24 resp. displaced in directions very roughly parallel to this axis (in the sense to the northwest) with regard to two other stations to the northeast of the first pair at a mutual distance of about 35 km, also on a line roughly parallel to the axis of Sumatra; the connecting line of the second pair is situated about 30 km northeast of the connecting line of the first pair of stations. The displacement of a fifth station in the area between both pairs over 0.64 m (roughly to the southeast) leads to the supposition that the faultplane was about parallel to the length-axis of Sumatra and located between the fifth station and the first pair; the displacement along this plane must have had a horizontal component of about 2.2 meters. For details the writer may refer to a paper by Muller in "Verhandelingen Kon. Nederl. Akad. v. Wetenschappen", Amsterdam, 1895.

For the strong earthquake in California of 1906 the faultplane was the San Andreas fault and the horizontal displacement was about 4 1/2 meters. The direction of this fault is about parallel to the coast. For this relative movement we find two strong indications of its being the continuation of a movement that has been going on during a long time already. In "Internal Constitution of the Earth" 2nd Edition, p. 2o7, Gutenberg gives a picture of the offsets of streams where they cross the San Andreas fault. In the second place the geology on both sides of this fault and of other parallel faults, slightly further inland, which have previously played, show a large offset. This has led to the conclusion that together a dextral displacement has taken place of at least 160 km. As in Indonesia, remeasuring of the triangulation has shown that not only regional deformation of the crust has occured, bat also a relative movement of great crustal blocks of thousands of kilometers dimension.

In the area of the Indonesian archipelago we have another possibility of obtaining data about the total displacement along the Sumatra fault or faults. A belt of strong negative gravity anomalies has been discovered of about - 130 mgals over a width of about 100 km, running outside Java and over Timor, Tanimbar islands and Kai islands. This belt is seismically disturbed and we may, therefore, be sure that the crust in this belt is subject to strong deformation. In Timor evidence ist found of folding and overthrusting and this points to the deformation having the character of horizontal compression of the crust.

These facts lead to the supposition that in this belt the crust is subject to plastic downbuckling and that the shortening in the direction of the compression amounts to about 40 km. For a more detailed study of the crustal phenomena, the writer may refer to different papers [1]. From these papers it follows that the total shear movement along the faults in Sumatra here discussed is of the order of 37 km, and that the whole crustal block of the archipelago has moved over 33 km to the SSE with regard to the surrounding crust. The Archipelago crustal block is bordered on the eastern side by faultplanes through the Philippine islands, which are well known to geologists. They found that the relative movements along these faults are of the sinistral kind, which fits to the general picture of movement of the Archipelago crustal block as we have here given; the movement along the Sumatra faults was, as it was mentioned, of the dextral kind.

In the eastern part of the archipelago the pattern of deformation shows a complication; the belt of strong negative anomalies, i.e. the belt of crustal downbuckling, runs from the Philippine islands to East Celebes where it ends at the west end of this arm of Celebes, while the belt running over Timor gets much weacker near the Kai islands and from there shows a weak connection to the former belt east of East Celebes. For the interpretation the writer may refer to the papers mentioned, but we must here draw attention to the Problem how it is possible that the belt of downbuckling ends at the west end of East Celebes. We can only understand this behaviour, if we assume a crustal faultplane running from there to the NNW, viz. in the direction of the uniaxial stress which causes the whole crustal deformation pattern in the archipelago; in this case the area to the ENE of this faultplane could move to the SSE for bringing about the compression causing the downbuckling in East Celebes.

However, no such faultplane has been found and so this puts us before a problem which can only be solved if we assume that also relative movements of two crustal blocks can occur in a zone between both blocks, where the limit of elasticity is exceeded and where pseudo-flow takes place. Gutenberg has already made mention of such a behaviour in the Buena Vista Oil Field in California [2]; he publishes a picture of bent pipe lines in an area where no earthquakes take place and where no fault can be found, but where unmistakably faulting movements of a gradual kind, in time and place, occur.

It is hardly necessary to stress the importance of this possibility of crustal behaviour for our problem. Its occurrence probably requires a value of the shear-stress in the mobile zone that is high enough to exceed the elastic limit. But the absence of earthquakes which can reveal such a zone makes it difficult to surmise whether and where it can be present. This leads to uncertainties where such mobile zones exist, and, consequently, where remeasuring of triangulation-nets may be needed. It is probable that at least one condition has to be fulfilled for their occurence, viz. the presence of strong deviatoric stresses in the crust, and so we may perhaps say that they will usually occur in areas which are seismically active. However observations are needed for making sure about this point.

In Japan many geodetic determinations have been made of the crustal movements, which are particularly strong in that area. Their results indicate the existence of several crustal blocks, which are moving with regard to each other with velocities of one or a few centimeters per year. There is a tendency of movement towards the Japan Trench. The presence of this trench, which no doubt can be attributed to crustal downbuckling, makes it likely that the total amount of this movement is at least equal to the value found for the Indonesian archipelago, i.e. 33 km.

A similar conclusion may be drawn for the Caribbean area. Here also the Puerto Rico Trench points to a total relative movement of about 30-40 km of two great crustal blocks. This is confirmed by the results of the gravity survey which indicates that the crustal deformations in this area caused by uniaxial compression in the direction SSW-NNE. The southern block comprises the South American Continent. The block is probably bounded by a deformation belt of the Buena Vista type, running from the SW-NE directed part of the Isthmus to the island of Santo Domingo, and along which the block moves to the NNE with regard to the crustal block to the WNW of this belt. As in the Indonesian archipelago, this belt is seismically inactive though the Caribbean as a whole is subject to many earthquakes. So the movement along this belt is probably continuous.

The curious twist in the isthmus, comprised in the mobile belt, seems to point to a relative displacement of the two blocks of a much greater amount than mentioned above, viz. about 200 km. It is likely that this last figure has regard to the whole period of existence of the South American continent and the isthmus, while the value of 30-40 km dates from the time the Puerto Rico trench started to form.

Up to now we have dealt with the degree of mobility of the mobile areas of continents and island-arcs, where orogenic phenomena are taking place. However, there are probably other continental areas that are not quite stable. If the crust is subject to horizontal tension, we must expect that, according to the theory about faulting of Anderson, King Hubbert and others, tilted faultplanes will come into being under an angle of hade of 25^{0} to 30^{0}. The tendency towards isostatic equilibrium will bring about that crustal blocks between two such tilted faultplanes will subside if the faultplanes converge downwards and will rise if they diverge downwards. This leads to the development of "graben" and "horsts". If through some disturbing effect the angles of hade of the faultplanes are somewhat different, the two statements remain true.

The vertical movements mentioned are only possible if the corresponding horizontal movements are possible and this is obviously not always true; if we examine geological profiles we see that often isostatic equilibrium is not attained. In that case we may expect that, if the crust is subject to tension and if, therefore, some movement becomes possible, vertical as well as horizontal movements will occur of the two blocks on both sides of the tilted faultplane. There is no doubt that such movements do take place; they have recently been occurring in the southeastern provinces of the Netherlands. Such movements will often not be accompanied by earthquakes; the presence of horizontal tension in the crust corresponds to small energies involved in these isostatic adjustment effects.

From what has been said it follows that the horizontal movements of this kind are smaller than the vertical ones; the ratio depends on the angle of hade of the faultplane along which movement occurs. Still, though the shifts of this type are in general much smaller than those along faultplanes in orogenic areas, they are not negligible for the geodesist, and as they may occur in areas which are not subject to earthquakes, the geodesist must be careful not to overlook this possibility.

In the above paper the writer has not taken up the problem of the cause of the considerable horizontal movements, of the horizontal compression in some parts of the crust and of the horizontal tension elsewhere. He believes that it is difficult to explain them otherwise than by half-turn convection currents in the mantle, which exert a strong drag on the rigid crust. He has advocated this hypothesis since 1930. For the most recent statement about it he may refer to two papers of 1961 [3]. The currents are supposed to make only a half turn, during which they bring down the upper part of the mantle which lost heat towards the outside, and bring up the lower part. This half turn is expected to last 50 to 100 million years, during which the crust is deformed by the drag exerted on it. A long period of rest ensues of several hundred million years, during which cooling acts on the upper mantle and heating by the core on the lower mantle, and then a new half-turn current comes into being. During the present period we are probably living in the second half of a half-turn current system. In other words, the orogenic period is still continuing.

L i t e r a t u r

[1] a) Vening Meinesz, F.A.: Indonesian Archipelago, a geophysical study. Bull. G.S.A., Feb. 1954

b) Plastic Buckling of the Earth's Crust: The Origin of Geosynclines. G.S.A. Spec. Paper 62, "The Crust of the Earth" pp. 319 - 330, 1955

c) The Geophysical History of a Geosyncline. Proc. Kon. Ned. Akad. v. Wetensch., Ser. B, 60, No. 2, pp. 126 - 140, 1957

[2] Gutenberg, B.: Internal Constitution of the Earth. 2nd. Ed., Dover Publ., p. 403,1951

[3] a) Vening Meinesz, F.A.: Convection-currents in the Mantle of the Earth;

b) Continental and Ocean-Floor Topography, etc.; both papers: Proc. Kon. Ned. Akad. v. Wetensch. Ser. B, 64, No. 4, 1961

Vertical movements of the earth's crust, temporal changes of gravity field and shift of subcrustal material

von

E. N. L u s t i k h , V. A. M a g n i t s k y (Moskau)

S u m m a r y

In the article are discussed relations of the gravity field to the vertical movements of the crust and to the transfer of the matter along the Earth's surface.

The gravity field of the stable platforms, disturbed areas and the areas of recent glaciation is considered. Principal inferences are as follows.

Vertical movements of the crust are accompanied by the displacement of subcrustal material in direction suitable to the establishment of the equilibrium of the crust.

Nevertheless it seems the dominating cause of vertical movements is the change of volume of the subcrustal material. Only in the course of time the disturbances of gravity field due to this process can be eliminated by the transfer of material in the subcrustal weak layer (probably the low-velocity layer).

The results are in fair agreement with results obtained by the other geophysical investigations and with data from investigations of mineral transformations due to the temperature and pressure changes.

Z u s a m m e n f a s s u n g

Es werden Beziehungen zwischen dem Gravitationsfeld und den Vertikalbewegungen der Kruste sowie der Verlagerung der Materie entlang der Erdoberfläche diskutiert.

Es wird das Gravitationsfeld stationärer Ebenen, gestörter Gebiete und Gebiete rezenter Vereisungen betrachtet. Die Hauptstörungen sind folgende:

Vertikalbewegungen der Kruste sind mit Ortsänderungen von unter der Kruste liegendem Material in einer bestimmten Richtung verbunden, um den Gleichgewichtszustand der Kruste zu erreichen.

Nichtsdestoweniger scheint der Hauptgrund für die Vertikalbewegungen in der Volumenänderung des Materials unterhalb der Kruste zu liegen. Nur in dem Falle können die Störungen des Gravitationsfeldes infolge dieses Prozesses eliminiert werden durch den Transport von Material in der unter der Kruste liegenden weichen Schicht (wahrscheinlich die Schicht mit der geringen Geschwindigkeit).

Die Ergebnisse stimmen mit den Ergebnissen, die von anderen geophysikalischen Untersuchungen vorliegen, sowie mit den Untersuchungsdaten von Mineraltransformationen infolge von Temperatur- und Druckänderungen gut überein.

We believe the vertical movements of the Earth's crust are major ones since they persist everywhere continuously whereas folding and volcanism occur only locally and mainly in certain periods.

While examining the causes of vertical movements, gravity data may be useful. As a rule, such movement changes gravity, and both the sign and amount of the change indicate the mode of the movement. The cause of uplift or subsidence may be either shift of material under the Earth's crust (or in the crust), or change in volume of the material, or both. At the surface of the Earth either denudation or sedimentation may be superimposed in the same time. Considering the simplest case (i.e. of one dimension) one can count the value of change in gravity anomaly of any kind. In this case free air anomaly is identical with isostatic one. Let us denote

H - vertical displacement of the crust (positive upward),

α - a fraction, indicating the share of the vertical displacement on account of horizontal shift of deep-seated material,

D - density of this material,

h - thickness of the layer which was deposited (positive) or carried away (negative),

d - density of this layer, we suppose this density was adopted for Bouguer reduction,

f - gravitational constant.

The change in Bouguer anomaly is

$$2\pi f\ (d-\alpha D)H \qquad (1)$$

and the change in isostatic anomaly is

$$2\pi f\ (dh-\alpha DH) \qquad (2)$$

Only the movements of platform are to be discussed since processes in geosinclynes are much more complicated and changes of gravity field are mostly unknown there. All the platform areas we can divide into "calm" and "disturbed". The calm areas are low plains, they may have a thick cover of sediments or be much truncated by denudation, as a rule, indeed, both the cases indicate slow monotonous vertical movement during several geologic cycles. The disturbed areas are either highly lifted above sea level or submerged under sea water, commonly this proves they have been greatly moved vertically during Alpine tectonic cycle.

The most difficult task is to determine the change of gravity field. We can measure the present gravity field, but not the ancient one. The only possible method is to compare identical areas of different age and to admit that a sequence from the oldest structures to the youngest is the sequence of evolution.

As examples of disturbed areas we may point to the western part of the North American platform, the southern part of the Indian platform, Central Mexico and East African basin. The ancient (of precambrian consolidation) calm platform areas are the Russian platform, the eastern part of the North American platform, the northern part of the Indian platform. The Young (of paleozoic consolidation) calm areas are the North German lowland, the French lowland, West Siberia.

It is easy to see from gravity data that the calm areas of any age are in isostatic equilibrium, and since they are lowlands, Bouguer anomalies are close to zero. From this we may conclude that gravity on such areas does not change in spite of addition of some mass due to sedimentation or removal of some mass by denudation. It follows from (1) that Bouguer anomaly remains without change only if

$$\alpha = \frac{d}{D} \qquad (3)$$

and from (2) we may see that isostatic anomaly remains without change if

$$\alpha = \frac{dh}{DH} \qquad (4)$$

if $h = H$, which is just the case, this condition turns into (3). It is most probable that $d \leq 2.5$ g/sq.cm and $D \leq 3.3$ g/sq.cm, so from (3) $\alpha > 0.75$, i.e. the long slow monotonous vertical movements in such areas minimum 75 per cent are on account of horizontal shift of deep-seated material.

Disturbed areas are mostly overcompensated, i.e. the rised areas have negative isostatic anomalies and the submerged areas positive anomalies. It suggests such fast movements of short duration are mainly due to dilatation or contraction of material below.

Considering our problem, areas of recent glaciation become to be of special importance. The example of the upheaval of Scandinavia after disappearance of the ice cover is a common one. The uprising of Scandinavia was used by many authors not only in order to prove the theory of isostasy but also to deduce the viscosity of the subcrustal material.

Nevertheless the example of Scandinavia is not quite convincing. The uprising of this region may be caused by some other causes not only by the tendency toward established isostatic equilibrium.

Recent geophysical investigations in Greenland and Antarctica have obtained evident proofs in favour of outflow of subcrustal matter under the pressure of glacial load.

On Fig. 1 is given a profile across East Antarctica containing cross-section of the ice-shield and its bottom and the free-air gravity anomalies.

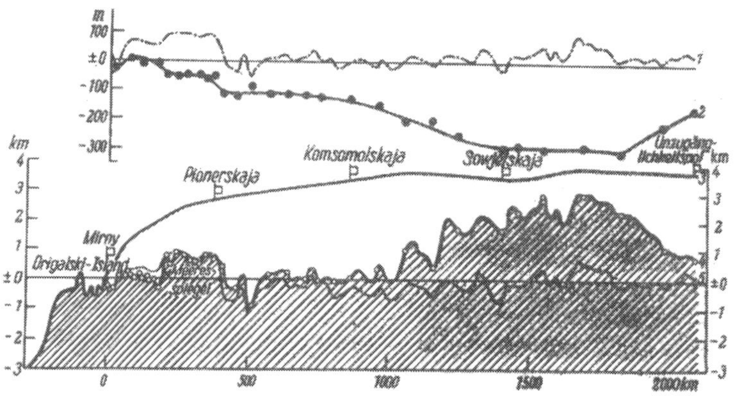

Fig. 1

Cross-section through the ice-shield and its bottom in East Antarctica and free-air gravity anomalies

One can see that the details of the free-air anomaly curve are connected with the correspondent details of the bottom of the ice-shield. On the contrary the average anomaly along every part of profile is very close to zero. This means proximity of the region in consideration to the isostatic equilibrium. As ice-covers in Greenland and Antarctica were superimposed in epoch not lower than tertiary one, we may conclude this interval to be sufficient for restoring the equilibrium due to outflow of subcrustal matter.

So we have studied the relations of the gravity field to the vertical movements of the crust and to the transfer of the matter along the Earth's surface.

Principal conclusions are as follows.

Vertical movements of the crust if there is sufficient interval of time are accompanied by the displacement of subcrustal material in direction suitable to the establishment of the equilibrium of the crust. This displacement takes place at the depth less than horizontal dimensions of moving blocks of the crust. Hence the gravity during such processes remains near to its normal value. But it does not mean that such a subcrustal displacement of material is the principal cause of the vertical movements in consideration. On the contrary it seems that the dominating cause of vertical movements is the change of volume of the subcrustal material. This consequence obviously follows from the gravity of newly displaced blocks. Changes of volume can occur not only in the upper parts of substratum (above the asthenosphere) but as well at the great depths. Only in the course of time the disturbances of gravity field due to this process can be eliminated by the transfer of matter in the subcrustal weak layer.

These results are in fair agreement with results obtained by the other geophysical investigations. The existence of a low-velocity layer at the depth of 50–200 km is established. The sources of volcanic activity are connected with this layer. The most probable explanation of this fact is supposititious that the temperature in that layer is near to the melting point and the material is there in the vitreous state. Thus the low-velocity layer can be considered as the layer in which the restoring displacements of subcrustal matter take place.

Fig. 2 shows a cross section through Central Mexico obtained by seismic investigations. In lower part of section there is a region of low velocities which is supposed to be the place of expansion of the matter causing the late upraising of Central Mesa.

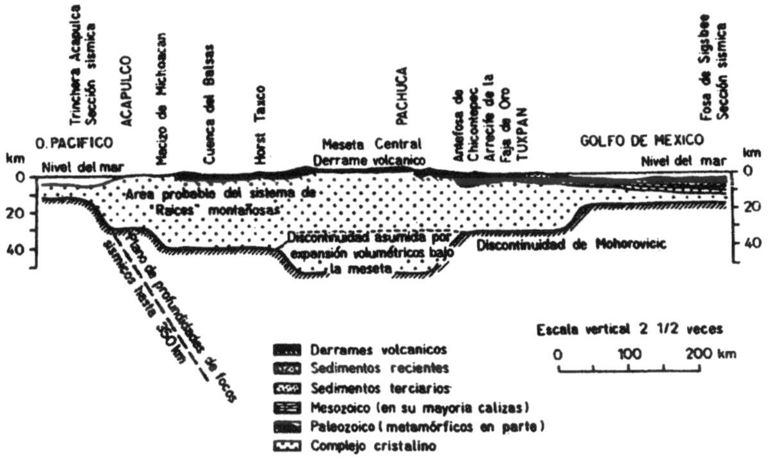

Fig. 2

Cross-section through Central Mexico obtained by seismic investigations

On Fig. 3 the transition curves of some mineral transformation due to the change of pressure and temperature are plotted. Comparison of these curves with temperature-pressure conditions in the Earth's interior plotted on the same figure confirms the possibility of the intense volume change in the subcrustal material.

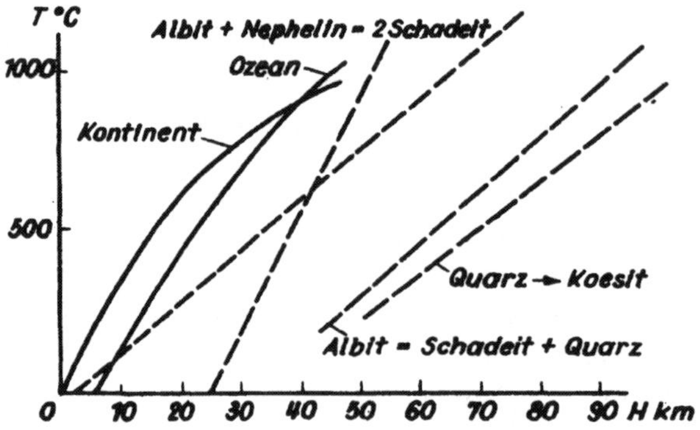

Fig. 3

Comparison of the transition curves of some mineral transformation due to the change of pressure and temperature with the temperature-pressure conditions in the Earth's interior

D i s k u s s i o n

D. I. G o u g h (Salisbury):

In congratulating Prof. Magnitsky on a most interesting discussion I should like to agree with him on the need for caution in assuming that the existence of isostatic anomalies is necessarily related to movement in the crust or mantle. In such deglaciated areas as Fennoscandia, uplift is combined with negative isostatic anomalies. In many other regions, however, large isostatic anomalies have persisted for times in excess of 10^7 years, and the corresponding loads have been supported by the strength of the crust.

Some geophysical problems in connection with recent and past crustal movements

von

L. E g y e d (Budapest)

S u m m a r y

The horizontal crustal movements bring about a mosaiclike structure on the Earth's surface. There are several stages in the evolution of this structure:

a) Fault systems of epirogenetic territories

b) Rift systems

c) Graben-like features

d) Continental drift.

The horizontal crustal movements cannot be separated from the vertical crustal movements as they are the results of the same stress system. The currently registrated crustal movements also show a graben-like system (the European crustal movements of the USSR). The development of the present-day crustal movements has begun in the geological past. An evidence thereof can be found in the linear connection between the quaternary territories and the level differences in Hungary. The area of the Little Hungarian Plain corresponds entirely to the stage of evolution of the Red Sea, which can be proved by geological and geophysical data, as well. However, the rate of sedimentation seems to be much quicker than in the area of the Red Sea. The subsidence of the area of the Little Hungarian Plain is going on since Paleozoic times. Most of the horizontal movements show a tensile character. An adequate explanation of the system of horizontal movements - also from the point of view of energetics - can be furnished only by the theory of general expansion. The mechanism of continental drift, however, cannot be derived exclusively from this theory. A magma current due to expansion must be assumed for the explanation of drifting away of a greater magnitude. The development of present-day forms of continents is shown on the results of several model experiments by the author.

Z u s a m m e n f a s s u n g

Die horizontalen Krustenbewegungen führen zu einer mosaikähnlichen Struktur an der Erdoberfläche. Es gibt mehrere Stufen in der Entwicklung dieser Struktur:

a) Verwerfungssysteme epirogenetischer Territorien

b) Spaltensysteme

c) Graben-ähnliche Gebiete

d) Kontinentalverschiebungen

Die horizontalen Krustenbewegungen können nicht von den vertikalen Krustenbewegungen getrennt werden, da sie die Ergebnisse desselben Spannungssystems sind. Die laufend registrierten Krustenbewegungen zeigen auch ein grabenähnliches System (die europäischen Krustenbewegungen der UdSSR). Die Entwicklung der jetzigen Krustenbewegungen hat in geologischer Vergangenheit begonnen. Ein Beweis dafür kann in dem linearen Zusammenhang der Quartärterritorien mit den Niveaudifferenzen in Ungarn gesehen werden. Die Fläche der Kleinen Ungarischen Ebene entspricht durchaus der Entwicklungsstufe des Roten Meeres, Das kann durch geologische wie auch geophysikalische Daten bewiesen werden. Jedoch scheint der Sedimentationsbetrag viel größer zu sein als in der Ebene des Roten Meeres. Die Senkung der Fläche der Kleinen Ungarischen Ebene schreitet seit paläozoischen Zeiten fort. Die meisten der horizontalen Bewegungen zeigen einen Dehnungscharakter. Eine hinreichende Erklärung des Systems der horizontalen Bewegungen - auch vom energetischen Standpunkt - kann nur durch die Theorie der allgemeinen Ausdehnung gegeben werden. Der Mechanismus der Kontinentalverschiebung kann jedoch nicht ausschließlich aus dieser Theorie abgeleitet werden. Für die Erklärung des Driftens größeren Ausmaßes infolge Expansion muß ein Magmastrom angenommen werden. Die Entwicklung der gegenwärtigen Formen der Kontinente wird vom Autor anhand von Ergebnissen mehrerer Modellversuche gezeigt.

All crustal movements can be reduced to stresses originating either in the upper part of the mantle, or the crust of the Earth, or in the sedimentary layers. The same stress system accounts for the horizontal and vertical movements as well. Crustal movements arise, generally, in places, where the stress surpasses the strength of the rocks or materials either on the surface or in the interior of the Earth.

Stress systems can be attributed to the energies acting in the interior of the earth. Generally fracture phenomena occur in the rocks, where the strength of the material is the smallest. The places of the least strength may be looked for in the areas of the sedimentary layers. Although the presence of fractures is not conspicuous in a sedimentary area, they reveal themselves in the formation of a drainage system. A drainage pattern shows the outlines of the fracture system developed in a given area covered by sediments. These statements are illustrated in the following Figs.

Fig. 1 Drainage pattern of the environs of Nipissing, after Hobbs

Fig. 1 is a known example by Hobbs showing the drainage map of the district of Nipissing representing the distribution of trajectories of the original stress system of the area. Our establishment is also corroborated by the river pattern in the area of Rockland (Fig. 2) and in the northern part of France (Fig. 3).

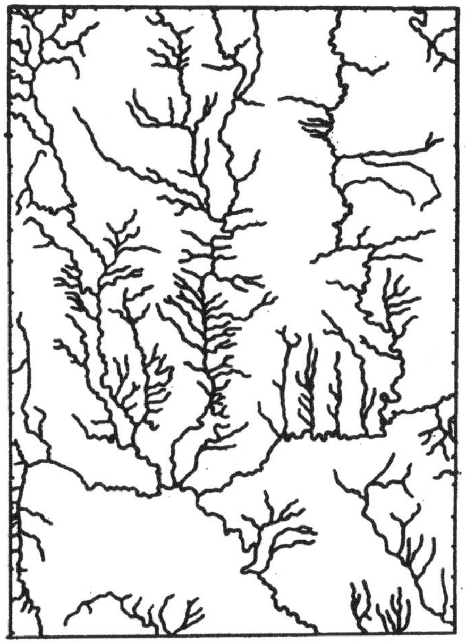

Fig. 2

Pattern of rivers in the area about Rockland, Michigan, after Hobbs

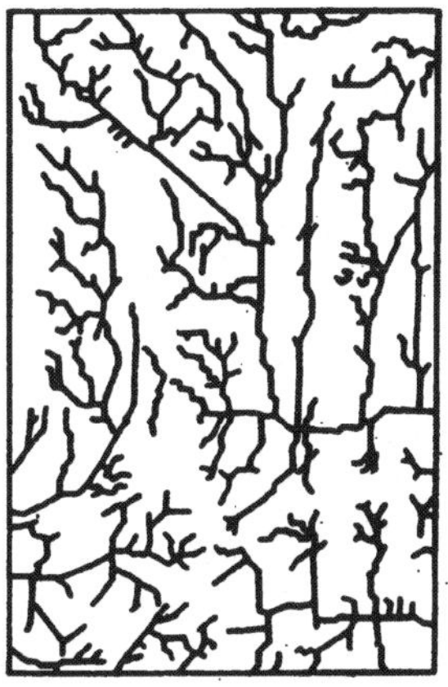

Fig.3

Drainage network in Northern France after Daubrée

Fig.4 shows the deep structure of the area of the Little Hungarian Plain. The course of the Danube coincides with a buried fault-line as proved by the isobaths of Pre-Tertiary sediments revealed by wells. Traces of another fracture system in close connection of the drainage pattern are indicated by the intrusions of the area. The fault line coinciding with the course of the Danube is seismically active displaying thus the presence of recent movements. Moreover, recent investigations have shown that the area is subsiding,i.e. recent movements seem to be continuations of past movements.

The drainage system of Western Hungary is a striking example of the fact that the fracture system of a given area is reflected by its drainage system.

The formation of rifts may be regarded as the second stage in the development of the crustal movements. At this stage faulting caused by increasing tensional stresses involves deeper parts of the crust and the separated parts of the fault line drift away forming a gap of some tens of km. In this respect, one of the most famous area is the area of the African rifts, however, this can be regarded only as part of a world-wide rift-system (Fig.) running along the crust of the oceanic ridges. There is almost no difference as to form or dimensions between the profile of the Tanganyika rift (Fig.5) and that of the rifts of the Midatlantic ridge. Both areas are characterized by a belt of shallow earthquakes and partly, by basaltic volcanism.

In areas of intense erosion the gaps of the rifts are filled up by sediments and thus, the striplike sinking tendency of the area refer to the presence of rifts.

The third stage in the evolution of fault systems displays itself in the formation of grabens. The distance between the separated parts of the fault-line amounts already to hundreds of kilometres and the role of the plastic deformation becomes more and more

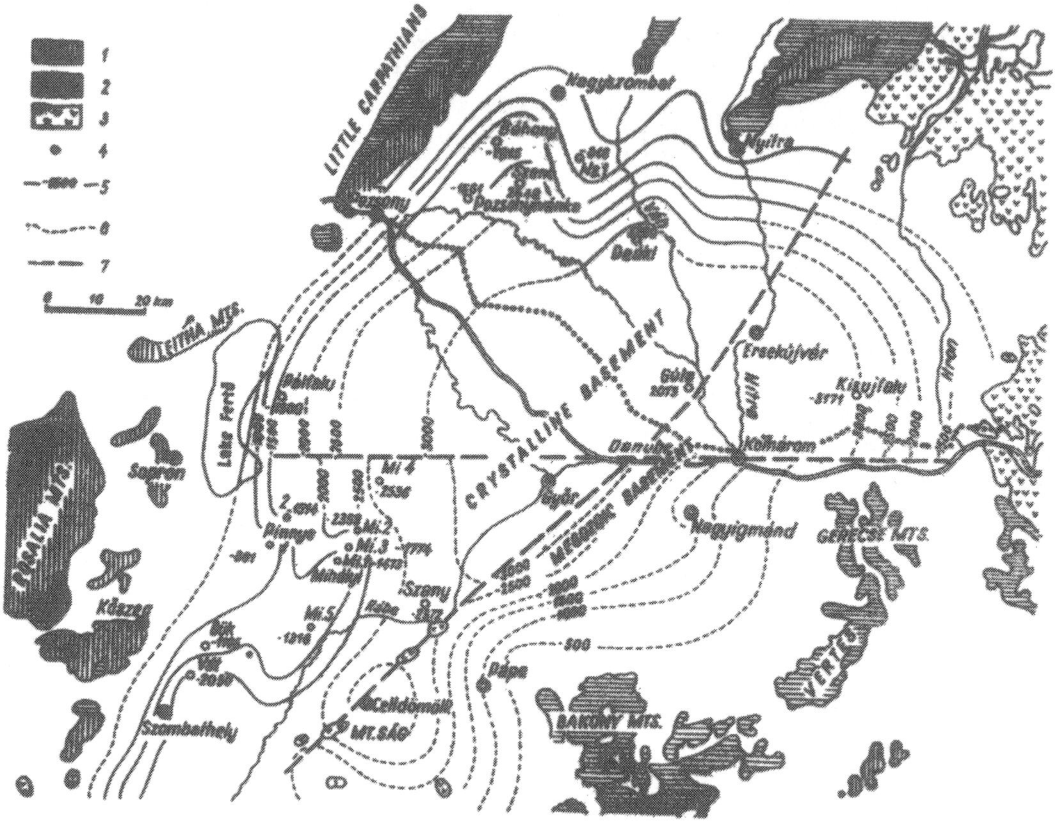

Fig. 4a

The Pre-Tertiary basement of the Little Hungarian Plain, after Kőrössy. Symbols: 1.Early Palaeozoic above-surface; 2. Pre-Tertiary deposits; 3. Neozoic volcanics; 4. Deep well; 5. Surface of Pre-Tertiary basement as revealed by deep-wells; 6. The same shown by geophysical data; 7. Structural lines in the interior of the basin

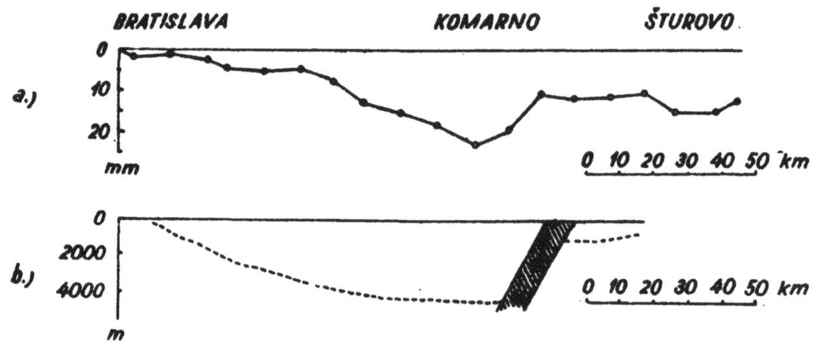

Fig. 4b

Precise levelling and structural profile across the Little Hungarian Plain

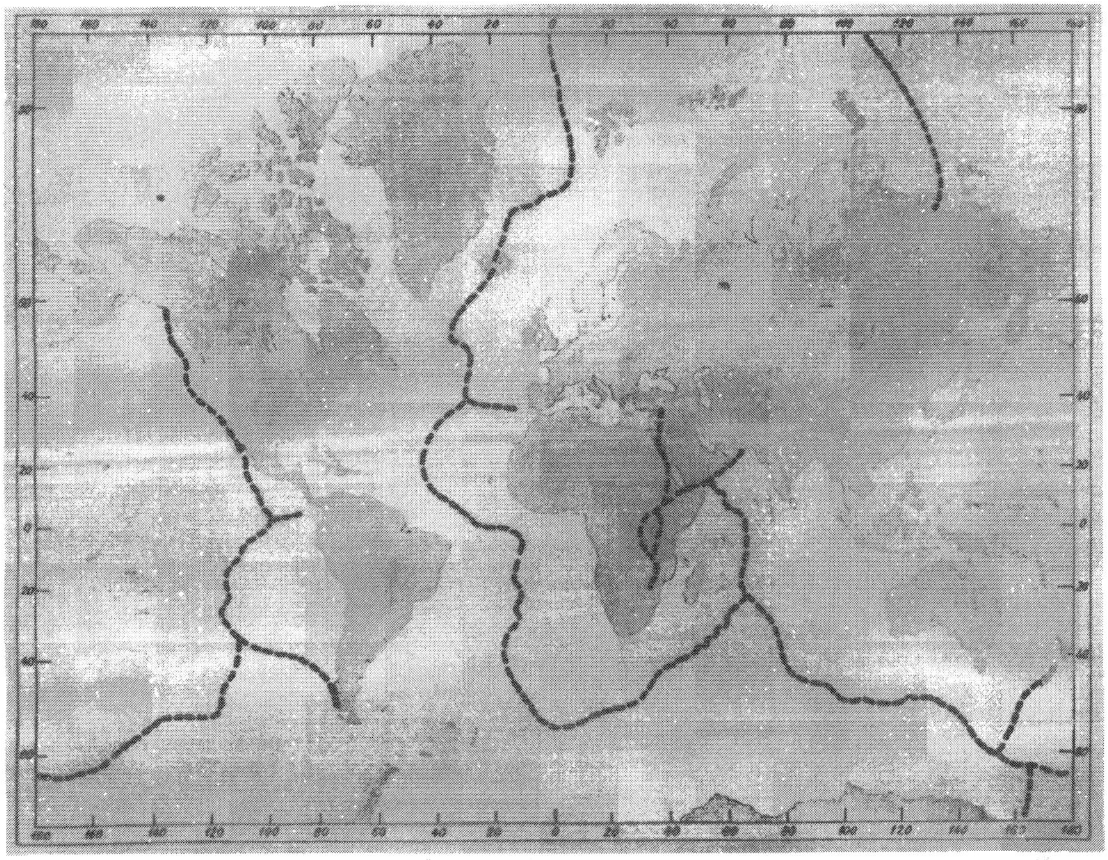

Fig. 5

Mid-Ocean rift-system

predominant. A continuous magmatic intrusion belt develops along the main fault and, sometimes, the graben stretches forth forming several rifts. At the same time the seismic activity of the area is reduced almost to minimum. The profiles running across the graben generally show in this case a gravity minimum, accompanied in the centre by a local maximum. The most characteristic representative of a graben seems to be the Red Sea (Fig. 7) displaying all the features mentioned above.

The dimensions of the Rhine-graben are about those of a rift. Its cross section, however, explored in detail, shows that its development is a result of tensional stresses.

Graben-like features may be observed also on the surface of the Moon. Fig. 8 exhibits the signs of drifting apart on the lunar surface.

The continental drift represents the largest horizontal movements of the Earth's surface. At the same time, however, very delicate questions arise as to the energies needed for large-scale horizontal movements. The difficulties resulted in opinions discarding the whole conception of continental drift. The increasing rate of the paleomagnetic data, however, corroborate the primary connection and continuity of the continents.

Solely the expansion of the Earth can be regarded as the cause of the tensional stresses, taking place in the upper part of the mantle and resulting in fracture systems,

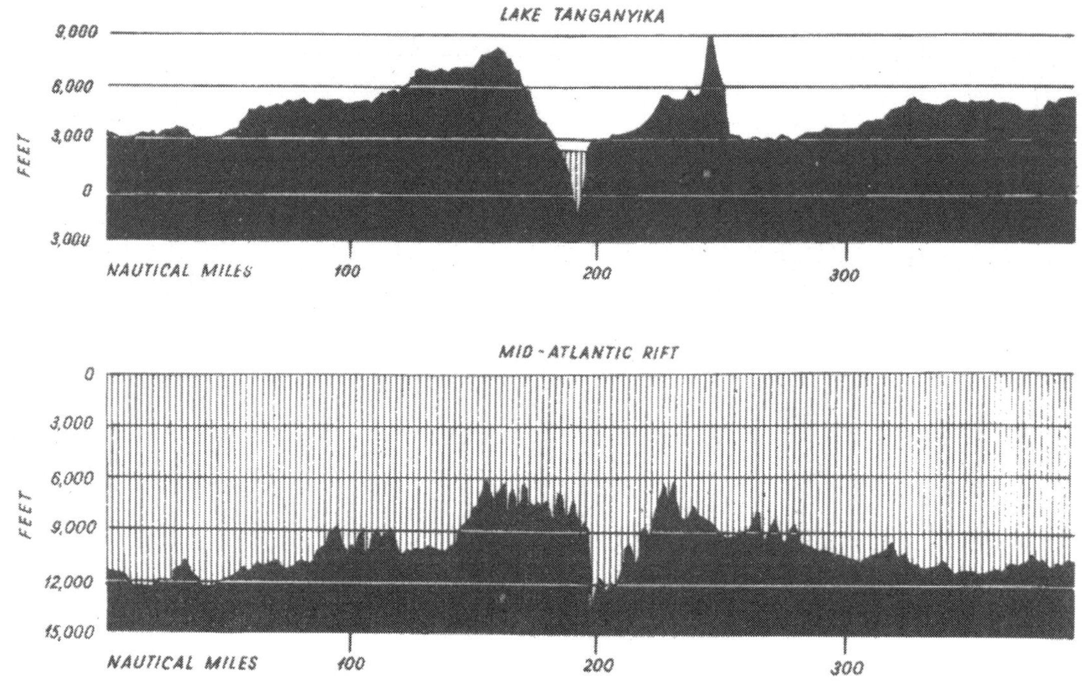

Fig. 6

Profiles across the African Rifts (Tanganyika Lake) and across the Mid-Atlantic ridge

Fig. 7

The structure of the Red-Sea.

Fig. 8

Graben-like formations on the Moon

rifts, grabens and the continental drift as well. The mechanism of the Earth's expansion can be reduced to the existence of irreversible degenerated phases in the interior of the Earth and to the monotonous decrease of the gravity coefficient. The energies released in the process of the expansion of the Earth are up to cover all requirements of large-scale horizontal and vertical movements.

In fact, continental drift cannot be regarded as a real drift, but as a result of the formation of ocean basins due to the expansions of the Earth. As to its mechanism, it seems to be a graben formation several times repeated.

Another question subject to the problem of continental drift and the formation of the Atlantic Ocean is the time of the disrupture of Pangea. According to Wegener this happened during the Carboniferous. The rate of expansion of the Earth contradicts to this establishment: the formation of the Atlantic Ocean due to expansion must have begun during the Late Precambrian or Early Cambrian. There is a possibility of larger continental displacements in case the heat developed during the disrupture and the formation of grabens is distributed by convections. In this case, the time necessary for the formation of an ocean will be shorter, however, the low-velocity channel must have become obliterated or disappeared altogether under the Atlantic Ocean. This supposition must be yet cleared up by seismological observations.

In the following it is attempted to show that the stresses due to the expansion of the Earth may produce fracture lines similar to the coastlines of the present Globe. The photo illustrations are results of the model experiments carried out by Prof. Kádár in Debrecen. He has put on a blown-up football bladder a gypsum layer and continued after solidification the blowing until the rupture of the gypsum layer. The rupture lines display forms similar to the coasts of India, South Africa and the Red Sea (Fig. 9a, b, c).

There is a final question: does a connection exist between recent crustal movements and those of the geological past? We shall try to prove that most of the recent crustal movements form a part of the movements which have begun in the geological past. To this purpose the vertical movements derived from precise levellings in Hungary will be applied. Undoubtedly, these data must be used with the utmost care. The tendency

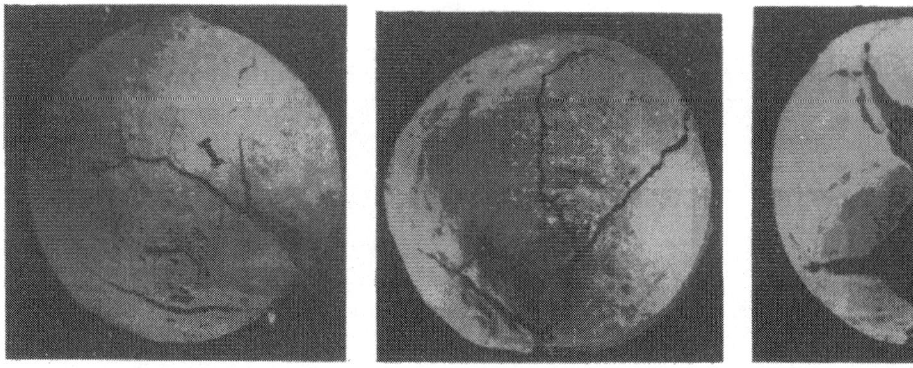

9a 9b 9c

Fig. 9
Fracture-lines produced on a gypsum layer by tensional stresses, after L. Kádár

of these vertical movements is, however, correct. Fig. 10 shows the thickness of the Pannonian and Pleistocene sediments of the Great Hungarian Plain, together with the isolines plotted by Gárdonyi's data of recent vertical movements. The figure proves that the largest subsidence may be observed in areas where the thickest sedimentary layers occur. The deviations may be attributed to errors in the measurements. Moreover, it is obvious from the figure that the isolines of the subsidence coincide with the fault lines. This supports also the reality of the vertical movements.

The problem is, however, whether the vertical movements derived from precise levellings have not been caused by the compaction of strata? It is known that the thickness of the Quaternary sediments amounts to 300 m, corresponding to an average subsidence of 0,3 mm/year, while the thickness of the Pliocene and Pleistocene strata is a little more than 3,000 m corresponding to the same yearly rate of sedimentation and subsidence. The compaction in time may be expressed as follows:

$$d\ (t) = a + be^{-\lambda t}$$

The thickness D of a layer composed during n-years by strata of a thickness of d can be expressed as follows:

$$D_n = na + b\ (1 + e^{-\lambda} + \ldots. + e^{-(n-1)}\lambda)$$

The same layer after compaction during one year

$$D_n = na + b\ (e^{-\lambda} + e^{-2\lambda} + \ldots. + e^{-n\lambda})$$

i.e. the yearly compaction of the layer will be

$$\Delta D = b\ (1-e^{-n\lambda})$$

Taking into account that $n = 10^6$ years in the case of Pleistocene, and 10^7 years in the case of Pliocene, the yearly compaction of the layer will be $\Delta D = b$, i.e. equal to the maximum compaction of the deposited sediments in a year. As to a compaction of 100 per cent, the maximum compaction can be only 0,3 mm/year. If recent vertical movements form part of the past movements, the yearly vertical movements observed by precise levellings can be 0,6 mm/year. On areas composed of elastic sediments the vertical movements can be considered real in the case only, if their rate exceeds the average rate of the yearly accumulation of sediments of the area during the geological past.

The yearly maximum subsidence of the Great Hungarian Plain is 2 mm/year according to Gárdonyi's data.

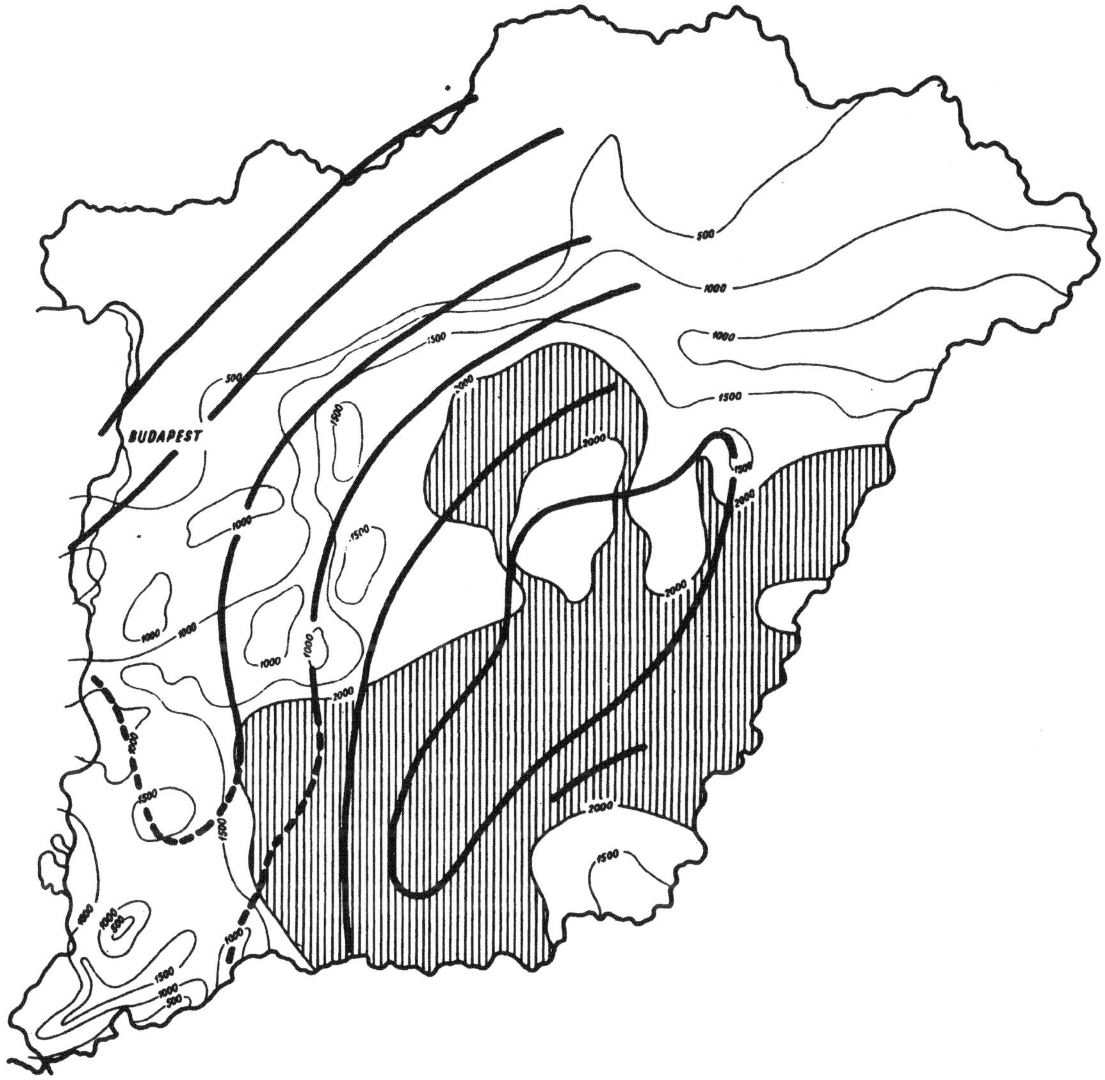

Fig. 10

Thickness of the Pannonian and Pleistocene sediments of the Great Hungarian Plain, compared with the isolines of recent vertical movements

D i s k u s s i o n

A. Z á t o p e k (Prag):

It is perhaps interesting to remark that the main NE-SW-line of the picture of W-Hungaria, shown by Prof. Egyed, is identical with the continuation of the main Czechoslovak seismic line on the Hungarian territory.

J. A. M e s c h e r i k o v (Moskau):

Is there a correlation between the position of the region of the maximum sedimentation in the Quaternary period and the region of recent subsidings?

What is the maximum thickness of Quaternary deposits?

What is the rate of recent subsiding?

L. E g y e d (Budapest):

The result of precise levelling is coinciding with the thickness of tertiary and quaternary sediments. The rate of subsidence derived from the sedimentation is 0,2 mm/year. The result from the precise levelling 2 - 3 mm/year. The difference can be partly results of compaction, partly of the inaccuracy of the measurements.

Recommendation for detecting and surveying the secular variation of the earth's gravity field

von

G. B a r t a (Budapest)

1. T h e c o n n e c t i o n b e t w e e n t h e t r i a x i a l i t y o f t h e E a r t h a n d t h e e c c e n t r i c i t y o f t h e g e o m a g n e t i c f i e l d

It is well known, that the geomagnetic field shows a permanent deformation, in so far as i t s c e n t r e d o e s n o t c o i n c i d e with the geometrical centre of the globe, but it h a s a n e c c e n t r i c s i t e t o w a r d s t h e d i r e c t i o n o f t h e M a r s h a l l - I s l a n d s [9. p. 651]. The amount of this eccentricity is far more greater, than the accuracy of the measurements and their processing, and it can be established not only on the basis of recent data, but also by using the coefficients of all the spherical functions calculated since Gauß; moreover, it can be followed back to 1550 in some extrapolated systems of data [5]. Therefore, its existence may be assumed as a firmly established fact of the physics of the Earth. B u t, i f i t i s s o, t h e n a s i m i l a r e c c e n t r i c i t y m u s t o c c u r i n t h e i n t e r n a l s t r u c t u r e o f t h e E a r t h, t o o.

It is interesting to note in this connection, that t h e d i r e c t i o n o f t h e e c c e n t r i c i t y o f t h e g e o m a g n e t i c f i e l d (in 1960, 32°W) c o i n c i d e s - within the limits of observational errors - w i t h t h e d i r e c t i o n o f t h e e q u a t o r i a l g r e a t e r a x i s o f t h e "t r i a x i a l E a r t h". Calculations show easily, that if the inner core of 2500 km diameter and of 17,0 density is embedded eccentrically with 300 kms in a surrounding outer core of 11,0 density, then the equatorial section of the geoid will be elliptical with a difference of 175 ms between the two axes, in full conformity with the geodetical informations [7].

Recently the ellipticity of the equator was computed on the basis of the perturbations of the motion of the Vanguard II. and Vanguard III. satellites. According to this determination, at about the middle of 1960 the following values have been valid [10]:

$$\lambda_a = -33{,}15^\circ \pm 0{,}53^\circ$$
$$a - b = 205 \pm 18 \text{ m}$$

These values are in better agreement with those computed from the magnetic eccentricity, than the geodetical and gravitational data obtained before. The orbit of an artificial satellite is influenced by the integrated effect of the Earth's mass; therefore, data obtained from its computation must be more accurate, than the gravitational or geodetical data founded on local measurements. Besides, the values obtained from the orbit of the satellites do not contain the secular change of the Earth's figure resulting from the shifting of the core either. Thus the investigation of the orbits of artificial satellites furnishes new and independent data for the support of the theory explained above.

T h e   m a g n e t i c   c e n t r e   h a s   b e e n   m o v i n g   in the last century towards WNW with an annual speed of 0,2°. This seems to indicate, that the eccentricity be not related to the continental structure of the Earth, but it must be the effect of some factors having their seat deeply under the crust and mantle, that is in the core itself. The longperiod character of the variation suggests a process of great inertia; thus, we can assume, that we have to do with a s h i f t i n g a n d s l o w l y d i s p l a c i n g o f i m m e n s e m a s s e s relatively to the surface. This displacement is brought about probably by the gravitational (tidal) effect exerted by the Moon and Sun upon the eccentric core [6].

In case of a movement of the inner core, the material of the outer core must show a divergence out of the point towards which the movement is directing. In fact, the vectors representing the magnetic secular variation show a s y m m e t r i c a l c o n f i g u r a t i o n around P a k i s t a n as centre, when projected orthogonally. The magnetic variation-vectors of the observatories lying in the middle of the projectional figur are converging around Pakistan, those of the observatories on the margine of the figur are diverging [2].

In the series of data of the m a g n e t i c o b s e r v a t o r i e s on the northern hemisphere - having been in function for a sufficiently long interval - one can state the existence of a w a v e w i t h a p e r i o d o f 50 y e a r s [1, 3, 12]. From this it is obvious, that the magnetic secular variation, beside its features of local nature possesses some large scale characteristics affecting the whole of the Earth, too. A period of about 50 years presents itself also in the rotational speed of the Earth as well as in the amplitude of the latitude variation [8, 11]; therefore one could suppose, that the magnetic secular variation is in connection with great mass movements, which may produce a change in the geoidal surface, too. Indeed, a certain large scale conformity can be discovered also in the o b s e r v a t i o n a l s e r i e s o f m a r e o g r a p h - s t a t i o n s being situated in great distances one from another and the 50 years' period can be established in these series of data, too [4].

B a s i n g o n t h e s e c o n s i d e r a t i o n s w e m i g h t s u p p o s e , t h a t t h e f o r m o f t h e g e o i d u n d e r g o e s s i g n i f i c a n t c h a n g e s i n t i m e a n d t h e g r a v i t a t i o n a l f i e l d m i g h t s h o w a s e c u l a r v a r i a t i o n o f s o m e 10 m g a l o r d r e o f m a g n i t u d e a s a c o n s e q u e n c e o f t h e c h a n g e s e x p e r i e n c e d i n t h e m a s s d i s t r i b u t i o n o f t h e E a r t h .

2. T h e p r o p o s a l a n d i t s m o t i v a t i o n

The above discussed connection between the triaxiality of the Earth and the eccentricity of the geomagnetic field suggests the possibility of a change in the figure of the Earth and of a secular variation of gravity. We wish to propose the establishment of an international survey, by which the expected variations could be most probably detected. By this way the high relative accuracy of modern new gravimeters could be better utilized for the study of the gravity field independently of any theory.

The proposal is

T h a t r e l a t i v e g r a v i m e t r i c o b s e r v a t i o n s o f a h i g h e s t o b t a i n a b l e p r e c i s i o n b e c a r r i e d o u t a t a c h a i n o f s t a t i o n s d i s t r i b u t e d a s u n i f o r m l y a s p o s s i b l e a r o u n d t h e e q u a t o r . The aim of the survey would be the detection of supposed mass m o v e m e n t i n t h e   E a r t h ' s   c o r e.

Preparatory to the projected survey, a network of about 20 stations, situated in a possibly uniform distribution in the vicinity of the equator, ought to be established and stabilized; between the gravity values of these stations, no significant differences should exist. (This requirement can be fulfilled with an accuracy of some tens of milligals by choosing the site and height of the stations accordingly. These preparations are necessary in order to make it possible to take fully the advantage afforded by the measuring accuracy of modern gravimeters.)

After having established such a stabilized net of 20 stations with nearly equal gravitational values, we should carry out - as second stage of the project - a survey of r e l a t i v e   g r a v i t a t i o n a l   m e a s u r e m e n t s around on the network with gravimeters of the m o s t   s u i t a b l e   t y p e (using about 8 - 10 instruments simultaneously).

Making the processing of the observations, only the lunisolar tidal corrections ought to be applied for reduction. Thus we could obtain a series of relative "g"-values around the equator, the differences of which would be known with an accuracy of at least a few hundred of milligal.

As for the next step of the project, t h e   o b s e r v a t i o n s   o u g h t   t o   b e   r e p e a t e d   s o m e   t i m e   a f t e r . If there is a mass movement in the Earth's core, we shall observe a difference against the preceding measurements. If the known shift of the magnetic centre of the Earth is identified with the displacement of the inner core of the Earth, then, after the lapse of ten years, we shall be able to establish in the differences - relatively to the values of the preceding measurements - a r o u n d   t h e   e q u a t o r   a   d o u b l e   w a v e   o f   o n e   m i l l i g a l   a m p l i t u d e . This amplitude is far more greater, than the measuring accuracy of relative gravimetric observations of today; therefore, the oscillation seems to be detectable in ten years or even in a shorter interval.

R e f e r e n c e s

[1] Barta, G.: On the 44 years' period of the secular variation of the terrestrial magnetic field. (In Hungarian, with English and Russian summaries). Geofizikai Közlemények III., 1. 1954

[2] Barta, G.: On the secular variation of the geomagnetic field. (In Hungarian with English summary). Geofizikai Közlemények VI., 1 - 2. 1957

[3] Barta, G.: Longitudinal and transversal effect of the secular variation of the geomagnetic field. (In Hungarian, with English summary). Geofizikai Közlemények VII., 1. 1958

[4] Barta, G.: On the secular variation of the level surface of gravity. Annales Universitatis Scientiarum Budapestinensis de Rolando Eötvös nominatae. Sectio Geologica. Tomus II. 1959

[5] Barta, G.: Über die Säkularbewegung des magnetischen Zentrums und der magnetischen Pole der Erde. Zeitschrift für Geophysik. Heft 4/5. Jahrgang 24

[6] Barta, G.: Connections between the secular variations of the earth's magnetic field and other phenomena. Annals of the New York Academy of Sciences. Vol. 95. Art. 1. Solar variations, climatic change, and related geophysical problems

[7] Barta, G.: The connection between the eccentricity of the geomagnetic field and the triaxiality of the Earth. Acta Technica. Tom. XXXVII. Fasc. 1 - 2

[8] Brouwer, D.: A new discussion of the changes in the Earth's rate of rotation. Proc. of the Roy. Soc. London, Ser. A., 243. 1950

[9] Chapman, S., and Bartels, J.: Geomagnetism II. Oxford

[10] Izsák, I.G.: A determination of the ellipticity of the Earth's Equator from the motion of two satellites. Smithsonian Institution Astrophysical Observatory. Special Report No. 56. 1961

[11] Munk, W.H., and Macdonald, G.J.F.: The rotation of the Earth. Cambridge University Press, 1960

[12] Slaucitajs, L.: Some notes on secular variation. Centr. of Baltic University, 6. Pinneberg, 1948

[13] Strasser, G.: Ellipsoidische Parameter der Erdfigur. Deutsche Geodätische Kommission, München, 1957

[14] Wolf, H.: Das von L. Tanni bestimmte Geoid und die Frage der Elliptizität des Erdäquators. Deutsche Geodätische Kommission, Frankfurt am Main, 1956

Diskussion

C. Morelli (Bari):

The I.A.G. is following with the greatest care this problem, owing to its fundamental interest. In the 1st Order World Gravity Net there exists a net of stations of about the same gravity values (so that calibration problems are ineffective), which have been connected in the past with many gravity-meters and high accuracy (± 0,1 mgals) and that have been surmounted this year. Since they are at about the same latitude (Rome - New York - S. Francisco - Tokyo - Beyrout - Rome), they constitute the first possibility for detecting such an effect. Calculations will be ready within few weeks. An enlargement of this net is possible to Buenos Aires, Capetown and Melbourne.

I. D. Boulanger (Moskau)

It does not seem very difficult to determine the changes of gravity of the order 0.1 mgal per year. However, the experiments carried out in this direction in the USSR do not confirm the suppositions given by Prof. Barta. The observations of gravity on the Caucasus during 30 years and in the Central part of the USSR during 10 years did not show noticeable changes in time.

Geomagnetische Informationen über die Bewegung der Erdkruste

von
G. F a n s e l a u (Potsdam)

S u m m a r y

The secular variation is the chief source of information of geomagnetic research with regard to recent movements of the earth's crust. As to this phenomenon a distinction must be made between the secular variation originating from the core of the earth and that which is to be localized in the earth's crust. As the two phenomena can hardly be separated from each other experimentally, we have to reflect on them in a theoretical-statistical manner. The secular variation taking its rise from the core of the earth has been researched by means of the parameters dependent on time of the calculations of potential, especially in regard to transformations of coordinates. The result has been a static homogeneous distribution with regard to varying positions of the axes of the coordinate system. With a certain approximation it is possible to separate the secular variation of the core of the earth from that of the earth's crust in virtue of computations of potential dependent on time. Moreover, indications of local anomalies of the secular variation with an extension not exceeding 100 km result from the research of the secular variation in Central Europe between 1935 and 1955. The attempt is made to assign these anomalies to the portion of the crust by a method designed for the separation of the main field into the portions of the core and the crust. Local processes of fusing in connection with tectonic movements may then be regarded as causes.

Z u s a m m e n f a s s u n g

Die hauptsächlichste Informationsquelle der geomagnetischen Forschung im Hinblick auf rezente Erdkrustenbewegungen ist die Säkularvariation. Bei dieser Erscheinung muß zwischen der Säkularvariation, die ihren Ursprung im Erdkern und jener, die ihren Sitz in der Erdkruste hat, unterschieden werden. Da beide Erscheinungen kaum auf experimentellem Wege voneinander getrennt werden können, müssen theoretisch-statistische Betrachtungen angestellt werden. Die Säkularvariation, die aus dem Erdkern stammt, wurde mit Hilfe der zeitabhängigen Parameter der Potentialberechnungen, besonders im Hinblick auf Koordinatentransformationen, untersucht. Es ergab sich eine statistische Gleichverteilung in bezug auf verschiedene Achsenlagen des Koordinatensystems. In gewisser Annäherung ist es möglich, auf Grund zeitabhängiger Potentialberechnungen die Säkularvariationen des Erdkerns von der der Erdkruste zu trennen. Bei einer Untersuchung der Säkularvariation zwischen 1935 und 1955 in Mitteleuropa ergeben sich fernerhin Hinweise auf lokale Anomalien der Säkularvariation mit einer Ausdehnung von höchstens 100 km. Es wird versucht, diese Anomalien mit einer für die Trennung des Hauptfeldes in Kern- und Krustenanteil entwikkelten Methode dem Krustenanteil zuzuordnen. Als Ursachen können dann örtliche Aufschmelzvorgänge im Zusammenhang mit tektonischen Bewegungen angesehen werden.

Die Untersuchung der zeitlichen Veränderlichkeit des geomagnetischen Feldes bietet die Möglichkeit, gewisse Aussagen über Bewegungszustände in der das Feld erzeugenden Materie zu machen. Als Grundlage für derartige Untersuchungen dienen die magnetischen Vermessungen und die Beobachtungen an den Observatorien.

Betrachtet man zunächst die Erde als Ganzes, so ist die Verteilung des geomagnetischen Feldes an der Erdoberfläche am zweckmäßigsten in einer Reihe von Kugelfunktionen darzustellen, deren numerische Koeffizienten - die Parameter - die Charakteristika für die Feldverteilung repräsentieren. Neben dem Feldverlauf selbst läßt sich dann auch die Verteilung der Säkularvariation in solchen Reihen erfassen, wenn die Potentialentwicklungen zu verschiedenen Epochen in irgendeiner analytischen Form zeitlich zusammengefaßt werden.

Darstellungen dieser Art sind wiederholt durchgeführt worden und ergeben besonders im Hinblick auf die Säkularvariation kein so einheitliches Gepräge, wie es etwa das Gesamtfeld in seiner Grundstruktur aufweist. Die Veränderungen in der Lage regionaler Anomalien der Verteilung der Säkularvariation sind untersucht worden und haben gewisse Hinweise auf Verschiebungen des Mantels zum Kern der Erde ergeben.

Eine Reihe wichtiger theoretischer Schlußfolgerungen konnte aus diesen Ergebnissen gezogen werden. Es ist nun möglich, noch einen anderen Weg einzuschlagen, der gewisse zusätzliche Erkenntnisse zu liefern imstande ist. Der Grundgedanke besteht darin, die zur Darstellung der Verteilung des geomagnetischen Feldes oder seiner Säkularvariation benutzten Kugelfunktionen nicht nur auf die Rotationsachse der Erde als Polarachse, sondern auf irgendwie anders im Erdkörper gelagerte Achsen zu beziehen, die aber alle durch den Erdmittelpunkt gehen sollen. Die Transformationen einer Kugelfunktionsreihe von einem Koordinatensystem auf das andere ist eine rein analytische Aufgabe, die von A. Schmidt gelöst wurde. Die Parameter, die auf die Rotationsachse der Erde bezogen sind, ändern nach Maßgabe der Pollage des neuen Koordinatensystems ihren Zahlenwert.

Untersuchungen in dieser Richtung wurden angestellt und haben ergeben, daß bei dem Hauptfeld ein deutlicher Einfluß der Pollage des Bezugssystems für die Kugelfunktionen vorhanden ist. Der numerische Anteil der zonalnahen Parameter - d.h. Parameter, die zu den Gliedern geringer Längenabhängigkeit gehören - nimmt ab, wenn man die Darstellung auf Polarachsen bezieht, die sich in ihrer Lage immer mehr von der Rotationsachse entfernen. Interessanterweise ergibt sich die Tatsache, daß dieser Einfluß für die Glieder gerader und ungerader Ordnung verschieden ist, und zwar in der Richtung, daß er nur bei den Gliedern ungerader Ordnung klar in Erscheinung tritt, bei denen gerader Ordnung dagegen verschwindet. Bei diesen Reihengliedern ist also eine Bevorzugung der zonalnahen Parameter für eine bestimmte Lage des Koordinatensystems nicht festzustellen.

Genau das gleiche numerische Ergebnis erhält man, wenn man die oben erwähnte Überlegung nicht auf das geomagnetische Feld, sondern auf die Darstellung der Höhen- und Tiefenverhältnisse nach Prey anwendet. Auch hier zeigt sich kein klares Hervortreten der Lage irgendeiner Bezugsachse. Bemerkenswerterweise ergibt sich nun, daß auch für die Säkularvariation ganz ähnliche Verhältnisse vorliegen wie etwa für die Verteilung der Höhen- und Tiefenverhältnisse auf der Erdoberfläche. Auch für dieses geophysikalische Phänomen zeigt sich, daß der Einfluß der Lage der Bezugsachse sehr gering ist. Die Anteile zwischen zonalnahen und zonalfernen Parametern sind für verschiedene Bezugsachsen ungefähr gleich. Das zeigt, daß die Verteilung der Säkularvariation und die damit in Verbindung stehenden Bewegungen innerhalb der Erde nicht mit der Rotationsachse der Erde in klar erkennbarem Zusammenhang stehen.

Die bisher betrachteten Untersuchungen beziehen sich im großen gesehen auf die Säkularvariation, die an der Grenze des Erdkerns entsteht. Diese zeitlichen Veränderungen des geomagnetischen Feldes umfassen jedoch nicht die gesamte Säkularvariation, vielmehr kommt noch ein Teil hinzu, der in der Kruste seinen Ursprung hat und in gewissem Umfang auf Bewegungen innerhalb der Erdkruste zurückgeführt werden kann.

In gewisser Näherung ist es möglich, in kleinen Teilgebieten der Erdoberfläche diese beiden Anteile der Säkularvariation, die im Erdinneren bzw. in der Erdkruste entstehen, voneinander zu trennen. Es wird dabei eine Methode herangezogen, die für die Trennung des Hauptfeldes selbst in Kern- und Krustenanteil von H. Kautzleben entwickelt worden ist. Das wesentliche Kennzeichen dieser Methode besteht dabei darin, daß ein geeignetes Normalfeld abgeleitet wird.

Unter bestimmten Voraussetzungen ist es möglich, die genannte Aufgabe in eine Trennung des Feldverlaufs nach geometrischen Konfigurationselementen zu überführen. Da nach Beobachtungen von Deel das Hauptfeld nur planetare und lokale Anomalien enthält, ist die Abspaltung des Kernanteils vom Hauptfeld mit genügender Annäherung möglich. In der für die quantitative Lösung erforderlichen Potentialentwicklung bedeutet das eine Trennung bei Reihengliedern etwa der Ordnung n = 50. Bei der Berechnung für die gesamte Erdoberfläche werden jedoch zu hohe Anforderungen an die Genauigkeit des Beobachtungsmaterials gestellt. Die Methode kann jedoch so umgeformt werden, daß für ein begrenztes Teilgebiet der Erdoberfläche die Lösung möglich wird. Man hat dabei lediglich zu beachten, daß der Anteil am Hauptfeld, der durch den ersten Teil der Reihenentwicklung erfaßt wird, sich in einem beliebigen vorgegebenem Punkt als lineare Kombination aller zur Berechnung herangezogenen Beobachtungswerte ergibt. Die Koeffizienten dieser Werte nehmen aber mit der Entfernung vom vorgegebenen Punkt ab, so daß der Einfluß dieser Beobachtungen sehr bald vernachlässigt werden kann. Demnach ist zur Berechnung des Kernanteils in einem begrenzten Teilgebiet ein dichtes Netz sehr genauer Beobachtungen lediglich in diesem Gebiet erforderlich, während außerhalb des Gebietes mit zunehmender Entfernung immer weniger Beobachtungen herangezogen werden müssen.

Bei praktischen Rechnungen kann der Formalismus der Reihenentwicklung nach Kugelfunktionen dann durch geeignete Mittelbildungen stark vereinfacht und durch eine Approximation mit Hilfe von Polynomen der Koordinatendifferenzen gegenüber einem Zentralpunkt ersetzt werden. Für das Gebiet der DDR wird man dabei in erster Näherung auf das von Rössiger vorgeschlagene Normalfeld für Mitteleuropa mit Zentralpunkt Niemegk geführt.

Wenn man diese Methode auf die Beschreibung der Säkularvariation anwendet, so hat man den Wert der Feldkomponente durch eine entsprechende Zeitfunktion zu ersetzen. Definiert man die Säkularvariation durch die Differenz der Jahresmittel von zwei Epochen, so ersetzt man den Wert der Feldkomponente einfach durch die entsprechende Differenz.

Bei der Ableitung lokaler Anomalien der Säkularvariation für ein begrenztes Gebiet muß man also für jeden Beobachtungspunkt mit Hilfe der beschriebenen Methode die Säkularvariation des Kernanteils (mit anderen Worten den normalen Wert der Säkularvariation) berechnen und diesen Wert darauf vom beobachteten Wert der Säkularvariation abziehen. Trägt man diese Differenz ($SV_{Beob.} - SV_{normal}$) in eine Karte des Gebietes ein, so muß man auf die Existenz von Anteilen der Kruste an der Säkularvariation schließen, wenn die Differenzen eine regionale Ordnung zeigen, die nicht durch die zufällige Verteilung der Beobachtungsfehler zu erklären ist.

Das beschriebene Verfahren wurde zur Analyse der Veränderung des Hauptfeldes im Gebiet der DDR zwischen 1935,0 und 1955,5 angewendet. Dazu standen Beobachtungen an 52 jeweils miteinander identischen Stationen der magnetischen Reichsaufnahme (reduziert auf 1935) und der magnetischen Vermessung der DDR (reduziert auf 1955,5) zur Verfügung. Die mittleren Fehler der Differenzen entsprechender Beobachtungswerte betragen dabei in D $\pm$ 2',5; in H $\pm 13\gamma$; in Z $\pm 23\gamma$.

Die Verteilung der oben eingeführten Differenzen ($SV_{Beob.} - SV_{normal}$) liefert Hinweise auf eine gewisse regionale Ordnung und damit auf krustenbedingte Anomalien der Säkularvariation. In D findet man ein Vorherrschen negativer Werte im Osten, in H ein Vorherrschen positiver Werte im Südwesten und in Z drei positive Zentren im Süden, im Nordwesten und im Nordosten.

Zur Deutung dieser Erscheinung könnte man etwa folgende Überlegung anstellen: Im

Normalfall wird der Hauptteil der Krustenmagnetisierung durch die magnetischen Gesteine oberhalb der Tiefe geliefert, in der die Curie-Temperatur erreicht wird. Vergrößert sich diese Tiefe (vielleicht im Zusammenhang mit tektonischen Vorgängen), so wird die Krustenmagnetisierung um einen Anteil zunehmen, der einer Schicht magnetischer Gesteine mit einer Mächtigkeit gleich der Differenz zwischen der neuen und der alten Tiefe der Curie-Temperatur entspricht. Demnach könnte eine Anomalie der Differenzen ($SV_{Beob.} - SV_{normal}$) durch den Feldanteil einer solchen Gesteinsplatte erklärt werden.

Die Beobachtungen im Gebiet der DDR lassen sich dann qualitativ so deuten, daß in den drei Gebieten, die den Anomalien in Z entsprechen, ein Vergrößern der Tiefe der Curie-Temperatur vor sich gehen müßte. Das Gebiet im Süden müßte dabei die größte Absenkung zeigen. Diese Deutung läßt sich mit den Beobachtungen in allen drei Feldgrößen D, H, Z qualitativ in Übereinstimmung bringen. Zur Ableitung von quantitativen Angaben sind jedoch noch Kenntnisse der Zusammensetzung und der Eigenschaften der Erdkruste in der fraglichen Tiefe erforderlich.

Geophysical investigations at Lake Kariba

von

D. I. Gough, W. I. Gough (Salisbury)

Summary

The Zambesi River was dammed in the Kariba Gorge in December 1958. The resulting lake is now in process of filling. In raising the water level 130 m above the bed of the gorge, the dam will eventually produce a lake some 250 km long storing 1.6 x 10^{11} m^3 of water. The resulting stress differences in the crust will be small compared with the strength of crystalline rock. If failure results from the water load, it will do so where the crust is already faulted or pre-stressed or both.

Two geophysical approaches are being used. Precise levels have been carried round the future shoreline, across the lake along three lines, and back to the primary levelling network. Re-levelling at a later date will enable deflections to be found. Three short-period vertical seismographs are in operation round the lake, and in conjunction with two permanent three-component stations, enable some of the tremors occurring to be located. Preliminary results on tremors located during 1961 are presented.

The investigation is a joint project of the Federal Department of Trigonometrical and Topographical Surveys, the University College of Rhodesia and Nyasaland, and the Federal Power Board.

Zusammenfassung

Der Sambesi wurde im Dezember 1958 in der Karibaschlucht aufgestaut. Der sich daraus ergebende See ist nunmehr im Stadium der Auffüllung. Durch die Erhöhung des Wasserstandes um 130 m über die Sohle der Schlucht wird die Staumauer eventuell einen See von ungefähr 250 km Länge und einem Wasserinhalt von 1,6 x 10^{11} m^3 erzeugen. Die resultierenden Spannungsunterschiede in der Kruste werden klein sein im Vergleich zu der Festigkeit der kristallinen Gesteine. Wenn sich eine Störung aus der Wasserbelastung ergibt, so wird dies dort geschehen, wo die Kruste schon gestört oder vorbeansprucht oder beides ist.

Zwei geophysikalische Methoden sind angewandt worden. Präzisionsnivellements sind rund um die künftige Küstenlinie, über den See entlang drei Linien und zurück zum Hauptnivellementsnetz ausgeführt worden. Ein Wiederholungsnivellement zu einem späteren Zeitpunkt wird es gestatten, die Abweichungen zu finden. Drei kurzperiodische Vertikalseismographen registrieren rund um den See und in Gemeinschaft mit den zwei permanenten Dreikomponentenstationen erlauben sie, einige der auftretenden Beben zu lokalisieren. Vorläufige Ergebnisse über Beben, die 1961 lokalisiert wurden, werden vorgelegt.

Die Untersuchung ist ein gemeinsames Projekt des Federal Department of Trigonometrical and Topographical Survey, des University College of Rhodesia and Nyasaland und des Federal Power Board.

Some 420 km downstream from the Victoria Falls the Zambezi River passes through the Kariba Gorge. The first contracts for the hydroelectric scheme sited in the Gorge were awarded in 1955, and in December, 1958, the river was closed. The dam wall was completed in June, 1959. The river channel at the wall is 150 m wide. The wall is 128 m high in the centre and contains some 750,000 m^3 of concrete. Six 100,000 kilowatt turbo-generators have been installed in an underground power station excavated in the right bank. At a later stage a second power station will be constructed in the left bank and will probably deliver 900,000 kilowatts. Transmission lines carry power at 330 kilovolts some 420 km north to the mines of the Copperbelt and a comparable distance south-east to the cities of Southern Rhodesia.

In raising the water level 130 m above the river bed, the wall will eventually produce a lake 250 km long storing 1.6×10^{11} m^3 of water. The formation of an artificial lake of these dimensions constitutes a major intervention in natural conditions in the area and affords corresponding research opportunities.

In a year or two, when the lake reaches its maximum extent, a load of some 1.6×10^{11} tons will have been added to the Earth's surface. For a mass distribution approximating to Lake Kariba, it appears that the stress difference will nowhere exceed about 10^6 dyne cm^{-2}, which is about 0.1 per cent of the strength of granite in the laboratory. However, the crust of the Earth contains many rocks weaker than granite, and the crust is extensively fractured along fault planes. Further, the fault zones themselves show that stress differences have existed sufficient to cause fracture and displacement of crustal blocks. It is therefore not improbable that stress differences exist in the crust which may be just too small, in places, to cause further displacement at some existing fault. The incremental stress differences caused by Lake Kariba might then cause movement.

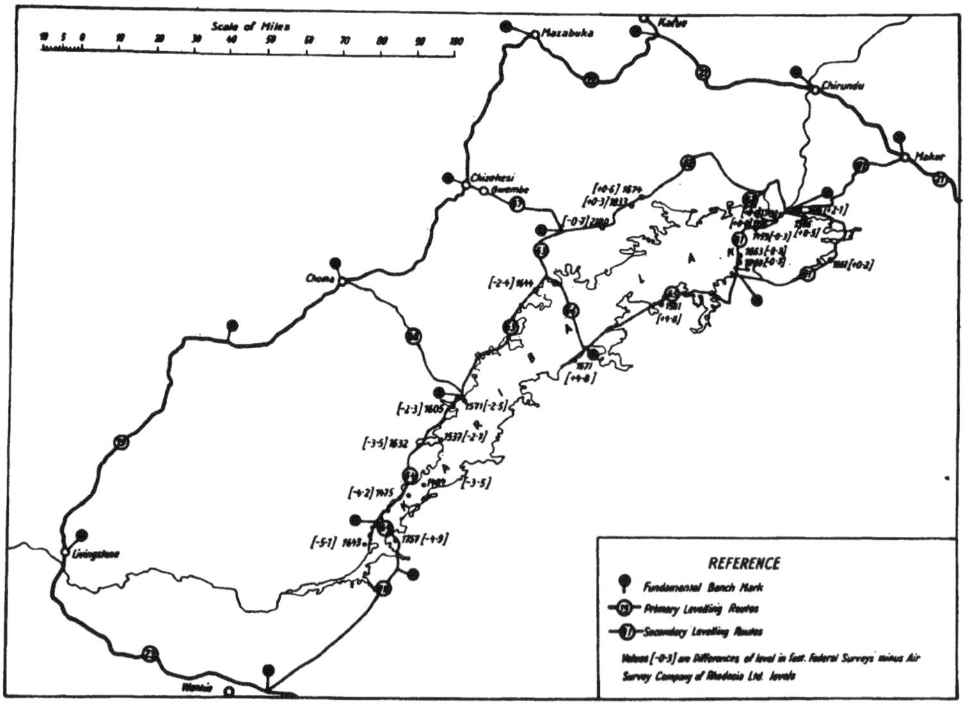

Fig. 1

The Department of Federal Surveys, under the direction of Brigadier M.O. Collins, has carried a precise levelling network around much of the perimeter of the future lake and across it along three lines. Special care was taken in the Sanyati Basin, the area where maximum depths will occur and where approximately one third of the total mass of water is concentrated. The network at the lake is tied to primary levelling routes. The immediate motive for the levelling survey was the need to locate the future shoreline at 484 m (1,590 ft) above sea level. The dense bush of the region has been cleared from the future shoreline to a depth of 30 m to allow the development of a trawl-fishing industry; and harbours and roads have been constructed. For all these operations the position of the shoreline was required. In addition to serving these immediate practical ends, the levelling survey will enable any subsidence of all or part of the lake area to be accurately determined, provided the survey can be repeated at a future date.

Two sections of repeat levelling were carried out late in 1960, the first in the Gwembe Valley in Northern Rhodesia. The second followed a route about 16 miles below the dam wall in Southern Rhodesia. In both cases the sections selected ran approximately at right angles to the lake shore. These established that at that time no appreciable tilt of the surface had occured along the lines repeated.

The positions of certain bench-marks are indicated in Figure 1, and it will be noticed that some of these lie in the lake. Some of these were sited on hills which will form permanent islands. At the others level steel tables have been sited. The depths of these will be measured from a boat in future re-levellings.

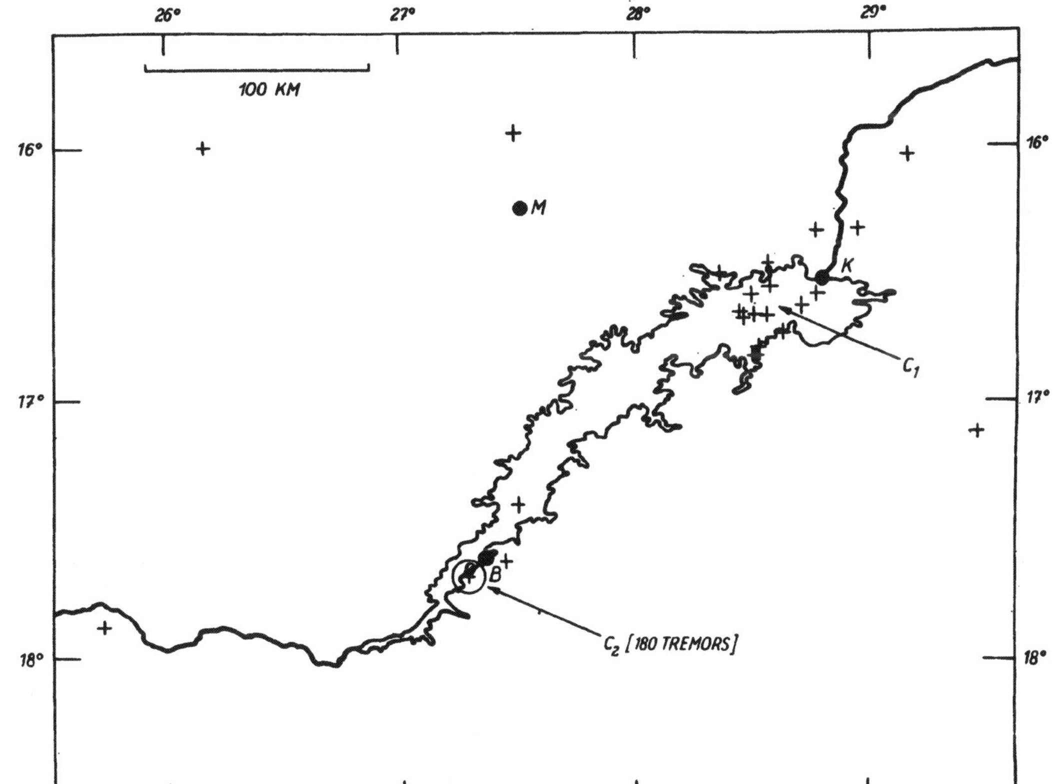

Fig. 2

The precise levelling approach will determine the total deflections over a period of years. The second geophysical approach is through the use of seismographs to locate individual fractures in time and place. Three short-period vertical-component Willmore seismographs from a local network round the Lake. These are supplemented by the permanent three-component Willmore seismographs at Broken Hill and Bulawayo where recording began in May and February 1959 respectively. The three local stations use quartz crystal clocks employing transistors, to produce a 50 c/s output derived from the 100 kc/s crystal oscillator frequency. The crystal clocks are rated against radio time signals recorded daily on the seismograms. At each station it has been possible to build a concrete slab on outcropping rock, and noise levels are low except at the Kariba station when the sluice-gates are open.

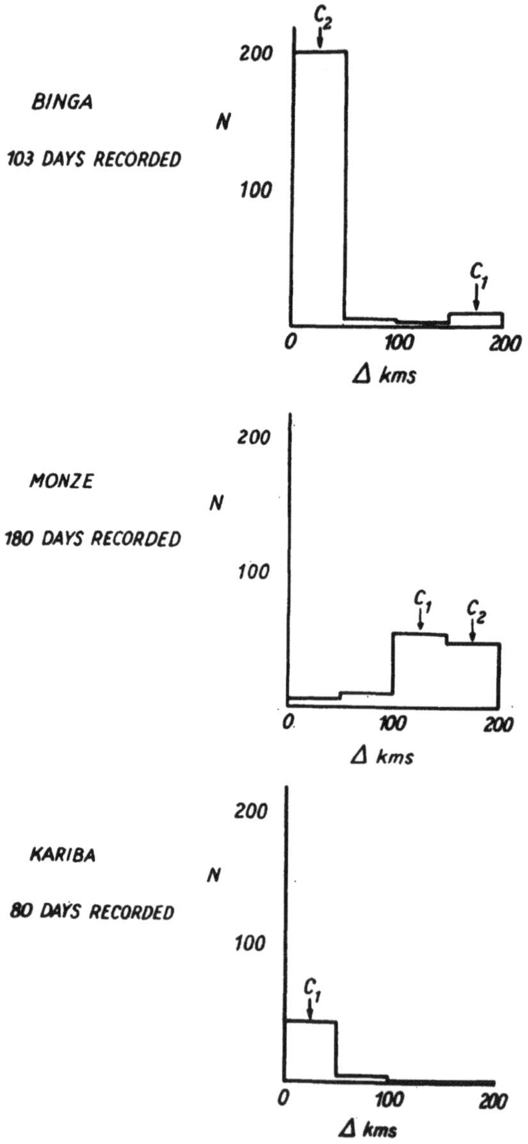

Fig. 3

Trial recordings were carried out at intervals through 1960 at the Kariba dam site, and full time recording at the three local stations began in January 1961. The lake was then rather more than half full. It would, of course, have been most desirable to have commenced recording much earlier. Ideally one would wish to have a control experiment extending over several years before the dam was built, in order to establish the seismicity in the area, so that any correlation between seismicity and load could be established. Resources of money and especially of scientific manpower are slight in Central Africa, and in spite of efforts by all concerned the recording did not start until the dates given. However, it is quite probable that a significant increase in tremor frequency will be observed when the observations of the next few years are complete.

It would, of course, be most desirable to add a fourth station to the "local" network, which is only relatively "local". The Kariba and Binga stations are some 200 km apart. Some tremors are too small to be recorded at stations other than the nearest, and these of course elude location. Furthermore, when seismographs and crystal clocks are operated at remote stations in tropical conditions many hundreds of kilometres from a seismologist, it is inevitable that a considerable fraction of recording time is lost through faults. Three stations are the minimum for an unique location, and when one station is inoperative location is possible only through the use of the Broken Hill records, when the tremor is large enough to record there. A fourth "local" station would raise location efficiency a great deal.

The seismological investigation at Lake Kariba is a joint undertaking of the Federal Government Department of Trigonometrical and Topographical Surveys and the University College of Rhodesia and Nyasaland. The servicing of the stations is done mainly by Surveys staff. The seismograms are read, and tremors located, by Mrs. W.I. Gough. In the first half of 1961, tremors located in the Lake area are shown in Fig. 3. There is a general tendency for tremors to occur in bursts lasting a few days or weeks, with epicentres in a restricted area. Thus in May 1961 most of the tremors shown at C_1 in (3) occurred, and in June 1961 some 180 tremors occurred in 80 hours on the Binga fault, at C_2 in (3). Most tremors of the latter burst were too small to be recorded elsewhere than at Binga. The tremors shown in the widest part of the Lake (C_3) do not follow any mapped fault. In July a number of tremors (not indicated in (3)) occurred along the Zambesi for some 60 km downstream from the dam.

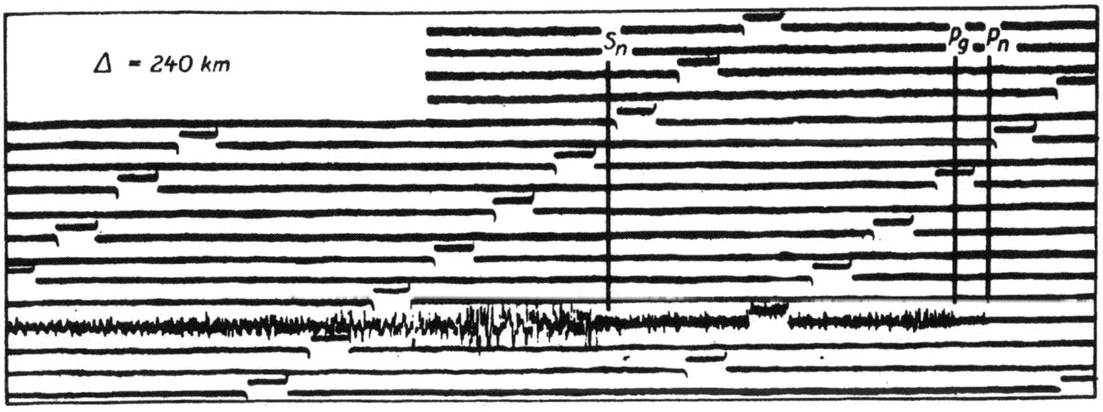

Fig. 4

Tremors which could not be located because they were recorded at less than three stations, are represented in the histograms of (4). These failed to record at three stations in some cases because they were too small, and in others because stations were inoperative. The histograms show their frequency distribution with distance from the individual stations. The distances from the stations to the centres C_1 and C_2 of seismic activity, indicated in Fig. (3), are shown in the histograms.

This is in the nature of a progress report. No answer is yet possible to the question whether the tremors are caused by the water load.

D i s k u s s i o n

T. S ϼ r g e n f r e i (Kopenhagen):

I ask whether the seismicity of the area had been investigated before the dam was built.

D. I. G o u g h (Salisbury):

The river was closed in December 1958. The seismograph station at Broken Hill commenced recording in May 1959. Unfortunately this constitutes our earliest record of seismicity in the area, and is effective only for the larger tremors.

T. S o r g e n f r e i (Kopenhagen):

I ask if there were faults along the borders of the lake basin where blocks could be displaced due to loosening on account of the water filling. The tremors might be related to such displacements or to genuine seismic tremors.

D. I. G o u g h (Salisbury):

The major shocks are of magnitude about 4, and many others are clearly aftershocks of these. This fact, and the distribution of epicentres, suggest that they represent genuine seismic tremors.

J. A. M e s c h e r i k o v (Moskau):

Dr. Gough, you have reported on an extremely interesting program of researches. In this connection I would like to ask you, how often do you intend to carry out relevelling?

D. I. G o u g h (Salisbury):

We cannot at present say when the relevelling will be done, as the cost over the great lengths of roads will be high, and financial support will have to be sought.

J. A. M e s c h e r i k o v (Moskau):

Could we consider the region of the Kariba lake as one of the polygones the network of which, according to the suggestions made by Prof. Boulanger and me, it is desirable to have on all continents?

D. I. G o u g h (Salisbury):

Yes. The local network may be regarded as a set of three polygons, having common sides along two routes, and the whole is tied to the national primary levelling network.

J. D. B o u l a n g e r (Moskau):

It is a very important project for geodetic and geophysical researches. The determination of the elastic properties of crust could be made with the help of these observations. However, during the studies it is necessary to take into account the hydro-geological conditions of the region, where the geodetic marks are set, because the changes of these conditions, as we have noted, cause the vertical displacements of the earth's surface.

O. M e i ß e r (Freiberg):

Have you in a wall-galery a special levelling arrangement? With our hose-level you can get an exactitude $\pm$ 0,005 mm.

D. I. G o u g h (Salisbury):

I think this is a valuable suggestion. At the moment strain gauges are used in the wall, and theodolite observations are taken on pegs mounted on it.

C. M o r e l l i (Trieste):

I would only give some more data to Prof. Sorgenfrei's question. In some artificial lakes in the Alps, a very strong seismicity has followed the first water filling of the lakes; it was extinguished exponentially only after years. Also the elastic characteristics of the rocks were found very different before, during and after the water filling.

Scheme of a complex investigation method of tectonic movements with special regard to mining areas

von

Z. Kowalczyk (Krakau)

Summary

The investigating of tectonic movements, necessary in mining areas, is hindered by the exploitation of deposits and its consequences. Rock stresses appearing in these areas and caused by tectonic movements probably affect the originating of mine shocks. This problem is associated with the need of establishing, which of the percussions are of tectonic, and which of mining origin. A fundamental part in these investigations should be attributed to seismic studies. The results of seismic observations should be analyzed in every respect. The seismic investigations should mainly aim at establishing the coordinates of percussion foci, in order to localize them. Since the percussion foci are situated in zones of tectonic disturbances, the geodetic observations (of vertical and horizontal displacements and of surface inclines) should also be concentrated in these places.

The establishing of percussion origin (whether tectonic or mining) may be helped by determining the percussion depth and the sign of the first seismic waves registered.

When investigating the genesis of percussions in mining areas, dynamic studies, being of no lesser importance, should be taken into account too. Geodetic studies will base foremost on the results of periodical precision levelling, executed for determining the magnitude of yearly subsidence, on those bench marks, which are exempt of the effect of mining exploitation, this being confirmed by calculations made e.g. with the author's method, and which do not remain in the reach of the depression funnel, occasioned by dewatering of tertiary or quaternary formations. Simultaneously with the observation of the subsidence of bench marks on the surface, observations of bench marks stabilized on lowest level bottom pillars should be made.

The results of seismic and geodetic investigations, conjointly with a geological and mining analysis, should find their confirmation in the presentday geomorphological phenomena, or should at least not be contradicted by them.

Zusammenfassung

Die in Bergbaugebieten notwendigen Untersuchungen von tektonischen Bewegungen werden durch den Abbau von Ablagerungen mit ihren Folgeerscheinungen behindert. Gebirgsspannungen, die in diesen Gebieten auftreten und durch tektonische Bewegungen verursacht sind, beeinflussen wahrscheinlich die Erzeugung von Gebirgsschlägen. Dieses Problem ist verbunden mit der Notwendigkeit, festzustellen, welche Erschütterungen tektonischen Ursprungs sind und welche vom Bergbau herrühren. Ein Großteil dieser Untersuchungen sollte den seismischen Arbeiten gewidmet werden. Die Ergebnisse der seismischen Beobachtungen sollten in jeglicher Hinsicht ausgewertet werden. Die seismischen Untersuchungen sollten hauptsächlich zur Feststellung der Koordinaten der Herdzentren ausgeführt werden, um dieselben zu lokalisieren. Da die Herdzentren in tektonischen Störungszonen gelegen sind, sollten auch die geodätischen Beobachtungen (von vertikalen und horizontalen Verrückungen und Oberflächenbewegungen) an diesen Stellen konzentriert werden.

Die Feststellung des Erschütterungsursprungs (ob tektonischer oder bergbaulicher Art) kann unterstützt werden durch die Bestimmung der Herdtiefe und den Einsatz der ersten seismischen Wellen. Bei der Untersuchung der Entstehung von Erschütterungen in Bergbaugebieten sind dynamische Studien von nicht geringer Wichtigkeit und sollten auch in Rechnung gestellt werden. Geodätische Arbeiten stützen sich vorerst auf die Ergebnisse periodischer Präzisionsnivellements, welche für die Bestimmung der Größe der jährlichen Absenkung an solchen Fixpunkten ausgeführt werden, die frei sind von den Einflüssen des Bergbaus (dies wurde durch Berechnungen z.B. nach einer Methode des Autors bestätigt) und die nicht im Bereich von Depressionstrichtern liegen, welche durch Entwässerung von Tertiär- und Quartärformationen entstanden sind. Gleichzeitig mit der Beobachtung der Absenkung der Fixpunkte auf der Oberfläche, sollten Beobachtungen von stabilisierten Fixpunkten auf tiefeingerammten Grundpfeilern gemacht werden. Die Ergebnisse seismischer und geodätischer Untersuchungen, zusammen mit einer geologischen und bergbaulichen Analyse, sollten ihre Bestätigung in den gegenwärtigen geomorphologischen Erscheinungen finden oder sollten ihnen wenigstens nicht widersprechen.

Introduction

The problem of present earth-crust movements has, besides its scientific aspect, an important practical meaning too. The earth-crust tectonic movements have been and are still occurring on the territories of Poland; this has been stated both by chronicles and by seismograph records taken by Polish and foreign stations. However, the investigations of tectonic movements on mining areas are greatly hindered by mining exploitation and its consequences. Nevertheless, investigating tectonic movements on mining exploitation areas is of great practical importance, since a lack of proper qualitative and quantitative recognition of the phenomenon discussed would account for an impossibility of discerning and mastering mine shocks. This problem is closely connected with the fundamental question, whether stresses, occasioned by tectonic movements, affect the originating and the course of mine shocks, and if so - what is the influence and what are the mining and geological conditions in which these shocks occur. A reply to this question aswell as to other questions, ensuing from it, may be given by a properly elaborated method of complex investigations.

A fundamental part in such investigations should be performed by the results of complex observations of seismic movements, thoroughly analyzed, as well as of periodically performed basic geodetical observations, compared with the results of geomorphic investigations, foremost. The sites of these investigations and the analysis of their results should be strictly coordinated with the tectonic geology of the place.

The purpose of seismic research

Seismic stations in mining areas are recording a great number of percussions, yet only some of them may be considered as being of tectonic origin. Tectonic percussion investigations in mining areas are greatly involved, foremost by percussions accompanying mine shocks, both of explosive strata and top rocks, as well as by percussions occasioned by caving or by blasting some of the shot holes in bed exploitation.

A fact, characterizing percussion of mining origin (mine shocks, cavings), is that the energy evolved very strongly diminishes as the distance increases. This is the reason why, when ascertaining the genesis of percussions in areas of underground mining exploitation, dynamic investigations should be taken into account too.

Dynamic investigations of percussions require computing the intensity of percussion in its focus and comparing the order of this intensity with that of the mine shock, simultaneously noted. In the case if both phenomena would prove to be of a similar order, and all the more if the intensity of the mine shock and its consequences in the mine itself should exceed the intensity of percussion computed by means of formulae (e.g. those of Gutenberg, Gibowicz or Karnik), this would suggest that the percussion was occasioned by a mine shock.

If, on the contrary, the energy of percussion distinctly exceeds the energy computed on the basis of consequences of the mine shock, it may be believed that it was the tectonic percussion that had originated the impulse of the simultaneous mine shock. One should however be very cautious in drawing the above conclusion as the determination of percussion intensity is being burdened with a considerable error, greatly depending on the intensity calculation method applied, as well as on the determination of the mine-shock area.

The amount of mine-shock energy depends on the weight of overburden and consequently on the works' depth, on the volume of the workings where the shock occurred, on the specific gravity of the rocks and their elastic characteristics. The labour necessary for crushing coal may be computed e.g. from the formula:

$$U = S \cdot H \cdot h \cdot \delta \text{ in kGm} \qquad (1)$$

whereas the energy in ergs is

$$U_c = 9.81 \cdot 10^7 \cdot U \text{ ergs}$$

where: S - the area of mine shock,
H - depth of mine shock,
h - height of overburden drop,
δ - specific weight of overburden rocks.

It should be nevertheless always taken into account, that a sudden mine shock does not make the whole energy being discharged. Moreover, in the case of strong and elastic sandstone, occurring above the workings undergoing a mine shock, energy will be liberated - not only due to the dynamic percussion of the overburden, but also owing to the sandstone plate breaking.

It should be also turned into account in such investigations that, as has been observed, in the case of mine shocks all the stations record the same sign of first wave deviation, that is either compressions or dilatations. Natural percussions, instead, have various signs of the first deviations on particular stations. Performing this kind of investigations requires the disposal of an adequately dense net of seismic stations. A considerable difficulty is encountered in the investigations discussed, namely that of ascertaining the proper sign of percussion, the records being but very little legible.

Percussions originated by mine shocks or cavings have, owing to their rapid decrease in intensity, but a small reach and are therefore often observed only by single nearest seismic stations. In order to classify the seismicity of mining fields, an adequate macroseismic service should be organized, for establishing curves of equal seismic oscillation intensity (isoseismals) in the case of stronger percussions; this would permit a future elaboration of the summary stresses for discretionary points of the region. It may be expected that comparing an isoseismal line map and a gravimetric map of the investigated region's anomalies would permit to establish whether an increased disturbance of isostatic equilibrium coincides with strongly seismic zones, since the investigations made till now have proved that the maximum disturbances of isostatic equilibrium occur in distinctly seismic zones, whereas in typical non-seismic zones there is a state approaching an equilibrium (isostasy).

Another purpose of seismological observations is establishing the coordinates (x, y, z) of the percussion foci. This remains in connection with the cardinal problem of accuracy attainable in establishing the focus location (x, y) and depth (z). In order to obtain sufficient accuracy (e.g. $\pm$ 1 km) in evaluating the x, y, z coordinates of the percussion focus, one should not only modernize the existing seismic stations, but still add some new ones. Besides that, there should be plotted a number of seismic profiles, obtained by deep sounding, in order to determine the rate of seismic wave advancement in the region investigated. When knowing with sufficient accuracy the coordinates (x, y, z) of the percussion foci, their location may be determined.

Since the percussion foci are usually located in zones of active tectonic disturbances, chiefly on a crossing of important faults, the geodetic, tensometric and inclinometric observations undertaken for investigating vertical and horizontal displacements and the ground incline should be concentrated in the percussion foci regions established.

When establishing the origin of percussion, that is when clearing whether the phenomenon is a tectonic percussion or else a mine shook, it is of great importance to determine with sufficient accuracy the focus (hypocentrum) depth. The fact being known that mine shocks occur at a depth of over 250 m and that percussion foci (on continental territories) are located at depths of over 10 km, an evaluation of the percussion depth with an accuracy of already 1 km would permit to establish the kind of percussion.

Hence the accuracy of determination of the foci coordinates (x, y, z) is affected, first - by all errors of evaluating the moments of the first impulses of particular seismic waves appearing, next - by the errors of identifying the place of first wave-impulse, by an improper determination of the wave character, by a wrongly evaluated mean velocity of particular seismic waves, an insufficiently accurate determination of the absolute time on the seismograph, a contraction of the seismograph strip and by other errors of lesser importance. Most of these errors might be eliminated by applying modern precision seismic apparatus.

Another very important problem, very unfavourable to the accuracy of determining the foci coordinates (x, y, z), is the variable seismic wave velocity and the heterogenous character of rocks. On the other hand, the exploitation of deposits in the presence of a very richly developed dislocation network causes the arising of numerous discontinuity planes, which refract and reflect the waves.

In order to calculate the coordinates with sufficient accuracy, at least two profiles of seismic sounding should be carried out through the area investigated, one of them in the West-East direction, the other in the North-South direction. These profiles will permit to elaborate hodographic curves and else to draw conclusions as to the geologic structure at greater depths.

The seismic percussion investigations in mining fields should base on seismographs, suitable in number and adequate as to accuracy and quality, of two types, i.e. of a higher and a lower frequency. High-frequency seismic stations with low oscillation amplifying would constitute a first-order net, consisting of at least three stations, while low-frequency stations with high oscillation amplifying would make a second-order net.

The necessity of having two orders of stations ensues from the fact that in the case when foci are more than thirty kilometres distant, i.e. for first-order stations, the periods of percussions remain in the range of 1.5 - 2 seconds, that is to say they are much longer than in the case of nearer foci, the periods of which remain in the range of 0.3 - 1 second. Similarly, the amplitude values of these two kinds of foci differ one from another. This explains why seismic stations of one order should be equipped with seismographs very sensitive and strongly damped, registering the wave impulses on a rapidly moving strip.

The choice of a proper calculation method is a problem of great importance, as it has an outstanding effect on the accuracy of evaluation of percussion foci coordinates, the more that the particular methods are characterized by different magnitudes of systematic errors, induced mainly by simplifications assumed.

It would also be most advisable to carry out seismoacoustic investigations in the mines themselves.

Geodetical investigations (of vertical and horizontal displacements)

The most universal method of investigating the magnitude of contemporary tectonic movements is the geodetical method, first of all as applied to investigations of ground surface vertical displacements. In regions where earth movements are slow, such being the case in Poland, where the magnitude of both vertical and horizontal displacements are but small, the measurement of horizontal displacements encounters special difficulties.

Investigations of tectonic movements in mining fields are, as has been already mentioned, strongly complicated by the consequences of mining exploitation. One should therefore be fully aware, which of the areas have been exploited in the lapse of time between two levelling observations (A), which of them have been exploited and are considered as completely stilled (B) and which of them remain under the effect of a current exploitation (C). The class (C) areas will be excluded from investigations of tectonic movements, until the ground surface becomes completely stilled. In order to do this, mine works maps (e.g. in the scale of 1 : 10 000) of the period investigated should be plotted with bench marks and one should next eliminate those bench marks, which are subjected to the effects of mining exploitation.

The aforesaid investigations could be accomplished by applying the author's method, published in the quarterly review "Geodezja i Kartografia" of the Polish Academy of Science, Warsaw 1960[x]. The magnitude of subsidence $\underline{z}$ displayed by the bench mark examined is shown by the formula:

$$z = a \cdot m \cdot e^{-\frac{a \cdot m \cdot i}{s^2 \cdot \delta \cdot h} \cdot \rho^2} = a \cdot m \cdot e^{-\frac{a \cdot m}{s^2 \cdot b} \cdot \rho^2} \qquad (2)$$

where: a – the coefficient of surface subsidence, depending on the exploitation system,

m – the bed thickness,

i – coefficient of the underlayer specific resistance,

δ – mean specific gravity of rock,

h – depth of exploitation,

$s = h \cdot \operatorname{ctg} \gamma$,

γ – angle of reach of noxious effects,

ρ – distance of the point examined from the centre of system, located in the range of the bed examined,

b – subsidence of top.

The magnitude of horizontal displacement is to be found from the formula:

$$p = \frac{z \cdot \rho \cdot r}{L \cdot h} \qquad (3)$$

where $r = L - s = L - h \cdot \operatorname{ctg} \gamma$

---

x) Z. Kowalczyk. Określenie wielkości odkształceń powierzchni na skutek eksploatacji górniczej metodą przekrojów pionowych (Evaluating the magnitude of surface deformation due to mining exploitation by a method of vertical cross-sections).

$$L = \sqrt{\frac{s^2 \cdot \delta \cdot h \cdot \log(1000\, a \cdot m)}{a \cdot m \cdot i \cdot \log e}} \qquad (4)$$

L is the reach of effect of the exploitation discussed.

In the case if the exploitation is accomplished in more than one place, one need establish the reach of particular exploitations and estimate the effect of all beds (applying in the calculations the principle of superposition) on the bench mark in question.

E x a m p l e 1

Bench mark No. 15, Table 1 is located in the vicinity of two exploitations, i.e. of bed 510, which approached in the investigated period 1947 - 1957 the exploitation centre O_1 by ρ_1 = 440 m, and the exploitation of bed 414 by ρ_2 = 150 m. The results of calculation are contained in Table 1. Considering that bench mark No 15 remains in the period examined under the effect of exploitation, it has to be eliminated from further investigations.

Table 1

Period of observation	Mean yearly sinking or raising mm	Distance ρ m	Bed denomination	Bed thickness m	Coefficient of subsidence a	Depth of exploitation l m	Cotangent of the angle of noxious effects ctg γ	Computed		Notes
								sinking z mm	horizontal displacement P mm	
1947 - 1957	-	440	510	5.0	0.5	400	0.6	245	189	Mark No 15
	-	150	414	3.0	0.15	300	0.6	281	94	

Summary sinking	526
Resultant horizontal displacement for both beds amounts to	195

After having eliminated bench marks underlying the effect of mining exploitation in the period examined, we proceed to evaluate the relative velocity of successive displacement of pairs of neighbouring reference bench marks

$$h_2 - h_1 = v \cdot t$$

hence

$$v = \frac{h_2 - h_1}{t} \qquad (5)$$

where: h_1 - first levelling excess of two neighbouring bench marks,

h_2 - second levelling excess of the same bench marks as above,

t - lapse of time between levellings.

In the case if a non-uniform and unimportant course of vertical movements should be found, a direct investigation of the area inclines by means of special declinators should be included. A manifold evaluation of the same bench mark's velocity by means of different levelling traverses permits to determine the most probable values of contemporary movements.

In mining fields where one of a pair of optional neighbouring bench marks may incur the effect of mining activity while the other not, one should calculate the values of vertical displacements of particular points in mm pro year of the period investigated, in relation to their known height, i.e. using formula (5), in which case:

h_2 - bench mark height in the subsequent levelling,

h_1 - bench mark height in the preceding levelling.

As has been already mentioned, these calculations have to be carried out in consecutive periods, eliminating for a particular bench mark those periods, where effects of mining exploitation have been stated.

We subject the computed velocities of contemporary tectonic movements of separate bench marks to an analysis, basing on a tectonic map on which they have been previously plotted. We draw on this map contour lines of yearly sinking and raising in the ranges of: 0 - 0.5 mm, 0.5 - 1 mm, 1 - 2 mm, 2 - 3 mm etc. When comparing the results of levelling, executed in different periods of time, one should take into account only results, basing on the same reductions. Similarly, the method of equalizing the whole network applied should be the same for all periodical levellings.

It should be endeavoured in the future that the levelling measurements be executed according to one single technical program, so that the results of levelling would be burdened by the same systematic errors. The movements of the earth surface have usually a continuous course and hence, for reducing the involved systematic error to minimum, the levelling observations should be executed in a possibly short time.

If the magnitudes of errors in the levelling of a certain part of the area investigated are of a similar order as the magnitudes of height excess in a given period, a new observation line should be established, for instance in a direction perpendicular to the first line and basing on a present-day tectonic map. The way of locating the observation lines in regard to the tectonic geology of the region investigated has a basic effect on the disclosing of present tectonic movements and this is the reason why the design of these lines should be elaborated basing on the aforesaid map.

The vertical displacements entail also horizontal displacements of the surface points. The calculation of horizontal displacements will be of special value for investigations made in regions where the differences of vertical movements of points are not uniform.

In order to make observations of horizontal displacements in regions with unimportant slow vertical movements, local microtriangulation nets should be laid - classical or uniform, built in the shape of closed nets with densely located triangulation points.

The skeleton of triangulation nets of a local character should consist of points, on which astronomical observations are to be made, viz. the measurement of astronomical azimuths, of geographical longitudes and latitudes, for reducing the astronomical into geodetical azimuth. This will permit to estimate the skeleton net geodetical azimuths with an accuracy up to 1". It would be very expedient to introduce a tellurometer into the investigations of horizontal displacements. Observations of these nets should be repeated simultaneously with precision levelling, hence not oftener than every 5 years. Post and target eccentricities should be avoided in these nets.

Independently of these observations, it is necessary to carry out, on certain posts, observations towards all visible triangulation points, as well as towards additionally laid permanent signals, doing it often. The choice of these points should be made on the basis of a tectonic map. The mean error of directions observed in these investigations should not exceed $\pm$ 2". Characteristic changes in the ground strains should be investigated by means of tensometers.

The tectonic movements of earth surface should be confirmed by simultaneously performed investigations of the vertical and horizontal displacements of points, stabilized on lowest mine levels in bottom pillars, but only in places where a squeeze of bottom or a lateral squeeze is not observed.

What pertains to earth tide investigations, it seems advisable to execute investigations in the lowest workings by means of suitably adapted gravimeters.

Besides basic geodetical observations, the observations of following items should be performed once a year:

1. single levelling transverses on the surface in regions with strong tectonic disturbances,
2. pencils of directions towards stabilized points (in characteristic tectonic regions) orientated towards a direction, the geodetical azimuth of which should every time be set, basing on astronomical observation,
3. vertical and horizontal displacements of fixed points in a mine, stabilized on the deepest level in a bottom pillar,
4. tensometric and inclinometric observations in places of the earth surface, considered as geodetically characteristic.

Determination of underground de-watering reach

When investigating tectonic movements in areas of mining exploitation, where marshy or peaty soils appear, the effect of dewatering cannot be overlooked. The problem concerns shallow subsoil water levels, i.e. water occurring in tertiary or quaternary formations.

As a result of this water being pumped away, the reach of the mine depression funnel, forming in the course of time around the mine workings, increases. The sinking of underground water levels in the reach of the mine depression funnel reduces the hydrostatic uplifting force, hence increases the specific gravity of the desiccated formations and, as consequence, causes their subsidence. This is why in the cases discussed, the radius R of the mine depression should be known; it may be computed e.g. according to Kusakin's formula:

$$R = 575 \cdot s \cdot \sqrt{k \cdot H} \qquad (6)$$

where: R - radius of depression, in m,
s - depression in the hole or in the centre of the field of workings in which the de-watering occurs, in m,
k - filtration coefficient of the water-bearing rock, in m.p.s.,
H - height of the natural water-level in respect to the substratum, in m.

The bench marks located in the reach of the depression funnel should also be eliminated from investigations, particularly in cases when the existing depression distinctly exceeds the order of tectonic subsidences in the area investigated.

In order of exactly determining the value of sinking caused by overburden de-watering, levelling observations should be made on the surface of bottom pillars and simultaneously on the lowest mine level. The points observed should be then connected with bench marks outside the region affected by mining exploitation. If the height difference between the bench marks set in the shaft collar and nearby and height marks fixed on the lowest level is constant, it proves that no sinking of the bottom pillar is occurring. It should be added that this might concern only such a bottom pillar which is not affected by an exploitation outside the pillar and there is no squeeze of bottom in the workings where the marks are fixed. However, if the height difference between fixed bench marks on the surface and fixed bench marks in the mine should be constant and at the same time the absolute height of the bench mark in the mine, determined by measuring the shaft depth with a precision exchangeable tape, should be variable, this would indicate that sinking of the entrance mark occurs; in the case if such a supposition should be excluded - it would mean, that a sinking of the rocks, caused by factors of a tectonic character, takes place.

Geomorphological investigations

The results of seismic and geodetic investigations, together with a geological and mining analysis, should find their confirmation in the present-day geomorphological phenomena or at least the possibility of their existing should not be excluded. It should be apprehended that geomorphological investigations in industrial regions and foremost in areas with mining exploitation encounter impediments resulting from the activity of the mining, as well as of other industries, which had caused disturbances in the picture of naturally proceeding geomorphological processes. For instance, phenomena like the modification of depth and of accumulation on short sections of rivers should in most cases be ascribed to man's activity. The sinking and variations of the river water levels in areas of mining exploitation are usually caused by diverting the water through fissures and faults into mine workings. Disturbances caused in the normal course of geomorphological processes by the industry might not only impede, but even make it impossible to state whether a geomorphological process is still proceeding.

Geomorphological investigations are foremost meant to state whether the configuration of the earth's surface and the present geomorphological processes, moulding it, indicate the existence of present-day tectonic sinking or raising movements. An answer to the aforesaid question may be given firstly by changes in the position of the river erosion basis, changes in the river gradients (eliminating the effect of their underlying stratum), the shape and depth of valleys and river beds, the size and structure of accumulative planes or a possible rejuvenating of smaller sidevalleys, as well as the arising of new sidehill valleys.

The aforesaid investigations should be conducted, basing on longitudinal profiles of river valleys and on their gradients, in order to determine the zones of increased gradients and those of equalized gradients. For instance, a sudden change of valley gradients in certain regions will give a confirmation of present-day tectonic movements.

Since the results of geomorphological investigations relate to prolonged tectonic processes, whereas those of geodetical investigations concern present-day processes, hence the correlation of both will be in many a case difficult. It is the reason why one should in doubtful cases carry out levelling of river beds and do boring work in order to execute thorough detailed studies, including the determining of conditions of ancient formations occurring in these valleys, taking advantage of the results of palaeontological examination. Soil scientific investigations might be helpful in some cases too.

Discussion

A. Zátopek (Prag):

1.) In der ČSSR werden Gebirgsschläge seit 1949 zunächst in der seismischen Station Prag, später auch in Pruhonice und in der letzten Zeit im Kohlengebiet von Kladnó und in Přibramer Bergwerken selbst verfolgt. Sehr viele Beobachtungen und Vergleichsuntersuchungen sind vorhanden. Es wurden in einer Anzahl von Fällen größere Epizentraldistanzen erhalten als der Abstand zwischen den Stationen und dem Bergbaugebiet ist, so daß man in diesen Fällen auf den tektonischen Ursprung der betreffenden Stöße schließen kann. In einigen Fällen haben sich die Energien wesentlich größer erwiesen, als es in der überwiegenden Mehrheit von Stößen der Fall war; man kann also diese Erschütterungen als tektonisch (wenigstens im Wesen) betrachten. Es ist zu bemerken, daß die angerichteten Schäden gar nicht der seismisch festgestellten Energie entsprechen müssen. Manchmal entspricht ein starker Effekt im Bergbaugebiet einer winzigen Energie, manchmal ist es umgekehrt. Es erscheint also die Zwischenwirkung zwischen den tektonischen und den durch menschliche Tätigkeit verursachten Anregungen in Bergbaugebieten sehr verwickelt.

2.) Zur Seismoakustik: Es wurden im Kohlengebiet von Kladnó unter Tage Versuche angestellt, den Herdpunkt oder das Herdgebiet möglichst genau mit einer größeren Anzahl von Geophonen zu bestimmen. Es wurde sogar eine Apparatur zusammengestellt, die das zeitliche energetische Regime liefern soll und somit eine Art von Vorhersage geben. Es zeigte sich dieser Weg als wenig geeignet.

Z. K o w a l c z y k (Kraków):

Ich möchte noch eine Bemerkung machen. Bei uns im Kohlenbecken stellen die Seismographen pro Monat durchschnittlich bis zu 400 Erschütterungen fest. Es ist sehr schwer und unmöglich, auf Grund einiger Faktoren den Ursprung der Erschütterungen zu ermitteln.

A. T á r c z y - H o r n o c h (Sopron):

Es ist interessant, daß die alten Beobachtungen von Erdkrustenbewegungen mit dem Bergbau verbunden waren und sehr praktische Ziele verfolgten: nachzuweisen, daß nicht alle Bauschäden durch unterirdische Abbaue verursacht worden sind und für diese der Bergbau nicht Schadenersatz zu leisten hat. Teilweise behandelte in diesem Zusammenhang die Schule von Prof. Wilski in Aachen vor mehr als 30 Jahren die sogen. "Sandgewandaufgabe" und untersuchte Prof. Niemczyk die Spalten auf Island, worüber er ein Buch veröffentlichte.

Bewegungsvorgänge der Erdkruste als Folge bergbaulicher Einwirkungen

von

K. N e u b e r t (Freiberg)

S u m m a r y

Differing hypotheses exist concerning the causes and the chronological course of movements of the earth's crust. Rockmovements and subsidences caused by mining, well explored through mining operations, can give indications as to the research programme. The laws found out are demonstrated by the effects on the surface and within the rocks.

Z u s a m m e n f a s s u n g

Über Ursache und zeitlichen Ablauf der Erdkrustenbewegungen bestehen unterschiedliche Annahmen. Hinweise für das Forschungsprogramm können die durch den Bergbau ausgelösten und gut untersuchten Gebirgs- und Bodenbewegungen liefern. Es werden die dabei erkannten Gesetzmäßigkeiten an den Auswirkungen über Tage und innerhalb des Gebirges aufgezeigt.

Epirogenetische und orogenetische Vorgänge in der Erdkruste sind aus mannigfaltigen Erscheinungsformen der Topographie sowie aus bergmännischen Aufschlüssen hinreichend bekannt. Über die Wirkung des Kräftespiels mit einfachen bis zu kompliziertesten Auswirkungen bestehen in den meisten Fällen klare Vorstellungen, während über Ursachen der dynamischen Vorgänge, sowohl Größenordnung als auch zeitlicher Ablauf der Kräfte bis zum Abschluß nach erreichtem Gleichgewichtszustand, unterschiedlichen Auslegungen ein weites Feld offen steht. Mit den heutigen Kenntnissen läßt sich die Entscheidung über Richtigkeit oder Abwegigkeit der Annahmen nur selten mit Bestimmtheit treffen. Der Durchdringung der Lithosphäre durch bergbauliche Eingriffe bis zu den tiefsten Bohrungen ist im Vergleich zum Erddurchmesser sehr bald ein Ende gesetzt. Hier bieten nur Methoden der Geophysik Aussicht auf weitere Erfolge. Über die Frage, ob eine Bewegung noch anhält und in welchem Sinne sie auftritt, darüber können nur Präzisionsmessungen höchster Feinheit Aufschluß geben. Auf diese Möglichkeit der Erfassung von Erdkrustenbewegungen wurde zur 1. Erdmessungskonferenz i. J. 1864 in Berlin hingewiesen. Diese Erkenntnis ist für die Erforschung großräumiger Erdkrustenbewegungen bei den letzten Generalversammlungen der Internationalen Union für Geodäsie und Geophysik zur beherrschenden Zielsetzung der Aufgabenstellungen geworden. Es gilt jedoch nicht nur die Höhenänderungen kennenzulernen. Von viel größerer Bedeutung ist die Horizontalverschiebung als 2. Komponente. Da es sich meist um kleine Größen auf beachtliche Entfernungen in Zeitintervallen von Jahren bis Jahrhunderten handelt, läßt sich nur in Ausnahmefällen kurzfristig eine Aussage machen, während erdweite Angaben über kontinentale Schollenbewegungen mühevolle Kleinarbeit von Generationen voraussetzen. Instrumente und Apparate zur Erfassung von Horizontalwanderungen mit der erforderlichen Genauigkeit und einem tragbaren Aufwand sind uns in neuer Zeit durch die Wellentechnik auf elektrooptischem und elektromagnetischem Gebiet an die Hand gegeben, so daß begründete Erfolgsaussicht besteht. Voraussetzung für die Durchführung dieser nicht nur interessanten, sondern auch zukunftsweisenden Aufgabe setzt gegenseitiges Vertrauen aller Völker und eine friedliche Entwicklung der Menschheit voraus.

Zur Vermeidung von Fehlplanungen dieser umfassenden Zielsetzung werden sicher Erkenntnisse Beachtung finden, die dem Bergbau infolge Hohlraumbildung und der damit ausgelösten Gleichgewichtsstörung des Deckgebirges bis zur Erdoberfläche praktisch und wissenschaftlich zu verdanken sind. Auf die Bedeutung dieses Analogieschlusses hat erstmalig Prof.Dr. Karl Lehmann mit einer i.J. 1919 in Bonn genehmigten Dissertation "Bewegungsvorgänge bei der Bildung von Pingen und Trögen" hingewiesen. Durch zahlreiche Abhandlungen und Tagungen auf dem Gebiet der Geomechanik, der Gebirgsdruckforschung und der Bergschadenkunde ist in den darauf folgenden Jahren ein umfangreiches Schrifttum entstanden. Exakte Messungen und Beobachtungen haben zu Gesetzmäßigkeiten geführt, mit deren Hilfe eine mit hohem Genauigkeitsgrad angestellte Vorausberechnung der zu erwartenden Auswirkungen vor Durchführung bergmännischer Eingriffe möglich ist.

Aus markscheiderischen Beobachtungen über bergmännischen Hohlräumen zog Lehmann in seiner Dissertation folgende Schlüsse:

1. Die Bildung eines Hohlraumes in der Erdrinde hat ein Nachsinken der darüber befindlichen Schichten zur Folge.
2. Der Bewegungsvorgang besteht aus der größeren Absenkung (Muldenbildung) über dem Hohlraum und der kleineren Überhebung (Aufwölbung) auf den Randgebieten.
3. Der Kräfteumsatz äußert sich in Pressungen und Zerrungen, die bei der Absenkung zugleich auftreten. In der Senkungsmulde entstehen Druckkräfte, die zu Pressungen und Stauchungen führen, während sich in den Randgebieten Zerrungen geltend machen, die Zerreißungen sowie Riß- und Spaltenbildungen bedingen.
4. Die Pressungen und Zerrungen sind proportional der Höhe des Hohlraumes und der Mächtigkeit der sinkenden Schichten und umgekehrt proportional der Spannweite des Beckens.

Im wesentlichen stimmen auch heute noch diese fundamentalen Erkenntnisse, die in den hinter uns liegenden 40 Jahren weiter ausgebaut, vervollkommnet und ergänzt wurden.

Umfangreiches Untersuchungsmaterial ergab, daß die Form des Senkungsprofiles über eine geradlinige lange Abbaufront der Gaußschen Glockenkurve sehr nahekommt (Bild 1).

$$f(x) = \frac{s_{max}}{r} \cdot e^{-\frac{\pi x^2}{r^2}}$$

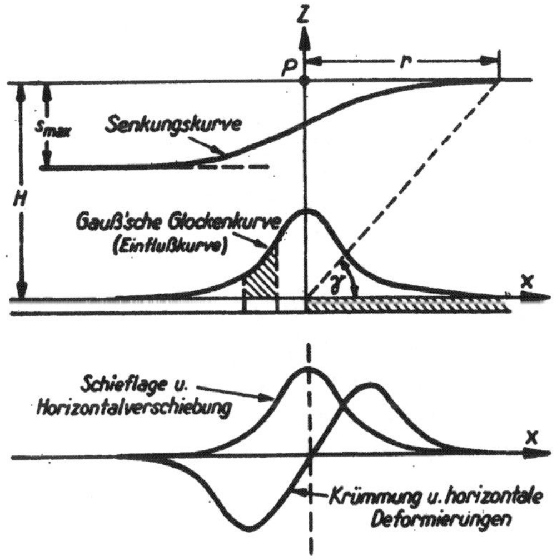

Abb. 1 Zusammenhänge von Bodenbewegungsvorgängen über Abbauhohlräumen

Für die Reichweite der Einflußzone $r = H \cdot \cot \gamma$ sind die Teufe und der Grenzwinkel maßgebend, während das Ausmaß der Senkung von der Mächtigkeit des Hohlraumes, dem Absenkungsfaktor, der Größe der abgebauten Fläche und dem Zeitfaktor abhängt. Umgekehrt kann von der Reichweite oder dem festgestellten Senkungsbetrag auf einen der übrigen Parameter geschlossen werden, ein für die vertikale Erdkrustenbewegung wichtiger Zusammenhang.

Ein Punkt der Erdoberfläche wird eine von der Größe der verhauenen Fläche abhängige Vertikalbewegung erfahren, die von der Teilfläche bis zur Vollfläche zunimmt. Mit der Erreichung des Maximums s_{max} vergrößert sie sich auch bei der Überfläche in ihrem Absolutbetrag nicht mehr, sondern weitet sich nur flächenhaft aus (Bild 2).

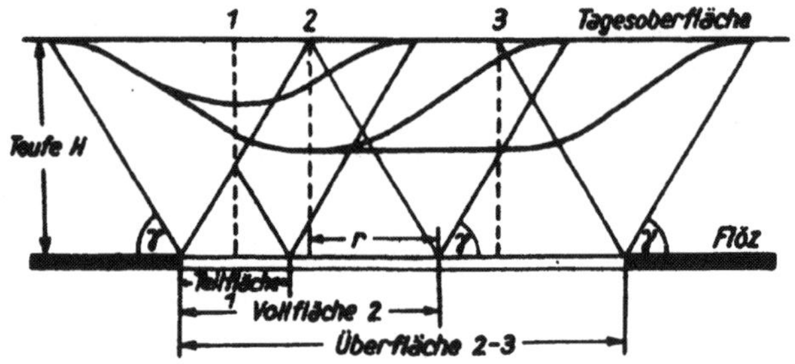

Abb. 2 Senkungsmulde bei Teil-, Voll- und Überfläche

Der allmähliche Übergang im zeitlichen Ablauf eines Punktes ist aus Bild 3 zu erkennen. Nach beschleunigter Anfangsbewegung der vertikalen Absinkung stellt sich das Maximum der Geschwindigkeit ein. Anschließend folgt eine verzögernde Bewegung bis zum Ruhezustand. Bild 4 zeigt den Vorgang bei einer Senkungsmulde. Bei gegebener Hohlraummächtigkeit wird der maximale Senkungsbetrag im Kraftzentrum der Vollfläche erreicht, unbeschadet der Teufe, in der sich der Hohlraum befindet. Mit zunehmender Teufe wächst jedoch die Hohlraumfläche quadratisch. Sind bei 250 m Teufe 10 ha als Einwirkungsvollfläche erforderlich, so werden in 1000 m Teufe 160 ha abgebaute Fläche benötigte (Bild 5).

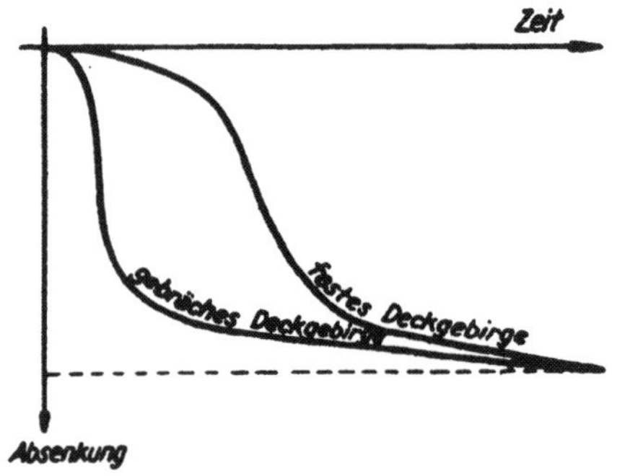

Abb. 3 Zeitlicher Ablauf des Senkungsvorganges eines Punktes über dem Abbau

Als erster Differentialquotient der Senkung wird die Schieflage s' am Senkungsrand, der von der konvexen Krümmung aus dem bergfreien Gebiet über den Wendepunkt zur konkaven Krümmung im Senkungstrog überleitet, ermittelt. Je geringer die Teufe, desto größer ist die Schieflage pro Längeneinheit. Die Linien gleicher Senkung (Isokatabasen) rücken um so dichter zusammen, je geringer die Teufe ist, in der der Abbau umgeht. Sie werden weitmaschiger mit zunehmender Teufe. Während bei geringen Teufen die Abbauwirkung an der Erdoberfläche sofort sichtbar ist, kann sie bei größeren Teufen nur meßtechnisch festgestellt werden. Infolge geringer Neigungsänderung ist sie äußerlich kaum erkennbar

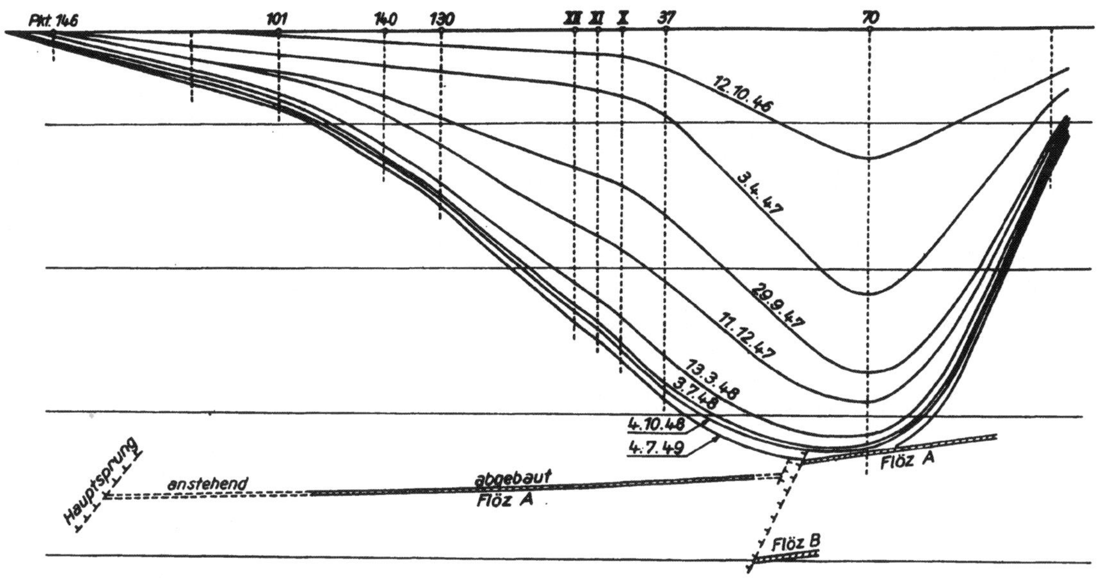

Abb. 4

Senkungskurven im zeitlichen Ablauf

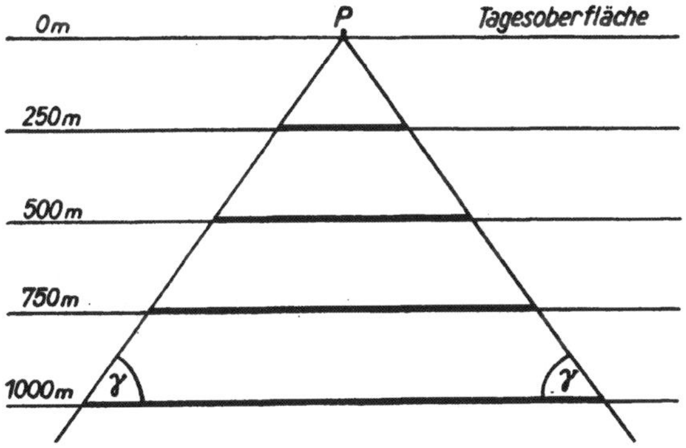

Abb. 5 Einwirkungsflächen für Punkt P in verschiedenen Teufen (Schnitt)

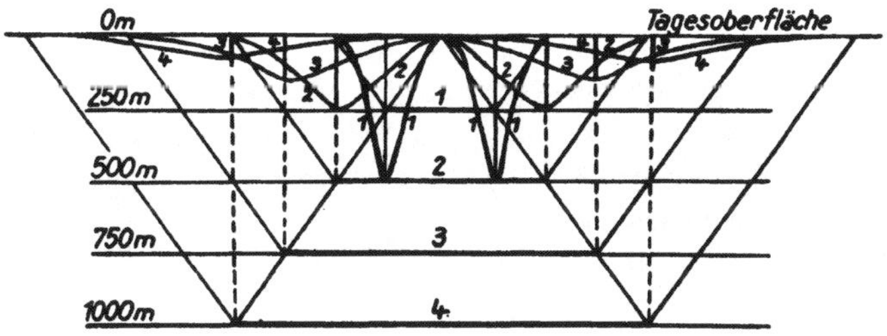

Abb. 6 Bodenneigungskurven bei verschiedenen Teufen

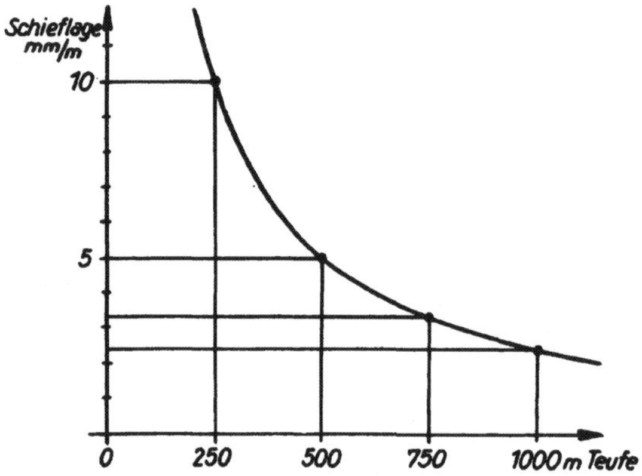

Abb. 7 Bodenneigungen bei Abbau in verschiedenen Teufen

(Bilder 6 und 7). Ein Analogon hierzu ist die Kippung von Schollen der Erdkruste, wie z.B. die Poebene, die norddeutsche, holländische Tiefebene sowie die französische Nordküste, Teile der britischen Südinsel, Kanada usw.

Tektonische Störungen wirken wie Trennwände bei der Auswirkung der Bodenbewegungen über Abbauhohlräumen. Emondts gibt in seinem Bericht [2] die Schieflage im Bereich eines Diagonalsprunges an (Bild 8).

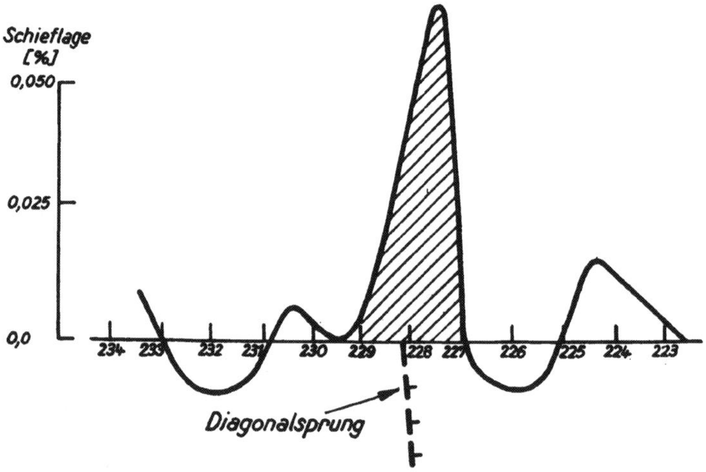

Abb. 8

Schieflage im Bereich eines Diagonalsprunges (8. Messung)

Der 2. Differentialquotient der Senkung ist die Krümmung s", die bei jedem Senkungsrand je ein Maximum im konkaven und im konvexen Abschnitt besitzt, während die Krümmung im Wendepunkt gleich Null ist. Mit zunehmender Teufe glättet sich die Kurve und kann nur meßtechnisch festgestellt werden. Der Krümmungsradius $\rho = \frac{1}{s''}$ nimmt große Beträge an.

Besitzt die Hohlraumfläche ein kleineres oder größeres Einfallen, so werden die Kurven immer unsymmetrischer (Bild 9).

Der zeitliche Ablauf der Vertikalbewegung ist von der Zusammensetzung des Deckgebirges, vom Zerrüttungsgrad und dem Abbaufortschritt abhängig. Aus dem Forschungsbericht von

Emondts [2] (Bild 10) sind die Veränderungen zwischen der 2. und 3. Messung dargestellt. Bild 3 zeigt die Auswirkung bei festem und gebrächem Deckgebirge, wobei die Grenzwinkel mit zunehmendem Zerrüttungsgrad anwachsen, während der zeitliche Ablauf auf eine viel kürzere Zeit zusammengedrängt ist.

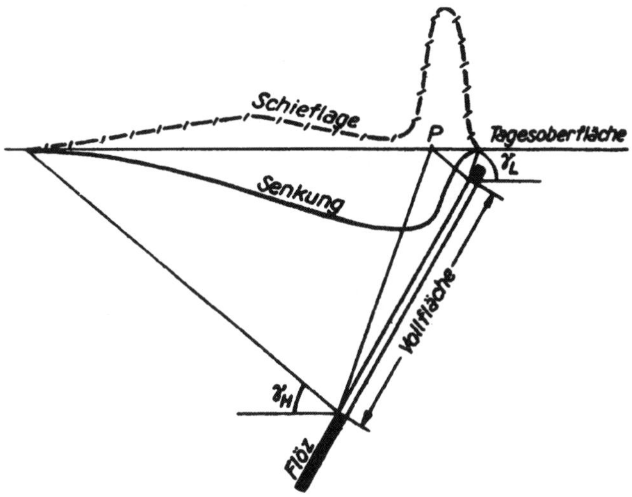

Abb. 9

Senkungs- und Schieflagekurve bei steilem Einfallen der Lagerstätte

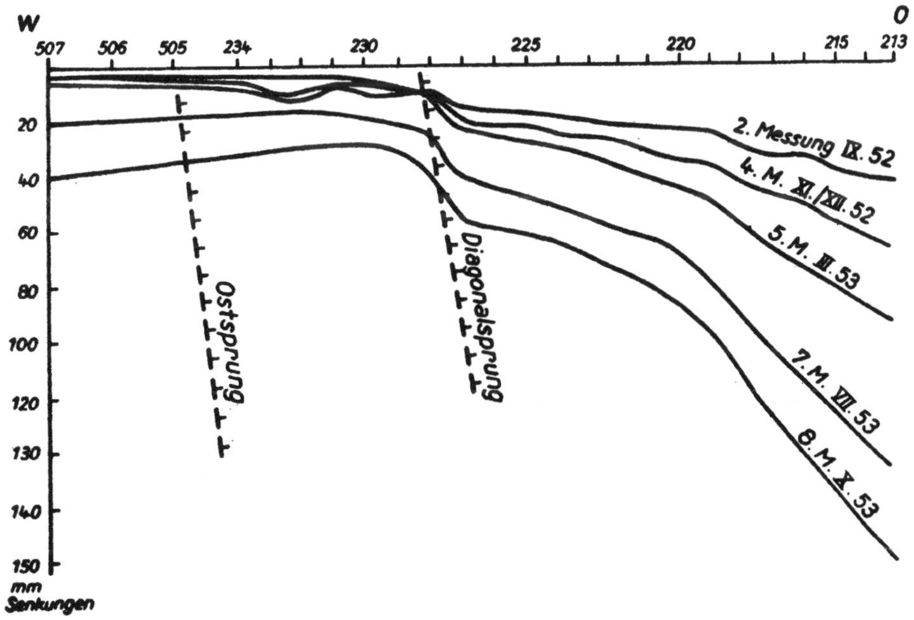

Abb. 10

Senkungskurven in der Beobachtunglinie 507-213

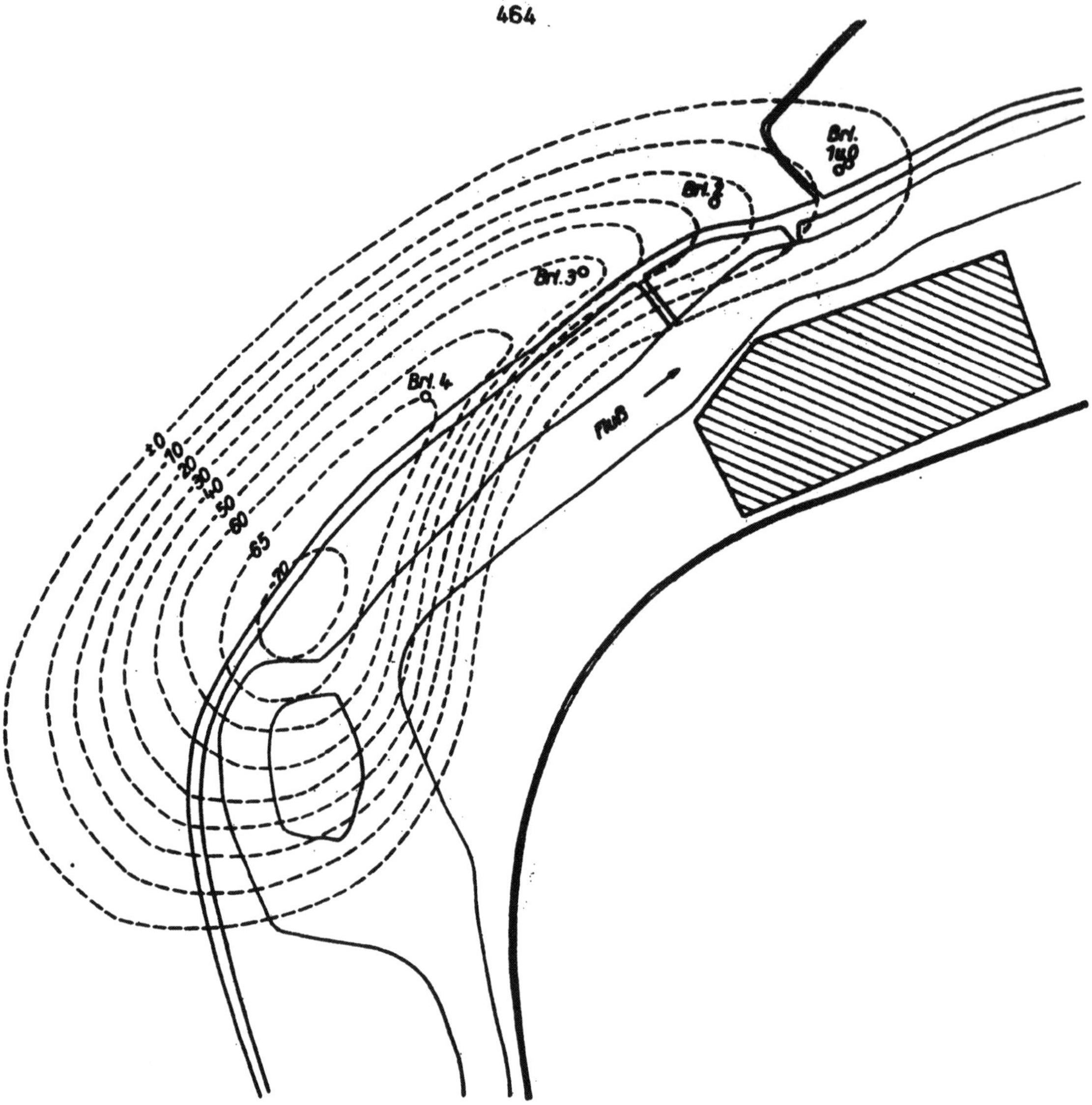

Abb. 11

Linien gleicher Senkungen über künstlichen Aussohlungsräumen, 1930 - 1957

Ähnliche Auswirkungen wie der Abbau rufen künstliche Aussohlungen hervor. Es bilden sich langsam vertiefende Senkungsmulden (Bilder 11, 12). Oberflächennahe Auslaugungen dagegen haben Erdfälle mit scharfen Abrissen zur Folge (Bild 13).

Auch im bergbauunbeeinflußten Gebiet sind die rezenten Vertikalbewegungen der Erdrinde in den einzelnen Ländern bereits jetzt z.T. gut beobachtet. Es fehlt jedoch noch der Zusammenhang über die Erstreckung der Kontinente hinweg und insbesondere der ozeanographischen Bodenveränderungen.

Wie bereits einleitend erwähnt, stellt die vertikale Punktwanderung nur eine Komponente der Abbaueinwirkung dar. Die 2. ist die horizontale, d.h. die lagemäßige Veränderung in der Abszissen- und Ordinatenrichtung, deren Kenntnis gerade für die bestehenden, z.T. voneinander sehr abweichenden geomorphologischen Hypothesen von besonderem Interesse sein kann.

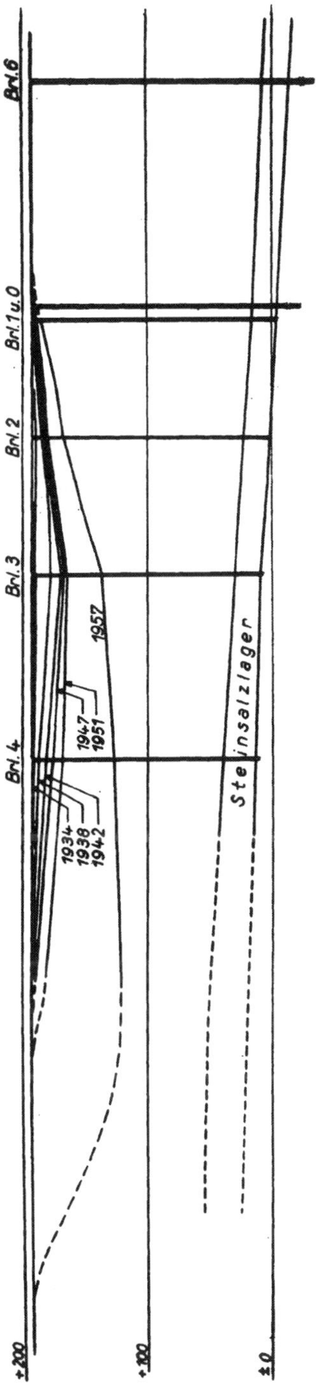

Abb. 12

Vertiefung der Senkungsmulde
über ausgesohlten Hohlräumen
1930-1957 (Schnitt)

Abb. 13

Erdfall natürliche Auslaugung

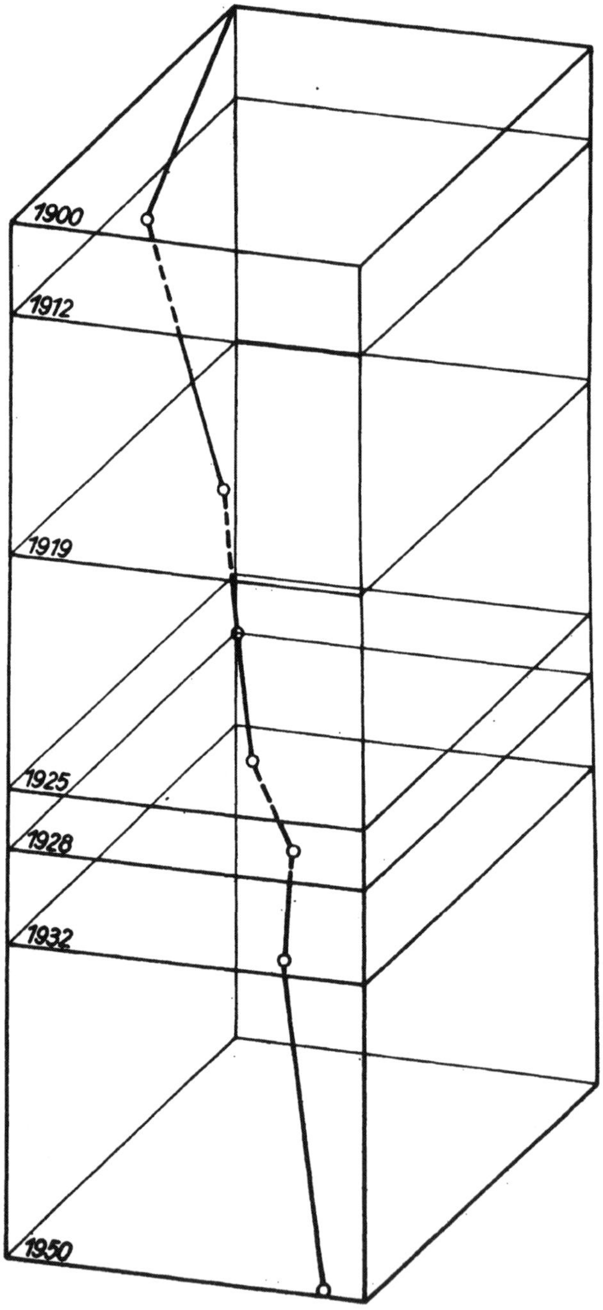

Abb. 14

Punktwanderung des Nagelschen Nullpunktes
von 1900 bis 1950
Monodimetrische Darstellung

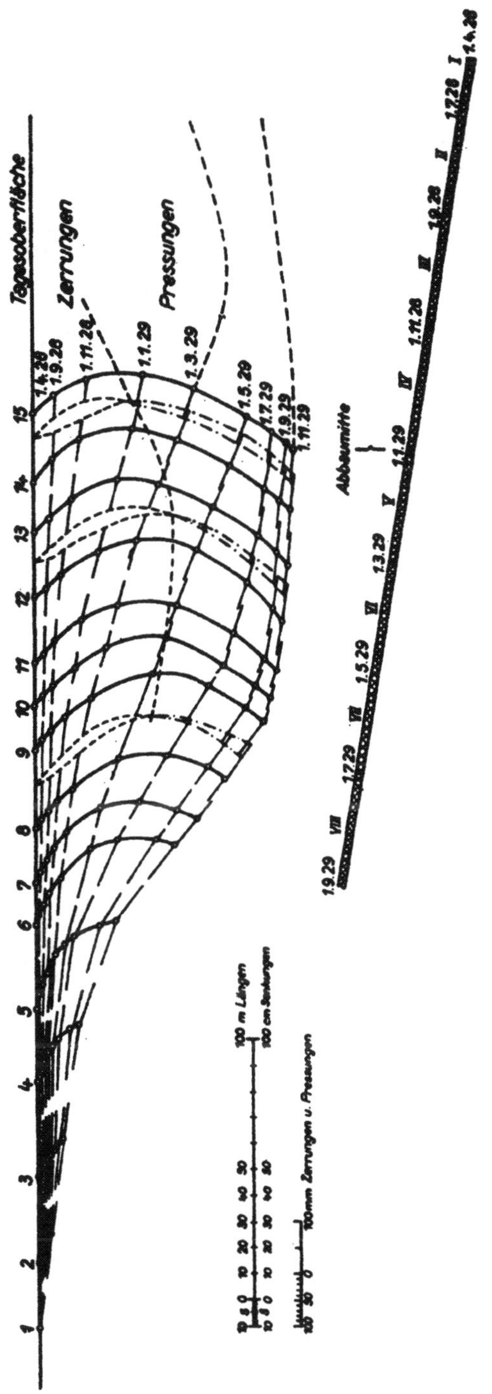

Abb. 15

Wanderung der Senkungen, Zerrungen und Pressungen bei fortschreitendem Verhieb

Betrachten wir einmal die räumliche Wanderung des Nagelschen Nullpunktes als Folge der Gewinnung im Zwickauer Steinkohlenbassin (Bild 14). Aus der anschaulichen Darstellung kann die Punktveränderung im Zeitintervall von 1900 bis 1950 verfolgt werden. Zeiten größerer vertikaler Wanderungen wechseln mit Abschnitten bedeutender horizontaler Bewegungen ab. Betrachtet man die zeitliche Folge der Auswirkungen über einem fortschreitenden Abbau im Schnitt, so erkennt man eine bankrechte Absenkung im Muldentrog, während an den Rändern der Senkungsmulde beachtliche Horizontalverschiebungen, dagegen geringe Vertikalbewegungen stattfinden (Bild 15) [3]. Richtung und Größe der Pfeile stellen in Bild 16 die von der Lage des Punktes abhängige Größe der Senkung und Horizontalwanderung

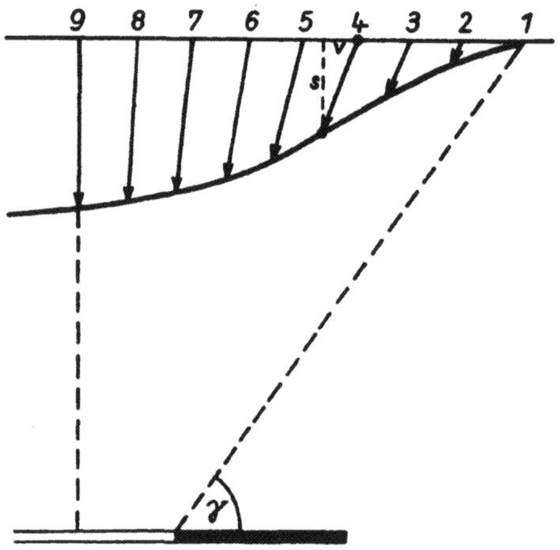

Abb. 16 Vektorielle Punktwanderung über Abbauhohlraum

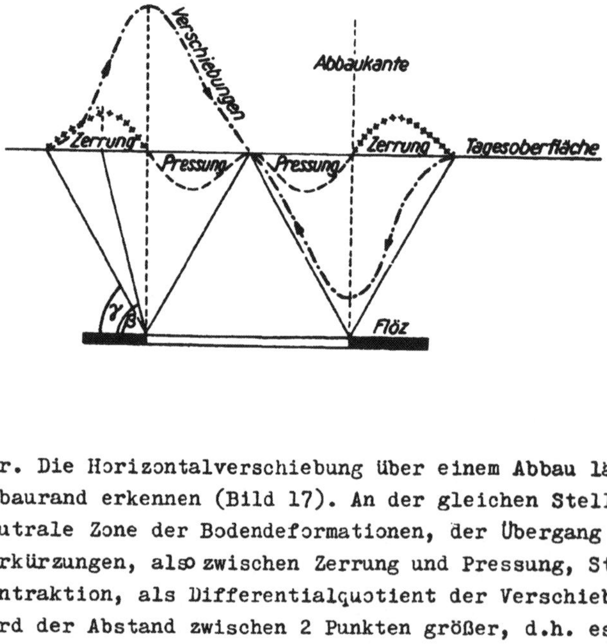

Abb. 17 Verschiebungen und Längenänderungen über einer Abbauvollfläche

dar. Die Horizontalverschiebung über einem Abbau läßt den maximalen Wert etwa über dem Abbaurand erkennen (Bild 17). An der gleichen Stelle dieses Maximums befindet sich die neutrale Zone der Bodendeformationen, der Übergang zwischen Längendehnungen und Längenverkürzungen, also zwischen Zerrung und Pressung, Streckung und Stauchung. Dilatation und Kontraktion, als Differentialquotient der Verschiebung. Am äußeren Rand der Senkungsmulde wird der Abstand zwischen 2 Punkten größer, d.h. es herrscht Zerrung. Das Maximum der Dehnung wird über dem freien Schenkel des Bruchwinkels erreicht. Auch hier gilt die Erkenntnis, daß die Größe der Verschiebung von der Teufe unabhängig ist, wenn die gesamte Einwirkungsfläche eines bestimmten Punktes der Tagesoberfläche unterbaut ist. Nur die

Verschiebungsdifferenz zwischen denselben Punkten bei größerer Teufe ist geringer, so daß die auftretenden Spannungen an der Tagesoberfläche mit wachsender Abbauteufe abnehmen (Bild 18). Exakte Messungen haben ergeben, daß die Summe der Zerrungen gleich der Summe der Pressungen ist (Bild 19).

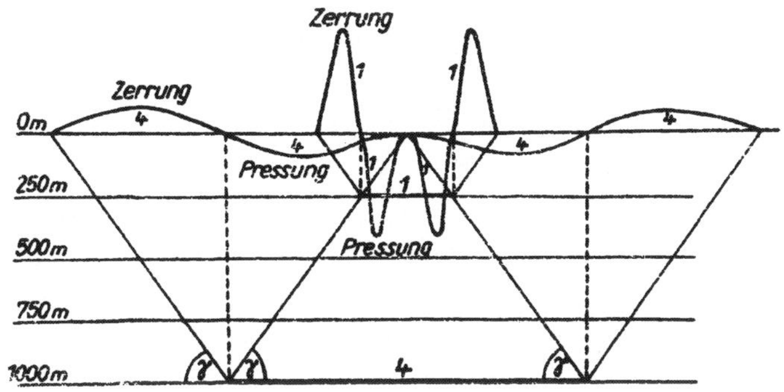

Abb. 18 Bodenspannungen bei Abbau in verschiedenen Teufen

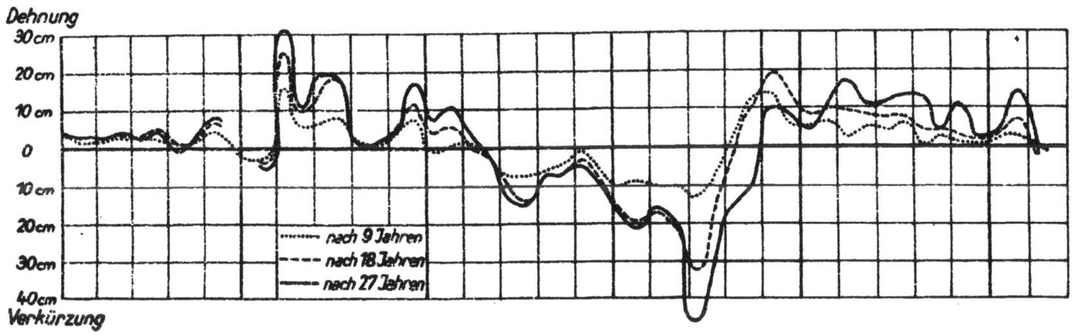

Abb. 19

Längenänderungskurven (auf 50 m Strecken bezogen)

Ähnliche Erscheinungen zeigen sich an den Rändern großer Braunkohlentagebaue mit beachtlichen horizontalen Punktwanderungen und Abrissen im Tagebauvorfeld. Bild 20 stellt den Tagebau B im Grundriß dar, und in Bild 21 sind horizontale Punktwanderungen des untrigonometrischen Punktes einer in Tagebaunähe befindlichen Kirche zu erkennen. Die unterschiedlichen Beträge sind die Folge einer mehr oder weniger intensiven Abbautätigkeit. Diese mit markscheiderischen Instrumenten festgestellten Veränderungen können auch mit einem hydraulischen Dilatometer ermittelt werden [4].

Für Vorausberechnungen werden markscheiderische Beobachtungen an einem umfangreichen Festpunktfeld über Tage mit einwandfreier Vermarkung und Bezugspunkten außerhalb des bergbaubeeinflußten Gebietes angestellt, um die vertikalen und horizontalen Punktwanderungen in bestimmten zeitlichen Abständen, die von der Intensität der Abbautätigkeit abhängen, kennenzulernen. Mit Hilfe dieser Unterlagen lassen sich folgende Größen ermitteln, um an anderen Teilen des Grubenfeldes unter Verwendung bergbaulicher Projektierungsarbeiten künftig eintretende Einwirkungen qualitativ und quantitativ angeben zu können:

1) Grenz- und Bruchwinkel (γ und β)

2) Absenkungsfaktor a

3) Zeitfaktor z

4) maximaler Senkungsbetrag s_{max}

5) maximaler Verschiebungsbetrag v_{max}.

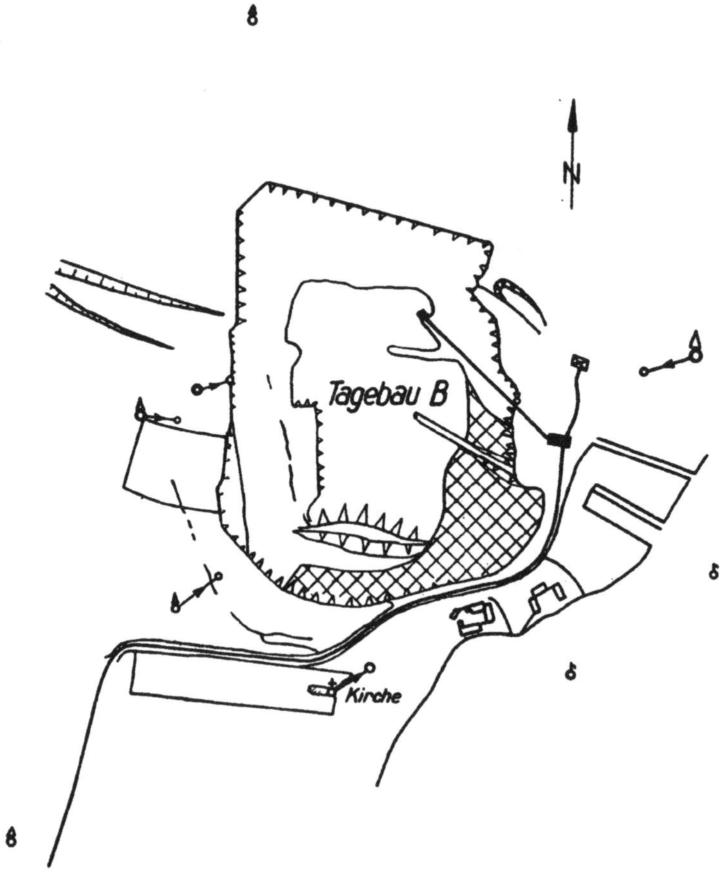

Abb. 20

Horizontale Punktwanderungen in Tagebaurandnähe

Es werden vor Beginn des Abbaues der voraussichtliche Verlauf der Isokatabasen, die Zonen der maximalen Schieflage und Bodenkrümmung sowie die Lage der Preßbruch- und Zerrbruchzonen ermittelt. Mit diesen Angaben muß entschieden werden, ob schädliche Auswirkungen an wertvollen Objekten der Tagesoberfläche durch Veränderung der Abbauplanung unterbunden oder verringert werden müssen. Hierfür ist der Umfang der zu erwartenden Schäden zu errechnen, die in Beziehung zur Verteuerung der Abbaumethode und des zu verändenden Abbauzuschnittes zu bringen sind. Auch Vorflutstörungen sowie Veränderungen des Grundwasserstandes spielen oft eine entscheidende Rolle.

Aber nicht nur die Beeinflussung der Tagesoberfläche infolge Abbau läßt Analogieschlüsse für Probleme der Erdkrustenbewegungen zu. Auch die Vorgänge innerhalb des Gebirgskörpers, die durch bergmännische Eingriffe ausgelöst werden, lassen Parallelen erkennen.

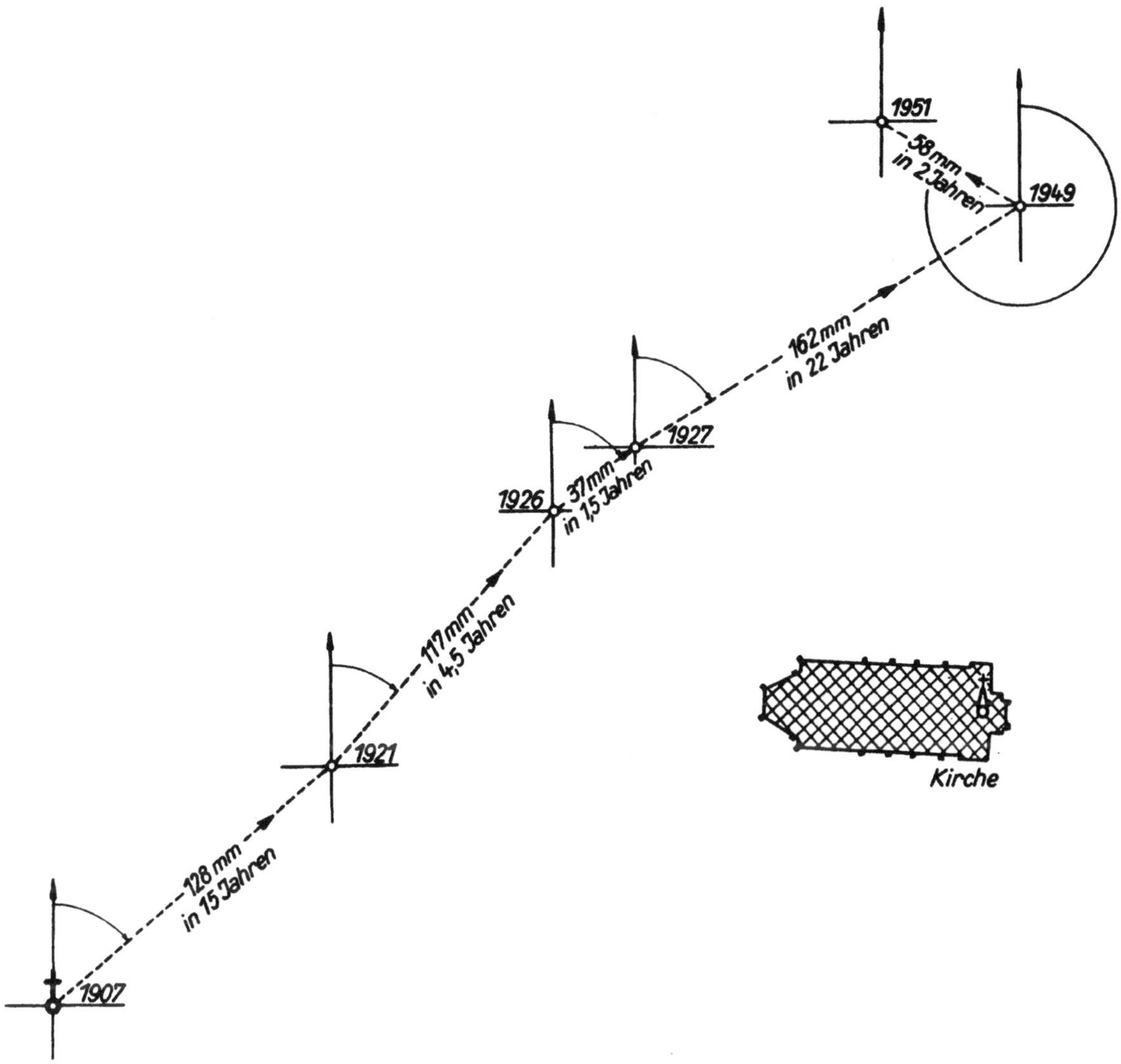

Abb. 21

Horizontale Punktwanderung eines Trig. Punktes in Tagebaunähe von 1907 - 1951

Gegenläufiger Abbaubetrieb in 2 verschiedenen Flözen ruft Firstsenkungen und Hebungen hervor (Bild 22). Gebirgsdruckunterschiede werden durch verschiedene Meßmethoden ermittelt. Hierzu dienen Messungen mit markscheiderischen Instrumenten und Geräten (Bild 23) oder mit Meßdosen einschließlich Gebern auf mechanischer oder elektrischer Grundlage. Zur Zeit sind wir nur bedingt in der Lage, den im anstehenden Gestein herrschenden Gebirgsdruck messen zu können. Um jeden bergmännischen Hohlraum bildet sich eine entspannte Trompetersche Zone, die nicht am Tragen des überlagernden Gebirges teilnimmt, so daß sich ein neuer Gleichgewichtszustand durch erhöhte Stützkraft in den Nachbarzonen einstellt. Bis zur Erreichung des Normaldruckes sind Entfernungen bis zu mehreren Hundert Metern im Vorfeld der Lagerstätte feststellbar.

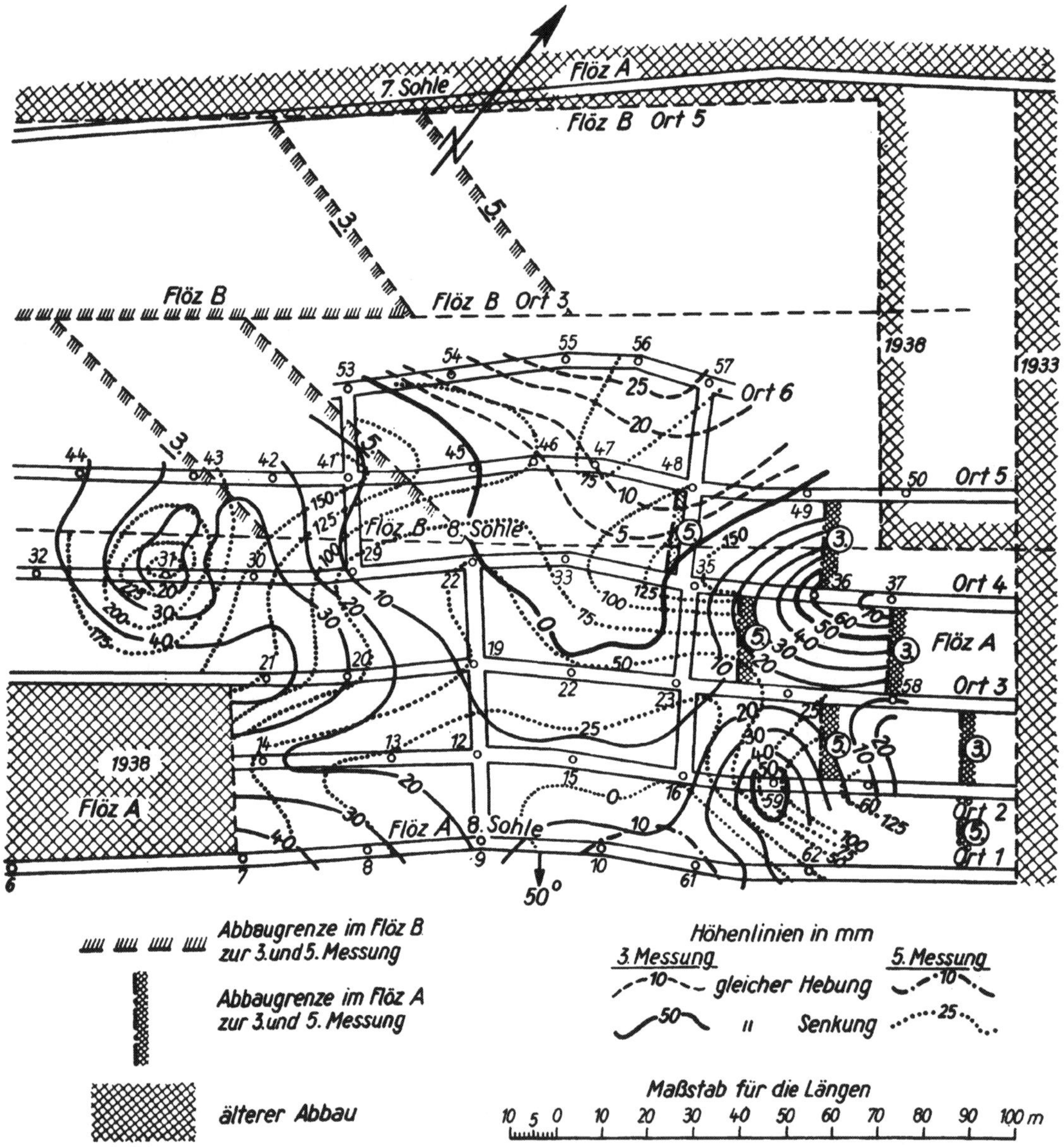

Abb. 22

Firstsenkungen bei gegenläufigem Verhieb

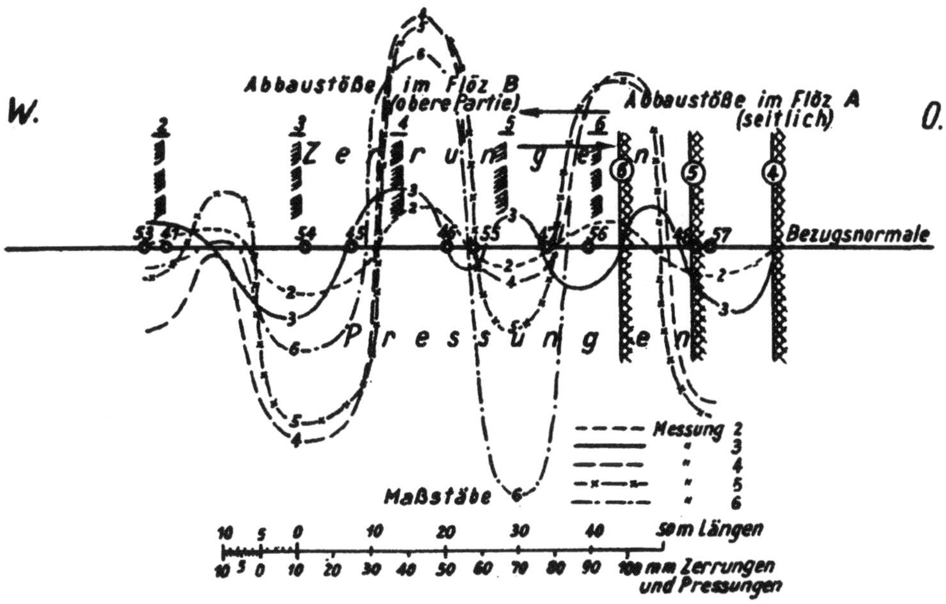

Abb. 23

Spannungsänderungen bei verschiedenen Abbaustellungen in zwei gegenläufig betriebenen Streben Ort 5 und 6

Mit zunehmender Teufe wächst der Überlagerungsdruck und damit das Verhalten der Gesteinspakete. Im deutschen Bergbau wird deshalb der Schwierigkeitsgrad unterirdischer Arbeiten in 3 Kategorien eingeteilt:

1. Oberflächennaher Abbau bis 300 m Teufe mit normalem Ausbau.
2. Abbau mittlerer Teufe von 300 bis 800 m bei erhöhtem Schwierigkeitsgrad. Druckkonzentrationen können zu Gebirgsschlägen führen.
3. Abbau bei Teufen über 800 m. Verschiedene Gesteine beginnen sich plastisch zu verformen. Die Offenhaltung der Strecken bereitet teilweise erhebliche Schwierigkeiten. Erhöhte Gebirgsschlaggefahr bei Gesteinen im unmittelbaren Hangenden, deren Elastizitätsgrenze in der Nähe der Bruchgrenze liegt.

In Tiefen, zu denen der Bergmann infolge untragbarer Temperaturen und nicht beherrschbaren Gebirgsdruckes nicht vorzudringen vermag, spielen sich die Ursachen der Erdkrustenbewegungen ab. Die Lithosphäre geht in einen plastischen Zustand über. Schollen beginnen auf dem leicht verformbaren Untergrund zu treiben.

Die Erscheinung der Scherwirkung zwischen den Gesteinsschichten besteht bereits in den durch den Bergbau beeinflußten Teilen der Gebirgsdecke. Dieser Horizontalkomponente kommt die Hauptbedeutung hoher Spannungsansammlung bis zur Auslösung eines Gebirgsschlages zu. Die Gefahr unzulässig hohen Druckanstieges durch Verspannung der Schichten gegeneinander, die durch unterschiedliche Eigenschaften der Gesteinspakete hervorgerufen wird, hat ihre Ursache in der Abbaumethode, im Zuschnitt, in den Lagerungsverhältnissen oder der Zusammensetzung der Gesteine. Stauchungen von Sandsteinen auf 10 mm pro Meter lösen bereits Gebirgsschläge aus. Besonders in größeren Teufen wird die gegenseitige Reibung der Gesteinsschichten überwunden, so daß intensive Verschiebungskräfte wirksam werden. Diese sind auch die Geburtsstätte der Erdkrustenbewegung. Der Hauptanteil der gesamten tektonischen Vorgänge geht auf Konto der horizontalen Beanspruchung, deren Ursache auch außerirdischen Kräften zuzuschreiben ist.

Durch die Abbauführung werden Kräfte ausgelöst, die Gebirgsbewegungen hervorrufen. Die Dynamik dieser Kräfte unter verschiedenen Bedingungen kennenzulernen, gehört zu den vornehmsten Aufgaben des Bergmannes. Die Lenkung des Abbaus muß derart erfolgen, daß die auftretenden Kräfte vom Feind zum Freund des Menschen werden, wie es Spackeler gelegentlich auszudrücken pflegte. Mit der Abbaudynamik beschäftigen sich unter den mannigfaltigsten Verhältnissen seit Jahrzehnten die Bergbauwissenschaftler und Praktiker fast aller bergbautreibenden Länder. Den Vorgang um den Abbauhohlraum im Gebirge stellte Labasse anschaulich dar (Bild 24) [5]. Der Kräftefluß läßt sich unter Veränderung der Abbaugeschwindigkeit und der Abbaumethode derart beeinflussen, daß eine Unterstützung der Ar-

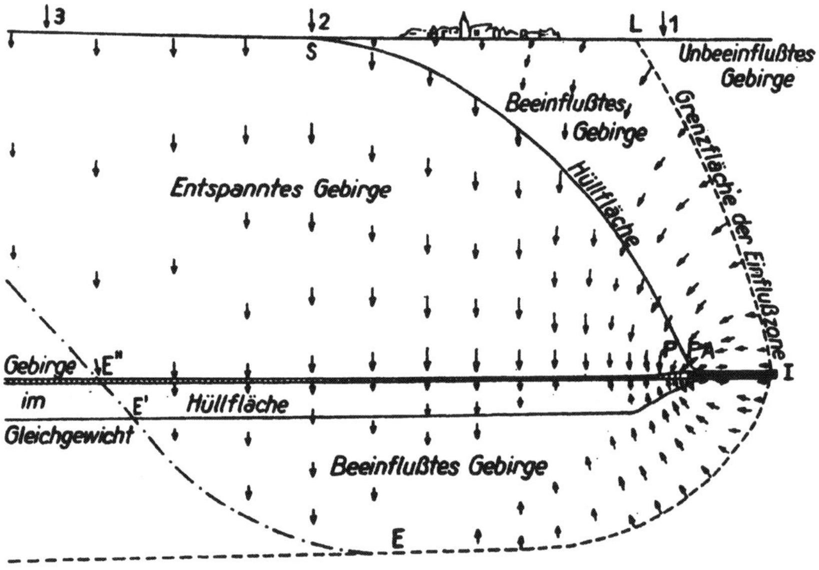

Abb. 24
Druckmechanik im Gebirge um den Abbauhohlraum

beitsleistung des Bergmanns zustande kommt. Einerseits muß das Streben darauf gerichtet sein, daß nicht geschossen zu werden braucht, um keine Unruhe im Gebirge hervorzurufen, andererseits sollen durch die ausgelösten Kräfte Lockerung und Zermürbung des Verbandes der zu gewinnenden Lagerstätte erfolgen, so daß nur geringe Kräfte zum Hereingewinnen aufzuwenden sind, ohne aber den Gebirgsverband im Hangenden über dem Abbauhohlraum gefahrbringend zu lockern. Brüche sind in den Versatzraum zu verlegen und können sich dort austoben. In den meisten Fällen ist diese Zielsetzung durch bergmännische Maßnahmen gelungen, wenn auch bisweilen in der Praxis andere Gesichtspunkte maßgebend sind. Plansoll und Zeitplan sind oft die alles beherrschenden motorischen Kräfte der Produktion.

Es ist manches erreicht. Es bleibt noch vieles zu tun. Wenn aber die wenigen Ausführungen für die weitere Erforschung der Vorgänge in der Erdrinde als Analogon zu den im Bergbau erkannten Gesetzmäßigkeiten anregend und fördernd wirken, so haben sich meine Hinweise gelohnt.

L i t e r a t u r

[1] Knothe, St.: Bodenbewegungen und Bodendeformierungen beim Streckentreiben bei geringer Teufe. Freib. Forsch.-H. A 57 (1957), S. 87 - 95

[2] Emondts, J.: Über Bodenverformungen bei stark gestörtem und mächtigem, wasserführendem Deckgebirge im Aachener Steinkohlengebiet. Forsch.Ber. d. Wirtschafts- u. Verkehrsministeriums Nordrhein-Westfalen, Nr. 123, Westdeutscher Verlag Köln und Opladen 1955

[3] Niemczyk, O.: Bergschadenkunde. Glückauf-Verlag, Essen 1949, S. 233

[4] Wilkening, W.: Bodenbewegungsmessungen mit dem hydraulischen Dilatometer. Bergbauwissensch. 9 (1962), S. 75 - 82

[5] Labasse, H.: Les propriétés mécaniques des bancs. Rev. de l'industrie minérale 39 (1957), S. 997

Vertikalbewegungen längs hauptsächlicher Linien des Präzisionsnivellements im Górnóslaskie Zaglebie (Oberschlesisches Kohlenbecken) im Zeitraum 1955 bis 1958

von

J. Niewiarowski (Warschau)

Summary

In order to determine vertical movements of bench marks on the chief lines of the precision levelling in Górnośląskie Zagłębie Weglowe, the results of measurement dating from 1955 and 1957/58 have been compared with each other on coinciding lines of the class II A.

The variations of differences of height between neighbouring bench marks as well as the summed up differences of height on the lines have been calculated. The canvas consisting of 12 closed polygons formed by the compared lines has been compensated.

For 1 km of compared lines we have got an error of ± 1,2 mm. The vertical displacements of bench marks represented graphically could be compiled in the final result of the compensation.

Zusammenfassung

Es wurden, zwecks Bestimmung der Vertikalbewegungen von Höhenfestpunkten auf den hauptsächlichen Linien des Präzisionsnivellements im Górnośląskie Zagłębie Weglowe, die Meßergebnisse aus den Jahren 1955 und 1957/58 auf sich überdeckenden Linien der Klasse II A miteinander verglichen.

Berechnet wurden die Differenzen von Höhenunterschieden zwischen benachbarten Höhenfestpunkten sowie die summierten Höhenunterschiede auf den Linien. Das Netz, das sich aus 12 geschlossenen, aus den verglichenen Linien gebildeten Polygonen zusammensetzt, wurde ausgeglichen.

Für 1 km verglichener Linien erhielt man einen Fehler von ± 1,2 mm. Im Endergebnis des Ausgleiches konnten die graphisch dargestellten Vertikalverschiebungen von Höhenfestpunkten erarbeitet werden.

Die Untersuchungen gegenwärtiger Vertikalbewegungen der Erdkruste beruhen grundsätzlich auf geologischen und tektonischen Voraussetzungen, durchgeführt werden dieselben mittels verschiedener Methoden.

Die hauptsächliche quantitative Methode ist die geodätische Methode eines wiederholten Präzisionsnivellements. Man geht dabei von der Voraussetzung aus, daß die gegenwärtigen Vertikalbewegungen der Erdkruste - hinsichtlich ihrer Kinematik - gleichmäßige und kontinuierliche Bewegungen sind. Abgesehen von katastrophenbedingten Bewegungen, werden die gegenwärtigen Vertikalbewegungen der Erdkruste durch eine Geschwindigkeit der Größenordnung Millimeter je Jahr gekennzeichnet.

In manchen Räumen wird das Auftreten derzeitiger Vertikalbewegungen durch den Einfluß menschlicher Tätigkeit gestört. Zu derartigen Räumen gehören in der Regel Gebiete intensiver Förderung von Mineralvorkommen und unter denselben vor allem Kohlenbecken.

Das Gesamtbild oberflächlicher Vertikalbewegungen von Erdmassen im Abbaugebiet eines Kohlenbeckens ist sehr kompliziert, da diese Bewegungen die Folge des Einwirkens einer Reihe von Ursachen sind. Zu diesen Ursachen gehören vor allem der das bestehende Gleichgewicht von Erdmassen an der Oberfläche störende Abbau von Kohlenflözen, die infolge der Kohlenförderung eintretenden Veränderungen im Stand und in den Bewegungen des Grundwassers u.ä.m. In derartigen Erscheinungen ist gleichfalls eine Komponente gegenwärtiger Vertikalbewegungen der Erdkruste zu erkennen.

Das größte Kohlenbecken Polens ist das sich in voller Entwicklung befindliche Zagłębie Górnośląskie (Oberschlesisches Kohlenbecken). Die im Raume dieses Beckens nicht nur mittels geodätischer Methoden, sondern gleichfalls auf Grund verschiedener Verfahren betriebenen und sich immer mehr entwickelnden Untersuchungen von Bewegungen tragen nicht nur den Charakter wissenschaftlicher Forschungen, mit denselben wird auch eine praktische Zielsetzung angestrebt. Die Kenntnis der Gesetzmäßigkeiten und des Verlaufs der Oberflächenbewegungen von Erdmassen im Bereich des Beckens, die Verknüpfung dieser Bewegungen mit bestimmten Ursachen und die Bestimmung ihrer Reichweite sind für den weiteren Abbau von hervorragender Bedeutung und können in hohem Grade den manchmal schädlichen Folgen einer gegebenen Nutzungsart vorbeugen.

Nach dem I. Weltkriege wurden im Gebiet eines Teiles des Zagłębie Górnośląskie eine Reihe von Präzisionsnivellementsvermessungen in den Jahren 1925, 1931, 1937 angestellt und nach dem II. Weltkrieg gelangten derartige Vermessungen bereits im Gebiete des gesamten Beckens in den Jahren 1955 und 1958 zur Durchführung.

Das Präzisionsnivellementsnetz des Jahres 1955 war mittels mit Planparallelplatten versehenen Wild N3-Nivellatoren sowie mittels Invarlatten vermessen worden. Ausgeglichen und berechnet wurde dasselbe in Anlehnung an Punkte des staatlichen Nivellements I. Klasse. Die Meßgenauigkeit wird durch die nachfolgenden, je 1 km zweifachen Nivellements (vor Ausgleichung) auftretenden Fehler gekennzeichnet:

einen zufälligen Fehler $\eta \leq \pm 1.5$ mm

" systematischen " $\sigma \leq \pm 0.4$ mm.

Der überwiegende Teil der Linien des in den Jahren 1957/58 vermessenen Präzisionsnivellementsnetzes des Zagłębie Górnośląskie deckt sich mit den Linien aus dem Jahre 1955. Die methodologische Verfahrensweise und die Meßgenauigkeit waren dieselben wie im Jahre 1955.

Das Netz v. J. 1957/58 war – wie bereits erwähnt – in Anlehnung an Punkte des staatlichen Präzisionsnivellements I. Klasse ausgeglichen worden, sein Bezugsniveau ist demnach das Niveau des Nullpegels von Kronstadt. Das Netz der Jahre 1957/58 ist mittels Einführung von die wirklichen Werte der Schwerkraftbeschleunigung berücksichtigenden Korrekturen in die Höhenunterschiede auf ein System von Normalhöhen berechnet. Im Südwesten verbindet sich das Präzisionsnivellementsnetz des Zagłębie Górnośląskie mit dem Nivellementsnetz des Zagłębie Ostrawsko-Karwińskie.

Für Forschungszwecke wurde im Institut für Geodäsie und Kartographie aus dem Nivellementsnetz des Zagłębie Górnośląskie entnommmenen Linien ein Netz der Klasse II A entworfen. Dieses Netz wurde bei einer Gesamtlänge von rd. 800 km in Anlehnung an drei säkulare Fundamentpunkte (Brudzowice, Racibórz, Koniaków) erarbeitet. Dasselbe ist in Höhenrichtung das grundsätzlich tragende Gerippe für die Untersuchung von Vertikalbewegungen im Gebiet des Zagłębie Górnośląskie. Die Vermessung dieses – sowie des mit demselben verbundenen Netzes Zagłębie Ostrawsko-Karwińskie soll je alle 4-5 Jahre wiederholt werden.

Als erste Etappe für die Bestimmung der Vertikalbewegungen oberflächlicher Erdmassen des Zagłębie Górnośląskie mittels geodätischer Methoden wurden vergleichende Ergebnisse der Präzisionsnivellements aus den Jahren 1955 und 1957/58 bearbeitet. Untereinander verglichen wurden vermessene Höhenunterschiede zwischen benachbarten Höhenfestpunkten längs der Linie. Als vermessene Höhenunterschiede wurde das bei Vermessung in beiden Richtungen aus Höhenunterschieden erhaltene Mittel angenommen, wobei eine die wirkliche Lattenlänge berücksichtigende Korrektur einzuführen war. Für jede Linie wurden die Differenzen zwischen den vermessenen Höhenunterschieden (r) sowie die summierten Differenzen der Höhenunterschiede (Σr) - vom Ausgangspunkt der Linie an - errechnet. In geschlossenen Figuren muß die Summe der so gebildeten Höhenunterschiede gleich Null sein, und die Nichterfüllung dieser Bedingung ist das Ergebnis von Nivellementsfehlern bzw. des Einflusses von Geländebewegungen während des Verlaufs der Linienmessung.

Das auf diese Weise entstandene Netz von 12 geschlossenen Polygonen wurde unter Voraussetzung der Unbeweglichkeit von auf dem Umfang des Netzes II A liegenden Verbindungspunkten bei Anwendung der Bedingungsmethode ausgeglichen. Als Ergebnis dieser Ausgleichung erhielt man die Korrekturen zu den summierten Höhenunterschiedsdifferenzen für 24 Linien. Diese Korrekturen wurden sodann - proportional zur Entfernung vom Ausgangspunkt der Linie - auf die Höhenunterschiedsdifferenzen zwischen den Höhenfestpunkten längs der Linie verteilt. Der 1 km verglichener Linien entsprechende Fehler hat die Größe $m_0 = \pm 1,2$ mm, er gibt die Genauigkeit der verglichenen Nivellements sowie den Einfluß der während des Messungsverlaufs vor sich gehenden Vertikalbewegungen wieder.

Die für einige besonders charakteristische Linien beigefügten Diagramme (Abb. 1, 2, 3, 4) stellen die Vertikalverschiebungen von Höhenfestpunkten längs dieser Linien im Zeitraum von 1955 - 1958 dar.

Die bisher errechneten Größen von Vertikalverschiebungen längs 24 Präzisionsnivellementslinien gestatten noch keine Schlußfolgerungen hinsichtlich gegenwärtiger Vertikalbewegungen zu ziehen, da hierfür der mittels wiederholter Nivellements erfaßte Zeitintervall zu kurz ist. Natürlich kann man in Fällen des Auftretens bedeutender und krasser Setzungserscheinungen an Höhenfestpunkten, was gleichbedeutend mit einem Setzen von an der Oberfläche befindlichen Erdmassen ist, mit vollkommener Sicherheit voraussetzen, daß dieselben eine direkte oder indirekte Folge von bergwerksmäßiger Nutzung des betroffenen Geländes sind.

Der vorliegende Rechenschaftsbericht enthält die Darstellung der ersten Untersuchungsetappe über Vertikalbewegungen mittels des Verfahrens wiederholter Präzisionsnivellements im Gebiet des Zagłębie Węglowe. Erst nach vergleichender Bearbeitung wiederholter Nivellements für einen längeren Zeitraum wird man bezüglich gegenwärtiger Vertikalbewegungen der Erdkruste innerhalb dieses Gebietes Schlußfolgerungen ziehen können.

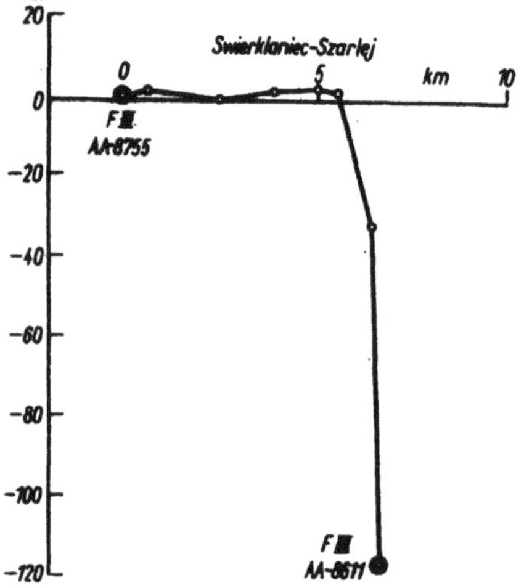

Abb. 1

Die Vertikalverschiebungen von Höhenfestpunkten längs 4 Linien im Zeitraum 1955 - 1958

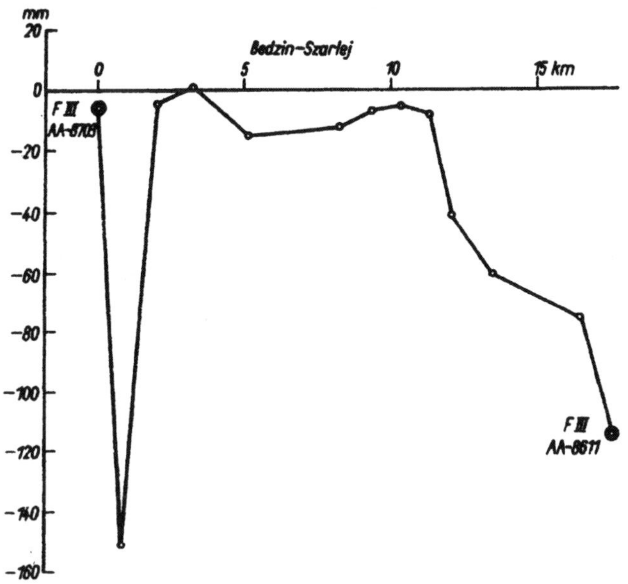

Abb. 2

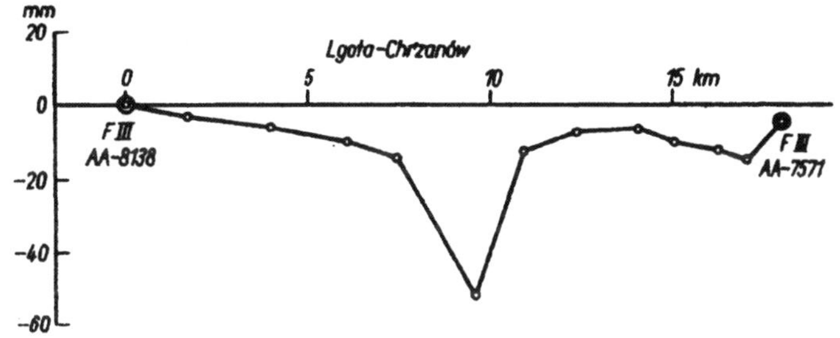

Abb. 3

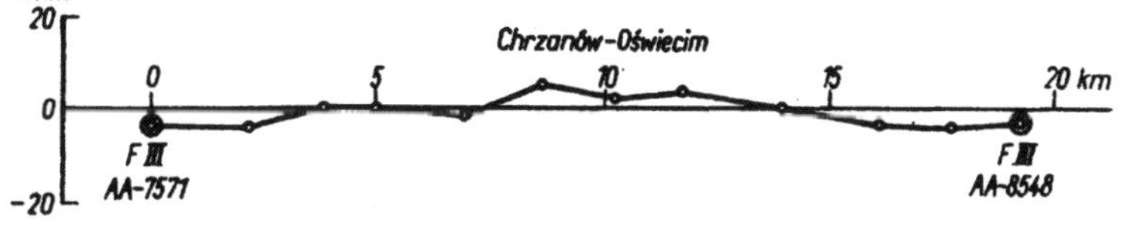

Abb. 4

About the measurements of active faults

von

J. A. B r o g g i (Lima)

S u m m a r y

It is evident that there are active faults that show relative displacements along their edges; but the dimensions in direction and depth are unknown, as in certain cases where their relations with superficial movements are not produced in rock formations perfectly responsive to deeper movements.

On the other hand, the measurement of displacements along the edge of a fault brings a conflict between desire and the possibility of determining them, due to the lack of funds and technically trained personnel in the areas where earthquakes have occurred and produced visible displacements of the crust.

The parts of the earth in which catastrophic earthquakes have occurred in recent years are in underdeveloped countries, that is to say, where the measurement of crustal movements is very difficult because of the lack of efficient technical personnel for the basic study of the geological phenomena of geofaults and their geodetic control. It is possible to do this in the principal countries of the Earth, but not in those where movements have occurred recently but no measurements of the geofault have been made. The difficulty increases in the case of faults tens of kilometers in length, where levels and triangulations have to be repeated at long intervals during which bench marks may change their position due to the fecund vegetative population growth of the present period and the effect of their work activities.

R e s u m e n

Es evidente que existen fallas activas o sea que muestran desplazamientos relativos según sus bordes; pero se ignoran sus dimensiones en dirección y profundidad como en ciertos casos sus relaciones con movimientos superficiales que no se producen en formaciones rocosas perfectamente responsables de movimientos de carácter más profundo.

De otro lado, la medición de desplazamientos de bordes de fallas es un asunto de conflictos entre deseos y posibilidades de determinarlos por razones de falta de medios económicos y técnicos de las diversas áreas en que se ha constatado terremotos en conexión con desplazamientos visibles de la corteza.

Las áreas terrestres en que se han constatado catástrofes sísmicas en los años recientes, son áreas de dominio político llamado de subdesarrollo ó sea de muy difícil control de movimientos corticales verdaderos tanto porque no hay técnicos eficientes capacitados para estudiar a fondo el fenómeno geológico de las geofallas como para medir sus movimientos o sea controlarlos. Es posible que en los principales países del mundo tal cosa pueda realizarse; pero no en aquellos donde recientemente se ha constatado sin medirse movimientos de los bordes de una geofalla. Tal cosa se dificulta cuando se trata sobretodo de fallas con longitud de decenas de kilómetros en que las nivelaciones y triangulaciones de precisión han de realizarse en tiempos distanciados durante los cuales es muy posible que los bench marks cambien de posición con el crecimiento vegetativo tan fecundo de la época actual y sus laboreos.

Zusammenfassung

Es ist erwiesen, daß es aktive Störungen gibt, die Relativverschiebungen entlang ihrer Bruchkante zeigen. Aber die räumlichen Ausmaße sind unbekannt wie in bestimmten Fällen, wo kein Zusammenhang zwischen oberflächlichen Bewegungen, die nicht durch Gebirgsbildung hervorgerufen worden sind, und tieferliegenden Bewegungen, vorhanden ist.

Andererseits ist die Messung von Verschiebungen entlang der Störungszonen eine Angelegenheit, bei der die Wünsche mit den Möglichkeiten wegen der fehlenden wirtschaftlichen und technischen Hilfsmittel in den Gebieten, in denen Erdbeben mit sichtbaren Verschiebungen der Erdkruste verbunden sind, nicht übereinstimmen.

Die Erdteile, in denen Erdbebenkatastrophen in den letzten Jahren stattgefunden haben, sind sogenannte unterentwickelte Länder, in denen die Messung von Krustenbewegungen sehr schwierig ist, weil es kein technisch versiertes Personal für das Grundlagenstudium der geologischen Phänomene der Geostörung und ihrer geodätischen Kontrolle gibt. Es ist möglich, diese Aufgabe in den meisten Ländern der Erde durchzuführen, aber nicht dort, wo in der letzten Zeit Bewegungen stattgefunden haben, aber keine Messungen der Geostörung erfolgt sind. Die Schwierigkeiten wachsen bei Störungen mit linearen Ausdehnungen von 10 km und mehr, wenn Nivellements und Triangulationen nur in großen Zeitintervallen wiederholt wurden, währenddessen die Fixpunkte ihre Lage infolge Pflanzenwachstum und Ackerbau in der gegenwärtigen Periode ändern konnten.

Condiciones del pronóstico sísmico para que sea eficaz

von

J. A. B r o g g i (Lima)

S u m m a r y

One of the most important tasks of our time is that of being able to foresee a destructive earthquake in order to avoid the catastrophes resulting from it. We are guided by previous minor earthquakes (Nanney) and changes of position of definite sectors of the earth's crust such as can be determined by extensometers, e.g. the extensometer designed by Benioff, or by horizontal pendulums. These instruments are very sensitive and are used in a depth of about 100 m in underground working (Nishimura), i.e. in fresh and solid rocks. Finally we are guided by the study of possible periodicities (Asada).

The results obtained till today render it possible to recognize the existence of such premonitory symptoms; nevertheless more observations are necessary to establish the precise relation between the records of the instruments and the time of the occurrence of earthquakes particularly as to their magnitude, intensity and the area affected by them.

At any rate an anti-seismic construction is to be recommended, as there will be no time for an opportune and sure forecast which will only be of any value if given at the right time with the precise fixation of the area affected. Even if a high percentage of personal fatalities could be eliminated, the damages caused in buildings would be inevitable by lack of time. Though an early warning is possible, the communications, however, are so defective in many cases that a great number of localities could not be protected from these fatalities. Our littoral reveals its seismic and catastrophic past but we don't know if this past began yesterday or a thousand years ago.

It is necessary to give the forecast in time, precisely, and with a sufficiently wide propagation that it may be useful and reliable - apart from being sure in the case of a destructive earthquake - in order to be effective for as large an area as possible.

R e s u m e n

Una de las mayores actividades de hoy, es la de poder prever un sismo destructivo para evitar las calamidades consiguientes, guiándose de microsismos previos (Nanney), cambios de posición de determinados sectores de la Costra terrestre los que podrían ser detectados por tensómetros como el de Benioff o inclinómetres (muy sensibles ubicados en labores subterráneas a más o menos 100 metros bajo la superficie (Nishimura) es decir en roca firme y fresca o por último mediante estudio de posibles periodicidades (Asada).

Los resultados obtenidos hasta hoy hacen ver la existencia de tales síntomas premonitores; pero faltan todavía más observaciones para establecer la relación precisa entre los registros de instrumentos y el tiempo de ocurrencia de sismos particularmente con su magnitud, intensidad y área afectada.

De todos modos la construcción antisísmica se impondría, pues no hay tiempo de dar un aviso previo y seguro de catástrofe por lo cual sino se pronostica con precisión con tiempo suficientemente largo y fijación del área por ser afedtada, no sería el pronóstico de valor; pues de otro modo si bien podría eliminarse un fuerte porcentaje de fatalidades personales, los daños a las construcciones serían inevitables por falta de tiempo para darles carácter de seguridad. Con todo, si bien es posible el aviso previo, en muchos casos las comunicaciones son tan defectuosas que este no libraría tampoco un buen número de poblaciones de las fatalidades. Nuestro litoral está revelando su sísmico y catastrófico pasado pero este pasado no se sabe si ocurrió ayer o hace mil años.

Para que un pronóstico sea pues útil y de fé, aparte de ser seguro para el caso de sismo destructivo, debe ser dado con la antelación, precisión y amplia propagación suficiente para que sea eficaz a la mayor área posible.

Zusammenfassung

Eine der wichtigsten Aufgaben besteht heute darin, ein gefährliches Erdbeben voraussehen zu können, um die daraus folgenden Katastrophen zu vermeiden. Man läßt sich dabei leiten von vorhergehenden kleinen Beben (Nanney) und Lageveränderungen bestimmter Sektoren der Erdkruste, wie sie durch Extensometer, z.B. den von Benioff, oder durch Neigungsmesser festgestellt werden können. Es sind sehr empfindliche Geräte, die in etwa 100 m Tiefe bei Arbeiten unter der Oberfläche aufgestellt werden (Nishimura), d.h. in festem und frischem Gestein. Schließlich geht man dabei auch vom Studium möglicher Periodizitäten aus (Aseda).

Die bis jetzt erzielten Ergebnisse lassen das Vorhandensein solcher warnender Anzeichen erkennen; aber es fehlen noch weitere Beobachtungen, damit man den genauen Zusammenhang zwischen den Registrierungen der Instrumente, der Herdzeit und besonders ihrer Magnitude und Intensität sowie dem betroffenen Gebiet feststellen kann.

Unter allen Umständen würde sich eine erdbebensichere Bauweise empfehlen, da keine Zeit zu rechtzeitiger, sicherer Katastrophenwarnung bleibt, die nur von Wert ist, wenn sie lange genug vorher mit einer genauen Angabe des betroffenen Gebietes erfolgt; denn andererseits wären, wenn sich auch ein hoher Prozentsatz von persönlichem Verhängnis vermeiden ließe, die Schäden an Gebäuden aus Zeitmangel unvermeidlich. Obwohl eine Vorwarnung möglich ist, sind doch in vielen Fällen die Verbindungen so mangelhaft, daß die Vorwarnung eine beträchtliche Anzahl von Ortschaften auch nicht vor diesen Katastrophen bewahren würde. Unser Küstengebiet offenbart seine Erdbeben- und Katastrophengeschichte, aber von dieser Geschichte weiß man nicht, ob sie gestern oder vor 1000 Jahren begann.

Damit die Vorhersage von Nutzen und zuverlässig ist, außerdem sicher für den Fall eines destruktiven Bebens, muß sie rechtzeitig, genau und hinreichend weit verbreitet werden, um für ein möglichst großes Gebiet wirksam zu werden.

Submeridionale Tiefenbrüche Mitteleuropas als Gebiete fossiler wie rezenter Krustenbewegungen

von

R. Lauterbach (Leipzig)

Die vom Institut für Geophysikalische Erkundung, Leipzig durchgeführten geophysikalischen und tektonischen Untersuchungen der Struktur der Kruste in Mitteleuropa zeigen folgende Ergebnisse:

1) Der europäische Kontinent wird von einem System Nord-Süd bis NNO - SSW streichender tektonischer Störungszonen durchzogen. Der Bereich rheinischer Tektonik der Mittelmeer-Mjösen-Zone gehört diesem größeren System an.

2) Diese tektonischen Zonen zeigen sich auch im Bild der geophysikalischen (gravimetrischen, magnetischen und seismischen) Regionalanomalien. In tektonischer Beziehung weisen sie ganz verschiedenen Charakter auf (Brüche, Überschiebungen, Förderzonen basischer Schmelze u.a.), sie queren tektonische Großschollen oder bilden gelegentlich auch ihre Grenze.

3) Die submeridionalen tektonischen Linien findet man besonders charakteristisch ausgeprägt, wenn sie als Tiefenbrüche entwickelt sind. Es handelt sich dann um lithosphärische Bruchzonen mit Hochstand der Gabbroschicht, magnetisch und gravimetrisch nachweisbaren basischen Intrusionen und stellenweise Extrusionen. Ein wichtiges Beispiel bildet in Mitteleuropa das Rhone-Rhein-Lineament, das seine tektonisch bedeutsamste, bisher weniger beachtete Fortsetzung nach Norden über die Kraichgausenke, die Rhön, das Thüringer Becken, die Sangenhäuser und Mansfelder Mulde nach dem Prignitz-Massiv und darüber hinaus findet.

4) Diese Lineamente spiegeln tiefkrustale Störungen wider, die sich im geologischen Befund als während der ganzen Erdgeschichte immer wieder in Bewegung begriffen nachweisen lassen. Auch die seismische Aktivität dieser failles lithosphériques (VECCHIA) weist darauf hin, daß sie für r e z e n t e K r u s t e n b e w e g u n g e n von großer Bedeutung sind. Für den Bereich der polnisch-deutschen Ostseeküste kann dies unmittelbar gezeigt werden.

Eine ausführliche Publikation ist im Druck, die einen Teil dieser Fragen behandelt (R. Lauterbach: Rhenotype Strukturen im Bilde geologisch-geophysikalischer Untersuchungsergebnisse Mitteleuropas. Berichte der Geologischen Gesellschaft in der Deutschen Demokratischen Republik 7. Band, 3. Heft, 1962). Den Teilnehmern des Symposiums über Erdkrustenbewegung wird ein Abdruck nach Erscheinen übersandt werden.

Über einige Grundsätze der Bestimmung von Unterschieden geographischer Längen

von

J. Radecki (Warschau)

S u m m a r y

Undoubtedly the results of the work of international services of time and latitudes are used for investigating the horizontal drift of continents and of parts of these continents. Nevertheless determinations of longitudes and perhaps of latitudes made at certain time intervals should be applied, too. The experiences made during numerous determinations of longitudes render it possible to establish differences of longitudes nowadays with an error not exceeding 0,002", which corresponds to a quantity of 60 cm in our latitudes.

During the determinations of longitudes made in the last years a great accuracy of these measurements could be attained by considering this principle and owing to the elimination - as complete as possible - of errors.

The following principal factors seem to play an important part which - when taken into consideration - ensure the possibility of eliminating quite a number of serious errors: the method of directly establishing differences in longitudes based on the measurement of the separating time intervals of the culminations of the same stars, the synchronization of the watches of different observation teams by receiving the same time signals, the use of the same types of instruments, identical operations when making observations and when handling appliances, identical methods of reduction and compensation. It seems to be possible to achieve a greater accuracy of precision of astronomical determinations during a certain time by the same method just mentioned.

Z u s a m m e n f a s s u n g

Zweifellos werden bei Untersuchungen der Horizontalverschiebung von Kontinenten und deren Teilen die Ergebnisse der Arbeit internationaler Zeit- und Breitendienste eingesetzt werden. Es sollten jedoch hierbei auch in bestimmten Zeitabständen durchgeführte Bestimmungen von geographischen Längen sowie eventuell geographischer Breiten zur Anwendung gelangen. Die während zahlreicher Längenbestimmungen gesammelten Erfahrungen ermöglichen es, gegenwärtig Unterschiede geographischer Längen mit einem 0,002" nicht überschreitenden Fehler zu bestimmen, was in unseren Breiten einer Größe von 60 cm entspricht.

Im Verlauf von Bestimmungen geographischer Längen, die während der letzten Jahre angestellt wurden, konnte dank weitgehender Berücksichtigung dieses Grundsatzes und einer möglichst vollständigen Ausschaltung von Fehlern eine hohe Genauigkeit dieser Messungen erzielt werden. Es scheinen folgende wesentliche Faktoren zu bestehen, die bei entsprechender Berücksichtigung die Möglichkeit des Ausschaltens einer ganzen Reihe von gefährlichen Fehlern gewährleisten: die Methode der unmittelbaren Bestimmung von Unterschieden geographischer Längen, die auf der Messung der trennenden Zeitintervalle von Kulminationen derselben Sterne beruht, der Vergleich der Uhren verschiedener Beobachtungskollektive mittels Empfang derselben Zeitzeichen, Einsatz derselben Gerätetypen, identische Verfahrensweise bei Beobachtung und Gerätebedienung, identische Methoden für Reduktion und Ausgleich. Es scheint, daß auf die eben angeführte Weise während einer bestimmten Zeit eine weitere Heraufsetzung der Präzision astronomischer Bestimmungen erreicht werden kann.

Očerk po neotektonike pol'ši (Preliminary outline of neotektonics in Poland)

von

K. S c h o e n e i c h (Warschau)

S u m m a r y

Neotectonics phenomena in Poland reveal as lowerings and uplifts of major areas - mostly structural units. These phenomena depend on tectonics.

The area of Prae - Europe was slightly uplifted from the beginning of Quarternary. The area of Caeno-Europe, in turn, was strongly uplifted. The Meso-Europe was subject of moderate lowering.

The course of neotectnoic phenomana differentiates further within each of the mentioned structural units.

In Northern Poland, from the beginning of Quarternary, synclinoriums relatively deepened against anticlinoriums and synclines - against anticlines.

In Southern Poland the Metacarpathian uplift and Carpathian folds were rised during the same period of time.

The Forecarpathian depression was lowered against the Metacarpathian uplift and Carpathian folds.

The neotectonic phenomena in Poland differ neither quantitatively, nor qualitatively from the tectonic phenomena of Pre-Quaternary time.

Z u s a m m e n f a s s u n g

Neotektonische Erscheinungen zeigen sich in Polen als Senkungen und Hebungen großer, meist strukturell einheitlicher Gebiete. Derartige Erscheinungen sind tektonisch bedingt.

Das Gebiet von Prä-Europa wurde vom Beginn des Quartärs an gering gehoben. Das Gebiet von Cäno-Europa wurde dann wieder kräftig gehoben. Meso-Europa war einer mäßigen Senkung unterworfen.

Der Ablauf der neotektonischen Phänomene unterscheidet sich ferner in jedem der erwähnten strukturell einheitlichen Gebiete.

Vom Beginn des Quartärs an vertieften sich in Nordpolen die Synklinorien im Vergleich zu den Anticlinorien und die Synklinen im Vergleich zu den Antiklinen.

In Südpolen stiegen während desselben Zeitabschnittes die mittelkarpatische Hebung und die karpatischen Falten.

Die vorkarpatischen Depressionen wurden erniedrigt im Vergleich zur mittelkarpatischen Hebung und zu den karpatischen Falten.

Die neotektonischen Phänomene in Polen unterscheiden sich weder quantitativ noch qualitativ von den tektonischen Phänomenen der vorquartären Zeit.

Тектоническая консолидация современной территории Польши окончилась на переломе палеогена и неогена превращением геосинклинали Тетис в карпатскую горную систему. Несмотря на эту консолидацию, на всей территории Польши так в неогене как и в четвертичном периоде в дальнейшем проявлялись тектонические явления с такой же силой, как и в древнейших геологических эпохах.

Вертикальные движения от начала неогена: В структурной провинции Кайно-Европы (Область образовавшаяся главным образом на протяжении кайнозойской эры), в Карпатах, судя по так называемой поверхности наибольших возвышенностей, пост-палеогеновое вынесение равнялось величине от + 200 до + 2200 м (Бескиды, Пенины) или даже + 1200 до + 2200 м (Татры). В Центральном Карпатском синклинории (центральная часть польских Карпат) и в Подгалянском синклинории (западная часть польских Карпат) вынесение было значительно меньше. Пост-палеогеновые смены уровней, образовавшиеся на территории Предкарпатского прогиба (Предкарпатский прогиб и Метакарпатский вал для удобства рассматриваем в пределах Кайно-Европы) фиксирует нам подошва миоценовых морских образований, залегающая ныне на глубине от 0 до - 2500 м н.у.м. Одновременно с погружением Предкарпатского прогиба, Метакарпатский вал, - судя по поверхности наибольших возвышенностей - вынесся на высоту от + 300 до + 450 м. Наиболее сильно выдвинулись те части вала, которые простираются на Центрально-Польском антиклинории, Предсудетской антеклизе, или на Пра-Европе. Более слабые выдвигающие движения отразились на части вала, интерферирующей с синклинорием Щецинско-Могиленско-Меховским и с Краевым синклинорием.

В области структурной провинции Мезо-Европы (без Метакарпатского вала и Предкарпатского прогиба) на пост-палеогеновые смены высот указывает залегание кровли морских отложений олигоцена или подошвы озерных осадков миоцена. В тех частях Мезо-Европы, в которых олигоцен не выступает, за поверхность отнесения принято кровлю старших пород. В этом случае вероятна большая ошибка.

Все эти данные указывают на незначительное пост-палеогеновое выдвижение Центрально-Польского антиклинория и значительно меньшее выдвижение могиленской части Щецинско-Могиленско-Меховского синклинория (разве реминисценция движений, выдвигающих Метакарпатский вал?). Одновременно, Щецинская часть синклинория Щецинско-Могиленско-Меховского и Краевой синклинории, поддавались сильному погружению. Южные части территории Мезо-Европы, поддавались тем тектоническим процессам, которые действовали на Метакарпатский вал, или на Предкарпатский прогиб.

В структурной провинции Палео-Европы, на территории судетских структур, реконструкцию пост-палеогеновых изменений уровней проведено путем реконструкции поверхности наибольших возвышенностей, которая в начале неогена была вынесена не более, чем несколько сот метров н.у.м. Дифференциация высот этой поверхности хотя в части может быть следствием разной отпорности на денудационные процессы - указывает на значительную вертикальную подвижность блоков, образующих Судетские структуры. Кажется, что наиболее сильно были выдвинуты батолит Карконоши, блоки гор Сових и Снежника. Относительному обнижению, в сравнении с соседними структурами, были подданы Грабен Нысы Клодзской и Еленегорская котловина. В общем Судетские структуры поднеслись в плоскости большого берегового сброса от 200 до 600 м, в отношении к Предсудетском антеклизе. В то время Предсудетская антеклиза погрузилась приблизительно до - 250 м н.у.м., о чем говорит залегание подошвы озерно болотных образований миоцена. Наиболее сильному погружению поддавалась при этом так называемая депрессия Одры - район отрицательных гравиметрических аномалий с неясной связью с тектоникой. Только юго-восточные части Предсудетской антеклизы принадлежащие в наложенным структурным единицам выдвинулись или погрузились, согласно с направлением движений, творящих эти структуры.

В структурной провинций Пра-Европы, - судя по глубине залегания подошвы озерно-болотных образований миоцена - наиболее поднеслась Мазурская антеклиза. Подлясская синеклиза не поддавалась вертикальным движениям, удерживаясь на прежнем уровне, сближением к уровню моря. О движении синеклизы Прибалтийской и антеклизы Лебы трудно сказать что либо определенного, потому что на переломе палеогена и неогена поверхность земли в этом районе была снесена денудацией. Быть может, что обнижение Прибалтийской синеклизы равняется около - 100 м. Наиболее южная часть Пра-Европы значительно выдвинулась как часть Метакарпатского вала.

В е р т и к а л ь н ы е д в и ж е н и я о т н а ч а л а ч е т в е р т и ч н о г о п е р и о д а

В области структурной провинции Кайно-Европы, карту смен уровней составлено опираясь на публикациях базирующих на подробном анализе морфологии речных долин и флювио-гляциальных наносов. Движения четвертичного периода в Карпатах тесно отражают план постпалеогеновых движений. Бескиды и Пенины на протяжении четвертичного периода выдвинулись от + 50 до + 100 м, Татры не более + 100 м. Не выключается, что по мере дальнейших исследований, принятые величины окажутся заниженными. Величины вертикальных движений в Предкарпатском прогибе и на Метакарпатском вале определено путем анализа глубины залегания дон раннечетвертичных долин. Опираясь на этой основе, принимаем, что уровни раннечетвертичной поверхности Предкарпатского прогиба погрузились на глубину - 100 до - 150 метров, а Метакарпатского вала немного выдвинулись. Кажется, что сильнее, в пределах выше + 50 м, была выдвинута восточная часть вала. Выдвижение западной части было, вероятно, незначительное.

Провинция Мезо-Европы, что следует из анализа глубины залегания дна раннечетвертичных долин, погрузилась максимально до - 200 м. Эти обнижающие движения были более сильные в обоих синклинориях, менее сильнее - на Центрально-Польском антиклинории. Они верно отражают план вертикальных движений от начала неогена. Южные части территории Мезо-Европы участвовали в движениях, выдвигающих Метакарпатский вал и погружающих Предкарпатский прогиб.

В структурной провинции Палео-Европы, Судетские структуры на протяжении четвертичного периода незначительно выдвинулись. В то же время, Предсудетская антеклиза, как вычислено на основе анализа глубины залегания раннечетвертичных речных долин, - довольно сильно погрузилась. Наиболее сильно погрузилась гравиметрическая депрессия Одры. Юговосточная часть Предсудетской антеклизы была выдвинута в пределах Метакарпатского вала.

В провинции Пра-Европы, поддавалась небольшому выдвижению (до + 50 м) Мазурская антеклиза. Синеклизы - Прибал тийская и Подлясская погрузились на величину - 200 м.

С о в р е м е н н ы е в е р т и к а л ь н ы е т е к т о н и ч е с к и е д в и ж е н и я

На целой территории Польши движения, измеряемые от начала четвертичного периода, являются почти непосредственным продолжением движений, измеряемых от начала неогена. При этом, и первые и другие, приспосабливаются к первоначальному тектоническому плану, который в нескольких районах начал формироваться в мезозое, или даже раньше. Этого плана придерживаются тоже современные движения поверхности земли. Современные движения являются, таким образом, продолжением древнейших тектонических движений.

В структурной провинции Кайно-Европы, на территории карпатских структур, нашлось несколько кратких линий повторной нивелировки. Из них следует, что карпатские структуры поддаются слабому выдвигающему движению, порядка + 0,5 мм на год. Центральный Карпатский синклинорий, выявленный одной линией повторной нивелировки, погружается со скоростью порядка - 0,5 мм в год. Быть может, что также, как в прошедших эпохах погружается и Подгалянский синклинории. Эта гипотеза однако требует подтверждения повторной нивелировкой. Также требует подтверждения область Татр, которые должны выдвигаться несколько большей скоростью от прилегающих карпатских структур. Предкарпатский прогиб был подтвержден только в его южной части с помощью двух кратких линий повторной нивелировки. В той зоне прогиба кажется нет никаких движений. В более центральных частях возможны погружающие движения прогиба. На Метакарпатском вале, почти все линии повторной нивелировки, указывают на существование здесь выдвигающих движений. Выделяются при этом два центра наиболее сильных движений; на востоке и на западе вала. В восточной части вала одной линией повторной нивелировки были выявлены движения, погружающие синеклизу Угруска.

Структурная провинция Мезо-Европы, довольно хорошо оконтурена линиями повторной нивелировки и долголетними померами мареографов. Центрально Польский антиклинорий выдвигается незначительно в отношении к уровню моря, а сильно выдвигается в отношении к двум соседним синклинориям. Щецинско-Могиленско-Меховский синклинорий и Краевой синклинорий погружаются со скоростью около - I до - 2 мм/год. Данные с повторной нивелировки почти идентичные с данными отчетов мареографов. Причина лежит в том, что Торунь, к которому навязаны линии повторной нивелировки, лежащий на переходе от погружающегося Краевого синклинория и выдвигающегося антиклинория, почти не поддается вертикальным тектоническим движениям, удерживаясь на одинаковом уровне.

В структурной провинции Палео-Европы, две краткие линии повторной нивелировки секущие судетские структуры, говорят о стабильности, или же о незначительных выдвигающих движениях этой территории. Одновременно, Предсудетская антеклиза в части находящейся вне Метакарпатского вала, погружается.

На территории Пра-Европы, судя по данным повторной нивелировки и многолетним измерениям мареографов, антеклиза Мазурская и антеклиза Влодавы выдвигаются со скоростью выше + I мм в год; Подлясская синеклиза и Прибалтийская синеклиза в место того незначительно погружаются.

Современные вертикальные движения углубляют неровности между отдельными структурными единицами и между синклиналями и антиклиналями, в пределах этих же структурных единиц. Можно их принять за вертикальную составляющую складкообразных орогенических движений. Несомненно, возможно тоже померить горизонтальную составляющую наблюдаемых тектонических движений. На территории карпатских структур например, которые на протяжении неполных 25 миллионов лет надвинулись несколько десятков километров на свое предгорье, - скорость горизонтального движения не может быть меньше чем I мм на год.

Одновременно со складкообразными современными движениями, существуют современные вертикальные движения в плоскостях существующих сбросов. Эти сбросы имеют региональный характер и отражаются иногда в морфологии района: например береговой сброс в Судетах, или же сброс, отделяющий восточную часть Метакарпатского вала от Предкарпатского прогиба.

В ы в о д ы: Территория Польши поддается теперь выдвигающим и погружающим движениям, тесно связанным с тектоникой. Выдвигаются карпатские структуры, а также, генетически связанный с Карпатами-Метакарпатский вал, наложенная структурная единица. Погружается Предкарпатский прогиб - другая наложенная структурная единица. На остальной территорий Польши, лежащей вне зоны интенсивных тектонических явлений, заходящих в связи с карпатским складкообразованием (на север от Метакарпатского вала) выдвигается антеклиза Лебы, Мазурская антеклиза, Центрально Польский антиклинорий и Судеты. Прибалтийская синеклиза, Подлясская синеклиза, Краевой синклинорий, Щецинско-Могиленский синклинорий и Предсудетская антеклиза - погружаются. В пределах Метакарпатского вала погружается тоже синеклиза Угруска.

Этот общий фон выдвигающих или погружающих движений, дифференцирует еще, в зависимости от размещения элементарных тектонических структур - синклиналей и антиклиналей.

Выявлены также вертикальные движения в плоскостях уже существующих сбросов.

Все современные тектонические движения можно проследить от начала неогена. Ритм их пробега является сложным, изменяющимся по силе, а временно и в направлении

Литература из неотектоники Польши

Boll, E. (1851): Erdbeben in Pommern. Archiv. d. Freund. d. Naturgesch. in Mecklenburg

Deecke, W. (1905): Das Skandinavische Erdbeben vom 23. Oktober 1904 und seine Wirkungen in den Südbaltischen Ländern. Jhrb. Geogr. Gesell., Greifswald

Halicki, B. (1930): Dyluwialne zlodowacenie północnych stoków Tatr. Sprawozdania PIG. t. 5, Warszawa

Halicki, B.; Olczak, T. (1953): Zlodowacenia czwartorzędu a anomalie grawimetryczne na niżu europejskim. Acta Geologica Polonica, z 3, Warszawa

Jahn, A. (1955): Dolina Kłodnicy i stratygrafia utworów plejstoceńskich pod Gliwicami. Biul. Inst. Geol. nr 97, Warszawa

Jahn, A. (1956): Wyżyna Lubelska. Warszawa

Janczewski, W. (1932): Ruchy sejsmiczne zauważone w Polsce w lutym 1932 roku. Pos. Nauk. PIG, nr 33, Warszawa

Janczewski, E. (1948): Zagadnienia wstrząsów podziemnych w Polskim Zagłębiu Węglowym i metody badania tych zjawisk. Biul. nr 1, Śląskiej Stacji Geofizycznej za rok 1948, Warszawa 1950

Janczewski, E. (1956): Trzęsienia ziemi na Górnym Śląsku. Archiwum Górnictwa, t. I, z. 4, Kraków

Jarmułowicz-Łozinska, H. (1961): Czwartorzędowe ruchy tektoniczne w Lesie Czyżowskim w północno wschodniej części Gór Świętokrzyskich. Biul. Geol. Uniw. Warsz. t. I, cz. 1, Warszawa

Kotański, Z.J. (1956): Budowa Geologiczna zachodniego brzegu Żuław. Biul. PIG nr 100, Warszawa

Kotański, Z.J. (1958): Preglacjalne i interglacjalne osady w Tatrach. Acta Geologica Polonica, vol. III, z. 2, Warszawa

Lencewicz, S. (1926): Czwartorzędowe ruchy epirogeniczne i zmiany sieci rzecznej w Polsce środkowej Przegląd Geograficzny, t. II, Warszawa

Laska, W. (1902): Die Erdbeben Polens, Wien

Laska, W. (1905): Jahresber. d. Geodynam. Observat. in Lemberg f. d. Jahr 1903 nebst Nachtr. d. Erdbeben Polens. Mitt. d. Erdbeben-Komiss. d.K. Akad. d.Wiss. Wien

Lewiński, J. (1924): Zaburzenia czwartorzędowe i morena dolinowa w pradolinie Wisły pod Włocławkiem. Spraw. PIG, t. II, Warszawa

Limanowski, M. (1922): O znaczeniu iłów wstęgowych Chełmna dla stratygrafii dyluwium Pomorza. Spraw. PIG, t. I, z. 4-6, Warszawa

Łoziński, W. (1921): Wpływy tektoniczne w rozwoju rzek Karpat fliszowych. Kosmos, z. 25, Lwów

Niemczyk, O. (1923): Die tektonische Absenkung des Beuthener Erz- und Steinkohlenbeckens und ihre Bedeutung für die Beurteilung von Bergschäden. Glückauf, Jahrg. 59, Essen

Niewiarowski, J.; Wyrzykowski, T. (1961): Wyznaczanie współczesnych pionowych ruchów skorupy ziemskiej na obszarze Polski przez porównanie wyników powtarzanych niwelacji precyzyjnych, Prace IGiK, z. 1 (17), t. VIII, Warszawa

Opolski, Z. (1930): Najmłodsze ruchy tangencjalne i małopromienne w Polskich Karpatach. Zbiór prac poświęcony Eugeniuszowi Romerowi pod redakcją H. Arctowskiego, Warszawa

Pawłowski, St. (1955): Przyczynek do poznania ruchów pionowych skorupy ziemskiej w Polsce. Przegląd Geologiczny z. 4, Warszawa

Pokorny, J. (1957): Współczesne ruchy pionowe skorupy ziemskiej w Europie. Czasopismo Geograficzne, t. XXVIII, z. 3 - 4, Warszawa

Premik, J. (1924): O zastoisku Widawskim. Spraw. PIG, t. II, z. 3-4, Warszawa

Romer, E. (1907): Kilka spostrzeżeń i wniosków nad utworami lodowcowymi między Przemyślem a Dobromilem, Kosmos, z. 32, Lwów

Romer, E. (1906): Kilka przyczynków do historii doliny Dniestru. Kosmos, z. 31, Lwów

Romer, E. (1929): Tatrzańska epoka lodowa. Prace Geograficzne t. 11, Lwów

Rosa, B. (1958): O postglacjalnej transgresji Bałtyku na polskim wybrzeżu. Czasopismo Geograficzne, t. 29, z. 3, Warszawa

Rosa, B. (1957): The Littorina Transgression and actual vertical movements of the earth's crust on the Polish coast. Résumés des communications Madrid-Barcelona, 1957, INQUA V Congrés

Rudnicki, J. (1958): Geneza jaskiń systemu Lodowego Źródła i ich związek z rozwojem Doliny Kościeliskiej . Acta Geologica Polonica, vol. VIII, z. 2

Sawicki, L. (1909): O młodszych ruchach górotwórczych w Karpatach Kosmos, t. 34, z. 5-6, Lwów

Samsonowicz, J. (1954): Sur l'âge de formation des dômes salifères sur l'anticlinorium de Couyavie. Bull. Acad. Pol. Sc. U III, vol. 2, nr 2, Warszawa

Schoeneich, K. (1961): Powstanie i rozwój dorzecza Wisły a tektonika północnej Polski. Zeszyty Naukowe Politechniki Szczecińskiej "Budownictwo", nr XX, z. 3, Szczecin

Schoeneich, K. (1961): Vertical tectonic movements on the North-West Poland area. Referat wygłoszony na VI Miedzynarodowym Kongresie INQUA w Warszawie

Schoeneich, K. (1962): Rozwój podwarszawskiego odcinka doliny Wisły w holocenie. Przegląd Geologiczny z 4, Warszawa

Schoeneich, K. (w druku): Żywe procesy tektoniczne w północnozachodniej Polsce, Poznań

Stahl, A. (1932): Das Relief des oberschlesischen Steinkohlen-Gebirges im Lichte der Paläogeographie. Jhrb. d. pr. geol. L.A., Berlin

Stenz, E. (1948): O domniemanej zależności wstrząsów górniczych od zmian ciśnienia atmosferycznego. Biul. nr 1 Sląskiej Stacji Geofizycznej za rok 1948, Warszawa 1950

Świderski, B. (1930): O młodych ruchach tektonicznych, erozji i denudacji Karpat. Rocznik P.T. Geol. t. 8, z. 1, Kraków

Wójcik, Z.; Zwoliński, S. (1959): Młode przesunięcia tektoniczne w jaskiniach tatrzańskich. Acta Geologica Polonica vol. IX, z. 2, Warszawa

Wyrzykowski, T. (1959): Zagadnienie współczesnych ruchów pionowych skorupy ziemskiej na obszarze Polski i ich badanie metodą powtarzalnej niwelacji precyzyjnej. Biul. IGiK, z. 2

Zierhoffer, A. (1925): Zagadnienie powierzchni poddyluwialnej na ziemiach Polski. Pokłosie Geograficzne, Lwów

Рис. I

СТРАТИГРАФИЧЕСКАЯ КАРТА

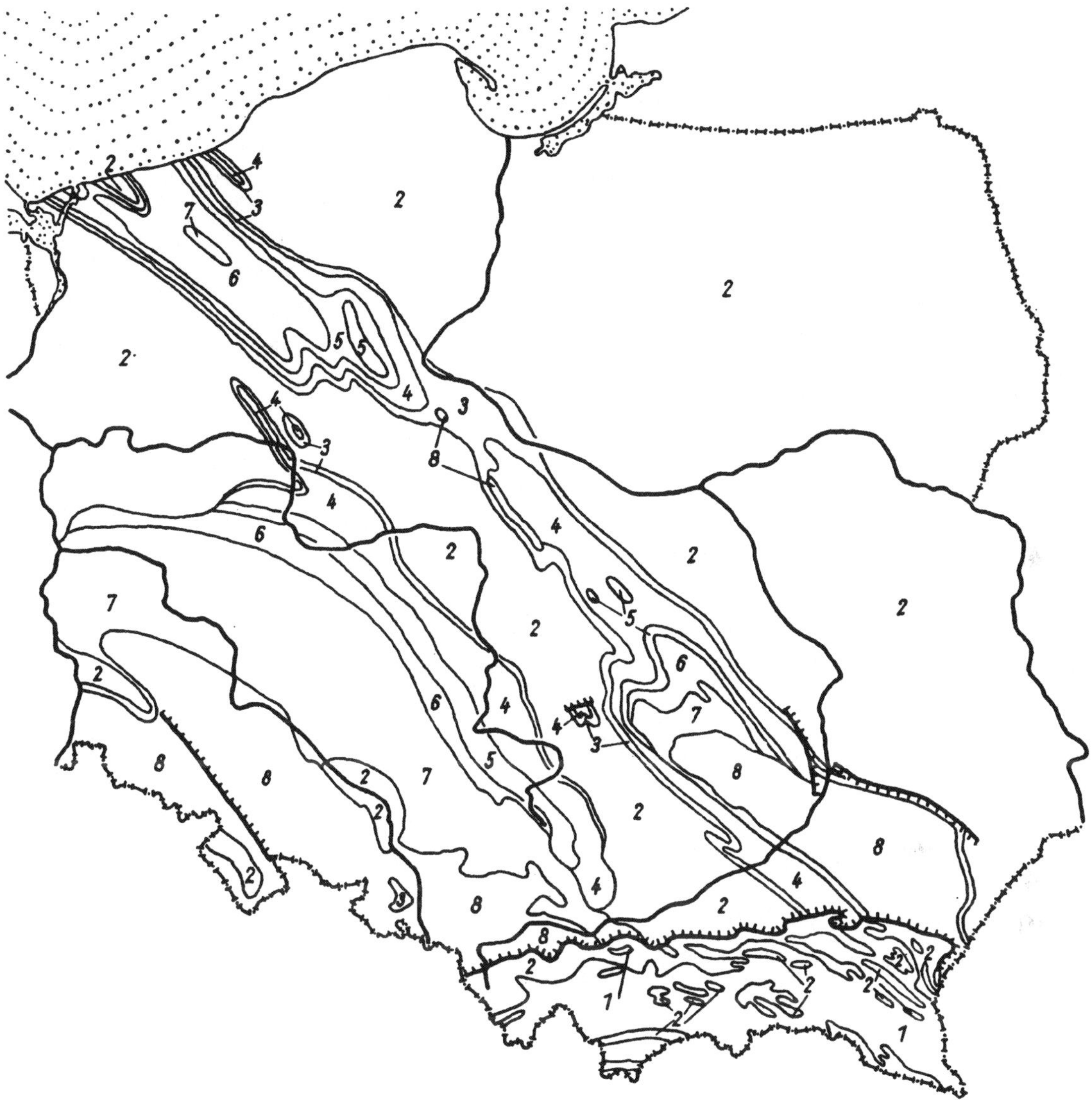

Обозначения

I. Палеоген (оставленный только в Карпатах)
2. Верхний мел
3. Нижний мел
4. Верхняя юра
5. Средняя юра
6. Нижняя юра
7. Триас
8. Палеозой и докембрий

разрывные нарушения

СТРУКТУРНАЯ КАРТА

Рис. 2

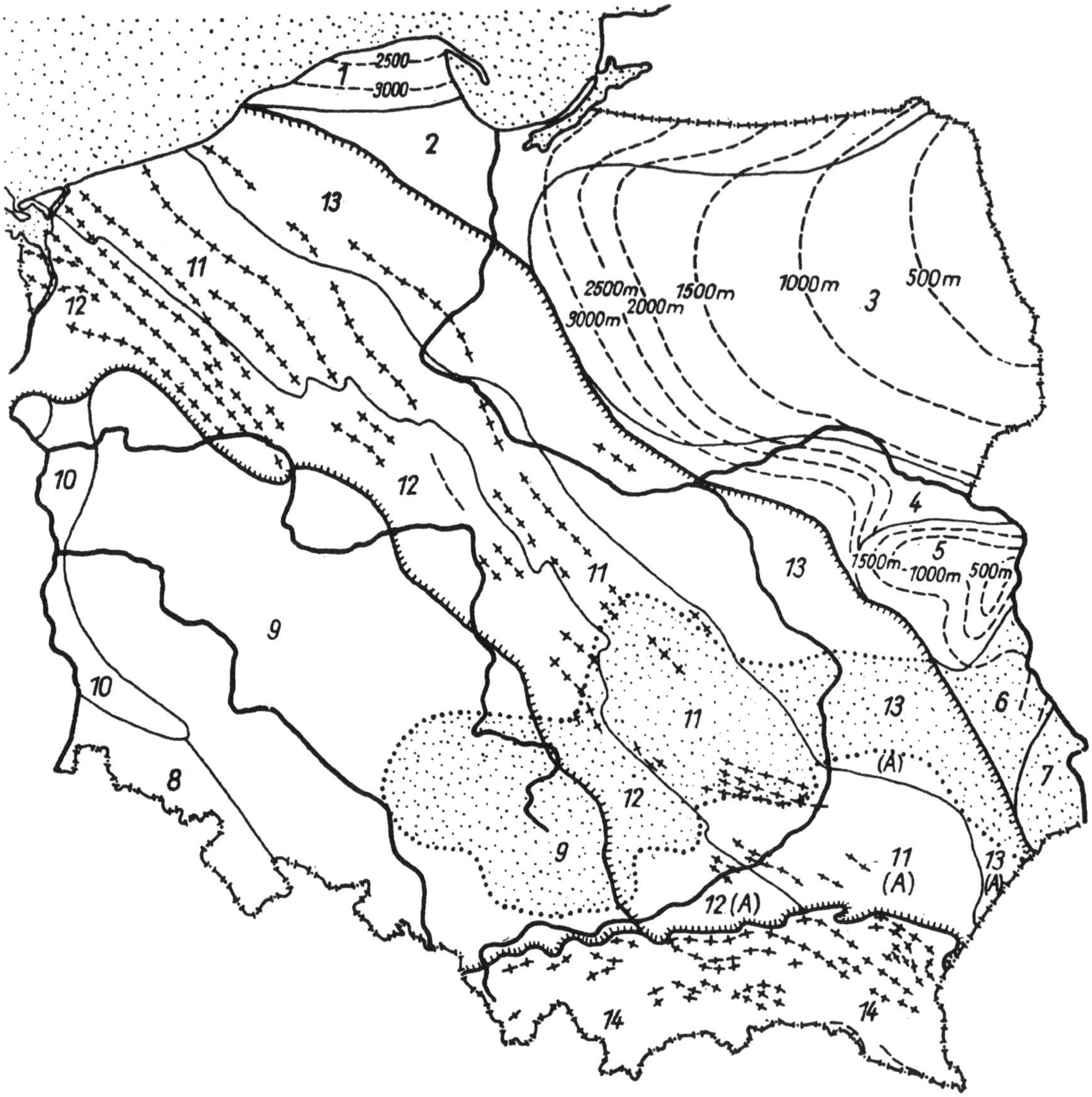

Обозначения

Провинция Пра-Европы

I. Антеклиза Лебы
2. Прибалтийская синекл.
3. Мазурская антеклиза
4. Подлясская синеклиза
5. Антеклиза Влодавы
6. Синеклиза Угруска
7. Антеклиза Дубенки

Провинция Палео-Европы

8. Судетские структуры
9. Предсудетская антеклизе
IO. Лужицкая синеклиза

Провинция Мезо-Европы

II. Центр. - Польский антиклин.
I2. Щецинско-Могиленско-Меховский синклинорий
I3. Краевой синклинорий

Провинция Кайно-Европы

I4. Карпатские структуры
(А) Предкарпатский прогиб
Метакарпатский вал

границы структурных провинций
антиклинали
изогипсы поверхности докембрия

КАРТА ИЗМЕНЕНИЯ ВЫСОТ ОТ НАЧАЛА НЕОГЕНА

Рис. 3

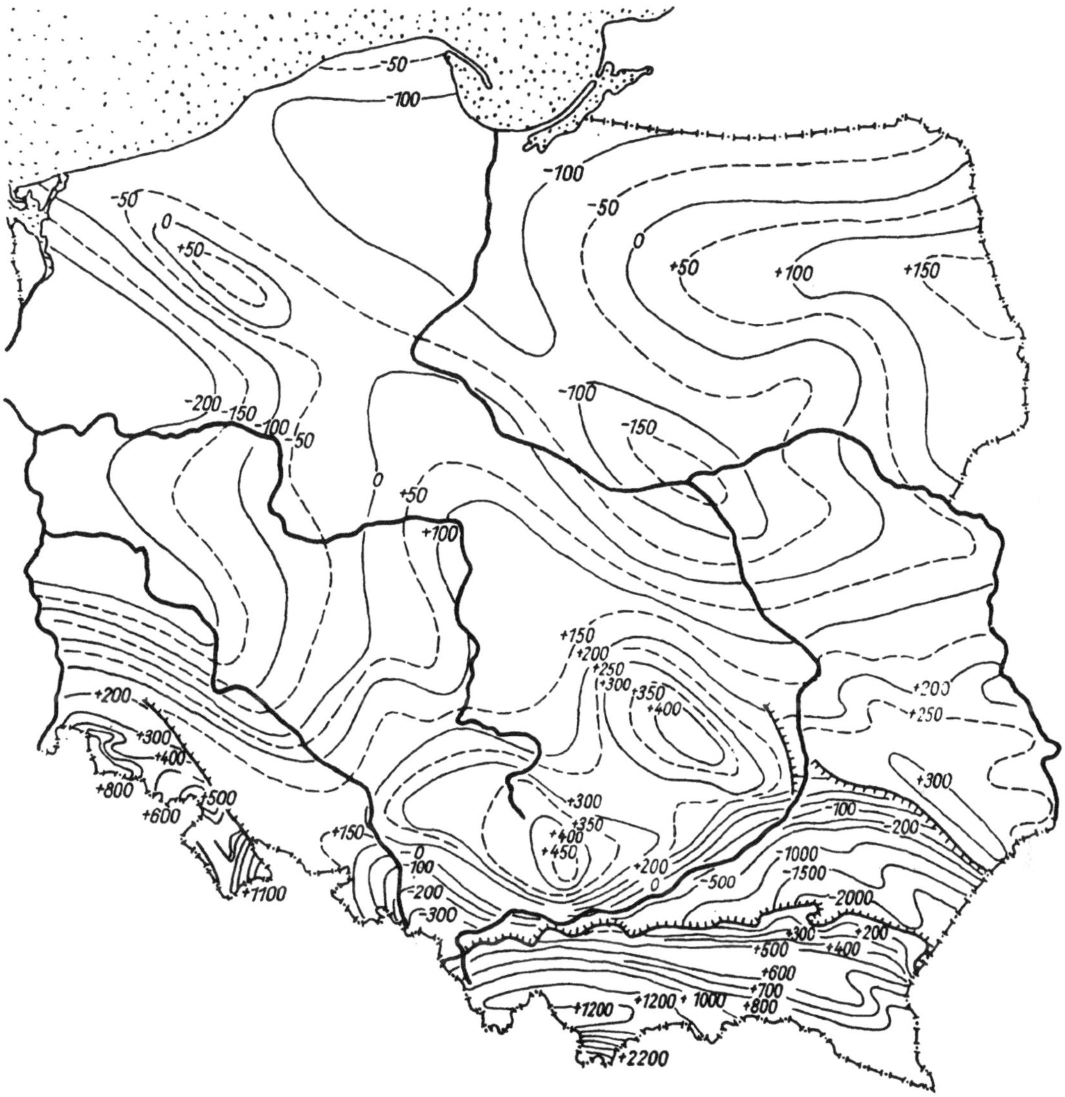

Обозначения

Разрывные нарушения

Изолинии изменения высот (в метрах)

КАРТА ИЗМЕНЕНИЯ ВЫСОТ ОТ НАЧАЛА ЧЕТВЕРТИЧНОГО ПЕРИОДА

Рис. 4

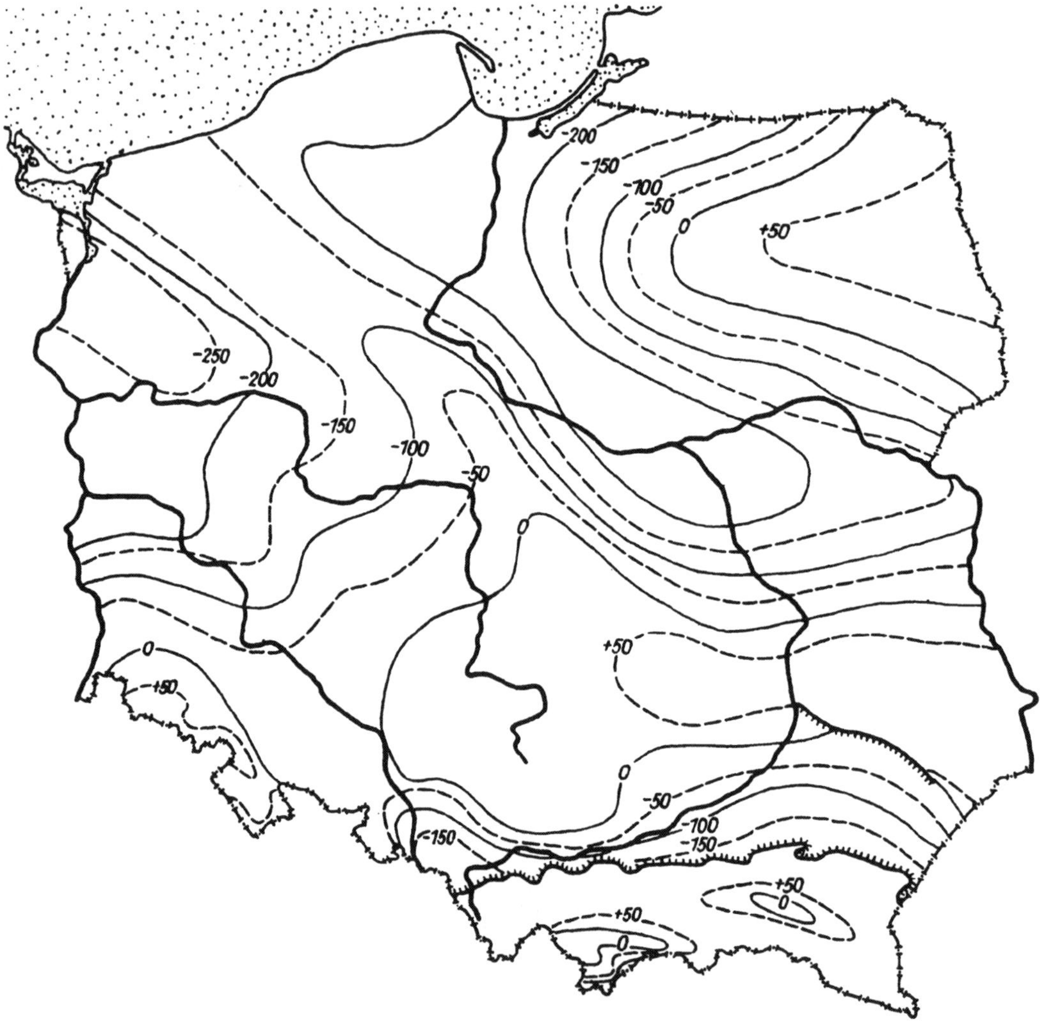

Обозначения

Разрывные нарушения

Изолинии изменения высот (в метрах)

Рис. 5

КАРТА СОВРЕМЕННЫХ ВЕРТИКАЛЬНЫХ ТЕКТОНИЧЕСКИХ ДВИЖЕНИЙ

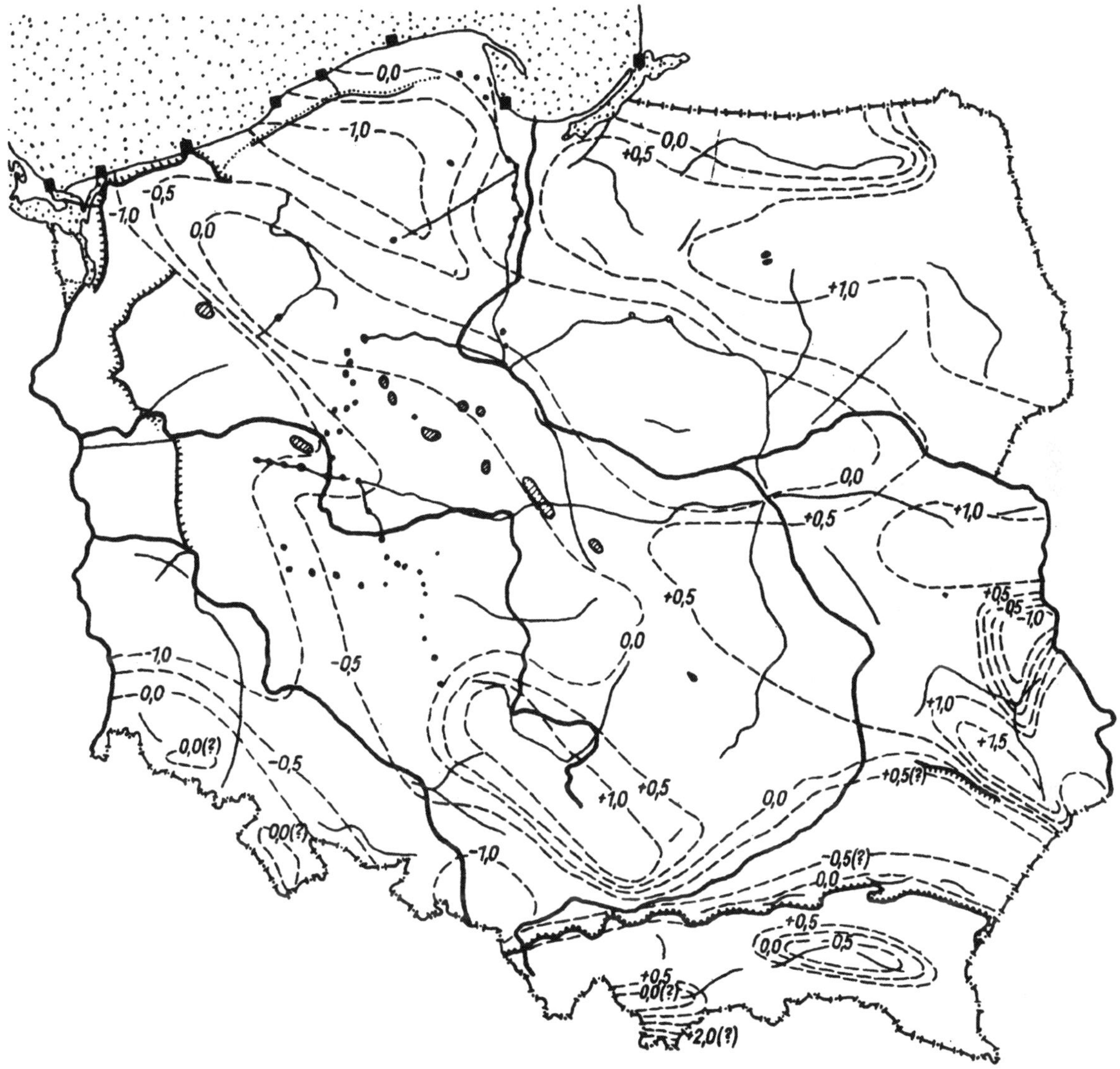

Обозначения

- Живые разрывные нарушения
- Живые соляные штоки
- Изолинии скоростей вертикальных тектонических движений (мм/год)
- Линии повторного нивелирования (по Ст. Павловскому, 1955)
- Линии повторного нивелирования (по И. Невяровскому и Т. Вырзыковскому 1961)
- Уровнемерные наблюдения (по Г. Валентину 1952)
- Линии повторного нивелирования (по автору 1962)

Zur Frage der Erdkrustenbewegungen in dem karpatischen Teil der ČSSR

von

K. Svoboda (Prag)

Erdkrustenbewegungen in dem karpatischen Teil der ČSSR hängen mit dem Bau der Karpaten eng zusammen. Die geologische Karte (Abb. 1) zeigt auf den ersten Blick, daß es in der ČSSR zweierlei voneinander ganz verschiedenartige strukturgeologische Gebiete gibt:

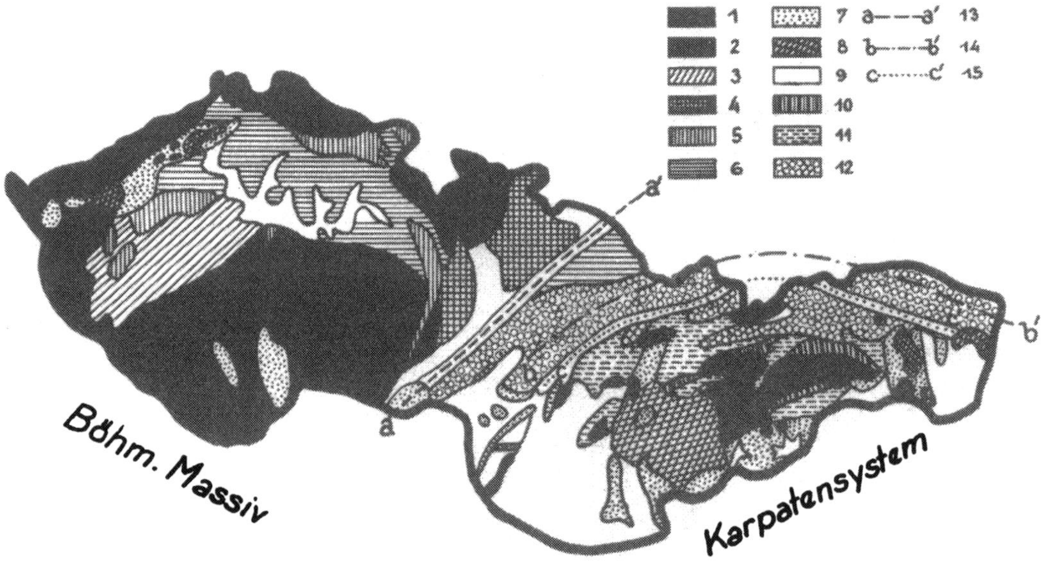

Abb. 1

Geologische Übersichtskarte der ČSSR. (1 - Kristallinikum, 2 - karpatische Kerngebirge, 3 - Barrandien, 4 - unteres Karbon, 5-- oberes Karbon und Perm, 6 = Kreide, 7 - Tertiär, 8 - tertiäre Vulkanite, 9 - Quartär, 10 - Karp. Devon, Karbon, Perm, 11 - Trias, Jura, untere Kreide, 12 - Flysch, 13 - karpatische Vortiefe (Lagerungszone), 14 - äußere Flyschzone, 15 - innere Klippenzone)

Das Böhmische Massiv, das zur varistischen Provinz gehört und die Karpaten, die zu der alpinischen Provinz gehören. Die beiden Gebiete werden durch die Karpatenvortiefe (Lagerungszone) voneinander abgesondert (Abb. 1 a - a').

Das Böhmische Massiv stellt einen Typ alter Gebirge dar, welche in den Hauptlinien konsolidiert sind, mit mäßigen Höhenunterschieden, und das durch niedrige seismische Aktivität mit maximalen Werten von 6° - 7° MCS in den Randteilen Böhmens charakterisiert ist (Abb. 2).

Die Karpaten hingegen sind ein verhältnismäßig junges Gebirge und die gebirgsbildende Wirkung in diesem Gebiete dauert noch an bzw. ist im Erlöschen. Das zeigt sich in den Bewegungen besonders an den Brüchen, die bereits auch ganz geringe junge Schichten zerstören. Dies bestätigen wohl auch zahlreiche karpatische Erdbeben, welche ein Maximum

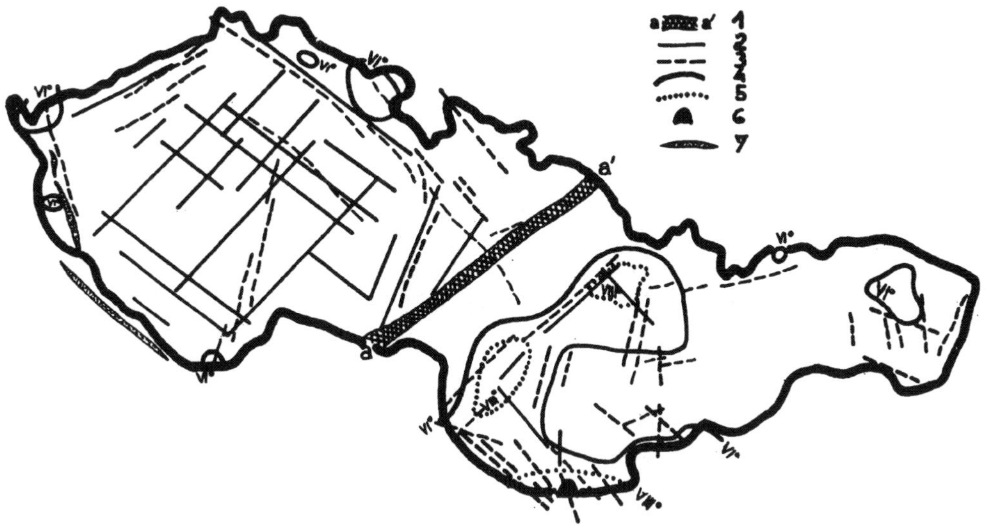

Abb. 2

Tektonische und seismische Karte der ČSSR (nach A. Dvořák 1956). 1 - (a - a') Karpatenvortiefe (Lagerungszone), 2 - statistisch festgestellte beweglichere Zonen nach A. Zátopek, 3 - schematischer Verlauf der jungen tektonischen Linien, 4 - seismische Gebiete ab VI°MCS, 5 - seismische Gebiete ab VIII°MCS, 6 - seismische Gebiete mit X°MCS, 7 - Quarzpfähle

von 8° - 9° MCS erreichen (Bild 2). In der Periode 1923 - 1938 wurden hier sogar 100 makroseismische Erschütterungen wahrgenommen.

Die Karpaten sind zum alpiden polyplikativen Typ zu zählen, d.h. sie weisen ein mehrmals intensiv gefaltetes Deckensystem auf. Sollten wir die Spuren von älteren Orogenen des vormesozoischen Zyklus (z.B. des jungpaläozoischen Faltungszyklus) verloren haben, die manchmal von weiteren Zyklen verdeckt wurden, so handelte es sich hierbei nach A. Andrusov (1960) wesentlich um eine zweifache mächtige Faltung, die von dem Aufkommen der Decken gefolgt wurde. Dies war vor allem die nordvergente vorgosauische Faltung (Voroberkreide) und teilweise die laramische (zwischen Kreide und Tertiär) in dem Gebiet der Westkarpaten und weiter die nordvergente mitteloligocäne Faltung, wahrscheinlich besonders die savsche (an der Grenze zwischen Oligocän und Miocän), im Flyschgebiet und bei dem Außenrand auch Intramiocän.

Das Falten im alpinischen Orogen bildete den Zonenbau der Karpaten, mit Hochgebirgs-Relief und mit dem charakteristischen bogenförmigen Verlauf rund um das große ungarische Becken (Bild 3). Der bogenförmige Verlauf der Karpaten entstand dadurch, daß sich mächtige Sedimente der karpatischen Geosynklinale beim Falten unter der Wirkung gewaltiger gebirgsbildender Drucke, welche in Richtung des Meridians wirkten, sich von Süden nach Norden verschoben, wo sie einerseits an das kompakte Böhmische Massiv und teils an die polnische Plattform anstießen.

Die Gliederung der karpatischen Zonen ist die folgende (Bild 1 - 3):

1. Karpatische Vortiefe mit Produkten der Sedimentation (Lagerungszone).
2. Äußerliche Klippenzone, welche die karpatische Vortiefe in getrennten Schollen von Kalkstein und Dolomiten umgrenzt.
3. Flyschzone, d.i. der Nordflügel der gesamten megastrukturellen Elevation der Westkarpaten, gefaltet im unteren Miocän, mit Sandsteinen und Schiefer der oberen Kreide und des Paläogens, die sich in der Reihe übereinander gefalteter Decken befinden.
4. Die innere Klippenzone, d.i. die labile Zone, deren Breite infolge wiederholter Faltungen des im Ursprung komplizierten tektonischen Baues stark verengt wurde. Die Fal-

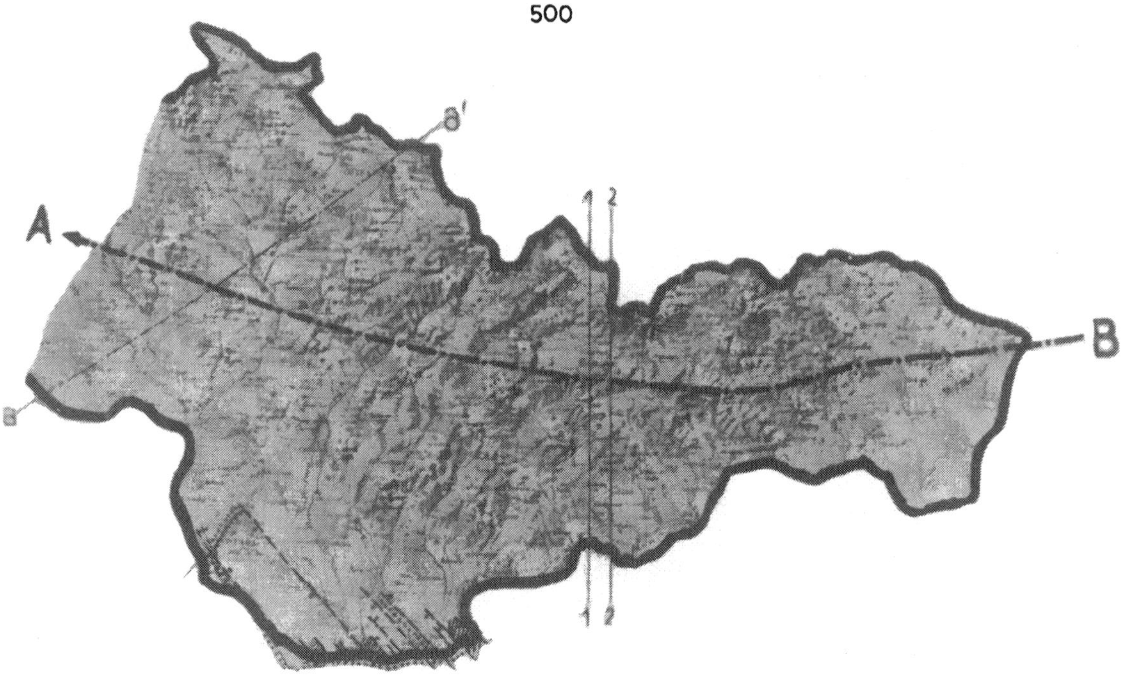

Abb. 3

Morphologisches Bild der slovakischen Karpaten. A - B Profil quer durch die ČSSR (siehe Bild 8), a - a' Karpatenvortiefe, 1 - 1 und 2 - 2 Nord-Süd Schnitte durch die Niedere Tatra (siehe Abb. 4), schräg schraffiert auf dem südl. Rand der Slovakei zwischen Bratislava und Štúrovo Horst-(steigende) Schollen, mit Beistrichen bezeichnet die Senkungsbrüche, voll mit schrägen Beistrichen und Überschiebungen , punktiert die Erdbebenlinien

tung verlief bereits in der Kreide und nach dem Paläogen und hinterließ in dieser Zone vereinzelte getrennte Schollen (Klippen) Kalk- und Dolomitgesteine, die aus den weicheren Sandsteinen und Mergeln emporragten.

5. Die Zentralkarpatenzone, die sich durch gegenseitigen Kontakt einiger tektonischer Einheiten und deren wiederholter tektonischer Superposition auszeichnet: Kerngebirge-Umschlagsserien - überschobene Decken.
   Die Gebirgszonen der Zentralkarpaten bilden sog. Kerngebirge, in deren Mittelpunkt Kristallinikum aus Graniten und stark metamorphierten Schiefer mit Umschlagsserien der erdmittelalterlichen Gesteine (hauptsächlich der Kalksteine und Dolomiten) vorkommt. Im Verlauf des Verfaltens kam es zu horizontalen Überschiebungen großen Umfanges und zur Bildung sog. Decken (Abb. 4). Der Deckenbau hat sich aber nicht nur auf die Schichten der Sedimentgesteine beschränkt, sondern erstreckte sich auch auf die Massen der Tieferuptivgesteine (Granit-Massive) und der kristallinen Schiefergesteine. Diese wurden mit deren sedimentären Umschlag gemeinsam verfaltet, was die hohe Plastizität der Gesteine bestätigt. Wir ersehen dies z.B. aus der Verfaltung des tatranschen Zentral-Granitmassives in der Falte des Rotberges und Giewont-Berges, u.a. aus dem Kern der Niederen Tatra. Gleichzeitig mit dem Ursprung gewaltiger linienförmiger Brüche in Richtung annähernd O-W entstanden ähnliche tiefe und mächtige Bruchzonen transversaler Richtungen N-S (Abb. 2). So tragen die Zentralkarpaten zur Bewegungsaktivität entscheidend bei.
6. Das Spisch-Gemer Erzgebirge ist von abweichender Bauart, mit überwiegend varistischen und kaledonischen - alpinotypisch metamorphierten Strukturen ohne Granitkern.
7. Tertiäre vulkanische Gebirge, welche sich kranzartig vom Kremnitzer-Stiavnitzer-Gebirge über die ungarische Mátra, Buchenberge und Tokayer Gipfel herumziehen und in

der ČSSR mit den Salzgipfeln und Vihorlat fortsetzen. Diese geologische Struktur beteiligt sich wesentlich an der Karpatenbewegungsaktivität, namentlich durch das Abklingen des großen Druckes, der von den subsequenten bis finalen Vulkaniten und deren Pyroklasiken bewirkt wird.

8. Im Süden wurden die Becken mit jungtertiären Sedimenten angefüllt, welche die Mächtigkeit von 1000 - 5000 m erreichten und die bei Schollenbewegungen eine bedeutende Rolle spielen.

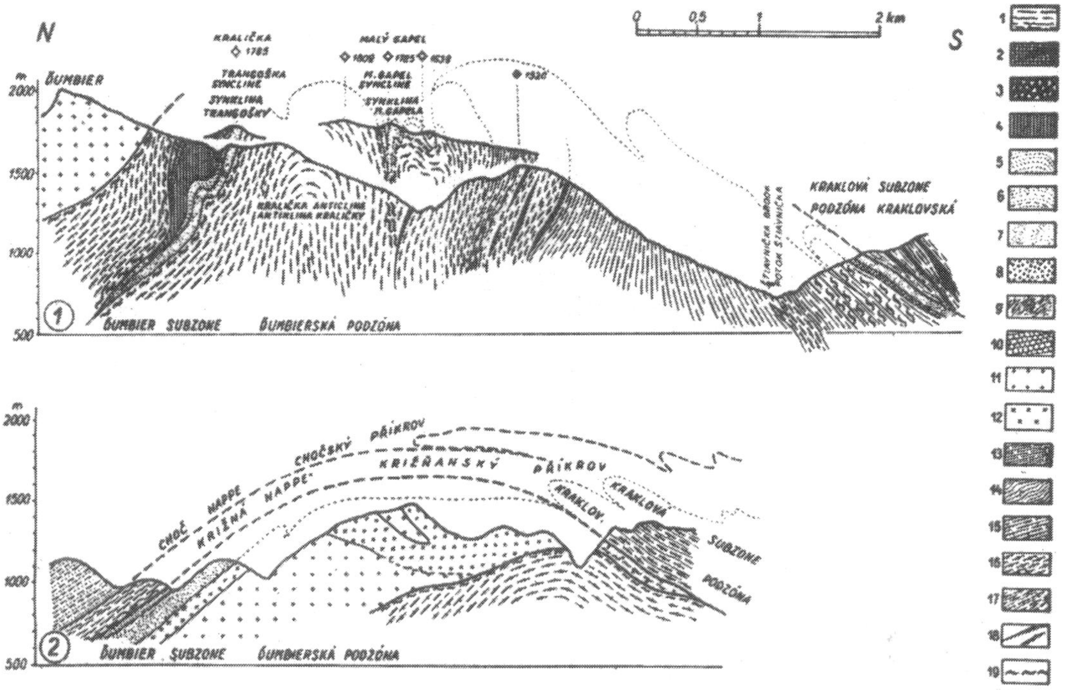

Abb. 4

Nord-Süd Schnitte durch die Niedere Tatra von Ďumbier; Achsenelevation (1) genommen zum Osten bis Achsendepression zwischen Čertovica und Velký Bok (nach V. Zoubek (1951))
1 - Neokom, 2 - obere Trias, 3 - Trias-Dolomite, 4 - mittlere Trias, 5 - Schiefer, Sandsteine, 6 - Werfenschichten, 7 - Quarzit, 8 - basales Quarzkonglomerat (5 bis 8 untere Trias) 9 - perm. Schiefer, 10 - perm. Arkosen, 11 - Ďumbiergranit: Quarzdiorit (Tatrit), 12 - Ďumbiergranit, 13 - Gneis und Migmatit, 14 - Gneis und diaphtor. Migmatit, 15 - Klammer, 16 - Granit-Gneis (Ďumbier Subzone), 17 - Granit-Gneis (Kraklov Subzone), 18 - Amphibolit, 19 - Mylonit

Die känozoische Revolution, deren Einflüsse im Karpatensystem solche merkbaren Spuren zurückließen, ist wahrscheinlich noch nicht ganz beendet. Nach G. Gamow (1941) dauert der zweite Ausbruch dieser Revolution, welcher im Miocän begann und in dem nachfolgenden Pliocän etwas mäßiger fortfuhr, wobei die alten Brüche, wie auch Dislokationen belebt wurden, bis heute an. Es ist sogar wahrscheinlich, daß die gegenwärtige, verhältnismäßig ruhige Periode, nur eine bloße Unterbrechung vor einer neuen Explosion umstürzender Wirkung darstellt. Die maximalen spezifischen Drucke, welche in der karpatischen Geosynklinale wirkten, waren größenordnungsmäßig hunderte kg/cm^2 größer als dies heute der Fall ist, so daß sie gebirgsbildende Werte erreichten. Die gewaltigen Auswirkungen der känozoischen Revolution milderten die Spannung in der Erdkruste. Die restliche Spannung in der Erdkruste und das Verhalten der verflossenen gebirgsbildenden Prozesse dauern noch an und zeigen sich, wie bereits erwähnt, in den merkbaren Bewegungen auf den aktiven Brüchen. Daraus folgt auch, daß die karpatischen Erdkrustenbewegungen vorwiegend orogenen und tektonischen Charakter haben.

Ein wichtiger Faktor der Bewegungsaktivität in den Karpaten sind die Großfalten (Grundfalten), das sind säkulare Undationen (Auswölbungen) der Erdkrustenteile. Nach R. Kettner (1955) entstehen sie in dem obersten Teil der Erdkruste - scheinbar durch den Einfluß bedeutender Materialversetzungen in der Tiefe der Erdkruste - gewöhnlich nach dem eigentlichen Prozeß der Faltung. So entstand auch die schlüssige Wölbung der Karpaten in der jüngsten Zeit, wo der Deckenbau bereits vollendet wurde. Bei dem Aufheben der Großfalten-Wölbungen entsteht oft auf ihrer Oberfläche ein bedeutender Zug, welcher die Ursache der sich öffnenden Spalten und damit gleichzeitig der Schollensenkung ist.

Zu weiteren säkularen Bewegungen kommt es durch die Wirkung von vertikalen und horizontalen Kräften, die durch aufgehäufte Massen der karpatischen Falten, Decken und vulkanischen Produkte hervorgerufen werden (sog. Erschwerung).

Die oben angeführten Bewegungen zeigten sich bedeutungsvoll schon bei der Erneuerung des alten österreich-ungarischen Präzisionsnivellements in den Jahren 1920 - 1939. So z.B. führt K. Srba in seinem "Bericht über das Präzisionsnivellement der Strecke Uherské Hradiště - Trenčianská Teplá" aus dem Jahre 1921 den Unterschied der Schlußhöhenmarke bei Trenčianská Teplá von den beiden Nivellements zu -97,1 mm an. Der Höhenunterschied, den das damalige Militär-Geographische Institut bestimmte, differierte im nördlichen Teil des Polygonzuges um 24,7 mm und im südlichen Teil um 144,6 mm. K. Srba führte diese bedeutenden Unterschiede auf Fehler in der ursprünglichen Messung und dem Setzen der Gebäude, in welchen die Höhenmarken bereits stabilisiert wurden, zurück.

Bei der Revision der Grundhöhenmarke Strečno wurde im Jahre 1922 eine Senkung um 78 mm festgestellt.

Gleichzeitig machte O. Matoušek (1926) auch auf die Differenzen der Höhen in der Umgebung vom slovakischen Erzgebirge aufmerksam unter Hinweis auf vertikale Erdkrustenbewegungen.

Zu interessanten Erkenntnissen führte die Vermessung, die im Jahre 1932 in der Slovakei durchgeführt wurde zwecks Verbindung der tschechoslovakischen und ungarischen Präzisionsnivellation. Der maximale Messungsunterschied war bei den tschechoslovakischen und ungarischen Gruppen zwischen zwei Marken in der Entfernung von 1,2 km, lediglich 1,8 mm auf der Strecke über die Donau bei der Markenentfernung von 0,5 km betrug der Unterschied 0,62 mm. Die Ergebnisse für drei gemeinsame Polygonzüge von Längen 235,0; 139,3 und 163,6 km waren + 13,1, - 5,2 und + 6,04 mm. Trotzdem zeigten sich beim Vergleich der neubestimmten Höhen, die aus den Grundhöhenpunkten Strečno und Nadap berechnet wurden, hohe Unterschiede in den Grenzen von - 78,7 bis - 100,3 mm. K. Srba (1935) hatte die Ursache dieser großen Unterschiede einerseits dem Einfluß der Messung und Ausgleichung dem alten Nivellement zugeschrieben, wie auch den Bewegungen der Objekte, auf welchen die Höhenmarken stabilisiert wurden, gab aber auch die Wirkung der Erdkrustenbewegungen unter Berücksichtigung der höheren Seismizität des Gebietes zu.

L. Čepek machte im Jahre 1938 auf die Bewegungen in dem Raum von Komárno aufmerksam. Beim Studium der Position von Beobachtungspunkten in der tektonischen Karte (Abb. 3) dieses Gebietes wurde von ihm wahrgenommen, daß diese auf Horstschollen liegen. Aus diesem Grunde hat er geschlußfolgert, es handle sich hierbei um Variationsbewegungen der Schollen: während sich die Grabenverfälle senken (Senkungsbrüche), heben sich die Horstschollen ihnen gegenüber. Zu diesem Zweck wurde ein Vergleich der genauen Nivellements aus den Jahren 1901 - 1908 - 1931 durchgeführt, aus denen folgte, daß die Variationen für die angeführte Periode die Werte von 3 - 64 mm erreichen und von Zeit zu Zeit das Vorzeichen ändern. Das Forschungsinstitut für Geodäsie, Topographie und Kartographie (FIGTK) hat in diesem Gebiet (1957) eine Kontrollvermessung durchgeführt und stellte in dem Komárno-Gebiet eine Senkung von ca. 20 mm in 5 Jahren fest. Die Kontrollmessungen werden fortgesetzt.

Elevations- und Depressionsstrukturen in dem ostslovakischen neogenen Becken studierte J. Janáček (1962).

Jüngere nachpliocäne Bewegungen im karpatischen Flysch erscheinen überwiegend in den Bruchzonen. Die Funktion dieser Brüche, welche nach T. Buday (1960) in der Zeit des unteren Tortons entstanden sind, zeigt sich an Stellen, wo auf der Grundlage der gravimetrischen und geomagnetischen Messungen die Beziehung dieser Dislokationen zur alten tektonischen Konfiguration der autochthonen Unterlage zum Vorschein kommt.

Neben den bereits angeführten säkularen Bewegungen regionalen Charakters und Schollenbewegungen sind die bedeutenden rezenten karpatischen Bodenbewegungen nicht weniger interessant. Diese haben einen großen Umfang und erscheinen an mehreren Stellen im Karpatengebiet. D. Andrusov und C.M. Paul beschreiben eine ganze Reihe von Rutschungen in Orava mit der durchschnittlichen Länge von 4 - 5 km. Diese Rutschungen sind während einer sehr langen Epoche entstanden, teilweise in Pleistocän, teilweise in Holocän und deren Bewegung dauert auch heute noch an (juvenile Rutschungen). In einigen Abschnitten der Waagtal-Klippenzone, z.B. in der Umgebung von Myjava, erscheinen zahlreiche subfossile Rutschungen senilen Typus.

Ende 1960 - Anfang 1961 kam es zu einer mächtigen Rutschung auf den "Ziegenrücken" bei Handlová, wo 25 Millionen m^3 Erdmasse in Bewegung gerieten (Abb. 5). Infolge dieser

Abb. 5
Rutschung Handlová (Blick auf die Stirn der Bodenlawine)
(Photo UGU)

Rutschung wurden 238 Familien obdachlos, sie setzte das Wasserbasin und die Wasserleitung außer Betrieb, vernichtete einige Verkaufsanlagen, zwei Schulgebäude und verursachte weiteren großen Schaden (Abb. 6). Zwecks Feststellung der Absonderungslinie und für die Sanationsarbeiten wurde das Rutschungsgebiet einigemale nacheinander stereophotogrammetrisch vermessen mit Auswertung auf dem Stereoautograph im Kartenmaßstab 1 : 2500. Diese Vermessung wurde von dem "Wissenschaftlichen Laboratorium für Photogrammetrie" der Slovakischen Technischen Hochschule durchgeführt.

Der Autor dieser Abhandlung machte bereits in den Jahren 1938 - 1939 den Versuch für eine systematische Erforschung der Erdkrustenbewegungen auf dem ganzen Gebiet der ČSSR. Wegen unzureichenden Materials und der Kriegszeit wurde jedoch diese Arbeit unterbrochen. Nach dem Krieg hat FIGTK in Prag die systematische Erforschung der Erdkrustenbewegungen durchgeführt. Als Ausgangsmaterial für diese Forschung der Bewegungsaktivität in der ČSSR lagen nach vorhergehender Ausgleichung die Ergebnisse des Präzisionsnivellements der österreichischen Nivellation aus den Jahren 1873 - 1896 (alte Nivellation) und des genauen tschechoslovakischen Nivellements aus den Jahren 1920 - 1939 (neue Nivellation) zu Grunde.

Abb. 6
Rutschung Handlová – Destruktion des Wohnungsviertels

Durch den Vergleich der identischen Punkte wurden für das ganze tschechoslovakische Gebiet Isobasen konstruiert und gleichzeitig wurde die Zunahme der wahrscheinlichen systematischen Fehler mit Rücksicht auf den Ausgangspunkt Lišov in Südböhmen eingetragen (Abb. 7).

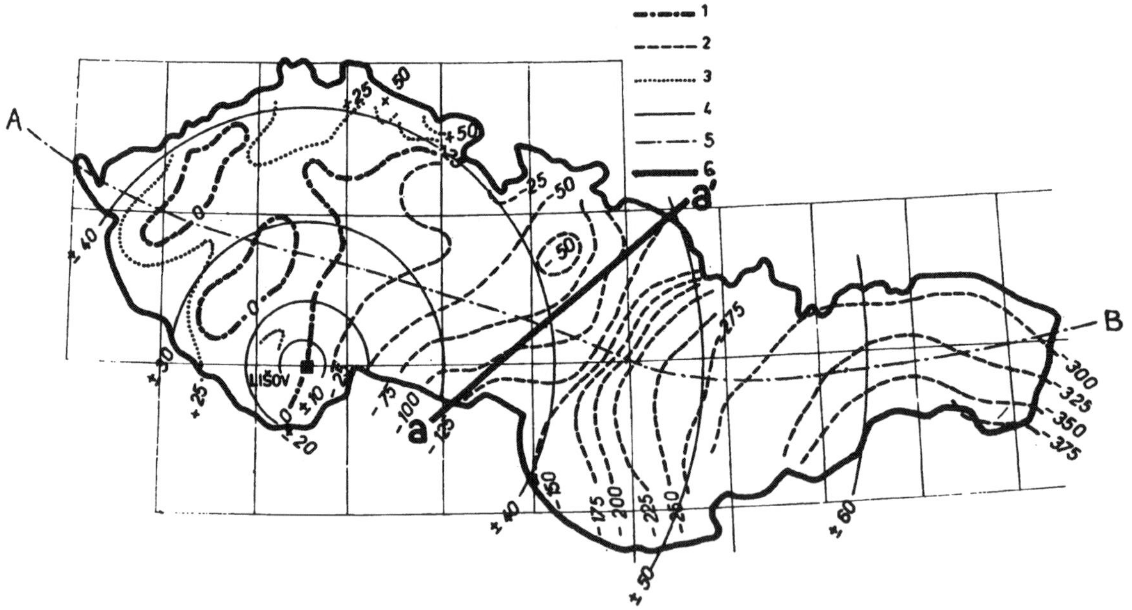

Abb. 7
Isobasenkarte der ČSSR, bezogen auf den Grundhöhenpunkt Lišov (teilweise nach FIGTK)
1 – Null-Isobase, 2 – Isokatabase (Senkungslinie), 3 – Isanabase (Hebungslinie), 4 – wahrscheinliche systematische Fehler vom Ausgangspunkt Lišov, 5 – Querprofil durch die ČSSR (siehe Abb. 8), 6 – Karpatenvortiefe (a – a')

Es ist allerdings festzustellen, daß den Vergleichsmaterialien die Voraussetzungen für die erwogenen wissenschaftlichen Ziele fehlten. So z.B. hatte die alte Nivellation ein Netz von großen Polygonzügen, die die Größe von 500 - 600 km erreichten. Große Flächen wiesen dabei keine Höhenmarken auf, und die Entfernung der Höhenpunkte erreichte 14 km. Namentlich in der Slovakei (Karpatensystem) wurde das Netz ungenügend ausgebaut, in einer Reihe von Nivellementspolygonen kam es zu hohen Unterschieden. Der Aufbau des Netzes wurde in verschiedenen Zeitintervallen und an verschiedenen Stellen des gesamten Gebietes durchgeführt, und die Nivellementszüge wurden fast ausschließlich längs der Eisenbahnlinien geführt. Das neue Nivellement wurde ebenfalls stufenweise in oft voneinander sehr entfernten Intervallen in der Periode von 20 Jahren durchgeführt und überwiegend längs den Landstraßen. Dadurch wurde die Anzahl der identischen Punkte für den Vergleich bedeutend reduziert, und manche von ihnen wurden noch infolge einer langsamen Durchführung des Nivellements mit Fehlern aus den säkularen Erdkrustenbewegungen beeinflußt, so daß die gesamte Plazierung, Dichte und Zuverlässigkeit der identischen Punkte kaum für die reale Konstruktion der Isobasen ausreicht. Dies gilt vor allem für das Karpatengebiet und in diesem namentlich für die Zentralkarpatenzone, mittelslovakisches als auch ostslovakisches vulkanisches Gebiet.

Trotz dieser Mängel stellen die Ergebnisse der Studienarbeiten von FIGTK wenigstens Orientierungs- und methodisches Material für künftige Arbeiten wissenschaftlichen Charakters und für eine vorläufige geologische und geophysikalische Interpretation der Bewegungsaktivität des Karpatensystems dar. Aus der Abb. 7 ersehen wir, daß östlich der karpatischen Vortiefe die Senkungen sehr rasch bedeutende negative Werte erreichen, wobei die Isokatabasen die bogenartige Form haben, die dem Bogenbau des Karpatensystems gleicht. Die folgende Abb. 8 zeigt die Hebungen und Senkungen im Profil quer durch die ČSSR in Beziehung zu morphologischen, geologischen, seismischen und den Schwereverhältnissen. Das morphologische Relief von den Zentralkarpaten ruft in diesem Profil die Vorstellung der Analogie zum Bodenumformen und den Bodendeformationen durch den Einfluß von deren Bau hervor. Unter diesem Gesichtspunkt - rein bodenmechanisch betrachtet - wäre es möglich, einen gewissen Anteil negativer Bewegungen (Senkungen) mit der Umformung der unteren Gesteine (dem Liegenden der Karpatendecken) und mit unterstützenden Drucken, namentlich der Zentralkarpaten und vulkanischen Effusiven zu erklären. Durch den Einfluß solcher Umformungen und Drucke muß es nicht unbedingt nur zu vertikalen Senkungen kommen, sondern eventuell auch zu Neigungen und waagerechten Verschiebungen, welche auch die Ursache für die Richtungsfehler bei Triangulationsmessungen sein können.

Interessant ist auch der Vergleich der geologischen Struktur des Karpatensystems mit der Bouguer-Isanomalenkarte (Abb. 9). In der Karpatenvortiefe bewegen sich die Werte $(g''_0 - \gamma'_0)$ um 0 mGal oder sie erreichen kleine positive Werte; lediglich im Raum Olomouc erlangen sie kleine negative Werte. In den ersten Karpatenzonen (äußere Klippenzone und Flysch) erreichen sie die Größe von - 20 bis - 45 mGal. Gleiche Werte erreichen sie in den Zentralkarpaten. Im Gebiet der Hohen Tatra wachsen sie bis auf - 60 mGal an. Den Ebenen gehören schwache Anomalien, in den Horst-(emporsteigenden) Schollen und tektonischen Elevationen bewegen sie sich zwischen 0 bis - 20 mGal (z.B. im Raum Komárno) und im Gebiet tektonischer Senkungsbrüche sind sie niedrig bis negativ. Auf diesen Umstand machte bereits im Jahre 1938 L. Čepek aufmerksam.

Schlußfolgerungen vom isostatischen Charakter der Bewegungen im Karpatensystem auf Grund dieses Vergleichs sind derzeit noch nicht möglich, weil die karpatischen Bewegungen überwiegend vom orogenen Typus sind und die Isostasie erst sekundär zum Vorschein kommt. Allgemein aber ist es möglich zu konstatieren, daß die karpatischen Bewegungen in enger Beziehung zu den Schwereverhältnissen, die das Gebiet charakterisieren, stehen. Weiter steht fest, daß in Gebieten der Jungtektonik, d.h. im ganzen Karpatensystem, die Bewegungsaktivität viel ausdrucksvoller und intensiver ist als auf den stabilen Schollen der Erdkruste (Böhmisches Massiv) und schließlich daß in seismisch aktiven Gebieten (Komárno-

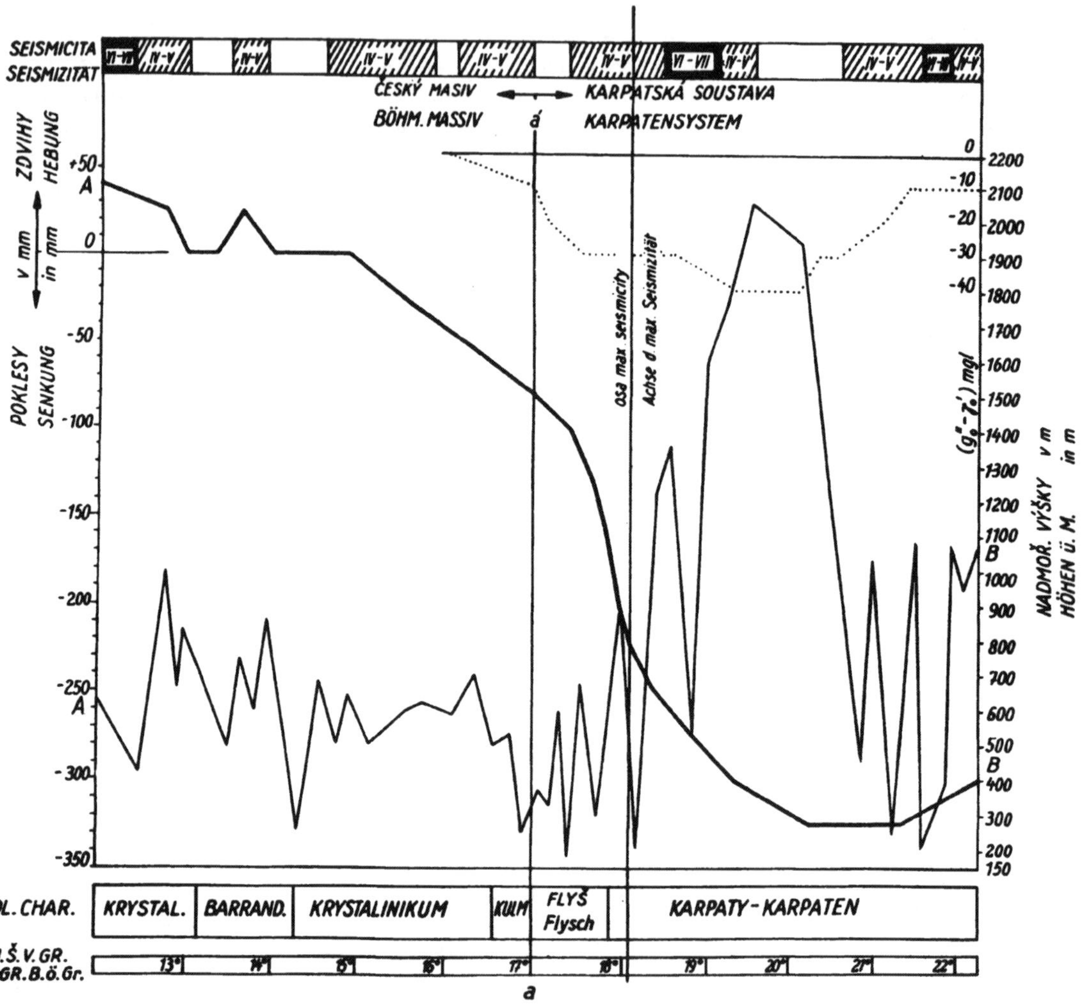

Abb. 8

Querschnitt durch die ČSSR. A – B mit Bezeichnung der Morphologie (schwache Linie), der Senkungen und Hebungen (starke Linie) und der Schwereverhältnisse (punktiert)

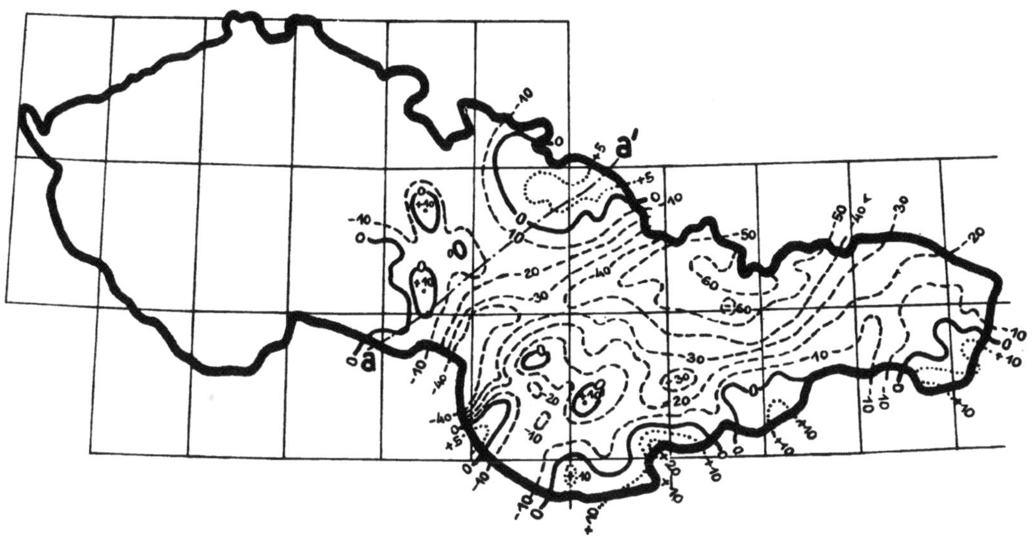

Abb. 9

Karte der Bouguer-Isanomalen ($g_0'' - \gamma_0'$) nach M. Wittinger (1956)
Angaben in mGal

Gebiet) sich die Richtung der Bewegungen im positiven und negativen Sinne ändert.

Für eine weitere wissenschaftliche Beobachtung dieser Probleme im Karpatensystem und in der ČSSR überhaupt wird es zweckmäßig sein, einige Grundnivellementszüge W - O quer durch die ČSSR und eine Reihe von Nivellementszügen in Richtung N - S aufzubauen. Eine spezielle Lösung wird die Beobachtung des Kontaktes der beiden geologischen Systeme: Böhmisches Massiv - Karpatensystem im Raum der Karpatenvortiefe erfordern. Es wird auch die Gründung eines neuen sogenannten "Indikations-Höhenpunktfeldes" erwogen und eine Verbindung der Grundpolygonzüge mit den stabilen Grundhöhenpunkten in benachbarten Staaten.

Das ist der Kern dieses Problemkreises: nur durch gemeinsame Arbeit der Forscher aller Nationen und auf Grund der einheitlichen geodätischen Grundnetze wird es möglich sein, zu befriedigenden Ergebnissen streng wissenschaftlicher Werte zu gelangen. Die um die Internationale Geodätische und Geophysikalische Union gruppierten Forscher sind aktive Vorkämpfer dieser Perspektiven.

Influence des mouvements systematiques verticaux de l'écorce terrestre sur les résultats et la détermination des grandeurs d'erreurs de nivellement de précision

von

T. Wyrzykowski (Warschau)

Summary

On the assumption of a uniform, but different vertical movement of the earth's crust as to its rate in various areas the influence of this movement on the differences of height between the extreme points of levelling lines as well as on the error caused by closing the levelling polygon has been analyzed. An analysis of the magnitude of the influence of the crustal movements on the results of levelling has been made as compared to accidental errors of the real levelling survey. The problem of the so-called mean moment is dealt with, i.e. of the common moment to which measurements of various lines of the levelling net, made at different points of time, can be reduced most advantageously.

Zusammenfassung

Es wurde - bei Voraussetzung einer gleichmäßigen, aber in bezug auf ihre Größe in verschiedenen Räumen unterschiedlichen Vertikalbewegung der Erdkruste - eine Analyse des Einflusses dieser Bewegung auf die Höhenunterschiede zwischen den Endpunkten von Nivellementslinien sowie auf den Fehler beim Abschließen des Nivellementspolygons durchgeführt. Der Analyse wurde die Größe des Einflusses der Erdkrustenbewegungen auf die Ergebnisse des Nivellements im Vergleich zu zufälligen Fehlern der eigentlichen Nivellementsvermessung unterzogen. Behandelt wird das Problem des sogenannten mittleren Moments, d.h. des gemeinsamen Momentes, auf welches sich zu verschiedenen Zeitpunkten ausgeführte Vermessungen verschiedener Linien im Nivellementsnetz am vorteilhaftesten reduzieren lassen.

FSC
www.fsc.org
MIX
Papier | Fördert gute Waldnutzung
FSC® C083411

Kolibri 360 GmbH
Ferdinand-Jühlke-Straße 7
99095 Erfurt, Deutschland
produktsicherheit@kolibri360.de